# Lecture Notes on Data Engineering and Communications Technologies

Volume 76

**Series Editor**

Fatos Xhafa, Technical University of Catalonia, Barcelona, Spain

The aim of the book series is to present cutting edge engineering approaches to data technologies and communications. It will publish latest advances on the engineering task of building and deploying distributed, scalable and reliable data infrastructures and communication systems.

The series will have a prominent applied focus on data technologies and communications with aim to promote the bridging from fundamental research on data science and networking to data engineering and communications that lead to industry products, business knowledge and standardisation.

Indexed by SCOPUS, INSPEC, EI Compendex.

All books published in the series are submitted for consideration in Web of Science.

More information about this series at http://www.springer.com/series/15362

Jude Hemanth · Tuncay Yigit ·
Bogdan Patrut · Anastassia Angelopoulou
Editors

# Trends in Data Engineering Methods for Intelligent Systems

Proceedings of the International Conference
on Artificial Intelligence and Applied
Mathematics in Engineering
(ICAIAME 2020)

 Springer

*Editors*
Jude Hemanth
Department of Electronics
and Communication Engineering
Karunya University
Coimbatore, Tamil Nadu, India

Bogdan Patrut
Department of Computer Science
Alexandru Ioan Cuza University
Iasi, Romania

Tuncay Yigit
Department of Computer Engineering,
Faculty of Engineering
Süleyman Demirel University
Isparta, Turkey

Anastassia Angelopoulou
Department of Computer Science
and Engineering
University of Westminster
London, UK

ISSN 2367-4512        ISSN 2367-4520   (electronic)
Lecture Notes on Data Engineering and Communications Technologies
ISBN 978-3-030-79356-2      ISBN 978-3-030-79357-9   (eBook)
https://doi.org/10.1007/978-3-030-79357-9

This Springer imprint is published by the registered company Springer Nature Switzerland AG
The registered company address is: Gewerbestrasse 11, 6330 Cham, Switzerland

# Foreword

It is a pleasure for me to write that Foreword for the 2020 edition of the Springer proceeding book for ICAIAME 2020: International Conference on Artificial Intelligence and Applied Mathematics in Engineering 2020. All readers are welcome to turn the pages for reading the latest advances and developments regarding engineering efforts of real-world-based problems.

As it is known, artificial intelligence has been widely applied in a wide variety of real-world problems associated with multiple engineering disciplines. From electronics, computer, or chemistry to biomedical, all engineering fields require intense efforts of advanced technologies to keep these fields balanced according to newly encountered problems. Actually, there is an unstoppable loop in terms of real-world problems experienced so there is always a critical need for developing innovative solutions. That can be achieved by computational approaches, as done by artificial intelligence and Applied Mathematics. In this book, there is a wide variety of chapters aiming to provide effective machine learning, deep learning, IoT, optimization, image processing, and even robotics-related solutions for problems of different engineering areas of biomedical engineering, electrical and electronics engineering, computer engineering, industrial engineering, civil engineering, mechanics engineering, and more. It is also remarkable that mathematical solutions coming from the subfields of applied mathematics are giving high level of value to the book content. I think scientists, professionals, experts, and even students from various degrees (BSc., MSc., PhD.) will be benefiting from that book greatly.

Since last year, the humankind is experiencing the massive COVID-19 issue so there is a great emergency state for especially field of medical. However, high-level usage of artificial intelligence, communication technologies, and even image processing has already given great importance to design innovative solutions sensitive to the data. In this book, readers will be able to figure out important clues about how to design such timely solutions by combining engineering knowledge/skills and the most recent high-technology components. After that, it will be possible to refer to COVID-19 targeting sources. So, this book is a timely contribution for a better technological future and a good starting point to understand exact mechanisms on the background of the latest high-technology tools.

I would like to thank dear ICAIAME 2020 staff to give me such a valuable opportunity to write these paragraphs, and finally, I wish all readers to enjoy all chapters of the book.

Deepak Gupta

# Foreword

Currently, we need advanced technologies to deal with a massive problem: COVID-19. It is a fatal virus type, so every fields of the modern life are hardworking for medical and public solutions. That is not a common thing that uncontrolled use of technology, population growth, and spending natural resources can easily trigger such problems. However, the most critical weapon that we can use for that is again advanced technology. We need that because we are able to use technology for performing early predictions and careful decision-making processes to catch important details of real-world problems.

As the humankind is experiencing a new century with high intense use of digital data, technological solutions require use of data and intelligent systems to have advantages against the massive issues we are facing. At this point, artificial intelligence has a great importance for present and the future world. As rising over the shoulders of mathematics, artificial intelligence has already started to shape the future at even its early steps in the middle of the twentieth century. Nowadays, advancements are exponentially growing and using numerical background to derive smart tools. So, it is a remarkable point to produce effective research at the intersection of artificial intelligence and mathematics.

The Second International Conference on Artificial Intelligence and Applied Mathematics in Engineering in 2020 (ICAIAME 2020) has been realized between October 9 and 11, 2020 after several updates of the dates, because of the COVID-19. Eventually, a hybrid mode event was run to host international authors as well as participants to enable them for sharing the most recent scientific knowledge. At this point, the scope targeted all engineering disciplines as they are intensively interested in use of mathematical solutions and intelligent systems to provide innovative developments. Engineering disciplines are too critical to support scientific developments with an applied perspective. In this context, use of advanced technology often requires employment of engineering knowledge and skills. So, ICAIAME 2020 has welcomed innovative research works from engineering disciplines. I enjoyed the parallel sessions as well as keynote speech sessions and felt mysterious synergy of science while discussing with participants. It was a pleasure for me to co-manage the whole event as the second time and perform them in both real and virtual environments.

This year, ICAIAME has come with more alternative publishing opportunities for the extended versions of the selected papers. Those included indexed journals as well as edited book studies, so the participants with top-level papers have found opportunity to raise the level of visibility for their papers. We will be doing our best to improve such opportunities in the future years. Because of the COVID-19, we have also canceled our workshop and specific seminar sessions during the event but we hope to realize them in next years.

This book is a great venue to see the most recent papers accepted and presented at ICAIAME 2020. Covering more than 70 papers, you will find different types of research works considering up-to-date use of intelligent systems and applied mathematics to solve problems of different engineering disciplines. All the papers have been carefully reviewed and revised in order to meet with high-quality standards and contribute to the state of the art. So, I kindly request all readers to enjoy their reading while diving deeper sides of advanced technologies.

As my final words, I would like to thank all event staff, authors, and the participants for their contributions to ICAIAME 2020.

Utku Kose

# Preface

As the continuation of the first event in 2019, the Second International Conference on Artificial Intelligence and Applied Mathematics (ICAIAME) was held in Antalya, Manavgat (Turkey), between October 9 and 11, 2020. On behalf of the organization and scientific committees of ICAIAME 2020, it is a great pleasure for us to provide this recent book. The objective of ICAIAME 2020 was to enable academia, industry, organizations, and governments in contributing to the actively improving knowledge regarding artificial intelligence and applied mathematics in engineering-oriented applications. In order to meet with that, ICAIAME ensures exchanging of ideas associated with machine/deep learning, algorithm design for intelligent systems, robotic systems, image processing, diagnosis applications, discrete mathematics, operations research, and even wide advanced technological applications of engineering fields. This year, ICAIAME 2020 was run in both face-to-face and online modes because of the COVID-19. However, the active role of knowledge sharing did not stop along the three-day-long experiences for the international audience.

In ICAIAME 2020, a total of 200 papers have been accepted to be presented in the context of 35 parallel face-to-face and online sessions. The selections were made according to the external reviewers as well as the members of the scientific committee. As it was done in 2019, the proceedings of the 2020 is again published by Springer under the Springer Series: Lecture Notes on Data Engineering and Communications Technologies. We are grateful for more than 70 high-quality papers that you are just a few pages away to start reading. In addition to the proceeding book opportunity, we have also selected high-quality papers to be suggested to internationally and nationally reputable journals (with SCI-E, ESCI, Scopus indexes) and book studies. When it is considered in international scope, ICAIAME 2020 hosted authors/participants from more than 10 different countries such as England, France, India, Kosovo, Kyrgyzstan, Mexico, Pakistan, Turkey, and USA. That has enabled the audience to share the most recent research findings with a wide international audience.

In ICAIAME 2020, a total of 8 invited keynote presentations were delivered by internationally reputable contributors in artificial intelligence and applied mathematics. The speech sessions included "Artificial Intelligence in Project Management" by Dr. Çetin Elmas (Gazi University, Turkey), "Project Management" by Timoty Jaques (International Project Management Agency, USA), "Poincare chaos in Hopfield type neural networks" by Dr. Marat AKHMET (Middle East Technical University, Turkey), "An Overview of Multiobjective Hyper-heuristics" by Dr. Ender ÖZCAN (University of Nottingham, England), "Cloud Computing Applications in Turkey" by Yusuf TULGAR (NetDataSoft CEO, Ankara, Turkey), "The Power of Data and The Things It Empowers" by Dr. Hüseyin SEKER (Staffordshire University, England), "Understanding The Mystery Behind Deep Learning-DEEP, DEEPER, DEEPEST" by Dr. Jude HEMANTH (Karunya University, India), and "Assistive Devices for Visually Impaired and Blind People in Smart City Infrastructure: A Retrospective Representation of an IoE Project" by Dr. Dmytro ZUBOV (University of Central Asia, Kyrgyzstan).

As like we always indicate, success of that event: ICAIAME 2020 could not be possible without the staff and international contributors. We thank everybody enrolled in the event to realize a good discussion environment and led such a resulting book work. We wish more success and wider effects of ICAIAME in the upcoming years. Thank you!

Jude Hemanth
Tuncay Yigit
Bogdan Patrut
Anastasia Angelopoulou

# Acknowledgement

As the editors, we would like to thank Lect. Çilem Koçak (Isparta University of Applied Sciences, Turkey) for her valuable efforts on pre-organization of the book content and the Springer team for their great support to publish the book.

# International Conference on Artificial Intelligence and Applied Mathematics in Engineering 2020

**Web:** http://www.icaiame.com
**E-Mail:** icaiame.umymk@gmail.com

## Briefly About

The International Conference on Artificial Intelligence and Applied Mathematics in Engineering (ICAIAME 2020) held within **October 9–11, 2020**, at the **Antalya, Manavgat (Turkey)**, which is the pearl of the Mediterranean, heaven corner of Turkey, and the fourth most visited city in the world.

The main theme of the conference, which will be held at Bella Resort & Spa with international participations along a three-day period, is solutions of artificial intelligence and applied mathematics in engineering applications. The languages of ICAIAME 2020 are English and Turkish.

## Scope/Topics

*Conference Scope/Topics (as not limited to):*
*In Engineering Problems:*

- Machine Learning Applications
- Deep Learning Applications
- Intelligent Optimization Solutions
- Robotics/Soft Robotics and Control Applications
- Hybrid System Based Solutions
- Algorithm Design for Intelligent Solutions
- Image/Signal Processing Supported Intelligent Solutions
- Data Processing Oriented Intelligent Solutions
- Prediction and Diagnosis Applications
- Linear Algebra and Applications
- Numerical Analysis

- Differential Equations and Applications
- Probability and Statistics
- Operations Research and Optimization
- Discrete Mathematics and Control
- Nonlinear Dynamical Systems and Chaos
- General Engineering Applications

## Honorary Chairs

İlker Hüseyin Çarikçi          Rector of Süleyman Demirel University, Turkey
İbrahim Diler          Rector of Isparta Applied Sciences University, Turkey

## General Chair

Tuncay Yiğit          Süleyman Demirel University, Turkey

## Conference Chairs

İsmail Serkan Üncü          Isparta Applied Sciences University, Turkey
Utku Köse          Süleyman Demirel University, Turkey

## Organizing Committee

Mehmet Gürdal          Süleyman Demirel University, Turkey
Anar Adiloğlu          Süleyman Demirel University, Turkey
Şemsettin Kilinçarslan          Süleyman Demirel University, Turkey
Kemal Polat          Bolu Abant İzzet Baysal University, Turkey
Okan Bingöl          Applied Sciences University of Isparta, Turkey
Cemal Yilmaz          Gazi University, Turkey
Ercan Nurcan Yilmaz          Gazi University, Turkey
Uğur Güvenç          Düzce University, Turkey
Yusuf Sönmez          Gazi University, Turkey
Hamdi Tolga Kahraman          Karadeniz Technical University, Turkey
Akram M. Zeki          International Islamic University Malaysia, Malaysia

Bogdan Patrut          Alexandru Ioan Cuza University of Iasi, Romania
Hasan Hüseyin Sayan          Gazi University, Turkey
Halil İbrahim Koruca          Süleyman Demirel University, Turkey
Ali Hakan Işik          Mehmet Akif Ersoy University, Turkey
Muhammed Maruf Öztürk          Süleyman Demirel University, Turkey
Osman Özkaraca          Muğla Sıtkı Koçman University, Turkey
Asım Sinan Yüksel          Süleyman Demirel University, Turkey
Mehmet Kayakuş          Akdeniz University, Turkey
Mevlüt Ersoy          Süleyman Demirel University, Turkey

| | |
|---|---|
| Gürcan Çetin | Muğla Sıtkı Koçman University, Turkey |
| Murat Ince | Süleyman Demirel University, Turkey |
| Gül Fatma Türker | Süleyman Demirel University, Turkey |
| Ferdi Saraç | Süleyman Demirel University, Turkey |
| Hamit Armağan | Süleyman Demirel University, Turkey |

## Secretary

Nilgün Şengöz     Mehmet Akif Ersoy University, Turkey

## Accommodation/Venue Desk

Çilem Koçak     Isparta Applied Sciences University, Turkey

## Travel/Transportation

Ferdi Saraç     Süleyman Demirel University, Turkey

## Web/Design/Conference Sessions

| | |
|---|---|
| Cem Deniz Kumral | Süleyman Demirel University, Turkey |
| Ali Topal | Süleyman Demirel University, Turkey |

## Scientific Committee

| | |
|---|---|
| Tuncay Yiğit (General Chair) | Süleyman Demirel University, Turkey |
| Utku Köse (Committee Chair) | Süleyman Demirel University, Turkey |
| Ahmet Bedri Özer | Fırat University, Turkey |
| Ali Öztürk | Düzce University, Turkey |
| Anar Adiloğlu | Süleyman Demirel University, Turkey |
| Aslanbek Naziev | Ryazan State University, Russia |
| Ayhan Erdem | Gazi University, Turkey |
| Çetin Elmas | Gazi University, Turkey |
| Daniela Elena Popescu | University of Oradea, Romania |
| Eduardo Vasconcelos | Goias State University, Brazil |
| Ekrem Savaş | Rector of Uşak University, Turkey |
| Eşref Adali | İstanbul Technical University, Turkey |
| Hüseyin Demir | Samsun 19 May University, Turkey |
| Hüseyin Merdan | TOBB Economy and Technology University, Turkey |
| Hüseyin Şeker | Staffordshire University, England |
| Igbal Babayev | Azerbaijan Technical University, Azerbaijan |
| Igor Litvinchev | Nuevo Leon State University, Mexico |
| İbrahim Üçgül | Süleyman Demirel University, Turkey |

| İbrahim Yücedağ | Düzce University, Turkey |
| Jose Antonio Marmolejo | Panamerican University, Mexico |
| Junzo Watada | Universiti Teknologi PETRONAS, Malaysia |
| Marat Akhmet | Middle East Technical University, Turkey |
| Marwan Bikdash | North Carolina Agricultural and Technical State University, USA |
| Mehmet Ali Akçayol | Gazi University, Turkey |
| Mehmet Gürdal | Süleyman Demirel University, Turkey |
| Melih Günay | Akdeniz University, Turkey |
| Mostafa Maslouhi | IbnTofail University, Morocco |
| Mustafa Alkan | Gazi University, Turkey |
| Nihat Öztürk | Gazi University, Turkey |
| Norita Md Norwawi | Universiti Sains Islam Malaysia, Malaysia |
| Nuri Özalp | Ankara University, Turkey |
| Oktay Duman | TOBB Economy and Technology University, Turkey |
| Ömer Akin | TOBB Economy and Technology University, Turkey |
| Ömer Faruk Bay | Gazi University, Turkey |
| Recep Demirci | Gazi University, Turkey |
| Resul Kara | Düzce University, Turkey |
| Reşat Selbaş | Applied Sciences University of Isparta, Turkey |
| Sabri Koçer | Necmettin Erbakan University, Turkey |
| Sadık Ülker | European University of Lefke, Cyprus |
| Sergey Bushuyev | Kyiv National University, Ukraine |
| Sezai Tokat | Pamukkale University, Turkey |
| Yusuf Öner | Pamukkale University, Turkey |
| Cemal Yilmaz | Gazi University, Turkey |
| Gültekin Özdemir | Süleyman Demirel University, Turkey |
| Ercan Nurcan Yilmaz | Gazi University, Turkey |
| Kemal Polat | Bolu Abant İzzet Baysal University, Turkey |
| Okan Bingöl | Applied Sciences University of Isparta, Turkey |
| Ali Keçebaş | Muğla Sıtkı Koçman University, Turkey |
| Erdal Kiliç | Samsun 19 May University, Turkey |
| Mehmet Karaköse | Fırat University, Turkey |
| Mehmet Sıraç Özerdem | Dicle University, Turkey |
| Muharrem Tolga Sakalli | Trakya University, Turkey |
| Murat Kale | Düzce University, Turkey |
| Ahmet Cüneyd Tantuğ | İstanbul Technical University, Turkey |
| Akram M. Zeki | Malaysia International Islamic University, Malaysia |
| Alexandrina Mirela Pater | University of Oradea, Romania |
| Ali Hakan Işik | Mehmet Akif Ersoy University, Turkey |
| Arif Özkan | Kocaeli University, Turkey |
| Aydın Çetin | Gazi University, Turkey |

| | |
|---|---|
| Samia Chehbi Gamoura | Strasbourg University, France |
| Ender Ozcan | Nottingham University, England |
| Devrim Akgün | Sakarya University, Turkey |
| Ercan Buluş | Namık Kemal University, Turkey |
| Serdar Demir | Muğla Sıtkı Koçman University, Turkey |
| Serkan Balli | Muğla Sıtkı Koçman University, Turkey |
| Serkan Duman | Düzce University, Turkey |
| Turan Erman Erkan | Atılım University, Turkey |
| Ezgi Ülker | European University of Lefke, Cyprus |
| Gamze Yüksel | Muğla Sıtkı Koçman University, Turkey |
| Hasan Hüseyin Sayan | Gazi University, Turkey |
| İlhan Koşalay | Ankara University, Turkey |
| İsmail Serkan Üncü | Applied Sciences University of Isparta, Turkey |
| J. Anitha | Karunya University, India |
| Jude Hemanth | Karunya University, India |
| M. Kenan Döşoğlu | Düzce University, Turkey |
| Nevin Güler Dincer | Muğla Sıtkı Koçman University, Turkey |
| Özgür Aktunç | St. Mary's University, USA |
| Ridha Derrouiche | EM Strasbourg Business School, France |
| Sedat Akleylek | Samsun 19 May University, Turkey |
| Selami Kesler | Pamukkale University, Turkey |
| Selim Köroğlu | Pamukkale University, Turkey |
| Tiberiu Socaciu | Stefan cel Mare University of Suceava, Romania |
| Ramazan Şenol | Applied Sciences University of Isparta, Turkey |
| Tolga Ovatman | İstanbul Technical University, Turkey |
| Ümit Deniz Uluşar | Akdeniz University, Turkey |
| Muhammed Hanefi Calp | Karadeniz Technical University, Turkey |
| Abdulkadir Karaci | Kastamonu University, Turkey |
| Ali Şentürk | Applied Sciences University of Isparta, Turkey |
| Arif Koyun | Süleyman Demirel University, Turkey |
| Barış Akgün | Koç University, Turkey |
| Deepak Gupta | Maharaja Agrasen Institute of Technology, India |
| Dmytro Zubov | University of Information Science and Technology St. Paul the Apostle, Macedonia |
| Enis Karaarslan | Muğla Sıtkı Koçman University, Turkey |
| Erdal Aydemir | Süleyman Demirel University, Turkey |
| Esin Yavuz | Süleyman Demirel University, Turkey |
| Fatih Gökçe | Süleyman Demirel University, Turkey |
| Gür Emre Güraksin | Afyon Kocatepe University, Turkey |
| Iulian Furdu | Vasile Alecsandri University of Bacau, Romania |
| Mehmet Kayakuş | Akdeniz University, Turkey |
| Mehmet Onur Olgun | Süleyman Demirel University, Turkey |
| Mustafa Nuri Ural | Gümüşhane University, Turkey |
| Okan Oral | Akdeniz University, Turkey |
| Osman Palanci | Süleyman Demirel University, Turkey |

| Paniel Reyes Cardenas | Popular Autonomous University of the State of Puebla, Mexico |
| S. T. Veena | Kamaraj Engineering and Technology University, India |
| Serdar Biroğul | Düzce University, Turkey |
| Serdar Çiftçi | Harran University, Turkey |
| Ufuk Özkaya | Süleyman Demirel University, Turkey |
| Veli Çapali | Uşak University, Turkey |
| Vishal Kumar | Bipin Tripathi Kumaon Institute of Technology, India |
| Bekir Aksoy | Applied Sciences University of Isparta, Turkey |
| Remzi Inan | Applied Sciences University of Isparta, Turkey |
| Anand Nayyar | Duy Tan University, Vietnam |
| Bogdan Patrut | Alexandru Ioan Cuza University of Iasi, Romania |
| Simona Elena Varlan | Vasile Alecsandri University of Bacau, Romania |
| Ashok Prajapati | FANUC America Corp., USA |
| Katarzyna Rutczyńska-Wdowiak | Kielce University of Technology, Poland |
| Mustafa Küçükali | Information and Communication Technologies Authority |
| Nabi Ibadov | Warsaw University of Technology, Poland |
| Özkan Ünsal | Süleyman Demirel University, Turkey |
| Tim Jaques | International Project Management Association, USA |

## Keynote Speaks

1- Çetin Elmas
(Gazi University, Turkey)
*"Artificial Intelligence in Project Management"*

2- Timoty Jaques
(International Project Management Agency, USA)
*"Project Management"*

3- Marat Akhmet
(Middle East Technical University, Turkey)
*"Poincare chaos in Hopfield type neural networks"*

4- Ender Ozcan
(University of Nottingham, England)
*"An Overview of Multiobjective Hyper-heuristics"*

5- Yusuf Tulgar
(NetDataSoft CEO, Ankara, Turkey)
*"Cloud Computing Applications in Turkey"*

6- Hüseyin Seker
(Staffordshire University, England)
*"The Power of Data and The Things It Empowers"*

7- Jude Hemanth
(Karunya Institute of Technology and Sciences, India)
*"Understanding The Mystery Behind Deep Learning-DEEP, DEEPER, DEEPEST"*

8- Dmytro Zubov
(University of Central Asia, Kyrgyzstan)
*"Assistive Devices for Visually Impaired and Blind People in Smart City Infrastructure: A Retrospective Representation of an IoE Project"*

# Contents

Prediction of Liver Cancer by Artificial Neural Network . . . . . . . . . . . .   1
Remzi Gurfidan and Mevlut Ersoy

Remarks on the Limit-Circle Classification of Conformable Fractional
Sturm-Liouville Operators . . . . . . . . . . . . . . . . . . . . . . . . . . . . .   10
Bilender P. Allahverdiev, Hüseyin Tuna, and Yüksel Yalçinkaya

Improving Search Relevance with Word Embedding Based Clusters . . .   15
Işılay Tuncer, Kemal Can Kara, and Aşkın Karakaş

Predicting Suicide Risk in Turkey Using Machine Learning . . . . . . . . .   25
Elif Şanlıalp, İbrahim Şanlıalp, and Tuncay Yiğit

Building an Open Source Big Data Platform Based on Milis Linux . . . .   33
Melih Günay and Muhammed Numan İnce

Machine Learning for the Diagnosis of Chronic Obstructive
Pulmonary Disease and Photoplethysmography Signal – Based
Minimum Diagnosis Time Detection . . . . . . . . . . . . . . . . . . . . . . . .   42
Engin Melekoğlu, Ümit Kocabıçak, Muhammed Kürşad Uçar,
Mehmet Recep Bozkurt, and Cahit Bilgin

Diagnosis of Parkinson's Disease with Acoustic Sounds by Rule
Based Model . . . . . . . . . . . . . . . . . . . . . . . . . . . . . . . . . . . . . . .   59
Kılıçarslan Yıldırım, Muhammed Kürşad Uçar, Ferda Bozkurt,
and Mehmet Recep Bozkurt

Text Classification Models for CRM Support Tickets . . . . . . . . . . . . .   76
Şevval Az, Uygar Takazoğlu, and Aşkın Karakaş

SMOTE-Text: A Modified SMOTE for Turkish Text Classification . . .   82
Nur Curukoglu and Alper Ozpinar

**Development of Face Recognition System by Using Deep Learning
and Face-Net Algorithm in the Operations Processes** . . . . . . . . . . . . . .   93
Ali Tunç, Mehmet Yildirim, Şakir Taşdemir, and Adem Alpaslan Altun

**Mobile Assisted Travel Planning Software: The Case of Burdur** . . . . . .  106
Ali Hakan Işık and Rifai Kuçi

**Text-Based Fake News Detection via Machine Learning** . . . . . . . . . . . .  113
Uğur Mertoğlu, Burkay Genç, and Hayri Sever

**A Recommendation System for Article Submission for Researchers** . . .  125
Seth Michail, Ugur Hazir, Taha Yigit Alkan, Joseph William Ledet,
and Melih Gunay

**Optimal Power Flow Using Manta Ray Foraging Optimization** . . . . . . .  136
Ugur Guvenc, Huseyin Bakir, Serhat Duman, and Burcin Ozkaya

**Optimal Coordination of Directional Overcurrent Relays Using
Artificial Ecosystem-Based Optimization** . . . . . . . . . . . . . . . . . . . . . . .  150
Ugur Guvenc, Huseyin Bakir, and Serhat Duman

**Performance Evaluation of Machine Learning Techniques on Flight
Delay Prediction** . . . . . . . . . . . . . . . . . . . . . . . . . . . . . . . . . . . . . . . . .  165
Irmak Daldır, Nedret Tosun, and Ömür Tosun

**Machine Breakdown Prediction with Machine Learning** . . . . . . . . . . . .  174
Hande Erdoğan, Ömür Tosun, and M. K. Marichelvam

**Prevention of Electromagnetic Impurities
by Electromagnetics Intelligence** . . . . . . . . . . . . . . . . . . . . . . . . . . . . . .  183
Mehmet Duman

**The Effect of Auscultation Areas on Nonlinear Classifiers
in Computerized Analysis of Chronic Obstructive
Pulmonary Disease** . . . . . . . . . . . . . . . . . . . . . . . . . . . . . . . . . . . . . . . .  190
Ahmet Gökçen and Emre Demir

**Least Square Support Vector Machine for Interictal Detection Based
on EEG of Epilepsy Patients at Airlangga University Hospital
Surabaya-Indonesia** . . . . . . . . . . . . . . . . . . . . . . . . . . . . . . . . . . . . . . .  198
Santi Wulan Purnami, Triajeng Nuraisyah, Wardah Rahmatul Islamiyah,
Diah P. Wulandari, and Anda I. Juniani

**Generating Classified Ad Product Image Titles
with Image Captioning** . . . . . . . . . . . . . . . . . . . . . . . . . . . . . . . . . . . . .  211
Birkan Atıcı and Sevinç İlhan Omurca

**Effectiveness of Genetic Algorithm in the Solution of Multidisciplinary
Conference Scheduling Problem** . . . . . . . . . . . . . . . . . . . . . . . . . . . . . .  220
Ercan Atagün and Serdar Biroğul

**Analyze Performance of Embedded Systems with Machine
Learning Algorithms** .......................................... 231
Mevlüt Ersoy and Uğur Şansal

**Classification Performance Evaluation on Diagnosis
of Breast Cancer** .............................................. 237
M. Sinan Basarslan and F. Kayaalp

**Effective Factor Detection in Crowdfunding Systems** .............. 246
Ercan Atagün, Tuba Karagül Yıldız, Tunahan Timuçin, Hakan Gündüz,
and Hacer Bayıroğlu

**Predictive Analysis of the Cryptocurrencies' Movement Direction
Using Machine Learning Methods** ............................. 256
Tunahan Timuçin, Hacer Bayiroğlu, Hakan Gündüz, Tuba Karagül Yildiz,
and Ercan Atagün

**Sofware Quality Prediction: An Investigation Based on Artificial
Intelligence Techniques for Object-Oriented Applications** ........... 265
Özcan İlhan and Tülin Erçelebi Ayyıldız

**A Multi Source Graph-Based Hybrid Recommendation Algorithm** .... 280
Zühal Kurt, Ömer Nezih Gerek, Alper Bilge, and Kemal Özkan

**Development of an Artificial Intelligence Based Computerized
Adaptive Scale and Applicability Test** ......................... 292
Mustafa Yagci

**Reduced Differential Transform Approach Using Fixed Grid Size
for Solving Newell–Whitehead–Segel (NWS) Equation** ............. 304
Sema Servı and Galip Oturanç

**A New Ensemble Prediction Approach to Predict Burdur
House Prices** ................................................. 315
Mehmet Bilen, Ali H. Işık, and Tuncay Yiğit

**The Evidence of the "No Free Lunch" Theorems and the Theory
of Complexity in Business Artificial Intelligence** ................... 325
Samia Chehbi Gamoura, Halil İbrahim Koruca, Esra Gülmez,
Emine Rümeysa Kocaer, and Imane Khelil

**Text Mining Based Decision Making Process
in Kickstarter Platform** ....................................... 344
Tuba Karagül Yildiz, Ercan Atagün, Hacer Bayiroğlu, Tunahan Timuçin,
and Hakan Gündüz

**Deep Q-Learning for Stock Future Value Prediction** .............. 350
Uğur Hazir and Taner Danisman

Effect of DoS Attacks on MTE/LEACH Routing Protocol-Based
Wireless Sensor Networks . . . . . . . . . . . . . . . . . . . . . . . . . . . . . . . . . . . . 360
Asmaa Alaadin, Erdem Alkım, and Sercan Demirci

On Connectivity-Aware Distributed Mobility Models for Area
Coverage in Drone Networks . . . . . . . . . . . . . . . . . . . . . . . . . . . . . . . . . . 369
Mustafa Tosun, Umut Can Çabuk, Vahid Akram, and Orhan Dagdeviren

Analysis of Movement-Based Connectivity Restoration Problem
in Wireless Ad-Hoc and Sensor Networks . . . . . . . . . . . . . . . . . . . . . . . 381
Umut Can Cabuk, Vahid Khalilpour Akram, and Orhan Dagdeviren

Design of External Rotor Permanent Magnet Synchronous Reluctance
Motor (PMSynRM) for Electric Vehicles . . . . . . . . . . . . . . . . . . . . . . . . 390
Armagan Bozkurt, Yusuf Oner, A. Fevzi Baba, and Metin Ersoz

Support Vector Machines in Determining the Characteristic
Impedance of Microstrip Lines . . . . . . . . . . . . . . . . . . . . . . . . . . . . . . . . 400
Oluwatayomi Adegboye, Mehmet Aldağ, and Ezgi Deniz Ülker

A New Approach Based on Simulation of Annealing to Solution
of Heterogeneous Fleet Vehicle Routing Problem . . . . . . . . . . . . . . . . . 409
Metin Bilgin and Nisanur Bulut

Microgrid Design Optimization and Control with Artificial
Intelligence Algorithms for a Public Institution . . . . . . . . . . . . . . . . . . 418
Furkan Üstünsoy, Serkan Gönen, H. Hüseyin Sayan,
Ercan Nurcan Yılmaz, and Gökçe Karacayılmaz

Determination of Vehicle Type by Image Classification Methods
for a Sample Traffic Intersection in Isparta Province . . . . . . . . . . . . . . 429
Fatmanur Ateş, Osamah Salman, Ramazan Şenol, and Bekir Aksoy

Prediction of Heat-Treated Spruce Wood Surface Roughness
with Artificial Neural Network and Random Forest Algorithm . . . . . . . 439
Şemsettin Kilinçarslan, Yasemin Şimşek Türker, and Murat İnce

Design and Implementation of Microcontroller Based Hydrogen
and Oxygen Generator Used Electrolysis Method . . . . . . . . . . . . . . . . . 446
Mustafa Burunkaya and Sadık Yıldız

ROSE: A Novel Approach for Protein Secondary
Structure Prediction . . . . . . . . . . . . . . . . . . . . . . . . . . . . . . . . . . . . . . . . . 455
Yasin Görmez and Zafer Aydın

A Deep Learning-Based IoT Implementation for Detection of Patients'
Falls in Hospitals . . . . . . . . . . . . . . . . . . . . . . . . . . . . . . . . . . . . . . . . . . . . 465
Hilal Koçak and Gürcan Çetin

Contents

**Recognition of Vehicle Warning Indicators** ..................... 484
Ali Uçar and Süleyman Eken

**Time Series Analysis on EEG Data with LSTM**.................. 491
Utku Köse, Mevlüt Ersoy, and Ayşen Özün Türkçetin

**Gaussian Mixture Model-Based Clustering of Multivariate Data Using
Soft Computing Hybrid Algorithm**........................... 502
Maruf Gögebakan

**Design Optimization and Comparison of Brushless Direct Current
Motor for High Efficiency Fan, Pump and Compressor Applications**... 514
Burak Yenipinar, Cemal Yilmaz, Yusuf Sönmez, and Cemil Ocak

**Stability of a Nonautonomous Recurrent Neural Network Model
with Piecewise Constant Argument of Generalized Type** ........... 524
Duygu Aruğaslan Çinçin and Nur Cengiz

**Dynamics of a Recurrent Neural Network with Impulsive Effects
and Piecewise Constant Argument** ............................ 540
Marat Akhmet, Duygu Aruğaslan Çinçin, and Nur Cengiz

**A Pure Genetic Energy-Efficient Backbone Formation Algorithm
for Wireless Sensor Networks in Industrial Internet of Things** ....... 553
Zuleyha Akusta Dagdeviren

**Fault Analysis in the Field of Fused Deposition Modelling (FDM)
3D Printing Using Artificial Intelligence**......................... 567
Koray Özsoy and Helin Diyar Halis

**Real-Time Maintaining of Social Distance in Covid-19 Environment
Using Image Processing and Big Data** ......................... 578
Sadettin Melenli and Aylin Topkaya

**Determination of the Ideal Color Temperature for the Most Efficient
Photosynthesis of Brachypodium Plant in Different Light Sources by
Using Image Processing Techniques** .......................... 590
İsmail Serkan Üncü, Mehmet Kayakuş, and Bayram Derin

**Combination of Genetic and Random Restart Hill Climbing
Algorithms for Vehicle Routing Problem** ...................... 601
E. Bytyçi, E. Rogova, and E. Beqa

**Reliability of Offshore Structures Due to Earthquake Load
in Indonesia Water**........................................ 613
Yoyok Setyo Hadiwidodo, Daniel Mohammad Rosyid, Imam Rochani,
and Rizqullah Yusuf Naufal

**Unpredictable Oscillations of Impulsive Neural Networks
with Hopfield Structure** ................................... 625
Marat Akhmet, Madina Tleubergenova, and Zakhira Nugayeva

**Performance Analysis of Particle Swarm Optimization and Firefly Algorithms with Benchmark Functions** .......................... 643
Mervenur Demirhan, Osman Özkaraca, and Ercüment Güvenç

**Optimization in Wideband Code-Division Multiple Access Systems with Genetic Algorithm-Based Discrete Frequency Planning** ......... 654
Sencer Aksoy and Osman Özkaraca

**Deep Learning Based Classification Method for Sectional MR Brain Medical Image Data** ......................................... 669
Ali Hakan Işik, Mevlüt Ersoy, Utku Köse, Ayşen Özün Türkçetin, and Recep Çolak

**Analysis of Public Transportation for Efficiency** .................. 680
Kamer Özgün, Melih Günay, Barış Doruk Başaran, Batuhan Bulut, Ege Yürüten, Fatih Baysan, and Melisa Kalemsiz

**Methods for Trajectory Prediction in Table Tennis** ............... 696
Mehmet Fatih Kaftanci, Melih Günay, and Özge Öztimur Karadağ

**Emotion Analysis Using Deep Learning Methods** ................. 705
Bekir Aksoy and İrem Sayin

**A Nested Unsupervised Learning Model for Classification of SKU's in a Transnational Company: A Big Data Model** ................. 715
Gabriel Loy-García, Román Rodríguez-Aguilar, and Jose-Antonio Marmolejo-Saucedo

**Identification of Trading Strategies Using Markov Chains and Statistical Learning Tools** ...................................... 732
Román Rodríguez-Aguilar and Jose-Antonio Marmolejo-Saucedo

**Estimation of the Stochastic Volatility of Oil Prices of the Mexican Basket: An Application of Boosting Monte Carlo Markov Chain Estimation** ................................................ 742
Román Rodríguez-Aguilar and Jose-Antonio Marmolejo-Saucedo

**Optimization of the Input/Output Linearization Feedback Controller with Simulated Annealing and Designing of a Novel Stator Flux-Based Model Reference Adaptive System Speed Estimator with Least Mean Square Adaptation Mechanism** .............................. 755
Remzi Inan, Mevlut Ersoy, and Cem Deniz Kumral

**Author Index** ................................................. 771

# Prediction of Liver Cancer by Artificial Neural Network

Remzi Gurfidan[1] and Mevlut Ersoy[2]([⊠])

[1] Isparta University of Applied Sciences, Isparta, Turkey
`remzigurfidan@isparta.edu.tr`
[2] Suleyman Demirel University, Isparta, Turkey
`mevlutersoy@sdu.edu.tr`

**Abstract.** In this study, a model was developed for the prediction of Hepatocellular Carcinoma, a liver tumor. Artificial Neural Network (ANN) technique was used in the developed model. ANN is created using the Python language. Hepatocellular Carcinoma data set was used in the study. Data set, *k*aggle.com (www.kaggle.com/mrsantos/hcc-dataset) there are 165 rows in the dataset consisting of 50 columns. There are 49 independent arguments in the dataset. Arguments were used as inputs of the ANN. The weights between the parameters used and the artificial nerve cells were not randomly distributed, but according to a specific algorithm. The Sigmoid and Relu functions are used as activation functions in ANN, which is formed in two layers. For the training of the Model, 100 repetitions of learning were performed. The dependent variable of ANN is given as a value of 1 or 0. The estimated success of the developed model was measured at 81%.

## 1 Introduction

Technology is a constantly developing concept. The center of all developing fields is human. The most important area where technology is used is the health sector. In particular, the use of software and hardware technological products in the detection of diseases is increasing day by day [1]. In the last few years, we see that cancer diseases have been increasing in the world. Early diagnosis of the disease is seen as important in these diseases. However, it becomes difficult to detect with imaging techniques in the first stage of the disease. All diagnostic and imaging systems become more effective with personal experiences depending on the infrastructure of the technology. However, advanced software systems can provide infrastructures to assist personal experiences.

In 2015, there were 8.8 million cancer-related deaths worldwide, and 1 out of every 6 deaths was caused by cancer [2,3]. In this study, we deepened our research on liver cancer, i.e. hepatocellular carcinoma. Liver cancer is the most common basic harmful tumor of the liver, caused by hepatocytes. The incidence is between ‰0.2 and ‰2. It is the fifth most common type of cancer. It ranks third in cancer-related deaths. Worldwide, it causes 250,000 to 1 million deaths per year [4]. Early diagnosis of a human life-threatening this disease will be

© The Author(s), under exclusive license to Springer Nature Switzerland AG 2021
J. Hemanth et al. (Eds.): ICAIAME 2020, LNDECT 76, pp. 1–9, 2021.
https://doi.org/10.1007/978-3-030-79357-9_1

important for treatment. For this reason, experimental studies were conducted using machine learning algorithms according to the measured values and characteristics of individuals for the diagnosis of liver cancer.

Data-based statistical research has been of interest to diagnose diseases. However, the large size of the data set and its heterogeneous structure is seen as a difficult problem in terms of predicting disease outcomes. Algorithms used in data mining may not produce correct results in terms of heterogeneous structure of the data. [5–7]. Therefore, evaluations can be made with training algorithms in order to examine the data sets in which they are in heterogeneous groups. In this context, one of the basic algorithms used for estimation is the Artificial Neural Network Model.

In this study, a prediction model was created for early diagnosis of disease diagnoses from blood samples taken from us liver cancer patients. A software tool has been produced in this proposed model. This software tool enables experts to generate a prediction by generating a prediction model to obtain a diagnosis of disease from patient blood results. In the developed study, Hepatocellular Carcinoma Dataset, an open data set obtained from 165 patients, was used [2].

## 2   Releated Works

In 2017, Han conducted a study on automatic classification of liver lesions using the deep convoluted neural networks method. In his work, the Convulation Neural Network (CNN) algorithm was used for image classification algorithms. He used a data set containing 130 images for model training and achieved a success rate of 70% [8]. Chaudhary and his team conducted a study that predicted liver cancer based on deep learning. 360 HCC data was used in the study. The Cancer Genome Atlas (TCGA) data was used in two phases. The first step was to obtain survival risk classes tags using the entire TCGA dataset. In another step, an SVM model was trained by dividing samples by 60% and 40% of training and test data. Five additional validation datasets were used to evaluate the prediction accuracy of the DL-based prognosis model. The developed model achieved 90%, 93% and 83% success in different tests [9]. Li and his team conducted a study on automatic classification using the deep neural network from CT images.sa They developed the model they designed using CNN algorithms. They compared data from 30 tomography images with CNN, Support Vector Machine and Random Forest algorithms. Success rates of 84%, 82% and 80% were achieved, respectively [10]. Hashem and Mabrouk conducted research on the diagnosis of liver disease using support vector machine algorithms. Diabetes diagnosis, Bupa liver disorders and Ilpd Indian liver dataset were used as data sets. Overall, a success rate of 88% was achieved [11]. Li and his team presented a machine learning approach to determine the size of liver tumors in CT images. Two contributions were identified in this study. First, adaboost learning was introduced to classify true boundary positions and false ones in one-dimensional density profiles. Secondly, the problem of how to find the two-dimensional tumor boundary was successfully solved by trained classifiers based on one-dimensional profiles [12].

# 3   Metodology

## 3.1   Artificial Neural Network

In order to transfer the capabilities of the human brain to other systems, the neurophysiological structure of the brain has been studied and mathematical models have been tried to be created. As a result of the studies, various artificial cell and network models have been developed. ANN is a parallel distributed processor capable of storing and generalizing information after the learning process [13]. ANNs are parallel and distributed information processing structures inspired by the human brain, and consist of processing elements, each with its own memory and connected to each other mainly through connections. In other words, ANNs are computer programs that mimic biological neural networks [14]. The structure of an artificial nerve cell is shown in Fig. 1.

Like biological neural networks, ANNs are made up of artificial nerve cells. Artificial nerve cells are also known as process elements in engineering science. As shown in Fig. 2, each process element has 5 basic elements. These are inputs, weights, collection function, activation function and outputs [15].

As a measure of how well the ANN provides the desired input-output properties, an error function is defined that is the sum of the squares of the error values of each neuron in the output layer of the ANN. The value of the neuron in the output layer of the ANN, its difference with the desired value from that neuron will show the error of that neuron. Error values are obtained by the following equations.

$$e_i = d_i - y_i(k) \tag{1}$$

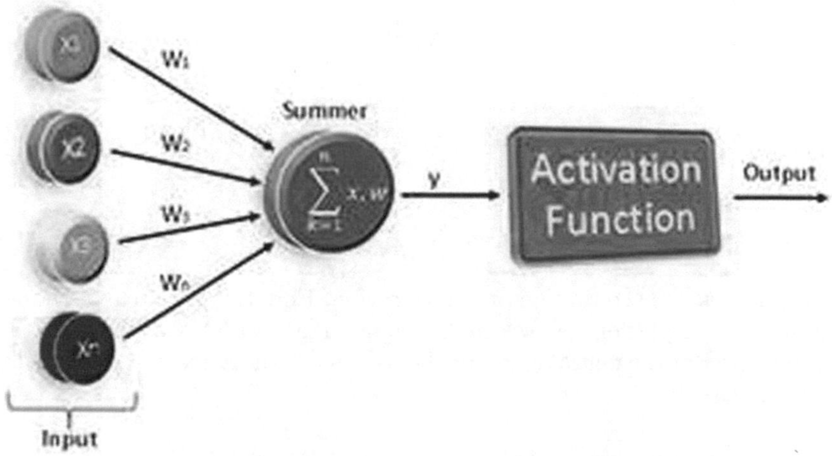

**Fig. 1.** Structure of artificial neural cell

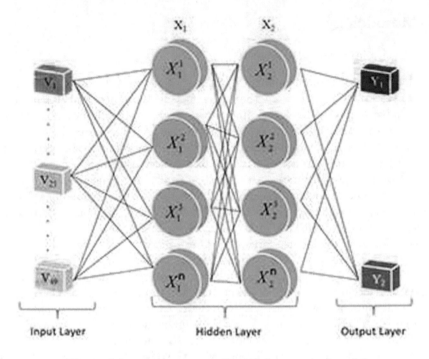

**Fig. 2.** Schematic representation of the proposed model

$$E = \frac{1}{2}(e_i^2 k) = \frac{1}{2}(d_i - y_i(k))^2 \tag{2}$$

$$\Delta\omega_{ij} = \eta\delta_j\gamma_i \tag{3}$$

$$\delta_j = e_j(k)f_j \tag{4}$$

$$\delta_j = f_j\sum\delta_m\omega_{mj} \tag{5}$$

Here, Eq. 1, i. is the error value of the iteration. Equation 2 indicates the adaptation function. Equation 3 shows each weight change in ANN. Equation 4 shows the output layer of the neural network. Equation 5 is the activation function used in hidden layers.

## 3.2   Training-Based Sampling

In this study, experimental studies were conducted to predict liver cancer disease using ANN according to the measurement values obtained from individuals.

**Table 1.** Arguments used as input data in ANN

| Parameters | Value |
| --- | --- |
| Gender | 1 = Male 0 = Female |
| Symptoms | 1 = Yes 0 = No |
| Alcohol | 1 = Yes 0 = No |
| Hepatitis B surface antigen | 1 = Yes 0 = No |
| Hepatitis B e antigen | 1 = Yes 0 = No |
| Hepatitis B core antibody | 1 = Yes 0 = No |
| Hepatitis C virus antibody | 1 = Yes 0 = No |
| Cirrhosis | 1 = Yes 0 = No |
| Endemic countries | 1 = Yes 0 = No |
| Smoking | 1 =Yes 0 = No |
| Diabetes | 1 =Yes; 0 = No |
| Obesity | 1 =Yes; 0 = No |
| Hemochromatosis | 1 = Yes; 0 = No |
| Arterial hypertension | 1= Yes; 0 = No |
| Chronic renal insufficiency | 1 = Yes; 0 = No |
| Human immunodeficiency Virus | 1 = Yes; 0 = No |
| Nonalcoholic steatohepatitis | 1 = Yes; 0 =No |
| Esophageal varices | 1 = Yes; 0 = No |
| Splenomegaly | 1 = Yes; 0 = No |
| Portal hypertension | 1= Yes; 0 = No |
| Portal vein thrombosis | 1 = Yes; 0 = No |
| Liver metastasis | 1 = Yes; 0= No |
| Radiological hallmark | 1 = Yes; 0 = No |
| Age at diagnosis | 0–100 |
| Grams of alcohol per day | 0–500 |
| Packs of cigarets per year | 0–510 |
| Performance status | 0;1;2;3;4;5 |
| Encephalopathy degree | 0–3 |
| Ascites degree | 0–3 |
| International normalised Ratio | 0.84-4.82 |
| Alpha-Fetoprotein (ng/mL) | 1.2–1810346 |
| Haemoglobin (g/dL) | 5–18.7 |
| Mean corpuscular volume (fl) | 69.5–119.6 |
| Leukocytes (G/L) | 2.2–13000 |
| Platelets | 1.71–459000 |
| Albumin (mg/dL) | 1.9-4.9 |
| Total Bilirubin (mg/dL) | 0.3–40.5 |
| Alanine transaminase (U/L) | 11–420 |
| Aspartate transaminase (U/L) | 17–553 |
| Gamma glutamyl transferase (U/L) | 23–1575 |
| Alkaline phosphatase (U/L) | 1.28–980 |
| Total Proteins (g/dL) | 3.9–102 |
| Creatinine (mg/dL) | 0.2–7.6 |
| Number of Nodules | 0–5 |
| Major dimension of nodule (cm) | 1–200 |
| Direct Bilirubin (mg/dL) | 0.1–29.3 |
| Iron | 0–244 |
| Oxygen Saturation | 0–126 |
| Ferritin (ng/mL) | 0–2230 |

Hepatocellular Carcinoma data set was used in the study. In the dataset, 49 arguments are defined as input data. The arguments and data types defined in the software are detailed in Table 1.

In this study, The HCC data set, which is available as open access over Kaagle, was used. The data set consists of 50 columns and 165 rows. There are 49 independent arguments in the dataset. The suitability of the data in the dataset was examined first and the appropriate independent variables were determined. The missing data in the data set was then examined and the missing data was completed with appropriate algorithms. The data types in the dataset are optimized for accurate training of data and accurate implementation of algorithms.

After the preparation of the training, the data is divided into training data and test data. Of all data, 78% was determined as training data and 22% as test data. After the split, we developed the first layer of the drape type. The number of neurons in the hidden layer was obtained as a result of experimental studies. While determining the initial value of the weights between layers, distribution was made according to a certain algorithm. Relu and Sigmoid activation functions were used because the process of our model was a prediction experiment. Since our model has a multicellular input, the Relu activation function is used first. This function allows some neurons to be active and an efficient computational load. The Sigmoid activation function is used, which converts values between 0 and $+\infty$ to values from 0 to 1. After defining the amount of data and error state parameters that the model will receive at once, the model is executed. Data were repeatedly trained to achieve more successful results. With each repeated training, the accuracy rate increased and the loss value decreased. The most optimized results were obtained in 100 iterations in different numbers of repeating models. Constant values of weights between layers have been reached since the training. The output of the system is the predictive result of whether the person suffers from liver cancer according to the values subjected to the algorithms.

## 4   Experimental Results

ANN results are shown with graphs of change in loss value and change in accuracy value according to number of iterations. As seen in Fig. 3, the loss value starting with 0.7922 was reduced to 0.4940 at the end of the training. As shown in Fig. 4, the accuracy value starting with 0.4902 increased to 0.8139 when the training ended.

Each time a new training is started, the loss value is completed with fewer errors than the results of the previous training. As the process progresses, the loss value remains at a constant value. The optimal number of training was accepted as this point and the training was completed again.

Figure 4 shows the change in accuracy value according to the number of training. The accuracy value was completed with each new training with higher values than the previous training results. Accuracy remained constant as the process progressed. The optimal number of training was accepted as this point and the training was completed again.

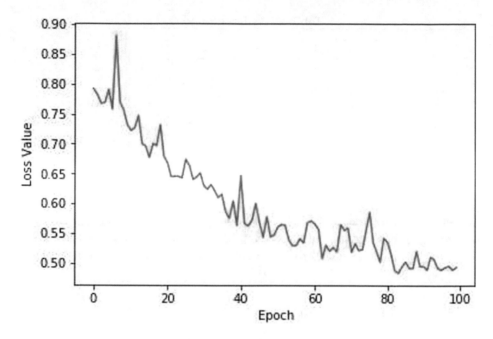

**Fig. 3.** The change of the lost value in repetitive education process.

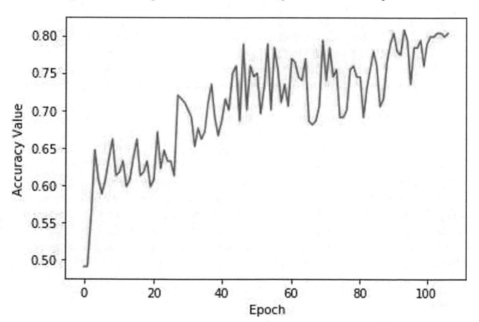

**Fig. 4.** The change of the lost value in repetitive education process.

**Table 2.** Confusion matrix

| Error matrix | Negative (0) | Positive (1) | Total |
| --- | --- | --- | --- |
| Negative (0) | 19 | 2 | 21 |
| Positive (1) | 1 | 22 | 23 |
| Total | 20 | 24 | 44 |

When the error matrix of the system is subtracted, the column section shows the actual values, and The Row section shows the estimated values. The table shows that the value of neg - neg is 19, the value of neg - pos is 2, the value of pos - neg is 1, and the value of pos - pos is 22. As a result of the system, the expected result is to determine whether an individual has liver cancer as a result of algorithms subjected to input values. 25% of the 165 data rows in the dataset are reserved as test data. Subjected to 100 learning cycles, the system achieved a successful prediction rate of 81% (Table 2).

## 5    Discussion and Conclusions

TIn this study, a basic artificial intelligence-based model is proposed for early diagnosis of liver cancer diseases. During the training process, both the loss values and the graphics of the accuracy values changed significantly. The main reason for this is that the number of input arguments of ANN is used too many times. After learning the model, a high prediction success of 81% was achieved according to the test results. Predicting these serious illnesses with high success based on various test results provides significant gains both temporally and financially.

n this study, experimental studies were conducted to predict liver cancer disease using ANN according to the measurement values obtained from individuals. Hepatocellular Carcinoma Data Sct data set was used in the study. In the dataset, 49 arguments are defined as input data. The expected result as a result of the system is to determine whether an individual has liver cancer as a result of algorithms subjected to input values. Of the 165 data lines in the dataset, 22% (44) are divided into test data and 78% (165) into training data. The estimated success rate of the developed model was 81%. In future studies, it is aimed to increase the amount of data in the data set, to observe the results by applying different algorithms and to increase the final success rate above 95%.

## References

1. Gürfidan R., Ersoy M., Yiğit T.: "Derin Öğrenme ile Kalp Hastalıkları Tahmini. 2. Uluslararası Türk Dünyası Mühendislik ve Fen Bilimleri Kongresi, Antalya (2019)
2. Santos, M.S., Abreu, H.P., Garcia-Laencina, P.J., Simao, A., Carvalho, A.: A new cluster-based oversampling method for improving survival prediction of hepatocellular carcinoma patients. J. Biomed. Inf. **58**, 49–59 (2015)

3. Alim, N.E.: Türkiye'de ve Dünyada Kanser Epidemiyolojisi. T.C, Halk Sağlığı Genel Müdürlüğü (2017)
4. Taş A.: Hepatosellüler Karsinom Tanı ve Tedavisi, Güncel Gastroenteroloji, **14**, 3145–3148, Ankara (2010)
5. Harrell, F., Lee, K., Mark, D.: Multivariable prognostic models: issues in developing models, evaluating assumptions and adequacy, and measuring and reducing errors. Stat. Med. **15**(4), 361–387 (1996)
6. Thongkam, J., Xu, G., Zhang, Y., Huang, F.: Toward breast cancer survivability prediction models through improving training space. Expert Syst. Appl. **36**(10), 12200–12209 (2009)
7. Andonie, R.: Extreme data mining: Interference from small datasets. Int. J. Comput. Commun. Control **5**(3), 280–291 (2010)
8. Han, X.: Automatic Liver Lesion Segmentation Using A Deep Convolutional Neural Network Method. Elektra Inc., St. Louis (2017)
9. Chaudhary, K., Poirion, O.B., Lu, L., Garmire, L.X.: Deep Learning Based Multi-Omics İntegration Robustly Predicts Survival İn Liver Cancer (2017). https://doi.org/10.1158/1078-0432.CCR-17-0853
10. Li, W., Jia, F., Hu, Q.: Automatic segmentation of liver tumor in CT images with deep convolutional neural networks. J. Comput. Commun. (2015). https://doi.org/10.4236/jcc.2015
11. Hashem, E.M., Mabrouk, M.S.: A Study of Support Vector Machine Algorithm for Liver Disease Diagnosis, Biomedical Engineering, p. 2015. Misr University for Science and Technology, MUST University (2015)
12. Li, Y., Hara, S., Sihimura, K.: A Machine Learning Approach for Locating Boundaries of Liver Tumors in CT Images. In: Proceedings of the 18th International Conference on Pattern Recognition (2006). 0-7695-2521-0/06
13. Ataseven, B.: Yapay Sinir Ağları ile Öngörü Modellemesi. Ocak. pp. 101-115 (2013)
14. Elmas, Ç.: Yapay Sinir Ağları (Kuram, Mimari, Eğitim, Uygulama). Seçkin Yayıncılık, Ankara (2003)
15. Öztemel, E. (2006). Yapay Sinir Ağları. 2. Baskı. Papatya Yayıncılı, İstanbul

# Remarks on the Limit-Circle Classification of Conformable Fractional Sturm-Liouville Operators

Bilender P. Allahverdiev[1], Hüseyin Tuna[2], and Yüksel Yalçinkaya[1(✉)]

[1] Department of Mathematics, Süleyman Demirel University, 32260 Isparta, Turkey
bilenderpasaoglu@sdu.edu.tr
[2] Department of Mathematics, Mehmet Akif Ersoy University, 15030 Burdur, Turkey

**Abstract.** In this work, we establish a criteria under which the conformable fractional singular Sturm-Liouville equation is of limit-circle case.

**Keywords:** Limit-circle case · Singular point · Conformable fractional Sturm-Liouville operator

## 1  Introduction

In a boundary value problem, if the differential expression and the coefficients are finite at each point in the defined range, the problem is regular; the problem is named as singular problem if the range is unlimited and at least one of the coefficients increases at least at one point of the range. In regular differential expressions, the boundary conditions are given directly by the values at that point, where as in the singular case these boundary conditions can not be given easily [1–3]. Therefore, it is difficult to solve singular problems. Another difficulty that arises in the analysis of singular problems is in which space element will be the solution of the problem. Weyl's analysis showed that a solution of the differential equation is necessarily second-order integrable. In the case that all solutions can be integrated quadratically, in the case of a borderline with differential expression; Otherwise this is called the limit point case. In the literature, there are many sufficient conditions for the differential equation to be in the limit-circle or limit-point case [2–8, 10–12].

The classification of the limit-circle and limit-point cases was first developed by Herman Weyl in the early 1900s [7]. Since then, such methods have become increasingly important thanks to their precise estimates of their potential form in solving a wide variety of singular Sturm-Liouville problems.

In a boundary value problem, if the range in which the differential expression is given is finite at each point in the range where the coefficients are defined, it is regular to the problem; If at least one of the coefficients is not summable in the range, or if the range is unlimited, the problem is called a singular problem. Although,

J. Hemanth et al. (Eds.): ICAIAME 2020, LNDECT 76, pp. 10–14, 2021.
https://doi.org/10.1007/978-3-030-79357-9_2

in the regular differential expressions, boundary conditions are given directly with the values at that point, these boundary conditions cannot be given easily in singular case. Therefore, it is difficult to solve singular problems. Another difficulty that arises in the analysis of singular problems is that which space element of the solution of the problem will be.

Initially, fractional-order differential equations are a subject of pure mathematics. They have found many applications today, thanks to rapid advances in computer technology and studies [13]. Examples of areas where fractional order differential equations are used, viscoelasticity, viscoelastic systems; diffusion, heat and moisture transfer phenomenon; economic and biological systems can be given.

Interest in fractional differential equations has been increasing in recent years, and the recently introduced definition of the appropriate fractional derivative includes a limit rather than an integral. Khalil et al. have redefined the definition of conformable fractional derivative using the classical derivative definition. In their work, they presented linearity condition, product rule, division rule and Rolle theorem for conformable fractional derivative [14]. Later in [15], Abdeljawad gave the definition of left and right conformable fractional derivatives, the definition of higher order fractional integral, fractional Grönwall inequality, chain rule and partial integration formulas for conformable fractional derivatives, fractional power series expansion and Laplace transformation. Conformable fractional derivative, aims to broaden the definition of classical derivative carrying the natural features of the classical derivative. In addition, with the help of the conformable differential equations obtained by the definition of derivative aims at a new look for differential equation theory [16]. In [17], the researchers investigated the Sturm-Liouville problem in the frame of the conformable fractional derivatives. In [18], Zheng et al. obtained two limit point condition criteria within the framework of compatible fractional derivatives, taking into account the Sturm-Liouville operator with a fraction of $2\alpha$ – degrees, and an example is presented.

In this article, we shall give a criterion that the conformable fractional Sturm-Liouville equation is in the limit-circle case. This construction follows [8,9].

## 2    Conformable Fractional Calculus

In this section, we provide some preliminaries for proving the main results.

**Definition 1** (see [15]). Given a function $x : (0, \infty) \longrightarrow (-\infty, \infty)$. Then, the conformable fractional derivative of order $\alpha$ of $x$ is defined by

$$T_\alpha x(t) = \lim_{\varepsilon \to 0} \frac{x\left(t + \varepsilon t^{1-\alpha}\right) - x(t)}{\varepsilon},  \tag{1}$$

where $t > 0$ and $0 < \alpha < 1$. The fractional derivative at zero, $(T_\alpha x)(0)$ is defined by

$$(T_\alpha x)(0) = \lim_{t \to 0^+} T_\alpha x(t).$$

**Definition 2** ([15]). The conformable fractional integral of a function $x$ of order $\alpha$ is defined by

$$(I_\alpha x)(t) = \int_0^t s^{\alpha-1} x(s)ds = \int_0^t x(s)d_\alpha s.$$

**Lemma 1** ([15]). *Let $x \in C(0,b)$. Then, we see that*

$$T_\alpha I_\alpha x(t) = x(t),$$

*for all $t > 0$.*

Let $L_\alpha^2(\mathcal{J})$ be a Hilbert space consisting of all complex-valued functions $x$ such that

$$\|x\| := \sqrt{\int_0^b |x(t)|^2 \, d_\alpha t} < \infty,$$

with the inner product

$$\langle x, y \rangle := \int_0^b \overline{y(t)} x(t) \, d_\alpha t,$$

where $0 < b \leq \infty$ and $\mathcal{J} = [0, b)$.

## 3    Main Result

Consider the following problem

$$\varrho[y] = -T_\alpha(p(t)T_\alpha)y(t) + q(t)y(t) = \lambda y, \text{ on } \mathcal{J}. \tag{2}$$

The coefficients $p(.)$ and $q(.)$ are real-valued functions on $\mathcal{J}$ and satisfy the conditions $\frac{1}{p(.)}, q(.) \in L_{\alpha,loc}^1(\mathcal{J})$. The following assumption will be needed throughout the paper: the endpoint $b$ is singular.

We will denote by $AC_{\alpha,loc}(\mathcal{J})$ the class of complex-valued functions which are absolutely continuous on all compact sub-intervals of $\mathcal{J}$.

Let

$$\mathcal{D} := \left\{ x \in L_\alpha^2(\mathcal{J}) : x, T_\alpha x \in AC_{\alpha,loc}(\mathcal{J}), \ \varrho[x] \in L_\alpha^2(\mathcal{J}) \right\}.$$

Then, for $x_1, x_2 \in \mathcal{D}$, we have

$$\int_0^b \varrho[x_1](t)\overline{x_2(t)}d_\alpha t - \int_0^b x_1(t)\overline{\varrho[x_2](t)}d_\alpha t = [x_1, x_2](b) - [x_1, x_2](0), \tag{3}$$

where $[x_1, x_2](t) = p(t)\left\{ x_1(t)\overline{T_\alpha x_2(t)} - T_\alpha x_1(t)\overline{x_2(t)} \right\}$ and $t \in \mathcal{J}$.

*Remark 1.* From (3), we see that $\lim_{t \to b^-} [x_1, x_2](t)$ exists and is finite for all $x_1, x_2 \in \mathcal{D}$. Furthermore, we know that $\varrho[.]$ is limit-point at $b$ if and only if $\lim_{t \to b^-} [x_1, x_2](t) = 0$ for all $x_1, x_2 \in \mathcal{D}$ (see ([8])).

**Theorem 1.** *We will make the following assumptions:*

*(i)*

$$T_\alpha p,\ T_\alpha q \in AC_{\alpha,loc}\,(\mathcal{J}),\ q \in C_\alpha\,(\mathcal{J})\,and\ T_\alpha^2 p, T_\alpha^2 q \in L_{\alpha,loc}^2\,(\mathcal{J}), \qquad (4)$$

*(ii)*

$$q(t) < 0\ \ and\ p(t) > 0\ for\ all\ t \in \mathcal{J}, \qquad (5)$$

*(iii)*

$$(-pq)^{-\frac{1}{2}} \in L_\alpha^2(\mathcal{J}), \qquad (6)$$

*(iv)*

$$T_\alpha\{pT_\alpha(pq)(-pq)^{-\frac{1}{2}}\} \in L_\alpha^2(\mathcal{J}), \qquad (7)$$

*Then, the conformable fractional Sturm-Liouville equation (2) is in the limit-circle case.*

*Proof.* We show that

$$\lim_{t \to b^-} [y, z](t) \neq 0, \qquad (8)$$

for one pair $y, z$ of elements of $\mathcal{D}$.

Now, we take $y = z$ and determine $y$ by

$$y(t) = \{-p(t)q(t)\}^{-\frac{1}{2}} \exp\left[i \int_0^t \left\{-\frac{q(s)}{p(s)}\right\}^{\frac{1}{2}} d_\alpha s\right], \qquad (9)$$

where $t \in \mathcal{J}$. A calculation shows that

$$T_\alpha y = \left[\frac{i(-q)^{\frac{1}{2}}}{p^{\frac{1}{2}}} + \frac{1}{4}\frac{T_\alpha(pq)}{(-pq)^{\frac{1}{2}}}\right]\exp[...]$$

and

$$T_\alpha^2 y = \left[\begin{array}{c}-\frac{(-q)^{\frac{1}{2}}}{p^{\frac{1}{2}}} + \frac{i}{4}\frac{T_\alpha(pq)}{p^{\frac{1}{2}}(-q)^{\frac{1}{2}}} - \frac{i}{4}\frac{T_\alpha p}{(-pq)^{\frac{1}{2}}} - \frac{3i}{4}\frac{(-q)^{\frac{1}{2}}T_\alpha p}{p^{\frac{1}{2}}} \\ +\frac{1}{4}\frac{T_\alpha^2(qp)}{(-pq)^{\frac{1}{2}}} + \frac{5}{16}\frac{\{T_\alpha(pq)\}^2}{(-pq)^{\frac{3}{2}}}\end{array}\right]\exp[...].$$

Thus we get

$$[y, y](t) = -2i, \qquad (10)$$

where $t \in \mathcal{J}$ and, with details of the calculation omitted,

$$\varrho[y] = -pT_\alpha^2 y - T_\alpha p y + q y = -\frac{1}{4}T_\alpha\{pT_\alpha(pq)(-pq)^{-\frac{1}{2}}\}\exp[...] \qquad (11)$$

It follows from (9), (5) and (6) that $y \in L_\alpha^2(\mathcal{J})$. From (9) and (4), we obtain $T_\alpha f \in AC_{\alpha,loc}(\mathcal{J})$. It follows from (11) and (7) that $\varrho[y] \in L_\alpha^2(\mathcal{J})$. Thus we have $y \in \mathcal{D}$. From (8) and (10), we conclude that the conformable fractional Sturm-Liouville equation (2) is in the limit-circle case at $b$. $\square$

# References

1. Naimark, M.A.: Linear Differential Operators II. Ungar, New York (1968)
2. Krein, M.G.: On the indeterminate case of the Sturm Liouvillemboundary problem in the interval (0,∞), Izv. Akad. Nauk SSSR Ser. Mat. **16**(4), 293–324 (1952). (in Russian)
3. Fulton, C.T.: Parametrization of Titchmarsh's $m(\lambda)$-functions in the limit circle case. Trans. Am. Math. Soc. **229**, 51–63 (1977)
4. Bairamov, E., Krall, A.M.: Dissipative operators generated by the Sturm-Liouville differential expression in the Weyl limit circle case. J. Math. Anal. Appl. **254**, 178–190 (2001)
5. Allahverdiev, B.P.: Nonselfadjoint Sturm-Liouville operators in limit-circle case. Taiwan. J. Math. **16**(6), 2035–2052 (2012)
6. Coddington, E.A., Levinson, N.: Theory of Ordinary Differential Equations. McGraw-Hill, New York (1955)
7. Weyl, H.: Über gewöhnliche Differentialgleichungen mit Singularitäten und die zugehörigen Entwicklungen willkürlicher Functionen. Math. Ann. **68**, 222–269 (1910)
8. Everitt, W.N.: On the limit-circle classification of second order differential expressions. Quart. J. Math. **23**(2), 193–196 (1972)
9. Everitt, W.N.: On the limit-point classification of second-order differential expressions. J. London Math. Soc. **41**, 531–4 (1966)
10. Dunford, N., Schwartz, J.T.: Linear Operators, Part II: Spectral Theory. Interscience, New York (1963)
11. Baleanu, D., Jarad, F., Ugurlu, E.: Singular conformable sequential differential equations with distributional potentials. Quaest. Math. **42**(3), 277–287 (2019)
12. Huaqing, S., Yuming, S.: Limit-point and limit-circle criteria for singular second-order linear difference equations with complex coefficients. Comput. Math. Appl. **52**(3–4), 539–554 (2006)
13. Podlubny, I.: Fractional Differential Equations. Academic Press, New York (1999)
14. Khalil, R., Al Horani, M., Yousef, A., Sababheh, M.: A new definition offractional derivative. J. Comput. Appl. Math. **264**, 65–70 (2014)
15. Abdeljawad, T.: On conformable fractional calculus. J. Comput. Appl. Math. **279**, 57–66 (2015)
16. Khalil, R., Abu-Shaab, H.: Solution of some conformable fractional differential equations. Int. J. Pure Appl. Math. **103**(4), 667–673 (2015)
17. Allahverdiev, B.P., Tuna, H., Yalçınkaya, Y.: Conformable fractional Sturm-Liouville equation. Math. Methods Appl. Sci. **42**(10), 3508–3526 (2019)
18. Zhaowen, Z., Huixin, L., Jinming, C., Yanwei, Z.: Criteria of limit-point case for conformable fractional Sturm-Liouville operators. Math. Methods Appl. Sci. **43**(5), 2548–2557 (2020)

# Improving Search Relevance with Word Embedding Based Clusters

Işılay Tuncer[✉], Kemal Can Kara, and Aşkın Karakaş

Kariyer.net Inc., Istanbul, Turkey
{isilay.tuncer,can.kara,askin.karakas}@kariyer.net

**Abstract.** The main purpose of web search engines is to provide the user with most relevant results based on the searched keywords. Finding content that meets the expectations of the user and the relevance of the content to these needs are very important for the success of the search engine. In this study; with the keywords written to the search engine, it is aimed to reach not only the results which contains these terms, but also the semantically related results. Words and phrases in all documents to be searched will be vectorized with Word2Vec model, then phrases will be clustered based on their similarity values. Finally, these outputs will be integrated into a Lucene based NoSQL solution at index time. The study will be used for Kariyer.net's job search engine. This study includes research and applications in word embeddings, machine learning, unsupervised learning.

**Keywords:** Word embeddings · Unsupervised learning · Elasticsearch

## 1 Introduction

The internet is the most important communication tool that we can achieve the results we want to reach by using search engines. Search engines track links to websites and pages in the database and provide a list of results that best match what the user is trying to find. Search engines have been diversified to date, starting with Archie, the first search engine developed by Alan Emtage in 1990. Search engines can be found inside websites or in devices such as operating systems.

The purpose of the web search engines is to lead the user to the results that have the best relationship with the search. In the Kariyer.net search engine, thousands of job text and skill searches are made for a day and it is an important mission for Kariyer.net that jobseekers reach the right job texts. If jobseekers reach only the results containing the keyword they are looking for in the search engine, they are deprived of some similar work related to their search. The purpose of the study is to create a step to eliminate this shortcoming and ensure that jobseekers reach the most suitable job texts for them.

In the study, Word2Vec and Fasttext, unsupervised learning and elasticsearch are used. The first step of the study is the process of clearing the texts to train Word2vec and Fasttext models. The texts used consist of job texts kept in Kariyer.net database. The cleaned texts are divided into words and phrases and converted into vector values with the help of Word2vec and Fasttext. In order to compare the accuracy of the vectors,

© The Author(s), under exclusive license to Springer Nature Switzerland AG 2021
J. Hemanth et al. (Eds.): ICAIAME 2020, LNDECT 76, pp. 15–24, 2021.
https://doi.org/10.1007/978-3-030-79357-9_3

Word2vec and Fasttext models trained with Wikipedia texts are used. Then, vectors are clustered with three different clustering algorithms, K-means, Spectral and Hdbscan.

Sets of word vectors and trained word embedding models are used in the query created by Elasticsearch. With the query, the job texts in the search engine results are accessed with priority in the searched word. Next, it shows job texts that include words from clusters and models and words that can be closely related.

## 2    Similar Studies

The role of search engines is important for the usefulness of the website. For this reason, many websites that use their own search engine need to improve the current system.

As a study similar to our study, Dice.com, a job search site in the field of information technologies, uses word vectors to conduct semantic search in its 2015 study [1]. It sets these vectors and integrates them into search queries with solr during indexing.

Woo-ju and Dong-he, who are working on document searching, say that keyword-based queries that use the vector space model, such as tf-idf, can create some problems in the search process and cannot distinguish between lexical and semantic differences. In their work, they combine central vectors with the Word2vec method to capture semantic similarity, instead of tf-idf vectors to represent query documents [2].

In order to provide more useful information to the user, KeyConcept was developed in 2004, a system containing information about the conceptual framework and keywords of the queries. KeyConcept is a search engine that receives documents based on a combination of keywords and conceptually close words [3]. During indexing, KeyConcept results confirm that conceptual matching significantly improves the accuracy of keyword search results [4].

Stokoe, Oakes and Tait say that the semantic ambiguity of words has a negative effect on web search systems in general. In their study, instead of the commonly used TF-IDF technique, research is conducted on a statistical word sense uncertainty system. The study analyzes the content of the text and presents the research of statistical approaches instead of the sense based vector space retrieval model [5].

Stating that the user has reached thousands of documents related to the subject with the increase of dynamic web search in recent years, Roul and Sahay argue that most of these documents are not related to the user's search. They are working on a more useful and efficient mechanism of clustering that can group similar documents in one place and sort the most relevant documents at the top. They propose a Tf-Idf based mechanism called Apriori for clustering web documents. Documents in each set are listed using the similarity factor of documents based on tf-idf and user query [6].

## 3    System Details

The data used in the study consist of job texts kept in Kariyer.net database. In order to educate Word2Vec and Fasttext models, the job texts published since January 2017 are based on. A total of 721,286 job texts are made. Apart from that, Word2vec and Fasttext models trained with Wikipedia data are tried to compare the accuracy of word vectors.

The study proceeds in 4 separate steps. The first step consists of cleaning and processing the data. Word2vec and fasttext word embedding algorithms are trained with the cleaned data. The word vectors are clustered with three clustering algorithms. The cluster and word embedding model that gives the best results is used in the query created with Elasticsearch.

## 3.1  Processing of Data

- First of all, in order for the models to produce correct results, all job texts are cleaned from HTML codes, junk words and special characters. Using the Python BeautifulSoup library, HTML and XML residues are removed from the job texts. Special characters such as #, '', / are also cleared while processing, but there are also words such as ASP.Net, C++, OS/390 that contain special characters in the text. There are 2 ways to avoid losing the characters in these words. Search is made among the most frequently used skills and the most searched words in Kariyer.net and if the word is included in this list, the special characters in the word is not deleted or changed.
- The n-gram method is used to identify common words in the text. N-gram is the method used to search for data, make comparisons and learn the number of repetitions of the searched expression. For each sentence in all data, unigram, bigram and trigrams are used to extract the words and phrases that will be used in the Word2Vec model.

As a result of the transaction, the word list obtained from all job text data is shown in Fig. 1.

```
['istatistiksel_analiz', 'inceleme', 'raporlama_becerisi', 'olan'],
['iyi_derecede', 'sözlü_yazılı', 'ingilizce', 'bilgiye_sahip'],
['sürekli_öğrenme', 'yetenek', 'isteği_olan'],
['veri_bilimci', 'çalışma_arkadaşı_arayışımız_bulunmaktadır'],
```

**Fig. 1.**  Word list created with n-gram

The words we want to cluster are the 1500 most frequently searched words in Kariyer.net search engine and 7112 words kept in the skill table in Kariyer.net database. A total of 8528 words and phrases are obtained by combining these two groups and cleaning the repeated words.

## 3.2  Training of Word Embedding Models

Word2Vec and Fasttext models are trained with n-grams obtained from cleaned text. Word2Vec and Fasttext models trained with Wikipedia texts are used to compare the accuracy of the models.

Word2Vec is a word embedding model developed by Tomas Mikolov and his team in 2013. Word2vec that used in this study, use the Python library gensim, whose purpose is natural language processing [7].

Fasttext is a word embedding model developed by Facebook AI Research in 2016. Like Word2Vec, Fasttext use the gensim library. It is an open source library based on N-gram attributes and size reduction, allowing the user to learn text representations and text classifiers [8, 9]. It gives the words in n-grams instead of giving them to the model individually.

After model training, the vector affinities of the "machine learning" word group are shown in the Fig. 2.

[('artificial_intelligence', 0.7436475157737732),

('data_science', 0.7397566437721252),

('natural_language_processing', 0.7383211851119995),

('deep_learning', 0.7260794639587402),

('big_data', 0.7243600487709045),

('ai', 0.7243313193321228),

('computer_vision', 0.7137678861618042),

('data_mining', 0.7135145664215088),

('machine_learning_deep_learning', 0.7120709419250488),

('artificial_intelligence', 0.7083033323287964)]

**Fig. 2.** Vector affinities of the "machine learning"

Each of the 8528 words we want to cluster is not included in models trained with job texts and Wikipedia texts. These words are separated before proceeding with the clustering step.

Of the 8528 words obtained,

- 5188 are in the Word2vec model trained with job text data,
- 4279 are in the Word2Vec model trained with wikipedia data,
- 4346 are in the Fasttext model trained with job text data,
- 5094 are in the Fasttext model trained with wikipedia data.

### 3.3 Clustering

Clustering is the process of grouping the data in the data set according to the relationship between the other data in the set [10]. Three different clustering algorithms, K-means, Spectral and Hdbscan, are used for clustering word vectors.

K-means is one of the first methods used in text mining as it is one of the simple and popular unsupervised machine learning algorithms. K-means takes a k value that determines the number of clusters as parameters. Elbow method and Silhouette score method are used to find the optimum k value.

Optimum number of clusters achieved by models for K Means clustering:

- With Word2vec model trained with job text data: 480 clusters
- With Word2vec model trained with Wikipedia data: 400 clusters

- With Fasttext model trained with job text data: 320 clusters
- With Fasttext model trained with Wikipedia data: 400 clusters

Spectral clustering performs better than many traditional clustering algorithms such as K-means. Spectral clustering is a clustering algorithm that is simple to implement and can be solved efficiently with standard linear algebra methods [11]. Like k-means, it takes the k value to determine the number of clusters as parameters, and the number of clusters is determined by elbow and silhouette score methods.

Optimal number of clusters achieved by models for Spectral Clustering:

- With Word2vec model trained with job text data: 120 clusters
- With Word2vec model trained with Wikipedia data: 120 clusters
- With Fasttext model trained with job text data: 320 clusters
- With Fasttext model trained with Wikipedia data: 150 clusters

HDBSCAN is a clustering algorithm developed by Campello, Moulavi and Sander, meaning "Hierarchical Density-Based Spatial Clustering of Applications with Noise" [12]. HDBSCAN works to create the best clusters by excluding this noise and outliers, even if there are different dimensions, different densities and noise in the data set. It is an algorithm that calculates automatically instead of trying to find the most suitable epsilon value for clustering [13].

Optimal number of clusters achieved by models for Hdbscan:

- With Word2vec model trained with job text data: 2638 words from 5188 words, 13 sets
- With Word2vec model trained with Wikipedia data: 1203 words from 4279 words, 11 sets
- With Fasttext model trained with job text data: 3819 words from 5094 words, 3 sets
- With Fasttext model trained with Wikipedia data: 3 sets of 3286 words from 4346 words

### 3.4 Elasticsearch

Elasticsearch is an Apache Lucene-based open source search and analysis engine used in content search, which can be used for all types of data such as textual, numerical, geographic [14]. Elasticsearch takes data in JSON format. It automatically stores documents and adds them as searchable documents to the cluster directory. Later, documents can be obtained using Elasticsearch API.

The search query of Kariyer.net, which is currently in use, provides the results of the word entered in the search engine and if a query containing more than one word is entered, it divides this word group and presents the results of the words separately. For example; If "machine learning" is searched, it separates the phrases "machine" and "learning" and feeds the query in this way.

In creating the query in the study, the query is expanded by adding the 5 closest words from the selected set and word2vec model to the query, in addition to the keyword or phrase itself. Weighting is done so that the weight of the keyword itself has the most weight and the weight of the close words added is lower. For example, part of the query that occurs when searching for "machine learning" is shown below.

```
{
  "multi_match": {
    "type": "phrase",
    "query": "machine learning",
    "fields": [
      "title^50",
      "positionName^10",
      "companyName^5",
      "searchKeywords^2",
      "qualifications^2"
    ],
    "boost": 50
  }
},
{
  "multi_match": {
    "type": "phrase",
    "query": "artificial intelligence",
    "fields": [
      "title^50",
      "positionName^10",
      "companyName^5",
      "searchKeywords^2",
      "qualifications^2"
    ],
    "boost": 0.5
  }
},
{
  "multi_match": {
    "type": "phrase",
    "query": "data science",
    "fields": [
      "title^50",
      "positionName^10",
      "companyName^5",
      "searchKeywords^2",
      "qualifications^2"
    ],
    "boost": 0.5
  }
}
```

With the searches made with this sample query, not only the results in the searched word are found, but also the job text with close words are also included.

## 4 Conclusion

K-means and Spectral clustering tries to import each data in the dataset into a cluster. It does not perform very well because it tries to cluster outliers. However, since HDBSCAN tries to aggregate in the best way, it excludes outliers and only aggregates data that is really relevant to each other.

1. By clustering the vectors produced by Word2Vec model trained with the job texts with HDBSCAN, the best results are obtained (Fig. 3).

```
0:['sanayii', 'seramik', 'yapı_kimyasalları', 'maden', 'demir_çelik', 'çelik', 'kauçuk', 'petrol', 'enjeksiyon', 'demir','kompozit',
1:['hyundai', 'volkswagen', 'bmw', 'renault', 'mercedes', 'toyota', 'oem', 'otokar', 'ford', 'honda']
2:['lojistik', 'dijital_pazarlama', 'satın_alma', 'tasarım', 'pazarlama', 'medya', 'tedarik_zinciri', 'proje', 'teknik', 'üretim',
3:['viaport', 'konya', 'pendik', 'akasya', 'akhisar', 'memorial', 'çorum', 'erzincan', 'rocher', 'soma', 'bergama', 'tekirdağ',
çekmeköy', 'aksaray', 'çatalca', 'yalova', 'beykoz', 'silivri', 'ordu', 'levent', 'bandırma', 'şişli', 'marmarapark', 'arnavutköy',
4:['garson', 'aşçı_yardımcısı', 'barista', 'aşçı', 'resepsiyonist', 'bellboy', 'komi', 'mutfak_şefi']
5:['kimya_mühendisliği', 'çevre_mühendisliği', 'makine_mühendisliği', 'elektrik_elektronik_mühendisliği', 'endüstri_mühendisliği',
6:['biyoloji_öğretmeni', 'acil_tıp_teknisyeni', 'ebe', 'psikolojik_danışman', 'rehber_öğretmen', 'okul_müdürü', 'diş_hekimi_asistanı
7:['inşaat_mühendisi', 'topoğraf', 'elektrik_teknisyeni', 'kimyager', 'gıda_mühendisi', 'makina_mühendisi', 'inşaat_teknikeri',
8:['fransızca', 'tercüman', 'almanca_bilen', 'arapça_bilen', 'almanca', 'italyanca', 'rusça_bilen', 'rusca', 'fransizca', 'türkçe',
9:['continuum', 'markt', 'das', 'bis', 'mit', 'deutsch', 'im']
```

**Fig. 3.** Clusters obtained byHdbscan

With the hdbscan algorithm, 2638 words are aggregated from 5188 words and a total of 13 sets are obtained.

2. Within the sets obtained, one set has a fairly large number of elements. 2343 words of 2637 words that can be clustered with hdbscan belong to this set. For this reason, this cluster is re-clustered in itself. When the vectors of the words in this set, which are obtained with the fasttext model, are re-clustered in itself, 35 sets are obtained in which 437 words can be clustered (Fig. 4).

```
0:['quick_learner', 'unsupervised_learning', 'positive_attitude', 'unsupervised', 'work_ethic']
1:['spanish', 'russian', 'chinese', 'french', 'japanese']
2:['directx', 'direct', 'active_directory', 'directors', 'director', 'directory']
3:['system_testing', 'regression_testing', 'functional_testing', 'unit_testing', 'integration_testing', 'automated_testing']
4:['technical_support', 'copywriting', 'technical_writing', 'report_writing', 'writing']
5:['uploading', 'reading', 'sequencing', 'mixing', 'loading', 'mirroring', 'scoping']
6:['technology', 'technologies', 'emerging_technologies', 'biotechnology', 'mobile_technology']
7:['design_documentation', 'project_documentation', 'documents', 'document', 'documentation', 'instrumentation']
8:['image_processing', 'process_mapping', 'signal_processing', 'processing', 'data_processing', 'order_processing']
9:['accounts_payable', 'accounting', 'accounts_receivable', 'accounts', 'receivable', 'account_reconciliation', 'cost_accounting', 'payables']
```

**Fig. 4.** Clusters obtained by re-clustering with Hdbscan

3. Although the hdbscan algorithm produces better clusters than other algorithms, it leaves a lot of outlier. 2551 outliers were determined from 5188 words given to the cluster. The vectors of Outlier that are obtained with Fasttext model are again clustered with Hdbscan algorithm. As a result, 44 clusters are obtained in which 637 words can be clustered (Fig. 5).

```
0:['ilişkiler', 'iliskiler', 'ilişkileri', 'halkla_ilişkiler', 'uluslararası_ilişkiler']
1:['öğretmen', 'anaokulu', 'öğretmeni', 'koleji', 'doğa_koleji', 'kolej', 'branş_öğretmeni']
2:['havaalanı', 'havalimani', 'hava', 'yeni_havalimanı', 'havalimanı']
3:['turk_telekom', 'telco', 'türk_telekom', 'telekom', 'türkcell']
4:['mimar', 'mimarlık', 'iç_mimar', 'mimarı', 'peyzaj_mimarı']
5:['endüstriyel_tasarım', 'grafiker', 'fotoğrafçı', 'moda_tasarım', 'grafik_tasarımcı', 'grafik_tasarım']
6:['field_service', 'managed_services', 'service', 'services', 'professional_services', 'shared_services']
7:['electric', 'general_electric', 'electrician', 'schneider_electric', 'electronic', 'electronic_warfare']
8:['özel_güvenlik', 'hizmetleri', 'eğitim_danışmanı', 'danışma', 'yer_hizmetleri', 'hasta_danışmanı', 'satış_danışmanı', 'güvenlik', 'danışmanı',
9:['asistan', 'doktor_asistanı', 'asistanı', 'ofis_asistanı', 'asistani', 'yönetici_asistanı', 'yönetici_adayı', 'yönetici']
```

**Fig. 5.** Clusters obtained by re-clustering of outliers

The five words that have the closest meaning from the resulting cluster and word2vec model are also added to the Elasticsearch query with low weight. When searching for "machine learning" in the search engine, the result is also shown in the Fig. 6.

| | |
|---|---|
| 5G Machine Learning Engineer | 5G Machine Learning Engineer |
| Software Engineer - Machine Learning | Software Engineer - Machine Learning |
| Machine Learning Engineer | Machine Learning Engineer |
| NLP Research Engineer | Learning & Development Intern |
| Data Scientist | Smart Start Digital Machinery |
| Kıdemli Java Developer | Data Scientist |
| Data Scientist | NLP Research Engineer |
| Yapay Zeka Mühendisi | Yapay Zeka Mühendisi |
| Mobile Operational Account Manager - Fresh Grad | Kıdemli Java Developer |
| Yapay Zeka Mühendisi | Data Scientist |
| Data Scientist ( Aı, Ml) | Makine Operatörü - Dağıtım Trafoları |
| Yapay Zeka Yazılım Uzmanı | Makine Operatörü |
| Software Engineer, Search | Makine Teknikeri |
| Tübitak 2244 Proje Bursiyeri | Makine Teknolojisi Öğretmeni |
| Smart Start Digital Machinery | Makine Operatörü |
| Yazılım Uzmanı | Makine Operatörü (Manisa) |
| E-Ticaret Data Analisti | Makine Ressamı |
| Ar-Ge Mühendisi | Makine Teknisyeni |
| Veri Bilimci | Makine Operatörü |

**Fig. 6.** Results from both queries for "machine learning" search

In the Fig. 6, the right column shows the results from the query we want to improve, and the left column shows the query results we have improved. As can be seen from the results, the results such as artificial intelligence, data scientist, software engineer, data analyst are encountered more frequently after the job texts that the "machine learning" phrases passes through the query we developed. On the other side, results such as machine operator and machine technician from different sectors and professions are presented.

In order to measure the success of the search engine, 10 search keywords from different fields are chosen among the 1500 most frequently searched words in Kariyer.net search engine. Results from both queries are tested and labeled by the expert as relevant (1) or not relevant (0) to the search keyword. Precision values are calculated with labeled data.

Precision tells us the percentage of our relevant results and is calculated by the formula shown in the Fig. 7 below.

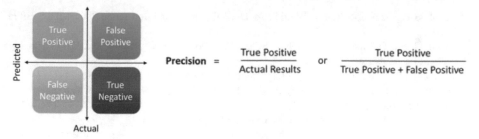

$$\text{Precision} = \frac{\text{True Positive}}{\text{Actual Results}} \quad \text{or} \quad \frac{\text{True Positive}}{\text{True Positive} + \text{False Positive}}$$

**Fig. 7.** Precision formula

The success of the search engine is calculated by averaging the precision values found using the formula in the Fig. 7. The results of this calculation are shown in the Table 1 for 10 different keywords belonging to different business fields.

**Table 1.** Results of average precision values for 10 different keywords

|  | Current query results | Improved query results |
|---|---|---|
| Bilgi İşlem | 0.975 | 1 |
| İnşaat Mühendisi | 0.925 | 0.975 |
| İç Mimar | 0.872 | 0.925 |
| Laboratuvar | 0.75 | 0.925 |
| Çevre Mühendisi | 0.75 | 0.818 |
| Makam Şoförü | 0.6 | 0.925 |
| Mechanical Engineer | 0.375 | 0.925 |
| Tekstil Mühendisi | 0.575 | 0.657 |
| Veri Bilimi | 0.225 | 0.666 |
| Türk Hava Yolları | 0.125 | 0.75 |

With close meaning words integrated into the query, the user is given much closer results to the results he wants to achieve. Thus, users achieve faster and more meaningful results.

# References

1. DiceTechJobs/ConceptualSearch (2015). https://github.com/DiceTechJobs/ConceptualSearch. Accessed 21 Jan 2020
2. Kim, W., Kim, D., Jang, H.: Semantic Extention Search for Documents Using the Word2vec (2016)
3. Gauch, S., Madrid, J.M., Induri, S., Ravindran, D., Chadalavada, S.: KeyConcept: A Conceptual Search Engine (2004)
4. Ravindran, D., Gauch, S.: Exploiting hierarchical relationships in conceptual search (2004)

5. Stokoe, C., Oakes, M., Tait, J.: Word Sense Disambiguation in Information Retrieval Revisited (2003)
6. Roul, R.K., Devanand, O.R., Sahay, S.K.: Web Document Clustering and Ranking using Tf-Idf based Apriori Approach (2014)
7. gensim 3.8.1 (2019). https://pypi.org/project/gensim. Accessed 11 Apr 2020
8. Joulin, A., Grave, E., Bojanowski, P., Mikolov, T.: Bag of Tricks for Efficient Text Classification (2016)
9. FastText (2020). https://fasttext.cc, Accessed 11 Apr 2020
10. Elankavi, R., Kalaiprasath, R., Udayakumar, R.: A fast clustering algorithm for high-dimensional data (2017)
11. Ulrike von Luxburg: A Tutorial on Spectral Clustering (2007)
12. Campello, R.J.G.B., Moulavi, D., Sander, J.: Density-based clustering based on hierarchical density estimates (2013)
13. İnsanların Kariyerlerindeki Bir Sonraki Pozisyonun Tahmin Edilmesi – 2 (2018). https://arge.kariyer.net/Makale/Ust-ve-Esdeger-Pozisyonlarin-Bulunmasi. Accessed 17 Apr 2020
14. What is Elasticsearch (2020). https://www.elastic.co/what-is/elasticsearch. Accessed 07 May 2020

# Predicting Suicide Risk in Turkey Using Machine Learning

Elif Şanlıalp[1], İbrahim Şanlıalp[2(✉)], and Tuncay Yiğit[3]

[1] Department of Computer Engineering, Ankara Yıldırım Beyazıt University, Ankara, Turkey
egul@ybu.edu.tr
[2] Department of Computer Engineering, Kırşehir Ahi Evran University, Kırşehir, Turkey
ibrahim.sanlialp@ahievran.edu.tr
[3] Department of Computer Engineering, Süleyman Demirel University, Isparta, Turkey
tuncayyigit@sdu.edu.tr

**Abstract.** Predicting suicide risk is a critical issue for the future of public health. Failure to accurately predict suicide risk limits solutions to this major public health problem. The aim of the presented study is to predict the risk of suicide by using the machine learning approach in Turkey. This study uses the Turkish Statistical Institute's public database for the prediction of suicide risk. The dataset consists of 30,811 patients committing suicide. Subject data includes all cities (81 cities) of Turkey and covers a 10-year period (2009–2018). Population information grouped by attributes in the data set is also taken from the Turkish Statistical Institute's public database (for all cities in Turkey). The structured patient's feature includes city, age-group, gender, and specific mortality rate. Multiple linear regression model is implemented and results indicate that age-group, gender, and city variables are promising success predictors of specific mortality rate in predicting future risk of suicide. (i.e., *MAE*: 0.0386959, *RMSE*: 0.0621640, $R^2$: 0.5648034). The findings are expected to help suicide prevention rehabilitation programs and to assist developers in Machine learning-based suicide risk assessment tools.

**Keywords:** Machine learning · Suicide risk · Linear regression · Prevention

## 1 Introduction

The World Health Organization (WHO) defines suicide as a serious global public health issue [1], as well as a deliberate action taken by the person concerned with the expectation of a fatal outcome [2]. In recent times, WHO divides the phenomenon of suicide into two groups: suicide action and suicide attempt. While true suicides involve events that result in death, suicide attempts involve all non-lethal voluntary attempts to commit suicide, which the individual attempts to harm, poison, or destroy himself [3].

Suicide is a phenomenon that has come from the past and there are economic, cultural, social, and psychological aspects of the suicide phenomenon seen in all societies throughout human history [4]. For this reason, suicide case is a subject to multidisciplinary approach.

© The Author(s), under exclusive license to Springer Nature Switzerland AG 2021
J. Hemanth et al. (Eds.): ICAIAME 2020, LNDECT 76, pp. 25–32, 2021.
https://doi.org/10.1007/978-3-030-79357-9_4

Suicide in the world is a major public health problem and there are about 25 million non-fatal suicide attempts each year [5]. In addition, suicide is responsible for over 800,000 deaths annually worldwide [10]. Studies in Turkey about suicide are substantially based on data received from the Turkish Statistical Institute (TURKSTAT) [4, 6–8]. According to a research in Turkey, between 2002 and 2014, approximately 67.4% of all suicides were male suicides, while the rate of female suicides was around 32.6% [7]. According to another study in Turkey, suicide has become a public mental health problem in Turkey and this can be explained by the increase in the number of deaths: in 1974, the suicide rate was 1.92% for males and 1.31% for females; in 2013 this had risen to 6.22% for males and 2.26% for females [9].

Researchers focused on empirical approaches to identify suicide risk [11, 12, 15]. In these empirical approaches, the plurality of regression-based models, which historically show slightly better accuracy than clinical judgments, comes to the fore [11]. The correct suicide risk prediction may require complex combinations of lots of risk factors. ML techniques can take into account complex factor combinations in number and type [12]. Traditional statistical techniques in clinical psychology remain poor for such combinations [12] and most research test isolated predictors are inaccurate [13, 14]. Therefore, machine learning (ML) techniques have many advantages in suicide risk prediction [12–15].

Predictive accuracy may vary across populations. For example, it can be relatively easy to accurately identify the risk in a general population sample [16]. Among many of the risk factors for suicide, such as gender, job, age, education, marital status, and history of suicide attempt, age, gender, and job were the most important risk factors for fatal suicide attempts [17]. Clinicians can also take a more comprehensive approach by evaluating additional demographic risk factors for suicide [18].

The purpose of this study is to predict the risk of suicide using age-group, gender, city, and specific mortality rate parameters by multiple linear regression models in Turkey. Exploring performance of the parameters of the ML approach for predicting suicide risk can provide a basis for future studies. In addition, the results are expected to assist those who develop suicide risk assessment and screening tools using ML for suicide prevention programs and to help suicide prevention rehabilitation programs. In the remainder of this paper, firstly section of Material and Method is presented. Then, Experimental Results section is presented. The Conclusions, Suggestions, and Future Work section is the last presented.

## 2 Material and Method

Sample data used in the research are taken from the website of TURKSTAT [19]. There is no need for a separate permission and ethical approval for the use of this data, which is published on the internet and available to everyone. Sample suicide dataset covers the number of patients (30,811) who commited suicide over city population according to age group, gender for the years 2009–2018. 30,811 patients who committed suicide consist of 22,650 (73.5%) men and 8161 (26.5%) women. Population information covering a 10-year period (2009–2018) grouped by attributes in the data set is also taken from TURKSTAT's public database (for all city in Turkey) [19]. Datasets of suicidal patients are arranged and a sample dataset is obtained. The variables of sample dataset is shown in Table 1.

**Table 1.** Variables of dataset

| Gender | City | Age-group | Rate |
|--------|------|-----------|------|
| Male | 81 Cities | −15 | Specific Mortality Rate (SMR) |
| Female | | 15–19 | |
| | | 20–24 | |
| | | 25–29 | |
| | | 30–34 | |
| | | 35–39 | |
| | | 40–44 | |
| | | 45–49 | |
| | | 50–54 | |
| | | 55–59 | |
| | | 60–64 | |
| | | 65–69 | |
| | | 70–74 | |
| | | 75+ | |

The specific mortality rate including suicide mortality is the output variable for multiple linear regression model and this variable values are obtained as:

$$SMR = \frac{No.\ of\ deaths\ of\ the\ relevant\ group\ in\ the\ ten\ year\ period\ \times\ 1000}{Population\ of\ the\ relevant\ group\ at\ the\ mid\ ten\ year\ period} \quad (1)$$

## 2.1  Data Representation

Data is modeled as a set of variables [21]. In the present study, output variable is specific mortality rate and input variables are age-group, city and gender. Also input variables are called predictors. The predictors are derived from the research literature on suicidality [7, 8, 11, 12, 15].

The sample data are transformed into forms appropriate for ML process. The data set is divided into two sub-groups: the training sample found the model and tested the resulting model of the test sample. Test and training samples are randomly generated among cases. The result obtained from the learning example (70% of cases) is then evaluated using the test sample (30% of cases).

Data is preprocessed using Python [22] with statistical analyses performed in Spider [23]. Relevant Python libraries are included the SciPy ecosystem, NumPy, Pandas, iPython [24, 25].

## 2.2  Normalization (z-Score)

Normalization is a pre-processing stage or scaling technique which is can find new ranges from an existing one range. It can be helpful for prediction or forecasting purposes a lot [26]. In this study z-score normalization is dealt with. For formula, let $A$ be a numeric attribute with $n$ observed values, $r_1$, $r_2$, ..., $r_n$. The values for an attribute, $A$, are normalized based on the standard deviation and mean of $A$. A value, $r_i$, of $A$ is normalized to $r_{izScored}$ by computing Eq. 2 [20].

$$r_{izScored} = \frac{\left(r_i - \overline{A}\right)}{\sigma(A)} \tag{2}$$

Where $\sigma$ and $\overline{A}$ are the standard deviation and mean, respectively, of attribute $A$ [20]. According to the information given, z-score normalization has applied to the input values in the data used in this study.

## 2.3  Linear Regression (LR)

Linear regression (LR) is one of the commonly applied prediction methods in health data analysis [27–30]. Once LR finds a line, it calculates the vertical distances of the points from the line and minimizes the sum of the vertical distance square [31]. Multiple linear regression is a kind of linear regression used to estimate the relationship between two or more explanatory variables and a response variable. The multiple linear regression model is expressed Eq. 3 [30]:

$$Y = b_0 + b_1\ X_1 + \ldots + b_n\ X_n + \varepsilon_n \tag{3}$$

Where $Y$ is predicted value, $b_0$ represent a value of parameter [30] called intercept, $b_1$ through $b_n$ are the vector of estimated regression coefficient, $X_1$ through $X_n$ are vector of explanatory variables [32], $\varepsilon_n$ are a random error term [32]. In this study, the multiple linear regression model is used for forecasting accuracy and analysis.

## 2.4  Measuring Analysis

Failure to accurately predict suicide risk limits solutions to this major public health problem. Therefore, model performance should be accurate and scalable on risk detection. Evaluating model predictions can provide valuable information about model performance. The risk factors derived from the research literature on suicidality comprise the model's input. The performance of the model is measured according to the coefficient of determination R-squared ($R^2$), Root Mean Squared Error ($RMSE$), and Mean Absolute Error ($MAE$). $R^2$ is a statistical metric. ($R^2$) metric is used to measure how much of the variation in outcome can be explained by the variation in input and to model validation. The coefficient of determination ($R^2$) is calculated using Eq. 4 [33]) and the error metrics are calculated using Eqs. 5 and 6 [34]:

$$R^2 = 1 - \left(\frac{\sum_{i=1}^{N}\left(y_i - \hat{y}_i\right)^2}{\sum_{i=1}^{N}(y_i - \bar{y}_i)^2}\right) \tag{4}$$

$$RMSE = \sqrt{\frac{1}{N} \sum_{i=1}^{N} (y_i - \bar{y}_i)^2} \tag{5}$$

$$MAE = \frac{1}{N} \sum_{i=1}^{N} |(y_i - \hat{y}_i)| \tag{6}$$

Where $\bar{y}$ is the mean value of the dependent variable $Y$, $\hat{y}$ is the predicted value of $Y$, $y$ is the actual value of $Y$, and $N$ is number of data samples.

For the training and testing process, Regression value $(R)$ [35] is used to find best correlations in combination trials at different rates ($70\% - 30\%$, $75\% - 25\%$, $80\% - 20\%$) for forecasting better accuracy.

## 3  Experimental Results

Performance of the Multi Linear Regression model depends on input parameter values of age-group, city, and gender. When a good predictive model is mentioned, out of the sample prediction result is of higher importance than the in-sample output [34]. Lower forecasting errors normally indicate a higher prediction ability of a predictor model [34]. In results, the degree of prediction of the dependent variable of the model to determining the training and testing data is measured up to $(R) = 0.752$ (p < 0.001). It is determined 70% of data as a training set and 30% as a testing set. The data are randomly separated to find the best model using Holdout Cross-Validation [36]. To evaluate the prediction accuracy ability of the model, $R^2$, $MAE$ and $RMSE$ is used. The best five forecasting values obtained from 20 iterations of the techniques are given in Table 2.

**Table 2.** Results from the multiple linear regression model.

| No. | $R^2$ | RMSE | MAE |
|-----|-------|------|-----|
| 1 | 0.5434789 | **0.0621640** | 0.0390860 |
| 2 | **0.4948002** | 0.0650729 | 0.0403089 |
| 3 | 0.5561675 | 0.0644416 | 0.0390073 |
| 4 | 0.5451537 | 0.0632218 | **0.0386959** |
| 5 | **0.5648034** | 0.0626744 | 0.0397366 |

It is seen that the best model's coefficient of determination explaining the variance in the dependent variable is $R^2 = 0.5648034$. It can be interpreted for suicide risk as follows: Fifty-six percent of the variance in the specific mortality rate variables can be explained by age-group, gender and city variables. The remaining 44 can be attributed to unknown or inherent variability. $RMSE$ and $MAE$, on the other hand, are low error, high performance measures inversely proportional to performance [38]. The best $RMSE$ and $MAE$ metric values are obtained 0.0621640 and 0.0386959, respectively.

## 4 Conclusions, Suggestions and Future Work

Suicide cases in the world are mostly seen in men, whereas women are attempting suicide more. In Turkey, both in terms of number of enterprises in terms of mortality and women constitute more suicides. The concentration of the 15–24 age range and suicides in Turkey are known to be in the same range in the world [3]. Suicidal behavior is a rather complex phenomenon that is influenced not only by personal but also by social and cultural variables [37]. In addition, the correct suicide risk prediction may require complex combinations of lots of risk factors [12]. In consequence, looking at error metrics and coefficient of determination, it can be said that the multiple linear regression model predicts the specific mortality rate values can explain to a certain extent well when studies on similar subject [37, 40] and regression model are examined [39]. The findings of the model indicate that age-group, gender, and city variables are promising success predictors of specific mortality rate in predicting future risk of suicide but not sufficient alone for very good prediction in predicting future risk of suicide with this model. It is thought that the results obtained will contribute greatly to the selection of effective factors to be selected for the examination of this health problem. For this reason, it is aimed to improve the forecast by addressing the findings carefully and expanding the scope of risk factors. Expanding the properties of the data set (business, education, city conditions etc.) and applying different methods (Support Vector Machines, Artificial Neural Network etc.) will contribute to the model.

## References

1. World Health Organization: Suicide and attempted suicide. In: Brooke, E. (ed.) Public Health Paper, vol. 58, pp. 1–128. World Health Organization, Geneva (1974)
2. World Health Organization: Primary Prevention of Mental, Neurological and Psychosocial Disorders. Suicide. WHO, Geneva (1998)
3. Harmancı, P.: Dünya'daki ve Türkiye'deki intihar vakalarının sosyodemografik özellikler açısından incelenmesi. Hacettepe Univ. Fac. Health Sci. J. 2(Suppl. 1), 1–15 (2015)
4. Özcan, B., Şenkaya, S., Ozdin, Y., Dinç, A.: Türkiye'deki intihar Vakalarının Çeşitli Kriterlere Göre_Istatiksel Olarak Incelenmesi. Sosyal Politika Çalışmalar_Dergisi
5. Buendía, J.A., Chavarriaga, G.J.R., Zuluaga, A.F.: Social and economic variables related with paraquat self-poisoning: an ecological study. BMC Public Health 20(1), 1–5 (2020)
6. Doğan, N., Toprak, D., Doğan, İ.: Influence of birth cohort, age and period on suicide mortality rate in Turkey, 1983–2013. Cent. Eur. J. Public Health 27(2), 141–144 (2019)
7. Ayas, S.: The infuence of unemployment and educational level on suicide: an empirical study on TUIK data. J. Adm. Sci. 14(28), 101–119 (2016)
8. Aşırdizer, M., Yavuz, M.S., Aydin, S.D., Tatlisumak, E.: Do regional risk factors affect the crude suicidal mortality rates in Turkey. J. Med. 23(2), 1–10 (2009)
9. Akyuz, M., Karul, C., Nazlioglu, S.: Dynamics of suicide in Turkey: an empirical analysis. Rom. J. Econ. Forecast. 20(3), 5–17 (2017)
10. Sikander, D., et al.: Predicting risk of suicide using resting state heart rate. In: Proceedings of the Asia-Pacific Signal and Information Processing Association Annual Summit and Conference (APSIPA), pp. 1–4. IEEE, December, 2016
11. Hill, R.M., Oosterhoff, B., Do, C.: Using machine learning to identify suicide risk: a classification tree approach to prospectively identify adolescent suicide attempters. Arch. Suicide Res. 24(2), 218–235 (2019). https://doi.org/10.1080/13811118.2019.1615018

12. Walsh, C.G., Ribeiro, J.D., Franklin, J.C.: Predicting risk of suicide attempts over time through machine learning. Clin. Psychol. Sci. **5**(3), 457–469 (2017)
13. Ribeiro, J.D., et al.: Self-injurious thoughts and behaviors as risk factors for future suicide ideation, attempts, and death: a meta-analysis of longitudinal studies. Psychol. Med. **46**(2), 225–236 (2016)
14. Franklin, J.C., et al.: Risk factors for suicidal thoughts and behaviors: a meta-analysis of 50 years of research. Psychol. Bull. **143**(2), 187 (2017)
15. Miché, M., et al.: Prospective prediction of suicide attempts in community adolescents and young adults, using regression methods and machine learning. J. Affect. Disord. **265**, 570–578 (2020)
16. Walsh, C.G., Ribeiro, J.D., Franklin, J.C.: Predicting suicide attempts in adolescents with longitudinal clinical data and machine learning. J. Child Psychol. Psychiatry **59**(12), 1261–1270 (2018)
17. Amini, P., Ahmadinia, H., Poorolajal, J., Amiri, M.M.: Evaluating the high risk groups for suicide: a comparison of logistic regression, support vector machine, decision tree and artificial neural network. Iran. J. Public Health **45**(9), 1179 (2016)
18. Poulin, C., et al.: Predicting the risk of suicide by analyzing the text of clinical notes. PLoS ONE **9**(1), e85733 (2014)
19. Turkish Statistical Institute (TURKSTAT) Homepage (2019). http://tuik.gov.tr/Start.do
20. Han, J., Pei, J., Kambel, M.: Data Mining: Concepts and Techniques, 3rd edn. Morgan Kaufman, Elsevier, New York (2012)
21. Buitinck, L., et al.: API design for machine learning software: experiences from the scikit-learn project. In: European Conference on Machine Learning and Principles and Practices of Knowledge Discovery in Databases, Praque (2013)
22. Python Homepage (2020). http://www.python.org
23. Spyder Homepage (2020). https://www.spyder-ide.org/
24. Pedregosa, F., et al.: Scikit-learn: machine learning in python. J. Mach. Learn. Res. **12**, 2825–2830 (2011)
25. Pérez, F., Granger, B.E.: IPython: a system for interactive scientific computing. Comput. Sci. Eng. **9**(3), 21–29 (2007)
26. Patro, S., Sahu, K.K.: Normalization: A Preprocessing Stage. arXiv:1503.06462 (2015)
27. Stebbings, J.H.: Panel studies of acute health effects of air pollution: II. A methodologic study of linear regression analysis of asthma panel data. Environ. Res. **17**(1), 10–32 (1978). https://doi.org/10.1016/0013-9351(78)90057-9
28. Speybroeck, N.: Classification and regression trees. Int. J. Public Health **57**(1), 243–246 (2012)
29. Ramli, A.A., Kasim, S., Fudzee, M.F.M., Nawi, N.M., Mahdin, H., Watada, J.: A revisited convex hull-based fuzzy linear regression model for dynamic fuzzy healthcare data. J. Telecommun. Electron. Comput. Eng. **9**(3–7), 49–57 (2017)
30. Nawi, A., Mamat, M., Ahmad, W.M.A.W.: The factors that contribute to diabetes mellitus in Malaysia: alternative linear regression model approach in the health field involving diabetes mellitus data. Int. J. Public Health Clin. Sci. **5**(1), 146–153 (2018)
31. Tomar, D., Agarwal, S.: A survey on data mining approaches for healthcare. Int. J. Bio-Sci. Bio-Technol. **5**(5), 241–266 (2013)
32. Fang, T., Lahdelma, R.: Evaluation of a multiple linear regression model and SARIMA model in forecasting heat demand for district heating system. Appl. Energy **179**, 544–552 (2016)
33. Menard, S.: Coefficients of determination for multiple logistic regression analysis. Am. Stat. **54**(1), 17–24 (2000)

34. Dash, R., Dash, P.K.: MDHS–LPNN: a hybrid forex predictor model using a legendre polynomial neural network with a modified differential harmony search technique. In: Handbook of Neural Computation, pp. 459–486. Elsevier, Amsterdam (2017). https://doi.org/10.1016/B978-0-12-811318-9.00025-9

35. Hanefi Calp, M., Ali Akcayol, M.: A novel model for risk estimation in software projects using artificial neural network. In: Jude Hemanth, D., Kose, U. (eds.) Artificial Intelligence and Applied Mathematics in Engineering Problems: Proceedings of the International Conference on Artificial Intelligence and Applied Mathematics in Engineering (ICAIAME 2019), pp. 295–319. Springer International Publishing, Cham (2020). https://doi.org/10.1007/978-3-030-36178-5_23

36. Koul, A., Becchio, C., Cavallo, A.: Cross-validation approaches for replicability in psychology. Front. Psychol. **9**, 1117 (2018)

37. Aradilla-Herrero, A., Tomás-Sábado, J., Gómez-Benito, J.: Associations between emotional intelligence, depression and suicide risk in nursing students. Nurse Educ. Today **34**(4), 520–525 (2014)

38. Wang, W., Xu, Z.: A heuristic training for support vector regression. Neurocomputing **61**, 259–275 (2004)

39. Nguyen, D., Smith, N.A., Rose, C.: Author age prediction from text using linear regression. In: Proceedings of the 5th ACL-HLT Workshop on Language Technology for Cultural Heritage, Social Sciences, and Humanities, pp. 115–123, June, 2011

40. Sharaf, A.Y., Thompson, E.A., Walsh, E.: Protective effects of self-esteem and family support on suicide risk behaviors among at-risk adolescents. J. Child Adolesc. Psychiatr. Nurs. **22**(3), 160–168 (2009)

# Building an Open Source Big Data Platform Based on Milis Linux

Melih Günay$^{(\boxtimes)}$ and Muhammed Numan İnce

Akdeniz University, Antalya, Turkey
mgunay@akdeniz.edu.tr

**Abstract.** With technology, all internet connected devices have become data producers and that lead to exponential growth in data. Data and processing are now increasingly taking place on Cloud while the cost of cloud services are decreasing and the capacity is increasing. This is because the high availability of the servers, their ease of creation, configuration and management. Open source Apache Hadoop and its add-on components are frequently used in cloud systems for the handling Big Data. The installation of the Hadoop infrastructure and its compatibility with other components is only possible through a proper deployment process. The package manager of the operating system in which it is hosted, ensures that the compatible versions of the Big Data frameworks and applications are installed and updated together.

At present, global companies that used to produce open source integrated solutions are also purchased, privatized and hence now closed sourced. However, the need of a Linux compatible big data platform based on open source Hadoop framework is still demanded. Therefore, in this paper, the components, the methodology and a case study based on novel big data platform that is high-performing, light-weight, scalable based on Milis Linux is presented.

## 1 Introduction

With the developing technology, the approaches to data processing and storage technologies have changed and as a result of the increasing data, the calculation factor is handled over big data. So if we define big data, the term of big data actually refers to the way that all of these data are generated in different formats such as log files, photo/video archives, social media shares, and other records that we continuously record.

Until recently, business data that are collected and processed include only the relational entities that are stored in disks. Organizations depended on this data and reports primarily to run their day-to-day operations. However, with the increased competition, that started to change as organizations recognize the value of uncollected, not-stored and/or not processed data especially to reduce cost and increase market share. Consequently, generated big data requires a special infrastructure and computational environment that is capable of handling large data sizes and able to scale as needed.

J. Hemanth et al. (Eds.): ICAIAME 2020, LNDECT 76, pp. 33–41, 2021.
https://doi.org/10.1007/978-3-030-79357-9_5

Although there are several options for the operating system of the underlying big data platform, Linux has been the mostly adapted solution. Linux's modular design supports many devices, architectures and use cases. Being open source, it enables 3rd parties to implement and deploy new techniques and algorithms for effective computation. In addition, Linux allows the management of third party open-source modules by sharing data and resources controlled through its flexible permission and rights architecture. Hereby, publications from industry and academia claim Linux is the sensible platform for cloud computing [3]. Linux based cloud services became the frequent choice for Big Data storage and processing. Major public cloud providers such as Microsoft Azure, Google Cloud Platform and Amazon Web Services (AWS) have various flavors of Linux available. About 30% of the virtual machines that Microsoft Azure uses are based on the Linux platform.

MILIS Linux [4] project is a Linux distribution project that started voluntarily in 2016 and implemented using LFS (Linux from Scratch) [5] techniques. Since 2019, the project is supported by the grant provided at Akdeniz University BAP unit [6]. MILIS Linux differs from other GNU/Linux projects with its independent base and own package manager. For the time being, a desktop-targeted operating system has being developed for x86-64 systems and also has being designed for RISC-V [7]. Target users of MILIS are, primarily government agencies, commercial organizations and private users. It has the goal of being a free, light-weight and easy to use operating system for individuals and organizations. Unlike other distributions, which is commonly forked from Debian, MILIS is not based on any existing Linux distribution. It has its own package manager and independent root file-system. It has open development platform for developers by providing core Linux development and user experience.

The research in the field of Big Data processing and management is relatively new. Therefore, the management of these data storage and processing platforms are left to the global cloud server providers. At present, global companies that produce open source integrated solutions for Big Data are also purchased and privatized. For example, RedHat, which provides infrastructure service for Linux operating system, was purchased by IBM [1]. Likewise, HortonWorks, the company that provided the big data frameworks and technologies, hence the ecosystem of Hadoop platform was purchased by its rival, Cloudera [2]. Considering the need of a Linux based open source Hadoop platform and software distribution, we developed a new distribution presented in this paper. We are motivated by the fact that; the market share of open source big data technologies are likely to increase exponentially and could be dominated by big companies.

## 2   Related Works

The introduction of cloud service providers such as Amazon Web Services has been instrumental in the growth of big data applications. The first services of Amazon Web Services started with S3 cloud storage, SQS (Simple Queue Service) and EC2 in 2006. Amazon supports big data analytics with integrated services using data lakes [8].

Google Cloud Platform (GCP) provides a distributed server infrastructure used by large companies such as YouTube. From simple to complex applications, different types of software may be developed through Google Cloud Platform.

Amazon and GCP have several psychical and virtual assets that are placed in data centers located across the globe. This distribution of assets comes with advantages, including improved recovery and low latency as a result of resource proximity to the client.

In 2011, several members of the Yahoo company in charge of the Hadoop project started up a company HortonWorks which quickly developed an open source big data platform based on Apache Hadoop distribution. HortonWorks was a major Hadoop contributor and charges for exclusively support and training. However at the end of 2018, Hortonworks was purchased by the rival Cloudera which sells software licences [9].

Cloudera Inc. was founded by big data experts used to work for Yahoo, Google, Oracle and Facebook in 2008. Cloudera was the first big data platform based on Apache Hadoop. Cloudera has the largest customer base including some major companies. Cloudera provides Cloudera Management for deployment and Cloudera Search for searching [10].

The MapR Converged Data Platform provides an alternative technology for the big data management Unlike Cloudera and Hortonworks, MapR Hadoop distribution has a distributed approach for storing metadata on the cluster nodes using MapR File System (MapR FS).

Microsoft Azure also plays a big role in meeting the big data needs. Microsoft Azure HDInsight is Microsoft's Apache Hadoop based Big Data solution. It is a fully managed cloud service that includes widely used Big Data open source frameworks such as Hadoop, Spark, Hive and R. HDInsight is competing with other competitors as it combines the integration and data storage services required for the Big Data solution with security support. It also allows you to use various development interface tools from Microsoft Visual Studio to Eclipse and IntelliJ and supports Scala, Python, R, Java and .Net platforms.

## 3   Methodology

### 3.1   MILIS Linux Building System

MILIS Linux base system is used as the foundation for the big data platform. Hadoop ecosystem reside on Linux userspace. MILIS, which is created from scratch, includes a package manager (MPS - MILIS Package System) that is used to manage applications in the user-space of Operating System. Base system of MILIS consists of approximately 100 packages including kernel, firm-wares, built tools, compression, text processing, network, C/C++ compiler, disk, startup, git, basic security and authorization tools. Graphics server and desktop modules are not be needed for a base system. The applications hosted by the base system is sufficient for access control and management. Like other big data platforms such as Hortonworks and Cloudera where components are managed using a custom GUI, in our system, MPS directly manages the components. Therefore, the proposed system is easier to setup. The implementation is broken down to 3 phases:

Build Preparation. First directives of selected packages are prepared. Next, package build scripts are collected at a different directive level. They are stored in a Git version track repository. By adding a new big data level to build environment, packages are assembled.

Installation and Configuration. After producing packages, packages are uploaded to our big data package repository instead of the base repository of MILIS. Configuration is done manually for stability.

Testing and Verification. Management is orchestrated by MPS but the dependency of the big data software modules is still needs to be tested. During the testing phase, components are tested individually and verified to work together. However, extensive testing at system level is still needed to ensure between module compatibility which may force repackaging or reconfiguration.

For each technology integrated into the proposed big data platform, the lifecycle included; the compilation from the source code, installation, and testing phases for optimum packaging. MILIS build process require each directive file (talimat) of the package to be committed into GIT repository. Components may need additional packages during the build process and/or runtime. MILIS build process which is managed by MPS is carried out in a container that is isolated from the host. Pre-installation and post-installation scripts setup the installation environment and configure the software afterwards. During the installation phase, packages are registered and uploaded to the Milis package repository. MPS eases installation, update and removal of the components of the big data platform. If an update is applied to a component, MPS manages to trigger an update on the dependent software.

## 3.2    Components of the Milis Big Data Platform

The criteria for choosing the big data modules include not only the functionality but also the module being light-weight in terms of computational resources. If a functionality requires a certain module to be deployed, then necessary optimizations are applied during the configuration phase even sometimes by stripping none essential features.

Since the proposed big data platform uses Hadoop ecosystem. HDFS is used as the base file system. However, it is important to select the proper version of Hadoop as the dependent technologies may not support the selection. Hadoop currently comes in 2 flavors version 2.x and version 3.x. Due to backward compatible with a number of dependent applications, 2.x software is still being deployed in production. HDFS 3.x has improved file system attributes therefore we base the proposed distribution on Hadoop version 3.2.1. Default Hadoop distribution loaded with unnecessary modules including cloud connections to various providers. In order to reduce package size, they are being removed from the installation ensemble. Table 1 shows package size after the build process.

**Table 1.** Hadoop package comparison

| MB | Official binary archive | Milis package |
|---|---|---|
| Archive size | 343 | 134 |
| Installation size | 369 | 209 |

Instead of Java console client for Hadoop client, gohdfs[11] which is written using Go programming language is installed for management. It performs file operations more efficiently then Java client. Table 2 shows comparison for both clients.

**Table 2.** HDFS client vs GoHDFS

| 1.5 GB data/s | Official HDFS client | GoHDFS |
|---|---|---|
| Put operation | 8.1 | 7.4 |
| Get operation | 5.6 | 2.1 |
| List operation | 2.1 | 0.153 |
| Disk usage operation | 1.907 | 0.107 |
| Disk usage operation-2 | 2.1 | 0.21 |

As cluster coordinator, ZooKeeper is a highly available coordination service for Hadoop ecosystem and be-came the de-facto software for coordination. It is originated in the Hadoop project allows one to choose or implement custom locking, fail over, leader election, group members-hip and other coordination software on top of it. We will use 3.5.6 version of Zookeeper.

We chose Apache Kafka as message queue and data collector due to its speed and high data processing capacity either in batch or real-time contributed to the large cache system on disk. Also state full processing with distributed joins and aggregations. In addition Kafka has wide support for a variety of programming languages. As version choosing we chose 2.4 version of Kafka [13] because of newer version. By the way, Kafka requires Apache Zookeeper to run and we provides Zookeeper instance with ours not other which packaged with Kafka.

As query execution engine; MapReduce, Tez and Spark are options. MapReduce comes as default, but is not preferred due to its disadvantage in speed in comparison to Tez and Spark. Tez mainly depends on Yarn resource manager. Spark can run standalone without Yarn. Both of them uses directed acyclic graphs while processing data [15]. Spark may be installed on any Hadoop cluster. Therefore, Spark instead of Tez is chosen as we wanted to lightly couple our components with each other. Consequently, we can change any components with a better one if needed or desired. Spark has also advantages over Tez due to its real-time and machine learning support. Spark version 2.4.x which is compatible with Hadoop 3.x is chosen.

We try to keep powerful integration between Kafka and Spark with direct stream approach [16] to favour of parallelism, for example it will create as many

RDD partitions as there are Kafka partitions to consume, with the direct stream, instead of creating multiple input Kafka streams and union them. So it can be read data from Kafka in parallel. Hence, it is a one-to-one mapping between Kafka and RDD partitions, which is easier to understand and tune.

We have to admit that data sources can come in different types where big data is available. Therefore, this type of data should be cleaned and collected categorically in a warehouse. So a data warehouse appears as a solution on top of distributed file system. Consequently data warehouse will become the main source of data for reporting and analysis. Also it can be used for ad-hoc queries, canned reports, and dashboards[17]. At this point, we chose Apache Hive for the data warehouse technology in our big data platform due to the following features:

- Built-in on top of HDFS as distributed architecture
- Includes SQL-like declarative language called HiveQL (HQL).
- Table structure seems like relational database
- Efficient data extract-transform-load (ETL) feature
- Supports variety of data formats. Also converting variety of format from to other within Hive

Doing personalizations, recommendations, and predictive insights on big data is quite hard with traditional methods. So we need a scalable and proper machine learning component. For this work, we select Apache Spark for machine learning stuff. Spark provides a general machine learning library that is designed for simplicity, scalability, and easy integration with other tools. The key benefit of MLlib is that it allows data scientists to focus on their data problems and models instead of solving the complexities surrounding distributed data such as infrastructure, configurations, and so on [12].

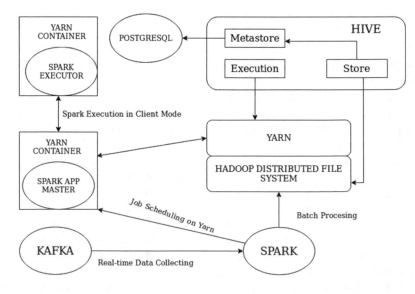

**Fig. 1.** Relationship between the big data components

Also in our big data platform design, Spark comes integrated with YARN. In particular, we can use it in two mode; YARN client mode or YARN cluster mode. At client mode, the driver of Spark is not managed as part of the YARN cluster only client program runs inside YARN container. Some of shell applications should be run in client mode such as PySpark and Spark Shell. At cluster mode, the driver program also runs on the Application Master, which itself runs in a container inside YARN cluster. The YARN client just synchronize status from the Application Master. In this mode, the client could exit after application submission. It is shown in Fig. 1.

Security includes authentication, authorization, audit tracking and data protection sections. For accomplishing to these steps we use default Linux applications. Authentication is handled by the Linux user management mechanism and ACL (Access Control Lists). Also, Kerberos provides an authentication protocol for secure network login. Kerberos works with the concept of tickets which are encrypted and can help reduce the amount of times passwords need to be sent over the network. Authentication is stronger with Kerberos. As authorization, the Polkit application can provide a mechanism for actions to be defined in XML files [14]. Encryption stuff handles by OpenSSL application. As a result, a security approach that blends the security tools in the system will be implemented. As a result of the above-mentioned selections, the MILIS Big Data Platform design is shown in the Fig. 2.

**Fig. 2.** Milis data platform

## 4   Conclusion

Big data is increasingly becoming an important area of research and development for businesses and organizations. The reflection of this to the software world is the development of big data components and the construction of platforms hosting these components. Construction based on open source software is the dominant approach for these platforms. Although, these platforms are open source, it does not mean that they are easy to deploy, adapt and maintain. Neither, they are completely free.

Being able to work from input to output in an integrated fashion brings the success of these big data platforms. Such integration however requires the right combination of components to be selected, installed and configured synchronously. This is not an easy task due to the configuration complexity of each component and system as a whole. Successful integration of each component into the platform, therefore requires distributed process management experience and programming capabilities. The purpose of this study is to build a free, light-weight big data platform based on independently developed MILIS Linux. Considering the needs of the big data processing, a minimal set of best performing, least resource hungry and complex modules are identified and became part of the final distribution image. Using the MILIS build-system, compatible and essential components of the Big Data platform are added to the package repository and verified to run after being individually tested under various conditions in capability and performance. Finally, during the deployment part of this effort inter-operability among these modules were ensured. Big Data distribution that is packaged is distributed with a prototype live image. The big data distribution developed in this study, is aimed to be the preferred software for a data scientist either local now or cloud in the future.

**Acknowledgement.** This work was supported by the Scientific Research Projects Coordination Unit of Akdeniz University Project Number: FBA-2018-4212.

## References

1. IBM: IBM to Acquire Redhat. https://www.redhat.com/en/about/press-releases/ibm-acquire-red-hat-completely-changing-cloud-landscape-and-becoming-worlds-1-hybrid-cloud-provider. Accessed 10 Apr 2020
2. Cloudera: Welcome Brand New Cloudera. https://hortonworks.com/blog/welcome-brand-new-cloudera. Accessed 10 Apr 2020
3. Bridgwater, A.: Why Linux is the powerhouse for big data. https://www.computerweekly.com/blog/Open-Source-Insider/Why-Linux-is-the-powerhouse-for-big-data. Accessed 10 Apr 2020
4. Milis Linux. https://mls.akdeniz.edu.tr. Accessed 10 Apr 2020
5. Linux From Scratch. https://www.linuxfromscratch.org/lfs. Accessed 10 Apr 2020
6. The Scientific Research Projects Coordination Unit. https://bap.akdeniz.edu.tr/english. Accessed 10 Apr 2020
7. Ince, M.N., Ledet, J., Gunay, M.: Building an open source Linux computing system on RISC-V. (2019). https://doi.org/10.1109/UBMYK48245.2019.8965559

8. Reeves, T.: Big Data on Amazon web services. http://www.npvstaffing.com/big-data-on-amazon-web-services. Accessed 10 Apr 2020
9. BusinessWire: Cloudera and Hortonworks announce merger to create world's leading next generation data platform and deliver industry's first enterprise data cloud. https://www.businesswire.com/news/home/20181003005869/en/Cloudera-Hortonworks-Announce-Merger-Create-World%E2%80%99s-Leading. Accessed 10 Apr 2020
10. Mindmajix: Cloudera vs Hortonworks. https://mindmajix.com/cloudera-vs-hortonworks. Accessed 10 Apr 2020
11. Github: Go HDFS Client. https://github.com/colinmarc/hdfs. Accessed 10 Apr 2020
12. Bradley, J.: Why you should use Spark for machine learning. https://www.infoworld.com/article/3031690/why-you-should-use-spark-for-machine-learning.html. Accessed 10 Apr 2020
13. Apache Kafka. https://kafka.apache.org. Accessed 10 Apr 2020
14. Wikipedia Contributors: Polkit. https://en.wikipedia.org/w/index.php?title=Polkit&oldid=936043697. Accessed 10 Apr 2020
15. Parker, E.: Spark vs. Tez: what's the difference? https://www.xplenty.com/blog/apache-spark-vs-tez-comparison. Accessed 10 Apr 2020
16. Spark: Spark Streaming with Kafka. https://spark.apache.org/docs/latest/streaming-kafka-integration.html. Accessed 10 Apr 2020
17. Gunay, M., Ince, M.N., Cetinkaya. A.: Apache hive performance improvement techniques for relational data (2019). https://doi.org/10.1109/IDAP.2019.8875898

# Machine Learning for the Diagnosis of Chronic Obstructive Pulmonary Disease and Photoplethysmography Signal – Based Minimum Diagnosis Time Detection

Engin Melekoğlu[1]([⊠])[ID], Ümit Kocabıçak[1][ID], Muhammed Kürşad Uçar[2][ID],
Mehmet Recep Bozkurt[2][ID], and Cahit Bilgin[3][ID]

[1] Faculty of Computer and Information Sciences, Computer Engineering,
Sakarya University, Sakarya, Turkey
`engin.melekoglu@ogr.sakarya.edu.tr`, `umit@sakarya.edu.tr`
[2] Faculty of Engineering, Electrical-Electronics Engineering, Sakarya University,
Sakarya, Turkey
`{mucar,mbozkurt}@sakarya.edu.tr`
[3] Faculty of Medicine, Sakarya University, Sakarya, Turkey
`cahitbilgin@sakarya.edu.tr`

**Abstract.** The main pathological characteristic of Chronic obstructive pulmonary disease (COPD) is chronic respiratory obstruction. COPD is a permanent, progressive disease and is caused by harmful particles and gases entering the lungs. The difficulty of using the spirometer apparatus and the difficulties in having access to the hospital, especially for young children, disabled or patients in advanced stages, requires to make the diagnosis process easier and shorter. In order to avoid these problems, and to make the diagnosis of COPD faster and then easier to track the disease, it is considered that the use of the Photoplethysmography (PPG) signal would be beneficial. PPG is a biological signal that can be measured anywhere on the body from the skin surface. The PPG signal, that is created with each heartbeat, is an easy measurable signal. The literature contains lots of information on the PPG signal of the body. In this study, a system design was made to use PPG signal in COPD diagnosis. The aim of the study is to determine: "Can COPD be diagnosed with PPG?" and "If it is possible, with the help of minimum how many seconds of signals this process can be completed?" In the line with this purpose, in average 7–8 h of PPG records was obtained from 14 individuals (8 COPD, 6 Healthy ones) for the study. The obtained records are divided into sequences of 2, 4, 8, 16, 32, 64, 128, 256, 512 and 1024 s. Studies were made for each group of seconds and it was tried to determine which second signals could diagnose with higher performance. The 8-h records of the sick individuals are divided into sequences of 2-s, and each sequence was given a patient tag. When the same procedure was done for the healthy individual, all parts were labeled as Healthy. Each signal group was firstly cleaned by 0.1–20 Hz numerical filtering method. Later on, 25 items feature extractions were made in time domains. Finally the formed data set (2, 4, 8 s) were classified by decision tree machine learning methods. According to the obtained results, the highest performance values were achieved with a 2-s data group, and 0.99 sensitivity, 0.99 specificity and 98.99% accuracy rate was obtained. Consistent with these results, it is assumed that COPD diagnosis could be made based on machine learning with PPG signal and biomedical signal processing techniques.

**Keywords:** Biomedical signal processing · Photoplethysmography signal · PPG · Decision tree · Chronic obstructive pulmonary disease · COPD

# 1   Introduction

Chronic Obstructive Pulmonary Disease (COPD) is a common respiratory disease. COPD is a common disease, characterized by airway and/or persistent respiratory symptoms and airway restriction caused by significant exposure to harmful particles or gases, that can be prevented and treated [1,2]. According to the World Health Organization (WHO), it is the fourth most common cause of death in the world, whereas in Turkey it is the third most common cause of death. Each year 2.9 million worldwide, whereas in Turkey 26 thousand people lose their lives due to COPD. The main pathological characteristic of COPD is chronic respiratory obstruction.

COPD is a permanent, progressive disease caused by harmful particles and gases entering the lungs Chronic cough and shortness of breath are the most important symptoms of the disease. Cigarette smoke is the main factors causing these symptoms. Gases and substances harmful to the breathing pipes and air vesicles are filled with cigarette smoke, over time, these harmful substances damage the structure of the bronchi and alveoli, resulting in the development and progression of COPD [3].

Since there is not enough information about the disease, diagnosis and thus the treatment of the disease are delayed. The diagnosis of the disease is made by the specialist doctor according to the report taken from the spirometer device, which is the standard method of diagnosis. This method can be applied only in hospitals with the help of a technician. However, monitoring the disease after diagnosis is also important for tracking the damages caused by the disease to the body. Therefore, it is vital to have an early and fast diagnosis process. As COPD is a progressive disease with permanent damages, the earlier it is diagnosed and treated, the less damage it causes to the individual. As with any other disease, monitoring of the disease during the treatment is possible with advanced equipment and only in hospitals. This process is very difficult and time consuming [2].

Today, COPD diagnosis can be made with a device called a spirometer. After measuring the difficult vital capacity ($FVC$) and the exhaled volume ($FEV1$) in the first second of the breath taken, the rate can be diagnosed by the specialist doctor by evaluating the $FEV1/FVC$ ratio. If $FEV1/FVC < 70\%$, the patient is considered with COPD [4,5].

The difficulty of using the spirometer device and the difficulties in accessing the hospital, especially for young children, disabled or patients in advanced stages of illness, necessitates facilitating and shortening the diagnosis process cite Er2008, Er2009. Due to these disadvantages, there is a need to develop systems that are easy to use and follow-up to make COPD diagnosis more practical [6, 7].

In this study, a machine learning based system with PPG signal was proposed to perform COPD diagnosis quickly and practically. Within the scope of the study, 25-time domain features were extracted from the PPG signal. Later, classification was made with decision trees by selecting and deselecting features. In order to increase the reliability of the study, the study performance values were supported with the Leave-One-Out method.

PPG is a biological signal that can be measured anywhere on the body near the heart. It is evaluated that according to the obtained study results, the electrocardiographic signal (ECG) contains important information about the body and the disease, and can be used in the diagnosis of COPD [6, 7].

There are many unique aspects of the study. These can be listed as follows. (1) Suggesting the use of PPG signal for COPD diagnosis. (2) PPG signal is a very easy signal to obtain. (3) As it is based on machine learning, it is a reliable system. (4) It is a system with a very low and fast process load. When all this is combined, the method to be developed will also create an infrastructure for the production of cost-effective and portable devices for the diagnosis of the disease.

## 2    Material and Method

The diagram flow in Fig. 1 was followed for the study. Together with the collection of data, preprocessing was carried out by making digital filtering. After filtering the data, the data was epoched in certain seconds (2–1024). Balancing process was done in the data set and divided into relevant qualification groups. The data groups created were sorted from the most relevant to the most irrelevant with the feature selection algorithm. After selecting a certain number of features, the classification process was carried out. Feature selection algorithm is optional and all features are classified separately.

### 2.1    Collection of Data

The data used in the study was obtained from Sakarya Hendek State Hospital Sleep Laboratory. The data in question; was examined by a specialist doctor and diagnosed according to COPD criteria, and it was labeled as ill or healthy. For diagnosis, a Vitalograph Alpha spirometer device was used. The studies were made on diagnosed patients, 8 sick and 6 healthy, 2 of them female and 13 male, 14 individuals in total. Demographic information and COPD registration information belonging to individuals were specified in Table 1.

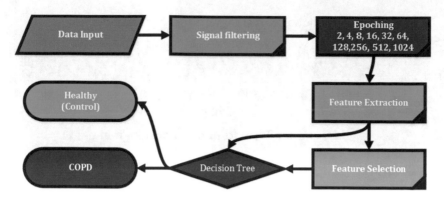

**Fig. 1.** Diagram flow

**Table 1.** Demographic information of individuals and distribution of records

|  | Female | | Male | | All individuals | |
|---|---|---|---|---|---|---|
|  | n1 = 2 | | n2 = 12 | | n = n1 + n2 = 14 | |
|  | ort | std | ort | std | ort | std |
| Age (Year) | 55.50 | ± 4.95 | 53.17 | ± 9.43 | 53.50 | ± 8.82 |
| Weight (kg) | 105.50 | ± 6.36 | 101.92 | ± 8.08 | 102.43 | ± 7.75 |
| Height (cm) | 170.00 | ± 7.07 | 173.42 | ± 6.52 | 172.93 | ± 6.43 |
| BMI (kg/m²) | 36.70 | ± 5.23 | 33.75 | ± 2.54 | 34.17 | ± 2.96 |
| Photoplethysmography time distribution record (sec) | | | | | | |
|  | ort | std | ort | std | ort | std |
| COPD group | – | ± – | 28643.50 | ± 11082.52 | 28643.50 | 11082.52 |
| Control group | 26041.00 | ± 4963.89 | 32611.00 | ± 5351.56 | 30421.00 | ± 5798.47 |

BMI Body Mass Index

## 2.2  Signal Pre-processing

By the filtering process performed during signal pre-processing phase, steps to derive HRV from PPG signal was achieved. For the purpose of clearing the noise on the PPG signal Chebyshev type II bandpass filter with a frequency between 0.1–20 Hz was applied, and then a noiseless PPG signal was obtained by applying the "Moving Average" filter [8]. Then the signal $T = [2\ 4\ 8\ 16\ 32\ 64\ 128\ 256\ 512\ 1024]$ was divided into seconds of epoch, and 25 features were obtained from each divided epoch in time domain. The obtained epoch data is shown in the Table 2.

In the Fig. 2 the PPG record and the Fast Fourier Transform with the Periodogram graph belonging to the COPD and Control groups is shown. As it is seen in the figure, there are differences between the signal amplitudes.

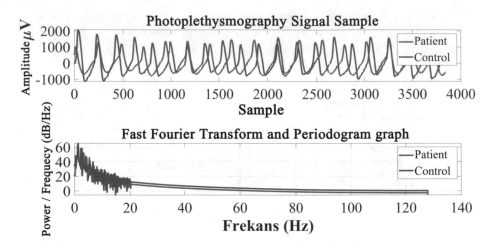

**Fig. 2.** Periodogram graph of Photoplethysmography signal

## 2.3  Feature Extraction

In the literature, many features have been used for PPG signal [9,10]. In this
study, a total of 25 features were extracted from the PPG signal in parallel
with the literature. The extracted properties are shown in 3 columns: property
number, property name and formula information in Table 3. The $x$ shown in the
formulas represents the signal.

**Table 2.** Epoch distribution

| Individual no | Group | Epoch (time - second) | | | | | | | | | |
|---|---|---|---|---|---|---|---|---|---|---|---|
| | | 2 | 4 | 8 | 16 | 32 | 64 | 128 | 256 | 512 | 1024 |
| 1 | COPD | 14323 | 7160 | 3578 | 1788 | 892 | 445 | 221 | 109 | 53 | 25 |
| 2 | | 22213 | 11105 | 5551 | 2774 | 1385 | 691 | 344 | 171 | 84 | 40 |
| 3 | | 2248 | 1122 | 560 | 278 | 138 | 67 | 32 | 15 | 6 | 1 |
| 4 | | 16228 | 8112 | 4055 | 2026 | 1011 | 504 | 251 | 124 | 60 | 29 |
| 5 | | 13978 | 6987 | 3492 | 1745 | 871 | 434 | 215 | 106 | 52 | 24 |
| 6 | | 15673 | 7835 | 3916 | 1956 | 977 | 487 | 242 | 119 | 58 | 28 |
| 7 | | 14428 | 7212 | 3605 | 1801 | 899 | 448 | 222 | 110 | 53 | 25 |
| 8 | | 13093 | 6545 | 3271 | 1634 | 815 | 406 | 202 | 99 | 48 | 23 |
| Total | | **112184** | **56078** | **28028** | **14002** | **6988** | **3482** | **1729** | **853** | **414** | **195** |
| 9 | Control | 17263 | 8630 | 4313 | 2155 | 1076 | 537 | 267 | 132 | 64 | 31 |
| 10 | | 19393 | 9695 | 4846 | 2421 | 1209 | 603 | 300 | 149 | 73 | 35 |
| 11 | | 15463 | 7730 | 3863 | 1930 | 964 | 480 | 239 | 118 | 57 | 27 |
| 12 | | 15463 | 7730 | 3863 | 1930 | 964 | 480 | 239 | 118 | 57 | 27 |
| 13 | | 11263 | 5630 | 2813 | 1405 | 701 | 349 | 173 | 85 | 41 | 19 |
| 14 | | 14773 | 7385 | 3691 | 1844 | 920 | 459 | 228 | 112 | 55 | 26 |
| Total | | **93618** | **46800** | **23389** | **11685** | **5834** | **2908** | **1446** | **714** | **347** | **165** |
| Overall total | | **205802** | **102878** | **51417** | **25687** | **12822** | **6390** | **3175** | **1567** | **761** | **360** |

## 2.4 Feature Selection

The number of the feature has both positive and negative effects on machine learning performance [11]. In order to isolate the negative effects, the feature selection process is used. With this process, a ranking is made from the relevant to the irrelevant according to the power effect of any feature in the label

**Table 3.** Properties of Photoplethysmography

| Line | Properties name | Formula |
|------|-----------------|---------|
| 1 | Kurtosis | $x_{kur} = \dfrac{\sum_{i=1}^{n}(x(i)-\bar{x})^4}{(n-1)S^4}$ |
| 2 | Skewness | $x_{ske} = \dfrac{\sum_{i=1}^{n}(x_i-\bar{x})^3}{(n-1)S^3}$ |
| 3 | Interquartile width | $IQR = iqr(x)$ |
| 4 | Coefficient of variation | $DK = (S/\bar{x})100$ |
| 5 | Geometric average | $G = \sqrt[n]{x_1 \times \cdots \times x_n}$ |
| 6 | Harmonic average | $H = n/(\frac{1}{x_1} + \cdots + \frac{1}{x_n})$ |
| 7 | Hjort activity coefficient | $A = S^2$ |
| 8 | Hjort mobility coefficient | $M = S_1^2/S^2$ |
| 9 | Hjort complexity coefficient | $C = \sqrt{(S_2^2/S_1^2)^2 - (S_1^2/S^2)^2}$ |
| 10 | Maximum | $x_{max} = max(x_i)$ |
| 11 | Median | $\tilde{x} = \begin{cases} x_{\frac{n+1}{2}} & : x \text{ tek} \\ \frac{1}{2}(x_{\frac{n}{2}} + x_{\frac{n}{2}+1}) : x \text{ ift} \end{cases}$ |
| 12 | Median absolute deviation | $MAD = mad(x)$ |
| 13 | Minimum | $x_{min} = min(x_i)$ |
| 14 | Moment, central moment | $CM = moment(x, 10)$ |
| 15 | Average | $\bar{x} = \frac{1}{n}\sum_{i=1}^{n} = \frac{1}{n}(x_1 + \cdots + x_n)$ |
| 16 | Average curve length | $CL = \frac{1}{n}\sum_{i=2}^{n}|x_i - x_{i-1}|$ |
| 17 | Average energy | $E = \frac{1}{n}\sum_{i=1}^{n} x_i^2$ |
| 18 | Average square root RMS value | $X_{rms} = \sqrt{\frac{1}{n}\sum_{i=1}^{n}|x_i|^2}$ |
| 19 | Standard error | $S_{\bar{x}} = S/\sqrt{n}$ |
| 20 | Standard deviation | $S = \sqrt{\frac{1}{n-1}\sum_{i=1}^{n}(x_i - \bar{x})^2}$ |
| 21 | Shape factor | $SF = X_{rms}/\left(\frac{1}{n}\sum_{i=1}^{n}\sqrt{|x_i|}\right)$ |
| 22 | Singular value decomposition | $SVD = svd(x)$ |
| 23 | 25% Trimmed mean value | $T25 = trimmean(x, 25)$ |
| 24 | 50% Trimmed mean value | $T50 = trimmean(x, 50)$ |
| 25 | Average teager energy | $TE = \frac{1}{n}\sum_{i=3}^{n}(x_{i-1}^2 - x_i x_{i-2})$ |

estimator. A researcher can include in his study as many of the features from the dataset that is sorted from the most relevant to the most irrelevant. Thus, by not using unnecessary data, he can get more accurate results and have a faster program cycle. In this study, Fisher feature selection algorithm is used. [12]. In each feature selection process, 20% of all features are selected. In other words, the best 5 features out of 25 features were selected.

## 2.5    The Decision Trees

The basic structure of the decision tree algorithm includes roots, branches, knots and leaves, and the basic structure of decision trees is shown in Fig. 3. While building the tree structure, each attribute is associated with a node. There are branches between the root and nodes. Through the branches each node passes to another node. The decision in the tree is made according to the final leaf. The last part in the tree is named the leaf and the top part is named the root [13]. The basic logic in creating a decision tree structure can be summarized as asking the relevant questions in each node reached and reaching the final leaf over the branches in the shortest path and time according to the given answers. Hence, it creates decision rules/models according to the answers obtained from the questions. The trained tree structure is tested for accuracy with different test data, and the model is used if it produces appropriate results.

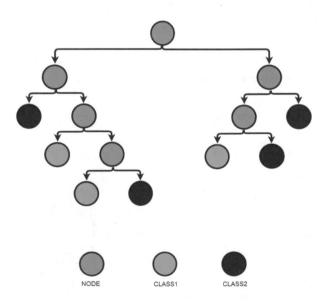

**Fig. 3.** Generic structure of decision tree

## 2.6     Performance Evaluation Criteria

Different performance evaluation criteria are used to test the accuracy rates of the proposed systems. These are accuracy rates, sensitivity, specificity, kappa coefficient (kappa value), Receiver Operating Characteristic - ROC, Area Under a ROC curve and k-fold is the cross-validation accuracy rate. These performance criteria are explained in detail in the subtitles below.

**Confusion Matrix, Kappa Coefficient, F-Measure and Decision Maker Efficiency.** In this study, for the purpose of performance evaluation, the accuracy ratios of classes, sensitivity and specificity values for each class were calculated and ROC curve analysis is performed. In addition, AUC, Kappa coefficient and F-criteria were calculated.

Sensitivity shows the ability of the test to distinguish patients among real patients. It ranges from 0 to 1. Sensitivity value of a diagnostic test is required to be 1. A test's sensitivity value of 1 indicates that test can accurately diagnose all patients. Specificity is the ability of the test to distinguish healthy ones from real healthy ones. It ranges from 0 to 1. It is used in situations where the disease needs to be confirmed. A test's specificity value of 1 indicates that test can accurately detect all healthy people. Sensitivity and specificity parameters were calculated as in Eq. 2 and 3. Accuracy rates in the study were calculated as in Eq. 1. TP, TN, FP and FN in Eq. 1, 2 and 3 are True Positive (TP), True Negative (TN), False Positive (FP), and False Negatives, (FN). In addition, a comparison matrix for accuracy, specificity and sensitivity is given in Table 4.

$$Accuracy = \frac{TP + TN}{TP + TN + FN + FP} \times 100 \tag{1}$$

$$Sensitivity = \frac{TP}{TP + FN} \times 100 \tag{2}$$

$$Specificity = \frac{TN}{FP + TN} \times 100 \tag{3}$$

**Table 4.** Confussion matrix

|  |  | Estimated | |
|---|---|---|---|
|  |  | P | N |
| Real status | P | TP | FN |
|  | N | FP | TN |

F-Measure is used to determine the effectiveness of the model created. The value obtained is the weighted Average of Sensitivity and Specificity values. The

F-measure is calculated as in Eq. 4. It takes a value ranging from 0 to 1. 1 indicates that the created model is perfect and 0 indicates that it is very bad.

$$F = 2 \times \frac{Specificity \times Sensitivity}{Specificity + Sensitivity} \tag{4}$$

The Kappa coefficient is a coefficient that gives information about reliability by correcting the "chance matches" that arise entirely due to chance. Regarding the degree of matching, different limit values have been defined in the literature for the Kappa coefficient [14]. In this study, the limit ranges in Table 5 are used.

**Table 5.** Kappa coefficients limit ranges

| Kappa coefficients | Explanation |
|---|---|
| 0.81–1.00 | Very good level of agreement |
| 0.61–0.80 | Good level of agreement |
| 0.41–0.60 | Moderate level of agreement |
| 0.21–0.40 | Low level of agreement |
| 0.00–0.20 | Poor level of agreement |
| < 0.00 | Very poor level of agreement |

In this study, it is classified according to calculation algorithms as 50% test and 50% test. Time-based training (50%) and test (50%) data set distribution for Machine Learning is shown in Table 6. Data distribution for sick and healthy was made between 2–1024 periods. Test and Testing data obtained from sick and healthy Patients was carried out by sick and healthy individual epoch process.

The success of the study results was reassessed with the Leave-one-out method in addition to all evaluation methods. Test and test data set for 8 sick and 6 healthy individuals are arranged as Table 7. The data set was divided into 6 groups and processed.

## 3    Results

In this study, machine learning based minimum diagnosis time was determined with PPG signal for COPD diagnosis. COPD operations were performed according to the flow diagram in Fig. 1 and results were obtained. First of all, the data obtained were made by numerical filtering and pre-processes were performed. For each data received from the patients, we performed epoch processing by distributing them in certain seconds. We created quality data groups by balancing data. Afterwards the features were selected. Then the data set was divided into training and test sections and classified with decision trees.

**Table 6.** Time related training (50%) and test (50%) data set distribution for machine learning

| Time | Training (%50) | | Test (%50) | | Total (%100) | |
|---|---|---|---|---|---|---|
| | Sick | Healthy | Sick | Healthy | Sick | Healthy |
| 2 | 56092 | 53408 | 56092 | 53407 | 112184 | 106815 |
| 4 | 28039 | 26699 | 28039 | 26698 | 56078 | 53397 |
| 8 | 14014 | 13343 | 14014 | 13343 | 28028 | 26686 |
| 16 | 7001 | 6666 | 7001 | 6666 | 14002 | 13332 |
| 32 | 3494 | 3328 | 3494 | 3328 | 6988 | 6656 |
| 64 | 1741 | 1659 | 1741 | 1659 | 3482 | 3318 |
| 128 | 864 | 825 | 865 | 824 | 1729 | 1649 |
| 256 | 426 | 408 | 427 | 406 | 853 | 814 |
| 512 | 207 | 198 | 207 | 198 | 414 | 396 |
| 1024 | 97 | 95 | 98 | 93 | 195 | 188 |

**Table 7.** Time based training and test data distribution for machine learning according to the Leave-one-out method

| Group | Training | | | | Test | | | |
| | SN | SE | HN | HE | SN | SE | HN | HE |
|---|---|---|---|---|---|---|---|---|
| 1 | 2–8 | 97861 | 2–6 | 76355 | 1 | 14323 | 1 | 17263 |
| 2 | 1, 3–8 | 89971 | 1, 3–6 | 74225 | 2 | 22213 | 2 | 19393 |
| 3 | 1–2, 4–8 | 109936 | 1–2, 4–6 | 78155 | 3 | 2248 | 3 | 15463 |
| 4 | 1–3, 5–8 | 95956 | 1–3, 5–6 | 78155 | 4 | 16228 | 4 | 15463 |
| 5 | 1–4, 6–8 | 98206 | 1–4, 6 | 82355 | 5 | 13978 | 5 | 11263 |
| 6 | 1–5, 7–8 | 96511 | 1–5 | 78845 | 6 | 15673 | 6 | 14773 |

Figures represent individuals numbers

SN Sick No, SE Sick Epoch, HN Healthy No, SE Healthy Epoch

In the results section, both training and test results of machine learning algorithms are given. However, it would be more appropriate to make success evaluations through the test.

The first application of the study includes the classification of the signals in the interval of 2–1024 s. The results of the analyzes performed for this purpose are shown in Table 8 and 9 together with the performance evaluation criteria. When the results were examined, the best performance was obtained with 2-s recordings (Table 9). Additionally, it has been found here that COPD diagnosis can be

**Table 8.** Training performance results for all features

| Second | Performance evaluation criteria | | | | | |
|---|---|---|---|---|---|---|
| | Sensitivity | Specificity | Accuracy rate | Kappa | F-Measure | AUC |
| 2 | 0.99 | 0.99 | 98.99 | 0.98 | 0.99 | 0.99 |
| 4 | 0.99 | 0.99 | 99.13 | 0.98 | 0.99 | 0.99 |
| 8 | 0.98 | 0.98 | 97.96 | 0.96 | 0.98 | 0.98 |
| 16 | 0.99 | 0.99 | 98.79 | 0.98 | 0.99 | 0.99 |
| 32 | 0.99 | 0.99 | 99.08 | 0.98 | 0.99 | 0.99 |
| 64 | 0.99 | 0.98 | 98.35 | 0.97 | 0.98 | 0.98 |
| 128 | 0.99 | 0.98 | 98.64 | 0.97 | 0.99 | 0.99 |
| 256 | 0.97 | 0.99 | 97.84 | 0.96 | 0.98 | 0.98 |
| 512 | 0.97 | 0.91 | 94.07 | 0.88 | 0.94 | 0.94 |
| 1024 | 0.89 | 0.99 | 93.75 | 0.88 | 0.94 | 0.94 |

**Table 9.** Test performance results for all features

| Second | Test | | | | | |
|---|---|---|---|---|---|---|
| | Performans Değerlendirem Kriterleri | | | | | |
| | Sensitivity | Specificity | Accuracy rate | Kappa | F-Measure | AUC |
| 2 | **0.94** | **0.97** | **95.31** | **0.91** | **0.95** | **0.95** |
| 4 | 0.94 | 0.91 | 92.36 | 0.85 | 0.92 | 0.92 |
| 8 | 0.92 | 0.95 | 93.76 | 0.88 | 0.94 | 0.94 |
| 16 | 0.92 | 0.88 | 89.97 | 0.80 | 0.90 | 0.90 |
| 32 | 0.92 | 0.87 | 89.53 | 0.79 | 0.89 | 0.89 |
| 64 | 0.92 | 0.88 | 90.00 | 0.80 | 0.90 | 0.90 |
| 128 | 0.90 | 0.93 | 91.53 | 0.83 | 0.92 | 0.92 |
| 256 | 0.84 | 0.84 | 84.03 | 0.68 | 0.84 | 0.84 |
| 512 | 0.86 | 0.85 | 85.93 | 0.72 | 0.86 | 0.86 |
| 1024 | 0.80 | 0.90 | 84.82 | 0.70 | 0.85 | 0.85 |

detected with a high accuracy rate with any length of PPG signal. Except for the accuracy rate, other performance criteria vary between 0–1 and 1 represents the best value. The accuracy rate is given over % and the highest value is 100.

The analyzes were repeated using also feature selection algorithm. (Table 10 and 11). Here, the best performance was obtained from 512 s recordings with 72.06% accuracy rate (Table 11).

**Table 10.** Training performance results for selected features performance evaluation criteria

| Second | Performance evaluation criteria | | | | | |
|--------|-------------|-------------|---------------|-------|-----------|------|
|        | Sensitivity | Specificity | Accuracy rate | Kappa | F-Measure | AUC  |
| 2      | 0.76        | 0.72        | 74.40         | 0.49  | 0.74      | 0.74 |
| 4      | 0.94        | 0.94        | 94.32         | 0.89  | 0.94      | 0.94 |
| 8      | 0.81        | 0.72        | 76.68         | 0.53  | 0.76      | 0.77 |
| 16     | 0.89        | 0.87        | 87.93         | 0.76  | 0.88      | 0.88 |
| 32     | 0.94        | 0.95        | 94.83         | 0.90  | 0.95      | 0.95 |
| 64     | 0.93        | 0.94        | 93.74         | 0.87  | 0.94      | 0.94 |
| 128    | 0.81        | 0.75        | 78.03         | 0.56  | 0.78      | 0.78 |
| 256    | 0.92        | 0.92        | 91.73         | 0.83  | 0.92      | 0.92 |
| 512    | 0.92        | 0.92        | 92.10         | 0.84  | 0.92      | 0.92 |
| 1024   | 0.92        | 0.54        | 72.92         | 0.46  | 0.68      | 0.73 |

The number of selected features is 5

**Table 11.** Test performance results for selected features performance evaluation criteria

| Second | Performance evaluation criteria | | | | | |
|--------|-------------|-------------|---------------|-------|-----------|------|
|        | Sensitivity | Specificity | Accuracy rate | Kappa | F-Measure | AUC  |
| 2      | 0.74        | 0.70        | 72.01         | 0.44  | 0.72      | 0.72 |
| 4      | 0.74        | 0.66        | 70.07         | 0.40  | 0.70      | 0.70 |
| 8      | 0.78        | 0.71        | 74.50         | 0.49  | 0.74      | 0.74 |
| 16     | 0.75        | 0.69        | 72.55         | 0.45  | 0.72      | 0.72 |
| 32     | 0.75        | 0.66        | 70.76         | 0.41  | 0.70      | 0.71 |
| 64     | 0.74        | 0.69        | 71.85         | 0.44  | 0.72      | 0.72 |
| 128    | 0.75        | 0.72        | 73.59         | 0.47  | 0.74      | 0.74 |
| 256    | 0.72        | 0.65        | 68.55         | 0.37  | 0.68      | 0.68 |
| 512    | **0.69**    | **0.82**    | **75.06**     | **0.50** | **0.75**  | **0.75** |
| 1024   | 0.86        | 0.53        | 69.63         | 0.39  | 0.65      | 0.69 |

The number of selected features is 5

Analysis performance results for other recording periods are also quite good. The sensitivity value is expected to be high for the diagnosis of the disease. Therefore, the 1024-s recordings with 0.86 sensitivity value can be useful for diagnosis of COPD (Table 11).

**Table 12.** Leave-one-out training performance results

| Group | Training | | | | | |
|---|---|---|---|---|---|---|
| | Performance evaluation criteria | | | | | |
| | Sensitivity | Specificity | Accuracy rate | Kappa | F-Measure | AUC |
| 1 | 0.99 | 0.99 | 99.28 | 0.99 | 0.99 | 0.99 |
| 2 | 0.99 | 0.99 | 99.29 | 0.99 | 0.99 | 0.99 |
| 3 | 0.98 | 0.97 | 97.89 | 0.96 | 0.98 | 0.98 |
| 4 | 0.99 | 0.99 | 98.98 | 0.98 | 0.99 | 0.99 |
| 5 | 0.99 | 0.99 | 99.33 | 0.99 | 0.99 | 0.99 |
| 6 | 1.00 | 1.00 | 99.51 | 0.99 | 1.00 | 1.00 |
| Average | 0.99 | 0.99 | 99.05 | 0.98 | 0.99 | 0.99 |

**Table 13.** Leave-one-test performance results

| Group | Test | | | | | |
|---|---|---|---|---|---|---|
| | Performance evaluation criteria | | | | | |
| | Sensitivity | Specificity | Accuracy rate | Kappa | F-Measure | AUC |
| 1 | 0.48 | 0.66 | 58.15 | 0.15 | 0.56 | 0.57 |
| 2 | 0.78 | 0.38 | 59.30 | 0.16 | 0.51 | 0.58 |
| 3 | 0.23 | 0.93 | 84.09 | 0.18 | 0.37 | 0.58 |
| 4 | 0.30 | 0.97 | 63.03 | 0.27 | 0.46 | 0.64 |
| 5 | 0.77 | 0.01 | 43.19 | −0.23 | 0.03 | 0.39 |
| 6 | 0.48 | 0.26 | 37.48 | −0.26 | 0.34 | 0.37 |
| Average | 0.51 | 0.54 | 57.54 | 0.05 | 0.38 | 0.52 |

In order to increase the reliability of the study, the classification process was repeated with the Leave-one-out method along with all the features. Training and test distribution according to the leave-one-out method is given in Table 7. Classification results for both training and testing are summarized in Tables 12 and 13. The training results are quite good. However, the average performance value of the test results is around 57.54%. This rate is an indication that the system needs improvement.

As a result of the analysis for COPD diagnosis using all the features, 0.94 sensitivity, 0.97 specificity and 99.8% accuracy rate with 2-s records was obtained, (Table 9). The decision tree structure for this performance value is shown in Fig. 4. The rules are complex, but they are computer-based applicable.

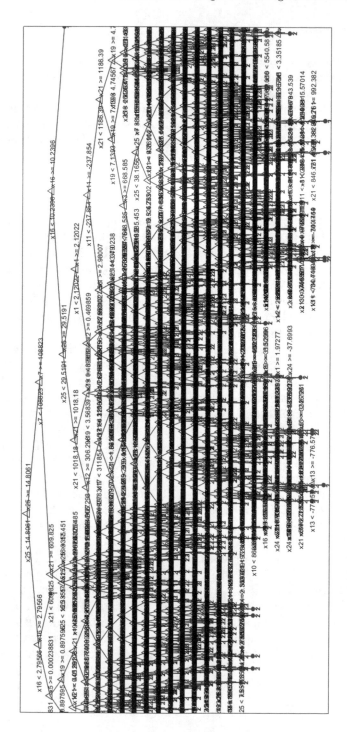

**Fig. 4.** The best decision tree structure for 2 s

## 4    Discussion

Because of day by day increasing air pollution, growing environmental pollution, tough city life and the use of tobacco products, chest diseases are one of today's biggest health problems [15,16].

COPD, a common preventable and treatable disease, is often characterized by progressive persistent airflow obstruction, associated with increased chronic inflammatory response of airways and lungs against harmful gases/particles. The emerging risk factors for COPD are; genetic status, cigarette smoke, organic and inorganic occupational dust and chemicals, indoor and outdoor air pollution, problems in the growth and development of the lungs, sex, age, respiratory infections, socio-economic level, chronic bronchitis, asthma/bronchial hyper-reactivity emerges as [17]. The symptoms are, chronic cough, phlegm production and shortness of breath. Despite being in the risk group and asymptomatic, the patients may neglect the symptoms of the disease until the appointment with a doctor. As moderate severity and airflow limitation become evident in COPD the patients often complain on shortness of breath, which affects their daily activities, Generally, at this stage, as symptoms become evident, a diagnosis of COPD is made by consulting a doctor. In mild COPD, the main symptoms are chronic cough and phlegm production.

In the literature review, until now no studies were encountered among already performed studies regarding the diagnosis or researches by collecting ECG signals belonging to the COPD patients [6,7].

In this study, a different sort of reflection of the ECG the PPG and artificial intelligence-based COPD diagnosis was proposed. Therefore, the proposed model is compatible with the literature.

In 2009, Er using Radial Based ANN tried to diagnose COPD, by asking the opinion of a specialist, he obtained 38 parameters from epicrisis reports and used them as a feature in practice. As a result, they concluded that ANN can be used in diagnosis with an accuracy rate of 90.20% in the diagnosis of COPD [15]. The model proposed in this study is ahead of the literature with an accuracy rate of 95.31%. Data collected from 14 real individuals were used for the proposed model.

In another study in the literature, it was concluded that PPG signal can be used to detect COPD. With PPG signal, it is thought that COPD can be diagnosed in real time and a short period like within 15 s [9]. However, PPG signals of different periods were not used in this study. With the applications in this study, it was determined that the diagnosis can be made with a minimum number of seconds. With the proposed model, it was concluded that the diagnosis of COPD disease can be detected with a 2-second record with the help of PPG signal.

# 5  Conclusion

According to the results of this study, it was concluded that COPD diagnosis can be made based on machine learning with PPG signal and biomedical signal processing techniques. In addition, it was determined how long a PPG record is needed. Just with only 2 s record the accuracy rate of 95.31% was obtained. These results are an indication that COPD diagnosis has practical diagnostic methods. COPD diagnosis is usually made with the help of spirometers. However, different alternatives are needed due to the difficulty of using these devices. The average market value of these devices is 2000 $. The proposed model has many advantages due to its technological infrastructure, rapid diagnosis, reliability, high accuracy rate and low cost. The proposed model can be added to simple oxygen saturation devices with simple software. Only with this economical dimension it provides a great innovation.

As a conclusion, the important parts of the study can be summarized as, (1) easy-to-use, (2) artificial intelligence-based, (3) very low-cost (4) reliable biomedical system that makes decisions based on signal measurement data has been developed. We hope this study will open up new horizons for COPD diagnosis.

# Information

Ethics committee report for the study was obtained from Sakarya University Faculty of Medicine Dean's Office with a letter numbered 16214662/050.01.04/70. In addition, data usage permission is given by T.C. Ministry of Health, Public Hospitals Authority Turkey Sakarya Province Public Hospitals Union General Secretariat of the 94556916/904/151.5815 numbered writing.

# References

1. Batum, M., Batum, Ö., Can, H., Kısabay, A., Göktalay, T., Yılmaz, H.: Evaluation of the severity of sleep complaints according to the stages of chronic obstructive pulmonary disease. J. Turkish Sleep Med. **3**, 59–64 (2015)
2. Zubaydi, F., Sagahyroon, A., Aloul, F., Mir, H.: MobSpiro: mobile based spirometry for detecting COPD. In: 2017 IEEE 7th Annual Computing and Communication Workshop and Conference (CCWC), pp. 1–4. IEEE, January 2017
3. Amaral, J.L.M., Lopes, A.J., Faria, A.C.D., Melo, P.L.: Machine learning algorithms and forced oscillation measurements to categorise the airway obstruction severity in chronic obstructive pulmonary disease. Comput. Methods Programs Biomed. **118**(2), 186–197 (2015)
4. Işık, Ü., Güven, A., Büyükoğlan, H.: Chronic obstructive pulmonary disease classification with artificial neural networks. In: Tıptekno 2015, Tıp Teknolojileri Ulusal Kongresi, Muğla, pp. 15–18 (2015)
5. Ertürk, E.: Chronic obstructive pulmonary disease (COPD) (2020)
6. Uçar, M.K., Moran, I., Altilar, D.T., Bilgin, C., Bozkurt, M.R.: Statistical analysis of the relationship between chronic obstructive pulmonary disease and electrocardiogram signal. J. Hum. Rhythm **4**(3), 142–149 (2018)

7. Uçar, M.K., Moran, İ., Altılar, D.T., Bilgin, C., Bozkurt, M.R.: Statistical analysis of the relationship between chronic obstructive pulmonary disease and electrocardiogram signal. J. Hum. Rhythm **4**(3), 142–149 (2018)
8. Kiris, A., Akar, E., Kara, S., Akdemir, H.: A MATLAB Tool for an Easy Application and Comparison of Image Denoising Methods (2015)
9. Uçar, M.K.: Development of a new method of machine learning for the diagnosis of obstructive sleep apnea. Ph.D. thesis (2017)
10. Sakar, O., Serbes, G., Gunduz, A.: UCI Machine Learning Repository: Parkinson's Disease Classification Data Set (2018)
11. Uçar, M.K., Nour, M., Sindi, H., Polat, K.: The effect of training and testing process on machine learning in biomedical datasets. Math. Probl. Eng. 1–17 (2020)
12. Duda, R.O., Hart, P.E., Stork, D.G.: Pattern Classification, 2nd edn. Wiley, New York (2001)
13. Çölkesen, I., Kavzoglu, T.: Classification of Satellite Images Using Decision Trees: Kocaeli Case. Technical report (2010)
14. Alpar, R.: Applied Statistic and Validation - Reliability. Detay Publishing (2010)
15. Er, O., Sertkaya, C., Tanrikulu, A.C.: A comparative study on chronic obstructive pulmonary and pneumonia diseases diagnosis using neural networks and artificial immune system diagnosis of chest diseases view project. J. Med. Syst. **33**(6), 485–492 (2009)
16. Er, O., Yumusak, N., Temurtas, F.: Diagnosis of chest diseases using artificial immune system. In: Expert Systems with Applications, vol. 39, pp. 1862–1868. Pergamon (2012)
17. Kocabaş, A., et al.: Chronic obstructive pulmonary disease (COPD) protection, diagnosis and treatment report. Official J. Turkish Thoracic Soc. **2**, 85 (2014)

# Diagnosis of Parkinson's Disease with Acoustic Sounds by Rule Based Model

Kılıçarslan Yıldırım[1] , Muhammed Kürşad Uçar[1(✉)] , Ferda Bozkurt[2] ,
and Mehmet Recep Bozkurt[1]

[1] Faculty of Engineering, Electrical-Electronics Engineering, Sakarya University,
Sakarya, Turkey
{mucar,mbozkurt}@sakarya.edu.tr
[2] Vocational School of Sakarya, Computer Programming, Sakarya University
of Applied Sciences, Sakarya, Turkey
fbozkurt@subu.edu.tr

**Abstract.** Parkinson's disease causes disruption in many vital functions such as speech, walking, sleeping, and movement, which are the basic functions of a human being. Early diagnosis is very important for the treatment of this disease. In order to diagnose Parkinson's disease, doctors need brain tomography, and some biochemical and physical tests. In addition, the majority of those suffering from this disease are over 60 years of age, make it difficult to carry out the tests necessary for the diagnosis of the disease. This difficult process of diagnosing Parkinson's disease triggers new researches. In our study, rule-based diagnosis of parkinson's disease with the help of acoustic sounds was aimed. For this purpose, 188 (107 Male-81 Female) individuals with Parkinson's disease and 64 healthy (23 Male-41 Female) individuals were asked to say the letter 'a' three times and their measurements were made and recorded. In this study, the data set of recorded 756 measurements was used. Baseline, Time, Vocal, MFCC and Wavelet that are extracted from the voice recording was used. The data set was balanced in terms of the "Patient/Healthy" feature. Then, with the help of Eta correlation coefficient based feature selection algorithm (E-Score), the best 20% feature was selected for each property group. For the machine learning step, the data were divided into two groups as 75% training, 25% test group with the help of systematic sampling method. The accuracy of model performance was evaluated with Sensivity, Specifitiy, F-Measurement, AUC and Kapa values. As a result of the study, it was found that the disease could be detected accurately with an accuracy rate of 84.66% and a sensitivity rate of 0.96. High success rates indicate that patients can be diagnosed with Parkinson's disease with the help of their voice recordings.

**Keywords:** Parkinson's disease · Acoustic sounds · Systematic feature selection · E-score · Decision tree

© The Author(s), under exclusive license to Springer Nature Switzerland AG 2021
J. Hemanth et al. (Eds.): ICAIAME 2020, LNDECT 76, pp. 59–75, 2021.
https://doi.org/10.1007/978-3-030-79357-9_7

# 1    Introduction

Parkinson's disease is the most common neurodegenerative disorder of the central nervous system after Alzheimer's disease that causes the partial or complete loss of brain-induced motor reflex, speech, behavior, mental process, and other vital functions [1]. This disease was first described by Doctor James Parkinson in 1817 as shaky paralysis and named after him [2]. Today, Parkinson's disease has become a universal neurological disease affecting more than 10 million people worldwide [3]. The incidence of this disease is 1 in 1000, while this rate increases to 1% in the age range of 60–85, 5% in the age range of 85 and above [4]. Since the course of the disease is mild in the first years, patients may not experience significant motor disorders. However, as it is a progressive disease, it leads to important health problems in later times. Although there is no definitive treatment for Parkinson's disease, medication is often used to reduce symptoms that affect patients' daily lives [5].

Early diagnosis of the disease plays an important role in increasing the effectiveness of the treatment. However, there is no known diagnostic method for Parkinson's disease. For the diagnosis of the disease, physicians require some physical tests to assess the functional competence of the legs and arms, muscle status, free walking and balance. Physicians may also request other diagnostic procedures, such as blood tests and neuroimaging techniques to diagnose [6]. Unfortunately, in the early stages of the disease as they may show similar symptoms to other neurological disorders such as Multi system atrophy (MSA) and Progressive supranuclear palsy (PSP) the symptoms are misleading and may cause misdiagnosis. [7]. In addition, the age of the patients to be diagnosed is usually 60 years and older, making these procedures even more difficult. All these facts have led to the need for an easier and more reliable method of diagnosing the disease [8–10].

One of the most common symptoms in the early stages of Parkinson's disease is vocal (speech and voice) problems [11–14]. Recently, speeches (voices) of individuals have been recorded and studies have been carried out on the diagnosis of Parkinson's disease [2,8–12,14–22]. Generally, two different data sets were used in these studies. The first of these is 195 sound measurements taken from 23 individuals with Parkinson's disease and 8 healthy individuals and the second data set was a public data set consisting of multiple speech records from 20 patients with Parkinson's disease and 20 healthy individuals [2,14–20]. Other new studies have also attempted to diagnose Parkinson's disease using small data sets [9–12,21,22]. Although the data sets in these studies have uneven distribution in terms of the number of people with Parkinson's disease and healthy individuals, the researches were carried out without any upper or lower sampling [9–12,14,22].

Numerous scientific studies have been conducted in recent years in order to make a reliable and rapid diagnosis of Parkinson's disease. The most important objective of these studies is to eliminate physical difficulties for patients, reduce the workload of clinical staff and provide support for physicians to diagnose [9–12,14,22].

In 2018–2019, many new machine-based systems for the diagnosis of Parkinson's disease were developed [10, 12, 22, 23]. The most recent approach among these is the model developed based on deep learning and predicting the severity of Parkinson's disease [22]. Developed Tensorflow-based system works with an accuracy rate of 62–81% using open UCI Machine Learning Repository data [22]. In a study conducted by Sadek et al. In 2019 based on artificial neural networks, they identified Parkinson's disease with a success rate of 93% [12]. 195 sound measurements from 31 individuals were used in the study. B. Karan et al. Established a model with support vector machines and random forest algorithm with 150 sound measurements taken from 45 people. 100% accuracy rate has been obtained from the model established with these small data sets [10]. However, when the training and test data are tested together, the machine moves towards memorization and makes biased decisions. When the data set is divided into training/test sets, the accuracy of the proposed model decreases effectively [10, 12, 14, 16, 19, 20]. High accuracy rates from such small data sets cannot be obtained from large data sets. In order to obtain more appropriate results, multiple data groups should be used in the larger data set and balancing operation must be performed in the data set [24–26].

Features extracted to use in machine learning; vocal basic frequency, the amount of variability in frequency, the amount of variability in amplitude, the ratio of components between noise and sound tone, non-linear dynamic values and non-linear basic frequency values include similar features [14, 16, 17]. These features help us achieve very high accuracy rates (98–99%) in distinguishing between patients and healthy individuals in the diagnosis of Parkinson's disease [12, 14, 16, 19, 20].

In this study, we aimed to diagnose Parkinson's disease by designing rule based models with the help of acoustic sounds. For this purpose, 756 measurements obtained from 252 individuals and data set containing 752 features were used. This data set is more extensive than the data sets used in other studies in the literature. Therefore, this data set will contribute to the creation of a more successful model.

The organization of our article is as follows. In Sect. 2, the data used in the study, data preprocessing operations, feature selection algorithm, decision trees (DT) and performance evaluation criteria were explained. The results obtained in Sect. 3 and the interpretation of results obtained in Sect. 4 were given. In the Sect. 5, future studies are mentioned in order to develop this study.

## 2  Materials and Methods

The study was carried out as shown in the flow diagram in Fig. 1. Accordingly, the data set was first balanced by systematic sampling method. Then, the data sets created were grouped according to feature extraction methods. Then, 20% of the properties of the data sets were selected with the help of Eta correlation coefficient based feature selection algorithm (E-Score). A rule-based model was created with 75% of these data sets and the system we designed with 25% of the data sets was tested. Finally, the performance of the system was evaluated.

## 2.1   Data Set

The data set used in this study was obtained from Machine Learning Repository (UCI) of Istanbul University Cerrahpaşa Faculty of Medicine Department of Neurology. The data set consisted of 756 measurements of 188 patients (107 males and 81 females) and 64 healthy individuals (23 males and 41 females). They were asked to say the letter 'a' three times and their measurements were made and recorded [27]. One for each measurement being a label, a total of 753 properties have been created. The label group consists of 1 (patient) and 0 (healthy). This data set is shown in Table 1. ID number in Table 1; is the number to specify the persons to be measured given in sequence starting from zero. Gender is shown as 1 (Male) and 0 (Female). 3 records were taken from each individual.

**Table 1.** Data set

| | | | Number of Data Attributes | | | |
|---|---|---|---|---|---|---|
| NM | ID | Gender | 1 2 . . . | 752 | | Label |
| 1 | 0 | 1 | . . . . . | . | | 1 |
| 2 | 0 | 1 | . . . . . | . | | 1 |
| 3 | 0 | 1 | . . . . . | . | | 1 |
| . | . | . | . . . . . | . | | . |
| . | . | . | . . . . . | . | | . |
| . | . | . | . . . . . | . | | . |
| 756 | 251 | 0 | . . . . . | . | | 0 |

NM Number of measurements, ID ID Number, Gender Female(0)-Male(1),
Label Patient(1)-Healthy(0)

## 2.2   Data Preprocessing

In the literature, there are some steps created by Han and Kamber (2006) to prepare the data set for analysis [28]. Data preprocessing steps performed in this study are described below.

**Dividing a Data Set into Feature Groups.** The 752 properties in the raw dataset consist of some property groups. These include Baseline features, Intensity Parameters, Formant Frequencies, Bandwidth Parameters, Vocal Fold, MFCC, and Wavelet Features. Intensity Parameters, Formant Frequencies and Bandwidth Parameters are combined as Time frequency features because they are close properties. Thus, 5 basic property groups to be processed and a total of 6 data groups consisting of all properties were created and these data groups are shown in Table 2.

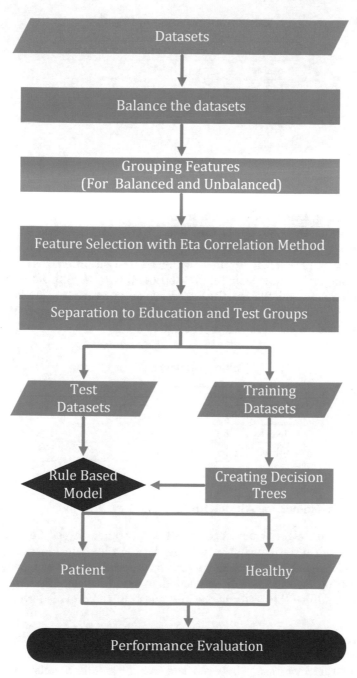

**Fig. 1.** Flow diagram

**Table 2.** Distribution of features

| Data property groups | Number of features |
|---|---|
| Baseline | 21 |
| MFCC | 84 |
| Time | 11 |
| Vokal | 22 |
| Wavelet | 614 |
| All | 752 |

**Balancing the Data Set.** If the number of label class values in the data set is not equal, the data set is unstable. If the data set used in the studies is unbalanced, the accuracy values may cause misleading decision making in performance evaluation [24]. To prevent this adverse situation, the data sets were balanced using systematic sampling method [25]. In the data set of this study, 756 measurements had 192 healthy (0) labels and 564 patient (1) labels. When we offset this unbalanced data set, we obtained 192 healthy and 192 patients. This balanced data set is shown in Table 3.

**Table 3.** Balanced data

|  | Raw data set | Balanced data set |
|---|---|---|
| Patient | 564 | 192 |
| Healthy | 192 | 192 |
| Total | 756 | 384 |

## 2.3  Feature Selection Algorithm

Feature selection algorithms are of great importance in the field of machine learning. Significantly reducing very large data is the main function of the feature selection algorithm. As data increases, more advanced and better performance selection algorithms will be needed. A good feature selection algorithm will further improve the performance and speed of the designed system [29].

**Eta Correlation Coefficient Based Feature Selection Algorithm.** Correlation coefficients are the criteria that give information about the strength and direction of the relationship between the variables. The correlation coefficient formula to be used varies according to the type of variables compared. In the field of machine learning, the overall data is continuous numerical data. In this study, correlation coefficient was calculated between qualitative and continuous numerical variables [30]. In this study, Eta correlation coefficient $(r_{pb})$ which is

suitable for these variables was calculated as shown in Eq. 1. Using these calculated values, properties were selected from the data set.

$$r_{pb} = \frac{\bar{Y}_1 - \bar{Y}_0}{s_y}\sqrt{p_0 p_1} \tag{1}$$

$\bar{Y}_0$ and $\bar{Y}_1$ is the average of data labeled 0 and 1, respectively. $s_y$ is the
standard deviation of data in both classes and is calculated from the expression
in Eq. 2.

$$s_y = \sqrt{\frac{\sum Y^2 - \frac{(\sum Y)^2}{n}}{n}} \tag{2}$$

$p_0$ ve $p_1$ is calculated by the expression in Eq. 3, wherein $N$, $N_0$ and $N_1$
represent the total number of labels, the number of elements labeled 0, and the
number of elements labeled 1, respectively.

$$p_0 = \frac{N_0}{N}, p_1 = \frac{N_1}{N} \tag{3}$$

The number of properties selected by performing the above steps is shown in
Table 4. Separate models were created for the whole data set shown in Table 4
and the best results were selected by calculating the performance values of the
system for each data set.

**Table 4.** Number of selected properties by the Eta correlation coefficient

| | Balanced | | Unbalanced | |
|---|---|---|---|---|
| | All | Eta | All | Eta |
| Baseline | 21 | 4 | 21 | 4 |
| MFCC | 84 | 17 | 84 | 17 |
| Time | 11 | 2 | 11 | 2 |
| Vocal | 22 | 4 | 22 | 4 |
| Wavelet | 614 | 123 | 614 | 123 |
| Total | 752 | 150 | 752 | 150 |

## 2.4  Classification

In our study, the classification process was performed with decision trees. Without classification, the data sets were divided into two groups, 75% training and
25% testing. The training and test datasets were shown in Table 5. This separation was carried out by systematic sampling method. Rule-based models were
established for each training data set. The performance of the system was evaluated by running the designed models with test data sets.

**Table 5.** Training (75%) and test (25%) data set distribution

| Label | Balanced | | | | Unbalanced | | | |
|---|---|---|---|---|---|---|---|---|
| | All features | | FSE | | All features | | FSE | |
| | Training | Test | Training | Test | Training | Test | Training | Test |
| Patient | 144 | 48 | 144 | 48 | 423 | 141 | 423 | 141 |
| Healthy | 144 | 48 | 144 | 48 | 144 | 48 | 144 | 48 |

FSE Features Selected with Eta

**Decision Tree.** The basic structure of the decision tree algorithm includes roots, branches, nodes and leaves, and the basic structure of decision trees is shown in Fig. 2. Each attribute is associated with a node when creating the tree structure. There are branches between the root and nodes. Each node switches to the other node through branches. The decision in the tree is made according to the final leaf [31]. The basic logic in forming a decision tree structure can be summarized as; asking the related questions in each node reached and reaching the final leaf in the shortest way and time according to the answers given. Thus, models are created according to the answers obtained from the questions. The performance of this trained tree structure is calculated with test data and the model is used if it produces appropriate results.

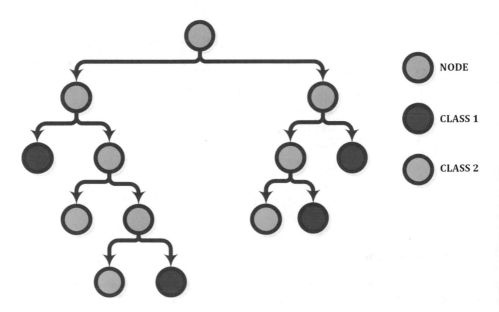

**Fig. 2.** Decision tree

## 2.5    Performance Evaluation Criteria

Different performance evaluation criteria were used to test the accuracy of the proposed systems. These are accuracy rates, sensitivity, specificity, F-Measurement, Area Under an ROC (AUC) and kappa coefficient value. These performance criteria are described in detail in the subheadings [32].

**Accuracy.** The accuracy is calculated as in Eq. 4. The expressions TP, TN, FP and FN in Eq. 4 are True Positives (TP), True Negatives (TN), False Positives (FP) and False Negative (FN) respectively. The accuracy rate of a system is intended to be 100. An accuracy of 100 means that the designed model answers all the questions correctly [32].

$$\text{Accuracy Rates} = \frac{TP + TN}{TP + TN + FN + FP} * 100 \tag{4}$$

**Sensitivity.** Sensitivity indicates the ability of the test to distinguish real patients within patients and is calculated as in Eq. 5. The sensitivity ranges from 0 to 1. A diagnostic test is required to have a sensitivity value of 1. A sensitivity value of 1 indicates that the test can correctly diagnose all patients [32].

$$\text{Sensitivity} = \frac{TP}{TP + FN} \tag{5}$$

**Specificity.** Specificity is the ability of the system to find what is actually healthy among the people who are actually healthy and is calculated as in Eq. 6. Specificity ranges from 0 to 1. Used in cases where the disease needs to be confirmed. A specificity value of 1 indicates that the test is able to correctly identify all healthy people [32].

$$\text{Specificity} = \frac{TN}{FP + TN} \tag{6}$$

**F-Measure.** F-Measure is used to determine the effectiveness of the model being created. The value obtained is the weighted average of sensitivity and specificity values. The F-measure is calculated as in Eq. 7. It gets a value ranging from 0 to 1. 1 indicates that the created model is excellent, and 0 indicates that it is very poor [32].

$$F = 2 * \frac{\text{specificity} * \text{sensitivity}}{\text{specificity} + \text{sensitivity}} \tag{7}$$

**AUC.** The AUC value is used to evaluate the performance of diagnostic tests used to diagnose a disease. When analyzing with the ROC curve, the curves of the different tests are drawn on top of each other and then the comparison is made. In each ROC curve given in the results, the ideal ROC curve is also given. In this way, it can be seen how close the designed system is to the ideal. The fact that the ROC curve is close to the left or top axis is an indication that it is better for diagnosing the location [32].

**Kappa.** The Kappa coefficient is a coefficient which provides information about reliability by correcting the "chance matches" that occur entirely depending on chance. Different limit values have been defined in the literature regarding the degree of agreement of the Kappa coefficient [32].

## 3    Results

In this study, a rule-based model was developed by using acoustic sounds for the diagnosis of Parkinson's disease. First, the data were balanced and then divided into groups according to feature types. E-Score based feature was selected for each feature group and decision tree based diagnostic system was created.

Table 6 shows the change in performance according to the distribution of the test data set in the decision tree process. The Average column represents the average of the accuracy rates for each row. The mean highest accuracy was obtained when the test data set was 25%. Therefore, in this study, the data set was divided into two groups as Training (75%) and Test (25%). The results obtained were obtained at this rate unless otherwise stated.

As shown in Table 7, the best test accuracy rate of unbalanced data was 81.48% from all properties. Sensitivity value is also 0.86. Although these values are good, the Specificity value is 0.67. Sensitivity values are much better, although the Specificity values are low, as shown in Table 7. The reason for this is that the number of healthy data in the unbalanced data set is less than the number of patient data. Therefore, this model does not work stable. As can be seen in Table 9, the accuracy ratio, specificity and sensitivity values obtained with the model generated with all unbalanced selected property data are 84.66%, 0.70 and 0.90 respectively. As it can be understood from here, the performance of the model designed by selecting feature with Eta correlation method is higher.

As shown in Table 8, the best test accuracy ratio of the balanced data was obtained from the Wavelet property with a rate of 75%. Specificity and sensitivity values of this property are 0.77 and 0.73, respectively. Although the accuracy rate (75%) obtained with the model created with balanced data is lower than the accuracy rate (81.48%) with the model created with unbalanced data, it is more stable. The reason for this is that we perform the balancing process and equalize the number of patients and healthy data. As we can see in Table 8, the performance evaluation criteria we obtained when we perform the balancing process are closer to each other. Therefore, this model works more stable. As

**Table 6.** Change in test ratio versus overall performance change

| Test % | U AR | US AR | B AR | BS AR | Average |
|--------|------|-------|------|-------|---------|
| 50.00 | 68.41 | 67.96 | 64.69 | 62.26 | 65.83 |
| 45.00 | 69.72 | 66.90 | 63.67 | 64.63 | 66.23 |
| 40.00 | 67.36 | 67.80 | 62.31 | 61.76 | 64.81 |
| 35.00 | 68.25 | 67.62 | 63.98 | 62.13 | 65.50 |
| 30.00 | 68.48 | 67.39 | 67.18 | 64.71 | 66.94 |
| 25.00 | 70.78 | 70.94 | 67.09 | 65.51 | 68.58 |
| 20.00 | 63.90 | 61.63 | 63.89 | 58.84 | 62.07 |
| 15.00 | 68.04 | 70.12 | 68.77 | 70.05 | 69.24 |
| 10.00 | 62.66 | 59.49 | 69.46 | 70.91 | 65.63 |

**BS Balanced Selection, US Unbalanced Selection,**
**U Unbalanced, B Balanced, AR Accuracy Rate**

**Table 7.** Classification results of unbalanced data

|  |  | AR | Specificity | Sensitivity | F-measure | AUC | Kappa |
|--|--|----|-------------|-------------|-----------|-----|-------|
| Baseline | Training | 82.89 | 0.53 | 0.93 | 0.68 | 0.73 | 0.51 |
|  | Test | 79.37 | 0.54 | 0.87 | 0.67 | 0.71 | 0.43 |
| MFCC | Training | 83.60 | 0.64 | 0.90 | 0.75 | 0.77 | 0.56 |
|  | Test | 80.42 | 0.61 | 0.87 | 0.72 | 0.74 | 0.47 |
| Time | Training | 80.78 | 0.50 | 0.91 | 0.65 | 0.71 | 0.45 |
|  | Test | 78.84 | 0.48 | 0.89 | 0.62 | 0.68 | 0.39 |
| Vocal | Training | 82.72 | 0.58 | 0.91 | 0.71 | 0.75 | 0.52 |
|  | Test | 72.49 | 0.41 | 0.83 | 0.55 | 0.62 | 0.24 |
| Wavelet | Training | 94.89 | 0.92 | 0.96 | 0.94 | 0.94 | 0.87 |
|  | Test | 75.13 | 0.61 | 0.80 | 0.69 | 0.70 | 0.38 |
| AF | Training | 97.88 | 0.97 | 0.98 | 0.97 | 0.97 | 0.94 |
|  | Test | 81.48 | 0.67 | 0.86 | 0.76 | 0.77 | 0.52 |

AF All Features, AR Accuracy Rate

shown in the Table 10, the highest accuracy rate in balanced selected properties was calculated as 81.25% in the model designed with the entire property data set.

Test performance is improved in models designed by applying feature selection process with Eta correlation method to all properties of both balanced and unbalanced data sets. While the test accuracy ratio of the model designed with "all balanced property data" shown in Table 8 is 66.67%, when we apply the feature selection process, this ratio increased to 81.25% as shown in Table 10. Similarly, while the test accuracy ratio of the model we designed with all

**Table 8.** Classification results of balanced data

|          |          | AR    | Specificity | Sensitivity | F-measure | AUC  | Kappa |
|----------|----------|-------|-------------|-------------|-----------|------|-------|
| Baseline | Training | 77.08 | 0.78        | 0.76        | 0.77      | 0.77 | 0.54  |
|          | Test     | 70.83 | 0.73        | 0.69        | 0.71      | 0.71 | 0.42  |
| MFCC     | Training | 73.96 | 0.68        | 0.80        | 0.73      | 0.74 | 0.48  |
|          | Test     | 66.67 | 0.63        | 0.71        | 0.66      | 0.67 | 0.33  |
| Time     | Training | 83.68 | 0.87        | 0.81        | 0.84      | 0.84 | 0.67  |
|          | Test     | 71.88 | 0.83        | 0.60        | 0.70      | 0.72 | 0.44  |
| Vocal    | Training | 70.49 | 0.79        | 0.62        | 0.69      | 0.70 | 0.41  |
|          | Test     | 69.79 | 0.79        | 0.60        | 0.69      | 0.70 | 0.40  |
| Wavelet  | Training | 90.97 | 0.90        | 0.92        | 0.91      | 0.91 | 0.82  |
|          | Test     | 75.00 | 0.77        | 0.73        | 0.75      | 0.75 | 0.50  |
| AF       | Training | 73.96 | 0.68        | 0.80        | 0.73      | 0.74 | 0.48  |
|          | Test     | 66.67 | 0.63        | 0.71        | 0.66      | 0.67 | 0.33  |

AF All Features, AR Accuracy Rate

**Table 9.** Classification results of unbalanced selected properties

|          |          | AR    | Specificity | Sensitivity | F-measure | AUC  | Kappa |
|----------|----------|-------|-------------|-------------|-----------|------|-------|
| Baseline | Training | 78.13 | 0.36        | 0.93        | 0.52      | 0.64 | 0.34  |
|          | Test     | 79.37 | 0.39        | 0.92        | 0.55      | 0.66 | 0.36  |
| MFCC     | Training | 85.19 | 0.58        | 0.95        | 0.72      | 0.76 | 0.57  |
|          | Test     | 80.95 | 0.50        | 0.91        | 0.65      | 0.70 | 0.44  |
| Time     | Training | 79.54 | 0.32        | 0.96        | 0.47      | 0.64 | 0.34  |
|          | Test     | 79.37 | 0.28        | 0.96        | 0.44      | 0.62 | 0.30  |
| Vocal    | Training | 77.78 | 0.36        | 0.92        | 0.51      | 0.64 | 0.33  |
|          | Test     | 71.96 | 0.28        | 0.86        | 0.43      | 0.57 | 0.16  |
| Wavelet  | Training | 93.30 | 0.84        | 0.97        | 0.90      | 0.90 | 0.82  |
|          | Test     | 84.13 | 0.74        | 0.87        | 0.80      | 0.81 | 0.59  |
| AF       | Training | 95.06 | 0.86        | 0.98        | 0.92      | 0.92 | 0.87  |
|          | Test     | 84.66 | 0.70        | 0.90        | 0.78      | 0.80 | 0.59  |

AF All Features, AR Accuracy Rate

unbalanced property data in Table 7 was 81.48%, this ratio increased to 84.66% when we applied feature selection process as shown in Table 9.

When we apply feature selection process to datasets with low number of Time, Baseline Vocal and MFCC features, the sensitivity value decreases while the specificity value of the system increases. In other words, while the patient detection ability of the system increases, its ability to determine healthy individuals decreases. This may be due to the fact that the E-Score features in the unbalanced data set have chosen more features that represent patient individuals.

**Table 10.** Classification results of balanced selected properties

|  |  | AR | Specificity | Sensitivity | F-measure | AUC | Kappa |
|---|---|---|---|---|---|---|---|
| Baseline | Training | 79.17 | 0.82 | 0.76 | 0.79 | 0.79 | 0.58 |
|  | Test | 71.88 | 0.73 | 0.71 | 0.72 | 0.72 | 0.44 |
| MFCC | Training | 79.51 | 0.88 | 0.72 | 0.79 | 0.80 | 0.59 |
|  | Test | 64.58 | 0.81 | 0.48 | 0.60 | 0.65 | 0.29 |
| Time | Training | 68.75 | 0.64 | 0.74 | 0.68 | 0.69 | 0.38 |
|  | Test | 62.50 | 0.63 | 0.63 | 0.63 | 0.63 | 0.25 |
| Vocal | Training | 65.97 | 0.56 | 0.76 | 0.64 | 0.66 | 0.32 |
|  | Test | 67.71 | 0.60 | 0.75 | 0.67 | 0.68 | 0.35 |
| Wavelet | Training | 95.49 | 0.93 | 0.98 | 0.95 | 0.95 | 0.91 |
|  | Test | 64.58 | 0.73 | 0.56 | 0.64 | 0.65 | 0.29 |
| AF | Training | 90.97 | 0.94 | 0.88 | 0.91 | 0.91 | 0.82 |
|  | Test | 81.25 | 0.88 | 0.75 | 0.81 | 0.81 | 0.63 |

AF All Features, AR Accuracy Rate

**Table 11.** Best results

|  |  |  | AR | Specificity | Sensitivity | F-Ölçümü | AUC | Kappa |
|---|---|---|---|---|---|---|---|---|
| US | W | Eğitim | 93.30 | 0.84 | 0.97 | 0.90 | 0.90 | 0.82 |
|  |  | Test | 84.13 | 0.74 | 0.87 | 0.80 | 0.81 | 0.59 |
| US | AF | Eğitim | 95.06 | 0.86 | 0.98 | 0.92 | 0.92 | 0.87 |
|  |  | Test | 84.66 | 0.70 | 0.90 | 0.78 | 0.80 | 0.59 |
| BS | AF | Eğitim | 90.97 | 0.94 | 0.88 | 0.91 | 0.91 | 0.82 |
|  |  | Test | 81.25 | 0.88 | 0.75 | 0.81 | 0.81 | 0.63 |
| U | AF | Eğitim | 97.88 | 0.97 | 0.98 | 0.97 | 0.97 | 0.94 |
|  |  | Test | 81.48 | 0.67 | 0.86 | 0.76 | 0.77 | 0.52 |
| U | MFCC | Eğitim | 83.60 | 0.64 | 0.90 | 0.75 | 0.77 | 0.56 |
|  |  | Test | 80.42 | 0.61 | 0.87 | 0.72 | 0.74 | 0.47 |
| B | W | Eğitim | 90.97 | 0.90 | 0.92 | 0.91 | 0.91 | 0.82 |
|  |  | Test | 75.00 | 0.77 | 0.73 | 0.75 | 0.75 | 0.50 |

BS Balanced Selection, US Unbalanced Selection, U Unbalanced, B Balanced,
AF All Features, W Wavelet, AR Accuracy Rate

Table 11 shows the performance values of the six best performing models. The best model is a model designed with the unbalanced selection of Wavelet data set. The decision tree of this model is shown in Fig. 3 and reached the result in three steps. In addition, as shown in Table 11, all three of the top three best models are feature-selective models. It was concluded that the models designed with Eta correlation method gave better results. In addition, it was concluded that the models designed with All Feature Data Set and Wavelet Feature Data

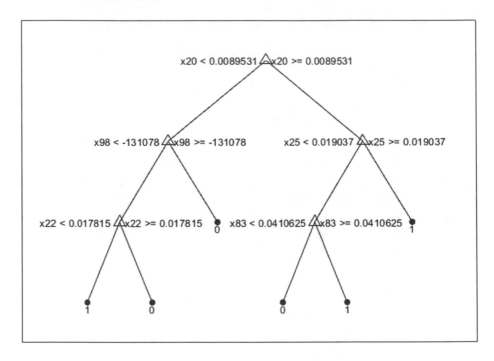

**Fig. 3.** Decision tree for best classification

Set from the data feature groups gave better results than the models designed with other data feature sets.

## 4   Discussion and Conclusion

There are many studies with high accuracy rate in the literature for the diagnosis of Parkinson's disease [2,9–12,14–22]. However, smaller data sets and less data characteristics were used in these studies. Therefore, they will not be able to obtain the high accuracy they obtain in large data sets. The data set and features we use are larger. For this reason, even though the accuracy rate of our designed system is lower than these studies, the system we designed is more stable. So the system we design gives more reliable results.

Many models designed in the literature have used unstable data sets [8–12,14,18–22]. In models created with unstable data, the system yields results close to the data in excess of quantity [25,26,28]. In this study, we used both balanced and unbalanced data sets. In addition, 20% of the characteristics of these datasets were selected by Eta correlation method. Thus, more specific data characteristics were used. In this study, we created models with data sets whose features were not selected. Thus, it was possible to compare the models created with Eta correlation method and normal models.

The results given in some studies in the literature are the average of training and test performances [16,33]. Therefore, the results of these studies are controversial. In this study, we performed the performance evaluation based on the test results. We also obtained the arithmetic average of all performance evaluation criteria while performing the performance evaluation. For this purpose, since the performance values of sensitivity, specificity, F-measurement, AUC and Kappa values are calculated over 1, we multiplied them by 100 and evaluated the performance on a scale of 100. Thus, the same performance evaluation size (100) as the Accuracy Ratio was obtained. After that, we obtained the arithmatic average of all performance evaluation criteria. The performance values obtained by these processes are more stable and more reliable than the other studies.

As a result, the model designed with this study requires only the voice recordings of the person to be diagnosed. In this way, the diagnostic process will be faster and at a lower cost. In addition, the workload of doctors will be reduced and patients will be provided with an easier diagnosis process.

## 5   Future Work

Our work in this area is ongoing and this work can be improved in many ways. Advances in signal processing applications occur with the development of signal processing techniques. The innovations for these steps can be listed as follows.

- The number of features can be increased with different processes.
- Different feature selection algorithms can be used to reveal useful features.
- Different machine learning algorithms can be used to increase the accuracy of decision making.
- The balancing process can be performed during and after the feature selection stage so that the location of the balancing process can be determined.
- Data groups for the diagnosis of gender-based Parkinson's disease can be classified as gender-based.

## References

1. Kurt, I., Ulukaya, S., Erdem, O.: Musical feature based classification of Parkinson's disease using dysphonic speech. In: 2018 41st International Conference on Telecommunications and Signal Processing, TSP 2018. Institute of Electrical and Electronics Engineers Inc., August 2018
2. Little, M.A., McSharry, P.E., Hunter, E.J., Spielman, J., Ramig, L.O.: Suitability of dysphonia measurements for telemonitoring of Parkinson's disease. IEEE Trans. Biomed. Eng. **56**(4), 1015–1022 (2009)
3. Tysnes, O.B., Storstein, A.: Epidemiology of Parkinson's disease, August 2017
4. Wood-Kaczmar, A., Gandhi, S., Wood, N.W.: Understanding the molecular causes of Parkinson's disease, November 2006
5. Jankovic, J.: Parkinson's disease: clinical features and diagnosis
6. Poewe, W., Scherfler, C.: Role of dopamine transporter imaging in investigation of parkinsonian syndromes in routine clinical practice. Mov. Disorders **18**(SUPPL. 7) (2003)

7. Perju-Dumbrava, L.D., et al.: Dopamine transporter imaging in autopsy-confirmed Parkinson's disease and multiple system atrophy. Mov. Disorders Offic. J. Mov. Disorder Soc. **27**(1), 65–71 (2012)
8. Okan Sakar, C., et al.: A comparative analysis of speech signal processing algorithms for Parkinson's disease classification and the use of the tunable Q-factor wavelet transform. Appl. Soft Comput. J. **74**, 255–263 (2019)
9. Mathur, R., Pathak, V., Bandil, D.: Parkinson disease prediction using machine learning algorithm. In: Rathore, V.S., Worring, M., Mishra, D.K., Joshi, A., Maheshwari, S. (eds.) Emerging Trends in Expert Applications and Security. AISC, vol. 841, pp. 357–363. Springer, Singapore (2019). https://doi.org/10.1007/978-981-13-2285-3_42
10. Karan, B., Sahu, S.S., Mahto, K.: Parkinson disease prediction using intrinsic mode function based features from speech signal. Biocybern. Biomed. Eng. (2019)
11. Tang, J., et al.: Artificial neural network-based prediction of outcome in Parkinson's disease patients using DaTscan SPECT imaging features. Mol. Imaging Biol. **21**, 1165–1173 (2019)
12. Sadek, R.M., et al.: Parkinson's Disease Prediction Using Artificial Neural Network (2019)
13. Torun, S.: Parkinsonlularda Konuşma Fonksiyonunun Subjektif ve Objektif (Elektrolaringografik) Yöntemlerle incelenmesi (1991)
14. Gürüler, H.: A novel diagnosis system for Parkinson's disease using complex-valued artificial neural network with k-means clustering feature weighting method. Neural Comput. Appl. **28**(7), 1657–1666 (2017)
15. Eskidere, Ö.: A comparison of feature selection methods for diagnosis of Parkinson's disease from vocal measurements. Technical report
16. Okan Sakar, C., Kursun, O.: Telediagnosis of Parkinson's disease using measurements of dysphonia. J. Med. Syst. **34**(4), 591–599 (2010)
17. Tsanas, A., Little, M.A., McSharry, P.E., Ramig, L.O.: Nonlinear speech analysis algorithms mapped to a standard metric achieve clinically useful quantification of average Parkinson's disease symptom severity. J. R. Soc. Interface **8**(59), 842–855 (2011)
18. Peker, M.: A decision support system to improve medical diagnosis using a combination of k-medoids clustering based attribute weighting and SVM. J. Med. Syst. **40**(5), 1–16 (2016). https://doi.org/10.1007/s10916-016-0477-6
19. Sakar, B.E., Serbes, G., Okan Sakar, C.: Analyzing the effectiveness of vocal features in early telediagnosis of Parkinson's disease. PloS one **12**(8), e0182428 (2017)
20. Peker, M., Şen, B., Delen, D.: Computer-aided diagnosis of Parkinson's disease using complex-valued neural networks and mRMR feature selection algorithm. J. Healthcare Eng. **6**(3), 281–302 (2015)
21. Pak, K., et al.: Prediction of future weight change with dopamine transporter in patients with Parkinson's disease. J. Neural Trans. **126**(6), 723–729 (2019)
22. Grover, S., Bhartia, S., Akshama, A.Y., Seeja, K.R.: Predicting severity of Parkinson's disease using deep learning. Procedia Comput. Sci. **132**, 1788–1794 (2018)
23. Wroge, T.J., Özkanca, Y., Demiroglu, C., Si, D., Atkins, D.C., Ghomi, R.H.: Parkinson's disease diagnosis using machine learning and voice. In: 2018 IEEE Signal Processing in Medicine and Biology Symposium, SPMB 2018 - Proceedings. Institute of Electrical and Electronics Engineers Inc., January 2019
24. Özen, Z., Kartal, E.: Dengesiz Veri Setlerinde Sınıflandırma (2017)
25. Alpar, R.: Reha Alpar. Spor Sağlık Ve Eğitim Bilimlerinden Örneklerle UYGULAMALI İSTATİSTİK VE GEÇERLİK GÜVENİRLİK. Detay Yayıncilik, 2018 edition (2018)

26. García, S., Ramírez-Gallego, S., Luengo, J., Benítez, J.M., Herrera, F.: Big data preprocessing: methods and prospects. Big Data Anal. **1**(1) (2016)
27. UCI. UCI Makine Öğrenimi Havuzu: Parkinson Hastalığı Sınıflandırma Veri Seti (2018)
28. Kartal, E.: Sınıflandırmaya Dayalı Makine Öğrenmesi Teknikleri ve Kardiyolojik Risk Değerlendirmesine İlişkin bir Uygulama. Ph.D. thesis (2015)
29. Kürşat Uçar, M.: Makine Öğrenimi için Eta Korelasyon Katsayısı Tabanlı Özellik Seçimi Algoritması: E-Skor Özellik Seçimi Algoritması - Semantik Bilgin (2019)
30. Alpar, R.: Applied statistic and validation-reliability (2010)
31. (John Ross) Quinlan, J.R., Ross, J.: C4.5: Programs for Machine Learning. Morgan Kaufmann Publishers, San Francisco (1993)
32. Kürşat Uçar, M.: OBSTRÜKTİF UYKU APNE TEŞHİSİ İÇİN MAKİNE ÖGRENMESİ TABANLI YENİ BİR YÖNTEM GELİŞTİRİLMESİ. Ph.D. thesis (2017)
33. Betul Erdogdu Sakar, M., et al.: Collection and analysis of a Parkinson speech dataset with multiple types of sound recordings. IEEE J. Biomed. Health Inf. **17**(4), 828–834 (2013)

# Text Classification Models for CRM Support Tickets

Şevval Az, Uygar Takazoğlu$^{(\boxtimes)}$, and Aşkın Karakaş

Kariyet.net Inc., Istanbul, Turkey
{sevval.az,uygar.takazoglu,askin.karakas}@kariyer.net

**Abstract.** With this study, a system will be implemented for the use of Kariyer.net Customer Solution Center, which estimates the category of the mails that comes from customers and job seekers. These e-mails sent to Customer Solution Center employees via an internal demand system will be automatically estimated by the system to which category they belong before the action is taken. And guidance will be provided for proper action to be taken. Technological expertise, such as natural language processing, classification algorithms, machine learning, user interface will be studied in the system.

**Keywords:** Natural language processing · Classification algorithms · Machine learning · Document classification

## 1 Introduction

The data set was taken from the biggest online recruitment company of Turkey, Kariyer.net, which was founded in 1999. The Ministry of Science, Industry and Technology assigned Kariyer.net as a research and development (R&D) company in 2014. The company uses the most updated technologies to let both job seekers easily look for jobs and companies find the employees fast. And, in order to use the big data effectively, it produces solutions that will take the technology to the next level.

Every day, an average of 350 e-mails are sent to Kariyer.net's CSC (Customer Solution Center) and this count was 85,793 for 2018. There is a large amount of labor and time used to examine and understand these incoming mails and to take action towards the demand to which they belong.

The aim of this system is to accelerate the solution process for the demand of customers and job seekers and to reduce the effort of the solution center employees for this process.

Classification algorithms and machine learning were utilized in the development of the demand classification system and in estimating the category to which the mail contents belong.

In text classification studies, in general, documents with specific labels are given to the system as input, as shown in Fig. 1. First, a pre-processing step is performed for inputs. In the preprocessing step, the information that is not important to the classifier, such as unnecessary characters, url, mail address, phone number and conjunction in the

J. Hemanth et al. (Eds.): ICAIAME 2020, LNDECT 76, pp. 76–81, 2021.
https://doi.org/10.1007/978-3-030-79357-9_8

text is cleared. The differences are eliminated by converting the letters to lowercase letters and a cleared text is created.

The feature vector space used in document classification studies is based on the frequency with which words are displayed in documents. The data, which is vectorized by various methods, is prepared to be given to the algorithm for the classification. Texts with different class labels are separated into two groups as train and test. Which data is selected for train and which data is selected for testing can change the success of the model and the accuracy of the model can be different in each selection. Training with the K Fold Cross Validation method is done over all data. Each data is included in the test set at least once and k − 1 is included in a training set once. Less attention is paid to how the data is separated.

The accuracy of the model is calculated by giving the separated data for training and testing to the selected classification algorithm. The basic concepts used when evaluating model success are accuracy, precision, recall and f measure [1].

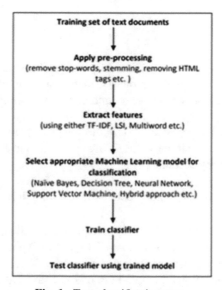

**Fig. 1.** Text classification steps.

## 2  Similar Works

The purpose of document classification is to estimate which document will be included in a certain number of predefined categories (the values we give as labels for the classification) based on its attributes [2].

When determining the features of the documents, it was seen that the bag of words method and tf-idf method were used most common methods when the studies in the literature were examined. In the bag of words method, the frequency of terms of a text document is important. The frequency of passing words and the positions of words are not important. In the study [3] of Bozkir et al., Spam emails were classified over hyperlinks using the n-gram and bag of words model. Doğan et al. also used the n gram method for author, genre and sex classification [4]. The tf-idf method, in which the frequency of passing the words in all documents is important, was found to be the most compared method with the bag of words model. The most commonly used methods when classifying over the feature extraction can be grouped under two main headings: traditional machine learning algorithms and deep learning algorithms [5, 6]. The most commonly used algorithms are Naive Bayes [7], Decision Trees [8], Random Forest, K-Nearest Neighbor Model K-NN [9], Support Vector Machines [10], Logistic Regression [11], Artificial Neural Networks [12, 13].

When comparing the successes of classification algorithms, it was seen that the confusion matrix mentioned in the previous section is used. As well as the accuracy of the model is important, the precision and recall values of the estimated classes are also important. These values become even more important when data is distributed unbalanced.

## 3 System Details

A total of 264 different sub-definitions are mailed to Kariyer.net Customer Solution Center under the main categories of requests, thanks and complaints from companies and job seekers. These e-mails can be short and clearly written mails such as;

- "I forgot the password to my account, I would like your help."
- "Can I get information about the status of my application to XXX Company?"
- "I want you to delete my account",

or these e-mails also be written in long text and take a long time to understand, such as;

- Hello, I've written to you before. I will try to explain in detail once again. An email was sent to my e-mail address on Monday, April 17, 2017. The mail said I had an unread message and the message was sent by the company "XXX". I'm adding the email so you can check it out. When I entered the inbox, I saw that there was no message. You said there was a technical problem and it would be fixed. When I forward this problem to you, you sent me the last message in my inbox. When I asked again, you forwarded the old message that XXX sent me on 14.03.2017. Please do not send me old messages or the last message in my inbox. Or don't say we couldn't bring it to your inbox because of a technical problem. I went to this company for a job interview. This message can be a question, an invitation to an interview, or a message stating that the process is negative. Please solve this problem.

In the first step of the study, e-mail contents that had already been sent to the MCM and whose categories were certain were collected. In terms of category, the four most common topics were selected. These topics are designated as;

- "Employer—Sending Password" (6675 mails)

- "Job Seeker—Application Tracking" (10,909 mails)
- "Job Seeker—Forgot My Password" (18,374 mails)
- "İşin Olsun—Membership Cancellation" (3031 mails)

The model is a multiple classification problem that will be established to produce an estimate under the heading "other" (38,518 mails) for the prediction of these four categories and for the mails that will come out of these categories. The collected mails were subjected to the pre-processing stage mentioned in the previous section. Single exist states of words in the e-mails and 2 n-grams of the words are prepared. Mails have been vectorized with Tf-idf Vectorizer method and Count Vectorizer method (for bag of words). The vectorized mail contents are separated for the model as 70% train and 30% test data. The selection of the training and test data has been made randomly, but the accuracy values of the models may be different over different data sets. In other words, the accuracy may vary in each selection, since the whole data set does not go through validation. For this reason, K-Fold Cross Validation method was used.

K-Fold Cross Validation divides the dataset into specific folds and performs iteration between these folds by shifting the test set at each iteration, it enables the entire data set to enter training and test sets. In this way, the average accuracy of the model can be calculated.

The accuracy of the models was compared by using the data separated as training and test together with the Logistic Regression, Random Forest and Support Vector Machine classification algorithms mentioned in the previous section. In the selection of these algorithms, we are also informed about the probability value of the data whose class is estimated and which class belongs to. It will also be used as an important metric in the launching of the model, in which category a mail enters with which probability value.

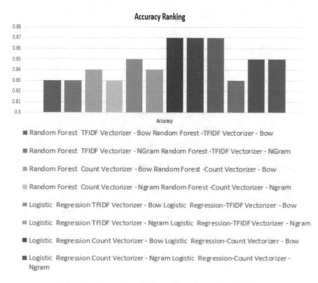

**Fig. 2.** Results of classification algorithms

When the methods used are compared, as can be seen in Fig. 2, the Logistics Regression algorithm gave the highest success result on a single word with a bag of words model. The results of classification algorithms are generally close to each other. In addition, the classification report of the model chosen as the most successful model is as in Fig. 3.

| | precision | recall | f1-score | support |
|---|---|---|---|---|
| 0 | 0.82 | 0.75 | 0.78 | |
| 1 | 0.83 | 0.72 | 0.77 | |
| 2 | 0.84 | 0.74 | 0.79 | |
| 3 | 0.86 | 0.93 | 0.89 | |
| 4 | 0.89 | 0.91 | 0.9 | |
| | | | | |
| accuracy | 0.87 | | | |

**Fig. 3.** Classification report of classification algorithms

In order to use the forecasting system by Kariyer.net CSC, a service was developed using Flask. The developed model produces a result indicating that the mails that are used by the service to the CSC are instantaneously one of the four most frequent classes or belong to a category other than these.

It is necessary to know the percentage of probable mails belonging to a class during the transition of the model to the live environment. According to a threshold value to be determined, the estimates that remain above this probability value will be directed directly to take action without the control of Kariyer.net CSC. In line with this need, the "predict_proba" function offered by the Logistic Regression model was used. The threshold has been determined as 0.80 and above. In the study, it was seen that 0.82 of the data remain above the threshold value. The most successful predicted class was found to be "Şifremi Unuttum" (Fig. 4).

| | Count of mails to be estimated | Correctly predicted above threshold | Percentage of accuracy |
|---|---|---|---|
| Şifre Gönderimi | 2003 | 1298 | 64% |
| Başvuru Takibi | 3273 | 2129 | 65% |
| Şifremi Unuttum | 5512 | 4991 | 90% |
| İO- Aday Üyelik İptali | 909 | 493 | 53% |
| Diğer | 11556 | 10140 | 87% |

**Fig. 4.** Accuracy distribution of classes

# References

1. Bhowmick, A., Hazarika, S.M.: Machine Learning for E-Mail Spam Filtering: Review, Techniques and Trends (2016)
2. Yilmaz, R.: Türkçe Dokümanların Sınıflandırılması (2013)
3. Bozkir, A., Sahin, E., Aydos, M., Sezer, E.A.: Spam E-Mail Classification by Utilizing N-Gram Features of Hyperlink Texts (2017)
4. Doğan, S.: Türkçe Dokümanlar İçin N-Gram Tabanli Siniflandirma: Yazar, Tür Ve Cinsiyet (2006)
5. Şeker, A., Diri, B., Balık, H.: Derin Öğrenme Yöntemleri ve Uygulamaları Hakkında Bir İnceleme (2017)
6. Bayrak, Ş., Takçı, H., Eminli, M.: N-Gram Based Language Identification with Machine Learning Methods (2012)
7. Kim, S.-B., Rim, H.-C., Yook, D., Lim, H.-S.: Effective methods for improving Naive Bayes text classifiers. In: Ishizuka, M., Sattar, A. (eds.) PRICAI 2002. LNCS (LNAI), vol. 2417, pp. 414–423. Springer, Berlin (2002). https://doi.org/10.1007/3-540-45683-X_45
8. Muh-Cherng, W., Lin, S.-Y., Lin, C.-H.: An effective application of decision tree to stock trading. Exp. Syst. Appl. **31**(2), 270–274 (2006). https://doi.org/10.1016/j.eswa.2005.09.026
9. Soucy, P., Mineau, G.W.: A Simple K-NN Algorithm for Text Categorization (2001)
10. Joachims, T.: Text categorization with support vector machines: learning with many relevant features. In: Nédellec, C., Rouveirol, C. (eds.) Machine Learning: ECML-98, pp. 137–142. Springer, Berlin (1998). https://doi.org/10.1007/BFb0026683
11. Sun Lee, W., Liu, B.: Learning with Positive and Unlabeled Examples Using Weighted Logistic Regression (2003)
12. Çalişkan, S., Yazicioğlu, S.A., Demirci, U., Kuş, Z.: Yapay Sinir Ağları, Kelime Vektörleri Ve Derin Öğrenme Uygulamaları (2018)
13. Levent, V., Diri, B.: Türkçe Dokümanlarda Yapay Sinir Ağları ile Yazar Tanıma (2014)

# SMOTE-Text: A Modified SMOTE for Turkish Text Classification

Nur Curukoglu[1] and Alper Ozpinar[2]([✉])

[1] Medepia IT Consulting, Istanbul, Turkey
[2] Istanbul Commerce University, Istanbul, Turkey
`aozpinar@ticaret.edu.tr`

**Abstract.** One of the most common problems faced by large enterprise companies is the loss of knowhow after employee's job replacements and quits. Creating a well-organized, indexed, connected, user friendly and sustainable digital enterprise memory can solve this problem and creates a practical knowhow transfer to new recruited personnel. In this regard, one of the problems that generated is the correct classification of documents that will be stored in the digital library. The most general meaning of text classification also known as text categorization is the process of categorizing text into labeled groups. A document can be related to one or more subjects and choosing the correct labels and classification is sometimes a challenging process. Information repository shows various distributions according to the company's business areas. For a good and successful machine learning based text classification requires balanced datasets related with the business and previous samples. Due to the lack of documents from minor business creates imbalanced learning dataset. To overcome this problem synthetic data can be created with some methods but those methods are suitable for numerical inputs not proper for text classification. This article presents a modified version of Synthetic Minority Oversampling Technique SMOTE algorithm for text classification by integrating the Turkish dictionary for oversampling for text processing and classification.

**Keywords:** SMOTE-text · Text classification · Machine learning · Imbalanced data sets · Oversampling

## 1 Introduction

Writing is one of the most basic and intelligent behavior of human beings. From Egyptian hieroglyphics to the tablets of the Sumerians; humans wrote about everything about their life, civilizations and formed the history as well as transferring the knowledge from past for future generations. This tradition continues in a more modern way with using technology, news, social networks, blogs, discussion forums wiki's and so on in public as well as private for companies. The new era of human evolution characterized by communication and use of technologies [1]. Rules of making successful business shaped how companies process and stores their knowledge and experience as an important organizational asset [2]. In order to protect and increase the usability of the this valuable asset; organizations creates knowledge management systems for storing and sharing them in

© The Author(s), under exclusive license to Springer Nature Switzerland AG 2021
J. Hemanth et al. (Eds.): ICAIAME 2020, LNDECT 76, pp. 82–92, 2021.
https://doi.org/10.1007/978-3-030-79357-9_9

the most appropriate ways [3]. There are lots of studies and researches made on this topic in the literature. Companies uses these systems for maintaining competitive advantages, reducing administrative costs, utilizing the resource-based view of the firm and dynamic capabilities and helping employees for taking decisions under situations and cases where SOP's not created [2, 4–7]. One common problem of knowledge management systems is lack of creating well classified, indexed and organized information library in other words classification and categorization of the information and knowhow [8]. After 1950's the era of digitalization started and information storing and processing using microcontrollers and computers started. IBM's 704 Data Processing System was one of the first computers used for this process both for encoding and analyzing [9]. The first practical and digital use of statistical probability for finding the meaning of a document, collecting similar documents, automatic encoding and creating a searching pattern for the stored information was made by the use of word frequency and distributions in order to extract a dimension for the significance of the document [10, 11]. The first auto extracted abstracts and summaries were also created. Nowadays text summarization mostly used for faster indexing and even for creating friendly SEO's which helps and improves the success rate of finding relevant information in big data, large document libraries or web pages [12]. There are various techniques used nowadays for text mining applications [13]. Especially after 2010 with the rise of cloud computing shared resources as well as combined use of hardware and software text mining algorithms improved significantly [14]. Lots of researchers and social media & network professionals focused on this area and started working afterwards [15]. Since sharing, collecting and storing the information have different ways of collaborating and information flow paths like web services, API's, microservices the amount of data stored is also started being treated as big data. Nowadays most of the organizations and companies already started recording and storing every piece of data without making a detailed analysis of possibility to become a meaningful information. So, soon there will be lots of unclassified data consuming the resources waiting for processed especially for classification. Classified data can be stored and reached in a more optimized ways; if the classification level for the data is top priority and urgently needed for high speed processing short term or working memory then fastest storage systems like in memory or flash drives like SSD's are used. These hardware's are the fastest on the other hand the most expensive way of doing the business, other alternatives like normal storage systems; energy efficient storage and etc. are more cheaper but less performance systems [16]. The goal is to classify all the information if possible, with supervised learning or doing an unsupervised learning, a more recent approach will be reinforcement learning. Typical machine learning works and learns from the data. A common expectation is the data is balanced by means of each class so it is more appropriate to learn from the original data. However sometimes the data is not balanced or not uniformly distributed there are dominant majority or poor minority. This paper organized as follows aiming to focus and explains working and learning on imbalanced classifying supervised text data with an improved SMOTE algorithm named as SMOTEText on Turkish Text. Section 2 introduces the machine learning, text classification and problem of imbalanced data. Section 3 reviews well known classification approaches for class imbalanced datasets. Section 4 explains the experiment data, data preprocessing and details of proposed algorithm. Finally, Sect. 5 concludes the outputs and explains the results with proposed future work.

## 2    Machine Learning

As can be seen from the historical developments; key findings, theories, developments in the algorithms are commonly after from key technology hives, mutations or big innovations in computers and information systems. Current developments in Superset of Artificial Intelligence have led to the hope that many useful applications and solutions to chronic and daily issues of human beings in the fields of similar behaviors of humans like reasoning, knowledge usage and storing, planning, learning, experiencing, understanding and generating with natural language processing (communication), perception and the ability to change, move and manipulate objects in the environment. Since artificial intelligence is the broader term including the algorithms without data as well as all the approached works with data namely machine learning introduced in our terminology in 1959 which an AI subset [17]; expanded with hybrid algorithms, big data and huge computational power as Deep Learning after 2010 [18]. Machine Learning is a subset of Artificial Intelligence. The universal set of Artificial Intelligence have to many different algorithms and solutions in use [19]. Digitization of services that supports and improves the life quality of societies from all over the world including also undeveloped countries. Data of a given problem represented by set of properties or attributes sometimes data is not in the form of direct attributes whereas in the raw data format needs to be preprocessed. If the data has already given with labels as classes or values named outputs for the given data learning is called supervised learning. If data is not labelled by a supervisor or system, then these types of learning called unsupervised learning. Text mining, text classification, text summarization and a broader term text analysis applications are commonly from supervised learning field with some exceptions [20].

### 2.1    Text Representation for Classification

Learning and extracting information from text documents in preliminary machine learning started in 1990's with implementation of back propagation with large unlabeled data [21] and then with statistical approaches later on [22]. Natural Language Processing NLP, Natural Language Understanding NLU and Natural Language Generation are three buzzwords which are extremely popular nowadays. Both deals with reading, writing, and understanding the language in text in order with a common property of working with corresponding numbers first. Reading includes the processes of simple filtering, sentiment detection, topic classification to aspects detection in various levels. These are basic requirements comparing with the expanded version as understanding which is a more complex modelling of the NLP. After understanding what is exactly asked from the machine the answer can be generated by querying, calculating, and expressed in the formal way of NLG. Starting from scratch to novice the initial step is to see the document in the machine perspective which are numbers.

Many recent studies have focused on NLP starts with classification of matching predefined category labels to freetext documents from various resources as mentioned before [23]. Next stage is extracting important content, keywords by automatic summarization [24]. Since machine learning based on mathematical calculations text cannot be processed so to process text in documents a conversion called text encoding should be done. This encoding converts the text in to datasets of vectors with direct coding or scoring of words and representing them with numbers [25]. Vectors forms the data input as a vector space for the learning domain [26].

Before starting encoding all words in the document should be simplified to roots in order to avoid duplications and repeating similar words by stemming or lemmatization algorithms and cleans the document with stop words to avoid useless words for understanding and classification with avoid punctuation, cases and misspelled ones. In order to do analyzes on sentences and sequences of words then n-grams models can be used [27]. Also all vectors and dimensions can be simplified with Latent Semantic Analysis/Indexing LSA using Singular Value Decomposition SVD [28].

The common representation for vectors are pure simple Boolean vectors like one-hot representation or encoding OHE which just gives 1/0 for the all words in the document without scoring them so each row in vector with one attribute representing the a word with a value of 1 and all the other features with value 0, if frequencies used then it will be the frequency based vectorization. Major disadvantages of this approach are the dimensions of the final vector may be excessively big and since word orders not stored understanding algorithms have difficulties in learning. To overcome this ordered OHE can be used then each document represented by a tensor and tensors of all documents formed the library document corpus [29]. Tensors are commonly used for Recurrent Neural Networks (RNNs) and Long Short Term Memory LSTM type of ANNs.

In index based encoding all words mapped to a unique index in the library and those indexes are filled into the vector with number of occurrences and common name for this approach is Bag of Words BoW [30]. The size of the input vector for each document is same and equal to the number of unique words in the document corpus.

To create a relation between the words in a document and label for the document it is important to identify and measure the influence of a word or ngram to the label. Especially in document classification based on statistical occurrence with calculating the Term Frequencies TF and Inverse Document Frequencies IDF which calculated by for a given document collection dataset or document corpus D, a word w and a sample individual document d from D;

$$w_d = TF*IDF$$

$$w_d = f_{w,d} * \log\left(\frac{|D|}{f_{w,D}}\right) \tag{1}$$

Where |D| is the number of documents in corpus, TF is fw,d as the number of times word appears in document d, fw, D is the number of documents have the word w. TF can be calculated with different formulas also can be normalized with the real frequency of the word in the document instead of counting. Normalization is useful especially document corpus has formed of different sized documents. And IDF helps the weight of rare words and lowers values for the most common words. This is useful especially in finding the rare but strong relation words in the documents. For example in document classification the name of an important scientist IDF value increase even occurs in one document but if the TF is great enough it makes meaningful results and learnings [31–33].

Another alternate approaches is Global Vectors for Word Representation GloVe which is an unsupervised learning algorithm for obtaining vector representations for words developed by MIT, where each row represents a vector representation and training from documents is performed on aggregated global word-word cooccurrence statistics and each word calculated by cooccurrence probabilities [34].

There are also other neural network based embeddings where more complex approaches can be used like Word2Vec [30, 35]. Which uses char n-grams focusing on rare words and Doc2Vec [36, 37] which built on the former one using Continuous bag of words CBoW for the context of a word and Skipgram for context words surrounding a specific word. These methods are more suitable for excavating the text for deep learning applications. These methods also learn the relationships between words/n-grams, such as synonymy, antonymy or analogies and paragraph representation can be done by using another layer.

## 2.2 Machine Learning Techniques for Text Classification

The study of statistical and computational approached has become an important aspect of text classification and pattern recognition, after numerical representation of documents for classification. The rest of the progress is not different then the well-known and applied machine learning classifications like Euclidean distance based nearest neighbor techniques, Bayesian learning, decision tree's, random forests, neural networks, ensemble learning techniques of boosting and bagging and support vector machines are some of them [38–41]. In this paper decision trees, boosting methods of AdaBoost and Gradient Boosting were used. The main concept of boosting is to add new learning models to the ensemble sequentially in order to improve the performance [42, 43]. These methods selected for their similarity and there are lots of good examples of these methods and their benchmarks for text classification in the literature [44–47] A multiclass study on the imbalanced dataset was made with decision trees, SVM and logistic regression [37].

## 3    Imbalanced Datasets in Machine Learning

In machine learning problems, the different datasets from different sources has been extensively studied in recent years which leads to different type of datasets collected with differences in prior class probabilities or distributions or in another words class imbalances can be observed so classification categories are not clearly represented. This may be due to lack of data, sensor problem or rare occurrence of one classes by the nature of the data [48]. Those unequal number of samples or documents for different classes, are said to be imbalanced. Fewer instances or samples from one class named as minority class namely irregular, exception or attractive examples and the larger number of samples named as majority class namely normal examples. These datasets can be seen in Machine Learning, Pattern Recognition, Data Mining, and Knowledge Discovery problems [49]. Since learning algorithms use the data for training the weights, selecting the neighbors, unbalanced data for learning and modelling minority class have some performance issues by standard learners.

### 3.1   Preparing Text Data for Machine Learning for Imbalanced Datasets

The methodologies or algorithms to overcome this problem can be solved by two approaches as data sampling and algorithmic modification of the imbalanced data [37]. There are different studies on the subject offering different methods and approaches to these approaches [49–52]. These approaches can be clustered in two diverse categories: one category corresponds to methods that operate on the dataset in a preprocessing step preceding classification as resampling methods on the other hand the other as cost-sensitive learning modifies the classification algorithm in order to put more emphasis on the minority class. Most of those resampling approaches focused on three methods as synthesis of new minority class instances like of cluster based, distance based, evolutionary based and other methods

- oversampling of minority class for eliminating the harms of skewed distribution by creating new minority class samples by cluster based, distance based, evolutionary based which this paper focused on,
- under-sampling of majority class for eliminating the harms of skewed distribution by discarding the intrinsic samples in the majority class,
- hybrid approach of combination of over-under sampling [51].

### 3.2   SMOTE Application of Synthetic Minority Oversampling Technique (SMOTe)

SMOTE was first introduced by Chawla [53] and one of the distance based methods using nearest neighbors judged by Euclidean Distance between data points in feature space. For each minority instance, k number of nearest neighbors are selected, and difference vectors found between the minority instance and each neighbor. The k difference vectors then multiplied by a random number selected uniformly between exclusively between 0 and 1. Those modified difference vectors added to the minority vector to synthetically generate a new oversampled minority class sample.

Due to the encoding of documents the difference of text classification can be categorized as high dimensional data. The researchers shown that high dimensional data SMOTE does not change the mean value of the SMOTE-augmented minority class, while it reduces its variance; the practical consequence of these results is that SMOTE hardly affects the classifiers that base their classification rules on class specific means and overall variances [54].

### 3.3   SMOTE-Text

SMOTE-Text is the modified version of SMOTE algorithm specially organized for TFIDF vectorization. The assumption of TFIDF calculations TF part can be sampled with the distance neighbors of the stem words, however IDF part modified for all the remaining dataset for updated version. So TFIDF data decomposed to TF and IDF values, and IDF values where N number of total documents updated, the number of documents with the given stems is not changed but the log value changes and some rare keywords becoming normal as a result of this.

# 4 Experiment

## 4.1 Data Collection and Preparation

Data for the experiment collected from the Turkish social media, blogs, news, wiki's, forums with Python Beautiful Soup and Request libraries. A collection of more than 1500 documents then manually classified and categorized with labels about daily topics like entrepreneurship, personal development, programming, information technologies, art and humanities, foreign languages, project management, time management and so on. Since the date collected from different resources their spelling, grammar and redaction were not perfect. Among those documents depending on the similarity of the document size, spelling, and grammar unity two topics time management and learning a foreign language was selected. Then the dataset normalized, cleaned, and isolated from all numbers, punctuation, cases and misspelled ones. A stop word list was also adapted to the algorithm. Since a Turkish language was selected as an experimental dataset an open source library Zemberek-NLP provides Natural Language Processing tools for Turkish was used for finding the roots, tokenization and normalization [55]. There are some other tools planned for implementing for future work [56]. Python was selected as a programming language and common machine learning libraries like Pandas, Scikit-Learn libraries used for reading data, finding TFIDF and machine learning algorithms like Decision Trees, Gradient Boosting and AdaBoosting.

Data is divided into training and test data in the beginning and test data reserved only for testing for all stages to avoid the random coincidence bias of test results. Original data consist of 150 data from each class of Foreign Language and Time Management a total of 300 data was used for training and 30 documents from each class a total of 60 documents was reserved for testing.

## 4.2 Experimental Results and Evaluation

Experiments divided into four stages for all stages the initial datasets are remained same. To evaluate the learning benchmarks for binary classification, confusion matrix outputs precision, recall, F1 score values compared. Among those indicators; precision shows the ratio of positive class predictions that actually belong to the positive class, recall shows the ratio of positive class predictions made out of all positive examples in the document corpus, F1 score provides a single score that balances both the concerns of precision and recall in one number to benchmark the results.

- S1: Balanced Stage.
  In the balanced stage experiments was conducted on the original data to test the pure benchmarks of the algorithms.
- S2: Imbalanced Stage.
  In the imbalanced stage experiments was conducted on the original data modified as one class is becoming imbalanced as 25 and 50 samples whereas the other class remains original as 150 sample documents to test the pure benchmarks of the algorithms.

- S3: Imbalanced Stage – Under sampling.

  In the imbalanced stage experiments was conducted on the original data modified as one class is becoming imbalanced as 25 and 50 samples whereas the other class also modified to under sampling to reduced balanced version of 25 and 50 samples amount the original as 150 sample documents to test the pure benchmarks of the algorithms.

- S4: Imbalanced Stage – SMOTE-Text.

  In the SMOTE-Text stage the imbalanced class samples oversampled by using the proposed SMOTE-Text algorithm to %200 oversampling.

Overall results can be observed from Table 1.

**Table 1.** Overall results

| Stage | Imbalance status | F1 DT $C_1$ | F1 DT $C_2$ | F1 AB $C_1$ | F1 AB $C_2$ | F1 GB $C_1$ | F1 GB $C_2$ |
|-------|------------------|-------------|-------------|-------------|-------------|-------------|-------------|
| Stage 1 | None | 0.86 | 0.84 | 0.9 | 0.9 | 1 | 1 |
| Stage 2 | 25/150 | 0.67 | 0.80 | 0.72 | 0.82 | 0.67 | 0.80 |
| Stage 2 | 150/25 | 0.70 | 0.57 | 0.79 | 0.73 | 0.76 | 0.65 |
| Stage 2 | 50/150 | 0.70 | 0.81 | 0.85 | 0.88 | 0.7 | 0.81 |
| Stage 2 | 150/50 | 0.73 | 0.61 | 0.82 | 0.77 | 0.81 | 0.75 |
| Stage 3 | 50/50 | 0.70 | 0.73 | 0.70 | 0.73 | 0.73 | 0.73 |
| Stage 4 | 50/150 → 100/150 | 0.72 | 0.82 | 0.80 | 0.83 | 0.68 | 0.79 |
| Stage 4 | 150/50 → 150/100 | 0.72 | 0.60 | 0.81 | 0.74 | 0.80 | 0.76 |

DT: Decision Tree; AB: Ada Boosting; GB: Gradient Boosting

# 5   Conclusion and Suggestions

In this study an improved version of an oversampling technique SMOTEText was used for imbalanced text classification problem in Turkish. As can be seen from the results first of all the selected machine learning algorithms were performed well for this type of a problem and performances from top to bottom for balanced data are Gradient Boosting, AdaBoosting and Decision Tree's however on the imbalanced datasets AdaBoosting performs better than the GradiendBoosting and for under sampling since the number of sample documents are too small decision trees and gradient boosting performs nearly same. With the F1 score of 1 in the original dataset Gradient Boosting performs perfect and as expected the imbalanced dataset performances decreased with the imbalance ratio increased. As seen under sampling approach was not successful as expected however SMOTEText performs slightly better than the imbalanced data in all techniques.

# References

1. Erickson, B.H.: Social networks and history: a review essay. Hist. Methods J. Quant. Interdiscip. Hist. **30**, 149–157 (1997)
2. Alavi, M., Leidner, D.E.: Knowledge management and knowledge management systems: conceptual foundations and research issues. MIS Q. 107–136 (2001)
3. Coakes, E.: Storing and sharing knowledge: supporting the management of knowledge made explicit in transnational organisations. Learn. Organ. Int. J. **13**, 579–593 (2006)
4. Becerra-Fernandez, I., Sabherwal, R.: Knowledge Management: Systems and Processes. Routledge, Milton Park (2014)
5. Maier, R., Hadrich, T.: Knowledge management systems. In: Encyclopedia of Knowledge Management, 2nd edn, pp. 779–790. IGI Global (2011)
6. Alavi, M., Leider, D.: Knowledge management systems: emerging views and practices from the field. In: Proceedings of the 32nd Annual Hawaii International Conference on Systems Sciences. HICSS-32. Abstracts and CD-ROM of Full Papers, pp. 8-pp. IEEE (1999)
7. Huber, G.P.: Transfer of knowledge in knowledge management systems: unexplored issues and suggested studies. Eur. J. Inf. Syst. **10**, 72–79 (2001)
8. Jacob, E.K.: Classification and categorization: a difference that makes a difference (2004)
9. Griffith, R.E., Stewart, R.A.: A nonlinear programming technique for the optimization of continuous processing systems. Manag. Sci. **7**, 379–392 (1961)
10. Luhn, H.P.: The automatic creation of literature abstracts. IBM J. Res. Dev **2**, 159–165 (1958). https://doi.org/10.1147/rd.22.0159
11. Luhn, H.P.: A statistical approach to mechanized encoding and searching of literary information. IBM J. Res. Dev. **1**, 309–317 (2010). https://doi.org/10.1147/rd.14.0309
12. Maybury, M.: Advances in Automatic Text Summarization. MIT Press, Cambridge (1999)
13. Ferreira, R., De Souza, C.L., Lins, R.D., et al.: Assessing sentence scoring techniques for extractive text summarization. Expert Syst. Appl. **40**, 5755–5764 (2013). https://doi.org/10.1016/j.eswa.2013.04.023
14. Svantesson, D., Clarke, R.: Privacy and consumer risks in cloud computing. Comput. Law Secur. Rev. **26**, 391–397 (2010). https://doi.org/10.1016/j.clsr.2010.05.005
15. Aggarwal, C.C., Zhai, C.X.: Mining Text Data, pp. 1–522. Springer, Boston (2013). https://doi.org/10.1007/978-1-4614-3223-4
16. Oberauer, K., Lewandowsky, S., Awh, E., et al.: Benchmarks for models of short-term and working memory. Psychol. Bull. **144**, 885 (2018)
17. Samuel, A.L.: Some studies in machine learning using the game of checkers. IBM J. Res. Dev. **3**, 210–229 (1959). https://doi.org/10.1147/rd.33.0210
18. Russel, S., Norvig, P.: Artificial Intelligence: A Modern Approach. Pearson Education Limited (2013)
19. Kotsiantis, S.B., Zaharakis, I., Pintelas, P.: Supervised machine learning: a review of classification techniques. Emerg. Artif. Intell. Appl. Comput. Eng. **160**, 3–24 (2007)
20. Ongsulee, P.: Artificial intelligence, machine learning and deep learning. In: 2017 15th International Conference on ICT and Knowledge Engineering (ICT&KE), pp. 1–6. IEEE (2017)
21. Rumelhart, D.E., Hinton, G.E., Williams, R.J.: Learning internal representations by error propagation. California Univ San Diego La Jolla Inst for Cognitive Science (1985)
22. Deerwester, S., Furnas, G.W., Landauer, T.K., Harshman, R.: Cara Hidup Susah.pdf. Kehidupan **3**, 34 (2015). https://doi.org/10.1017/CBO9781107415324.004
23. Zhang, X., Zhao, J., LeCun, Y.: Character-level convolutional networks for text classification. In: Advances in Neural Information Processing Systems, pp. 649–657 (2015)

24. Lund, K., Burgess, C.: Producing high-dimensional semantic spaces from lexical co-occurrence. Behav. Res. Methods Instruments Comput. **28**, 203–208 (1996). https://doi.org/10.3758/BF03204766
25. Nenkova, A., McKeown, K.: A survey of text sum-marization techniques. In: Aggarwal, C., Zhai, C. (eds.) Mining Text Data, pp. 43–76. Springer, Boston (2012). https://doi.org/10.1007/978-1-4614-3223-4_3
26. Lloret, E., Palomar, M.: Text summarisation in progress: a literature review. Artif. Intell. Rev. **37**, 1–41 (2012)
27. Brown, P.F., Desouza, P.V., Mercer, R.L., et al.: Class-based n-gram models of natural language. Comput. Linguist. **18**, 467–479 (1992)
28. Deerwester, S., Dumais, S.T., Landauer, T.K., et al.: Improving Information-Retrieval with Latent Semantic Indexing (1988)
29. Slonim, N., Tishby, N.: Document clustering using word clusters via the information bottleneck method. In: Proceedings of the 23rd Annual International ACM SIGIR Conference on Research and Development in Information Retrieval, pp. 208–215 (2000)
30. Bojanowski, P., Grave, E., Joulin, A., Mikolov, T.: Enriching word vectors with subword information. Trans. Assoc. Comput. Linguist. **5**, 135–146 (2017). https://doi.org/10.1162/tacl_a_00051
31. Ramos, J.: Using TF-IDF to determine word relevance in document queries. In: Proceedings of the First Instructional Conference on Machine Learning, Piscataway, NJ, pp. 133–142 (2003)
32. Yu, C.T., Lam, K., Salton, G.: Term weighting in information retrieval using the term precision model. J. ACM **29**, 152–170 (1982). https://doi.org/10.1145/322290.322300
33. Salton, G., Buckley, C.: Term-weighting approaches in automatic text retrieval. Inf. Process. Manag. **24**, 513–523 (1988). https://doi.org/10.1016/0306-4573(88)90021-0
34. Pennington, J., Socher, R., Manning, C.: Glove: global vectors for word representation. In: Proceedings of the 2014 Conference on Empirical Methods in Natural Language Processing (EMNLP), pp. 1532–1543. Association for Computational Linguistics, Stroudsburg (2014)
35. Choi, J., Lee, S.W.: Improving FastText with inverse document frequency of subwords. Pattern Recognit. Lett. **133**, 165–172 (2020). https://doi.org/10.1016/j.patrec.2020.03.003
36. Lau, J.H., Baldwin, T.: An empirical evaluation of doc2vec with practical insights into document embedding generation. In: Proceedings of the 1st Workshop on Representation Learning for NLP, pp. 78–86. Association for Computational Linguistics, Stroudsburg (2016)
37. Padurariu, C., Breaban, M.E.: Dealing with data imbalance in text classification. Procedia Comput. Sci. **159**, 736–745 (2019). https://doi.org/10.1016/j.procs.2019.09.229
38. Breiman, L., Friedman, J.H., Olshen, R.A., Stone, C.J.: Classification and Regression Trees. Routledge, Milton Park (2017)
39. Quinlan, J.R.: Induction of decision trees. Mach. Learn. **1**, 81–106 (1986). https://doi.org/10.1007/bf00116251
40. Hastie, T., Tibshirani, R., Friedman, J.: The Elements of Statistical Learning: Data Mining, Inference, and Prediction. Springer, New York (2009)
41. Breiman, L.: Random forests. Mach. Learn. **45**, 5–32 (2001)
42. Freund, Y., Schapire, R., Abe, N.: A short introduction to boosting. J.-Japan. Soc. Artif. Intell. **14**, 1612 (1999)
43. Friedman, J.H.: Greedy function approximation: a gradient boosting machine. Ann. Stat. 1189–1232 (2001)
44. Bloehdorn, S., Hotho, A.: Boosting for text classification with semantic features. In: Mobasher, B., Nasraoui, O., Liu, B., Masand, B. (eds.) WebKDD 2004. LNCS (LNAI), vol. 3932, pp. 149–166. Springer, Heidelberg (2006). https://doi.org/10.1007/11899402_10

45. Kudo, T., Matsumoto, Y.: A boosting algorithm for classification of semi-structured text. In: Proceedings of the 2004 Conference on Empirical Methods in Natural Language Processing, pp. 301–308 (2004)
46. Ikonomakis, M., Kotsiantis, S., Tampakas, V.: Text classification using machine learning techniques. WSEAS Trans. Comput. **4**, 966–974 (2005)
47. Aggarwal, C.C., Zhai, C.: A survey of text classifica-tion algorithms. In: Aggarwal, C., Zhai, C. (eds.) Mining Text Data, pp. 163–222. Springer, Boston (2012). https://doi.org/10.1007/978-1-4614-3223-4_6
48. Japkowicz, N., Stephen, S.: The class imbalance problem: a systematic study. Intell. Data Anal. **6**, 429–449 (2002)
49. García, V., Sánchez, J.S., Mollineda, R.A.: On the effectiveness of preprocessing methods when dealing with different levels of class imbalance. Knowl.-Based Syst **25**, 13–21 (2012). https://doi.org/10.1016/j.knosys.2011.06.013
50. Liu, J., Hu, Q., Yu, D.: A comparative study on rough set based class imbalance learning. Knowl.-Based Syst. **21**, 753–763 (2008). https://doi.org/10.1016/j.knosys.2008.03.031
51. Haixiang, G., Yijing, L., Shang, J., et al.: Learning from class-imbalanced data: review of methods and applications. Expert Syst. Appl. **73**, 220–239 (2017). https://doi.org/10.1016/j.eswa.2016.12.035
52. Loyola-González, O., Martínez-Trinidad, J.F., Carrasco-Ochoa, J.A., García-Borroto, M.: Study of the impact of resampling methods for contrast pattern based classifiers in imbalanced databases. Neurocomputing **175**, 935–947 (2016). https://doi.org/10.1016/j.neucom.2015.04.120
53. Chawla, N.V., Bowyer, K.W., Hall, L.O., Kegelmeyer, W.P.: SMOTE: synthetic minority over-sampling technique. J. Artif. Intell. Res. **16**, 321–357 (2002). https://doi.org/10.1613/jair.953
54. Blagus, R., Lusa, L.: SMOTE for high-dimensional class-imbalanced data. BMC Bioinform. **14**, 106 (2013). https://doi.org/10.1186/1471-2105-14-106
55. Akın, A.A., Akın, M.D.: Zemberek, an open source NLP framework for Turkic languages. Structure **10**, 1–5 (2007)
56. Gan, G., Özçelik, R., Parlar, S., et al.: Türkçe Doğal Dil İşleme için Arayüzler (2019)

# Development of Face Recognition System by Using Deep Learning and Face-Net Algorithm in the Operations Processes

Ali Tunç[1], Mehmet Yildirim[1(✉)], Şakir Taşdemir[2(✉)], and Adem Alpaslan Altun[2(✉)]

[1] Kuveyt Türk Participation Bank Konya R&D Center, Konya, Turkey
{ali.tunc,mehmet-yildirim}@kuveytturk.com.tr
[2] Computer Engineering of Technology Faculty, Selcuk University, Konya, Turkey
{stasdemir,altun}@selcuk.edu.tr

**Abstract.** Biometry is a branch of science that studies the physical and behavioral characteristics of living things. Every human being has its physical characteristics. These features are unique features that distinguish people from others. Biometric systems are the systems established by examining the biometric properties of people to determine their identity. Facial recognition systems, a type of biometric system, using the facial features of individuals, identify snow-bodies, identify information at the distance between the eyes, nose, cheekbones, jawline, jaw and so on and individual. The real purpose here is to confirm for whom did the operation. With the development of technology, many organizations use biometric data as an authentication tool. Within the scope of our study, to check the accuracy of customers who make transactions by identification method during operational transactions. In this study, if the level of accuracy in the face detection system is above a certain rate, no process blocking is performed. In cases where the similarity is low, alert information is sent to the operator and re-authentication with different data is want to be required. This study has been proposed to overcome the difficulties arising from a new convolutional neural network (CNN) framework, matrix completion, and deep learning techniques. FaceNet algorithm, which is one of the deep face recognition algorithms that can be used in the recognition of face images, is the algorithm used in the study. FaceNet is a neural network that learns to map distances from face images to a compact Euclidean space that corresponds to the face similarity measure. In this study, we will explore how to develop a face detection system using the FaceNet algorithm to identify people from photographs. The picture of the customer taken during an operational transaction is converted into a matrix expression using a trained system using this algorithm. If the matrix value obtained is greater than the threshold in the learned system, the definition is considered positive. If the obtained matrix value is less than the threshold value in the learned system, the definition is considered negative. Our data and success rates show the applicability of the study.

**Keywords:** Face recognition · Deep learning · Convolutional Neural Network (CNN) · FaceNet

© The Author(s), under exclusive license to Springer Nature Switzerland AG 2021
J. Hemanth et al. (Eds.): ICAIAME 2020, LNDECT 76, pp. 93–105, 2021.
https://doi.org/10.1007/978-3-030-79357-9_10

# 1  Introduction

The introduction of concepts such as big data into our lives, the collection of data is very important for the evaluation and analysis process. The transformation of the data obtained into information has become one of the areas of work in which many companies and scientists spend the most intensive time in today's conditions. The analysis of computer-aided images and the transformation of these images into information has been carried to high levels, especially in parallel with the development of artificial intelligence and image processing approaches. Image processing algorithms that enable feature extraction to identify and determine the dominant and meaningful characteristics of existing images and videos with various machine-learning algorithms and led to the development of systemic classification or processes.

In today's engineering applications, modeling the human thinking system, thinking like a human, learning from the behavior done and exhibiting behaviors such as human beings are emphasized applications. This approach is known as naming machine learning, which is used to take place in engineering applications of a human phenomenon in consideration [1, 2]. In the field of machine learning, artificial neural networks have often been used to solve many problems. In the early 2000s, artificial neural networks, which began to become a favorite area again, are the rapid development of technology and With its development in GPUs, it has migrated from neural networks to deep neural networks in many areas. With this approach, it has been successfully used in a wide range of applications, from image processing to data identification, natural language processing to healthcare applications. One of the most important factors in the expansion and development of Deep Learning algorithms is training with a high level of labeled data resulting from the use of technology at a high level is to withstand.

Studies on facial recognition have been conducted in many areas such as facial expression analysis, gender, and race classification. Face recognition, in other words, is the process of identifying the faces in a given picture or video and determining whose face belongs to. The topic, although every semester is attractive, has become quite popular today with the expansion of technology and the increase of the internet and mobile applications can be used in all areas in society. Analysis of still images can be performed using image processing methods and extracting meaningful information from those images. Face detection and face recognition, one of the most important working elements of digital image processing applications, has been one of the topics that have been studied for many years. Different algorithms and methods for face detection and face recognition have been developed. These processes provided solutions consist of different stages such as data entry, data preprocessing and stages, data attribute extraction and attribute selection, identification and solution to these stages with the help of different algorithms.

# 2  Face Recognition

People's face expressions play an important role in the digital environment, to give information about their identity and character. It is effective to be used in a wide variety of areas such as safety, education, health, law, entertainment. In the last twenty years, the subject of biometric recognition systems has been examined in many ways, using various techniques, efficient algorithms and methods for different aspects and details of the subject Several solutions have been put forward for many problems on the subject [3]. Face expression recognition is an active field of research in various fields such as human-machine interaction, face animation and robotics [4]. Face recognition is defined as identification studies conducted taking into account the physical and behavioral characteristics of persons. Biometric systems are divided into 4 groups according to the mode of operation and the state of an application. They are listed as follows [5].

1. Recording mode (enrollment)
2. Confirmation mode (verification)
3. Recognition mode (identification)
4. Monitoring mode (screening)

They all have the same working principle, but they have some differences in the way of application and the area of use. The process of face recognition is detecting certain faces (face detection) on a complex background, the localization of these faces and the removal of features from the face areas and finally recognition and verification operations can be parsed into basic 3 steps [6]. The face records of people must first be included in the system for the face recognition system. People have registered in the system by measuring one or more physical/behavioral characteristics in one or more time zones. Then when the data of the face to be recognized come, the face features are filtered, the noise is purified, the data is evaluated statistically, and the data classifiers are trained, attempting to integrate multiple pieces of data or decision mechanisms. Raw data is pretreated, such as removal of lighting effects, and the introduction of different scales of faces to the same paint. In this way, attribute vectors are extracted from the normalized faces, simplified and intensified, classifiers are trained, and validity tests are completed.

Each person at this stage is defined by a face stencil. Integration takes advantage of systems that use multiple data representations, multiple templates, or decision mechanisms. A new template is created, and the test template is tried to recognize it by comparing the templates previously obtained in the template library. Recognition accuracy must be high, as a valid person can not be recognized or rejected by the system creates a rather negative experience. In this regard, the reliability of the face recognition system requires both a low probability of false rejection and a low probability of misconception. This is defined as "Equal Error Rate". Face recognition software typically determines 80 points on a human face. In this regard, the dots are used to measure their variables, such as the length or width of the nose in a person, the depth of the eye sockets and the shape of the cheekbones. The system works by capturing data according to the points determined in the digital image of a person's face and saving the obtained data to be the front face. This recorded data is then used as the basis for comparing with data obtained from faces in an image or video [7]. The important point in order to achieve a successful result is to work on a picture with a low noise level, which will reveal the contours of the face. There is a correct ratio between the success of recognition and the quality of the selected picture. Face-identification points are shown in Fig. 1 and Fig. 2.

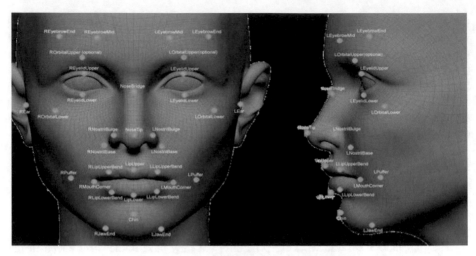

**Fig. 1.** Face recognition points on the face [8]

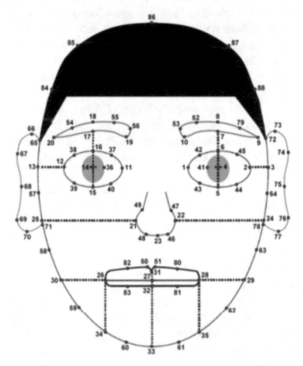

**Fig. 2.** Face recognition points on the face [9]

There are many studies in the literature in the field of face recognition. Hakan Çetinkaya and Muammer Akçay have worked on Face Recognition systems and application areas [10]. Baback Moghaddam et al. propose a new technique for direct visual matching of images for the purposes of face recognition and image retrieval, using a probabilistic measure of similarity, based primarily on a Bayesian (MAP) analysis of image differences [11]. Guodong Guo, S.Z Li and Kapluk Chan Support vector machines (SVM) have been recently proposed as a new technique for pattern recognition. SVM with a binary tree recognition strategy are used to tackle the face recognition problem [12]. Imran Naseem et al. present a novel approach of face identification by formulating the pattern recognition problem in terms of linear regression. Using a fundamental concept that patterns from a single-object class lie on a linear subspace, they develop a linear model representing a probe image as a linear combination of class-specific galleries [13]. O. Yamaguchi, K. Fukui and K.-i. Maeda present a face recognition method using image sequence. As input they utilize plural face images rather than a ldquosingle-shotrdquo, so that the input reflects variation of facial expression and face direction [14]. Priyan Malarvizhi Kumar et al. have worked on a subject that can make voice notification using face recognition system for blind people [15]. Naphtali Abudarham, et al. applied a reverse engineering approach to reveal which facial features are critical for familiar face recognition. In contrast to current views, they discovered that the same subset of features that are used for matching unfamiliar faces, are also used for matching as well as recognition of familiar faces [16].

## 3   Method

In this study, with one of the deep-learning algorithms, convolutional neural network algorithms, the face recognition problem was studied by extracting the features of the face. In this context, The FaceNet model that performs face recognition by google using CNN deep learning algorithm, an advanced version of artificial neural networks learning has been studied. The algorithms used and the work done has been given information about the results and success rates have been shared.

### 3.1   Deep Learning

The concept of deep learning is a class of machine learning from computer algorithms. In 2006 Geoffrey Hinton designed a multilayer network with deep learning machine learning algorithms has been one of the most popular research topics among. The principle of processing information between consecutive layers using many non-linear processing unit layers to make property extraction and conversion by properties is based on. Each layer became receives input from the previous layer's output [17]. Algorithms can be either supervised (classification) or unsupervised (audio/video analysis) properties. In deep learning, there is a structure based on learning more than one level of property or representations of data. These structures, developed form of artificial neural networks, are used for interlayer property extraction. A hierarchical representation is created by deriving top-level properties from down-level properties. This representation learns multiple levels of representation corresponding to different levels of abstraction [18]. Deep learning is based on the system-learning of the information in the represented data. When you call representation for an image, pixel values or density values per pixel can be evaluated as vector or edge sets, or defined properties. Some of these features can represent data better than other features. Here, feature extraction algorithms are used to reveal better representing attributes. As an advantage at this stage, deep learning methods are effective algorithms for hierarchical property extraction that best represent data rather than handcrafted features uses [19].

The distinctive aspect of Deep Learning algorithms from machine learning algorithms is the ability to process very large amounts of data and is the work of the device hardware with very high computing power. Deep learning algorithms have become very popular with the increasing use of the internet and mobile phones in recent years, the number of labeled data has increased, especially in the field of image processing, and the major technological advances in the hardware sector have provided wide working area. With deep learning methods, the best-known levels of achievement (State of the art) have been far higher [20]. On the one hand, financial and R&D technology companies (Apple, Google, Amazon, Facebook, Microsoft, etc.) are using this technology and integrating it into their products. These firms opening the algorithms developed to developer communities support the rapid progress in this area. Many models have been developed for deep learning. The most basic deep learning models are Convolutional Neural Network (CNN), Recurrent Neural Network (RNN), Sparse Auto Encoder (SAE) and Deep Belief Networks (DBN). These models are the most commonly used. We conducted our work on the Convolutional Neural Network (CNN) model.

When researching the literature, Ganin and Kononenko have conducted a study that changed the views of people in the picture using deep learning algorithms [21]. Isola and Zhu have done a study that translates the drawn objects from the map draft into a real picture [22]. Assael and Shillingford have performed to create voices by looking at the lip movements of people in a silent video [23]. Hwang and his colleagues have done a study on the coloring of images [24]. Graves worked on a study that translates normal writing into handwriting [25]. Companies such as Google and Facebook have developed different systems based on labeling and analyzing images. Cao Simon has conducted a study with predictions on what people would do according to their posture and movements [26]. Hinton et al. have conducted various studies on voice recognition [27], and Lanchantin and his friends have studied medical using deep learning algorithms [28].

## 3.2 Convolutional Neural Network

Convolution Neural Network (CNN) is a type of multi-layer sensor (Multi-Layer Perceptron-MLP). It is a structure based on filtration based on an approach based on animal vision system [29]. The specified filter makes the image's attributes distinct and uses different sizes and values to reveal low dominance properties [30]. CNN is one of the most commonly used deep learning models in image analysis processes and computerized vision studies in recent years. CNN begins with the implementation of the convolution process to the matrix obtained from the pixel values of the image at its initial stage. Convolution allows the subtraction of low and medium and high level attributes in the initial stages according to the type of filter in which it is applied. Learning low-level attributes in the first layers, the model transfers the most dominant pixels and attributes learned from the first layer to subsequent layers. This step has given 'feature learning' or 'transfer learning' names.

CNN algorithm consists of certain stages an introduction, convolution (Convolution-Conv), linearized linear unit (Rectifiedlinearunit-RELU), pooling, flattening, fully connected layers (Fully connected layer-FC) layers and activation function. Data is taken at the entrance stage. In the Convolution phase, the conversion process is based on the process of moving a particular filter over the entire image, and filters are an integral component of the layered architecture. Filters create output data by applying convolution to images from the previous layer. The activation map (Feature map) is generated as a result of this convolution process. (Rectified Linear Units Layer (ReLu)) follows the convolution layers and is known as the rectifier unit that is activated for the output of CNN neurons. The effect on the input data is that it draws negative values to zero. The pooling layer is usually placed after the ReLu layer. Its main purpose is to reduce the size of the input for the next convolution layer. This does not affect the depth size of the data system. The operation performed on this layer is also known as "Downsampling." The CNN architecture follows the concomitant convolution, ReLu, and the repositing layer, followed by a Fully Connected Layer. This layer depends on information from all areas of the previous layer. In different architectures, the number of this layer may vary. CNN also performs network memorization because of the training process with big data. The DropOut Layer is used to prevent the memorization of the network. The basic logic applied in this layer is the removal of some nodes of the network. The classification Layer follows the fully linked layer. The classification process is carried out in this layer of deep learning architecture. The output value of this layer is equal to the number of objects to be classified as a process result. Generally, a CNN model is depicted in Fig. 3.

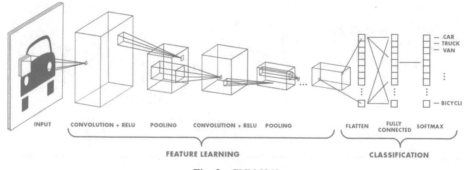

**Fig. 3.** CNN [31]

## 3.3  Face-Net

FaceNet provides a unified placement for face recognition, verification, and clustering tasks. Learning a direct mapping from face images, FaceNet is mapped to a compact Euclidean space where distances correspond to the measure of direct face similarity. This method uses a deep evolutionary network to optimize the embedding itself rather than a middle layer of congestion. Florian Schroff et al. given a face image develop FaceNet, so it will release high-quality features and these features are called face recessed 128 element vector representation is the foreseen face recognition system [32]. The model is a deep evolutionary neural network trained with a triple loss function while encouraging vectors of the same identity to be more similar (located at a smaller distance). The vectors of different identities located at a larger distance are expected to become less similar. This system is used to extract high-quality features from faces called face embedding, which can be used to train the face recognition system. The main difference between FaceNet and other techniques is that it learns the map from images and creates embedded systems instead of using any narrow throat layer for recognition or verification tasks.

All other operations, such as validation and recognition, are then performed using the standard techniques of that custom field, using newly created embedded systems as property vectors. Model structure Fig. 4 is also shown. Another important aspect of FaceNet is the loss function. Figure 5 show the Triple Loss function. To calculate the triple loss, the anchor needs 3 images - positive and negative. The system is trained as an image of one face ("anchor"), another image of the same face ("positive sample") and the image of a different face ("negative sample").

**Fig. 4.** Triplet model structure [32]

**Fig. 5.** Triplet loss [32]

## 4   Study and Results and Discussion

The main purpose of the study is to check customer accuracy by using an additional face recognition system to support operators who control identity during operational operations is to help in making. It is the similarity ratio of a person registered in the system by comparing the face picture taken during the transaction with the images of that person in the database. To do this, the customer's information must be stored in the system first. In the process of registering each customer to the system, three images of the customer are recorded in the database with the basic information of the customer. The purpose of taking three pictures here is to obtain pictures of the customer according to different angles or situations of view. The images taken are sent to the service on the virtual server located on the Google Cloud platform, shown in Fig. 8 and Fig. 9. The 512 array byte value returned from the service is saved in the database. In this way, the information of the customers recorded in the system will be compared with the images that will be taken later, and the pictures that are above a certain threshold are considered correct. Basic personal data and images taken are kept in tables prepared in the MSSQL Server database. Customers' images are also kept in the database in the form of a byte array. In this way, the pictures will be taken later and will be compared over the values and results of the bytes will be showing according to the threshold value and accuracy levels. For the accuracy and success rate of the system to be high in the picture shooting stage; characteristics of the camera taking the picture, having a single person's facial expression in the picture, ambient light, shooting quality of the pictures is very important that the noise data on the pictures are minimal. In Fig. 6, a screenshot showing the stage of recording customers is shown.

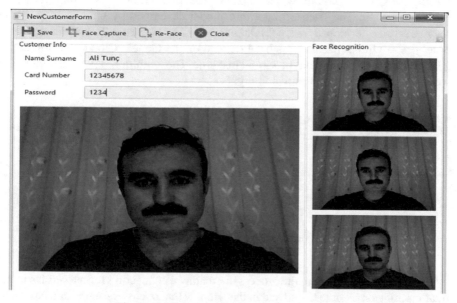

**Fig. 6.** Registration of customers in the system

After registering customers in the system, when the customer applies to the operator to perform an operational transaction, the customer's identity information is checked by the operator. In addition to this process, with the form screen in the application we developed, 3 pictures of the customers are taken within 10 s. The captured images are compared with those of the customer previously saved with the help of face recognition algorithms. The result of this face recognition process is used for customer accuracy in addition to the identification information. If the result from the face recognition algorithm is successful, the system will print the successful information. If it is unsuccessful, the operator is informed by an additional warning message that the check operation has failed. The forming screen we developed for the Validate screen is shown in Fig. 7 also shown.

**Fig. 7.** Validation of customers

To mention how the customer's face images are verified during operation, first of all, using the CNN algorithm, which is one of the deep learning algorithms as a face recognition system The Facenet algorithm model developed by Google is available through a service developed by PYTHON programming language. To provide access to this service, a virtual machine has been defined on the Google Cloud platform and it has been enabled to operate on the virtual machine. The definition representation of the respective virtual machine is shown in Fig. 8.

**Fig. 8.** Setting up the virtual machine on Google Cloud

A new window opens with the process of creating a connection for the service in the virtual machine on Google Cloud and in the window that opens username and password of the admin user of our server information is entered. Then, in turn, enter the directory in which the service is located, the necessary commands for running the python service are entered. After entering the corresponding directory, the "python3 restapi.py" command is provided to run the service that will convert the values of the face recognition points into the byte directory. When the service is running, it listens to request requests from the respective screens and sends the expressions it detected here as a response statement by translating them to the byte index according to the facenet model. The screenshot of the console screen on which the service is running is shown in Fig. 9.

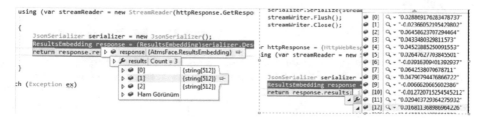


**Fig. 9.** Running the service on the virtual machine on Google Cloud

In this study, we can summarize how face recognition performs. Images taken on the Validate screen are accessed with IP information to the service on the virtual server defined on Google Cloud, sending processed images in the service. This Python service translates incoming images into a 512 element byte array according to the facenet algorithm model and returns them as a string. On the Validate screen, this incoming information is compared with the pictures taken at the customer's registration stage. The comparison is made according to a threshold value. According to this threshold value, the result is shown as successful or unsuccessful. Figure 10 shows the 512 element series where the service of the customer's picture is returned.

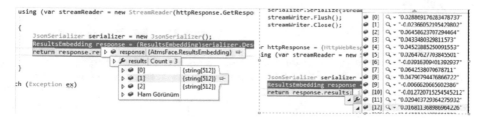

**Fig. 10.** Improving array values for images

To determine the threshold value, a study was conducted through my image. The images I took by the system were sent to the algorithm on the service and recorded the returned array values. Then the pictures taken from the validate screen are again sent to the algorithm on the service. Returning array values and values in the database were matched with a value nuance of 0.7. This value of 0.7 was considered a threshold value, and this threshold value was used for all subsequent image matches. Successful results were obtained when the study was evaluated in general. To make accurate face recognition, the most important points that affect success are the quality of the cameras used, the ambient light is appropriate and the noise effects on the pictures are minimal. summarized. More precise picture verifications can be made based on the arrangement of the threshold value.

## 5 Conclusions and Future Works

With the expansion of virtual server services through Cloud Technologies and the introduction of deep learning algorithms developed by large companies to developers studies will continue to increase day by day. Especially large data sets had opened to developers anonymously, the spread of algorithms and models trained with this data, the success rates of developers by experimenting with different will lead new studies that will achieve the chance of benchmarking and increasing. The difference that distinguishes our work from other studies is that the facenet algorithm trained data set can be applied to any field that requires authentication. The applicability and success rates in the study have always been tried to be high by updating the threshold values learned by the system. With the help of FaceNet is a deep convolutional network designed by Google algorithm, we have done our work using a system trained with images of hundreds of well-known people in terms of authentication has been attempted to support operator users as an auxiliary application. Moving success rates to high values and further shortening of transaction times are the two most important issues that are targeted to be done in the next work of the project. Further studies will be carefully studied on these issues, ensuring that face recognition is used for verification purposes in all Operational services. Furthermore, the choice of the right person on screens with multiple face images and the ability to identify only the face image of that person is also among future studies.

## References

1. Goldberg, D.E., Holland, J.H.: Genetic algorithms and machine learning. Mach. Learn. **3**(2), 95–99 (1988)
2. Quinlan, J.R.: Induction of decision trees. Mach. Learn. **1**(1), 81–106 (1986)
3. Zhao, W., Chellappa, R., Phillips, P.J., Rosenfeld, A.: Face recognition: a literature survey. ACM Comput. Surv. **35**, 399–459 (2003)
4. Hariri, W., Tabia, H., Farah, N., Benouareth, A., Declercq, D.: 3D facial expression recognition using kernel methods on Riemannian manifold. Eng. Appl. Artif. Intell. **64**, 25–32 (2017)
5. Jain, A.K., Ross, A., Pankanti, S.: Biometrics: a tool for information security. IEEE Trans. Inf. Forensics Secur. **1**(2), 125–143 (2006)
6. Cevikalp, H., Neamtu, M., Wilkes, M., Barkana, A.: Discriminative common vectors for face recognition. IEEE Trans. Pattern Anal. Mach. Intell. **27**(1), 4–13 (2005)
7. https://searchenterpriseai.techtarget.com/definition/facial-recognition
8. https://vpchothuegoldenking.com/tr/face-recognition-how-it-works-and-what-will-happen-next/
9. Özkaya, N., Sagiroglu, S.: Parmakizinden yüz tanima. Gazi Üniv Müh. Mim. Fak. Der. Cilt **23**(4), 785–793 (2008)
10. Çetinkaya, H., Akçay, M.: Yüz Tanıma Sistemleri ve Uygulama Alanları, Akademik Bilişim'12 - XIV. Akademik Bilişim Konferansı Bildirileri 1-3 Şubat 2012 Uşak Üniversitesi
11. BabackMoghaddam, T.: AlexPentland, Bayesian face recognition. Pattern Recogn. **33**(11), 1771–1782 (2000)
12. Guo, G., Li, S.Z., Chan, K.: Face recognition by support vector machines. In: Proceedings Fourth IEEE International Conference on Automatic Face and Gesture Recognition (Cat. No. PR00580), 06 August 2002. Print ISBN 0-7695-0580-5
13. Naseem, I., Togneri, R., Bennamoun, M.: Linear regression for face recognition. IEEE Trans. Pattern Anal. Mach. Intell. **32**(11), 2106–2112 (2010)

14. Yamaguchi, O., Fukui, K., Maeda, K.-I.: Face recognition using temporal image sequence. In: Proceedings Third IEEE International Conference on Automatic Face and Gesture Recognition (1998). Print ISBN: 0-8186-8344-9

15. Kumar, P.M., Gandhi, U., Varatharajan, R., Manogaran, G., Jidhesh, R., Vadivel, T.: Intelligent face recognition and navigation system using neural learning for smart security in internet of things. Cluster Comput. **22**(4), 7733–7744 (2019)

16. Abudarham, N., Shkiller, L., Yovel, G.: Critical features for face recognition. Cognition **182**, 73–83 (2019)

17. Deng, L., Yu, D.: Deep learning: methods and applications. Found. Trends® Signal Process. **7**(3–4), 197–387 (2014)

18. Bengio, Y.: Learning deep architectures for AI. Found. Trends® Mach. Learn. **2**(1), 1–127 (2009)

19. Song, H.A., Lee, S.-Y.: Hierarchical representation using NMF. In: Lee, M., Hirose, A., Hou, Z.-G., Kil, R.M. (eds.) ICONIP 2013. LNCS, vol. 8226, pp. 466–473. Springer, Heidelberg (2013). https://doi.org/10.1007/978-3-642-42054-2_58

20. https://devblogs.nvidia.com/parallelforall/mocha-jl-deep-learning-julia. Accessed 19 Dec 2019

21. Ganin, Y., Kononenko, D., Sungatullina, D., Lempitsky, V.: DeepWarp: photorealistic image resynthesis for gaze manipulation. In: Leibe, B., Matas, J., Sebe, N., Welling, M. (eds.) ECCV 2016. LNCS, vol. 9906, pp. 311–326. Springer, Cham (2016). https://doi.org/10.1007/978-3-319-46475-6_20

22. Isola, P., Zhu, J.-Y., Zhou, T., Efros, A.A.: Image-to-image translation with conditional adversarial networks. arXiv preprint arXiv:1611.07004 (2016)

23. Assael, Y.M., Shillingford, B., Whiteson, S., de Freitas, N.: LipNet: End-to-End Sentence-level Lipreading (2016)

24. Hwang, J., Zhou, Y.: Image Colorization with Deep Convolutional Neural Networks (2016)

25. Graves, A.: Generating sequences with recurrent neural networks. arXiv preprint arXiv:1308.0850 (2013)

26. Cao, Z., Simon, T., Wei, S.-E., Sheikh, Y.: Realtime multi-person 2D pose estimation using part affinity fields. arXiv preprint arXiv:1611.08050 (2016)

27. Hinton, G., et al.: Deep neural networks for acoustic modeling in speech recognition: the shared views of four research groups. IEEE Signal Process. Mag. **29**(6), 82–97 (2012)

28. Lanchantin, J., Singh, R., Wang, B., Qi, Y.: Deep Motif Dashboard: Visualizing and Understanding Genomic Sequences Using Deep Neural Networks. arXiv preprint arXiv:1608.03644 (2016)

29. Hubel, D.H., Wiesel, T.N.: Receptive fields and functional architecture of monkey striate cortex. J. Physiol. **195**(1), 215–243 (1968)

30. Fukushima, K., Miyake, S.: Neocognitron: a self-organizing neural network model for a mechanism of visual pattern recognition. In: Amari, S., Arbib, M.A. (eds.) Competition and Cooperation in Neural Nets, pp. 267–285. Springer, Heidelberg (1982). https://doi.org/10.1007/978-3-642-46466-9_18

31. https://towardsdatascience.com/a-comprehensive-guide-to-convolutional-neural-networks-the-eli5-way-3bd2b1164a53

32. Schroff, F., Kalenichenko, D., Philbin, J.: FaceNet: A Unified Embedding for Face Recognition and Clustering. Computer Science. Computer Vision and Pattern Recognition (2015)

# Mobile Assisted Travel Planning Software: The Case of Burdur

Ali Hakan Işık$^{(\boxtimes)}$ ⓘ and Rifai Kuçi

Department of Computer Engineering, Burdur Mehmet Akif Ersoy University, Burdur, Turkey
ahakan@mehmetakif.edu.tr

**Abstract.** Tourism activities requires different alternatives such as history, culture, nature and shopping. In this context, people want to visit different tourism destinations according to their travel preferences. Therefore, they need user friendly mobile applications. In this study, mobile application was developed for tourism purposes. In the developed application, users are able to take information about the location and investigate desired location that is related to other location on Google map, and also create a personal travel route after the information is obtained. Calculation of the shortest route for preferred places is realized with genetic algorithm. The optimum sightseeing route of the preferred places is created with the help of mapbox and displayed visually on the mobile map. Accommodation and eating places close to the obtained route are visualized on the map. In this study, unlike the other studies in the literature, there are walking and cycling options when calculating the route. In the presented study, the developed mobile application provides information about tourism destinations, find the shortest route, save expenses, and provide information about eating and accommodation places on the appropriate route.

**Keywords:** Genetic algorithm · Travelling salesman problem · Android · Route planning

## 1 Introduction

Due to the busy pace of work, the density of the city, the stress of work and the boredom of the same places and the need to find different activities such as traveling and walking are increasing day by day. Nowadays, people travel for different reasons in terms of their reasons. Visiting places consist of different alternatives such as coastal edges, cultural places, ancient cities, natural areas, entertainment places, flavors belonging to different regions. Each person has a different understanding of traveling. A person may want to feel the city by walking around the streets and talking to people. Another one may like historical places, nature and works of art. Therefore, travel plan should be done in private at the request of the person travel plans. In the developed mobile application, users are able to take information about the location, can see desired location that is related to other location on Google map, and also create a personal travel route after these informations are obtained. Travel plans can be created on Google map, and the

© The Author(s), under exclusive license to Springer Nature Switzerland AG 2021
J. Hemanth et al. (Eds.): ICAIAME 2020, LNDECT 76, pp. 106–112, 2021.
https://doi.org/10.1007/978-3-030-79357-9_11

users can be informed about the most suitable accommodation and eating places with the user comments with the google services. In addition, application provides the contact information of the accommodation and eating places for the users so that they can book accommodation and eating places before coming to the relevant city. Thus, the users can create a more comfortable route about places. The trip route can be determined by the user in the shortest way through the genetic algorithm and presented to the user with the Google and Mapbox map.

In literature, there are a lot of studies regarding to the traveling salesman problem optimization with genetic algorithm. In the study of Narwadi etc. all, genetic algorithm is used to find the relevant route between 5 and 20 selected cities. Helshani solved the problem of travelling salesman problem using android mobile application with web services. The routing information sent from the mobile device to the server. The genetic algorithm based software runs on the server. Suitable route is displayed with the mobile android device via web services and route also displayed on the map. Kulkarni realized the vehicle routing problem with Python programming language using Google map and Google API. Dwivedi etc. all solved the general purpose traveling salesman problem using a new cross method in genetic algorithm. In another study, Helshani solved the problem of traveling salesman using simulated annealing algorithm with the help of RESTFul web service and Google services. İlkucar and his colleague solved the problem of routing with the help of the genetic algorithm for the tourism purposes [1–7].

In this study, the shortest travel plan was calculated using the genetic algorithm. The required data is determined by the user on Android-based devices and visualized in Google and Mapbox maps. In the application, the location of the user is requested. Thus, it is possible to reorganize the trip plan. All operations are done on the mobile device.

## 2   Travelling Salesman Problem

The simplest and most famous traveler route problem is known as travelling salesman problem (TSP). In this problems, a seller 2, …. , n should visit the city and this journey should start at 1. Starting from city 1, seller should visit each city only once in a certain order and eventually return to the starting city. The aim is to determine the order of the visit, giving the minimum cost. Typical salesman problem chart is shown in Fig. 1. Based on this explanation, we can describe TSP briefly as follows: The problem of traveling salesman is the problems that search the route that will give the shortest total distance to visit each point in a certain cluster. Having a wide range of applications in problems such as distribution, planning, logistics, routing and planning, the travelling salesman problem is also a problem in the field of optimization, in the NP-hard class, which has been studied by researchers for many years [3, 7]. This problem can be solved with the genetic algorithm.

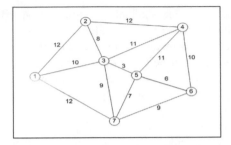

**Fig. 1.** Traveling salesman problem chart

Genetic Algorithm (GA) is a heuristic algorithm that imitates living things' survival skills and investigates more than one possible solution at the same time. In the genetic algorithm, the most successful solutions can be transferred to the next solution chromosome thanks to the selection, crossover and mutation functions inspired by the evolutionary processes in nature. In addition, more successful solutions can be obtained by exchanging parameters between solution sets. Thanks to the mutation function, which is one of its strongest feature, GA can include parameters that do not fit a certain pattern and cannot be predicted into the solution set. Thus, different solutions can be investigated without relying only on local solutions [7]. The flow diagram in which the components of the GA are shown together with the working principle is presented in Fig. 2.

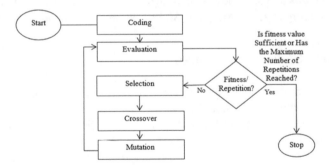

**Fig. 2.** Flow diagram of a traditional GA

## 3  Findings (Study)

In the study, a mobile application has been developed to find the shortest route on Google maps. In the first part of the application, what is included in the application is explained to the users as a form of slider. Thus, the user is informed about the places to visit in Burdur. This information helps the user to choose the more appropriate route to visit. Two examples of sliders that we developed are shown in Fig. 3.a and Fig. 3.b.

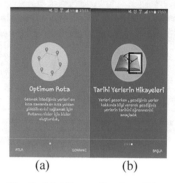

(a)    (b)

**Fig. 3.** Mobile application sliders

After making the necessary explanations in the sliders, the main part of our application was started. In the main part of the application, important places in Burdur are marked with a marker in different colors. It is shown in in Fig. 4.a. For example; Ulu Mosque, Hacılar Höyük, İncir Inn places are shown in green color as they are historical places, Burdur Lake, Salda Lake and Gölhisar Lake are in shown Turquoise color. In this way, the user can determine the route by taking into account the connection of places to visit. The markers can be used to get information where the user wants to go. This is shown in Fig. 4.b. The user can listen to the written information by clicking the sign on the top of the Slider. After getting information about places to visit, the route starts to be created. In the list that opens, the places to be chosen are selected It is shown in Fig. 4.c. Then the optimum route is found with genetic algorithm.

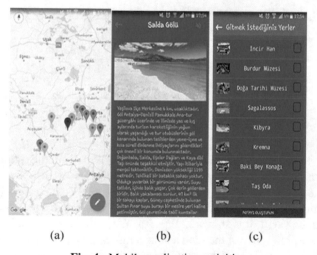

(a)    (b)    (c)

**Fig. 4.** Mobile application activities

For example, the distances from each other were calculated as shown in Table 1 by taking the latitude and longitude information of the places in Table 2. Google API is used in this calculation.

**Table 1.** Latitude-longitude information of travel places

| Trip point | Place name | Latitude, Longitude |
|---|---|---|
| 1 | İncir han | (37.478628, 30.533486) |
| 2 | Salda lake | (37.552650, 29.672964) |
| 3 | Burdur lake | (37.735082, 30.170443) |
| 4 | Sagalassos | (37.677311, 30.516952) |
| 5 | Taş Oda | (37.717124, 30.288768) |

The distance is based only on the distance between two places. Alternatively, information such as arrival time, distance traveled by walking and cycling, etc. can be obtained and the most optimum route can be found with the genetic algorithm. As seen in Table 2, when there are 5 travel destinations in total, the size of the chromosome in the genetic algorithm should also be appropriate. Since each individual represents a travel tour, it becomes an element of the solution set.

**Table 2.** Distance information of travel places

|  | İncir han | Sagalassos | Taş Oda | Burdur lake | Salda lake |
|---|---|---|---|---|---|
| İncir han | - | 36.3 km | 46,7 km | 53,6 km | 123,0 km |
| Salda lake | 36,2 km | - | 39,5 km | 46,3 km | 116,0 km |
| Burdur lake | 46,7 km | 39,5 km | - | 7,6 km | 77,0 km |
| Sagalassos | 53,4 km | 46,2 km | 8,7 km | - | 73.7 km |
| Taş Oda | 123,0 km | 116,0 km | 78,2 km | 73,6 km | - |

According to the travel route we gave as an example in our study, the user; Starting from İncir Han, it will go to Sagallassos, from Sagalassos to Taş Oda, from Taş Oda to Burdur Lake and Burdur Lake to Salda Lake, the total distance it will cover will be 157.1 km. With the help of the work we have made the user's route, it was drawn with the help of Mapbox API. It is shown in Fig. 5.a. While the user is visiting these places, depending on the preference of the food and accommodation places next to the route, the food icon at the bottom right of the activities and the accommodation icon can be seen. These are shown in Fig. 5.b and Fig. 5.c. We marked in different colors to distinguish between eating places and accommodation. Accommodations are shown with a green marker while meals are shown with a blue marker. According to the obtained route, the closest eating and accommodation places are shown on the map. For example; The user will be able to find accommodation and eating places close to the route we have specified above. In Fig. 5.b, Salda lake hotel can be seen as the accommodation in our example. By clicking on the marker on the Salda lake hotel, traveler able to access the contact information of the accommodation and make the right decision when choosing the accommodation according to the google comment points we obtained from google services. The same information can be obtained for Serenler hill restaurant as an eating place. Our application offers eating place and accommodation only as a suggestion. If the user does not want to see eating place and accommodation, he can disable the icons in the bottom right slider in the application [8].

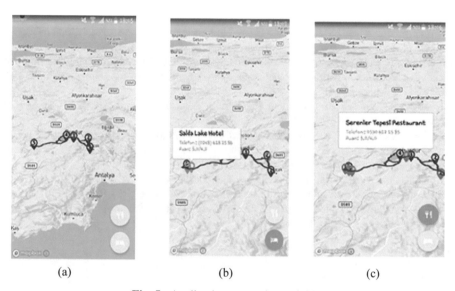

|       (a)       |       (b)       |       (c)       |

**Fig. 5.** Application promotion activities

## 4 Conclusions

In our study, a mobile application has been developed to increase the demand for tourism. Thus, while traveling in any city, it will be possible to make visits in a shorter time by saving a great deal of cost and time. In the developed mobile application, additional information regarding to the accommodation, eating and drinking places are given with the help of sliders. Information about the relevant places can be obtained by pressing and holding on the places indicated by the markers on the Google map and the information can also be listened. A special preference screen has been prepared in order to prefer places to visit. After creating the trip route, the most optimum travel route was shown on the map with the help of Google and Mapbox maps through the genetic algorithm. The location of the user can be taken from GPS system and the distance to the route can be seen. According to the obtained route, the closest eating and accommodation places are shown on the map. Our application was made with the support of Turkish language during the testing phase. In addition to these languages, it is aimed to add world languages such as English, German and Chinese during the development period.

# References

1. Helshani, L.: An android application for Google map navigation system, solving the traveling salesman problem, optimization throught genetic algorithm. In: Proceedings of FIKUSZ Symposium for Young Researchers, pp. 89–102 (2015)
2. Helshani, L.: Solving the traveling salesman problem using google services and simulated annealing algorithm. Eur. Acad. Res. **4**(3), 2321–2330 (2016)
3. lkuçar, M, Çetinkaya, A.: Mobil telefon ve google harita destekli yerel seyahat rotasi optimizasyonu: burdur örneği. Mehmet Akif Ersoy Üniversitesi İktisadi ve İdari Bilimler Fakültesi Dergisi **5**(1), 64-74 (2018)
4. Narwadi, T., Subiyanto: An application of traveling salesman problem using the improved genetiv algorithm on android google maps. In: AIP Conference Proceedings, pp. 020035-1–020035-11, March 2017
5. Kulkarnı, S.: Vehicle routing problem solver. Int. J. Eng. Res. Technol. (IJERT) **5**(12), 226–229 (2016)
6. Dwivedi, V., Chauhan, T., Saxena, S., Agrawal, P.: Travelling salesman problem using genetic algorithm. In: IJCA Proceedings on Development of Reliable Information Systems, Techniques and Related Issues (DRISTI 2012), pp. 25–30 (2012)
7. Bilen, M., Işik, A.H., Yiğit, T.: İnteraktif ve Web Tabanlı Genetik Algoritma Eğitim Yazılımı. Süleyman Demirel Üniversitesi Fen Bilimleri Enstitüsü Dergisi **21**(3), 928–934 (2017)
8. T.C. Burdur Governorship Culture and Tourism City Burdur (2020). http://www.burdur.gov.tr/kultur-ve-turizm-sehri-burdur

# Text-Based Fake News Detection via Machine Learning

Uğur Mertoğlu[1]([⊠]), Burkay Genç[1], and Hayri Sever[2]

[1] Department of Computer Engineering, Hacettepe University, Ankara, Turkey
umertoglu@hacettepe.edu.tr
[2] Department of Computer Engineering, Çankaya University, Ankara, Turkey

**Abstract.** The nature of information literacy is changing as people incline more towards using digital media to consume content. Consequently, this easier way of consuming information has sparked off a challenge called "Fake News". One of the risky effects of this notorious term is to influence people's views of the world as in the recent example of coronavirus misinformation that is flooding the internet. Nowadays, it seems the world needs "information hygiene" more than anything. Yet real-world solutions in practice are not qualified to determine verifiability of the information circulating. Presenting an automated solution, our work provides an adaptable solution to detect fake news in practice. Our approach proposes a set of carefully selected features combined with word-embeddings to predict fake or valid texts. We evaluated our proposed model in terms of efficacy through intensive experimentation. Additionally, we present an analysis linked with linguistic features for detecting fake and valid news content. An overview of text-based fake news detection guidance derived from experiments including promising results of our work is also presented in this work.

**Keywords:** Fake news detection · Machine learning · News credibility · Word2Vec · Text-based features

## 1 Introduction

Although there are many definitions in the literature, fake news can be simply defined as a kind of sloppy journalism, or misinformation that involves deliberate/nondeliberate deception spreading mainly through conventional media, popular media outlets and social media.

Recently, many issues related with research topics such as journalism, politics, health, sports, and information technologies etc. are directly related to or effected by "fake news". And this became a norm, because within the last decade we have witnessed many global crisis and important events some of which are shown in Table 1 and "fake news" is shown as the triggering factor for them. It later became a subject of interest within the field of Information Technology (IT). It is most likely that different types of fake news (e.g. click-baits, propaganda, misinformation, satire, junk-science) has different motivations, but almost all of them have severe effects. Consequently, it is seen that the fake news may take on various faces in different fields.

© The Author(s), under exclusive license to Springer Nature Switzerland AG 2021
J. Hemanth et al. (Eds.): ICAIAME 2020, LNDECT 76, pp. 113–124, 2021.
https://doi.org/10.1007/978-3-030-79357-9_12

**Table 1.** An overview of some important events associated with fake news over the past decade.

| Fake news with the related events |
| --- |
| Weapons of mass destruction (WMD) reports and the growing belief that Iraq would attack many countries [1] |
| Triggering violence against ethnic minorities in countries such as Sri Lanka and Myanmar [2] |
| Power elites-sponsored fake news causing the Arab spring uprisings [3] |
| Ukraine&Russia conflict [4] |
| The misinformation circulated in social media in the aftermath of bombing at the Boston Marathon [5] |
| Fake news stories associated with US Presidential Elections [6] |
| Pizzagate, a discredited conspiracy theory that went viral [7] |
| Computational propaganda of the Brexit debate [8] |
| Disinformation and social bot operations associated with French presidential election [9] |
| The case of Cambridge Analytica&Facebook: Misinformation causing damage to almost all of the actors [10] |
| Many fake stories spreading online about refugees causing polarization |
| Despite the lack of any scientific evidence, disinformation spreading online about the dangers of 5G technology regarding COVID-19 |

The examples are much more than presented in Table 1, including urban legends, junk-science, disinformation etc. The accepted opinion about fake news in all these events is that even if fake news is not the root factor at all, but it makes the situation much more chaotic. Once it was understood that interdisciplinary approaches, collaborative effort, and computational solutions were required, researchers particularly from IT departments started to study on the topic. But the truth verification also known as fact-checking is a difficult problem. So far, even fact-checking organizations, pioneering technology companies or big media platforms have not been able to fully cope with it.

In this paper, we focus on textual content-based fake news detection, but it should not be forgotten that, as Sina et al. [11] made clear, credibility of the news is not limited to the correctness of textual content, and other components such as images and videos have to be evaluated as well.

Moreover, there is a lack of studies in Turkish in this field. In this manner, the main purpose of this paper is to develop a text-based automated fake news detection system in Turkish, which tries to distinguish between valid or fake (fabricated) news. We offer a machine learning based approach to reveal whether a news text is worth to be considered as valid or not.

Our contribution in this paper is to present a text-based fake news detection workflow in Turkish aiming to be a guidance for the researchers interested in the field, particularly for the agglutinative languages. With motivating results obtained, we believe our work has the potential to be a basis for embodying of a sophisticated tool for determining the news credibility.

The rest of the paper is structured as follows: Sect. 2 presents a review of the relevant literature. We explain our workflow briefly together with the features and the techniques used, in Sect. 3. Then, we discuss experimental results and some other findings relatively in Sect. 4. Finally, in Sect. 5 we provide conclusions and propose possible future study topics.

## 2 Relevant Literature

Just as the popularity of the word "fake news" rose after some claims related with politics in 2016 [12], it can be said that after the influential study of Rubin et al. [13] in 2015, frequency of scholarly studies on fake news started to increase. In that study, Rubin et al. looked at fake news on a conceptual level. Soon, social scientists such as Nagi and Lazer [14, 15], started to study the philosophy and science of fake news and the impacts on society in the same period. And in a similar way, researchers especially from IT related departments has been trying to find computational solutions in recent years.

It can easily be understood from the studies in the field that challenges in this topic start from the very beginning: existence of suitable datasets. Seeing the lack of the datasets on the topic, researchers has concentrated on building fake news datasets. For example, Perêz-Rosas et al. [16] benefited from the study [17] presenting the guidelines to build a proper corpus which suggests some requirements for constructing fake news datasets that best suit the real world.

From the point of the linguistic analysis in Natural Language Processing (NLP), researchers have used different approaches utilizing several methodologies such as syntactic, grammatical, morphological, rhetorical. The scientific basis for fake news detection studies based on the deception detection studies. In the early studies [18, 19] of deception detection, some linguistic features are shown to be good indicators of falsehood. Thus, Zhou et al. used Linguistic Based Cues (LBC) and showed explicitly that linguistic information could be useful with regards to detection of deception [19]. Accordingly, Ott et al. [20] regard linguistic approaches as complementary tool to their approach.

In some machine learning studies, a variety of topics such as click-bait detection [21], spam detection [22], hoax/fraud detection [23], satire detection [24] etc. also utilized textual features. Besides linguistic analysis, network analysis, social media analysis, source analysis, semantic analysis and other hybrid analysis were also used in the literature and the studies mostly focused on English texts. Therefore, as we have mentioned before, our study differs in that we focus on the Turkish language which is structurally different than the English language.

## 3 Model

Our purpose of fake news detection is to classify Turkish news into predefined binary labels: "Fake" and "Valid". This classification is a supervised learning task [25] because the learning is based on training data with known class labels.

## 3.1   Data and Methodology

Before explaining the methodology, it should be noted the lack of fake news datasets is one prominent reason for the difficulties of fake news detection which discriminates this task from other deception detection tasks and it is same for Turkish language.

We collected the news belonging to the year 2019. The dataset consists of three topics; politics, sports and health. The statistics of the data is shown in Table 2. The whole collection of data is done automatically. We collected the first set of the data from Global Database of Events, Language and Tone (GDELT) [26] which publishes data on broadcast and news articles on the WWW. The GDELT data contain information automatically extracted from online news media around the world. This set is composed of news content from verified state and private news agencies. Moreover, we cross-checked all automatically collected news from GDELT with fact-checking organizations' databases to make sure that there is no content that is known to be fake but labelled automatically as valid in our dataset. As distinct from this verification, our research team manually verified the set taken from popular media outlets online.

Cleaning the text data such as removing punctuation, tokenization, removing stop-words, lowering cases etc. was handled next and was a challenging part of the study. This is not only because Turkish is an agglutinative language but also due to the fact that the machine learning library scikit-learn in Python or any other NTLK libraries needed extra customizations for Turkish.

**Table 2.** News data statistics.

| Collection source | Class | News topic | Count |
|---|---|---|---|
| GDELT | Valid | POLITICS | 3412 |
| | Fake | | 84 |
| | Valid | HEALTH | 845 |
| | Fake | | 52 |
| | Valid | SPORTS | 2216 |
| | Fake | | 79 |
| Popular News Sites | Valid | POLITICS | 812 |
| | Fake | | 64 |
| | Valid | HEALTH | 530 |
| | Fake | | 74 |
| | Valid | SPORTS | 740 |
| | Fake | | 99 |

The pre-processing work was done using the Turkish NLP engine Zemberek[1]. And afterwards we utilised machine learning techniques based on Word2Vec [27–29] to

---

[1] Zemberek https://code.google.com/p/zemberek.

capture the difference between fake and valid texts. We used skip-gram algorithm having a window-size of two, for the pairs to be trained. The logical model is shown in Fig. 1. Similar studies using Word2Vec in Turkish language is relatively sparse compared to English. The reason for this is in line with our previous experiences that in larger corpora TF-IDF generally gives better results than using raw Word2Vec in Turkish. One other reason can be the technique is only able to convert words/phrases into vector space representation causing a lack of finding the importance of a word in corpus. Hence, to make efficient use of Word2Vec, we did some minor adaption as weighting word vectors according to TF-IDF. We concatenated the vectors with other features selected. Contrary to our expectation, having a size of 250, the most efficient size of the representation is low.

Our data is composed of two classes: the majority (valid) class and the minority (fake) class. We used imblearn Python package to tackle the unbalanced data problem. Synthetic Minority Oversampling Technique[2] here helped us to synthesize the fake samples.

| Text<br>Önemli bilgi: Yarından itibaren tüm dünyada yeni iletişim kuralları yürürlüğe girecek.<br>(Important information: From tomorrow on, new communication rules will come into force all over the world.) | | | | | | Word pairs to be trained |
|---|---|---|---|---|---|---|
| Önemli | bilgi | Yarından | itibaren tüm dünyada yeni iletişim kuralları yürürlüğe girecek. | | | [önemli, bilgi]<br>[önemli, yarından] |
| Önemli | bilgi | Yarından | itibaren | tüm dünyada yeni iletişim kuralları yürürlüğe girecek. | | [bilgi, önemli]<br>[bilgi, yarından]<br>[bilgi, itibaren] |
| Önemli | bilgi | Yarından | itibaren | tüm | dünyada yeni iletişim kuralları yürürlüğe girecek. | [yarından, önemli]<br>[yarından, bilgi]<br>[yarından, itibaren]<br>[yarından, tüm] |

**Fig. 1.** Skip-gram model to predict the context of the word in the corpus. In the example, one sentence belongs to a fake news telling a story about bitcoin shown with its English translation.

## 3.2  Feature Selection

In this section, we will explain the features which we found more effective in terms of linguistic clues. All the features used are shown in Table 3. Some other features were excluded due to their low contribution to performance.

---

[2] SMOTE, https://imbalanced-learn.readthedocs.io/en/stable/index.html.

**Table 3.** All of the features those have been used for training.

| Feature set | Details |
|---|---|
| Punctuation marks | Number of periods, commas, question marks, exclamation marks, brackets, quotes, dashes, ellipsis |
| Text characteristics | Number of letters, capitals, abbrevations, words, function words, sentences, paragraphs |
| Part of speech | Number on nouns, adjectives, verbs, adverbs |
| News characteristics | Number of links, words in header |

### 3.2.1  Sentences Count

The expressions in fake news are mostly short, but remindful. The manner of telling the story shows some characteristics such as influencing the readers in a sense of sharing the information. When considered from this view, the anonymous quote commonly attributed to Mark Twain that "a lie can travel halfway around the world while the truth is still putting on its shoes" seems quite a pertinent remark. Because, fake news are commonly more interesting and people find it alluring. Even the number of sentences is a not full indicator alone, but gives us some clues of poor journalism. And to increase user's retention the fake news are designed short and catchy.

We used MinMaxScaler scaling method of scikit-learn to revalue features in more digestible form for the machine learning algorithms we used. The values are drawn to the scale of 0–1, preserving the shape of the original distribution. Because, the method doesn't reduce the importance of outliers, we excluded some of the outliers having extreme values. Some other features such as word count of words and paragraphs showed closer correlations with the sentences count feature, so we didn't add these features in optimized feature set.

### 3.2.2  Part of Speech (POS)

Generally, "Part of Speech" tags are defined as the words belong to classes depending on the role they assume in the sentence. In Turkish language, POS related with text classiffications is used by several NLP studies. We conducted an approach of Bag of POS as Robinson did [30], seeing many volatile counts of several POS isn't as efficient as expected. But this feature still needs some improvement stemming from the morphological disambiguation of Turkish which is discussed in detail in some of the studies [31, 32].

### 3.2.3  Punctuation Marks

Previous work on the fake news detection of Rubin et al. [17] suggests that the use of punctuation might be useful to differentiate between deceptive and truthful texts. We measured the Pearson's correlation between the style markers and the class. The prominent features extracted (normalized) from the writing-style analysis are as follows:

- Count of question marks
- Count of exclamation marks
- Count of ellipsis

### 3.2.4 Function Words

Mostly used for linking the words, function words have ambiguous meaning and are context independent. The frequencies of these words are used as a feature vector in this work and although we have a relatively small set, this feature is an effective one. Some of the most frequently used function words, such as işte (here/here it is), neden (why), belki (may be), nasıl (how), bile (even), daha (yet/else), etc. are good indicators of fake news in our corpus.

Apart from these features, we did some exploratory data analysis along with some statistical analysis which is discussed in Sect. 4.

### 3.3 Recurrent Neural Networks (RNN)

There are artificial neural network models which are adaptable for text modeling such as convolutional neural networks [33], recurrent neural networks [34] and recursive neural networks. Recurrent neural networks (RNNs) are a class of neural networks that allow previous outputs to be used as inputs while having hidden states. In this study, we used Long-Short Term Memory (LSTM) which is a special form of RNNs and is powerful when the chain of input-chunks becomes longer as in our problem. In our case, the main input is a string (processed news context) with the other features as well and the output a two-element vector indicating whether the news is fake or valid. For each timestep t, $a^t$ is the activation and $y^t$ is the output. The shared coefficients are $W_{ax}$, $W_{aa}$, $W_{ya}$, $b_a$, $b_y$ and $g_1$, $g_2$ are the activation functions. In Eq. (1, 2) the corresponding expressions are shown:

$$a^t = g_1\left(W_{aa}a^{t-1} + W_{ax}x^t + b_a\right) \tag{1}$$

$$y^t = g_2\left(W_{ya}a^t + b_y\right) \tag{2}$$

## 4 Experimental Results and Discussion

To assess the performance of different models, we compared results from three classification models: Random Forest, Support Vector Machine (SVM), Gradient Boosting Classiffier. And we also trained Long Short-Term Memory (LSTM) networks. We feed the classification models and RNN with several combinations of the extracted text features. Then, we evaluated the model with Precision, Recall, Accuracy and F-score metrics. Here, the metrics precision and recall stand for performance of the model prediction of the fake class.

Considering both Precision and Recall, F-score shows the model's quality. We got the highest F-score in the Sports domain with neural networks. The second highest score was obtained by the Gradient Boosting Classifier. In Table 4 and Table 5, we presented some of the best scores we obtained. We used an optimal set of features which refers to combination of the features explained in Sect. 3.2.

We used Keras, a Python-based neural networks library which is easy to use. The parameter tuning for RNN-LSTM is different from the other machine learning classification models. We carried out experiments (parameter tuning) to choose right parameters for number of hidden layers, drop-out rate, number of hidden units per layer etc. We used Rectified Linear Unit (ReLU) activation function to have the sparse representations.

**Table 4.** Evaluation results of Word2Vec and Optimised set of features combined together with the three classifiers trained.

| Model | Domain (Word2Vec+Optimized Features) | Precision | Recall | Accuracy | F1 Score |
|---|---|---|---|---|---|
| *Gradient Boosting* | POLITICS | 89,71 | 92,48 | 91,20 | 91,07 |
| | SPORTS | **93,24** | **93,66** | **93,46** | **93,45** |
| | HEALTH | 92,69 | 93,55 | 93,15 | 93,12 |
| | ALL | 91,97 | 93,14 | 92,59 | 92,55 |
| *SVM* | POLITICS | 85,95 | 88,96 | 87,64 | 87,43 |
| | SPORTS | 89,30 | 90,54 | 89,98 | 89,91 |
| | HEALTH | 86,51 | 89,60 | 88,23 | 88,03 |
| | ALL | 87,29 | 89,67 | 88,62 | 88,46 |
| Random Forest | POLITICS | 89,29 | 91,24 | 90,36 | 90,25 |
| | SPORTS | 91,95 | 92,62 | 92,31 | 92,28 |
| | HEALTH | 89,70 | 92,06 | 90,98 | 90,87 |
| | ALL | 90,64 | 92,13 | 91,45 | 91,38 |

**Table 5.** Evaluation results of RNN-LSTM.

| Model | Domain | Precision | Recall | Accuracy | F1 Score |
|---|---|---|---|---|---|
| Recurrent Neural Network- LSTM | POLITICS | 93,57 | 93,75 | 93,66 | 93,66 |
| | SPORTS | 95,07 | 98,15 | 96,64 | 96,59 |
| | HEALTH | 95,20 | 94,39 | 94,77 | 94,79 |
| | ALL | 94,24 | 95,26 | 94,77 | 94,75 |

We evaluated the cross-domain classification performance for the three news domains as politics, sports, and health. It can easily be seen from the results; the sports domain results are slightly better than others. We believe this is because of the sensational and attractive sort of expressions used in this domain. The results show that fake news in politics can not easily be predicted compared to other domains. Some semantic approaches should be used in this domain. In health news, where so-called science stands out, we believe better scores can be achieved with a long-term data. The results are promising for future work.

The training is slightly slower compared to other machine learning algorithms. It is normal considering the computation feeds from the historical information. And we also got slightly better results from recurrent neural networks. This shows that the difficulty of the task when approached with traditional machine learning algortihms. Experimental results show that our model can be improved when used with other features which is not directly related with text itself.

## 4.1 Exploratory Data Analysis

In this work, besides the experimental results, we also observed some interesting findings which we think worth to discuss briefly. Thus, we explored the main characteristics of the data by exploratory data analysis via visualization and statistical methods.

In fact, it can be observed that the statements which add seriousness to the news, ironically lose their meaning when used together in the fake news. Some of the interesting examples are shown in Table 6. We suggest that the numbers of examples can be increased with larger corpus. The main characteristic of these expressions is to attract people at first sight.

**Table 6.** Statements.

| Statements | Intent of Author&Truth |
| --- | --- |
| Son dakika! (Newsbreak!) | To make people feel like they will give a very surprising news. In fact, there is nothing worth to read it or it is a click-bait kind of a news |
| Çok önemli (Very Important) | To make people feel like they will learn about an important event. But there is nothing special |
| İddiaya/İddialara göre (According to the claim/s) | To support the story as if there is reasonable background of the story |
| Bunları biliyor muydunuz? (Do you know about these?) | To influence the reader as if reader will learn something quite new. But generally some confusing information is presented as junk science |
| Sürpriz Gelişme (Surprise) | An unexpected news or astonishing claims. Especially in sports news (e.g. illusive transfer rumors) |

We also examined the subject headings of news, some of our interesting findings are as follows:

- In sports domain, the transfer news has almost the same style, only the authors change the actors of the news from time to time.
- Health news, the junk science and the urban legends can easily be detected.
- In politics, especially in the certain periods of the year, approaches with identification of news actors can be helpful in term of fact-checking.
- The harmony or the relation between the header (title) and text may be caught with a semantic approach. We suggest that the title of fake news also differs as in their text from the valid ones.

## 5   Conclusion and Future Work

In this study, we applied machine learning techniques to detect fake news in Turkish using some linguistic feature vectors trained in three classification models and neural network. Our goal was predicting reliability of the news via detecting whether it is fake or valid. News content verification, in other words to verify the information in the text along with the video, photo and other components belonging the news is quite challenging when the text used only. Nevertheless, the results are quite encouraging.

Experimenting on a relatively small corpus including approxiamately 9000 news, may be criticized in our work considering in an agglutinative language such as Turkish has huge number of potential words, to be more representative the corpus should be larger. We plan to have a larger corpus soon. We believe combination of psycholinguistic features, rhetorical features, n-grams for words/characters, social context features etc. may give better performance in a big corpus, especially when used with deep learning networks.

## References

1. Adams, J.S.: Internet Journalism and Fake News. Cavendish Square Publishing, LLC (2018)
2. Taub, A., Fisher, M.: Where Countries are Tinderboxes and Facebook is a match. New York Times, April 21 (2018)
3. Douai, A.: Global and Arab media in the post-truth era: globalization, authoritarianism and fake news. IEMed: Mediterranean Yearbook 2019, pp. 124–132 (2019)
4. Khaldarova, I., Pantti, M.: Fake news: the narrative battle over the Ukrainian conflict. J. Pract. **10**(7), 891–901 (2016)
5. Starbird, K., et al.: Rumors, false flags, and digital vigilantes: misinformation on twitter after the 2013 boston marathon bombing. In: IConference 2014 Proceedings (2014)
6. Barbera, P.: Explaining the spread of misinformation on social media: evidence from the 2016 US presidential election. In: Symposium: Fake News and the Politics of Misinformation. APSA (2018)
7. Gillin, J.: How Pizzagate went from fake news to a real problem for a DC business. PolitiFact (2016)
8. Narayanan, V., et al.: Russian involvement and junk news during Brexit. Computational Propaganda Project, Data Memo 2017.10 (2017)
9. Ferrara, E.: Disinformation and social bot operations in the run up to the 2017 French presidential election. arXiv preprint arXiv:1707.00086 (2017)
10. Ireton, C., Posetti, J.: Journalism, fake news & disinformation: handbook for journalism education and training. UNESCO Publishing (2018)

11. Mohseni, S., Ragan, E., Hu, X.: Open issues in combating fake news: interpretability as an opportunity. arXiv preprint arXiv:1904.03016 (2019)
12. Bovet, A., Makse, H.A.: Influence of fake news in Twitter during the 2016 US presidential election. Nat. Commun. **10**(1), 1–14 (2019)
13. Rubin, V.L., Chen, Y., Conroy, N.J.: Deception detection for news: three types of fakes. Proc. Assoc. Inf. Sci. Technol. **52**(1), 1–4 (2015)
14. Nagi, K.: New social media and impact of fake news on society. In: ICSSM Proceedings, pp. 77–96, July 2018
15. Lazer, D.M.J., et al.: The science of fake news. Science **359**(6380), 1094–1096 (2018)
16. Perez-Rosas, V., et al.: Automatic detection of fake news. arXiv preprint arXiv:1708.07104 (2017)
17. Rubin, V.L., et al.: Fake news or truth? Using satirical cues to detect potentially misleading news. In: Proceedings of the Second Workshop on Computational Approaches to Deception Detection (2016)
18. Buller, D.B., Burgoon, J.K.: Interpersonal deception theory. Commun. Theory **6**(3), 203–242 (1996)
19. Zhou, L., et al.: Automating linguistics-based cues for detecting deception in text-based asynchronous computer-mediated communications. Group Dec. Negotiation **13**(1), 81–106 (2004)
20. Ott, M., Cardie, C., Hancock, J.T.: Negative deceptive opinion spam. In: Proceedings of the 2013 Conference of the North American Chapter of the Association for Computational Linguistics: Human Language Technologies (2013)
21. Potthast, M., Köpsel, S., Stein, B., Hagen, M.: Clickbait detection. In: Ferro, N., Crestani, F., Moens, M.-F., Mothe, J., Silvestri, F., Di Nunzio, G.M., Hauff, C., Silvello, G. (eds.) ECIR 2016. LNCS, vol. 9626, pp. 810–817. Springer, Cham (2016). https://doi.org/10.1007/978-3-319-30671-1_72
22. Akbari, F., Sajedi, H.: SMS spam detection using selected text features and boosting classifiers. In: 2015 7th Conference on Information and Knowledge Technology (IKT). IEEE (2015)
23. Sadia Afroz, Michael Brennan, and Rachel Greenstadt. Detecting hoaxes, frauds, and deception in writing style online. In: 2012 IEEE Symposium on Security and Privacy (SP), pages 461–475. IEEE (2012)
24. Reganti, A.N., et al.: Modeling satire in english text for automatic detection. In: 2016 IEEE 16th International Conference on Data Mining Workshops (ICDMW). IEEE (2016)
25. Caruana, R., Niculescu-Mizil, A.: An empirical comparison of supervised learning algorithms. In: Proceedings of the 23rd International Conference on Machine Learning (2006)
26. Global Database of Events, Language and Tone (GDELT) Project. https://www.gdeltproject.org/
27. Wensen, L., et al.: Short text classification based on Wikipedia and Word2vec. In: 2016 2nd IEEE International Conference on Computer and Communications (ICCC). IEEE (2016)
28. Word2Vec. https://code.google.com/archive/p/word2vec/
29. Mikolov, T., et al.: Efficient estimation of word representations in vector space. arXiv preprint arXiv:1301.3781 (2013)
30. Robinson, T.: Disaster tweet classification using parts-of-speech tags: a domain adaptation approach. Doctoral dissertation, Kansas State University (2016)
31. Görgün, O., Yildiz, O.T.: A novel approach to morphological disambiguation for Turkish. In: Gelenbe, E., Lent, R., Sakellari, G. (eds.) Computer and Information Sciences II, pp. 77–83. Springer, London (2011). https://doi.org/10.1007/978-1-4471-2155-8_9
32. Oflazer, K., Kuruöz, I.: Tagging and morphological disambiguation of Turkish text. IN: Proceedings of the Fourth Conference on Applied Natural Language Processing. Association for Computational Linguistics (1994)

33. Kalchbrenner, N., Grefenstette, E., Blunsom, P.: A convolutional neural network for modelling sentences. arXiv preprint arXiv:1404.2188 (2014)
34. Liu, P., et al.: Multi-timescale long short-term memory neural network for modelling sentences and documents. In: Proceedings of the 2015 Conference on Empirical Methods in Natural language processing. 2015.

# A Recommendation System for Article Submission for Researchers

Seth Michail, Ugur Hazir, Taha Yigit Alkan, Joseph William Ledet, and Melih Gunay[✉]

Akdeniz University, Antalya, Turkey
{josephledet,mgunay}@akdeniz.edu.tr

**Abstract.** Finding the best publication for article submission is a challenge because of the need to ensure that the publication is relevant. The relevancy of the publication is important because an article published in a less relevant publication will have less exposure to the intended target audience.

Publications can be characterized by a "finger print" or "signature" that can be a composition of the abstracts of articles already accepted by these publications. In turn, these abstracts can be characterized by vector representations obtained through the methods of natural language processing. Using publication abstract and the obtained finger print to search the publication space for the best fit is the aim of this research. There is a lack of substantial research related to discovering an recommending appropriate journals to those seeking to publish research papers. In this research, we extend the current state of the art and propose a robust and generic article matching tool for all publishers, using Web of Science data.

## 1 Introduction

With increasing frequency, performance of academic institutions and researchers are ranked by several organizations and reported routinely. Among the major factors affecting the performance are the quality of academic publication and its impact, which are often measured by the citations [3]. In the meantime, with the increase of online publications and ease of publishing through the internet, the number of publications has increased dramatically [8]. With this quantitative increase, the impact of the article drops, unless it is published in a relevant journal and brought to the attention of an audience substantially interested in the topic.

Some problems are well suited for specific solution applications, such as machine learning. One example is choosing an appropriate publication for an author to submit their work to be published. The choice of appropriate publication can be difficult, since there are a large number of publications, and researchers have limited time to read articles needed to gain background for their research. Recommendation systems exist that are not appropriate for this task due to focusing on recommending articles to a researcher for reading, rather than recommending a publication for the submission of an article. Yet finding

© The Author(s), under exclusive license to Springer Nature Switzerland AG 2021
J. Hemanth et al. (Eds.): ICAIAME 2020, LNDECT 76, pp. 125–135, 2021.
https://doi.org/10.1007/978-3-030-79357-9_13

the best journal to which to submit an article for publication is an even more difficult task due to increasing specialization, resulting in a very large number of publications covering related topics.

Selecting the best publication for article submission is made challenging by the need to ensure that the publication is relevant. The relevancy of the publication is important because an article published in a less relevant publication will have less exposure to the intended target audience. As time progresses, more publications are being started, each with its own specialty or subfield. Since these specialties and subfields are increasingly more closely related, the task of finding an appropriate candidate publication for submission of work becomes increasingly more difficult.

Publications can be characterized by a "signature" that can be a composition of the conceptual content of the abstracts of articles already accepted by these publications. In turn, these abstracts can be characterized by vector representations obtained through the methods of natural language processing. Humans can, with ease, understand the meaning of a correctly written block of text. As of the time of this writing, natural language understanding is still far from feasible. Therefore, an approximation or alternative measures must be used. This is the reason for the vectorization of blocks of texts, such as abstracts.

While using a simple keyword search can be useful, it can also easily fail to properly reflect the relevancy of a publication. This should be immediately clear on the basis that keywords are single words, and keyphrases are short arrangements of words that include one or more keywords. If an article could be written entirely in keywords, a keyword search would be enough to efficiently and effectively find relevant candidate publications for submission. However, natural language is not composed of just keywords, a significant portion of meaning is contextual. It is this context dependent meaning that is a critical part of natural language use that leads to keywords being insufficient to adequately distinguish appropriate publications in which to publish. The increasing specialization, together with increasing relatedness, and the inability of keywords to sufficiently distinguish publications, makes the approach outlined here necessary.

As stated above, finding the best journal to submit an article for publication is an even more difficult task due to increasing specialization. A solution for this task exists[6], but is proprietary, and only available for publications owned or operated by a single specific publishing company. The aim of this paper is to describe a solution that is accurate, easy to use, and independent of specific publishers.

This paper is organized in the following manner. Section 2 gives the background for the task outlined, as well as background for the methods employed. Section 3 gives a detailed description of the methodology of how the task was performed. Section 4 gives results and analysis, to include issues encountered in performing the task. Section 4 is a summary of the work done and results generated. We conclude the paper in Sect. 5 with some final notes.

## 2    Background

There is little work directly related to this particular application. An extensive search of available literature revealed only one extant application [6], which is proprietary, and therefore not available for examination. The ability to examine the application may be irrelevant however, since the application appears to recommend only those publications owned by the publishing company.

In 2013, Milikov et al. [12] first discuss continuous bag-of-words and skip-gram models, and later Mikolov et al. [10] introduce `word2vec` with great influence on future approaches. They introduce frequent word subsampling and an alternative to hierarchical softmax they call negative sampling. They describe frequent word subsampling as discarding words according to the probability given by $P(w_i) = 1 - (t/f(w_i))^{1/2}$, where $f(w_i)$ is the frequency of word $w_i$ in the document and $t$ is a threshold (typically $\sim 10^{-5}$). They define their negative sampling as

$$log\,\sigma({v'_{w_O}}^\top v_{wI}) + \sum_{i=1}^{k} \mathbb{E}_{w_i} \sim P_n(w) \left[log\,\sigma(-{v'_{w_i}}^\top v_{wI})\right]$$

which they describe as a simplified version of Noise Contrastive Estimation. Here, $\sigma$ is the sigmoid function, $P_n(w)$ is the noise distribution, and $-v'_{w_i}$ is the negative sample vector for word $w$. They also consider bigrams, weighting according to $score(w_i, w_j) = \frac{count(w_i, w_j)}{count(w_i) \times count(w_j)} - \delta$. A later refinement by Horn et al. [7] is given by

$$score(w_i, w_j) = \frac{count(w_i, w_j)}{max\{count(w_i), count(w_j)\}}$$

which is 1 if the words only appear as a co-occurrence, and a threshold is usually chosen to be less than one.

In 2014, Le and Mikolov [13] introduced `doc2vec`, based on, and extending, the previous work on `word2vec`. Like `word2vec`, it captures the meaning associated with word order, which is lost when using TF-IDF or Continuous-Bag-of-Words models. The principle is to concatenate a paragraph vector with several word vectors, and both forms of vectors are trained using stochastic gradient decent and backpropagation. The purpose is to have vector representations of words and documents that have some semantic similarity. They describe their overall approach, which can be used with common machine learning techniques, during which they introduce two versions of their paragraph vector concept. Paragraph vector distributed memory (PV-DM), which preserves word ordering, and paragraph vector distributed bag-of-words (PV-DBOW), which does not preserve word ordering. They encourage using both in combination.

In 2014, Pennington et al. [14] proposed a weighted least squares regression model defined by

$$J = \sum_{i,j=1}^{V} f(X_{ij}) \left(w_i^T w_j + b_i + \tilde{b}_j - logX_{ij}\right)^2$$

where $f(x_{ij})$ is a weighting function, $X_{ij}$ is the $i, j^{th}$ entry in the word-word co-occurrence matrix, $V$ is the size of the vocabulary, $w_i$ and $\tilde{w}_j$ are context word vectors from $W$ and $\widetilde{W}$ respectively.

Lau and Baldwin [9] give a rigorous evaluation of doc2vec in 2016. They ask 4 questions, the answers of which, they conclude, are that doc2vec performs well in different task settings, that PV-DBOW is a better model than PV-DM, that careful hyperparameter optimization can lead to overall performance improvements, and that similarly to word2vec, the basic unmodified version of doc2vec works well.

Zhu and Hu [17] propose an enhanced version IDF weights that they call context aware as an efficiency improvement on doc2vec. For the weighting they use a global temperature softmax for a normalization function expressed as

$$\psi(\mathbf{w}_j^i) = \frac{|W| e^{\mathbf{w}_j^i}}{\sum_{\mathbf{w}_j^i \in W} e^{\mathbf{w}_j^i} / T}$$

and they call this model w-dbow.

In 2017, Horn et al. [7] develop a method to be used in text classification that is based on the notion of relevancy of words in a body of text for the purpose of representing the content of the text. Their relevancy scores are presented as $r_c\_diff(t_i) = max\{TPR_c(t_i) - FPR_c(t_i), 0\}$ where

$$TPR_c(t_i) = \frac{|\{k : y_k = c \wedge \mathbf{x}_{ki} > 0\}|}{|\{k : y_k = c\}|}$$

and

$$FPR_c(t_i) = mean(\{TPR_l(t_i) : l \neq c\})$$
$$+ std(\{TPR_l(t_i) : l \neq c\})$$

Here, $\mathbf{x}_{ki} = tf(t_i) \cdot idf(t)_i)$ and $y_k = c$ means all documents belonging to class $c$, and $l \neq c$ means all classes other than $c$. Additionally, the rate quotient is given as

$$r_c quot(t_i) = \frac{min\{max\{z_c(t_i), 1\}, 4\} - 1}{3}$$

where

$$z_c(t_i) = \frac{TPR_c(t_i)}{max\{FPR_c(t_i), \epsilon\}}$$

and finally, $r_c\_dist(t_i) = 0.5(r_c\_diff(t_i) + r_c quot(t_i))$ So the rate distance is the average of the rate difference and rate quotient , where the rate difference is the measure of words occurring within a class compared to occurring in other classes, and the rate quotient captures words that would otherwise be lost.

Lilleberg, Zhu, and Zhang [11] indicate in 2015 that word2vec combined with tf-idf performs better than either alone, supporting the usefulness of tf-idf despite its drawbacks.

In 2017, Vaswani et al. [16] introduced the Transformer architecture, which uses attention mechanism alone. Their approach made RNNs and LSTMs nearly obsolete for use on NLP tasks. This was accomplished by encoding the position information in the state, sharing the states across time steps, and including this information when passing the states to the decoder stack. The Transformer architecture now forms the basis for the majority NLP task applications, despite compute requirements scaling quadratically with input sequence length.

Devlin et al. [5] developed an approach in 2018 that applies the encoder part of the encoder-decoder stack of a Transformer to generate encoder representations of text. This representation is generated by masking a random portion of the text and setting a model to the task of predicting the masked tokens. Additionally, their approach was applied bidirectionally, meaning that future tokens were considered when predicting the masked or next token. Their model architecture extends the compute demands of the Transformer architecture, thus motivating the search for alternative model architectures with greater efficiency, lower compute requirements, and comparable performance.

## 3   Methodology

The primary source for the dataset for this research came from Web of Science database [1]. As a case study, we choose Computer Science as a sub field, not only because of our experience in the field, but also Computer Science being quite interdisciplinary [2].

The dataset for this paper was maintained in a database containing 444 publications related to computer science, with a total of 23,060 article abstracts. This dataset has significant class imbalance, and to address this issue, a second, reduced copy of the dataset was made for further comparison. The necessary data was retrieved and then divided into training and testing sets. A doc2vec model was implemented using Gensim[1] The input into Gensim was abstracts, since Gensim includes its own methods for splitting and cleaning the text, the only manual preprocessing of the data that was necessary was reading in and storing to memory the abstract text, and to remove a few special characters. The abstracts were tokenized and a vector representation generated by doc2vec, and then placed in a matrix. Initially, word2vec was investigated, but ultimately abandoned due to training and testing accuracy below 1%.

Gensim outputs a vocabulary of word vectors. Using this vocabulary each abstract was again processed using functions included with Gensim to generate abstract vectors as the sum of the word vectors of the abstract. This is shown in the pseudocode below.

---

[1] https://radimrehurek.com/gensim/index.html.

---

**Algorithm 1.** Continuous space vector representation of words

---
1: **procedure** EMBEDDING($a, b$)
2:     Load Model
3:     Load Dataset
4:     Create WordsList
5:     Create empty feature vector, $X$
6:     **for** $i <$ size of *dataset* **do**
7:         Create list of words, $g$, tokenized from abstract
8:         Append $g$ to WordsList
9:         **for** $j <$ size of $g$ **do**
10:             **if** $g[j]$ in vocabulary of *docVectors* **then**
11:                 Create word vector same size as $X$ from $g[j]$
12:                 **for** $k <$ size of $X; k + +$ **do**
13:                     Append each component of word vector to $X$
        return X;

---

A Gensim model with output length 1200 was considered but not implemented in order to maintain greater comparability with the BERT[2] model. The Gensim model created was based on the need to predict on 392 classes, and a requirement that the feature vector length be limited to 256 features. All other parameters were the same, and are given as follows. Distributed Bag-of-Words versions of `doc2vec` was used with $window = 4$, $\alpha = 0.006$, $min\_alpha = 2.5 \times 10^{-6}$, $min\_count = 3$, $sample = 10^{-3}$, $negative = 7$, $ns\_exponent = -0.7$, $workers = 8$, $epochs = 30$ $dbow\_words = 0$, $hs = 0$. Respectively, these parameters are as follows [15]:

- number of vector components used to represent each abstract
- maximum distance between the current and predicted word within a sentence
- the initial learning rate of the model
- minimum number of occurrences of a word needed for the word to be used in constructing the abstract vector
- threshold frequency for downsampling high frequency words
- number of "noise" words used in negative sampling
- shape of the negative sampling probability distribution
  - if negative, samples low-frequency words more than high-frequency words
- number of worker threads used for training the model
- whether skip-gram is used
  - if 0, skip-gram is not used
- hierarchical softmax
  - if 0, default softmax is used
  - must be 0 for negative sampling

After training, the `doc2vec` model was used to generate the feature vectors used as input to an ANN. Several architectures were tested for the model, and the model architecture that gave the best performance is described here. Class

---

[2] https://github.com/huggingface/transformers.

imbalance was addressed in two ways. First, by discarding all samples from publishers having less than a threshold number of articles published. After minor preprocessing, the threshold was set to 16 articles. This resulted in the datasets having 392, which leads to the problem of class interference. The ANN was first trained with approximately 16.000 ($\approx$80%) abstracts for training dataset, with the remainder evenly split for validation and test datasets. The relative sizes of the datasets were due to the need to balance having a large enough test and validation datasets for testing and validation, and large enough training dataset so that the model could learn each class. Second, by splitting into test, validation, and training datasets, and then making the probability distribution for the training sets uniform by duplicating samples, such that each publisher class was represented by the same number of samples.

The vector representations of the abstracts were then used to train the ANN to classify the abstracts. The ANN used 2 layers, the first being 300 neurons, and the second containing 100 neurons. In the usual way, the output had a number of neurons corresponding to the number of classes, in this case, 392. Deeper architectures were investigated, and the performance was found to degrade with depth. None of the architectures investigated used any additional techniques, such as convolution, recurrence, or gating. The dropout, which was set to 20% for the first layer, and 10% for the second layer during training, improved the learning by extending the time over which learning occurred. The performance of the model was measured using cross entropy softmax loss function, as provided by TensorFlow. During training, the loss function for the model was optimized using Adam Optimizer.

The predictions were obtained by TensorFlow after training of ANN, where the predictions were in the form of the predicted publication ID for a given abstract being tested. This predicted publication ID is used in a lookup table to find the name of the publisher. The number of correct predictions was counted and divided by the total number of predictions to get the overall prediction accuracy. For class based accuracy, an array with length being the number of classes was used, with each publication ID mapped to each array index. When a correct prediction was made, the value at the corresponding array index was incremented. The 'best-of-$n$' accuracy was determined by taking IDs corresponding to the $n$ highest softmax outputs, and if the correct publication was among these $n$, the prediction was considered successful. The confidence for each correct prediction was also recorded, and was determined as the difference between the first and second highest softmax values.

## 4   Result and Discussion

The bodies of the existing articles are not included due to the possibility of influencing the development of a "signature", especially since the result of this work is intended to allow only the submission of abstracts. The justification for this restriction is the need of significant computation required by the natural language processing methods, as well as the summarative nature of abstracts leading to possible reduction of confounding.

**Table 1.** Example output

| Doc2Vec | Publication |
|---------|-------------|
| 17.2% | COMPUTER LANGUAGES SYSTEMS AND STRUCTURES |
| 12.5% | SCIENCE OF COMPUTER PROGRAMMING |
| 8.7% | JOURNAL OF WEB SEMANTICS |
| 5.5% | ARTIFICIAL INTELLIGENCE |
| 4.6% | INFORMATION SYSTEMS |
| **BERT** | Publication |
| **10.9%** | **IEEE WIRELESS COMMUNICATIONS** |
| 7.1% | IEEE NETWORK |
| 2.7% | IEEE COMMUNICATIONS MAGAZINE |
| 1.9% | JOURNAL OF THE INSTITUTE OF TELECOMMUNICATIONS |
| 1.7% | INTERNATIONAL JOURNAL OF SENSOR NETWORKS |

As a sanity check, a randomly selected sample from the test dataset was given to the best performing of each model. The outputs are shown in the Table 1, along with the corresponding softmax probability, in descending order (for brevity, only the top 5 are shown). The ground truth label was 'IEEE WIRELESS COMMUNICATIONS'.

**Table 2.** Best of 10 accuracy

|  | Model name | First | Second | Third | 80 |
|--|-----------|-------|--------|-------|-----|
| Doc2Vec | None | 3.3% | 3.3% | 3.0% | 6.0% |
| Doc2Vec | None | 5.3% | 11.4% | 13.3% | 43.6% |
| BERT | None | 3.7% | 4.7% | 4.3% | 5.0% |
| BERT | 'bert-base-cased' | 32.0% | 45.4% | 52.5% | NA |

In Table 2, the first 3 entries are the embeddings as feature vectors given to a vanilla neural network. In the first and third entries, the initial learning rate was set to $1.5 \times 10^{-3}$, the second and fourth are for initial learning rate of $1.5 \times 10^{-5}$, and in all cases, the learning rate was decayed by an order of magnitude for each of the first 3 passes over the training dataset. In the fourth case, the embedding feature vectors were given as input to a pre-trained BERT model, the output of which was given to a vanilla neural network with the same topology and parameter settings as the previous 3.

The difficulty of text classification can be quantified, according to Collins et al. [4], though only a qualitative assessment is given here.

The relative class balance was considered by comparing each publication on the basis of the number of article abstracts they contained. It was found that some publications contained only one article's abstract, while many others contained 76 article abstracts, with a range of number of article abstracts in between. This indicates that some publications have significantly fewer abstracts, meaning they are underrepresented, and thus will lead to increased misclassification. The class interference was moderate in significance, due to initially having 455 classes, and this directly leads to increased difficulty in classification, but noise was not a consideration as the data were already examined for missing or misplaced data. Overall, the difficulty of this classification task was ameliorated partly by the dataset size not being very large; the total number of words being estimated less than 25 million.

The initial results gave an accuracy of ≈0.4% in training and testing using TensorFlow native methods. The performance of the models previously developed purport to range from about 40% to about 90%. After adjusting the doc2vec model parameters, the training accuracy rose to 99%, suggesting overfitting.

Several runs were conducted, varying (one at a time) only learning rate, dropout, or number of nodes in the first layer. The figures below show the loss vs iteration (Fig.1).

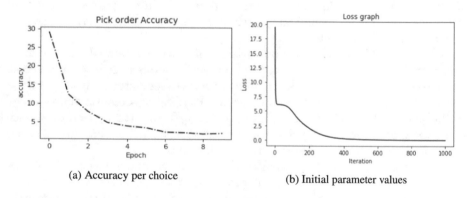

(a) Accuracy per choice          (b) Initial parameter values

**Fig. 1.** Parameters and accuracy

## 5    Conclusion

Many architectures and configurations were explored, and the best performance was obtained using two hidden layers, with 300 neurons and 100 neurons in the first and second hidden layers, respectively, along with other factors listed previously. The other architectures and configurations are not reported because they performed poorly by comparison.

It was observed that by increasing the number of recommendations, the accuracy can be improved, and that the outcome of the model reduces the number of possible publications to be considered, based on the relevance of such publications to the work of the researcher. Since many other factors, such as the language of the publication, or the publications impact factor, may be just as significant to the researcher in making a decision, the output of the model presented here, can be, at best, suggestive.

Additionally, some of the factors affecting choice of publication to submit to are author fees, whether the publication is open access or pay-walled, likelihood of submission being accepted, and reputation of the publication. This list of factors is clearly not exhaustive. Unfortunately, the appropriateness of a publications' area of coverage is often not considered, or is poorly considered, due to the difficulty of judging such appropriateness and the 'good enough' compromise, partly motivated by time constraints on authors. The problem is made worse by both multidisciplinary publications, and some similar publications varying in subtle but important ways. In the first case, articles can reach a wider readership, but may also be missed by specialists in the appropriate field. In the second case, articles may narrowly miss the intended readership.

The proposed solution may work in tandem with existing solutions, and together these solutions can lead to more efficient distribution of conducted research.

Finally, the writing styles of humans can vary in subtle ways that can be difficult for humans, especially those without specialized training, to detect. Machine learning techniques can detect and amplify subtle patterns, and thus distinguish publications small variations in subject matter and writing style of authors, among other factors.

Solutions such as recommender systems that suggest related articles for a researcher to read may suffer from low accuracy in terms of relevance or scope. These recommendation systems employ some form of semantic analysis, yet these systems are not sufficient for several reasons. The field of semantic analysis may not be sufficiently developed, or existing tools may be limited by resources, such as available data for training models. On the other hand, SA may be a mature NLP tool that is simply not put to this application.

The main benefits of the proposed solution is recommendation of publications an author may not have been previously aware of, recommendation of publications that are more relevant to the topic of an article being submitted, and reducing the number of publications that would need to be considered, thus reducing the effort of finding an appropriate publication to which to submit for publication.

In conclusion, this work was of an exploratory nature, and can be further extended by examining the impact of other approaches, such as TF-IDF, which can be used in conjunction with relevancy scores. Primary Component Analysis (PCA) may be used for dimensionality reduction, which can, in turn, improve accuracy by reducing the number of classes, and therefore improve the distinguishability of the classes. Singular Value Decomposition (SVD) may lead to some improvement in performance by reducing the number of data points. Crossover, mutation, fitness can be used for optimizing hyperparameters.

It is the hope of the authors that this work be used as one component of a more comprehensive solution for assisting others in making their submission decisions.

# References

1. Aggarwal, A.: Overview and description. ISI web of knowledge. Thomson Reuters (2010)
2. Alkan, T.Y., Gunay, M., Ozbek, F.: Clustering of scientist using research areas at Akdeniz university. In: 2019 4th International Conference on Computer Science and Engineering (UBMK), pp. 1–5 (2019)
3. Alkan, T.Y., Özbek, F., Günay, M., San, B.T., Kitapci, O.: Assessment of academic performance at Akdeniz University. In: Hemanth, D.J., Kose, U. (eds.) ICAIAME 2019. LNDECT, vol. 43, pp. 982–995. Springer, Cham (2020). https://doi.org/10.1007/978-3-030-36178-5_87
4. Collins, E., Rozanov, N., Zhang, B.: Evolutionary data measures: Understanding the difficulty of text classification tasks. arXiv preprint arXiv:1811.01910 (2018)
5. Devlin, J., Chang, M.W., Lee, K., Toutanova, K.: Bert: pre-training of deep bidirectional transformers for language understanding. In: NAACL-HLT (2019)
6. Elsevier: Journal finder. https://journalfinder.elsevier.com/
7. Horn, F., Arras, L., Montavon, G., Müller, K.R., Samek, W.: Exploring text datasets by visualizing relevant words. arXiv preprint arXiv:1707.05261 (2017)
8. Larsen, P.O., Ins, M.V.: The rate of growth in scientific publication and the decline in coverage provided by science citation index. Scientometrics **84**, 575–603 (2010)
9. Lau, J.H., Baldwin, T.: An empirical evaluation of doc2vec with practical insights into document embedding generation. arXiv preprint arXiv:1607.05368 (2016)
10. Le, Q., Mikolov, T.: Distributed representations of sentences and documents. In: Proceedings of the 31st International Conference on Machine Learning, vol. 32 (2014)
11. Lilleberg, J., Zhu, Y., Zhang, Y.: Support vector machines and Word2vec for text classification with semantic features. In: Proceedings of 2015 IEEE 14th International Conference on Cognitive Informatics and Cognitive Computing (2015). http://dx.doi.org/10.1109/ICCI-CC.2015.7259377
12. Mikolov, T., Chen, K., Corrado, G., Dean, J.: Efficient estimation of word representations in vector space. arXiv preprint arXiv:1301.3781 (2013)
13. Mikolov, T., Sutskever, I., Chen, K., Corrado, G., Dean, J.: distributed representations of words and phrases and their compositionality. In: Proceedings of the 26th International Conference on Neural Information Processing Systems, vol. 2, pp. 3111–3119 (2013)
14. Pennington, J., Socher, R., Manning, C.D.: GloVe: global vectors for word representation. In: Proceedings of the 2014 Conference on Empirical Methods in Natural Language Processing (EMNLP), pp. 1532–1543 (2014). https://doi.org/10.3115/v1/D14-1162
15. Rehurek, R.: Doc2vec paragraph embeddings. https://radimrehurek.com/gensim/models/doc2vec.html
16. Vaswani, A., Shazeer, N., Parmar, N., Uszkoreit, J., Jones, L., Gomez, A.N., Kaiser, L., Polosukhin, I.: Attention is all you need. In: Advances in Neural Information Processing Systems, pp. 5998–6008 (2017)
17. Zhu, Z., Hu, J.: Context aware document embedding. arXiv preprint arXiv:1707.01521 (2017)

# Optimal Power Flow Using Manta Ray Foraging Optimization

Ugur Guvenc[1] , Huseyin Bakir[2(✉)] , Serhat Duman[3] , and Burcin Ozkaya[4]

[1] Duzce University, Duzce, Turkey
ugurguvenc@duzce.edu.tr
[2] Dogus University, Istanbul, Turkey
hbakir@dogus.edu.tr
[3] Bandirma Onyedi Eylul University, Balıkesir, Turkey
sduman@bandirma.edu.tr
[4] Isparta University of Applied Sciences, Isparta, Turkey
burcinozkaya@isparta.edu.tr

**Abstract.** The optimal power flow (OPF) stands for the problem of specifying the best-operating levels for electric power plants in order to meet demands given throughout a transmission network, usually with the objective of minimizing operating cost. Recently, the OPF has become one of the most important problems for the economic operation of modern electrical power plants. The OPF problem is a non-convex, high-dimensional optimization problem, and powerful metaheuristic optimization algorithms are needed to solve it. In this paper, manta ray foraging optimization (MRFO) was used to solve the OPF problem which takes into account the prohibited operating zones (POZs). The performance of the MRFO was tested on IEEE 30-bus test system. The results obtained from the simulations were compared with well-known optimization algorithms in the literature. The comparative results showed that the MRFO method ensures high-quality solutions for the OPF problem.

**Keywords:** Power system optimization · Optimal power flow · Manta ray foraging optimization · Prohibited operating zones

## 1 Introduction

The solution to the OPF problem has gained more importance in recent years owing to increasing energy demand and a decrease in energy resources. Also, fossil resources are exhausted and their environmental damage is getting larger day by day [1, 2]. The OPF problem is expressed as finding the optimal value of the objective function that is subject to various constraints of equality and inequality. In the OPF problem, state variables are consist of generator reactive power, slack bus generator active power, load bus voltage, and transmission line apparent power. Control variables involve the active power of generators, generator bus voltages, reactive power of shunt VAR compensators, and tap settings of tap regulating transformers [3, 4].

© The Author(s), under exclusive license to Springer Nature Switzerland AG 2021
J. Hemanth et al. (Eds.): ICAIAME 2020, LNDECT 76, pp. 136–149, 2021.
https://doi.org/10.1007/978-3-030-79357-9_14

Recently, researchers have studied the OPF problem with several optimization algorithms. In [4], the gravitational search algorithm (GSA) was used to overcome OPF problem. The effectiveness of the method used in relevant paper has tested in different test systems. In [5], the authors recommended a genetic algorithm (GA) for the solution of optimal reactive power dispatch problem. In the study, the power systems with 51-bus and 224-bus were used to show the success of GA method. In [6], symbiotic organisms search (SOS) algorithm was used for OPF problem which consideration the valve point effect (VPE) and POZs. The success of the suggested method was simulated on IEEE 30-bus power system. In [7], a backtracking search algorithm (BSA) was suggested to overcome OPF problem. The success of BSA method was confirmed by different test cases. In [8], authors used teaching-learning based optimization (TLBO) algorithm to solve OPF problem. In optimization problem, several objective functions were taken into consideration. IEEE 30-bus and 57-bus test systems were used for simulation studies. In [9], a chaotic artificial bee colony (CABC) algorithm was preferred to overcome OPF problem which considers transient stability and security constraints. The success of recommended algorithm was tested on both New England 39-bus and IEEE 30-bus power system. In [10], enhanced self-adaptive differential evolution (ESDE) algorithm was used for solution of multi-objective OPF problem. Algeria 59-bus and IEEE 30-bus test systems were used for simulation studies. In [11], particle swarm optimization (PSO) was propound to solve integrated OPF problem with FACTS devices. In [12], hybrid dragonfly algorithm with aging particle swarm optimization (DA-APSO) was applied to OPF problem. Several objective functions such as fuel cost, voltage profile deviation and actual power losses were used in optimization problem. The success of DA-APSO algorithm was confirmed by different case studies in IEEE 30-bus power system. In [13], the authors presented a hybrid shuffle frog leaping algorithm (SFLA) algorithm for OPF that considering VPE and POZs. In [14], harmony search algorithm (HSA) was used to overcome OPF problem.

In this paper, MRFO algorithm was used to solve OPF problem which considering POZs of thermal generation units. Simulation studies were performed on IEEE-30 bus power system. The success of the recommended algorithm was tested in two different scenarios. The results achieved from MRFO algorithm were compared with different optimization algorithms in the literature.

The remainder of the article is organized as follows: formulation of the OPF problem was introduced in Sect. 2. The structure and pseudo-code of MRFO algorithm are explained in Sect. 3. In Sect. 4, simulation results are given to show validity and effectiveness of the MRFO algorithm. Finally, conclusions were given.

## 2 OPF Formulation

The aim of the OPF problem is to obtain the optimal settings of control variables for minimizing the selected objective function while satisfying various equality and inequality constraints. Mathematically, the form of OPF problem can be formulated as follows:

$$Minimize \quad f(x, u)$$
$$subject\ to: \quad g(x, u) = 0$$

$$h(x, u) \leq 0 \tag{1}$$

The state variable vector of the OPF problem is defined as follows:

$$x^T = \begin{bmatrix} P_{G1}, & V_{L1} \dots V_{LNPQ}, & Q_{G1} \dots Q_{GNG}, \dots S_{l1} \dots S_{lNTL} \end{bmatrix} \tag{2}$$

where $P_{G1}$ is the active power of the slack generator, $VL$ is voltage magnitudes of the load buses, $QG$ is the reactive power of the thermal generators, and $Sl$ is the apparent power of transmission lines. $NPQ$ and $NG$ are the numbers of load buses and generator units. $NTL$ is the number of transmission line number.

OPF problem control variables can be expressed as follows:

$$u^T = [P_{G2} \dots P_{GNG}, \ V_{G1} \dots V_{GNG}, \ T_1 \dots T_{NT}] \tag{3}$$

where $PG$ is the active power of the generator units (except slack bus), $VG$ is voltage magnitudes of the generators, $T$ is tap settings of transformers. $NG$ and $NT$ are numbers of generator units and tap regulating transformers.

## 2.1 Objective Function

The OPF problem objective function was considered as the minimization of the fuel cost (FC). The FC function which includes the generation units in different characteristics can be formulated as follows [15]:

$$Min \ FC \ (x, u) = \sum_{i=1}^{NG} \left( a_i P_{Gi}^2 + b_i P_{Gi} + c_i \right) \tag{4}$$

where $P_{Gi}$ denotes active power value of the i[th] generator (MW), $a_i$, $b_i$, and $c_i$ are cost coefficients of the i[th] generator, $NG$ is total generator number.

### 2.1.1 Prohibited Operating Zones (POZs)

Since physical constraints of power system constituents, POZs occur in thermal generation units. Vibrations of machines or shaft bearings may cause POZs to form in practical operation of generators. In order to minimize FC in power systems, the POZs of generators should be avoided. The POZs are expressed mathematically as follows [6, 13]:

$$
\begin{aligned}
P_{Gi}^{\min} &\leq P_{Gi} \leq P_{Gi,1}^{low} \\
P_{Gi,k-1}^{up} &\leq P_{Gi} \leq P_{Gi,k}^{low} \qquad k = 2, .., mi \\
P_{Gi,mi}^{up} &\leq P_{Gi} \leq P_{Gi}^{\max}
\end{aligned}
\tag{5}
$$

where $mi$ is the total number of POZs. $k$ is the number of POZ, $P_{Gi,k}^{low}$ and $P_{Gi,k-1}^{up}$ are the lower and upper bound of the $(k-1)$[th] POZ of the i[th] generator. The fuel cost curve in which the POZs effect is taken into account was shown in Fig. 1. [6].

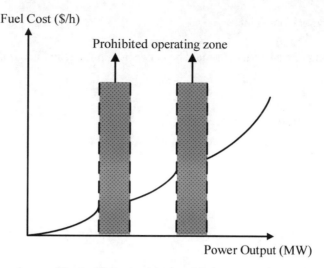

**Fig. 1.** Fuel cost curve considering POZs

## 2.2 Equality Constraints

For all buses, power flow equations should be evaluated.

$$P_{Gi} - P_{Di} - V_i \sum_{j=1}^{N} V_J \left[ G_{ij} \cos(\delta_i - \delta_j) + B_{ij} \sin(\delta_i - \delta_j) \right] = 0 \tag{6}$$

$$Q_{Gi} - Q_{Di} - V_i \sum_{j=1}^{N} V_J \left[ G_{ij} \sin(\delta_i - \delta_j) - B_{ij} \cos(\delta_i - \delta_j) \right] = 0 \tag{7}$$

where $N$ represents total bus number, $V_i$ and $V_J$ denote voltage of $i^{th}$ and $j^{th}$ bus, respectively. $\delta$ is voltage angle [4].

## 2.3 Inequality Constraints

In the optimization problem, there are three inequality constraints.

I. Generator Constraints

In the PV buses generator voltage, generator active, and reactive power output should be restricted as follows [6].

$$V_{Gi}^{min} \leq V_{Gi} \leq V_{Gi}^{max} \qquad i = 1, \ldots, NG \tag{8}$$

$$P_{Gi}^{min} \leq P_{Gi} \leq P_{Gi}^{max} \qquad i = 1, \ldots, NG \tag{9}$$

$$Q_{Gi}^{min} \leq Q_{Gi} \leq Q_{Gi}^{max} \qquad i = 1, \ldots, NG \tag{10}$$

## II.  Transformer Constraints

The tap setting of transformers should be between minimum and maximum limits.

$$T_i^{min} \leq T_i \leq T_i^{max} \qquad i = 1, \ldots, N_T \tag{11}$$

## III.  Security Constraints

The voltage of load buses must be between upper and lower limits. For each transmission line Eq. (13) should be evaluated.

$$V_{Li}^{min} \leq V_{Li} \leq V_{Li}^{max} \qquad i = 1, \ldots, N_{PQ} \tag{12}$$

$$S_{Li} \leq S_{Li}^{max} \qquad i = 1, \ldots, N_{TL} \tag{13}$$

Finally, considering the penalty conditions, the objective function is defined mathematically as follows:

$$J = f(x, u) + \lambda_P \left(P_{Gslack} - P_{Gslack}^{lim}\right)^2 + \lambda_V \sum_{i=1}^{NPQ} (V_{Li} - V_{Li})^2$$

$$+ \lambda_Q \sum_{i=1}^{NG} (Q_{Gi} - Q_{Gi}^{lim})^2 + \lambda_S \sum_{i=1}^{NTL} (S_{li} - S_{li}^{lim})^2 \tag{14}$$

where $\lambda_P, \lambda_V, \lambda_Q$ and $\lambda_S$ are penalty factor terms. If the value of state variables is lower or higher than limit values, the value of these variables is set to limit. Constraint violation states are expressed in following equations [6]:

$$P_{Gslack}^{lim} = \begin{cases} P_{Gslack}^{min} & if \ P_{Gslack} < P_{Gslack}^{min} \\ P_{Gslack}^{max} & if \ P_{Gslack} > P_{Gslack}^{max} \end{cases} \tag{15}$$

$$V_{Li}^{lim} = \begin{cases} V_{Li}^{min} & if \ V_{Li} < V_{Li}^{min} \\ V_{Li}^{max} & if \ V_{Li} > V_{Li}^{max} \end{cases} \tag{16}$$

$$Q_{Gi}^{lim} = \begin{cases} Q_{Gi}^{min} & if \ Q_{Gi} < Q_{Gi}^{min} \\ Q_{Gi}^{max} & if \ Q_{Gi} > Q_{Gi}^{max} \end{cases} \tag{17}$$

$$S_{li}^{lim} = S_{li}^{max} \quad if \ S_{li} > S_{li}^{max} \tag{18}$$

# 3   Manta Ray Foraging Optimization

Manta ray foraging optimization (MRFO) [16] is a novel optimizer [25] that imitates three intelligent foraging strategies of the manta rays which are chain foraging strategy, cyclone foraging strategy, and somersault foraging strategy. The pseudo-code of the MRFO algorithm was shown in Fig. 2.

## 3.1   Chain Foraging

Manta rays can follow up on the location of the plankton and swim towards it. In the swarm, whichever manta ray has more plankton concentration, it is the best position. Manta rays form a food search chain. Those who are out of the first move towards both the food and the food in front of it. Thus, each individual is updated on each iteration with the best solution ever. The mathematical formulation of this foraging is as follows [16]:

$$x_i(t+1) = \begin{cases} x_i(t) + r(x_{best}(t) - x_i(t)) + \alpha(x_{best}(t) - x_i(t)), & i = 1 \\ x_i(t) + r(x_{i-1}(t) - x_i(t)) + \alpha(x_{best}(t) - x_i(t)), & i = 2, \ldots, N \end{cases} \tag{19}$$

$$\alpha = 2.r.\sqrt{|\log(r)|} \tag{20}$$

where $x_i(t)$ is the position of the $i^{th}$ manta ray at time t, r is the random value in the interval of [0, 1], $\alpha$ is the weight coefficient, and $x_{best}(t)$ is the individual with the highest plankton concentration at time t.

## 3.2   Cyclone Foraging

Manta ray swarms in line developing a spiral perform foraging. The mathematical formulation is as follows:

$$x_i(t+1) = \begin{cases} x_{best} + r(x_{best} - x_i(t)) + \beta(x_{best} - x_i(t)), & i = 1 \\ x_{best} + r(x_{i-1}(t) - x_i(t)) + \beta(x_{best} - x_i(t)), & i = 2, \ldots, N \end{cases} \tag{21}$$

$$\beta = 2e^{r_1 \frac{T-t+1}{T}}.\sin(2\pi r_1) \tag{22}$$

where $\beta$ is the weight coefficient, T denotes the maximum iteration number and $r_1$ is the random number in the interval of [0, 1].

## 3.3   Somersault Foraging

In the present stage, the location of the food is considered a pivot. Each individual is tasked to swim around the pivot and take a new position. For this reason, they at all times update their positions around the best available position. This behaviour is expressed mathematically as:

$$x_i(t+1) = x_i(t) + S * (r_2 * x_{best} - r_3 * x_i(t)) \tag{23}$$

where S is the somersault factor and, $r_2$ and $r_3$ a random numbers in interval [0, 1].

---

**Start**
Define the number of population size (N) and maximum iteration number($T_{max}$)
Initialize random position of the manta rays
$x_i(t) = x_l + rand\,(x_u - x_l)$ for i=1,…,N  and t=1
Compute the fitness of each individual $f_i = f(x_i)$
Obtain the best solution $x_{best}$
**WHILE** stop criterion is not satisfied  **DO**
**FOR** i=1 to N   **DO**
**IF** rand<0.5 THEN      //**Cyclone foraging**
**IF** (t/$T_{max}$) < rand   THEN
$x_{rand} = x_l + rand\,(x_u - x_l)$
$x_i(t+1) = \begin{cases} x_{rand} + r(x_{rand} - x_i(t)) + \beta(x_{rand} - x_i(t)) & i = 1 \\ x_{rand} + r(x_{i-1}(t) - x_i(t)) + \beta(x_{rand} - x_i(t)) & i = 2, …, N \end{cases}$
**ELSE**
$x_i(t+1) = \begin{cases} x_{best} + r(x_{best} - x_i(t)) + \beta(x_{best} - x_i(t)) & i = 1 \\ x_{best} + r(x_{i-1}(t) - x_i(t)) + \beta(x_{best} - x_i(t)) & i = 2, …, N \end{cases}$
**END IF**
**ELSE**           //**Chain foraging**
$x_i(t+1) = \begin{cases} x_i(t) + r(x_{best} - x_i(t)) + \alpha(x_{best} - x_i(t)) & i = 1 \\ x_i(t) + r(x_{i-1}(t) - x_i(t)) + \alpha(x_{best} - x_i(t)) & i = 2, …, N \end{cases}$
**END IF**
Compute the fitness of each individual $f(x_i(t+1))$
**IF** $f(x_i(t+1)) < f(x_{best})$ THEN
$x_{best} = x_i(t+1)$
**FOR** i=1 to N    **DO**  //**Somersault foraging**
$x_i(t+1) = x_i(t) + S * (r_2 * x_{best} - r_3 * x_i(t))$
Compute the fitness of each individual $f(x_i(t+1))$
**IF** $f(x_i(t+1)) < f(x_{best})$  THEN
$x_{best} = x_i(t+1)$
**END FOR**
**END WHILE**
Return the best solution found so far $x_{best}$

---

**Fig. 2.**  Pseudocode of the MRFO algorithm

## 4  Simulation Results

In this study, the MRFO optimization algorithm is used to solve the OPF problem on IEEE 30-bus test system. Two different cases have been tested with a view to evaluating the success of the proposed algorithm. Test system data was taken from Ref. [4, 17]. In the test system, there are six generators and four transformers and also total load demand is 283.4 MW. The lower and upper voltage limits of load buses are set as 0.95–1.05 p.u. Figure 3 shows the IEEE 30-bus power system. Test system generator data were obtained from Ref. [6] and given in Table 1.

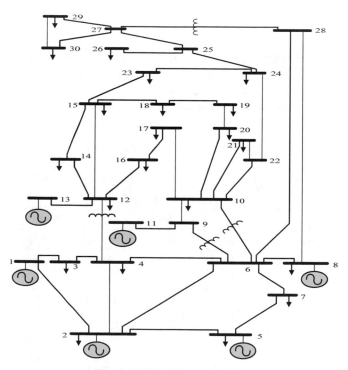

**Fig. 3.**  IEEE-30 bus test system

**Table 1.** IEEE-30 bus generator data

| Unit | $a$ | $b$ | $c$ | POZs | $Q_{Gi}^{min}$(MVAR) | $Q_{Gi}^{max}$(MVAR) |
|------|-----|-----|-----|------|------|------|
| $P_{G1}$ | 0.00375 | 2 | 0 | [55–66], [80–120] | −20 | 150 |
| $P_{G2}$ | 0.0175 | 1.75 | 0 | [21–24], [45–55] | −20 | 60 |
| $P_{G5}$ | 0.0625 | 1 | 0 | [30–36] | −15 | 62.5 |
| $P_{G8}$ | 0.0083 | 3.25 | 0 | [25–30] | −15 | 48.7 |
| $P_{G11}$ | 0.025 | 3 | 0 | [25–28] | −10 | 40 |
| $P_{G13}$ | 0.025 | 3 | 0 | [24–30] | −15 | 44.7 |

## 4.1    Case 1: Classical OPF Problem

In this case, VPE and POZs were ignored and the basic OFP problem was solved. Table 2 presents the value of the control variables and the objective function value obtained with the MRFO algorithm. It is clear that the control variable values are maintained within the lower and upper limits after the optimization process.

**Table 2.** Simulation results for test cases

| Control Variables | Limits | | Case 1 | Case 2 |
|------|------|------|------|------|
| | Minimum | Maximum | | |
| $P_{G1}(MW)$ | 50 | 250 | 177.1230 | 179.2218 |
| $P_{G2}(MW)$ | 20 | 80 | 48.9210 | 44.9994 |
| $P_{G5}(MW)$ | 15 | 50 | 21.2910 | 21.5448 |
| $P_{G8}(MW)$ | 10 | 35 | 21.1612 | 22.4382 |
| $P_{G11}(MW)$ | 10 | 30 | 11.9995 | 12.3016 |
| $P_{G13}(MW)$ | 12 | 40 | 12.0194 | 12.0007 |
| $V_{G1}(pu)$ | 0.95 | 1.1 | 1.0807 | 1.0816 |
| $V_{G2}(pu)$ | 0.95 | 1.1 | 1.0625 | 1.0625 |
| $V_{G5}(pu)$ | 0.95 | 1.1 | 1.0322 | 1.0321 |
| $V_{G8}(pu)$ | 0.95 | 1.1 | 1.0366 | 1.0369 |
| $V_{G11}(pu)$ | 0.95 | 1.1 | 1.0773 | 1.0842 |
| $V_{G13}(pu)$ | 0.95 | 1.1 | 1.0648 | 1.0623 |
| $T_{6-9}(pu)$ | 0.9 | 1.1 | 1.0357 | 1.0204 |
| $T_{6-10}(pu)$ | 0.9 | 1.1 | 0.9410 | 0.9664 |
| $T_{4-12}(pu)$ | 0.9 | 1.1 | 0.9812 | 0.9762 |
| $T_{28-27}(pu)$ | 0.9 | 1.1 | 0.9632 | 0.9610 |
| $FC(\$/h)$ | – | – | **800.7680** | **801.0305** |
| $P_{loss}(MW)$ | – | – | 9.1150 | 9.1066 |

As can be clearly seen from Table 2, the FC value obtained from the MRFO algorithm for Case 1 is 800.7680 ($/h). The results obtained from the MRFO algorithm and other optimization algorithms considered in this study are tabulated in Table 3. According to comparative results in Table 3, MRFO algorithm is more successful than other methods. The convergence curve of the MRFO algorithm for Case 1 is shown in Fig. 4. Accordingly, MRFO rapidly converged to the global optimum for Case 1, thanks to its powerful exploration and balanced search capabilities.

**Table 3.** Comparative results in terms of the FC ($/h) parameter for Case 1

| Methods | FC ($/h) | Methods | FC ($/h) |
|---|---|---|---|
| GA [18] | 804.10 | EGA [30] | 802.06 |
| SA [18] | 804.10 | FGA [32] | 802.00 |
| GA-OPF [19] | 803.91 | MHBMO [28] | 801.985 |
| SGA [20] | 803.69 | SFLA [13] | 801.97 |
| EP-OPF [19] | 803.57 | PSO [31] | 801.89 |
| EP [21] | 802.62 | Hybrid SFLA-SA [13] | 801.79 |
| ACO [22] | 802.57 | MPSO-SFLA [31] | 801.75 |
| IEP [23] | 802.46 | ABS [29] | 801.71 |
| NLP [24] | 802.4 | BSA [29] | 801.63 |
| DE-OPF [25] | 502.39 | SOS [6] | 801.5733 |
| MDE-OPE [25] | 802.37 | HHO [33] | 801.4228 |
| TS [26] | 802.29 | HHODE/best/2 [33] | 801.3732 |
| MSFLA [27] | 802.287 | HHODE/current-to-best/2 [33] | 800.9959 |
| HBMO [28] | 802.211 | MRFO | 800.7680 |

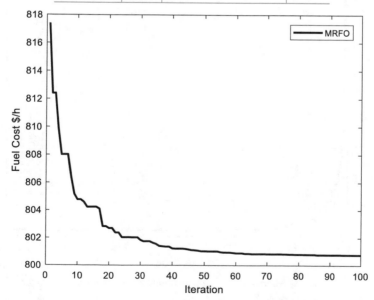

**Fig. 4.** Convergence curve of the MRFO algorithm for Case 1

## 4.2  Case 2: OPF Problem Considering POZs

In the present case, OPF problem was solved by considering POZs of generators. Simulation results obtained by the MRFO for Case 2 was tabulated in Table 2. As shown in Table 2, total FC value obtained by the MRFO algorithm is 801.0305 ($/h). The comparative results in Table 4 make it clear that the proposed method to solve OPF problem is more successful than the algorithms considered in this study. The convergence curve of the MRFO algorithm for Case 2 is shown in Fig. 5.

**Table 4.** Comparative results in terms of FC ($/h) parameter for Case 2

| Methods | FC ($/h) | Methods | FC ($/h) |
| --- | --- | --- | --- |
| GA [13] | 809.2314 | BSA [29] | 801.85 |
| SA [13] | 808.7174 | SOS [6] | 801.8398 |
| PSO [13] | 806.4331 | HHO [33] | 801.8058 |
| SFLA [13] | 806.2155 | HHODE/best/2 [33] | 801.4898 |
| Hybrid SFLA-SA[13] | 805.8152 | HHODE/current-to-best/2 [33] | 800.2161 |
| ABC [29] | 804.38 | MRFO | 801.0305 |

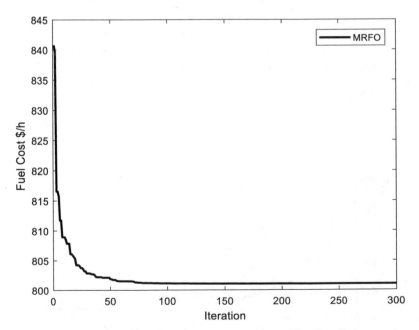

**Fig. 5.** MRFO algorithm convergence curve for Case 2

Figures 6 and 7 show the voltage magnitudes of load buses on the IEEE 30-bus test system. It is clear that the voltage magnitudes (p.u) of the load buses are maintained within the lower and upper voltage limits after the optimization process.

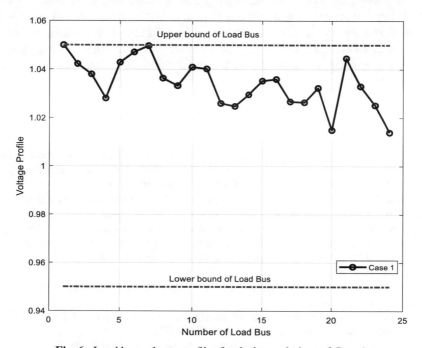

**Fig. 6.** Load bus voltage profiles for the best solutions of Case 1

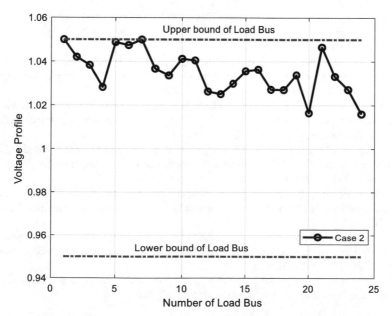

**Fig. 7.** Load bus voltage profiles for the best solutions of Case 2

# 5    Conclusion

In this paper, a recently developed MRFO algorithm has been employed for the solution of a non-convex, highly-dimensional OPF problem. A number of case studies in the IEEE-30 bus test system were investigated to evaluate the performance of the proposed algorithm. The obtained numerical results from the MRFO algorithm are compared with those of well-known optimization algorithms in the literature. The comparative results showed that the proposed method obtained better solutions for the OPF problem than its competitors.

# References

1. Niu, M., Wan, C., Xu, Z.: A review on applications of heuristic optimization algorithms for optimal power flow in modern power systems. J. Mod. Power Syst. Clean Energy 2(4), 289–297 (2014). https://doi.org/10.1007/s40565-014-0089-4
2. Güvenç, U., Özkaya, B., Bakir, H., Duman, S., Bingöl, O.: Energy hub economic dispatch by symbiotic organisms search algorithm. In: Jude Hemanth, D., Kose, U. (eds.) Artificial Intelligence and Applied Mathematics in Engineering Problems: Proceedings of the International Conference on Artificial Intelligence and Applied Mathematics in Engineering (ICAIAME 2019), pp. 375–385. Springer International Publishing, Cham (2019). https://doi.org/10.1007/978-3-030-36178-5_28
3. Özyön, S., Yaşar, C.: Farklı Salınım Barası Seçimlerinin Optimal Güç Akışı Üzerindeki Etkilerinin İncelenmesi
4. Duman, S., Güvenç, U., Sönmez, Y., Yörükeren, N.: Optimal power flow using gravitational search algorithm. Energy Convers. Manage. 59, 86–95 (2012)
5. Iba, K.: Reactive power optimization by genetic algorithm. IEEE Trans. Power Syst. 9(2), 685–692 (1994)
6. Duman, S.: Symbiotic organisms search algorithm for optimal power flow problem based on valve-point effect and prohibited zones. Neural Comput. Appl. 28(11), 3571–3585 (2016). https://doi.org/10.1007/s00521-016-2265-0
7. Chaib, A.E., Bouchekara, H.R.E.H., Mehasni, R., Abido, M.A.: Optimal power flow with emission and non-smooth cost functions using backtracking search optimization algorithm. Int. J. Electr. Power Energy Syst. 81, 64–77 (2016)
8. Ghasemi, M., Ghavidel, S., Gitizadeh, M., Akbari, E.: An improved teaching–learning-based optimization algorithm using Lévy mutation strategy for non-smooth optimal power flow. Int. J. Electr. Power Energy Syst. 65, 375–384 (2015)
9. Ayan, K., Kılıç, U., Baraklı, B.: Chaotic artificial bee colony algorithm based solution of security and transient stability constrained optimal power flow. Int. J. Electr. Power Energy Syst. 64, 136–147 (2015)
10. Pulluri, H., Naresh, R., Sharma, V.: An enhanced self-adaptive differential evolution based solution methodology for multiobjective optimal power flow. Appl. Soft Comput. 54, 229–245 (2017)
11. Naderi, E., Pourakbari-Kasmaei, M., Abdi, H.: An efficient particle swarm optimization algorithm to solve optimal power flow problem integrated with FACTS devices. Appl. Soft Comput. 80, 243–262 (2019)
12. Shilaja, C., Ravi, K.: Optimal power flow using hybrid DA-APSO algorithm in renewable energy resources. Energy Procedia 117, 1085–1092 (2017)

13. Niknam, T., Narimani, M.R., Azizipanah-Abarghooee, R.: A new hybrid algorithm for optimal power flow considering prohibited zones and valve point effect. Energy Convers. Manage. **58**, 197–206 (2012)

14. Sivasubramani, S., Swarup, K.S.: Multi-objective harmony search algorithm for optimal power flow problem. Int. J. Electr. Power Energy Syst. **33**(3), 745–752 (2011)

15. Kenan Dosoglu, M., Guvenc, U., Duman, S., Sonmez, Y., Tolga Kahraman, H.: Symbiotic organisms search optimization algorithm for economic/emission dispatch problem in power systems. Neural Comput. Appl. **29**(3), 721–737 (2016). https://doi.org/10.1007/s00521-016-2481-7

16. Zhao, W., Zhang, Z., Wang, L.: Manta ray foraging optimization: an effective bio-inspired optimizer for engineering applications. Eng. Appl. Artif. Intell. **87**, 103300 (2020)

17. IEEE 30-bus test system data. http://www.ee.washington.edu/research/pstca/pf30/pg_tca30bus.htm

18. Ongsakul, W., Bhasaputra, P.: Optimal power flow with FACTS devices by hybrid TS/SA approach. Int. J. Electr. Power Energy Syst. **24**(10), 851–857 (2002)

19. Sood, Y.R.: Evolutionary programming based optimal power flow and its validation for deregulated power system analysis. Int. J. Electr. Power Energy Syst. **29**(1), 65–75 (2007)

20. Bouktir, T., Slimani, L., Mahdad, B.: Optimal power dispatch for large scale power system using stochastic search algorithms. Int. J. Power Energy Syst **28**(2), 118 (2008)

21. Yuryevich, J., Wong, K.P.: Evolutionary programming based optimal power flow algorithm. IEEE Trans Power Syst **14**(4), 1245–1250 (1999)

22. Slimani, L., Bouktir, T.: Economic power dispatch of power system with pollution control using multiobjective ant colony optimization. Int. J. Comput. Intell. Res. **3**(2), 145–154 (2007)

23. Ongsakul, W., Tantimaporn, T.: Optimal power flow by improved evolutionary programming. Electric power components and systems **34**(1), 79–95 (2006)

24. Alsac, O., Stott, B.: Optimal load flow with steady-state security. IEEE Trans. Power Appar. Syst. **3**, 745–751 (1974)

25. Sayah, S., Zehar, K.: Modified differential evolution algorithm for optimal power flow with non-smooth cost functions. Energy Convers. Manage. **49**(11), 3036–3042 (2008)

26. Abido, M.A.: Optimal power flow using tabu search algorithm. Electric Power Compon. Syst. **30**(5), 469–483 (2002)

27. Niknam, T., rasoul Narimani, M., Jabbari, M., & Malekpour, A. R. : A modified shuffle frog leaping algorithm for multi-objective optimal power flow. Energy **36**(11), 6420–6432 (2011)

28. Niknam, T., Narimani, M.R., Aghaei, J., Tabatabaei, S., Nayeripour, M.: Modified honey bee mating optimisation to solve dynamic optimal power flow considering generator constraints. IET Gener. Transm. Distrib. **5**(10), 989–1002 (2011)

29. Kılıç, U.: Backtracking search algorithm-based optimal power flow with valve point effect and prohibited zones. Electr. Eng. **97**(2), 101–110 (2014). https://doi.org/10.1007/s00202-014-0315-0

30. Bakirtzis, A.G., Biskas, P.N., Zoumas, C.E., Petridis, V.: Optimal power flow by enhanced genetic algorithm. IEEE Trans. Power Syst. **17**(2), 229–236 (2002)

31. Narimani, M.R., Azizipanah-Abarghooee, R., Zoghdar-Moghadam-Shahrekohne, B., Gholami, K.: A novel approach to multi-objective optimal power flow by a new hybrid optimization algorithm considering generator constraints and multi-fuel type. Energy **49**, 119–136 (2013)

32. Saini, A., Chaturvedi, D.K., Saxena, A.K.: Optimal power flow solution: a GA-fuzzy system approach. Int. J. Emerg. Electric Power Syst. **5**(2) (2006)

33. Birogul, S.: Hybrid Harris Hawks Optimization Based on Differential Evolution (HHODE) Algorithm for Optimal Power Flow Problem. IEEE Access **7**, 184468–184488 (2019)

# Optimal Coordination of Directional Overcurrent Relays Using Artificial Ecosystem-Based Optimization

Ugur Guvenc[1] , Huseyin Bakir[2(✉)] , and Serhat Duman[3]

[1] Duzce University, Duzce, Turkey
ugurguvenc@duzce.edu.tr
[2] Dogus University, Istanbul, Turkey
hbakir@dogus.edu.tr
[3] Bandirma Onyedi Eylul University, Balıkesir, Turkey
sduman@bandirma.edu.tr

**Abstract.** Optimal directional overcurrent relays (DOCRs) coordination aims to find the optimal relay settings in order to protect the system, where, the primary relays are operated in the first to clear the faults, then the corresponding backup relays should be operated in case of failing the primary relays. DOCRs coordination problem is a non-convex and high dimensional optimization problem and it should be solved subject to operating constraints. The objective function for optimal coordination of DOCRs aims to minimize total operation time for all primary relays without violation in constraints to maintain reliability and security of the electric power system. This paper proposes the artificial ecosystem-based optimization (AEO) algorithm is for the solution of the DOCRs coordination problem. Simulation studies were carried out in IEEE 3-bus and IEEE 4-bus test systems to evaluate the performance of the proposed algorithm. The simulation results are compared with differential evolution algorithm (DE), opposition based chaotic differential evolution algorithm (OCDE1and OCDE2), and three real coded genetic algorithms (RCGAs) namely: Laplace crossover power mutation (LX-PM), Laplace crossover polynomial mutation (LX-POL), bounded exponential crossover power mutation (BEX-PM). The results clearly showed that the proposed algorithm is a powerful and effective method to solve the DOCRs coordination problem.

**Keywords:** Power system optimization · Directional overcurrent relays · Artificial ecosystem-based optimization

## 1 Introduction

Continuous expansion of the power system network increases complexity of power system operation [1]. Overcurrent relays play a major role in the reliable operation of power systems. These relays operate with the principle of activation when the predetermined current value is exceeded [2, 3]. DOCRs are widely used in power systems. Because they are more economical and simple to implement [4].

© The Author(s), under exclusive license to Springer Nature Switzerland AG 2021
J. Hemanth et al. (Eds.): ICAIAME 2020, LNDECT 76, pp. 150–164, 2021.
https://doi.org/10.1007/978-3-030-79357-9_15

In the DOCRs coordination problem, the goal is to optimize relay control parameters. Each relay has two control variables: time dial setting (TDS) and pickup current setting (PCS). These parameters directly affect the relay operation time [5, 6]. Also, in the DOCRs coordination problem, coordination time interval (CTI), which provides the reliability of the protection system, should be taken into account [2, 7].

Recently, the researchers have studied the DOCRS coordination problem with different algorithms. Some of these algorithms are: particle swarm optimization (PSO) [8, 9], genetic algorithm (GA) [10, 11], ant colony optimization (ACO) [12], nondominated sorting genetic algorithm-II (NSGA-II) [13], seeker optimization algorithm (SOA) [14], teaching learning-based optimization algorithm (TLBO) [15], adaptive differential evolution (ADE) algorithm [16], opposition based chaotic differential evolution (OCDE) algorithm [17], biogeography-based optimization (BBO) [18], modified electromagnetic field optimization algorithm (MEFO) [19], grey wolf optimizer (GWO) [20], harmony search algorithm (HSA) [21], modified water cycle algorithm (MWCA) [2], stochastic fractal search algorithm (SFSA) [22], firefly algorithm (FA) [23], artificial bee colony (ABC) [24], flower pollination algorithm (FPA) [25], symbiotic organisms search (SOS) [26], improved invasive weed optimization (IIWO) [27], hyper-sphere search (HSS) algorithm [28]. Also, the problem has been solved with hybrid algorithms such as hybrid whale optimization algorithm and grey wolf optimizer (HWGO) [29], hybrid whale optimization algorithm (HWOA) [30], hybrid gravitational search algorithm-sequential quadratic programming (GSA-SQP) [31], hybrid particle swarm optimization-differential evolution (PSODE) [32]. Although there are many meta-heuristic algorithms used to solve the DOCRs optimization problem in the literature, powerful optimization algorithms are needed to obtain better results than the previous ones. In this study, the AEO algorithm was used to solve the challenging DOCRs coordination problem on IEEE 3-bus and IEEE 4-bus test systems.

The paper is organized as follows: the problem formulation is presented in Sect. 2. After that, a meta-heuristic algorithm inspired by nature, called AEO, is explained in Sect. 3. Thereafter the simulation results and comparative results are given in Sect. 4. Finally, the main conclusions are provided in Sect. 5.

## 2    Formulation of Problem

In order to minimize the operating time of all primary relays and to avoid miscoordination between backup/primary relay pairs, the decision variables should have optimal values [3]. The decision variables in DOCRs coordination problem are TDS and PCS. The operation time of the relay depends on $TDS$, $PCS$ and fault current ($If$) and it is expressed by the nonlinear equation as shown below [7]

$$T = \frac{\alpha \; .TDS}{\left( \dfrac{I_f^i}{PCS \; .CT_{pr\_rating}} \right)^{\beta} - \gamma} \tag{1}$$

where $\alpha$, $\beta$ and $\gamma$ fixed values that show off the relay characteristic. In accordance with IEC Standard 60255–151, these constants are 0.14, 0.02 and 1.0 respectively. $CT_{pr\_rating}$ is the current transformer's primary rating and relay current ($I_{relay}$) is expressed as follows [3, 17].

$$I_{relay} = \frac{I_f}{CT_{pr\_rating}} \tag{2}$$

A fault occurring close to a relay is known as the close-in fault for that relay and a fault occurring at the other end of the line is known as a far-bus fault for the same relay, shown in Fig. 1.

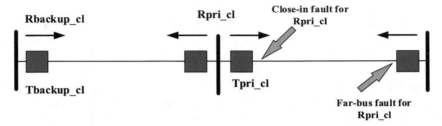

**Fig. 1.** Close-in and far-bus faults for relay $R_{Pri\_cl}$ [17]

### 2.1 Objective Function

The objective function (OF) is defined as minimizing operating times of all primary relays [17].

$$\begin{array}{c} \min \\ \text{TDS, PCS} \end{array} \text{OF} = \sum_{i=1}^{N_{cl}} T^i_{pri\_cl\_in} + \sum_{j=1}^{N_{far}} T^j_{pri\_far\_bus} \qquad (3)$$

where $N_{cl}$ and $N_{far}$ are close-in and far-bus relay numbers respectively. $T^i_{pri\_cl\_in}$ and $T^j_{pri\_far\_bus}$ represent relay response time to clear the close-in fault and far-bus fault respectively and formulated as:

$$T^i_{pri\_cl\_in} = \frac{0.14 \,.TDS^i}{\left(\dfrac{I^i_f}{PCS^i \,.CT^i_{pr\_rating}}\right)^{0.02} - 1} \qquad (4)$$

$$T^j_{pri\_far\_bus} = \frac{0.14 \,.TDS^j}{\left(\dfrac{I^j_f}{PCS^j \,.CT^j_{pr\_rating}}\right)^{0.02} - 1} \qquad (5)$$

### 2.2 Constraints

#### 2.2.1 Decision Variable Limits

TDS decision variable value should be limited within a certain bounds.

$$TDS_i^{\min} \leq TDS_i \leq TDS_i^{\max} \qquad i = 1,\ldots\ldots, N_{cl} \qquad (6)$$

where $TDS_i^{\min}$ and $TDS_i^{\max}$ are minimum and maximum limits of the TDS. The limit values considered as 0.05 and 1.1 respectively.

The bounds of the PCS decision variable can be expressed by

$$PCS_j^{\min} \leq PCS_j \leq PCS_j^{\max} \qquad j = 1,\ldots\ldots, N_{far} \qquad (7)$$

where $PCS_j^{\min}$ and $PCS_j^{\max}$ are minimum and maximum limits of PCS. The limit values considered as 1.25 and 1.5 respectively.

### 2.2.2 Relay Operation Time Limits

The operating time of each primary relay should be between 0.05 and 1 s.

### 2.2.3 Coordination Criteria

In order to prevent mal-operation, the backup relay should be set up to activate only when the primary relay fails. Coordination criteria are required to keeping the selectivity of relay pairs [3]. CTI is expressed as follows:

$$CTI = T_{backup} - T_{primary} \tag{8}$$

$$CTI \geq CTI_{min} \tag{9}$$

All primary/backup relay pairs must meet the selectivity criteria. The $CTI_{min}$ value in Eq. (9) is considered to be 0.3 s. Figure 2 depicts the coordination between primary and backup relays for F1 and F2 faults.

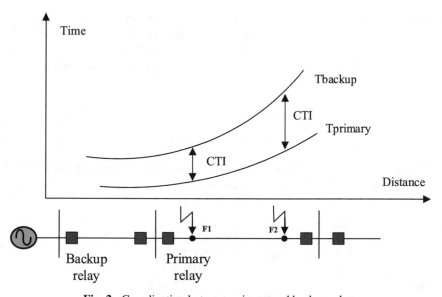

**Fig. 2.** Coordination between primary and backup relays

## 3   Artificial Ecosystem-Based Optimization Algorithm

The artificial ecosystem-based optimization (AEO) [33] is a powerful and robust meta-heuristic search algorithm with strong fundamentals inspired by the energy flow in the ecosystem. The algorithm imitates the production, consumption, and decomposition behavior of living organisms. The pseudocode of the AEO algorithm is depicted in Algorithm 1.

## 3.1 Production

This operator randomly generates a new individual. This operator sets balance between the algorithm's local and global search strategy. The production operator is expressed mathematically as follows [33]:

$$x_1(t+1) = (1-\alpha)x_n(t) + \alpha\, x_{rand}(t)$$
$$\alpha = \left(1 - \frac{t}{T}\right)r_1 \tag{10}$$
$$x_{rand} = r\,(U - L) + L$$

where, $U$ and $L$ are the upper and lower bounds of search spaces, respectively. $r_1$ and $r$ represent random values in the range between 0 and 1. a is the weight coefficient, $n$ denotes population size. The $t$ and $T$ are the current iteration and maximum iteration numbers, respectively.

## 3.2 Consumption

The consumption operator enables the exploration of the algorithm to be improved. Levy flight is a mathematical operator that simulates the food search process of animals. Levy flight prevents the algorithm's ability to trapped in local optima. It also contributes to the algorithm's explore of the entire search space. Levy flight can be expressed as follows [33]:

$$C = \frac{1}{2}\frac{v_1}{|v_2|}$$
$$v_1 \sim N(0, 1),\, v_2 \sim N(0, 1) \tag{11}$$

The consumption factor allows each consumer to search for food with distinct hunting strategies. Consumers are classified into three groups:

*Herbivore*
An herbivore type consumer can only eat the producer

$$x_i(t+1) = x_i(t) + C\,(x_i(t) - x_1(t)) \qquad i = 2, \ldots, n \tag{12}$$

*Carnivore*
The carnivore consumer can only eat the high-energy consumer.

$$\begin{cases} x_i(t+1) = x_i(t) + C\left(x_i(t) - x_j(t)\right) & i = 3, \ldots, n \\ j = randi\,([2 \quad i - 1]) \end{cases} \tag{13}$$

*Omnivore*
The omnivore consumer can eat both a producer and a consumer with a higher energy level.

$$\begin{cases} x_i(t+1) = x_i(t) + C\,(r_2\,(x_i(t) - x_1(t)) + (1 - r_2)\left(x_i(t) - x_j(t)\right) \\ j = randi\,([2 \quad i - 1]) \end{cases} \tag{14}$$

## 3.3 Decomposition

At this stage, the nutrients required for the growth of the producer are provided. [33].

$$x_i(t+1) = x_n(t) + D(e.x_n(t) - h.x_i(t)) \qquad i = 1, \ldots, n$$
$$D = 3u, \qquad u \sim N(0, 1)$$
$$e = r_3. \, randi \, ([1 \quad 2]) - 1 \tag{15}$$
$$h = 2. \, r_3 - 1$$

---

**Algorithm 1.** Pseudo-code of AEO

*1.* **Input:** Ecosystem size ($n$), Maximum iteration number (T)
*2.* **Output:** $x_{best}$
*3.* **Begin**
*4.*      $x$: randomly create an ecosystem of individuals
*5.*      **for** i=1: $n$ **do**
*6.*          $F$: evaluate the fitness for each individual
*7.*      **end**
*8.*      **while** *search process lifecycle: up to termination criteria* **do**
*9.*          *// Production //*
*10.*             For individual $x_1$ , update its location using Eq. (10)
*11.*          *// Consumption //*
*12.*             For individual $x_i$, (*i=2, ..., n*)
*13.*                 • **Herbivore**
*14.*                 *if rand<1/3 then* update its location using Eq. (12*)*
*15.*                 • **Omnivore**
*16.*                 *else if 1/3 < rand <2/3 then* update its location using Eq. (14)
*17.*                 • **Carnivore**
*18.*                 *else* update its location using Eq. (13)
*19.*                 *end if*
*20.*             For each individual calculate the fitness
*21.*             Update the best solution $x_{best}$
*22.*          *// Decomposition //*
*23.*             Update the location of each individual using Eq. (15).
*24.*             For each individual calculate the fitness
*25.*             Update the best solution $x_{best}$.
*26.*      **end while**
*27.*      return $x_{best}$
*28.* **End**

---

# 4 Simulation Results

## 4.1 Experimental Settings

The success of the AEO algorithm in the optimization of the DOCRs coordination problem has been tested with simulation studies performed on the IEEE-3 and IEEE-4 bus test systems. The schematic diagrams of the test systems are showed in Figures 3 and 4. The data of the test systems were taken from [7]. Also, the ecosystem size and the maximum number of iterations are considered as 50 and 10000, respectively.

## 4.2   Model I: IEEE-3 Bus Test System

IEEE 3-bus test system consists of 1 generator, 3 lines, 6 DOCRs, and 8 P/B relay pairs. In this case, the best value of 12 decision variables (TDS1-TDS6 and PCS1-PCS6) is sought. The values of $I_f$, $CT_{pr\_rating}$ parameters, and fault current values for the primary/backup relay pairs belonging to Model I are provided in Tables 1 and 2 [7].

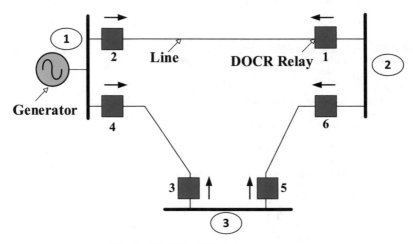

**Fig. 3.**  Model-I: IEEE 3-bus test system

**Table 1.**  Values of $I_f$ and $CT_{pr\_rating}$ for Model-I with $T_{pri\_cl\_in}$ and $T_{pri\_far\_bus}$

| $T^i_{pri\_cl\_in}$ | | | $T^j_{pri\_far\_bus}$ | | |
|---|---|---|---|---|---|
| $TDS^i$ | $I_f{}^i$(p.u.) | $CT^i_{pr\_rating}$ | $TDS^j$ | $I_f{}^j$(p.u.) | $CT^j_{pr\_rating}$ |
| $TDS^1$ | 9.46 | 2.06 | $TDS^2$ | 100.63 | 2.06 |
| $TDS^2$ | 26.91 | 2.06 | $TDS^1$ | 14.08 | 2.06 |
| $TDS^3$ | 8.81 | 2.23 | $TDS^4$ | 136.28 | 2.23 |
| $TDS^4$ | 37.68 | 2.23 | $TDS^3$ | 12.07 | 2.23 |
| $TDS^5$ | 17.93 | 0.80 | $TDS^6$ | 19.20 | 0.80 |
| $TDS^6$ | 14.35 | 0.80 | $TDS^5$ | 25.90 | 0.80 |

Table 3, presents the optimal value of the decision variables obtained by using different optimization algorithms in the literature as well as the AEO algorithm. It shows that the optimal value of the control variables is within the acceptable range. The comparison of the CTI values for Model I are reported in Table 4. It can be reported that the DE, OCDE1, and OCDE2 algorithms violate the coordination criteria expressed in Eq. (9). The obtained results can be defined as the solution with miscoordination between B/P relay pairs: R5/R1 for the DE algorithm, the solution with miscoordination between B/P relay pairs: R6/R3, R4/R5, R2/R6 for OCDE1 algorithm, and the solution with miscoordination between B/P relay pairs: R6/R3, R4/R5 for the OCDE2 algorithm.

**Table 2.** Fault currents for backup/primary relay pairs for Model-I

| $T^i_{backup}$ | | | $T^j_{primary}$ | | |
|---|---|---|---|---|---|
| Relay | $I_f^i$(p.u.) | $CT^i_{pr\_rating}$ | Relay | $I_f^j$(p.u.) | $CT^j_{pr\_rating}$ |
| R5 | 14.08 | 0.8 | R1 | 14.08 | 2.06 |
| R6 | 12.07 | 0.8 | R3 | 12.07 | 2.23 |
| R4 | 25.90 | 2.23 | R5 | 25.90 | 0.8 |
| R2 | 14.35 | 0.8 | R6 | 14.35 | 2.06 |
| R5 | 9.46 | 0.8 | R1 | 9.46 | 2.06 |
| R6 | 8.81 | 0.8 | R3 | 8.81 | 2.23 |
| R4 | 19.20 | 2.06 | R5 | 19.20 | 0.8 |
| R2 | 17.93 | 2.23 | R6 | 17.93 | 0.8 |

**Table 3.** Optimal TDS and PCS for Model I

| Item | DE [17] | OCDE1 [17] | OCDE2 [17] | LX-PM [34] | LX-POL [34] | BEX-PM [7] | AEO |
|---|---|---|---|---|---|---|---|
| TDS1 | 0.0500 | 0.0500 | 0.0500 | 0.0501 | 0.0500 | 0.0500 | 0.0500 |
| TDS2 | 0.2194 | 0.1976 | 0.1976 | 0.2044 | 0.2083 | 0.2002 | 0.1976 |
| TDS3 | 0.0500 | 0.0500 | 0.0500 | 0.0500 | 0.0503 | 0.0500 | 0.0500 |
| TDS4 | 0.2135 | 0.2090 | 0.2090 | 0.2140 | 0.2118 | 0.2108 | 0.2091 |
| TDS5 | 0.1949 | 0.1812 | 0.1812 | 0.1936 | 0.1951 | 0.1814 | 0.1812 |
| TDS6 | 0.1953 | 0.1806 | 0.1806 | 0.1949 | 0.1935 | 0.1814 | 0.1806 |
| PCS1 | 1.250 | 1.250 | 1.250 | 1.2507 | 1.2508 | 1.250 | 1.250 |
| PCS2 | 1.250 | 1.499 | 1.500 | 1.4354 | 1.3829 | 1.467 | 1.500 |
| PCS3 | 1.250 | 1.250 | 1.250 | 1.2509 | 1.2501 | 1.250 | 1.250 |
| PCS4 | 1.460 | 1.499 | 1.500 | 1.4614 | 1.4933 | 1.478 | 1.500 |
| PCS5 | 1.250 | 1.500 | 1.500 | 1.2915 | 1.2513 | 1.496 | 1.500 |
| PCS6 | 1.250 | 1.499 | 1.500 | 1.2590 | 1.2878 | 1.487 | 1.500 |

Comparative results of AEO and well-known methods in the literature for Model 1 are presented in Table 9. DE, OCDE1, and the OCDE2 algorithms have an infeasible solution because they do not meet the CTI constraints. According to comparison with results in the literature, the minimum relay operation time obtained with the proposed algorithm is 4.7803 s, for Model I. In other words, AEO is 1.1108%, 1.1681%, and 0.2004% less than the results of LX-PM, LX-POL, and BEX-PM respectively. From the comparison results in Table 9, it is observed that the performance of the proposed method is better than other methods in the literature.

**Table 4.** Comparison of the CTI values for Model I (Tbackup-Tprimary)

| Fault | Relay | | | DE | OCDE1 | OCDE2 | LX-PM |
|---|---|---|---|---|---|---|---|
| | Backup | Primary | | | | | |
| Near-end | R5 | R1 | CTI1 | 0.2998 | 0.3000 | 0.3000 | 0.3024 |
| | R6 | R3 | CTI2 | 0.3000 | 0.2996 | 0.2998 | 0.3004 |
| | R4 | R5 | CTI3 | 0.3001 | 0.2997 | 0.2999 | 0.3005 |
| | R2 | R6 | CTI4 | 0.3791 | 0.3898 | 0.3901 | 0.3915 |
| Far-end | R5 | R1 | CTI5 | 0.3281 | 0.3362 | 0.3362 | 0.3324 |
| | R6 | R3 | CTI6 | 0.3141 | 0.3207 | 0.3209 | 0.3147 |
| | R4 | R5 | CTI7 | 0.4020 | 0.4000 | 0.4003 | 0.4022 |
| | R2 | R6 | CTI8 | 0.3001 | 0.2998 | 0.3000 | 0.3013 |
| Fault | Relay | | | LX-POL | BEX-PM | AEO | |
| | Backup | Primary | | | | | |
| Near-end | R5 | R1 | CTI1 | 0.3004 | 0.3000 | 0.3000 | |
| | R6 | R3 | CTI2 | 0.3002 | 0.3001 | 0.3000 | |
| | R4 | R5 | CTI3 | 0.3018 | 0.3007 | 0.3002 | |
| | R2 | R6 | CTI4 | 0.3865 | 0.3883 | 0.3901 | |
| Far-end | R5 | R1 | CTI5 | 0.3289 | 0.3361 | 0.3362 | |
| | R6 | R3 | CTI6 | 0.3152 | 0.3209 | 0.3209 | |
| | R4 | R5 | CTI7 | 0.4061 | 0.4000 | 0.4007 | |
| | R2 | R6 | CTI8 | 0.3001 | 0.3000 | 0.3000 | |

## 4.3   Model II: IEEE-4 Bus Test System

IEEE 4-bus test syste contains 2 generators, 4 lines, 8 relays, and 9 P/B relay pairs. Model II has 16 decision variables (TDS1–TDS8 and PCS1–PCS8). The values of $I_f$, $CT_{pr\_rating}$ parameters, and fault current values for the primary/backup relay pairs belonging to Model II are provided in Tables 5 and 6 [7].

For the IEEE-4 bus system, the optimized TDS and PCS decision variables are tabulated in Table 7. It is understood from the table that all decision variables are within acceptable limits.

The comparison of CTI values for Model II is given in Table 8. It is clear that some relay pairs do not meet the selectivity criteria for DE, OCDE1, OCDE2, and BEX-PM algorithms. The obtained results can be defined as the solution with miscoordination between B/P relay pairs: R7/R3,R2/R6,R4/R8 for the DE algorithm, the solution with miscoordination between B/P relay pairs: R5/R1, R7/R3, R2/R6,R4/R8 for OCDE1 algorithm, the solution with miscoordination between B/P relay pairs: R5/R1, R7/R3, R2/R6, R4/R8 for the OCDE2 algorithm, and the solution with miscoordination between B/P relay pairs: R5/R1,R2/R6,R4/R8 for the BEX-PM algorithm These algorithms have an infeasible solution because they violate selectivity constraints. LX-PM, LX-POL

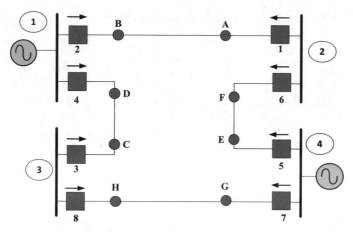

**Fig. 4.** Model-II: IEEE 4-bus test system

**Table 5.** Values of $I_f$ and $CT_{pr\_rating}$ for Model-II with $T_{pri\_cl\_in}$ and $T_{pri\_far\_bus}$

| $T^i_{pri\_cl\_in}$ | | | $T^j_{pri\_far\_bus}$ | | |
|---|---|---|---|---|---|
| $TDS^i$ | $I_f^i$(p.u.) | $CT^i_{pr\_rating}$ | $TDS^j$ | $I_f^j$(p.u.) | $CT^j_{pr\_rating}$ |
| $TDS^1$ | 20.32 | 0.4800 | $TDS^2$ | 23.75 | 0.4800 |
| $TDS^2$ | 88.85 | 0.4800 | $TDS^1$ | 12.48 | 0.4800 |
| $TDS^3$ | 13.6 | 1.1789 | $TDS^4$ | 31.92 | 1.1789 |
| $TDS^4$ | 116.81 | 1.1789 | $TDS^3$ | 10.38 | 1.1789 |
| $TDS^5$ | 116.70 | 1.5259 | $TDS^6$ | 12.07 | 1.5259 |
| $TDS^6$ | 16.67 | 1.5259 | $TDS^5$ | 31.92 | 1.5259 |
| $TDS^7$ | 71.70 | 1.2018 | $TDS^8$ | 11.00 | 1.2018 |
| $TDS^8$ | 19.27 | 1.2018 | $TDS^7$ | 18.91 | 1.2018 |

and AEO algorithms provide a feasible solution for relay coordination problem in the IEEE 4-bus system. When these algorithms are compared according to the objective function value, the best result belongs to the AEO algorithm and its value is 3.6757 s. It is understood from the comparison results in Table 9 suggested method is better than other algorithms considered in this study.

**Table 6.** Fault currents for backup/primary relay pairs for Model-II

| $T^i_{backup}$ | | | $T^j_{primary}$ | | |
|---|---|---|---|---|---|
| Relay | $I_f^i$(p.u.) | $CT^i_{pr\_rating}$ | Relay | $I_f^j$(p.u.) | $CT^j_{pr\_rating}$ |
| R5 | 20.32 | 1.5259 | R1 | 20.32 | 0.4800 |
| R5 | 12.48 | 1.5259 | R1 | 12.48 | 0.4800 |
| R7 | 13.61 | 1.2018 | R3 | 13.61 | 1.1789 |
| R7 | 10.38 | 1.2018 | R3 | 10.38 | 1.1789 |
| R1 | 116.81 | 0.4800 | R4 | 116.81 | 1.1789 |
| R2 | 12.07 | 0.4800 | R6 | 12.07 | 1.5259 |
| R2 | 16.67 | 0.4800 | R6 | 16.67 | 1.5259 |
| R4 | 11.00 | 1.1789 | R8 | 11.00 | 1.2018 |
| R4 | 19.27 | 1.1789 | R8 | 19.27 | 1.2018 |

**Table 7.** Optimal TDS and PCS for Model II

| Item | DE [17] | OCDE1 [17] | OCDE2 [17] | LX-PM [34] | LX-POL [34] | BEX-PM [7] | AEO |
|---|---|---|---|---|---|---|---|
| TDS1 | 0.0500 | 0.0500 | 0.0500 | 0.0500 | 0.0500 | 0.0500 | 0.0504 |
| TDS2 | 0.2248 | 0.2122 | 0.2121 | 0.2249 | 0.2249 | 0.2196 | 0.2122 |
| TDS3 | 0.0500 | 0.0500 | 0.0500 | 0.0500 | 0.0500 | 0.0500 | 0.0500 |
| TDS4 | 0.1515 | 0.1515 | 0.1515 | 0.1637 | 0.1637 | 0.1634 | 0.1524 |
| TDS5 | 0.1264 | 0.1262 | 0.1262 | 0.1318 | 0.1318 | 0.1320 | 0.1267 |
| TDS6 | 0.0500 | 0.0500 | 0.5000 | 0.0500 | 0.0500 | 0.0500 | 0.0500 |
| TDS7 | 0.1337 | 0.1337 | 0.1337 | 0.1371 | 0.1371 | 0.1356 | 0.1338 |
| TDS8 | 0.0500 | 0.0500 | 0.0500 | 0.0500 | 0.0500 | 0.0500 | 0.0500 |
| PCS1 | 1.2734 | 1.2500 | 1.2500 | 1.2859 | 1.2860 | 1.2846 | 1.2730 |
| PCS2 | 1.2500 | 1.4999 | 1.5000 | 1.2502 | 1.2502 | 1.3475 | 1.5000 |
| PCS3 | 1.2500 | 1.2500 | 1.2500 | 1.2501 | 1.2501 | 1.2500 | 1.2500 |
| PCS4 | 1.4997 | 1.4999 | 1.5000 | 1.2501 | 1.2501 | 1.2500 | 1.5000 |
| PCS5 | 1.4997 | 1.500 | 1.5000 | 1.3757 | 1.3756 | 1.3656 | 1.5000 |
| PCS6 | 1.2500 | 1.2500 | 1.2500 | 1.2503 | 1.2503 | 1.2500 | 1.2500 |
| PCS7 | 1.5000 | 1.5000 | 1.5000 | 1.4304 | 1.4304 | 1.4600 | 1.5000 |
| PCS8 | 1.2500 | 1.2500 | 1.2500 | 1.2523 | 1.2523 | 1.2500 | 1.2500 |

**Table 8.** Comparison of the CTI values for Model II (Tbackup-Tprimary)

| Fault | Relay | | | DE | OCDE1 | OCDE2 | LX-PM |
|---|---|---|---|---|---|---|---|
| | Backup | Primary | | | | | |
| Near-end | R5 | R1 | CTI1 | 0.3000 | 0.2999 | 0.2999 | 0.3006 |
| | R7 | R3 | CTI2 | 0.2997 | 0.2997 | 0.2997 | 0.3003 |
| | R2 | R6 | CTI3 | 0.3002 | 0.2848 | 0.2848 | 0.3005 |
| | R4 | R8 | CTI4 | 0.3225 | 0.3259 | 0.3257 | 0.3227 |
| | R1 | R4 | CTI5 | 0.3973 | 0.3973 | 0.3973 | 0.3862 |
| Far-end | R5 | R1 | CTI6 | 0.4002 | 0.4002 | 0.4002 | 0.3954 |
| | R7 | R3 | CTI7 | 0.3495 | 0.3495 | 0.3495 | 0.3484 |
| | R2 | R6 | CTI8 | 0.2998 | 0.3001 | 0.2999 | 0.3000 |
| | R4 | R8 | CTI9 | 0.2997 | 0.2998 | 0.2998 | 0.3006 |
| Fault | Relay | | | LX-POL | BEX-PM | AEO | |
| | Backup | Primary | | | | | |
| Near-end | R5 | R1 | CTI1 | 0.3005 | 0.2999 | 0.3002 | |
| | R7 | R3 | CTI2 | 0.3003 | 0.3000 | 0.3000 | |
| | R2 | R6 | CTI3 | 0.3005 | 0.3001 | 0.3028 | |
| | R4 | R8 | CTI4 | 0.3227 | 0.3238 | 0.3259 | |
| | R1 | R4 | CTI5 | 0.3862 | 0.3853 | 0.4007 | |
| Far-end | R5 | R1 | CTI6 | 0.3954 | 0.3941 | 0.4006 | |
| | R7 | R3 | CTI7 | 0.3484 | 0.3487 | 0.3499 | |
| | R2 | R6 | CTI8 | 0.3000 | 0.2999 | 0.3001 | |
| | R4 | R8 | CTI9 | 0.3006 | 0.2999 | 0.3024 | |

**Table 9.** Comparison results in the literature (Model I- II)

| Algorithm | IEEE 3-bus | IEEE 4-bus |
|---|---|---|
| DE [17] | $4.8422^{\gamma}$ | $3.6774^{\gamma}$ |
| OCDE1 [17] | $4.7806^{\gamma}$ | $3.6674^{\gamma}$ |
| OCDE2 [17] | $4.7806^{\gamma}$ | $3.6674^{\gamma}$ |
| LX-PM [34] | 4.8340 | 3.7029 |
| LX-POL [34] | 4.8368 | 3.7034 |
| BEX-PM [7] | 4.7899 | $3.6957^{\gamma}$ |
| AEO | **4.7803** | **3.6757** |

Infeasible solutions $\gamma$.

## 5 Conclusion

DOCRs coordination including many operational constraints is a highly complex real-world engineering problem that is difficult to solve using MHS algorithms. In this paper, the AEO algorithm with strong exploration and balanced search capabilities is used to solve the DOCRs coordination problem. The performance of the proposed method was tested on IEEE-3 and IEEE-4 power systems. The results obtained from the simulation studies were compared with that of DE, OCDE1, OCDE2, LX-PM, LX-POL, and BEX-PM algorithms. The simulation results showed that the AEO provided better solutions to the DOCRs coordination problem compared to other methods considered in the study.

## References

1. Güvenç, U., Özkaya, B., Bakir, H., Duman, S., Bingöl, O.: Energy hub economic dispatch by symbiotic organisms search algorithm. In: Jude Hemanth, D., Kose, U. (eds.) Artificial Intelligence and Applied Mathematics in Engineering Problems: Proceedings of the International Conference on Artificial Intelligence and Applied Mathematics in Engineering (ICAIAME 2019), pp. 375–385. Springer International Publishing, Cham (2019). https://doi.org/10.1007/978-3-030-36178-5_28
2. Korashy, A., Kamel, S., Youssef, A.R., Jurado, F.: Modified water cycle algorithm for optimal direction overcurrent relays coordination. Appl. Soft Comput. **74**, 10–25 (2019)
3. Radosavljevic, J.: Metaheuristic Optimization in Power Engineering. Institution of Engineering and Technology (2018). https://doi.org/10.1049/PBPO131E
4. Singh, M., Panigrahi, B.K., Abhyankar, A.R., Das, S.: Optimal coordination of directional over-current relays using informative differential evolution algorithm. J. Comput. Sci. **5**(2), 269–276 (2014)
5. Zeineldin, H.H., El-Saadany, E.F., Salama, M.M.A.: Optimal coordination of overcurrent relays using a modified particle swarm optimization. Electric Power Syst. Res. **76**, 988–995 (2006)
6. Mahari, A., Seyedi, H.: An analytic approach for optimal coordination of overcurrent relays. IET Gener. Transm. Distrib. **7**(7), 674–680 (2013)
7. Thakur, M., Kumar, A.: Optimal coordination of directional over current relays using a modified real coded genetic algorithm: a comparative study. Int. J. Electr. Power Energy Syst. **82**, 484–495 (2016)
8. Zeineldin, H.H., El-Saadany, E.F., Salama, M.M.A.: Optimal coordination of over-current relays using a modified particle swarm optimization. Electr. Power Syst. Res. **76**, 988–995 (2006)
9. Mansour, M.M., Mekhamer, S.F., El-Kharbawe, N.E.: A modified particle swarm optimizer for the coordination of directional overcurrent relays. IEEE Trans. Power Deliv. **22**(3), 1400–1410 (2007)
10. So, C.W., Li, K.K., Lai, K.T., Fung, K.Y.: Application of genetic algorithm for overcur-rent relay coordination. In: International Conference of Developments in Power System Protection, pp. 66–69 (1997)
11. Razavi, F., Abyaneh, H.A., Al-Dabbagh, M., Mohammadi, R., Torkaman, H.: A newcomprehensive genetic algorithm method for optimal overcurrent relays coordination. Electr. Power Syst. Res. **78**, 713–720 (2008)
12. Shih, M.Y., Enríquez, A.C., Trevino, L.M.T.: On-line coordination of directional overcurrent relays: performance evaluation among optimization algorithms. Electr. Power Syst. Res. **110**, 122–132 (2014)

13. Moravej, Z., Adelnia, F., Abbasi, F.: Optimal coordination of directional overcurrent relays using NSGA-II. Electr. Power Syst. Res. **119**, 228–236 (2015)
14. Amraee, T.: Coordination of directional overcurrent relays using seeker algorithm. IEEE Trans. Power Deliv. **27**(3), 1415–1422 (2012)
15. Singh, M., Panigrahi, B.K., Abhyankar, A.R.: Optimal coordination of directional over-current relays using teaching learning-based optimization (TLBO) algorithm. Electr. Power Energy Syst. **50**, 33–41 (2013)
16. Moirangthem, J., Krishnanand, K.R., Dash, S.S., Ramaswami, R.: Adaptive differential evolution algorithm for solving non-linear coordination problem of directional overcurrent relays. IET Gener. Transm. Distrib. **7**(4), 329–336 (2013)
17. Chelliah, T.R., Thangaraj, R., Allamsetty, S., Pant, M.: Coordination of directional overcurrent relays using opposition based chaotic differential evolution algorithm. Electr. Power Energy Syst. **55**, 341–350 (2014)
18. Albasri, F.A., Alroomi, A.R., Talaq, J.H.: Optimal coordination of directional overcurrent relays using biogeography- based optimization algorithms. IEEE Trans. Power Deliv. **30**(4), 1810–1820 (2015). https://doi.org/10.1109/TPWRD.2015.2406114
19. Bouchekara, H.R.E.H., Zellagui, M., Abido, M.A.: Optimal coordination of directional over-current relays using a modified electromagnetic field optimization algorithm. Appl. Soft Comput. **54**, 267–283 (2017)
20. Kim, C.H., Khurshaid, T., Wadood, A., Farkoush, S.G., Rhee, S.B.: Gray wolf optimizer for the optimal coordination of directional overcurrent relay. J. Electr. Eng. Technol. **13**(3), 1043–1051 (2018)
21. Rajput, V.N., Pandya, K.S.: Coordination of directional overcurrent relays in the interconnected power systems using effective tuning of harmony search algorithm. Sustain. Comput.: Inf. Syst. **15**, 1–15 (2017)
22. El-Fergany, A.A., Hasanien, H.M.: Optimized settings of directional overcurrent relays in meshed power networks using stochastic fractal search algorithm. Int. Trans. Electr. Energy Syst. **27**(11), e2395 (2017)
23. Zellagui, M., Benabid, R., Boudour, M., Chaghi, A.: Application of firefly algorithm for optimal coordination of directional overcurrent protection relays in presence of series compensation. J. Autom. Syst. Eng. 92–107 (2014)
24. Hussain, M.H., Musirin, I., Abidin, A.F., Rahim, S.R.A.: Solving directional overcurrent relay coordination problem using artificial bees colony. Intern. J. Electr. Electron. Sci. Eng **8**(5), 705–710 (2014)
25. El-Fergany, A.: Optimal directional digital overcurrent relays coordination and arc-flash hazard assessments in meshed networks. Int. Trans. Electr. Energy Syst. **26**(1), 134–154 (2016)
26. Saha, D., Datta, A., Das, P.: Optimal coordination of directional overcurrent relays in power systems using symbiotic organism search optimisation technique. IET Gener. Transm. Distrib. **10**(11), 2681–2688 (2016)
27. Srinivas, S.T.P.: Application of improved invasive weed optimization technique for optimally setting directional overcurrent relays in power systems. Appl. Soft Comput. **79**, 1–13 (2019)
28. Ahmadi, S.A., Karami, H., Sanjari, M.J., Tarimoradi, H., Gharehpetian, G.B.: Application of hyper-spherical search algorithm for optimal coordination of overcurrent relays considering different relay characteristics. Int. J. Electr. Power Energy Syst. **83**, 443–449 (2016)
29. Korashy, A., Kamel, S., Jurado, F., Youssef, A.R.: Hybrid whale optimization algorithm and grey wolf optimizer algorithm for optimal coordination of direction overcurrent relays. Electric Power Compon. Syst. **47**, 1–15 (2019)
30. Khurshaid, T., Wadood, A., Farkoush, S.G., Yu, J., Kim, C.H., Rhee, S.B.: An improved optimal solution for the directional overcurrent relays coordination using hybridized whale optimization algorithm in complex power systems. IEEE Access **7**, 90418–90435 (2019)

31. Radosavljević, J., Jevtić, M.: Hybrid GSA-SQP algorithm for optimal coordination of directional overcurrent relays. IET Gener. Transm. Distrib. **10**(8), 1928–1937 (2016)
32. Zellagui, M., Abdelaziz, A.Y.: Optimal coordination of directional overcurrent relays using hybrid PCSO-DE algorithm. Int. Electr. Eng. J. **6**(4), 1841–1849 (2015)
33. Zhao, W., Wang, L., Zhang, Z.: Artificial ecosystem-based optimization: a novel nature-inspired meta-heuristic algorithm. Neural Comput. Appl. **32**(13), 9383–9425 (2019). https://doi.org/10.1007/s00521-019-04452-x
34. Thakur M.: New real coded genetic algorithms for global optimization Ph.D. Thesis. India: Department of Mathematics, Indian Institute of Technology Roorkee (2007)

# Performance Evaluation of Machine Learning Techniques on Flight Delay Prediction

Irmak Daldır[1], Nedret Tosun[2], and Ömür Tosun[3(✉)]

[1] Department of International Trade and Logistics, Faculty of Applied Sciences,
Akdeniz University, Antalya, Turkey
irmakdaldir@akdeniz.edu.tr
[2] Department of Econometrics, Institute of Social Sciences, Akdeniz University,
Antalya, Turkey
tosunn@baib.gov.tr
[3] Department of Management Information Systems, Faculty of Applied Sciences,
Akdeniz University, Antalya, Turkey
omurtosun@akdeniz.edu.tr

**Abstract.** Machine learning has emerged as a solution to different variety of problems. It is commonly used where big data exist and solve the problem which is normally hard with traditional methods. Within the scope of this study, whether a given flight scheduled to be late or on time will be estimated with various machine learning methods. For this approach; XGBoost, Support Vector Machines, Random Forest, Artificial Neural Networks, and CatBoost algorithms have been used for the problem. Based on the results, it has been determined that the method that put forward the best solution for this problem and the data set is logistic regression.

**Keywords:** Flight delay prediction · Machine learning · Performance evaluation

## 1 Introduction

Flight delays are an important problem for the sector and passengers. Federal Aviation Administration estimated the annual cost of delays for the 2018 to be $28 billion [1]. Cost component items included direct cost both for airlines and passengers, lost demand and indirect costs. Despite the advantages of air transport like high speed, broad service range, unpredictable cost components can also be considered as major disadvantages.

In order to eliminate the delay problems, before 2016, many of the studies are used modelling and simulation techniques [2–5]. These methods have also their own pros and cons. In the context of this study, XGBoost, Support Vector Machines, Random Forests, Artificial Neural Networks, and CatBoost algorithms are used to predict whether a flight will be on time or not. The machine learning methods were preferred to find out patterns in the data set. The data consists of 13319 flights which belongs to a Turkish airport from 2018. The data gathered from the airport was combined with weather data to include the weather impacts on the flight delays.

© The Author(s), under exclusive license to Springer Nature Switzerland AG 2021
J. Hemanth et al. (Eds.): ICAIAME 2020, LNDECT 76, pp. 165–173, 2021.
https://doi.org/10.1007/978-3-030-79357-9_16

In Sect. 1, a brief information on flight delay is presented and the purpose of the study is expressed. In Sect. 2, Detailed literature review on flight delays with machine learning is discussed. In Sect. 3, all methods that are used for the problem are introduced briefly. Application results are given in Sect. 4. Finally, the results and future work propositions are discussed at the Sect. 5.

## 2 Literature Review on Machine Learning and Flight Delays

Machine learning and flights delay started to be studied not long ago. The reason for this is the development of machine learning methods and the fact that big data operations can be done easily with machine learning. Since the topic has vital importance in air traffic control, airline decision making processes, and ground operations, the topic have been studied from various perspectives. Very first study is done by [6]. Flight delays that are caused because of weather is forecasted with de domestic flight data and the weather data is used from 2005 to 2015. [7] also worked on two step machine learning model with Recurrent Neural Networks and Long Short Term Memory. The model predicts whether the aircraft will be delayed, and then gives results on how long they will be late within 15 min of categorical time periods. Data of ten major airports in the U.S. have been collected and the model's accuracy is between 85% and 92%. The US Domestic Airline On-Time Performance data and weather data (World Weather Online API) from the year 2012 to 2016 are used. [8] build a model to determine the delay time of aircrafts. They differently used Gradient Boosted Decision Tree and find out that the method gives better accuracy results then other methods with Coefficient of Determination of 92.31% for arrival delays and d 94.85% for departure delays. Again, the US data used the busiest 70 airports in the USA within April-November 2013 time period. Similarly, a comparison of methods study is done by [9]. Decision tree, logistic regression and artificial neural network models are developed. The results of the models are compared and it is shown that they are almost the same. The data set is collected from Kaggle. Another study which again based on the Kaggle dataset is done [10]. Artificial neural network and deep belief network are used for forecasting of whether the aircraft on time or late. The model gives the 92% accuracy. On average delays between Beijing Capital airport and Hangzhou Xiaoshan International Airport is also studied [11]. Delays which are less than 15 min are predicted with Long Short-Term Memory (LSTM). Following study is focused on delays for connected flights. Bureau of Transportation Statistics are used with Multi-Label Random Forest and approximated delay propagation model [12].

Differently in this study, five different methods are applied to determine to if the delay will occur for the selected flight or not problem and the methods are compared with each other.

## 3 Methods

### 3.1 Artificial Neural Network

The ANN is a machine learning tool, which is prevalently used for fitting, pattern recognition, clustering and prediction. The main concept of the tool is mimicking the human brain procedures. It consists of interconnected group of linked artificial neurons. Multi layered neural network with supervised learning is implemented for the model. Before 2006, only one hidden layer was used, but after Hinton and et al. suggested the deep learning, it is common to prefer more hidden layers if the data is suitable [13]. There are multiple layers in a feed forward neural network algorithm. The network consists of an input layer, hidden layers and output layer [14]. In these types of neural networks, input layer contains the external data. Each data means a neuron and these are connected to following hidden layer. The layer which gives the result represents the output layer. In the hidden layer each neuron has an activation and a sum function. First the received data is multiplied by weight that the link at the network has and then added to each other. Followingly, a bias is added to this calculation and the obtained result is evaluated is evaluated according to activation function and the result will be used as an input for the next neuron.

## 3.2  Extreme Gradient Boosting

XGBoost algorithm depends on the gradient boosting which is proposed by [15]. Boosting technique introduce new models to lower the error that existent models have made. Models are introduced sequentially until the best possible improvement are made. Gradient boosting is a method where new models are generated to predict the residuals or errors of previous models, and then used to make the final prediction together. It is called gradient boosting, since it uses an algorithm of gradient descent to minimize the loss when introducing new models. The improved version, XGBoost, is developed by [16]. The key factor behind XGBoost's performance is its scalability in all scenarios and scales to billions of examples in distributed or memory-limited settings are ten times faster than current common solutions on one computer [17].

## 3.3  Logistic Regression

Logistic regression models are used for only binary classification problems. The logistic regression algorithm's coefficients have to be derived from the training data. Maximum likelihood estimation is a popular learning algorithm which is used also at this stage of logistic regression. The best coefficients would lead to a model to predict which class the given data belong. The theory for maximum likelihood of logistic regression aims to reduce the error in the probabilities predicted by the model [18]. Alternatively, Gradient descent can be applied to minimize the cost value of the logistic regression model.

## 3.4  Random Forest

Random Forests consist of different tree classifiers' combination where each of them using a random vector that sampled independently from the input, and each tree has a unit weight to classify an input vector in the optimum class [19]. The unit weight of trees means that each tree has equal chance while sampling is done. The main idea behind this approach is the trees should have the low correlation between them to get best results. Bootstrap Aggregation (Bagging) and feature randomness are the key methods to not too correlated.

### 3.5  CatBoost

CatBoost also a gradient boosting library which's working method improved by [20]. The name come from the "category" and "boosting". There are several advantages of the method, which are mainly performance, handling categorical features automatically, robust, and easy to use. A comparison of XGBoost, Light BGM and Catboost methods are done and the tuning times are respectively, 500, 200, and 120 min [21]. CatBoost also allows the usage of categorical columns without encoding. The method reduces the chance of overfitting. It also gives opportunity to tune parameters like number of trees, learning rate, tree depth, and others.

### 3.6  Support Vector Machines (SVM)

This method is another machine learning model that is originally used to solve binary problems [22]. SVMs aim is to find a line that separates two classes while maximizing the margins from both classes. It can be applied to nonlinear data too. The tuning parameters for SVM are regularization, gamma, kernel and margin. Regularization parameters is used to determine to avoid misclassifying of each training example. Kernel is used to make faster calculations. Low gamma considers only training data that nearby the separation line and high gamma acts vice versa. Finally, good margin should place the separation line equidistant possible for both class.

## 4   Implementation of Machine Learning Methods on Flight Data

In this study flight delay prediction of a Turkish international airport will be predicted by using the flight data of 2018. There are 13319 samples in the data set. The problem is designed as a classification problem in which the aim is to predict if a given flight will be delayed or not. If the time difference between the actual flight time and scheduled flight time is bigger than 15 min, that flight is labeled as delayed.

Through a comprehensive literature survey, features in Table 1 are used in the analysis.

**Table 1.** Variable explanations

| Variable name | Definition | Data type |
| --- | --- | --- |
| Month | Month of the year | Categorical |
| Day of the month | Day of the month | Categorical |
| Day of the week | Day of the week | Categorical |
| Cargo | If the flight has a cargo or not | Categorical |
| Pax | Number of passengers on the given flight | Continuous |
| STD | Scheduled time of the flight | Continuous |

*(continued)*

**Table 1.** (*continued*)

| Variable name | Definition | Data type |
|---|---|---|
| ETD | Estimated time of the flight | Continuous |
| Start_up | Start up time of the engine of the given flight | Continuous |
| ATD | Actual time of the flight | Continuous |
| T | Air temperature of the airport at or near the flight time | Continuous |
| P0 | Atmospheric pressure measured at weather station of the airport at or near the flight time | Continuous |
| U | Relative humidity of the airport at or near the flight time | Continuous |
| DD | Mean wind direction measured of the airport at or near the flight time | Categorical |
| FF | Mean wind speed measured of the airport at or near the flight time | Continuous |
| C | Total cloud cover of the airport | Categorical |
| VV | Horizontal visibility measured at the airport (km) | Continuous |

Except from the Cargo, DD, C and date features, all the variables are collected as continuous numbers. The categorical variables are combined to numerical values using "one-hot-encoding" transformation technique, in which each unique observation of the variables is transformed to binary variables. There are 2 unique values in Cargo, 17 unique values in DD and 283 in C. Also 7 for the day of the week, 12 for the month and 31 for the day of the month variables. Therefore, after the transformation the data set has 362 variables.

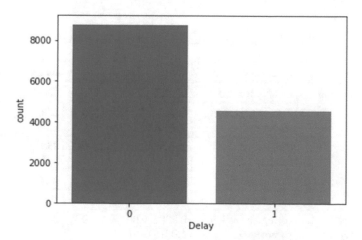

**Fig. 1.** Number of delayed and on time flights

If the output of the data analyzed, there are 8754 on-time flights and 4565 delay flights. Also it is shown in Fig. 1.

Before beginning the analysis, the data is randomly divided into two parts. 70% of the observations are used for training of the machine learning models and 30% for the test phase.

Performance of each model is measured using accuracy, recall, Receiver Operating Characteristic (ROC) Score, and Cohen's Kappa score. Performace ratios usually based on the confusion matrix. In a binary classification model, observations can be classified as either positive or negative. Based on this, results can be: true positive which is observations that are correctly classified as positive, false positives are incorrectly classified as positive, true negatives refer to negative observations that are correctly labeled, and false negatives that are incorrectly labeled as negative [23]. The accuracy rate is calculated by multiplying the portion of the correctly categorized observations to all observations by 100. Accuracy rate is suitable when the classification classes are equal and if the misclassification of minor class observations are high then it could lead misevaluation. Recall is the proportion of real positive cases that are correctly predicted as positive. In other words, the recall ratio is suitable to apply when it is aimed to minimize false negative classifications. ROC graph includes true positive (Y axis) and false positive rates (X axis). ROC score is the area that is under the ROC curve. Higher ROC score means a better model, because it shows that model's capability of distinguish the classes. Finally, Cohen's Kappa score is a quantitative measure of interrater reliability. The value that Cohen's Kappa can get is change between 0–1. 0 means there is random agreement and 1 means complete agreement between the rates. Higher score means correct negative negatives, positive positives classifications. Before using each model, a parameter optimization is applied with a 10-fold cross-validation.

For each machine learning algorithm, different control parameters are selected to optimize based on the previous studies in the literature. Experimental analysis results are given in Table 2:

**Table 2.** Experimental analysis results of the methods

| Algorithm | Parameter | Values | Best value found |
|---|---|---|---|
| Logistic regression | Penalty function | L1, L2 | L2 |
| | C (regularization parameter) | 0.001, 0.01, 0.1, 1, 10, 100, 1000 | 1000 |
| XGBoost | Number of estimators | 150, 200, 250, 300, 350 | 350 |
| | Learning rate | 0.05, 0.01, 0.15, 0.1, 0.2 | 0.15 |
| | Max depth | 5, 6, 7, 8, 9 | 8 |
| Support vector machines - Linear | C (regularization parameter) | 0.0001, 0.001, 0.01, 0.1, 1.0, 10.0, 20 | 20 |

(continued)

**Table 2.** (*continued*)

| Algorithm | Parameter | Values | Best value found |
|---|---|---|---|
| Support vector machines - rbf | C (regularization parameter) | 0.0001, 0.001, 0.01, 0.1, 1.0, 10.0, 20 | 0.0001 |
| | Gamma (kernel coefficient) | 0.1, 0.2, 0.4, 0.6, 0.8, 0.9 | 0.1 |
| Random forest | Number of estimators | 150, 200, 250, 300, 350 | 150 |
| | Max depth | 5, 6, 7, 8, 9 | 9 |
| | Min samples leaf | 5, 6, 7, 8, 9 | 6 |
| Artificial neural networks | Number of hidden layers | 1, 2 | 2 |
| | Hidden layer size | 5, 10, 15, 20, 25 | 5–15 |
| | Transfer function | Relu, tanh | Tanh |
| CatBoost | Learning rate | 0.005, 0.03, 0.1,0.3 | 0.3 |
| | Depth | 4, 6, 8, 10, 12 | 8 |
| | L2 leaf reg (L2 regularization parameter) | 1, 3, 5, 7, 9 | 7 |

In Table 3 the results of the algorithms are given.

**Table 3.** Algorithms results

| | Accuracy | | Recall | | ROC score | | Cohen's Kappa | |
|---|---|---|---|---|---|---|---|---|
| | Train | Test | Train | Test | Train | Test | Train | Test |
| Logistic regression | 0.996 | 0.987 | 0.993 | 0.971 | 0.996 | 0.984 | 0.993 | 0.971 |
| XGBoost | 1 | 0.973 | 1 | 0.937 | 1 | 0.964 | 1 | 0.940 |
| SVM – Linear | 0.979 | 0.970 | 0.945 | 0.921 | 0.971 | 0.958 | 0.953 | 0.931 |
| SVM – RBF | 0.653 | 0.666 | 0.945 | 0.921 | 0.819 | 0.556 | 0.676 | 0.128 |
| Random forest | 0.715 | 0.702 | 0.218 | 0.163 | 0.599 | 0.568 | 0.239 | 0.169 |
| ANN | 0.942 | 0.890 | 0.923 | 0.836 | 0.938 | 0.836 | 0.874 | 0.753 |
| CatBoost | 0.999 | 0.954 | 0.999 | 0.906 | 0.999 | 0.942 | 0.999 | 0.895 |

Table 3 summarizes of all algorithms results based on the determined performance measures. For this particular problem correct classification of the true negative is important, since it is aimed to determine to which flights are the delayed ones. Accordingly, relying only accuracy scores could be deceptive. According to test results logistic regression gives the best results, based on all performance measures; accuracy, recall, ROC score and Cohen's Kappa. XGBoost seems better as train results, but test results are lower than logistic regression and high train and low test results can be explained as overlearning of the model. Because of this reason, XGBoost seems weaker than logistic regression. Followingly, SVM –Linear results are better than radial basis function (RBF). Even the recall results are good of the algorithm other performance measurements are very low. Random forest algorithm has relatively high accuracy, but the other performance measures are very low to select this model for forecasting. ANN has reasonable results within the algorithms. nevertheless, CatBoost algorithm has better performance results than ANN.

## 5    Conclusions and Future Work

From this study, it is shown that some machine learning methods can be fit better than others. In order to find out this difference seven different algorithms are tried and results are compared. Logistic regression gives the best solution not only by the perspective of recall but also with accuracy, ROC score, and Cohen's Kappa score. Furthermore, high performance results of algorithm results are promise better predictions for delayed flights. As a result, it can be seen that, in case of actual implementation at airports, algorithms can give good answers to the question of which flights may be late. According to these estimates, it will be possible to make preliminary studies on the flights that are likely to be late. Thus, cost items can be reduced or customer dissatisfaction can be prevented. For further studies it is proposed to extend the data by increasing inputs to get more stable results. Moreover, delay time period can be forecasted.

**Acknowledgments.** This work was supported by The Scientific Research Projects Coordination Unit of Akdeniz University. Project Number: SBG-2020–5120.

## References

1. Airlines for America: U.S. Passenger Carrier Delay Costs. At the date of 2020 Airline for America:is taken from (2017). https://www.airlines.org/dataset/per-minute-cost-of-delays-to-u-s-airlines/
2. Kim, Y.J., Pinon-Fisher, O.J., Mavris, D.N.: Parallel simulation of agent-based model for air traffic network. A. M. Technologies, p. 2799 (2015)
3. Campanelli, B., et al.: Modeling reactionary delays in the European air transport network. In: Proceedings of the Fourth SESAR Innovation Days. SeSAR INnovation Days, Madrid (2014)
4. Gao, W., Xu, X., Diao, L., Ding, H.: SIMMOD based simulation optimization of flight delay cost for multi-airport system. In: 2008 International Conference on Intelligent Computation Technology and Automation (ICICTA), vol. 1, pp. 698–702. IEEE (2008)

5. Lee, L.H., Lee, C.U., Tan, Y.P.: A multi-objective genetic algorithm for robust flight scheduling using simulation. Eur. J. Oper. Res. **177**(3), 1948–1968 (2007)
6. Choi, S., Kim, Y.K., Briceno, S., Mavris, D.: Prediction of weather-induced airline delays based on machine learning algorithms. In: 35th Digital Avionics Systems Conference, pp. 1–6. IEEE (2016)
7. Kim, Y. J., Choi, S., Briceno, S., Mavris, D.: A deep learning approach to flight delay prediction. In: IEEE/AIAA 35th Digital Avionics Systems Conference (DASC), pp. 1–6. IEEE (2016)
8. Manna, S., Biswas, S., Kundu, R., Rakshit, S., Gupta, P., Barman, S.: A Statistical approach to predict flight delay using gradient boosted decision tree. In: International Conference on Computational Intelligence in Data Science (ICCIDS), pp. 1–5. IEEE (2017)
9. Kuhn, N., Jamadagni, N.: Application of Machine Learning Algorithms to Predict Flight Arrival Delays, pp. 1–6 (2017)
10. Venkatesh, V., Arya, A., Agarwal, P., Lakshmi, S., Balana, S.: Iterative machine and deep learning approach for aviation delay prediction. 4th IEEE Uttar Pradesh Section International Conference on Electrical, Computer and Electronics (UPCON), pp. 562–567. IEEE (2017)
11. Yu, B., Guo, Z., Asian, S., Wang, H., Chen, G.: Flight delay prediction for commercial air transport: A deep learning approach. Transp. Res. Part E **125**, 203–221 (2019)
12. Chen, J., Li, M.: Chained Predictions of Flight Delay Using Machine Learning. AIAA SciTech Forum, pp. 1–25. American Institute of Aeronautics and Astronautics, Inc., San Diego (2019)
13. Hinton, G. E., Osindero, S., Teh, Y.-W.: A fast learning algorithm for deep belief nets. Neural Comput. **18**(7), 1527–1554 (2006)
14. Hornik, K., Stinchcombe, M., White, H.: Multilayer feedforward networks are universal approximators. Neural Netw. **2**(5), 359–366 (1989)
15. Friedman, J.H.: Greedy function approximation: a gradient boosting machine. Ann. Statist. 1189–1232 (2001)
16. Chen, T., He, T.: Xgboost: extreme gradient boosting. R package version 0.4–2, 1–4 (2019)
17. Chen, T., Guestrin, C.: XGBoost: a scalable tree boosting system. KDD '16: Proceedings of the 22nd ACM SIGKDD International Conference on Knowledge Discovery and Data Mining, pp. 785–794. Association for Computing Machinery, NY (2016)
18. James, G., Witten, D., Hastie, T., Tibshirani, R.: An Introduction to Statistical Learning: with Applications in R (Cilt 1). Springer (2013). https://doi.org/10.1007/978-1-4614-7138-7
19. Breiman, L.: Random forests. UC Berkeley TR567 (1999)
20. Dorogush, A.V., Ershov, V., Gulin, A.: CatBoost: gradient boosting with categorical features support. arXiv:1810.11363 (2018)
21. Swalin, A.: CatBoost vs. Light GBM vs. XGBoost. at the date March 30, 2020 (2018). https://towardsdatascience.com/: taken from https://towardsdatascience.com/catboost-vs-light-gbm-vs-xgboost-5f93620723db
22. Lin, C.-F., Wang, S.-D.: Fuzzy support vector machines. IEEE Trans. Neural Networks **13**(2), 464–471 (2002)
23. Davis, J., Goadrich, M.: The relationship between Precision-Recall and ROC curves. ICML '06: Proceedings of the 23rd international conference on Machine learning, pp. 233–240 (2006)

# Machine Breakdown Prediction with Machine Learning

Hande Erdoğan[1](✉), Ömür Tosun[2], and M. K. Marichelvam[3]

[1] Department of Management Information Systems, Faculty of Applied Sciences,
Akdeniz University, Antalya, Turkey
handeaktan@akdeniz.edu.tr
[2] Department of Management Information Systems, Faculty of Applied Sciences,
Akdeniz University, Antalya, Turkey
omurtosun@akdeniz.edu.tr
[3] Department of Mechanical Engineering, Mepco Schlenk Engineering College,
Sivakasi, Tamilnadu, India
mkmarichelvamme@gmail.com

**Abstract.** Industry 4.0 and artificial intelligence can help to generate predictive maintenance models for the better usage of equipment by maximizing its lifetime and operational efficiency. The decision of shutting down an equipment for maintenance can have strategic value, because an error in the timing of the maintenance can produce a loss in the company's competitiveness. By using small sensors to collect infinite amount of data, decisions makers can create mathematical models with the help of machine learning algorithms. In this study, using different machine learning algorithms, a predictive maintenance model will be introduced to better predict the status of an equipment as working or breakdown.

**Keywords:** Predictive maintenance · Machine learning · Industry 4.0 · Sensors

## 1 Introduction

Traditional production systems, which are widely used in many manufacturing companies, have static and hierarchical structures that try to adapt their production policies and product portfolios in accordance with the needs of the market. However, this situation can prevent companies from responding quickly to changes and cause the costs of them to increase. But now the reality has changed. In today's business environment, such as fierce competition, increasing customer demands on price, delivery, diversity, etc., obligation to be flexible, companies are faced with quick response and decision-making in order to be more efficient. Now the customers demand to buy a smaller amount of more diverse products at less cost, while demanding more flexible solutions with a high level of mass customization [1]. Because of many companies using traditional production systems could not keep up with this obligatory and necessary change, Industry 4.0 revolution is needed. With this revolution led by Germany, it has been aimed to transform into highly automated production processes that have the capacity to respond and adapt to the needs and demands of the market quickly by using cyber-physical systems [2].

J. Hemanth et al. (Eds.): ICAIAME 2020, LNDECT 76, pp. 174–182, 2021.
https://doi.org/10.1007/978-3-030-79357-9_17

The first and second industrial revolution was realized by the invention of the steam engine, which greatly changed sense of production, and by using assembly lines that carry out mass production, respectively. Information and digitalization using programmable logic control (PLC) was pioneered the third industrial revolution. The fourth industrial revolution, in which the internet is everything and everything can be connected via the internet, and where any data collected in the industry can be used for various analyzes, was put forward by the German National Academy of Science and Engineering in 2011 and became currently the last stage of the industrial revolution [3].

In today's conditions, it is inevitable for the production to be smart and to have a systematic approach in order to realize the production with a continuous and almost zero downtime [4]. Industry 4.0 transforms a factory into a smart factory by applying future-oriented techniques such as advanced sensor technology, computer science, internet, communication technology, big data, artificial intelligence technology (AI) [5]. Industry 4.0 introduced the "Smart Factory" approach by changing the traditional factory concept in every field. Machines, components and systems in these factories are connected by smart common network channels and can control each other autonomously along the entire value chain [3].

In the smart factory of Industry 4.0, physical processes can be monitored with the help of cyber-physical systems, a virtual image of the physical world can be created, decentralized decisions can be made, large amounts of data can be produced, the collected data can be represented in real-time using large data processing technologies, and can be controlled and managed efficiently [5, 6].

Industry 4.0 technologies, which offer great advantages such as improving system performance, achieving zero downtime, providing predictive maintenance and more [5], are the most important element of smart factories. They enable IoT to communicate and collaborate with each other and people in real-time [6]. The most important reason for the existence of smart factories is to ensure that production is not interrupted by preventing failures caused by human or equipment in production. This can be achieved by predictive maintenance most effectively. The predictive maintenance strategy stands out among other maintenance strategies in the Industry 4.0 due to its ability to optimize the use and management of equipment [7].

In recent years, with the increasing demand for both quality and product variety, the equipment used in production has also become more complicated and complex. The early detection and prediction of a possible failure or breakdown has become critical by following the interactions between this equipment so that companies can fulfill these demands to meet the desired expectations and avoid losing money and trust. With such an attitude, breakdowns can be avoided and/or prevented before they occur. For this reason, with the continuous monitoring of the process in today's industry, failure prediction and diagnosis have gained importance [6].

Reactive systems based on monitoring the condition of the equipment can only recognize this situation when a failure occurs in the processes. Due to longer downtimes than expected and the possibility of damaging to the machines, these systems cost too much to companies. The way to deal with this situation is to increase the transparency and predictability of the working conditions of the production systems. Production systems can cope with this situation by using cyber-physical systems with the understanding of

Industry 4.0 [8]. Using the cyber physical systems, the physical world is transferred to the virtual environment for communication, measurement, analysis and decision making. In such environments, predictive maintenance -intelligent maintenance- is preferred to achieve goals such as minimizing maintenance costs, performing zero defects or reducing downtime [6]. Developing and implementing a smart maintenance strategy that allows to determine the status of in-service systems is essential to reduce maintenance costs, increase production efficiency and predict when maintenance should be performed [9].

Predictive maintenance approach that aims to plan the maintenance of the equipment before the failure occurs, to monitor the operating conditions of the equipment, to predict when the equipment failure will occur and thus to reduce maintenance times and maintenance costs must be adopted [6]. Using the technological infrastructure, predictive maintenance which is also called as condition-based maintenance based on the correct prediction of the working conditions of an equipment [8] aims to prevent breakdown and protect the operating conditions of the system.

In the realization of Industry 4.0, predictive maintenance comes to the fore in equipment maintenance, service and after-sales services, while artificial intelligence (AI) also offers central and data-oriented methods [10]. AI, one of the knowledge-based systems, is a new programming approach and methodology that is widely used for maintenance management of industrial machinery for fault detection and troubleshooting for effective plant maintenance management. AI, which guides operators in rapid fault detection, is used in a wide range from failures occurring in the maintenance of general-purpose industrial machines to rarely occurring emergencies [11].

In this study, a machine learning based decision support system will be proposed. Machine learning models of linear regression, artificial neural networks, random forest, XGBoost and CatBoost algorithms are used to predict the failure time of an equipment. A proper decision for a maintenance will provide efficiency for the management. In traditional maintenance policy, an equipment must be shut down for the inspection, whether there is a failure or not. With integrating machine learning with preventive maintenance right time for the maintenance can be predicted which provide both time and cost advantage for the system.

## 2    Predictive Maintenance

In the simplest form, breakdown is any change or abnormality in a system [12]. Breakdowns can be ranged from quick and simple replacement of an inexpensive part to high cost ones that can lead to high value loss of production, injury or pollution [6].

Companies have to maintain their existing systems in order to deliver their customers the products they want at the required time at the required quality. Therefore, they may have to bear the maintenance costs corresponding to approximately 15%–70% of their total production costs. Enterprises have to choose the most suitable one among the maintenance management strategies in order to realize the production at the desired level and not to bear the very high maintenance/breakdown costs [8]. The purpose of maintenance, expressed as all actions required to maintain or renew a system to perform its intended function; on the one hand, to control the cost and potential loss of production caused by maintenance activities, on the other hand, to ensure that the system maintains its capacity and functionality [12].

Maintenance management is to manage increasing the time between failures that will occur with regular monitoring of process, machine, material or product conditions, thereby minimizing the number and cost of unplanned breakdowns that affect the efficiency of production processes, product quality and overall effectiveness [12, 13].

The widely accepted approach among the maintenance management strategies classified as reactive, preventive and proactive is the corrective maintenance strategy, which is the reactive strategy [9]. This strategy, which tends to cause serious harm to people and the environment due to delays in production or unexpected failures, is used to eliminate the fault occurring while production continues and to sustain production [8, 9]. In progress of time, the increase in both production quantities and complexity of production systems has transformed the corrective maintenance strategy which is difficult to estimate maintenance resources such as human resources, tools and spare parts [9] required for repairs [8].

In order to overcome the shortcomings of corrective maintenance, which is an acceptable strategy in larger profit margins, another strategy - preventive maintenance strategy - has emerged in which maintenance prevents faults. With this strategy, planned activities based on periodic programs are carried out in order to keep the system running continuously by looking at the historical data and the experiences of the experts. Preventive maintenance strategy in order to reduce the costs caused by the faults and prevent them from occurring, a device can be maintained at certain time intervals [8]. Preventive maintenance allows for more consistent and predictable maintenance programs and, unlike reactive maintenance, requires more maintenance activities [9]. This strategy is not very successful in dynamic systems, often leading to excessive or less maintenance that causes unnecessary waste or inefficiency in production processes [8]. Predictive maintenance, which is a proactive approach to eliminate the deficiencies of preventive maintenance related to efficiency and effectiveness, is an alternative strategy in which maintenance actions are planned according to equipment performance or conditions rather than time [9]. Reality of today shows that production systems can no longer achieve the desired efficiency with reactive and preventive maintenance strategies. For this reason, reliability-oriented predictive maintenance strategy based on continuous monitoring system, also called condition-based maintenance strategy, has been developed [8].

In [14], it's stated that approximately 30% of industrial equipment does not use predictive maintenance techniques, and a large majority of companies use periodic maintenance to detect any abnormalities or faults in the components of their systems. When preventive and predictive maintenance is compared, it is seen that predictive maintenance contributes more to the efficiency of companies and reduces the amount of breakdown even more. Periodic maintenance based on periodic revisions is not successful in providing the desired results, since it is not always possible to know when a component of an industrial process will fail based on a certain time period. With the development of technology, predictive maintenance techniques have also improved as a result of collecting information about the performance of industrial machines using wireless sensors and central control and data collection systems.

Predictive maintenance requires an infrastructure that can make a maintenance decision to prevent the failures effectively, to provide equipment and personal safety, and to reduce the economic loss caused by failures [6]. With the implementation of a comprehensive predictive maintenance management, the maintenance manager can provide real data about the actual mechanical condition and the operational efficiency of each device, so that it can plan maintenance activities with real data [13]. With predictive maintenance, companies can use fault diagnosis, performance evaluation of the level of deterioration/wear, and error prediction models to achieve near-zero fault performance and increase efficiency [6].

The main purpose of predictive maintenance, which can be applied to any industrial system or equipment, is to determine the best time to perform the required maintenance operations from both component condition and new diagnostic perspectives [15]. Predictive maintenance uses real production data as a combination of the most cost-effective tools (vibration monitoring, process parameter monitoring, thermal imaging, friction technology, visual inspection) to achieve the actual operating conditions of critical equipment, materials and systems in optimizing production processes. Thus, it can program all maintenance activities as required. Taking into account all of these, predictive maintenance optimizes the availability of equipment, greatly reduces maintenance costs and improves production quality [13].

## 3    Predictive Maintenance with Artificial Intelligence

With the adoption of the Industry 4.0 approach in production, it is possible to make a failure and error estimation before a failure occurs with the predictive maintenance [10] which takes advantage of the opportunities that arise in the field of artificial intelligence. Predictive maintenance can be examined in three main categories such as statistical approaches that can monitor equipment conditions for diagnostic and prognostic purposes, artificial intelligence approaches and model-based approaches. Artificial intelligence approaches are increasingly used in predictive maintenance applications, since model-based approaches require mechanical knowledge and theory of the equipment to be monitored, and statistical approaches require a strong mathematical infrastructure [7]. Artificial intelligence used in predictive maintenance not only helps companies to diagnose equipment failures, but is also used to predict them and suggest repair strategies [11].

When predictive maintenance and artificial intelligence applications are examined together, some of the studies published in this field are as follows. Modeling dynamic nonlinear industrial processes via artificial neural networks, identifying and explaining both quantitative information from historical data through artificial neural networks and qualitative information from operational experts through expert systems, performing dynamic multi-purpose nonlinear optimization with constraints by genetic algorithms, stated that artificial intelligence and modeling techniques are sufficient to reach the basic goals of the predictive maintenance strategy [15]. And in their study, they mentioned the intelligent maintenance system based on artificial intelligence to monitor the online conditions of a wind turbine's gearbox. [9] examined the ability to predict tool wear with the random forests-based prognostic method. They then compared the performance of random forests with feed-forward back propagation FFBP, ANNs and SVR. Using AI techniques for effective plant maintenance, displacement, speed, acceleration, frequency, dominant vibration direction, lubrication properties, machine speed, etc., have developed information-based intelligent machine troubleshooting/maintenance software for general purpose industrial machines for the diagnosis of general-purpose industrial machines [11].

## 4 Application

In this study, a predictive maintenance model as a decision support system is proposed. The aim of the model is to predict the "time to failure" of an equipment. In a real-life scenario, an operator must disassemble an engine to start its scheduled maintenance. Therefore, a wrong decision to start a maintenance can increase the costs and reduce the operational efficiency of the machine.

A sample dataset from the literature is used for this purpose. Observing health and condition of different sensors, status of an engine can be measured. In the dataset there are 100 engines, 19 sensors and two different operational settings of the engines. All the variables in the dataset are continuous type. A total of 20631 observations are given in the dataset. 70% of the data is used for the training phase and the rest for the test phase.

Five machine learning algorithms are selected for the study. They are linear regression, artificial neural networks (ANN), random forest algorithm, XGBoost and CatBoost. In regression analysis a simple linear relation between inputs and the output is assumed. Whereas, non-linear relations are used in the remaining machine learning algorithms. ANN is the one of the widely used algorithm in both machine learning and deep learning. Random forest, XGBoost and CatBoost are the part of the family known as ensemble methods. In these techniques, the main idea is that multiple models give better performance than a single model. In this point of view, XGBoost is the most popular algorithm in machine learning environment.

For each machine learning model, parameter optimization is applied with 10-fold cross-validation. Using the best-found values, performance of each algorithm is measured with $R_2$, MAE (mean absolute error) and R_MSE (root of mean squared error). In Table 1, design of experiment for the parameter optimization is given.

**Table 1.** Parameter values used in the optimization test

| Algorithm | Parameter | Values | Best value found |
|---|---|---|---|
| XGBoost | Number of estimators | 150, 200, 250, 300, 350 | 350 |
| | Learning rate | 0.05, 0.01, 0.15, 0.1,0.2 | 0.15 |
| | Max depth | 5, 6, 7, 8, 9 | 5 |

(*continued*)

**Table 1.** (*continued*)

| Algorithm | Parameter | Values | Best value found |
|---|---|---|---|
| Random Forest | Number of estimators | 150, 200, 250, 300, 350 | 150 |
| | Max depth | 5, 7, 9, 11, 13, 15 | 15 |
| | Min samples leaf | 3, 5, 7, 9, 11 | 3 |
| | Min samples split | 5, 7, 10, 12, 15 | 5 |
| Artificial neural networks | Number of hidden layers | 1, 2 | 2 |
| | Hidden layer size | 5, 10, 15, 20, 25 | 10 – 10 |
| | Transfer function | Relu, tanh, logistic | Tanh |
| CatBoost | Learning rate | 0.005, 0.03, 0.1, 0.15 | 0.15 |
| | Depth | 4, 7, 10, 13 | 7 |
| | Iterations | 250, 300, 350, 400, 450 | 450 |
| | L2 leaf reg (L2 regularization parameter) | 1, 3, 5, 7, 9 | 5 |

The results are given in Table 2. As seen from the Table 2, linear regression gives the worst solution for the dataset. Assuming a linear relationship between the variables can be the weak point of this method. So more complicated relations are needed to better understand the dataset. For this result, first ANN is used. Its performance is between the linear regression and ensemble-based methods. All three ensemble-based methods give similar solution quality. CatBoost produce a little better than both XGBoost and Random Forest. Also based on the similar results between train and test sets, it can be interpreted that there is no over-fitting in the methods.

**Table 2.** Best values found for each algorithm

| | $R_2$ | | MAE | | R_MSE | |
|---|---|---|---|---|---|---|
| | Train | Test | Train | Test | Train | Test |
| Linear regression | 0.670 | 0.668 | 30.588 | 30.252 | 39.695 | 39.333 |
| XGBoost | 0.823 | 0.810 | 21.859 | 21.017 | 29.523 | 31.114 |
| Random Forest | 0.996 | 0.986 | 3.434 | 5.855 | 4.589 | 7.953 |
| ANN | 0.986 | 0.936 | 6.396 | 12.243 | 9.133 | 17.221 |
| CatBoost | 0.993 | 0.987 | 4.126 | 5.368 | 5.346 | 7.064 |

# 5   Conclusions and Future Work

The companies, which act with the understanding of Industry 4.0, prefer the predictive maintenance management among the maintenance strategies and can handle this maintenance strategy together with artificial intelligence, equipment, etc. to prevent economic losses and increase their performance. It can expeditiously enable them to effectively prevent malfunctions in all systems and to determine the best time for troubleshooting.

In this study, a decision support system is proposed to estimate the failure time of an equipment using machine learning, which is defined as a sub-discipline of artificial intelligence. With the use of machine learning and predictive maintenance together, it is possible to make predictions in a way that provides time and cost advantage by intervening the system before a failure occurs. For this purpose, in this study, linear regression, artificial neural networks, random forest, XGBoost and CatBoost algorithms has been compared and the results has been interpreted. When the results of the study have been examined, it has seen that the linear regression algorithm gives the worst result in the algorithms. This strengthens the assumption that there is no linear relationship between the variables. The ANN algorithm has been first used to make sense of the complex relationships between the dataset. It has seen that the results of this algorithm take a value between the results of linear regression and ensemble-based methods. The results of the remaining three ensemble-based method methods were obtained very similarly and the analysis results have demonstrated that that there is no overfitting between train and test sets.

In this study, it has been shown that machine learning techniques can be used to estimate maintenance of an enterprise. The data set was developed for the use of machine learning. As this data set is expanded, machine learning will deepen and the findings will improve. With this study, it has been aimed to highlight the fact that using the predictive maintenance strategy by using machine learning algorithms, managing the processes of preventing, defining and defining the fault before it occurs, can provide the enterprise with additional time and cost advantages, as well as additional benefits. The scope of this research can be improved using different parameters. Monitoring accuracy with this developed model may be take part among the future studies.

# References

1. Diez-Olivan, A., Ser, J.D., Galar, D., Sierra, B.: Data fusion and machine learning for industrial prognosis: trends and perspectives towards industry 4.0. Inf. Fus. **50**, 92–111 (2019)
2. Bressanelli, G., Adrodegari, F., Perona, M., Saccani, N.: Exploring how usage-focused businss models enable circular economy through digital technologies. Sustainability **10**(639), 1–21 (2018)
3. Zolotova, I., Papcun, P., Kajati, E., Miskuf, M., Mocnej, J.: Smart and cognitive solutions for operator 4.0: laboratory H-CPPS case studies. Comput. Ind. Eng. **139**, 1–15 (2020)
4. Bagheri, B., Yang, S., Kao, H.A., Lee, J.: Cyber-physical systems architecture for self-aware machines in industry 4.0 environment. IFAC-PapersOnLine **48**(3), 1622–1627 (2015)
5. Yan, J., Meng, Y., Lu, L., Li, L.: Industrial big data in an industry 4.0 environment: challenges, schemes, and applications for predictive maintenance. IEEE Access **5**, 23484–23491 (2017)

6. Li, Z., Wang, Y., Wang, K.-S.: Intelligent predictive maintenance for fault diagnosis and prognosis in machine centers: Industry 4.0 scenario. Adv. Manuf. **5**(4), 377–387 (2017). https://doi.org/10.1007/s40436-017-0203-8

7. Carvalho, T.P., Soares, F.A.A.M.N., Vita, R., Francisco, R.P., Basto, J.P., Alcala, S.G.S.: A systematic literature review of machine learning methods applied to predictive maintenance. Comput. Ind. Eng. **137**, 1–10 (2019)

8. He, Y., Han, X., Gu, C., Chen, Z.: Cost-oriented predictive maintenance based on mission reliability state for cyber manufacturing system. Adv. Mech. Eng. **10**(1), 1–15 (2017)

9. Wu, D., Jennings, C., Terpenny, J., Gao, R.X., Kumara, S.: A comparative study on machine learning algorithms for smart manufacturing: tool wear prediction using random forests. J. Manuf. Sci. Eng. **139**, 1–9 (2017)

10. Rodseth, H., Schjolberg, P., Marhaug, A.: Deep digital maintenance. Adv. Manuf. **5**(4), 299–310 (2017)

11. Nadakatti, M., Ramachandra, A., Kumar, A.N.S.: Artificial intelligence-based condition monitoring for plant maintenance. Assem. Autom. **28**(2), 143–150 (2008)

12. Lee, C.K.M., Cao, Y., Ng, H.K.: Big Data Analytics for Predictive Maintenance Strategies, Supply Chain Management In The Big Data Era, Advances in Logistics Operations and Management Science (ALOMS) Book Series, chapter 4, 50–74 (2017)

13. Spendla, L., Kebisek, M., Tanuska, P., Hrcka, L.: Concept of predictive maintenance of production systems in accordance with industry 4.0. In: IEEE 15th International Symposium on Applied Machine Intelligence and Informatics, pp. 405–409. 26–28 January 2017, Herl'any, Slovakia (2017)

14. Canizo, M., Onieva, E., Conde, A., Charramendieta, S., Trujillo, S.: Real-time predictive maintenance for wind turbines using big data frameworks. IEEE Int. Conf. Progn. Health Manage. **2017**, 1–8 (2017)

15. Garcia, M.C., Sanz-bobi, M.A., Pico, J.: SIMAP: Intelligent system for predictive maintenance: application to the health condition monitoring of a windturbine gearbox. Comput. Ind. **57**(6), 552–568 (2006)

# Prevention of Electromagnetic Impurities by Electromagnetics Intelligence

Mehmet Duman[(✉)]

Faculty of Enginering, Department of Electrical and Electronics Engineering, Düzce University, Düzce, Turkey
mehmetduman@duzce.edu.tr

**Abstract.** Nowadays, electromagnetic pollution has an increasing rate. 4G and 5G battles of base stations, 3G, GPRS and radio broadcasts, incredible increase in bluetooth modules, the huge volume of electromagnetic waves generated by wireless communication while talking about smart systems and IOT, infrared systems, NFC, wireless internet networks. Even if it has not been proven that they cause human health or diseases such as infertility, why should we not work to minimize this pollution? Why not protect especially our children from such factors? We may want to take advantage of these important wireless blessings of technology despite electromagnetic impurities, but it would be more appropriate to prevent our children from being exposed to these impurities, at least in some time periods, such as sleep. Within this study; preventing electromagnetic waves from reaching children's beds, car seats, playgrounds; measuring the frequency and amplitude values of the waves, followed by automatic generation or selection of band stop filters will be done with frequency selective surface analysis method according to these values by artificial intelligence. The decision of preventing which of the transmissions measured with the help of frequency meters and computer programs will be made by artificial intelligence, and then the information such as which electromagnetic waves are propagated more during which hours will be obtained by electromagnetics intelligence to prevent the impurities. According to the information obtained, the use of band stop filters in various fields may become more common.

**Keywords:** Electromagnetic impurities and pollution · Band stop filter · Electromagnetics intelligence · Frequency selective surface analysis

## 1 Introduction

The measurement of the frequency of electromagnetic waves (EMW) can be done quite easily today. In order to perform this process, a circuit can be realized by combining various electronic components or a frequency meter device that can operate at appropriate threshold frequencies can be obtained. Optical, laser, sound et cetera techniques can be used according to the frequency value [1]. While using these techniques, attention should be paid to the stability and required standards (such as IEEE standards) when measuring the EMW period or frequency [2]. It is also possible to make frequency meter

© The Author(s), under exclusive license to Springer Nature Switzerland AG 2021
J. Hemanth et al. (Eds.): ICAIAME 2020, LNDECT 76, pp. 183–189, 2021.
https://doi.org/10.1007/978-3-030-79357-9_18

measurements in microwave and millimeter wavelengths. These measurements can also be carried out simultaneously [3]. The frequency range to be measured and the magnitude of the frequency are very important. In some cases, while only one test equipment of the manufacturers is sufficient [4, 5], sometimes when the frequency values are too high, frequency can also be found by measuring the EM field values in order to obtain a precise result [6]. Using spectrum analyzer or analytical calculation with FFT method is another method used for frequency measurement. Computer programs or mobile applications are also available for Wi-Fi and mobile data signals. In order to perform radar, jamming, communications, EM pollution measurement etc. applications, frequency measurement methods, which are the first application to be performed, have been transferred with various methods [7]. The use of frequency counting is common not only in the EMW area, but in control and power systems [8, 9].

In this study, EMW propagation frequency values were found by using various methods and studies were conducted on inhibition or reduction of EMW emitted at the frequency found. Frequency selective surface (FSS) analysis techniques have been used for this prevention and reduction. By means of FSS analysis, band pass-stop structures, low or high pass filters can be established [10, 11]. While designing FSS; triangle, star, square, hexagon, anchor, cross, spiral etc. structures which are showed in Fig. 1 can be used during design. This usage varies by application or structure. Array method and cell method are the most commonly used methods among the others [12]. FSS structures are also available as multilayer [13, 14]. In this work, since EMW propagation frequencies will be different, FSS structures created in previous studies [15, 16] or already existing studies will be activated automatically according to the value measured by the frequency meter. In this way, some kind of frequency breaker application and EMW shielding [17] will occur and EM pollution will be prevented. After deciding which EMW shield to be used automatically, control can be achieved with a relay system, the subject of this study is not mechanical. The frequency of high propagation is learned and propagations at that frequency will be prevented.

In this work where EMW shielding is carried out intensely, it can be predicted by artificial intelligence at which time of the day or on which days of the week, in which frequency the strongest EMW is spread. Because artificial intelligence will take the information recorded by the frequency meter from the database and will be able to generate new estimates based on this information. I called this system, which uses EM and artificial intelligence together, as Electromagnetics Intelligence. Electromagnetics Intelligence will soon decide which signal (which radiated signal at which frequency) is stronger and will shield that frequency without the need for a frequency meter.

If this system is installed around the baby seats in the vehicles, around the cradles in the houses or in the places such as children's nurseries (kindergartens), the elders' dormitory, these people will be protected from the potential damage caused by EMW pollution [18] and the exposure to the warming in human tissue [19]. To remind; EMW damage to human health has not been proven precisely; in accordance with the principle of protective approach [20], precautions are required to be taken in this study.

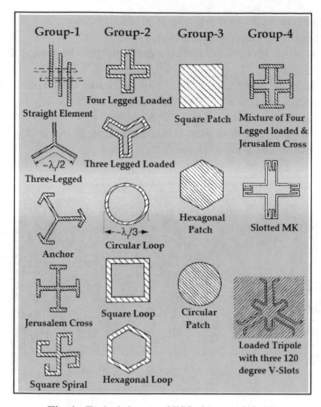

**Fig. 1.** Typical shapes of FSS elements [10, 12]

## 2 Materials and Methods

In this study, CST Design Environment program was used for performing FSS techniques and EM shield production. Since the aim is to make a band-stop filter at various frequencies and to block the frequency where EMW is the most intense, the designed shields can be realized later and can be activated automatically with a relay system. In Fig. 2, some of the designed shields can be seen. The structure in the middle is in a bandpass style, the opposite must be taken for the band stop. First one works at 1.3 GHz and the opposite form of the second one can work at 13 GHz (10 times bigger) frequencies. In the last, the input reflection graphic acts as a band stop filter at 11 GHz. This means that it is band pass for parameter S21 at 11 GHz. It is possible to design for each frequency value or band such as X band [22, 23].

Various WEB-mobile applications and programs have been used for frequency measurement. In addition, a frequency meter device that can measure up to 2.4 GHz frequency is used. It is necessary to be able to measure around the frequencies of 5 GHz for 4.5 G and 60 GHz for 5 G. For this reason, a device capable of wireless measurement up to 60 GHz will be provided for future work. Currently there are programs that show 5 GHz signals. There is no need to place a frequency meter device in the final product because artificial intelligence will decide by itself after a few measurements with instruments which frequency and EMW intensity are. If it is to be used only in moving products (such as cars, stroller), a frequency meter is needed.

Various forecasting applications can be used for the predictive algorithm of artificial intelligence.

In this article, as mentioned in the introduction, filtering will be done for use in items such as baby cradles and nurseries places. If the materials used are flexible [22, 24], it can also be designed as clothing. Here, it does not need to be flexible.

One of the technical devices used is the network analyzer (NA). Produced FSS models in previous studies were tested with NA.

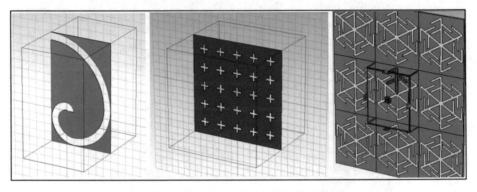

**Fig. 2.** Shield examples [15, 16, 21].

## 3    Discussions and Suggestions

Combination of EMW sources, EMW, frequency meter, FSS structure types, FSS structure frequencies (or multilayer), artificial intelligence (EM intelligence) and control mechanism (relay system) respectively is described in Fig. 3.

It can be used some estimation tools or forecasting programs for predicting by artificial intelligence. Markov chain algorithm is one of these programs. In this way, it is possible to estimate the future EMW in the environment where the EMW values are measured at certain times for several weeks. Thus, frequency meter need is not always required. The FSS needed is operated by the last relay mechanism.

In this system, designed and fabricated FSS layers can be used as multilayers without mounting any control system or writing any codes about artificial intelligence process to be protected from EMW in the environment. But in this case, it does not differ from the Faraday cage! (Fig. 4)

If sufficient measurements are made, there is no need for more measurement devices. It can procees directly to the decision of artificial intelligence. If it is a non-mobile system and at when it is known which EMW is spread, filtering process can be started without the need for artificial intelligence as measurements.

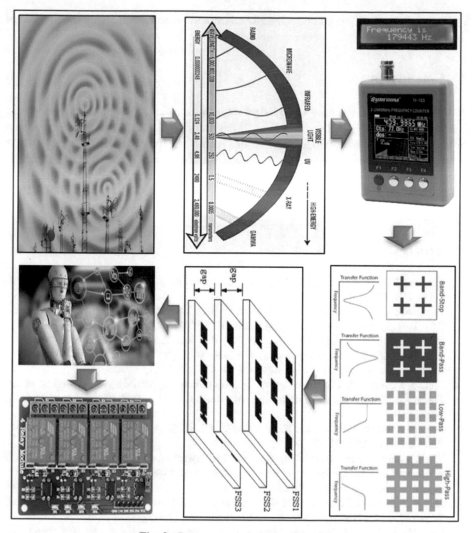

**Fig. 3.** Recommended system [25–28].

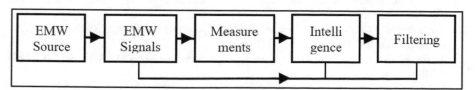

**Fig. 4.** Schematic representation of the proposed system.

For circuits that are measured and which frequency range is decided to operate, eventually; FSS filters that band stop [29], broadband [30] or ISM band [31] can be designed and created. These filters can be selected and implemented with artificial intelligence at the end.

## 4 Conclusion

As a result, various filters are required to prevent possible damage due to electro-magnetic waves. The structure of these filters and their frequency ranges are different. In this project, frequency measurement with various measuring instruments and which filter is used according to the frequency are explained. Artificial intelligence can decide which filter will be used, as well as decide after several measurements without taking measurements. It is easy to do so, because it knows which wave has spread at which time intervals.

## References

1. Jennings, D.A., Evenson, K.M., Knight, D.J.E.: Optical frequency measurements. Proc. IEEE **74**(1), 168–179 (1986)
2. Lombardi, M.: The mechatronics handbook, chapter: 17 fundamentals of time and frequency. CRC Press, IEEE Microwave Mag. **4**(3), 39–50 (2002)
3. Shi, J., Zhang, F., Ben, De., Pan, S.: Simultaneous radar detection and frequency measurement by broadband microwave photonic processing. J. Lightwave Technol. **38**(8), 2171–2179 (2020). https://doi.org/10.1109/JLT.2020.2965113
4. Frekans Metreler (2020). https://www.tekniktest.com/kategori/fre-kans-sayici-frekansmetre
5. Frekans ve Zaman Ölçerler (2020). https://www.netes.com.tr/u-runler/frekans-ve-zaman-olc erler
6. Measuring electric fields from 100 MHz to 60 GHz (2020). https://www.narda-sts.us/pdf_files/Datasheet_EF6092_EN.pdf
7. The Mystery of Measuring Microwave and Millimeter-Wave Frequencies (2020). https://www.mwrf.com/technologies/test-mea-surement/article/21849053/the-mystery-of-measuring-microwave-and-millimeterwave-frequencies
8. Luiten, A.N. (ed.): Frequency Measurement and Control: Advanced Techniques and Future Trends. Springer Berlin Heidelberg, Berlin, Heidelberg (2001)
9. Chih-Hung, L., Chi-Chun, H., Men-Shen, T.: Curve fitting approach for fundamental frequency measurement for power systems. Sens. Mater. **32**(1), 357–373 (2020)
10. Munk, B.A.: Frequency Selective Surfaces Theory and Design. Wiley Interscience, Hoboken (2000)
11. Vardaxoglou, J.C.: Frequency Selective Surfaces: Analysis and Design. Somerset, England, Research Studies Press, Taunton (1997)
12. Anwar, R.S., Mao, L., Ning, H.: Frequency selective surfaces: a review. Appl. Sci. **8**(9), 1689 (2018)
13. Erdemli, Y.E., Sertel, K., Gilbert, R.A., Wright, D.E., Volakis, J.L.: Frequency-selective surfaces to enhance performance of broad-band reconfigurable arrays. IEEE Trans. Antennas Propag. **50**(12), 1716–1724 (2002)
14. Ranga, Y., Matekovits, L., Esselle, K.P., Weily, A.R.: Design and analysis of frequency-selective surfaces for ultrawideband applications. In: IEEE EUROCON - International Conference on Computer as a Tool, Lisbon, pp. 1–4 (2011)

15. Duman, M.: Frekans Seçici Yüzey Analizi Yöntemiyle Gerçekleştirilmiş 1.35 GHz Frekanslı Bant Durduran Süzgeç Tasarımı. Dicle Üniversitesi Mühendislik Fakültesi Mühendislik Dergisi. **10**(3), 807–813 (2019)
16. Duman, M.: Designing metal band pass filter plate with the golden ratio rule and effect of golden ratio rule to the filter. In: 24th Signal Processing and Communication Application Conference (SIU), Zonguldak, pp. 213–216 (2016)
17. Katoch, K., Jaglan, N., Gupta, S.D.: A Review on Frequency Selective Surfaces and its Applications, International Conference on Signal Processing and Communication (ICSC), NOIDA, India, pp. 75–81 (2019)
18. Özen, Ş, Uskun, E., Çerezci, O.: Üniversite Öğrencileri Arasında Cep Telefonu Kullanımı ve Elektromanyetik Kirlilik Üzerine Bir Çalışma. SAU Fen Bilimleri Enstitüsü Dergisi **6**(2), 153–159 (2002)
19. Özdinç, P.L., Çömlekçi, S.: Elektromanyetik Alan Maruziyetinin Kas Dokusunda Oluşturduğu Etkinin Modellenmesi ve Analizi. Mühendislik Bilimleri ve Tasarım Dergisi **7**(3), 498–504 (2019)
20. Sevgi L.: Teknoloji, Toplum ve Sağlik: Cep Telefonları ve Elektromanyetik Kirlilik Tartışmaları (2013). http://www.emo.org.tr/ekler/e73a9a0d37ef-b96_ek.pdf
21. Menekşe M.: Frekans Seçici Yüzey ile Mikrodalga Soğurucu Tasarımı. Kocaeli Üniversitesi, Elektronik ve Haberleşme Mühendisliği Bölümü Lisans Bitirme Çalışması (2015)
22. Bilal, M., Saleem, R., Abbasi, Q.H., Kasi, B., Shafique, M.F.: Miniaturized and flexible FSS-based EM shields for conformal applications. IEEE Trans. Electromag. Compatibil. **62**(5), 1703–1710 (2020). https://doi.org/10.1109/TEMC.2019.2961891
23. Kanchana, D., Radha, S., Sreeja, B.S., Manikandan, E.: Convoluted FSS structure for shielding application in X-band frequency response. IETE J. Res. 1-7 (2019).https://doi.org/10.1080/03772063.2019.1691062
24. Sivasamy, R., Kanagasabai, M.: Design and fabrication of flexible FSS polarizer. Int. J. RF Microw. Comput. Aid. Eng. **30**, e22002 (2020)
25. Elektromanyetik Dalgaların Oluşumu (2020). http://www.za-mandayolculuk.com/html-3/elektromanyetik_dalgalar.htm
26. Graduate Research School Selective Transmission of RF Signals through Energy Saving Glass for smart buildings (2011). https://slideplayer.com/s-lide/10308146/
27. Segundo, F.C.G.S., Campos, A.L.P.S., Gomes, N.A.: A design proposal for ultrawide band frequency selective surface. J. Microw. Optoelectron. Electromag. Appl. **12**(2), 398–409 (2013)
28. Yapay Zekâ Hakkında Bilmemiz Gereken 10 Madde (2019). https://magg4.com/yapay-zeka-hakkinda-bilmemiz-gereken-10-madde/
29. Hong, T.Y., et al.: Study of 5.8 GHz band-stop frequency selective surface (FSS). Int. J. Integr. Eng. **11**(4), 244–251 (2019)
30. Xiaoming, L., et al.: Broadband FSS for millimeter radiometer. In: IEEE Cross Strait Quad-Regional Radio Science and Wireless Technology Conference, Xuzhou, China, pp. 1–4 (2018)
31. Can, S., Kapusuz, K.Y., Yilmaz, A.E.: A dual-band polarization independent FSS having a transparent substrate for ISM and Wi-Fi shielding. Microw. Opt. Technol. Lett. **59**, 2249–2253 (2017)

# The Effect of Auscultation Areas on Nonlinear Classifiers in Computerized Analysis of Chronic Obstructive Pulmonary Disease

Ahmet Gökçen$^{(\boxtimes)}$ and Emre Demir

Department of Computer Engineering, Iskenderun Technical University, İskenderun 31200, Hatay, Turkey
ahmet.gokcen@iste.edu.tr

**Abstract.** Today, with the rapid development of technology, various methods are being developed for computer assisted diagnostic systems. Computer assisted diagnostic systems allow an objective assessment and help physicians to diagnose. Pulmonary sounds are effective signals in the diagnosis of Chronic Obstructive Pulmonary Disease (COPD). In this study, multichannel respiratory sounds obtained by auscultation method were analyzed and the effect of auscultation points on nonlinear classifiers was investigated. In computerized analysis, EWT (Empirical Wavelet Transform) method was applied to 12-channel lung sounds which obtained from different auscultation points. Then, statistical properties based on extracted frequency components were calculated. COPD and healthy lung sounds were classified using nonlinear classifiers for each channel. At the end of the study, the effects of channels on classifiers were compared.

**Keywords:** COPD · Pattern recognition · Biomedical signal processing · Auscultation · Empirical wavelet transform

## 1 Introduction

With the development of science and technology, studies on computer-aided diagnostic systems have been affected greatly. In computer-aided diagnostic systems, the aim is to provide an objective interpretation to the diagnosis by developing systems that are fast, reliable and low cost, and to help experts diagnose by reducing the margin of error. Evaluation of biomedical signals used in computer-aided diagnostic systems can vary according to many different parameters, especially the expert's experience. Computer-aided diagnostic systems can help physicians reduce their margin of error by providing objectivity to evaluation.

In this study, we focused on Chronic Obstructive Pulmonary Disease (COPD), which is a global health problem. The aim of the study to detect the most significant auscultation point and investigating the effect of channels on classifiers. COPD, which is on the list of diseases that have become the biggest causes of mortality and morbidity, is a chronic disease that causes blockage of air sacs in the lung. It is one of the most fatal respiratory disorders that occur and proceeds due to prolonged exposure to harmful gases and

J. Hemanth et al. (Eds.): ICAIAME 2020, LNDECT 76, pp. 190–197, 2021.
https://doi.org/10.1007/978-3-030-79357-9_19

particles, especially cigarette smoke, and it imposes a heavy burden on economies economically. Therefore, early diagnosis and treatment is important for these diseases [1–5]. Auscultation method is the most widely used diagnostic method for respiratory diseases. Auscultation is a procedure performed with a stethoscope to listen to the sounds produced by the body, such as inhalation, whisper from the chest wall. The stethoscope, one of the most important examination tools, was invented in 1816 by René-Théophile-Hyacinthe Laennec. Adaptive methods aim to extract meaningful modes that represent the signal based on the meaningful information in the signal. Empirical Wavelet Transform (EWT) method was developed by combining the advantages of Wavelet Transform (WT) and Empirical Mode Decomposition (EMD). EWT is a signal decomposition method based on the design of an adaptive wavelet filter bank and to extract different modes of the signal [1, 2, 6].

When the literature is examined, there are many studies on respiratory diseases using computer assisted analysis methods, but the use of EWT method in biomedical analysis is limited. Altan et al. [7] used Deep Belief Network (DBN) and multiple machine learning algorithms for early diagnosis of COPD using multi-channel lung sounds. In the DBN classification results, they obtained 91%, 96.33% and 93.67% rates for sensitivity, specificity and accuracy, respectively. Amaral et al. [8] developed a diagnostic model for the diagnosis of COPD using Artificial Neural Networks (ANN). Nabi et al. [9] studied the classification of asthma by severity. Hossain and Moussavi [10] conducted a study on the analysis techniques that are meaningful in the diagnosis of asthma and its connection with the flow rate. Fernandez-Granero et al. [11] studied computer aided analysis of COPD exacerbations. İçer and Gengeç [12] conducted a study on pathological lung sounds to distinguish crackles and rhonchi sounds from normal lung sounds. They used the Support Vector Machine (SVM) method at the classification stage. Huang et al. [13] conducted a study on emotional prediction from EEG signals using the EWT method. Maharaja and Shaby [14] studied the diagnosis of glaucoma disease from digital fundus images using the EWT method. Babu et al. [15] used the EWT method for automatic detection of lung sounds from phonocardiogram signals.

There are some points that are considered standard for auscultation. In this study, after performing the necessary pretreatments on 12-channel lung sounds obtained by auscultation method, the EWT method was applied and the feature vector was created by the EWT-based attributes. COPD was detected by using nonlinear classification algorithms widely used on the obtained feature vectors. Finally, the effect of preferred auscultation zones on nonlinear classifiers used in this study was investigated and performance analysis was performed. While analyzing auscultation sounds, 5-s recordings which is very short duration when compared with the literature ones were used.

## 2  Materials and Methods

### 2.1  Database

In this study, RespiratoryDatabase@TR database created using Littmann 3200 digital stethoscope, which has many capabilities for computerized analysis of lung sounds, was used. RespiratoryDatabase@TR database enables analysis with signal processing techniques on multichannel respiratory sounds. The RespiratoryDatabase@TR database contains chest X-Rays, St. George's Respiratory Questionnaire (SGRQ-C) responses and Pulmonary Function Test measurements, primarily digitized auscultation sounds consisting of 12-channel lung sounds and 4-channel heart sounds [16]. Auscultation sounds were recorded simultaneously from left and right foci with the help of two digital stethoscopes from 16 different channels. Auscultation areas used for the chest and back are shown in Fig. 1.

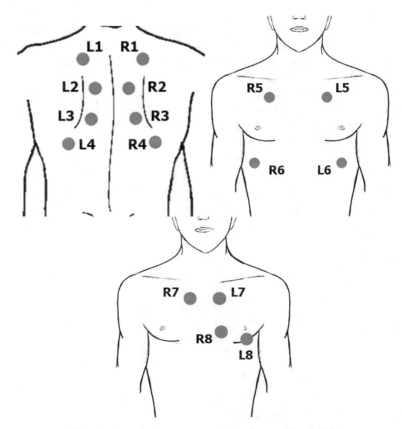

**Fig. 1.** Auscultation areas used for chest and back [16]

RespiratoryDatabase@TR Data of patients with COPD and healthy individuals were tagged in Antakya State Hospital after the necessary evaluations were made by pulmonologists. There is also ethical committee approval by Mustafa Kemal University [7]. Lung sounds in the RespiratoryDatabase@TR database are passed through a 7.5 Hz (1st Level Butterworth filter) filter to remove DC offset, and a 10 kHz (8th Grade Butterworth filter) low pass filter to reduce high frequency noise [17]. The length of the signal is also an important factor in biomedical analysis. In this study, the signal length is set to be 5 s in order for the application to run faster, to show the success of the developed system and to be easily applied on the hardware in future studies.

## 2.2  Empirical Wavelet Transform

Adaptive methods used in digital signal processing applications aim to extract the modes representing the signal based on the meaningful information inside the signal and provide a suitable basis for the analysis of the signal. The Empirical Mode Decomposition (EMD) method, which is the most common of these methods, aims to separate the signal into multiple IMFs and select the appropriate IMF to generate the envelope spectrum using transformations. The EMD method still causes many problems due to the lack of mathematical theory. Another method used to separate the signal into different modes is Empirical Wavelet Transform (EWT). EWT was developed by combining the advantages of Wavelet Transform (WT) and EMD. The basic principle of this method used in the signal separation and analysis process is based on the design of an adaptive wavelet filter bank [1, 6]. It does this by creating adaptive wavelets that make it adaptable based on the information content of the given signal. In the study, EWT was applied on the data obtained after necessary pretreatment.

## 2.3  Support Vector Machine

Support Vector Machine (SVM) is a classification algorithm based on statistical learning theory that can be applied successfully in many fields. Developed by Vapnik for pattern recognition and classification problems, SVM is one of the most used and highly effective classification methods in the literature [18]. SVM was originally designed for classification of linear data. However, since the problems generally encountered are of the type that cannot be separated linearly, they were later developed for the classification of nonlinear data. The basic principle of the algorithm is to define the most appropriate hyperplanes that can separate the two classes. In this study, the classification process was performed by applying SVM for each channel on the feature vectors created as a result of the previous stages. Based on the classification results with the help of SVM, the performance analysis of the algorithm and the performance effect of each auscultation channel on the SVM classifier were examined.

## 2.4  Multilayer Perceptron

Artificial neural network (ANN) is a mathematical modeling method developed inspired by the work of the human brain. It consists of nerve cells. ANN is applied in many areas such as pattern recognition and signal processing. There are several ANN models in the literature. The most common of these is Multilayer Perceptron (MLP). Effective and successful results can be obtained in systems developed by applying MLP. MLP basically consists of an input layer, one or more hidden layers and an output layer. Hidden layers consist of nodes and there is no strict rule regarding the number of layers. In this study, the MLP neural network model was used on the feature vectors obtained as a result of the previous stages. The most suitable parameters were determined by experimental comparisons. According to these results, the classification process was carried out with the help of MLP model, which includes 9 inputs, 2 hidden layers with 5 neurons and 2 output elements. Also, sigmoid activation function was used in all layers.

## 3    Experimental Results

In this study, the effect of auscultation channels on nonlinear classifiers was investigated by using EWT-based attributes for computer-aided COPD diagnosis. Multichannel lung sounds from each subject were used to diagnose computer-assisted COPD. The length of the signals is determined to be 5 s. After necessary pretreatment, EWT method was applied to the lung sound signals. As a result of the EWT method, a large number of IMFs are obtained. Signals of the same length can have a number of IMFs. The feature extraction process, which is frequently used in studies such as pattern recognition and signal processing applications, is one of the most important stages, and by extracting some meaningful features in the information, complex data is simplified and operation is performed on clean data. Correct feature extraction will directly affect the system's success and performance. In the study, necessary mathematical calculations were made on the IMFs and statistical feature extraction was applied for each IMF. After the feature extraction and selection phase, the classification process was made with MLP and SVM algorithms using the feature vector created. Confusion matrix is often used to evaluate the performance of the system. In this study, confusion matrix was used and sensitivity (Eq. 1), specificity (Eq. 2) and accuracy (Eq. 3) performance criteria were calculated depending on the predictive state. In this study, 10-part cross-validation method was applied.

$$Sensitivity = \frac{TP}{TP + FN} * 100 \tag{1}$$

$$Specificity = \frac{TN}{TN + FP} * 100 \tag{2}$$

$$Accuracy = \frac{TP + TN}{TP + TN + FN + FP} * 100 \tag{3}$$

The classification process of respiratory sounds using EWT-based features distinguished COPD patients from healthy subjects. Performance evaluation of each channel was performed on all classifiers used to evaluate the effects of the channels on the classifiers by evaluating the performance obtained after the classification phase. Performance effects were analyzed by comparing the results obtained by applying different classification algorithms for each channel. As a result, the results regarding the evaluation of classifiers according to auscultation channels are shown in Table 1.

**Table 1.** Evaluation of classifiers according to auscultation channels

| Channels | Multilayer perceptron | | | Support vector machine | | |
|----------|-------------|-------------|----------|-------------|-------------|----------|
|          | Sensitivity | Specificity | Accuracy | Sensitivity | Specificity | Accuracy |
| L1       | 87.50       | 70.00       | 80.77    | 80.00       | 90.00       | 85.00    |
| L2       | 90.00       | 70.00       | 80.00    | 80.00       | 80.00       | 80.00    |

(*continued*)

**Table 1.** (*continued*)

| Channels | Multilayer perceptron | | | Support vector machine | | |
|---|---|---|---|---|---|---|
| | Sensitivity | Specificity | Accuracy | Sensitivity | Specificity | Accuracy |
| L3 | 80.00 | 80.00 | 80.00 | 80.00 | 90.00 | 85.00 |
| L4 | 80.00 | 70.00 | 75.00 | 80.00 | 90.00 | 85.00 |
| L5 | 60.00 | 90.00 | 75.00 | 60.00 | 100 | 80.00 |
| L6 | 50.00 | 80.00 | 65.00 | 70.00 | 80.00 | 75.00 |
| R1 | 80.00 | 70.00 | 75.00 | 80.00 | 90.00 | 85.00 |
| R2 | 90.00 | 70.00 | 80.00 | 100 | 60.00 | 80.00 |
| R3 | 80.00 | 70.00 | 75.00 | 90.00 | 50.00 | 70.00 |
| R4 | 80.00 | 70.00 | 75.00 | 90.00 | 80.00 | 85.00 |
| R5 | 90.00 | 60.00 | 75.00 | 60.00 | 90.00 | 75.00 |
| R6 | 80.00 | 70.00 | 75.00 | 90.00 | 90.00 | 90.00 |

As a result of the application of the MLP algorithm, the graph about sensitivity, specificity and accuracy values for each auscultation channel is shown in Fig. 2.

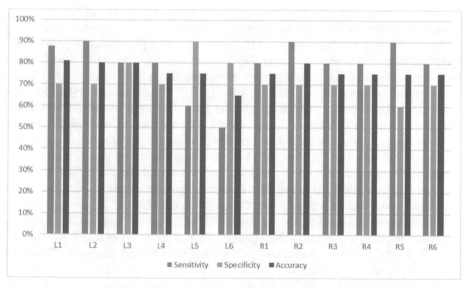

**Fig. 2.** The success of channels on MLP algorithm

As a result of the implementation of the SVM algorithm, the graph about sensitivity, specificity and accuracy values for each auscultation channel is shown in Fig. 3.

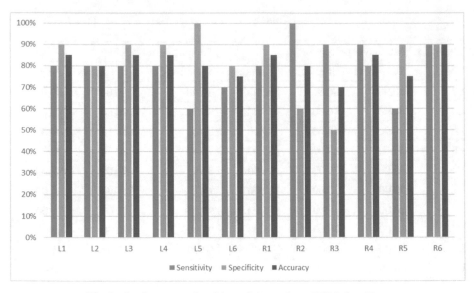

**Fig. 3.** Performance algorithm of channels on SVM algorithm

## 4 Conclusion

COPD is a global health problem that negatively affects the quality of life. It is very important to prevent and control the progression of the disease. Lung sounds are very important signals in the diagnosis of COPD. Rapidly developing technology has made it possible to work on computer aided diagnostic systems. In this study, the effects of auscultatory points on the classifier were investigated comprehensively using a system developed based on artificial intelligence. The EWT method was applied for each channel after the necessary pretreatment on the signals received from the Respiratory-Database@TR database. Then, by using statistical feature extraction methods, feature extraction and selection process is done and feature vector is created. The classification process was performed on MLN and SVM algorithms on the last feature, and each channel was analyzed separately. Thus, the effects of the channels on the applied classifier were analyzed by analyzing the lung sounds received from different channels. According to the data obtained as a result of these studies, when the effect of the channels on the overall performance is analyzed based on classifiers, it is seen that L1, L2, L3 and R2 channels are close to each other for MLP and L6 channel shows lower results than others. For SVM, R6 channel shows more successful results than other channels and R3 channel shows lower results than others. As a result, the effects of the classifiers were examined and it was observed that the performance rates of the channels were different for each classifier when both classifiers were compared. Since the effects of the channels on each classifier are different, it is seen that a generalization about their effects cannot be made.

# References

1. Demir, E., Gökçen, A.: Empirical wavelet transform analysis and classification of multichannel lung sounds. In: 2nd International Mersin Symposium, vol. 6, pp. 218–226 (2019)
2. Demir, E., Gökçen, A., Altan, G., Kutlu, Y.: The effect of nonlinear classifiers on the diagnosis of respiratory diseases using empirical wavelet transform based features. In: 3nd International Academic Studies Conference, pp. 178–185 (2019)
3. WHO: The top 10 causes of death (2018). https://www.who.int/news-room/fact-sheets/detail/the-top-10-causes-of-death
4. Erdinç, E., et al.: Türk toraks derneği kronik obstrüktif akciğer hastalığı tanı ve tedavi uzlaşı raporu 2010. Türk Toraks Dergisi Mayıs 11(1), 1–66 (2010)
5. Gunadyn, F.E., Kalkan, N., Günlüoglu, G., Aktepe, E.N., Demirkol, B., Altin, S.: The relationship between serum levels of surfactant protein D in COPD exacerbation severity and mortality. Turk. J. Med. Scı. 49(3), 888–893 (2019). https://doi.org/10.3906/sag-1809-6
6. Gilles, J.: Empirical wavelet transform. IEEE Trans. Signal Process. 61(16), 3999–4010 (2013)
7. Altan, G., Kutlu, Y., Allahverdi, N.: Deep learning on computerized analysis of chronic obstructive pulmonary disease. IEEE J. Biomed. Health İnf. 24, 1344–1355 (2019)
8. Amaral, J.L., Faria, A.C., Lopes, A.J., Jansen, J.M., Melo, P.L.: Automatic identification of chronic obstructive pulmonary disease based on forced oscillation measurements and artificial neural networks. In: 2010 Annual International Conference of the IEEE Engineering in Medicine and Biology, pp. 1394–1397. IEEE (2010)
9. Nabi, F.G., Sundaraj, K., Lam, C.K.: Identification of asthma severity levels through wheeze sound characterization and classification using integrated power features. Biomed. Signal Process. Control 52, 302–311 (2019)
10. Hossain, I., Moussavi, Z.: Finding the lung sound-flow relationship in normal and asthmatic subjects. In: 26th Annual International Conference of the IEEE on Engineering in Medicine and Biology Society, 2004. IEMBS 2004, vol. 2, pp. 3852–3855. IEEE (2004)
11. Fernandez-Granero, M.A., Sanchez-Morillo, D., Leon-Jimenez, A.: An artificial intelligence approach to early predict symptom-based exacerbations of COPD. Biotechnol. Biotechnol. Equip. 32(3), 778–784 (2018)
12. İçer, S., Gengeç, Ş: Classification and analysis of non-stationary characteristics of crackle and rhonchus lung adventitious sounds. Digital Signal Process. 28, 18–27 (2014). https://doi.org/10.1016/j.dsp.2014.02.001
13. Huang, D., Zhang, S., Zhang, Y.: EEG-based emotion recognition using empirical wavelet transform. In 2017 4th International Conference on Systems and Informatics (ICSAI), pp. 1444–1449. IEEE (2017)
14. Maharaja, D., Shaby, S.M.: Empirical wavelet transform and GLCM features based glaucoma classification from fundus image. Int. J. MC Square Sci. Res. 9(1), 78–85 (2017)
15. Babu, K.A., Ramkumar, B., Manikandan, M.S.: Empirical wavelet transform based lung sound removal from phonocardiogram signal for heart sound segmentation. In ICASSP 2019–2019 IEEE International Conference on Acoustics, Speech and Signal Processing (ICASSP), pp. 1313–1317. IEEE (2019)
16. Altan, G., Kutlu, Y., Garbi, Y., Pekmezci, A.Ö., Nural, S.: Multimedia respiratory database (respiratorydatabase@ TR): auscultation sounds and chest x-rays. Nat. Eng. Sci. 2(3), 59–72 (2017)
17. Altan, G., Kutlu, Y., Pekmezci, A.Ö., Nural, S.: Deep learning with 3D-second order difference plot on respiratory sounds. Biomed. Signal Process. Control 45, 58–69 (2018)
18. Vapnik, V.N.: The Nature of Statistical Learning Theory. Springer, New York (1995). https://doi.org/10.1007/978-1-4757-2440-0

# Least Square Support Vector Machine for Interictal Detection Based on EEG of Epilepsy Patients at Airlangga University Hospital Surabaya-Indonesia

Santi Wulan Purnami[1]([✉]), Triajeng Nuraisyah[1], Wardah Rahmatul Islamiyah[2], Diah P. Wulandari[3], and Anda I. Juniani[4]

[1] Department of Statistics, Institut Teknologi Sepuluh Nopember Surabaya, Surabaya, Indonesia
santi_wp@statistika.its.ac.id
[2] Department of Neurology, Universitas Airlangga Surabaya, Surabaya, Indonesia
[3] Department of Computer Engineering, Institut Teknologi Sepuluh Nopember Surabaya, Surabaya, Indonesia
[4] Design and Manufacture Study Program, Politeknik Perkapalan Negeri Surabaya, Surabaya, Indonesia

**Abstract.** Epilepsy is a chronic disease characterized by recurrent seizures. Epileptic seizures occur due to central nervous system (neurological) disorders. Around 50 million people worldwide suffer from epilepsy. The diagnosis of epilepsy can be done through an electroencephalogram (EEG). There are two important periods to consider in EEG recording, the interictal period (clinically no seizures) and ictal (clinically seizures). Meanwhile, visual inspection of EEG signals to detect interictal and ictal periods often involves an element of subjectivity and it requires experience. So that automatic detection of interictal periods with classification method is badly needed. In this study, Least Square Support Vector Machine (LS SVM) method for classification of interictal and ictal was used. Data preprocessing process was carried out using Discrete Wavelet Transform (DWT). The result showed that the classification using LS SVM with kernel RBF method achieved an accuracy under curve (AUC) of 96.3%.

**Keywords:** Square Support Vector Machine · Interictal · Ictal · EEG signal

## 1 Introduction

It is around 50 million people worldwide suffer from epilepsy. This causes epilepsy becomes one of the most common neurological diseases globally. Nearly 80% people suffering from epilepsy live in low and middle-income countries [1]. Epilepsy affects central nervous system. Usually epilepsy is characterized by recurring seizures for no reason that manifests itself in some variations, including emotional or behavioral disorders, seizure movements, and loss of consciousness [2]. People with epilepsy will experience recurrent seizures when they relapse. Seizures can occur due to an electrical

© The Author(s), under exclusive license to Springer Nature Switzerland AG 2021
J. Hemanth et al. (Eds.): ICAIAME 2020, LNDECT 76, pp. 198–210, 2021.
https://doi.org/10.1007/978-3-030-79357-9_20

impulse occurs in the brain that exceeds normal limit. According to WHO [1], people with epilepsy have a three times higher risk of early death than people in normal condition. Abnormal signs of brain can be detected visually by scanning electroencephalogram (EEG) signal. Electroencephalography is an electrophysiological technique for recording electrical activity arising from human brain [3]. In EEG recording process, the patient will be given several stimuli that trigger a seizure. Sleep condition and lack of sleep are two activation procedures that are very useful for stimulating seizures. Stimulus given in EEG recording process will cause EEG recording process is able to detect signs of neuronal dysfunction, both in the ictal period (clinically seizures) and interictal (clinically no seizures) [4]. Basically, the ictal and interictal periods need to be considered. This is because both periods show abnormal signals on the results of EEG recording. However, it is just a clinically different. Abnormal signals in the interictal periods occur within milliseconds, so the patient does not experience seizures. While abnormal signals in the ictal periods occur longer, in seconds or even minutes. In the ictal period the patient experiences clinical seizures. Therefore, detection is needed to determine the signal part of EEG recording included in the ictal and interictal periods as the first step in the diagnosis of epilepsy.

Meanwhile, the element of subjectivity is often involved in scanning EEG record. In addition, scanning EEG record also requires considerable experience and time. Given this difficulty, automatic detection of ictal and interictal signals is proposed using classification method. Support Vector Machine (SVM) method is appropriate for detecting epileptic seizures automatically. This is because EEG is a nonlinear and non-stationer signal [5]. Research on the comparison of various SVM methods had been done by Lima & Coelho [6]. Research by Lima & Coelho aimed to compare various SVM methods to classify EEG signals from normal patients and epilepsy patients. The results of the study showed that the classification using SVM and LS SVM had more consistent accuracy than Smooth SVM (SSVM), Lagrangian SVM (L SVM), Proximal SVM (P SVM), and Relevance Vector Machine (RVM).

The development of SVM methods in classifying EEG signals had been carried out a lot. Research on the detection of epileptic seizures using LS SVM was conducted by Ubeyli [7], Bajaj & Pachori [8], Sharma & Pachori [5], and Li, Chen & Zhang [9]. Research conducted by Ubeyli [7], Sharma & Pachori [5], Li, Chen, & Zhang [9] produced an accuracy of 99.56%, 98.67%, and 99.40–99.60% respectively. Research on the classification of epileptic seizures using SVM had also been undertaken by Sihombing [10] and Purnami, et al. [11]. Their studies identified EEG signals of normal patients and epileptic patients. In a study conducted by Sihombing [10], the data used were public data with details of five normal patients and five epileptic patients. Using K-Nearest Neighbor (KNN) method, research conducted by Sihombing [10] achieved an accuracy of 98.4%. Purnami et al. [11] used both public data and Airlangga University Hospital (RSUA) data. The classification methods used included KNN, SVM, and Naive Bayes. The result showed that accuracy with RSUA data was generally lower than accuracy with public data. But in both studies conducted by Sihombing [10] and Purnami, et al. [11], there were a very clear difference between signals originating from normal patients and epilepsy patients in the process of deepening the characteristics of the data. So that the classification process was considered to be less necessary to do.

In this study, ictal and interictal classification were carried out through EEG signals using SVM and LS SVM methods. The data pre-processing process was done before classification. First, the EEG signal was decomposed into theta, alpha, and beta sub-bands by the DWT method with the Daubecchies 7 (Db7) level 7 mother wavelet (morlet). Then, the signals in each sub-band were extracted into statistical features. Classification results were compared based on higher performance. In this study, the epilepsy patient data from Airlangga University Hospitalwere used. This research was expected to help the medical staff to distinguish ictal and interictal in epilepsy patients, so that the treatment of epilepsy patients could be more quickly and appropriately carried out by the medical team.

## 2   Theoretical Basis

### 2.1   Discrete Wavelet Transform (DWT)

Wavelet transform is a time frequency decomposition of a signal into a set of basis wavelet functions. Discrete Wavelet Transform (DWT) is derived from Continuous Wavelet Transform (CWT) through signal discretization $\psi_{(a,b)}(t)$. The equation of CWT can be written as follows [12]:

$$CWT(a, b) = \int_{-\infty}^{\infty} x(t)\psi_{(a,b)}^{*}(t)dt \tag{1}$$

with,

x (t): analyzed signal.

a, b: scale factor (coefficient of dilation/compression).

$\psi_{(a,b)}^{*}(t)$: wavelet on a scale factor a to b.

The most common discretization process is dyadic discretization with Eq. (2).

$$\psi_{(j,k)}(t) = \frac{1}{\sqrt{2^j}}\psi\left(\frac{t - 2^j k}{2^j}\right) \tag{2}$$

where a has been substituted with $2^j$ and b has been substituted with $2^j k$ [12].

One morlet that can be used in applying the DWT method is Daubechies. The Daubechies wave is an orthogonal wave family that defines DWT and is characterized by a maximum number of disappearing moments for some of the support provided. Feature extraction is the determination of features or feature vectors from vector patterns [13]. There are several features resulting from the extraction process, among others are as follows (Table 1);

**Table 1.** Statistical features

| Features | Formula |
|----------|---------|
| Variance | $\frac{\sum_{i=1}^{n}(x_i-\bar{x})^2}{n-1}$ |
| Energy | $\sum_{i=1}^{n} x_i^2$ |
| Maximum | $\max(x_i)$ |
| Minimum | $\min(x_i)$ |
| Entropy | $\sum_{i=1}^{n} x_i^2 \log\left(x_i^2\right)$ |

## 2.2 Support Vector Machine (SVM)

SVM is one of the classification methods used for linear and non-linear data. SVM finds hyperplane using support vector and maximal margins (which are defined from support vector). SVM approach to classification is to maximize the marginal hyperplane which is the distance between class [14]. In standard SVM formulations, in general the optimal separation of hyperplane w is determined by minimizing functions,

$$\min_{w,\, \xi_i} J(w, \xi_i) = \frac{1}{2} w^T w + C \sum_{i=1}^{N} \xi_i^2 \tag{3}$$

with,

w: normal hyperplane vector (n × 1).

C: Cost.

$\xi_i$: non-negative slack vector in the ith sample, i = 1,..., N (m × 1).

Optimization in constraints is shown by the following inequality.

$$y_i \left| w^T \varphi(x_i) + b \right| \geq 1 - \xi_i \tag{4}$$

Based on Eqs. (3) and (4), the results of quadratic programming (QP) are derived into two-dimensional space with the following equation.

$$\max_{\alpha} J(\alpha) = \sum_{i=1}^{N} \alpha_i - \frac{1}{2} \sum_{i=1}^{N} \sum_{i=1}^{N} \alpha_i \alpha_j y_i y_j \varphi(x_i)^T \varphi(x_j) \tag{5}$$

where $\sum_{i=1}^{N} \alpha_i y_i = 0$ and $0 \leqslant \alpha_i \leqslant C$, for i = 1, ..., N. To get $\varphi(\mathbf{x_i})^T \varphi(\mathbf{x_j})$ in (5), $\varphi(\mathbf{x_i})$ and $\varphi(\mathbf{x_j})$ need to be explicitly calculated. Instead, a kernel K matrix can be designed [6].

In non-linear separate data, some kernel functions can be used, among others are Radial Basis Function (RBF), Linear, Polynomial, and Sigmoid. The RBF kernel is increasingly being used in SVM nonlinear mapping. The RBF kernel function is shown by the following equation [15].

$$K(x_i x_j) = \exp\left(-\gamma x_i - x_j^2\right) \tag{6}$$

## 2.3  Least Square Support Vector Machine (LS SVM)

LS SVM classifier is close to Vapnik SVM formulation but solving a linear system is not a quadratic problem (QP) [16]. LS SVM was first introduced by Suykens and Vandewalle. The main concept of LS SVM is to take the least amount of error squares from an objective function, with the constraint of equation which is not inequality as in SVM [6]. The optimal separation of hyperplane in LS SVM method is shown by Eq. (7).

$$\min_{w, \, \xi_i} \quad J(w, b, \, e) = \frac{1}{2} w^T w + \frac{1}{2} \tau \sum_{i=1}^{N} e_i^2 \tag{7}$$

with,

w: normal vector to hyperplane (n × 1).

τ: parameter regulation.

$e_i$: least square error in sample ith, i = 1,..., N (m × 1),

depending on the constrain as follows.

$$y_i \left| w^T \varphi(x_i) + b \right| = 1 - e_i \tag{8}$$

with,

b = bias term

i = 1,..., N

For this reason, Lagrange multiplier formula is converted into the dual problem form of Eqs. (7) and (8) as follows:

$$L(w, \, b, \, e; \, \alpha) = J(w, b, \, e) - \sum_{i=1}^{N} \alpha_i \left\{ y_i \left( w^t \varphi(x_i) + b \right) - 1 + e_i \right\} \tag{9}$$

With, $\alpha_i = (\alpha_1, \ldots, \alpha_N)^t$ is Lagrange multipliers [16]. Optimal conditions are obtained by deriving equation (x) with respect to $w$, $b$, $\alpha_i$, $and$ $e_i$ and equating the result of equation with 0 [7].

## 2.4  Evaluation of Classification Performance

K-fold cross validation is one way to evaluate the model that is widely used and is considered effective. This approach divided the set of observations into groups k, with the same or nearly the same size. The fold 1 is used as testing set, the rest k − 1 as training data. The first misclassified observation (Err1), then it was calculated based on observations on the group used. This procedure was repeated k times. Each time, a different observation group was treated as a validation set. This process resulted in an error rate k of the test error, $Err_1$, $Err_2$, ..., $Err_k$. The k-fold Cross Validation (CV) error rate is calculated by the average value as follows,

$$.CV_{(k)} = \frac{1}{k} \sum_{i=1}^{k} Err_i \tag{10}$$

with $Err_i = I(y_i \neq \hat{y}_i)$ [17].

The measure of classification quality is made from a confusion matrix. Table 2 presents the confusion matrix for binary classification [18].

**Table 2.** Confusion matrix

| Actual/Prediction | Positive | Negative |
|---|---|---|
| Positive | TP (True positive) | FN (False negative) |
| Negative | FP (False positive) | TN (True negative) |

The classification performance can be evaluated by the following measures:

1.  1. Accuracy
    The equation that can be used to calculate accuracy is as follows.

$$Accuracy = \frac{TP + TN}{TP + FP + FN + TN} \tag{11}$$

2.  Sensitivity
    The sensitivity value can be calculated using the following equation [9].

$$Sensitivity = \frac{TP}{TP + FN} \tag{12}$$

3.  Specificity
    The calculation of specificity values can use the following equation [9].

$$Specificity = \frac{TN}{TN + FP} \tag{13}$$

In addition to using the three measures of goodness above, an Area Under Curve (AUC) area was also used. AUC was one of the criteria that could be used to measure the quality of the results of classification method [19].

AUC calculation is shown by Eq. (14).

$$AUC = \frac{1}{2}(Sensitivity + Specifisity) \tag{14}$$

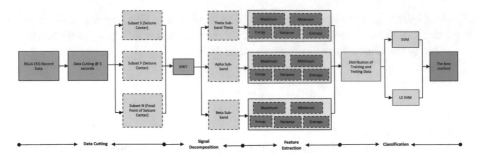

**Fig. 1.** Research flow chart.

## 3   Methodology

In this study secondary data were used from Airlangga University Hospital (RSUA), Surabaya Indonesia. The data obtained were recorded data of human brain EEG signals of five epilepsy patients who came to the neurology clinic. Recording in each patient used 25 channel electrodes taped to his scalp (extra cranial), of which 22 channels were for recording the brain signals (EEG), 2 channels for recording the eye (EOG), 1 channel for recording the heart (ECG). This research would only use EEG recording data.

The recording process at RSUA was carried out for 30 min with a sampling rate of 500 samples per second. The recording duration taken for this study was 5 s, resulting in 2500 samples in each channel. The cutting data were taken based on the doctor's recommendation at RSUA, considering the average length of the ictal and interictal periods. Data obtained from EEG recording was referenced in the Cz channel (channel in the center of brain), causing the resulting signal to fall into a single channel. Three data subsets consisting of N, F, and S were obtained from 5 epilepsy patients. Subset S was a subset with signal data originating from epilepsy patients who experienced an ictal period (clinical with seizures). While signal data on subsets N and F were obtained from epilepsy patients who experienced interictal peri-od (clinical without seizures). The data in subset F were obtained from the center of the seizures, while the data in subset N was obtained from the point farthest from the center of seizures. Waves in the interictal period were called sharp wave. The taking of EEG signal pieces in the ictal period was done in all channels, while in the interictal period the taking of EEG signal pieces was only conducted in the central channel of the occurrence of sharp wave and the farthest point from the center of the occurrence of sharp wave. Table 3 shows the details of data used in this study.

The analysis steps carried out can be described by Fig. 1.

**Table 3.** Patient information and data subset

|  | Patient | | | | |
| --- | --- | --- | --- | --- | --- |
|  | 1 | 2 | 3 | 4 | 5 |
| Electrode installation | Rule 10–20, extracranial | | | | |
| The period used | ictal; interictal | ictal; interictal | interictal | interictal | interictal |
| Subset data generated | S; F and N | S; F and N | F and N | F and N | F and N |
| Number of data Signal ictal; interictal | 220; 18 | 132; 6 | −; 6 | −; 6 | −; 21 |

## 4  Result and Suggestions

### 4.1  Signal Characteristics

The result of identifying the characteristics of the data based on their amplitudes is presented in Fig. 2. Figure 2 shows one-time series plot of the signal cut in each subset of F, N, and S of one patient. The number of samples produced per second was 500 samples, so that the signal cut taken for 5 s had a total sample (point) of 2500. The amplitude ranges of each of the subsets F, N, and S of patient 1 shown in Fig. 2 is different. Subset S (ictal period) had the widest range of amplitudes, whereas subset F (interictal period) had the narrowest amplitude range.

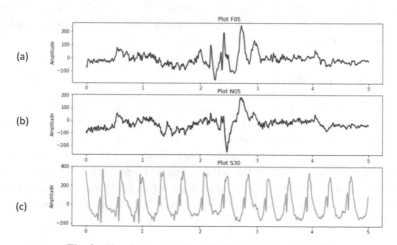

**Fig. 2.** Plot time series data (a) F5, (b) N5, and (c) S30

### 4.2  Sub-band EEG Signal

After describe the signal characteristics in each subset, the next step was to decompose the EEG signal. Signal decomposition was undertaken using DWT method with Daubechies

7 (Db7) level 7 morlet. In addition to decomposing the signal, the DWT method was also used to reduce the influence of physiological artifacts in the stages of ictal and interictal classification. In this study, theta, alpha, and beta sub-bands were taken. The results of signal decomposition using the Daubechies 7 level 7 morlet are shown in Fig. 3.

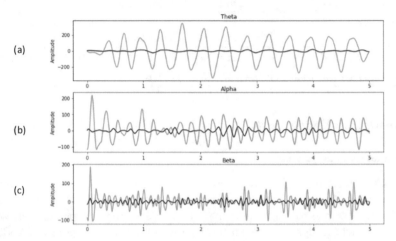

**Fig. 3.** Comparison of subset F and S in sub-band (a) Beta, (b) Alpha, and (c) Theta

Signals in the ictal period in Fig. 3 are shown in light blue, whereas signals in the interictal period represented by subset F are shown in red. In Fig. 3 shows significant differences in the amplitude range between signals in the ictal period (subset S) and those interictal (subset F).

### 4.3  Signal Feature Extraction

The next step taken in this research was extracting each sub-band on each signal into 5 statistical features, including energy features, maximum, minimum, variance, and entropy. The results from feature extraction will be explained as follows:

Based on Fig. 4, it can be seen that in general the signal in the ictal period (subset S) has a difference with the signal in the interictal period (subsets F and N). In the energy, maximum, variance, and entropy features, the period ictal signal has a greater value than the interictal period signal. As for the minimum features, the ictal period signal has a smaller value.

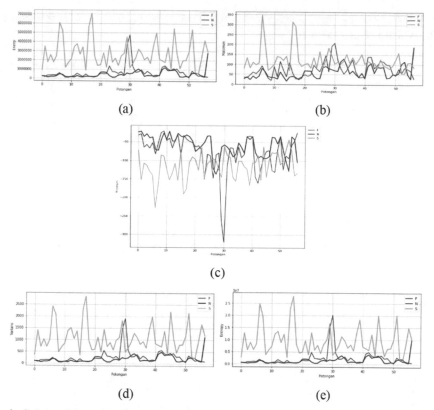

**Fig. 4.** Sub-band theta plot (a) Energy, (b) Maximum, (c) Minimum, (d) Variance, dan (e) Entropy

## 4.4 Classification

After obtaining 15 statistical features from data pre-processing stage, the classification stage was carried out. Statistical features were used as predictor variables. In this study the RBF kernel was used for both SVM and LS SVM methods. The 10-fold cross validation was used in this study. Data of 466 signal cuts consisted of 114 signals in the interictal period and 352 signals in the ictal period were divided randomly into 10 folds. The hyperparameter tuning stage was then performed. In both methods, parameters C, $\gamma$ and $\tau$ were tested with a value of 0.0001; 0,001; 0.01; 0.1; 1; 10 and 100. The selection of the best parameters based on the highest accuracy value. The effect of parameters used on classification accuracy using SVM and LS SVM visually is shown in Fig. 5. The difference in the value of $\gamma$ in Fig. 5(a) is shown by the difference in color. Increasing the value of C at the same value will cause the classification accuracy to increase. Visually it can be seen that the parameters that produce the highest accuracy are C = 100 and $\gamma = 0.1$. The classification accuracy with these parameters is 96.80%. In Fig. 5(b), the color difference of the lines shows the difference in value $\tau$. Based on Fig. 5(b), the same $\tau$ value causes the classification accuracy decreases as the value of $\gamma$ increases. Whereas with the same $\gamma$ value, the highest accuracy value is generally generated when

$\tau = 0.0001$. The highest accuracy value is generated when using $\tau = 0.0001$ and $\gamma = 0.01$, which is equal to 96.16%.

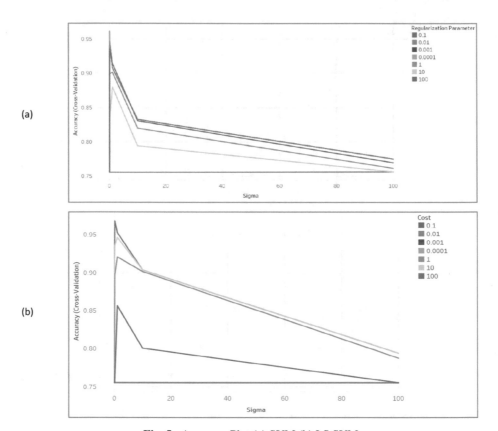

(a)

(b)

**Fig. 5.** Accuracy Plot (a) SVM (b) LS SVM

The selection of the best classification method was based on the average and standard deviation of the optimum performance of the resulting SVM classification and LS SVM. The results of the comparison of SVM and LS SVM methods based on the average and standard deviation of their performance are presented in Table 4.

**Table 4.** Comparison performance of SVM and LS SVM

| Method | Accuracy | Sensitivity | Specificity | AUC |
|--------|----------|-------------|-------------|-----|
| SVM | **0,96795** | 0,94848 | **0,97444** | 0,96146 |
| LS SVM | 0,96161 | **0,96591** | 0,96032 | **0,96311** |

In Table 4 it can be seen that the SVM and LS SVM achieve almost the same average performance. The average performance has exceeded the determined threshold of 76%, so the classification stage uses the 10-fold cross validation method to share training and testing data can be said to be successful. In this study the data used is imbalance data. Therefore, the basic for selecting the best method, AUC value was used in this study. So it could be concluded that the LS SVM method with RBF kernel was slightly better than the SVM RBF kernel method in classifying signals in the ictal and interictal periods originating from epilepsy patients at RSUA.

## 5 Conclusion

In this study, the classification of interictal and ictal signals was carried out using machine learning based on the SVM method. The results showed that LSSVM produced slightly better performance than SVM. The AUC of LSSVM reaches 96.3%. This results still needs to be improved, one of which is by dealing with the problem of imbalanced data.

The results of this study are very useful, especially in helping neurologists to identify interictal and ictal signals, especially at Airlangga University Hospital Surabaya, Indonesia. Furthermore, with automatic detection it can be early warning (identify interictal) before a seizure occurs (ictal).

**Acknowledgement.** This work is financially supported by Indonesian Ministry of Research, Technology, and Higher Education, through DRPM Grant "Penelitian PDUPT" under contract number 1181/PKS/ITS/2020.

## References

1. WHO: Epilepsy (2018). http://www.who.int/mediacentre/factsheets/fs999/en/. Accessed 3 Apr 2008
2. Mormann, F., Andrzejak, R.G., Elger, C.E., Lehnertz, K.: Seizure prediction: the long and winding road. Brain **130**, 314–333 (2007)
3. S.L.K., Louis, Frey: Electroencephalography. American Epilepsy Society, Chicago (2016)
4. Bowman, J., Dudek, F.E., Spitz, M.: Epilepsy. Encyclopedia of Life Sciences, pp. 1–7 (2001)
5. Sharma, R., Pachori, R.B.: Classification of epileptic seizures in EEG signals based on phase space representation of intrinsic mode function. Elsevier Expert Syst. Appl. **42**, 1106–1117 (2014)
6. Lima, C.A.M., Coelho, A.L.V.: Kernel machines for epilepsy diagnosis via EEG signal classification : a comparative study. Artif. Intell. Med. **53**, 83–95 (2011)
7. Ubeyli, E.D.: Least square support vector machine employing model-based methods coefficients for analysis of EEG signals. Expert Syst. Appl. **37**, 233–239 (2010)
8. Bajaj, V., Pachori, R.B.: Classification of seizure and nonseizure EEG signals using empirical mode decomposition. Trans. Inf. Technol. Biomed. **16**(6), 1135–1142 (2012)
9. Li, M., Chen, W., Zhang, T.: A novel seizure diagnostic model based on kernel density estimation and least squares support vector machine. Biomed. Signal Process. Control **41**, 233–241 (2017)
10. Sihombing, C.H.: Diagnosa Kejang Epilepsi Berdasarkan Sinyal EEG Menggunakan Metode K-Nearest Neighbor. ITS, Surabaya (2018)
11. Purnami, S.W., et al.: Comparison of nonparametric classifications approaches for epileptic seizure detection based on electroencephalogram signals. In: ISMI ICTAS, Department of Mathematical Sciences, UTM, Malaysia (2018)

12. Saravanan, N., Ramachandran, K.I.: Incipient gear box fault diagmosis using Discrete Wavelet Transform (DWT) for feature extraction and classification using Artificial Neural Network (ANN). Expert Syst. Appl. **37**, 4168–4181 (2010)
13. Cvetkovic, D., Ubeyli, E.D., Cosic, I.: Wavelet transform feature extraction from human PPG, ECG, and EEG signal responses to ELF PEMF exposures : a pilot study. Digital Signal Process. **18**, 861–874 (2008)
14. Han, J., Kamber, M., Pei, J.: Data Mining Concepts and Techniques. 3rd penyunt. Elsevier, USA (2012)
15. Shunjie, H., Qubo, C., Meng, H.: Parameter Selection in SVM with RBD Kernel Function. World Automation Congress (2012)
16. Suykens, J.A.K., Van Gestel, T., De Brabanter, J., De Moor, B., Vandewalle, J.: Least Squares Support Vector Machines. World Scientific, Singapore (2002). https://doi.org/10.1142/5089
17. James, G., Witten, D., Hastie, T., Tibshirani, R.: An Introduction to Statistical Learning with Applications in R. Springer Science+Business Media, New York (2013)
18. Sokolova, M., Japkowicz, N., Szpakowicz, S.: Beyond Accuracy, F-score and ROC: A Family of Discriminant Measures of Performance Evaluation. Semantics Scholar (2006)
19. Huang, J., Ling, C.X.: Using AUC and accuracy in evaluating learning algorithms. Trans. Knowl. Data Eng. **17**(3), 299–310 (2005)
20. Reynolds, E.H.: Epilepsy: The Disorder. In: Milestones The History of Epilepsy, Epilepsy Atlas, pp. 15–27 (2005)

# Generating Classified Ad Product Image Titles with Image Captioning

Birkan Atıcı$^{(\boxtimes)}$ and Sevinç İlhan Omurca

Department of Computer Engineering, Kocaeli University, Kocaeli, Turkey
{130201088,silhan}@kocaeli.edu.tr

**Abstract.** Image captioning is a field of study to describe images by automatically generating semantic sentences in accordance with natural language rules. In recent years, studies in the literature of challenging problem of image captioning have gained momentum with the advances in deep learning. The reason why these studies are challenging is they use both image processing and natural language processing methods together. The subject of this study is generating the most suitable title for the product images in the shopping category of classified advertising by using state of the art deep learning image captioning methods. This is the first study which generates titles to classified ad product images in Turkish in the literature. The results of the experimental study, conducted with approximately 1.8 million images and titles on the sahibinden.com website, are fairly successful for describing product images.

**Keywords:** Image captioning · Image processing · Natural language processing

## 1 Introduction

Image captioning is a field of automatically generating captions according to the content of the images, in accordance with natural language rules. In image captioning studies should make sense from the image and form a sentence from the object, action and space in the image using both image processing and natural language processing. For these reasons, the studies are considered challenging in the literature.

Before deep learning, these studies were done with template based and retrieval based [1]. With the increase of the deep learning researches, image captioning studies are also focused on deep learning techniques. In the deep learning based image captioning methods, caption generation is done by using the encoder decoder structure, which is basically used in deep learning based machine translation. Decoder uses Convolutional Neural Network (CNN) and generates a fixed-length vector representation for feature map of the image. Encoder generates sentences on Recurrent Neural Networks (RNN) using the vector representation of image which is already created before.

Image captioning studies can be used in many fields, such as getting images in the database of images with text search, automatically generating subtitles of images shared on the social media or labelling images. The purpose of this study is to provide that

© The Author(s), under exclusive license to Springer Nature Switzerland AG 2021
J. Hemanth et al. (Eds.): ICAIAME 2020, LNDECT 76, pp. 211–219, 2021.
https://doi.org/10.1007/978-3-030-79357-9_21

people who want to sell their second-hand items can post classified ads without losing time by simply uploading images of the products.

In this study, the subject of automatically generating titles from product images was discussed using approximately 1.8 million image-title pairs on the 'sahibinden.com' web site. The architecture in the Show, Attend and Tell [9] study, which has shown the best performance in recent years, has been used. As a result of the experimental study, Bilingual Evaluation Understudy (BLUE), Metric for Evaluation of Translation with Explicit ORdering (METEOR) and Consensus-based Image Description Evaluation (CIDER) scores are given to evaluate the experimental results.

## 2   Related Work

In the literature, two different traditional machine learning based methods have been used before the deep learning techniques for studies of automatically generating caption from images. These are retrieval-based and template-based image captioning methods.

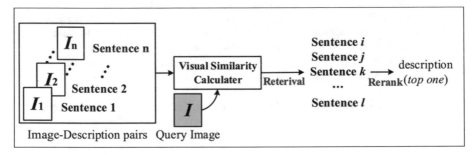

**Fig. 1.**  Retrieval based image captioning architecture [8].

In the retrieval based image captioning methods [2–4], firstly image-caption pairs pool is created. Similar image-caption pairs of the given image are retrieved from the previously created image-caption pairs pool. The caption with the best rank score from the captions sorted with different rank methods is selected as the caption of the given image (Fig. 1).

In template-based image captioning studies [5–7], templates consisting of a certain number of gaps are created. The caption generation is performed by filling the gaps in the template with detected features such as scene, action and object on the image (Fig. 2).

Although these traditional machine learning methods produce captions of images in accordance with the language rules, they form a very limited number of different sentences and are inadequate to fully define the image.

The first study in the field of image captioning with deep learning techniques made by Kiros et al. [10]. Unlike the previous studies, Kiros et al. proposed an image captioning method that operates on a multimodal space which does not depend on any template, structure or constraint. This method simply consists of 4 parts: image encoder, language encoder, multimodal space and language decoder. The image encoder extracts the features of the image using CNN layers. The language encoder creates embeddings

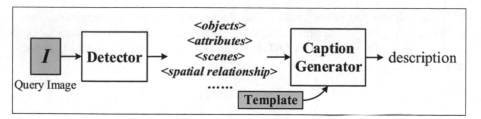

**Fig. 2.** Template based image captioning architecture [8].

of the words with an RNN model. Image feature vectors and word embeddings, which are extracted from the image and text, are given to the multimodal space and image and text are learned jointly. The language decoder performs the generation of the caption for the given image with learned information [1].

Vinyals et al. [11] in their study, inspired by neural machine translation, applied the encoder-decoder architecture to the image captioning (Fig. 3). In the neural machine translation, the texts to be translated are given Long Short-Term Memory (LSTM) model as an input. Translated texts are expected as LSTM model output. Similar to the neural machine translation, Vinyals et al. in their work, they gave the vector representing the feature of the image as the input of the model and the word representation of the caption as the model output. Equation 1 is used as the loss function of the deep learning model. In this equation, aim is to maximize the probability using $\theta$ model parameters.

$$\theta^* = \arg\max_\theta \sum_{(I,\,S)} \log p(S|I; \theta) \tag{1}$$

$\theta$: model parameters, I: image vector representation, S: image caption.

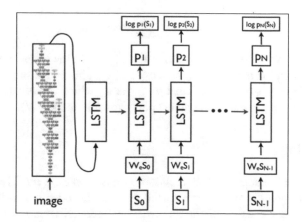

**Fig. 3.** LSTM based encoder-decoder architecture [11].

In the encoder part of the model in Fig. 3, Inception v3 also known GoogLeNet, a pre-trained convolutional neural network model, is used to represent the feature vector of the image. In the decoder part, LSTM, which is a recurrent neural network model, is used.

$$x_{-1} = CNN(I) \tag{2}$$

$$x_t = W_e S_t, \quad t \in \{0 \ldots N - 1\} \tag{3}$$

$$p_{t+1} = LSTM(x_t), \quad t \in \{0 \ldots N - 1\} \tag{4}$$

The model in Fig. 3 can express with the Eqs. (2–4). First, the image is given to the CNN model. The resulting features vector of the CNN model is shown as $x_{-1}$. The vector $x_{-1}$ is given as an input to LSTM and used to start the encoder-decoder model. $S_0$ shows the beginning of the sentence and $S_N$ shows the end of the sentence. $S_t$ shows the one-hot vector of each word, the $W_e$ word embedding matrix, and the word probability vector shown as $p_{t+1}$ produced from the model at the time of the $t + 1$. The highest probability vector determines the next word for the generated caption. Xu et al. in their study [9], added a visual attention mechanism to the encoder-decoder architecture. They used the convolutional neural networks to extract vectors representing different parts of the image as well as to extract the features vector of the image. The convolutional neural network model produces vectors that are LxD dimension shown as the Eq. (5).

$$a = \{a_1, \ldots, a_L\}, \quad a_i \in \mathbb{R}^D \tag{5}$$

$$\sum_{i=1}^{L} a_{ti} = 1 \tag{6}$$

$$z_t = \sum_{i=1}^{L} a_{ti} a_i \tag{7}$$

Each $a_i$ vector in the Eq. (5) represents different parts of the image. For each location $i$ in Eq. (6), the attention mechanism produces an $a_i$ weight and the sum of these weights is equal to 1. The context vector created with the $a_i$, which represents the part of the image, and $a_i$ weight corresponding to this part shown in the Eq. (7).

## 3 Methodology

In this study, to automatically caption classified ads images, a model was created by applying visual attention mechanism to the encoder-decoder mechanism with deep learning. The model consists of three networks: Encoder, attention mechanism and decoder.

## 3.1 Model Architecture

In this study, encoder-decoder architecture with attention mechanism proposed in the Show, Attend and Tell [9] article, which is one of the most successful image captioning studies of recent years, is used (Fig. 4).

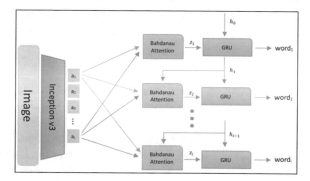

**Fig. 4.** Image captioning architecture with attention mechanism.

### 3.1.1 Encoder

Inception v3, an CNN model trained on the Imagenet dataset, is used as an encoder. The softmax layer at the end of the Inception v3 model is discarded and the transfer learning is applied to extract features from images.

### 3.1.2 Attention Mechanism

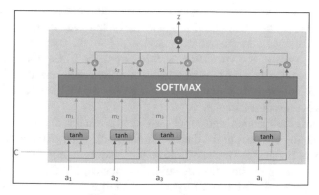

**Fig. 5.** Attention mechanism.

In the decoder part of the model, attention mechanism proposed by Bahdanau et al. [12], is used. The attention mechanism ensures that the focus is on different parts of the image at every step while the next word is generated for the title. The attention mechanism takes $a_1, \ldots, a_i$ vectors which are representing the different parts of the image and the context vector C generated from the sequence of hidden states of the model as input. As the output of the attention model, it creates the $z_t$ vector, which represents the summary of the image that it focuses at time t. General architecture of the attention mechanism is given in the Fig. 5.

### 3.1.3  Decoder

In the decoder part of the image captioning model a Recurrent Neural Networks model is used. As distinct from the feed forward neural networks, the outputs feed the inputs backwards in the RNN models. In RNN model, vanishing gradient problem is encountered during back propagation. Vanishing gradient is a problem of gradients getting too small during training. This causes learning to stop. Gated Recurrent Units (GRU) which is a customized RNN model is used to avoid vanishing gradient problem in the decoder part. As shown in Fig. 4, the decoder provides to generate a new word for the title using the context vector z created from the attention mechanism and hidden states created from previous steps.

## 4  Experiments

### 4.1  Data Preprocessing

In machine learning algorithms, data quality is one of the most important factors that affect the success rate of the model. Therefore, the data preprocessing and data filtering are very important in this study as in all machine learning studies. Our dataset consists of about 5.6 million image-title pairs obtained from the "sahibinden.com" which is the largest, the most visited online classified ads and shopping platform through which people buy and sell in many categories such as real estate, vehicles, shopping products and services in Turkey. In the dataset, dirty data is quite high. There were approximately 230000 unique words in the vocabulary created from the titles. 5000 words with the highest frequency were taken from the vocabulary. Data with less than 3 words in the titles were filtered out in the dataset because such titles do not have enough information content to make sense of the image.

Two different classification algorithms have been created to classify images and titles to filter out meaningless data in the dataset. In the multi-class deep learning algorithm created to classify images, the last layer shows the maximum probability of the category of the image. In order to filter the meaningless images, the data with the last layer probability values less than 0.9 were removed from the dataset. In the same way, the titles are given to the multi-class classification algorithm. Data whose probability value of the last layer is lower than 0.75 is removed from the dataset. After the preprocessing steps, details of the dataset are given in Table 1.

**Table 1.** Details of the dataset.

| Average of title length | Unfiltered unique word count | Filtered unique word count | Total unfiltered image-title pair count | Total unfiltered image-title pair count |
| --- | --- | --- | --- | --- |
| 9.4 | 232025 | 5000 | 5591956 | 1816146 |

## 4.2  Experimental Results

In this study, we made automatic Turkish title generation from the product images of second-hand advertisements with attention mechanism on the encoder-decoder method. Trained with large scale data 1.8 million, this model has fairly successful results.

In image captioning studies, different methods such as BLUE score, METEOR and CIDEr are used to measure the success of the model. These methods generally measure the success of the model by comparing the reference titles with the titles produced by the model. The experimental results with different optimizers of the model are given in Table 2.

**Table 2.**  Scores of experimental results.

| Optimizer | B@1 | B@2 | B@3 | B@4 | Meteor | Cider |
| --- | --- | --- | --- | --- | --- | --- |
| SGD | 0.567 | 0.440 | 0.330 | 0.231 | 0.114 | 0.452 |
| Adam | 0.687 | 0.623 | 0.564 | 0.494 | 0.130 | 0.634 |
| RMSProp | 0.442 | 0.374 | 0.320 | 0.258 | 0.125 | 0.372 |

When we compare the results in Table 2 with the image captioning studies results in the literature, it is seen that it is fairly successful. The most successful of the results in Table 2 with different optimizers is Adam optimizer. In Table 3, some of the successful, partially successful and failed results of the titles produced from the model are given respectively.

**Table 3.** Examples of generated titles from the model (TR: Turkish, EN: English).

| Successful | Partially successful | Failed |
|---|---|---|

TR: az kullanılmış dji mavic pro 1 drone
EN: slightly-used dji mavic pro 1 drone

TR: jamo audio satılık hoparlör
EN: jamo audio speaker for sale

TR: pro gaming denon koltuk
EN: pro gaming denon chair

TR: asus rog gl vw oyun laptopu
EN: asus rog gl vw gaming laptop

TR: sıfır gibi apple watch yeni sezon
EN: new season apple watch like new

TR: oyun bilgisayarı sıfır
EN: new gaming laptop

TR: sahibinden temiz çamaşır makinesi
EN: slightly-used washing machine from owner

TR: panasonic g su hatasız
EN: spotless panasonic g su

TR: ucuz kanvas tablo
EN: heap canvas painting

## 5   Conclusion

This is the first study to generate titles from the product image on Turkish in the literature. The reason that this study is not fully successful is our dataset is very dirty. In future studies, the data set can be collected with more meaningful titles and images and, different architectures other than the attention mechanism can be tried.

**Acknowledgments.** In this study, we would like to thank sahibinden.com company for providing image and title data, as well as for supporting GPU usage on the cloud.

# References

1. Hossain, M.D., Sohel, F., Shiratuddin, M.F., Laga, H.: A comprehensive survey of deep learning for image captioning. ACM Comput. Surv. (CSUR) **51**(6), 118 (2019)
2. Farhadi, A., et al.: Every picture tells a story: Generating sentences from images. In: Daniilidis, K., Maragos, P., Paragios, N. (eds.) ECCV 2010. LNCS, vol. 6314, pp. 15–29. Springer, Heidelberg (2010). https://doi.org/10.1007/978-3-642-15561-1_2
3. Mason, R., Charniak, E.: Nonparametric method for data-driven image captioning. In: Proceedings of the 52nd Annual Meeting of the Association for Computational Linguistics (Volume 2: Short Papers), pp. 592–598, June 2014
4. Kuznetsova, P., Ordonez, V., Berg, T.L., Choi, Y.: Treetalk: composition and compression of trees for image descriptions. Trans. Assoc. Comput. Linguist. **2**, 351–362 (2014)
5. Yang, Y., Teo, C.L., Daumé III, H., Aloimonos, Y.: Corpus-guided sentence generation of natural images. In: Proceedings of the Conference on Empirical Methods in Natural Language Processing, pp. 444–454. Association for Computational Linguistics, July 2011
6. Mitchell, M., et al.: Midge: generating image descriptions from computer vision detections. In: Proceedings of the 13th Conference of the European Chapter of the Association for Computational Linguistics, pp. 747–756. Association for Computational Linguistics, April 2012
7. Ushiku, Y., Yamaguchi, M., Mukuta, Y., Harada, T.: Common subspace for model and similarity: phrase learning for caption generation from images. In: Proceedings of the IEEE International Conference on Computer Vision, pp. 2668–2676 (2015)
8. Wang, Y., Xu, J., Sun, Y., He, B.: Image captioning based on deep learning methods: a survey. arXiv preprint arXiv:1905.08110 (2019)
9. Xu, K., et al.: Show, attend and tell: neural image caption generation with visual attention. In: International Conference on Machine Learning, pp. 2048–2057, June 2015
10. Kiros, R., Salakhutdinov, R., Zemel, R.: Multimodal neural language models. In: International Conference on Machine Learning, pp. 595–603, January 2014
11. Vinyals, O., Toshev, A., Bengio, S., Erhan, D.: Show and tell: a neural image caption generator. In: Proceedings of the IEEE Conference on Computer Vision and Pattern Recognition, pp. 3156–3164 (2015)
12. Bahdanau, D., Cho, K., Bengio, Y.: Neural machine translation by jointly learning to align and translate. arXiv preprint arXiv:1409.0473 (2014)

# Effectiveness of Genetic Algorithm in the Solution of Multidisciplinary Conference Scheduling Problem

Ercan Atagün$^{(\boxtimes)}$ and Serdar Biroğul

Department of Computer Engineering, Duzce University, Duzce, Turkey
{ercanatagun,serdarbirogul}@duzce.edu.tr

**Abstract.** Multidisciplinary conferences are the types of conferences that allow the presentation of studies in different disciplines such as naturel science, social sciences, health and arts sciences etc. In these conferences, determining the days, halls and sessions and making presentations according to the related main-scope and sub-scopes, are an important limited optimization problem. The fact that there are presentations from different disciplines in same session during the conference is a big problem for the conference participants. In this study, the solution of the scheduling problem of multidisciplinary conferences with Genetic Algorithm approach is discussed. The basic concepts of Genetic Algorithm are given and conference scheduling schemes and the elements to be considered in scheduling are indicated. In this study, two different multidisciplinary conference datasets have been used. An application has been developed with Genetic Algorithm in C# language under some constraints of the different days, different sessions and different rooms. As a result of the study, it is seen that the solutions obtained with Genetic Algorithm are generally close to optimum solutions.

**Keywords:** Genetic Algorithm · Optimization · Scheduling

## 1 Introduction

Scheduling problems have been a subject of research for many years. Scheduling problems are the carrying out of processes by arranging them to be done in order. There are different types of scheduling, such as workshop scheduling [1], nurse scheduling [2, 3], curriculum scheduling [4, 5], and CPU scheduling [6]. This study deals with the conference scheduling problem. Conference scheduling involves the presentation of conference papers in specific halls and sessions within the specified number of days of the conference. The International Conference on Automated Planning and Scheduling (ICAPS) is the first conference held on scheduling and it provided the opportunity to discuss a number of new algorithms [7]. In his study, Aktay emphasized that these conferences, which are scientific activities, primarily aim to enable the production and sharing of information. He also stated that by sharing research results and experiences during information sharing, all participants, including listeners and speakers, are provided with information [8]. Many threads need to be utilized in the planning of conferences. In their

© The Author(s), under exclusive license to Springer Nature Switzerland AG 2021
J. Hemanth et al. (Eds.): ICAIAME 2020, LNDECT 76, pp. 220–230, 2021.
https://doi.org/10.1007/978-3-030-79357-9_22

study, Andlauer et al. emphasize the importance of scientific, organizational and human skills in order to hold a productive conference, the scientific board, the topic of the conference, the schedule of the conference, the sessions and the session chairs must be determined properly [9]. As conference scheduling involves many constraints, classical optimization methods fall short. Nicholls proposed a heuristic method for the conference scheduling problem in which quick planning is made by the chair of the conference organizing committee [10]. Tanaka et al. developed conference scheduling in which sessions are grouped according to the keywords in the conference paper by using the Self Organization Map algorithm [11]. Vangerven et al. assigned appropriate sessions to speakers in a personalized fashion in conference scheduling. An integer program model was used for this process [12]. Zulkipli et al. led the authors to specific sessions by assessing the conference venue capacity in conference scheduling [13]. Kim et al., by bringing a different approach to the problem, have developed an interface named Cobi that shows visual feedback which takes into account the preferences and constraints of the organizers, participants, and the authors who will make presentations at the conference [14]. Ibrahim et al., for conference scheduling, have developed a combinatorial design model through which three concurrent sessions can be held [15]. Stidsen et al. brought together speeches on similar topics in EURO-k, one of the most important operations research conferences in the world [16]. Eglese and Rand practiced conference scheduling using annealing algorithm, which is a heuristic approach, in the study, the presentations of the participants were arranged before the conference by taking their expectations into consideration [17].

## 2  Problem Definition

Some definitions need to be made for the conference scheduling problem. Multidisciplinary conference is a term suggesting that multiple scientific fields are utilized harmoniously. Multidisciplinary conferences are conferences in which studies and presentations regarding several science fields are made. In this study, the Genetic Algorithm is proposed for the solution of conference scheduling problem. To solve the conference scheduling problem, presentations must be assigned on the basis of conference days, halls, sessions and times. The definitions of gene and chromosome structures, which are the basis of the Genetic Algorithm, are needed for problem definition.

### 2.1  Using Genetic Algorithm in Constricted Optimization Problems

The biggest difficulty encountered in constricted optimization problems is data groups that do not comply with restrictions. In order to overcome this difficulty, the penalty function has been defined. Genes that do not comply with the restrictions on the purpose of the problem are multiplied by the defined penalty score [18, 19]. In solving this problem, it is aimed to reach a low penalty score as possible after the application.

## 2.2  Problem Constraints

Problem constraints are divided into two categories: hard and soft constraints. Hard constraints:

1. An author cannot be present in a concurrent session.
2. The primary subfield of all studies in a session cannot be different.
3. The secondary subfield of all studies in a session cannot be different.
4. The tertiary subfield of all studies in a session cannot be different.

   Soft constraints:

1. In the scheduling process, as little vacancies as possible are left in sessions. In addition, this enables the bringing together of the conference sessions that have vacancies.

**Table 1.**  Penalty coefficients.

| Problem constraint | Penalty coefficient |
|---|---|
| An author is in in a concurrent session | 10000 |
| There are different primary subfields in a session | 500 |
| | 300 |
| There are different secondary subfields in a session | 20 |
| There are different tertiary subfields in a session | 1 |

## 2.3  Genetic Algorithm

Genetic Algorithm is based on the process of evolution in natural life and was defined by John Holland in the 1970s [20].

```
Genetic Algorithm
{
  Create Initial Population
  Set Criterias
        while(! (End of iteration) or ! (Healing stopped))
            {
            Evaluate chromosomes according to fitness function
            Send chromosomes to match pool
            Get the best chromosome with Elitism
            Apply Crossover operator to chromosomes
            Apply Mutation operator to chromosomes
            Apply Repair operator to each chromosomes
            Apply Roulett Wheel method according to
                        the fitness value of each chromosomes
            Create new population from previous population
            }
        Get the best chromosome as a result
}
```

**Fig. 1.** Genetic algorithm pseudo code [18].

Figure 1 shows the pseudo-code of the Genetic Algorithm.

### 2.3.1 Chromosome

The chromosome comes into existence by the coming together of genes.

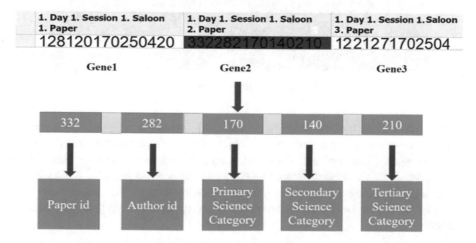

**Fig. 2.** Gene structure.

In this study, information regarding genes, which are the most fundamental units of the chromosome, is given in Fig. 2. Within the gene, a 3-byte part represents the conference paper number, a 3-byte part represents the primary field of the paper, a 3-byte part represents the secondary field of the paper, and a 3-byte part represents the tertiary field of the paper. In the admin panel of the developed software, the user should type the user study name and author name, and then should choose the topic of the study from the list on the screen. Science fields can be found in the database prepared for the program. There are 13 primary discipline fields registered in the database. There are 258 secondary disciplines that belong to these 13 scientific fields. There are 1904 tertiary disciplines that belong to these primary and secondary disciplines. After selecting the primary discipline area of the chosen subject secondary areas belonging to this primary area are listed. After the secondary areas are listed, the relevant one should be selected. After selecting the secondary field, the tertiary fields belonging to this secondary field are listed. For example, one of the fields such as Engineering, Law and Health Sciences should be selected as the primary discipline. After Engineering is chosen as the primary discipline, a secondary discipline field such as Computer Science and Engineering, Biomedical Engineering, or Electrical and Electronic Engineering should be chosen as the secondary discipline. When the secondary discipline is selected as Computer Science and Engineering, a tertiary discipline such as Cloud Computing, Information Security and Cryptology, Artificial Intelligence should be selected (Fig. 3).

**Fig. 3.** Paper registration screen.

After the related tertiary field is selected, all information related to the report is included in the system after clicking the "Save" button. Gene production is achieved when this process is completed. Through the 15-byte gene, the software stores all the information. In this way, genes come together to form a structure in which each conference paper is represented by a gene. Genes come together to form the chromosome. In the Genetic Algorithm process, chromosomes are the information that represents the solution.

## 2.4  Crossover

Crossover is one of the most basic operators of the Genetic Algorithm. The process of replacement of genes in order to generate individuals with better characteristics than the existing population is done based on a specific probability ratio. In the literature, the crossover rate varies between 50% and 90% [18]. In this study, the crossover rate was determined as 90%. There are different types of crossover, namely single-point crossover, two-point crossover, multi-point crossover, uniform crossover, and sequential crossover. In this study, the multi-point crossover was chosen due to the structure of the genes. Two chromosomes are required for crossover. These two chromosomes are selected from the population formed after the reproduction operator.

## 2.5 Mutation

Mutation is the random changes within the chromosome. Since the crossover operator, one of the genetic operators, is applied in every generation, individuals that resemble each other very much are formed in the next generations. The purpose of the mutation operator is to provide hereditary diversity by replacing genes when the similarity increases too much. In the literature, the mutation operator has been reported to be 0.1% and 0.15% [18]. In this study, multipoint mutation was used and the rate of application was determined as 0.1%.

## 2.6 Penalty Function and Fitness Value

In the Genetic Algorithm, it is an important problem that constraints can be expressed in problem-solving steps. To overcome this problem, the penalty function definition was used. Penalty function checks whether the restrictions are met. Each gene that does not comply with the problem constraint is multiplied by a predefined penalty coefficient. In this way, the penalty value of each chromosome is obtained. In this problem, the penalty values of the chromosomes are determined as follows:

$$\sum_{a=0}^{x} \sum_{b=0}^{y} \sum_{c=0}^{z} P_a * C_{bc} \tag{1}$$

y = number of chromosomes, z = number of genes, x = number of constraints, $C_{bc}$ = b. The gene in the chromosome with penalty value c. Gene, $P_a$ = a. penalty value of the constraint.

The fitness function is the function that determines how suitable a chromosome is for the solution. After the initial population is generated in a completely random fashion, the penalty value is determined by using the penalty function. Then, with the help of these penalty values, fitness values are calculated using the fitness function. In this study, fitness value is calculated as:

$$f = \frac{1}{1 - \sum_{a=0}^{x} \sum_{b=0}^{y} \sum_{c=0}^{z} P_a \, X \, C_{bc}} \tag{2}$$

# 3 Assessment of the Results

What kind of outcomes the Genetic Algorithm provides in different situations of conference scheduling problems are examined in this section. In this section, different values of "Day Number", "Session Number", and "Hall Number" were assigned to the conference scheduling problem to which the Genetic Algorithm was applied, and the results of and changes in these different experiments were shown. Graphs and tables created

**Table 2.** Scheduling of the conference with 200 papers.

| 2 days | 4 session | | | 6 session | | |
|---|---|---|---|---|---|---|
| | 4 saloon | 5 saloon | 6 saloon | 4 saloon | 5 saloon | 6 saloon |
| First penalty value | XXX | 17896 | 20130.9 | 16307.0 | 26376.4 | 65566.4 |
| Last penalty value | XXX | 2745 | 2269 | 2343 | 2099 | 1672 |
| Fitness value | XXX | 0.9932 | 0.9967 | 0.9966 | 0.9971 | 0.9973 |
| Number of iterations | XXXX | 300000 | 300000 | 300000 | 300000 | 300000 |

| 3 days | 4 session | | | 6 session | | |
|---|---|---|---|---|---|---|
| | 4 saloon | 5 saloon | 6 saloon | 4 saloon | 5 saloon | 6 saloon |
| First penalty value | 32172.2 | 60428.8 | 122774.7 | 78856.9 | 192882.4 | 382878.2 |
| Last penalty value | 3939 | 2132 | 1557 | 1579 | 1236 | 1084 |
| Fitness value | 0.9942 | 0.9961 | 0.9971 | 0.9967 | 0.9972 | 0.9973 |
| Number of iterations | 300000 | 300000 | 300000 | 300000 | 300000 | 300000 |

**Table 3.** Scheduling of a conference with 300 papers.

| 2 days | 4 session | | | 6 session | | |
|---|---|---|---|---|---|---|
| | 4 saloon | 5 saloon | 6 saloon | 4 saloon | 5 saloon | 6 saloon |
| First penalty value | XXX | XXX | XXX | XXX | 25887.0 | 52137.9 |
| Last penalty value | XXX | XXX | XXX | XXX | 8149 | 5769 |
| Fitness value | XXX | XXX | XXX | XXX | 0.9948 | 0.9971 |
| Number of Iterations | XXX | XXX | XXX | XXX | 300000 | 300000 |

| 3 days | 4 session | | | 6 session | | |
|---|---|---|---|---|---|---|
| | 4 saloon | 5 saloon | 6 saloon | 4 saloon | 5 saloon | 6 saloon |
| First penalty value | XXX | 52291.6 | 82760.2 | 66528.1 | 232700.8 | 552858.6 |
| Last penalty value | XXX | 8136 | 5941 | 5886 | 4641 | 3848 |
| Fitness value | XXX | 0.9961 | 0.9952 | 0.9972 | 0.9968 | 0.9973 |
| Number of Iterations | XXX | 300000 | 300000 | 300000 | 300000 | 300000 |

based on the changes in the Penalty Value and Fitness Value and the results obtained are presented.

Table 2 shows the results of the problem solving based on different numbers of the day, session and hall in a situation where there are 200 conference papers. According to the results shown in this table, it was observed that as the number of days, halls and sessions of the conference increased, the Penalty Function decreased and the Fitness Function was more converged to 1.

\* XXX: Indicates that the scheduling could not be performed because of the insufficiency of the number of days, sessions, and halls of this conference.

Table 3 shows the results of the problem solving based on different numbers of the day, session and hall in a situation where there are 300 conference papers. According to the results shown in this table, it was observed that as the number of days, halls and sessions of the conference increased, the Penalty Function decreased and the Fitness Function was more converged to 1. Of course, having 300 papers that need to be scheduled increases the difficulty of the problem. Considering the results of Table 2 and Table 3, it is seen that the final penalty value of the 300-paper-problem was higher than the final penalty value of the 200-paper-problem (Fig. 4 and Fig. 5).

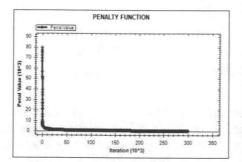
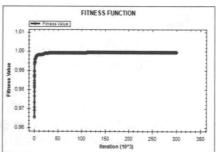

**Fig. 4.** 200 papers 2 days, 6 sessions, 6 halls; penalty function and fitness function.

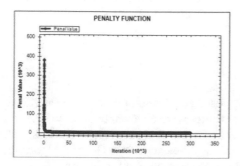
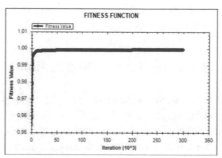

**Fig. 5.** 200 papers 3 days, 6 sessions, 6 hall; penalty function and fitness function.

When Fig. 6 and Fig. 7 are compared, it is seen that the change in the penalty value in Fig. 7 is quite high. In addition, Fig. 7 converged to a lower Penalty value and reached a higher fitness value.

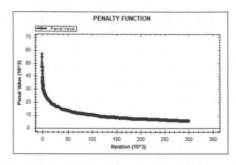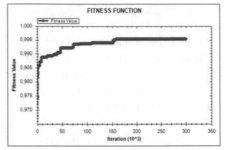

**Fig. 6.** 300 papers 2 days, 6 sessions, 4 hall; penalty functions and fitness function.

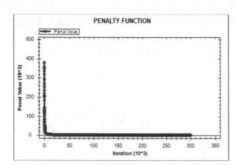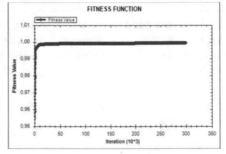

**Fig. 7.** 300 papers 3 days, 6 sessions, 6 hall; penalty function and fitness function.

The developed software was run with the determined values and the changes in the Penalty Value and Fitness Value in the process of three hundred thousand iterations were observed in the graphs. A conference with 200 papers was scheduled for 2 days, 6 sessions, 6 halls and 3 days, 6 sessions, 6 halls. It was observed that the lowest penalty value was obtained when the scheduling was for 3 days, 6 sessions and 6 halls. A conference with 300 papers was scheduled for 3 days, 6 sessions, 4 halls and 3 days, 6 sessions, 6 halls. The lowest penalty value was obtained when the scheduling was for 3 days, 6 sessions and 6 halls.

## 4   Conclusions and Suggestions

In this study, the Genetic Algorithm is proposed for the solution of the conference scheduling problem. Data of 2 different conferences were obtained in order to perform scheduling. The conference scheduling problem was handled based on parameters such as the number of days, sessions and halls, and the efficiency of the Genetic Algorithm was examined. In this study, the Genetic Algorithm is proposed for the solution of the conference scheduling problem. In order to make the conference scheduling in the most optimal way, it was aimed to achieve scheduling through which papers on the same topic will be presented in each session and the authors will not be allowed to be in a concurrent session. Data of 2 different conferences were obtained in order to perform scheduling. The conference scheduling problem was handled based on parameters such as the number of days, sessions and halls, and the efficiency of the Genetic Algorithm was examined. The efficiency of the Genetic Algorithm was examined by changing the number of days, sessions, and halls required for scheduling the conference. The changes in the Penalty Value and Fitness Value were observed by changing the number of days, sessions and halls. While examining the effect of the number of days on the scheduling, the number of sessions and halls were kept constant. While examining the effect of the number of sessions on the scheduling, the number of days and halls were kept constant. While examining the effect of the number of halls on the scheduling, the number of days and sessions were kept constant. The lowest penalty value was obtained when the scheduling was for 3 days, 6 sessions and 6 halls. In addition, when the graphs are examined, it is seen that the lowest penalty values were obtained when the number of days, sessions and halls were high. The situations where the changes in the penalty value are the most frequent and the most progress was made in the process of reaching the optimum result were the situations where the number of days, sessions, and halls were the highest.

# References

1. Pezzella, F., Morganti, G., Ciaschetti, G.: A genetic algorithm for the flexible job-shop scheduling problem. Comput. Oper. Res. **35**(10), 3202–3212 (2008)
2. Aickelin, U., Dowsland, K.A.: An indirect genetic algorithm for a nurse-scheduling problem. Comput. Oper. Res. **31**(5), 761–778 (2004)
3. Zanda, S., Zuddas, P., Seatzu, C.: Long term nurse scheduling via a decision support system based on linear integer programming: a case study at the University Hospital in Cagliari. Comput. Ind. Eng. **126**, 337–347 (2018)
4. Wang, Y.Z.: Using genetic algorithm methods to solve course scheduling problems. Expert Syst. Appl. **25**(1), 39–50 (2003)
5. Hossain, S.I., Akhand, M.A.H., Shuvo, M.I.R., Siddique, N., Adeli, H.: Optimization of university course scheduling problem using particle swarm optimization with selective search. Expert Syst. Appl. **127**, 9–24 (2019)
6. Kadhim, S.J., Al-Aubidy, K.M.: Design and evaluation of a fuzzy-based CPU scheduling algorithm. In: Das, V.V., et al. (eds.) BAIP 2010. CCIS, vol. 70, pp. 45–52. Springer, Heidelberg (2010). https://doi.org/10.1007/978-3-642-12214-9_9
7. Zilberstein, S., Koehler, J., Koenig, S.: The fourteenth international conference on automated planning and scheduling (ICAPS-04). AI Mag. **25**(4), 101–101 (2004)
8. Aktay, S.: Sempozyum Web Sitesi Yönetim Paneli İncelemesi. Sosyal ve Beşeri Bilimler Araştırmaları Dergisi **18**(39), 133–146 (2017)
9. Andlauer, O., Obradors-Tarragó, C., Holt, C., Moussaoui, D.: How to organize and manage scientific meetings. Psychiatry Pract.: Edu. Experience Expertise **10**(50), 97 (2016)
10. Nicholls, M.G.: A small-to-medium-sized conference scheduling heuristic incorporating presenter and limited attendee preferences. J. Oper. Res. Soc. **58**(3), 301–308 (2007)
11. Tanaka, M., Mori, Y., Bargiela, A.: Granulation of keywords into sessions for timetabling conferences. In: Proceedings of Soft Computing and Intelligent Systems (SCIS 2002), pp. 1–5 (2002)
12. Vangerven, B., Ficker, A.M., Goossens, D.R., Passchyn, W., Spieksma, F.C., Woeginger, G.J.: Conference scheduling—a personalized approach. Omega **81**, 38–47 (2018)
13. Zulkipli, F., Ibrahim, H., Benjamin, A.M.: Optimization capacity planning problem on conference scheduling. In: 2013 IEEE Business Engineering and Industrial Applications Colloquium (BEIAC), pp. 911–915. IEEE, April 2013

14. Kim, J., et al.: Cobi: A community-informed conference scheduling tool. In: Proceedings of the 26th Annual ACM Symposium on User İnterface Software and Technology, pp. 173–182, October 2013
15. Ibrahim, H., Ramli, R., Hassan, M.H.: Combinatorial design for a conference: constructing a balanced three-parallel session schedule. J. Discrete Math. Sci. Crypt. **11**(3), 305–317 (2008)
16. Stidsen, T., Pisinger, D., Vigo, D.: Scheduling EURO-k conferences. Eur. J. Oper. Res. **270**(3), 1138–1147 (2018)
17. Eglese, R.W., Rand, G.K.: Conference seminar timetabling. J. Oper. Res. Soc. **38**(7), 591–598 (1987)
18. Timucin, T., Birogul, S.: Implementation of operating room scheduling with genetic algorithm and the ımportance of repair operator. In: 2018 2nd International Symposium on Multidisciplinary Studies and Innovative Technologies (ISMSIT), pp. 1–6. IEEE (2018)
19. Timuçin, T., Biroğul, S.: Effect the number of reservations on ımplementation of operating room scheduling with genetic algorithm. In: Hemanth, D., Kose, U. (eds.) ICAIAME 2019. LNDECT, vol. 43, pp. 252–265. Springer, Cham (2020). https://doi.org/10.1007/978-3-030-36178-5_20
20. Holland, J.H.: Adaptation in Natural and Artificial Systems: An İntroductory Analysis with Applications to Biology, Control, and Artificial İntelligence. MIT Press, Cambridge (1992)

# Analyze Performance of Embedded Systems with Machine Learning Algorithms

Mevlüt Ersoy[1] and Uğur Şansal[2(✉)]

[1] Suleyman Demirel University, Isparta, Turkey
mevlutersoy@sdu.edu.tr
[2] Iskenderun Technical University, Hatay, Turkey
ugur.sansal@iste.edu.tr

**Abstract.** This article aims to analyze the performance of embedded systems using various machine learning algorithms on embedded systems. For this, data set were analyzed using Python programming language and related libraries. Using different machine learning algorithms, it has been useful in terms of giving us an idea about how these cards are sufficient for solving deep learning and artificial intelligence problems or how close they will be to the desired performances. Raspberry pi and jatson nano embedded system boards were used for performance measurement and they were compared with notebook performance.

**Keywords:** Machine learning · Deep learning · Artificial intelligence · Embedded systems

## 1 Introduction

When it was first started to produce embedded systems, it served as a single purpose. They are used in cell phones, air defense systems, medical devices and many more. Overtime, as well as single-purpose production, multi-purpose operations, programmable embedded system boards began to be produced [1].

It is widely used in research and development activities due to its low power consumption and very small size. Embedded systems can communicate with existing computers as well as communicate and control resources simultaneously. Examples of this are access to embedded systems over the internet, connecting security cameras and monitoring them. Many applications have been written on embedded systems with the support of programming languages such as Java and Python. It has been an important factor in choosing these applications with embedded library support [2].

Today, applications in the field of artificial intelligence are progressing rapidly. Artificial intelligence applications on embedded systems have started to gain importance. With the advantage of both its size and its increasing features, applications have been realized with embedded system boards. Machine learning algorithms, one of the subfields of artificial intelligence technology, give us ideas in terms of implementing this in embedded systems and making an idea about their performance.

J. Hemanth et al. (Eds.): ICAIAME 2020, LNDECT 76, pp. 231–236, 2021.
https://doi.org/10.1007/978-3-030-79357-9_23

In this application, after passing the necessary data set, machine learning algorithms such as K-Nearest Neighbor (KNN), Support Vector Machines (SVM), Normal Bayesian Classifier (NBC), Random Trees (RT), Trees Gradient Boosted (GBT) and and Artificial Using Neural Networks (ANN), performances between Raspberry pi 3B+, Jetson Nano and a notebook are evaluated [5]. The software is also implemented with the Python programming language and library via Ubuntu operating system.

## 2   Used Algorithms

Obtained data should form clustering and classification models with unattended and supervised machine learning algorithms.

### 2.1   Unsupervised Learning Algortihm

k-Means Algorithm;
The subject used in the k-means algorithm consists of k sets. Algorithm is similar to each other when the data is taken into the same set of features. The similarity variable in the algorithm is determined by the distance between the data. The smaller the distance, the higher the similarity [3]. In the cluster analysis, the similarities of the data are determined by determining the distances between them. Different distance measurement methods are used for this purpose [3] (Fig. 1).

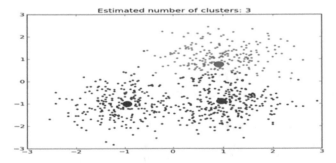

**Fig. 1.**  $K = 3$ for K-means algorithm

### 2.2   Supervised Learning Algorithms

In supervised learning, the training data we send to the algorithm includes the solutions we want, called labels. The most typical supervised learning task is classification. Another typical example of supervised learning is regression. In the regression problem, we try to estimate a numerical value from the available data. Some regression algorithms can also be used for classification problems. Logistic regression is one of the commonly used techniques for classification. In our performance analysis application, the following 3 classification algorithms are used [4].

- k-Nearest Neighbors (KNN)
- Naive Bayes Classification (NBC)
- Decision Tree Classification (CART)

k-Nearest Neighbors-KNN;

The KNN algorithm is one of the easiest algorithms in Machine learning. K means the number of neighbors. K data is classified by 3 closest data around K data. To find the distance between two points [4] (Fig. 2 and 3).

$$d(\mathbf{p}, \mathbf{q}) = d(\mathbf{q}, \mathbf{p}) = \sqrt{(q_1 - p_1)^2 + (q_2 - p_2)^2 + \cdots + (q_n - p_n)^2}$$

$$= \sqrt{\sum_{i=1}^{n}(q_i - p_i)^2}.$$

**Fig. 2.** Finding distance between 2 points

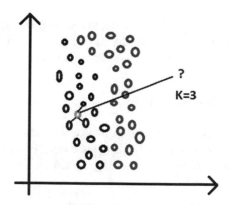

**Fig. 3.** K classification with KNN algorithm

The K value is counted from the highest class.

Naive Bayes Classification (NBC);

The work of the Naive Bayes Algorithm calculates the probability of all states for an element and finds the one with the highest value and classifies it accordingly.

$$P(A/B) = \frac{P(A/B) * P(A)}{P(B)}$$

A represents the given target and B represents the properties. Naive Bayes gives the product of all conditional probabilities [5].

In Bayesian network algorithm, graphical models are used to determine the relationships between variables in various events [6]. In Bayesian networks, each variable is represented by nodes. They represent the interaction between variables with conditional probabilities and graphically. This graph is used as a user interface and visually reflects the interaction between the variables of the modeled problem [7] (Fig. 4).

Decision Tree Classification (CART);

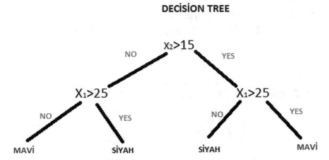

**Fig. 4.** Decision Tree Structure

Decision Tree Algorithm is a classification method consisting of decision and tree structure of leaf nodes. This algorithm allows the data set to be divided into small pieces. A decision node can contain segments with one or more branches [8].

## 3   Application of Algorithms

First, the installation of raspberry pi, Jatson nano and Anaconda Spyder for notebook has been completed. By using Python language and library as programming language, both visualization of data and machine learning algorithms have been performed.

### 3.1   Importing Data and Visualizing the Data Set

(See Fig. 5).

| Index | pelvic_incidence | pelvic_tilt numeric | lumbar_lordosis_angle | sacral_slope |
|---|---|---|---|---|
| 0 | 0.355688 | 0.5199 | 0.22918 | 0.250857 |
| 1 | 0.124501 | 0.296783 | 0.0985783 | 0.144629 |
| 2 | 0.411666 | 0.513932 | 0.322995 | 0.307661 |
| 3 | 0.416151 | 0.557414 | 0.27126 | 0.289436 |
| 4 | 0.227272 | 0.289479 | 0.128129 | 0.247022 |
| 5 | 0.136009 | 0.365744 | 0.0995589 | 0.119943 |
| 6 | 0.26315 | 0.400439 | 0.207316 | 0.22396 |

**Fig. 5.** Normalization

### 3.2  Train and Test Data

Separated data using the Sklearn library were train and tested. 80% of the data was trained and 20% was separated as test.

from sklearn.model_selection import train_test_split.

x_train,x_test,y_train,y_test=train_test_split(x,y,test_size=0.20,random_state=1).

### 3.3  Classification/Prediction

According to the visualization process of our data set, appropriate classification procedures were selected and ML algorithms were applied to the data set. As a result, predicted values were noted.

## 4  Experiments

A dataset consisting of 6 columns with 310 samples was used in our dataset. 248 of them were used as train and 62 of them were used as test (Table 1).

**Table 1.**  Systems features

| Notebook and embedded system | Parameters | | |
| --- | --- | --- | --- |
| | Core/Bit | Cpu | Memory |
| Asus Notebook | Core 4 X64 | İntel İ3 1.80 ghz | 6 GB |
| Raspberry Pi 3B+ | Core 4 X64 | ARM cortex-A53 | 1 GB |
| Jatson Nano | Core 4 X64 | ARM® cortex-A57 | 4 GB |

**Table 2.**  Elapsed time for machine learning algorithms of systems

| Notebook and embedded system | Classification/Prediction delay for algorithms | | | |
| --- | --- | --- | --- | --- |
| | KNN ms | Naive Bayes ms | Decision Tree ms | k-Means ms |
| Asus Notebook | 44 ms | 52 ms | 46 ms | 203 ms |
| Raspberry Pi 3B+ | 1554 ms | 1468 ms | 1378 ms | 2622 ms |
| Jatson Nano Developer | 1383 ms | 1749 ms | 1090 ms | 1562 ms |

## 5 Conclusion

As a result of the data obtained in Table 2, our application, which was created by using a data set at a level that does not exhaust the performance of the systems, has quite sufficient features and performances for such algorithms. The next applications should be done to cover the data that occurs in the image sets. Because GPU support has different features between the cards. Especially, Jatson Nano stands out with its GPU feature. I believe that by trying these, we will make even better judgments about the performance of the embedded system boards.

## References

1. Şansal, U., Temurtaş, F.: Gömülü Sistem Tabanlı Ev Otomasyon ve Güvenlik Sistemi Tasarımı. Bozok Üniversitesi Fen Bilimleri Enstitüsü, Mekatronik Müh. Anabilim Dalı Yüksek Lisans Tezi, Yozgat
2. İlhan, H.O.: Gömülü Sistemlerde Kablosuz Haberleşme Protokolü ile Görüntü ve Video Aktarımı. Yalova Üniversitesi, Fen Bilimleri Enstitüsü, Bilgisayar Müh. Anabilim Dalı Yüksek Lisans Tezi, Yalova
3. https://medium.com/@k.ulgen90/python-ile-k%C3%BCmeleme-algoritmalar%C4%B1-mak ine-%C3%B6%C4%9Frenimi-b%C3%B6l%C3%BCm-8-8204ffa702f2. Accessed 12 Jan 2020
4. https://makineogrenimi.wordpress.com/2017/05/25/makine-ogrenmesi-teknikleri/. Accessed 12 Jan 2020
5. Zidek, K., Pitel, J., Hošovský, A.: Machine learning algorithms ımplementation into embedded systems with web application user ınterface. In: INES 2017 21st International Conference on Intelligent Engineering Systems, Larnaca, Cyprus, 20–23 October 2017 (2017)
6. Dünder, E., Cengiz, M.A., Koç, H., Savaş, N.: Bayesci ağlarda risk analizi: Bankacılık sektörü üzerine bir uygulama. Erzincan Üniversitesi Sosyal Bilimler Enstitüsü Dergisi 6(1), 1–14 (2013)
7. Bayer, H., Çoban, T.: "Web İstatistiklerinde Makine Öğrenmesi Algoritmaları ile Kritik Parametre Tespiti" (Electronic Journal of Vocational Colleges- Special Issue: The Latest Trends in Engineering), 28 Aralık 2015
8. https://erdincuzun.com/makine_ogrenmcsi/decision-tree-karar-agaci-id3-algoritmasi-classi fication-siniflama/. Accessed 11 Jan 2020

# Classification Performance Evaluation on Diagnosis of Breast Cancer

M. Sinan Basarslan[1(✉)] and F. Kayaalp[2]

[1] Doğuş University, İstanbul, Turkey
mbasarslan@dogus.edu.tr
[2] Düzce University, Düzce, Turkey

**Abstract.** Cancer, which has many different types such as breast, pleural, and leukemia, is one of the common health problems of today. Most of them cause pain and treatment processes are so challenging. Medical authorities report that the diagnosis of cancer at early stages has a positive effect on medical treatments' success. On the way to design a computer-aided cancer diagnosis system about breast cancer to support the decisions of doctors about medical treatments, classification performances of six classifiers are investigated in this study. For this purpose, classifier models have been created with machine learning algorithms such as Support Vector Machine (SVM), Naïve Bayes (NB) and Random Forest (RF); and deep learning algorithms such as Recurrent Neural Network (RNN), Gated Recurrent Unit (GRU), and Long Short-Term Memory Network (LSTM). Two different open-source breast cancer datasets were used namely Wisconsin and Coimbra on experiments. Accuracy (Acc), Sensitivity (Sens), Precision (Pre), F-measure (F) were used as performance criteria. As a result of the tests, Acc values between 85% and 98% were obtained in the Coimbra breast cancer dataset; while Acc values were obtained between 92% and 97% in Wisconsin. According to the results obtained, it is seen that deep learning algorithms (RNN, GRU, and LSTM) are more successful than machine learning algorithms (SVM NB and RF). Among the deep learning algorithms, LSTM is more successful.

**Keywords:** Breast cancer · Deep learning · Machine learning

## 1 Introduction

Cancer is one of the diseases that affect people's health. There are many types of cancer. One of these women is cancer that negatively affects their lives.

Cancer disease is the uncontrolled division of cells as a result of DNA damage in cells [1]. While the World Health Organization reports the number of people with cancer as 18 million in 2018, it claims that this number will be 40 million in 2040 [2].

There are many types of cancer such as skin and lung. Early diagnosis is important for cancer patients to live more comfortably and overcome this disease with medical treatments.

Data analysis aims to help decision-makers to make accurate and precise decisions, which includes many areas from data collection to data and to achieve the end result. Data

J. Hemanth et al. (Eds.): ICAIAME 2020, LNDECT 76, pp. 237–245, 2021.
https://doi.org/10.1007/978-3-030-79357-9_24

analysis in areas related to people, especially health, makes life easier for people. The aim of this study is to evaluate doctors who will use this model by creating models with a computer breast cancer solution system and finally with widely used class algorithms to help make important decisions about their patients. Bit dataset Breast cancer is Coimbra. The classification algorithms used in the study are SVM, RF, RNN, and LSTM.

It is seen that cancer studies have been carried out with deep learning algorithms in recent years. Osman et al. offered a new method for breast cancer diagnosis using deep learning algorithms [3].

Patricio et al. used Decision Tree (DT), SVM, and Logistic Regression(LR) for prediction breast cancer. Their Acc performances in classifier models, respectively; It is 87% in SVM, 91% in DT and 83% in LR [4].

Ahmad et al., found SVM algorithm to achieve 95.7% Acc rate when predicting the occurrence of breast cancer, using records from 1189 patients collected in the Iranian Center for Breast Cancer [5].

RF gave the best performance in models established by Li and Chen on Coimbra and Wisconsin breast cancer datasets with various machine learning algorithms [6].

In a study on Parkinson's disease-related to early diagnosis, feature selection with Aspect Importance and Recursive Feature Removal, and classification models were created with Regression Trees, Artificial Neural Networks (ANN), and SVM. He gave a model created with SVM and Recursive Feature Elimination with an accuracy of 93.84% [7].

A study has been conducted on the early diagnosis of mesothelioma, a cancer type, with machine learning algorithms. In this study, Gradient Enhanced Trees yielded 100% ANN in both Acc and Sens [8].

In the second section, the dataset used in the study, the classification algorithms and the performance criteria of the models created with these algorithms are explained. In the third section, the experimental results of our study are shown, and in the fourth section, the conclusion and suggestions of the study are included.

## 2   Method

This section explains the datasets, classification algorithms, and performance criteria used in the study.

### 2.1   Datasets

Information on the two different breast cancer datasets studied are given in Table 1 (Breast Cancer Coimbra) [4] and Table 2 (Breast Cancer Wisconsin) [9].

**Table 1.** Attributes of breast cancer coimbra dataset.

| Attribute name | Attribute datatype |
|---|---|
| Adiponectin | Numerical |
| Age | Numerical |
| BMI | Numerical |
| Glucose | Numerical |
| HOMA | Numerical |
| Insulin | Numerical |
| Leptin | Numerical |
| MCP-1 | Numerical |
| Resistin | Numerical |
| Class | Categorical |

Ten real-valued properties are calculated for each cell nucleus, these values are shown in Table 2 as 1–10.

**Table 2.** Attributes of breast cancer wisconsin dataset.

| Attributes | Attribute datatype | Domain |
|---|---|---|
| Sample code number | Numerical | id number |
| Clump Thickness | Numerical | 1–10 |
| Uniformity of Cell Size | Numerical | 1–10 |
| Uniformity of Cell Shape | Numerical | 1–10 |
| Marginal Adhension | Numerical | 1–10 |
| Single Epithelial Cell Size | Numerical | 1–10 |
| Bare Nucleoli | Numerical | 1–10 |
| Mitoses | Numerical | 1–10 |
| Class | Categorical | 2: bening, 4: malign |

## 2.2 Classification Algorithms

Six different classification algorithms, SVM, RF, NB, RNN, GRU and LSTM, were used in the breast cancer Coimbra dataset.

### 2.2.1   Support Vector Machines Algorithm

It is a machine learning method that sets a boundary between any point in training data and the other furthest point [10]. One feature of SVM is inherent risk minimization in statistical learning theory [11].

### 2.2.2   Random Forest Algorithm

It was introduced by Leo Breiman. Researchers were able to get good results in both regression and classification without the need for hyper-parameters. It is an algorithm that aims to train in random trees instead of training in a single tree [12].

### 2.2.3   Naïve Bayes Algorithm

This algorithm is based on the statistical Bayes theorem. It is a classification algorithm that shows the relationship between independent properties and the target property [13].

### 2.2.4   Recurrent Neural Network

RNN has begun to be used again thanks to the recently developing technology. RNN is a neural network model developed to learn existing patterns by taking advantage of sequential information [14]. In RNN, each output is determined by the continuous processing of the same task on each instance of the array. Output is determined by previous calculations. In RNN, the length of time steps is determined by the size of the input. Let $x_t$ denote the input to the architecture at time step t and let $s_t$ denote the hidden state at time step t. The current hidden state $s_t$ has been computed as given by (1), by taking the current input and the hidden state for the former timestamp [15, 16]:

$$s_t = f(U_{x_i} + W_{s_{t-1}})  \tag{1}$$

In (1), $f$ denotes the activation function, which is usually taken as, tanh or ReLU function. U and W corresponds to the weights that are shared across the time [16].

### 2.2.5   Gated Recurrent Neural Network

The GRU is a type of RNN similar to LSTM. LSTMs and GRUs have been developed as a solution to the vanishing gradient problem in RNN. The difference from LSTM is that it has a reset and an update gate. The update gate acts similarly to the forget and entry gate of an LSTM, deciding what information to keep, discard, or add new information [17].

### 2.2.6   Long Short-Term Memory

LSTM is an artificial RNN architecture used in deep learning. Unlike standard feed forward neural networks, LSTM has feedback links. The LSTM unit consists of a cell, an entrance gate, an exit gate and a forget gate, and three types of gates. Based on the open-closed state of the gates, the cells determine what information should be preserved and when to access the units. The LSTM transition was performed as in the time in the equations below [16]:

$$i_t = \sigma(W^{(i)}x_t + (U^{(f)}h_{t-1} + b^{(i)})$$ (2)

$$ft = \sigma(W^{(f)}x_t + U^{(o)}h_{t-1} + b^{(f)})$$ (3)

$$ot = \sigma(W^{(o)}x_t + U^{(i)}h_{t-1} + b^{(o)})$$ (4)

$$ut = tanh(W^{(u)}x_t + U^{(u)}h_{t-1} + b^{(u)})$$ (5)

$$c_t = it^{\circ}u_t + ct^{\circ}c_{t-1}$$ (6)

$$h_t = ot^{\circ}tanh(ct)$$ (7)

where $x_t$ denotes the input vector to the LSTM unit, $ft$ denotes the activation vector for the forget gate, it denotes the activation vector for the input gate, $ot$ corresponds to the activation vector for the output gate, $h_t$ denotes the hidden state vector and $c_t$ denotes the cell state vector. In this model, we correspond to the weight matrices and b corresponds to bias vector parameters.

## 2.3  Performance Criteria

The models created in the study were evaluated by using confusion matrix. In the binary classification, we can numerically see the examples classified correctly and incorrectly according to the class with the confusion matrix. In the confusion matrix $(t_p)$ shows true positive, $(f_p)$ shows false positive, $(t_n)$ shows true negative, $(f_n)$ shows false negative, numbers [18].

Acc, Sens, Pre, F performance criteria obtained from the confusion matrix are given between Eq. (8) and Eq. (11).

$$Acc = \frac{t_p + t_n}{t_p + t_n + f_p + f_n}$$ (8)

$$Sens = \frac{t_p}{t_p + f_n}$$ (9)

$$Pre = \frac{t_p}{t_p + f_p}$$ (10)

$$F = \frac{2 * Pre * Sens}{Pre + Sens}$$ (11)

## 3   Experimental Results

Performance evaluations regarding the results of the models created on Coimbra and Wisconsin datasets with SVM, NB, RF, LSTM GRU, and RNN classifier algorithms are given in Fig. 1, Fig. 2, Fig. 3, and Fig. 4. In these figures, Acc, Sens, Pre, and F performance criteria are used for model performance evaluation. The training and test set seperations were made by using 10 fold cross-validation.

As its seen in Fig. 1 that SVM performed better than other classifier models NB, and RF.

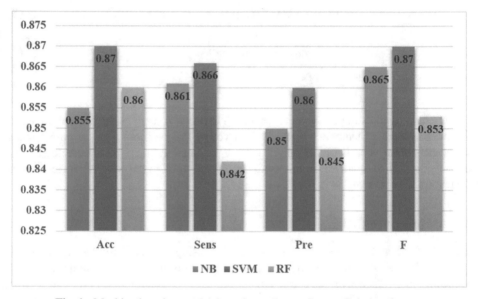

**Fig. 1.** Machine learning models' results on Breast Cancer Coimbra Dataset

As it seen in Fig. 2 that LSTM performed better than other deep learning models RNN, and GRU.

As its seen in Fig. 3 that NB performed better than other classifier models SVM, and RF.

As it seen in Fig. 4 that LSTM performed better than other deep learning models RNN, and GRU.

Accuracy value of NB is 0.01 higher than accuracy of SVM on Wisconsin (Fig. 1); whereas accuracy of SVM is 0.02 higher than accuracy of NB on Coimbra (Fig. 3). These results show that SVM and NB algorithms have close performances on both datasets.

Among the Deep Learning algorithms, the classifier models created with LSTM performed better than RNN and GRU, whereas RNN and GRU performed fairly close (Fig. 4).

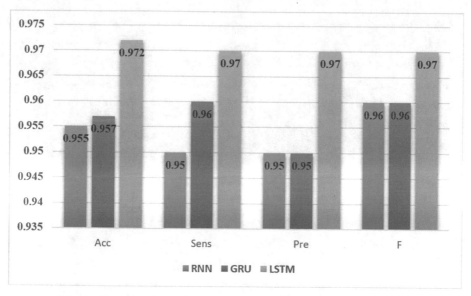

**Fig. 2.** Deep learning models' results on Breast Cancer Coimbra Dataset

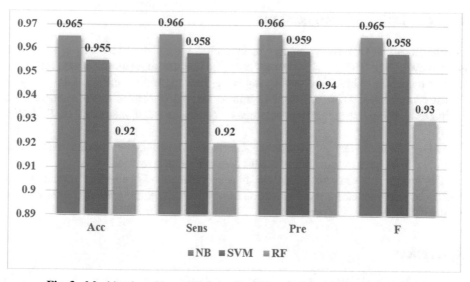

**Fig. 3.** Machine learning models' results on Breast Cancer Wisconsin Dataset

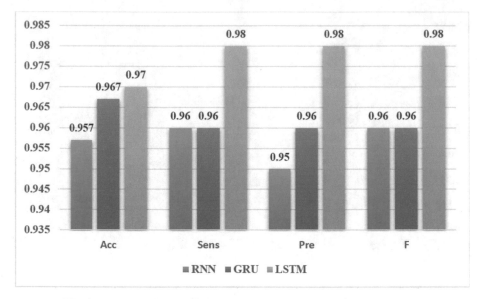

**Fig. 4.** Deep learning models' results on Breast Cancer Wisconsin Dataset

## 4   Conclusion and Suggestions

In order to investigate and compare the classification performances of deep learning algorithms (RNN, GRU, LSTM) versus machine learning algorithms (SVM, NB, RF), various models have been designed by applying these algorithms on two datasets about Breast cancer called Wisconsin and Coimbra. The performances of the models were evaluated by using five parameters obtained by confusion matrix namely Acc, Sens Pre, and F.

In the light of the results given above, LSTM has shown the best performance and it is advised to be used in this kind of health decision support system.

The results also show that deep learning algorithms have better performance than traditional machine learning classification algorithms. Especially, the more samples in the datasets, the better the results.

## References

1. Siegel, R.L., Miller, K.D., Jemal, A.: Cancer statistics. CA: A Cancer J. Clin. **70**(1), 7–30 (2020)
2. Bray, F., Ferlay, J., Soerjomataram, I., et al.: Global cancer statistics 2018: GLOBOCAN estimates of incidence and mortality worldwide for 36 cancers in 185 countries. CA: A Cancer J. Clin. **68**(6), 394–424 (2018)
3. Osman, A.H., Aljahdali, H.M.A.: An effective of ensemble boosting learning method for breast cancer virtual screening using neural network model. IEEE Access **8**, 39165–39174 (2020)
4. Patrício, M., et al.: Using Resistin, glucose, age and BMI to predict the presence of breast cancer. BMC Cancer **18**(1) (2018). 1(8)

5. Ahmad, L.G., Eshlaghy, A.T.: Using three machine learning techniques for predicting breast cancer recurrence. J. Heal. Med. Inform. **04**(02) (2013). 2(4)
6. Li, Y., Chen, Z.: Performance evaluation of machine learning methods for breast cancer prediction **7**(4), 212–216 (2018)
7. Senturk, Z.K.: Early diagnosis of Parkinson's disease using machine learning algorithms. Med. Hypotheses **138**, 109603 (2020)
8. Senturk, Z.K., Çekiç, N.: A machine learning based early diagnosis system fo r mesothelioma disease. Düzce Üniversitesi Bilim ve Teknoloji Dergisi **8**(2), 1604–1611(2020)
9. Blake, C.L., Merz, C.J.: UCI repository of machine learning databases (1998)
10. Elmas, Ç.: Yapay Sinir Ağları (Kuram, Mimari, Eğitim, Uygulama). Seçkin Yayıncılık, Ankara (2003)
11. Song, O., Hu, W., Xıe, W.: Robust support vector machine with bullet hole ımage classification. IEEE
12. Pattanayak, S.: Pro Deep Learning with TensorFlow. Apress, New York (2017). ISBN 978-1-4842-3095-4, 153-278
13. Senturk, Z.K., Kara, R.: Breast cancer diagnosis via data mining: performance analysis of seven different algorithms. Comput. Sci. Eng. **4**(1), 35 (2014)
14. Sherstinsky, A.: Fundamentals of recurrent neural network (RNN) and long short-term memory (lstm) network. Physica D: Nonlinear Phenomena, 132306 (2020)
15. Hochreiter, S., Schmidhuber, J.: Long shortterm memory. Neural Comput. **9**(8), 1735–1780 (1997)
16. Rojas-Barahona, L.M.: Deep learning for sentiment analysis. Lang. Linguist. Compass **10**(12), 701–719 (2016)
17. Onan, A.: Sentiment analysis on massive open online course evaluations: a text mining and deep learning approach. Computer Applications in Engineering Education (2020)
18. Basarslan, M.S., Kayaalp, F.: A hybrid classification example in the diagnosis of skin disease with cryotherapy and immunotherapy treatment. In: 2018 2nd International Symposium on Multidisciplinary Studies and Innovative Technologies (ISMSIT), pp. 1–5. IEEE, October 2018

# Effective Factor Detection in Crowdfunding Systems

Ercan Atagün[⊠], Tuba Karagül Yıldız, Tunahan Timuçin, Hakan Gündüz,
and Hacer Bayıroğlu

Department of Computer Engineering, Duzce University, Duzce, Turkey
{ercanatagun,tubakaragul,tunahantimucin,hakangunduz,
hacerbayiroglu}@duzce.edu.tr

**Abstract.** Crowdfunding is a platform that brings together entrepreneurs without their own budgets and investors who do not have to be related to each other. Crowdfunding, a model that enables the entrepreneurs to get the support they need to realize their projects, is increasingly used in the world and in our country as well. This study aims to open new horizons for entrepreneurs about what they should pay attention to get support for their projects. The crowdfunding system is important to prevent wasting of potential projects that may be beneficial to the society when they are implemented due to lack of resources. The fact that the project is not expressed in a way that attracts the attention of investors makes it difficult to use the system effectively. Our study answers the question of how to use this system more effectively by applying machine learning methods on a public data set. Also, the outcome of the study is expected to benefit both entrepreneurs and investors.

**Keywords:** Crowdfunding · Machine learning · Data mining · Funding success factors

## 1  Introduction

Crowdfunding consists of crowd and fund concepts. Crowd means "a group of people gathered and gathered together". A fund is called money to finance an institution or to spend it when necessary to ensure that a particular business is run. Crowdfunding is that the masses come together and collect funds by giving small amounts [1]. It is a method of financing that has been common in many parts of the world especially in recent years. Crowdfunding systems allow entrepreneurs to find finances in virtual environments due to lack of capital, lack of institutional identity, and lack of efficiency from existing financial resources [2]. With the developing technology, businesses and markets have become increasingly competitive. This situation caused the creation of financing instruments such as securities, forfaiting, factoring, leasing. These methods have benefited greatly. However, most of the time, they are systems that provide certain collateral to large-scale enterprises and have no contribution to small businesses. Therefore, new financing methods such as venture capital investment fund, venture capital fund, angel

J. Hemanth et al. (Eds.): ICAIAME 2020, LNDECT 76, pp. 246–255, 2021.
https://doi.org/10.1007/978-3-030-79357-9_25

investor have emerged for small businesses and individual entrepreneurs [3]. However, while these methods provided support for a very small number of entrepreneurs, they were insufficient to support a large number of entrepreneurs [4]. Angel investors and entrepreneurs are investors who provide funding support during the initial establishment phase. However, the low number of angel investors and the inability of angel investors to reach entrepreneurs is another problem. Such problems have led to the emergence of a different financing model, including small investors, with the financial support needed by small businesses and individual entrepreneurs. Competition is increasing in a globalized world. This situation causes the companies in the markets to be inadequate while looking for the necessary financial resources for their investments, and especially the newly established enterprises, start-ups are financially difficult. The difficulties experienced by the startups while procuring loans from banks, the insufficient available financial resources have made new financing models mandatory. It contributes to the formation of the necessary funds by providing small amounts of support to investors who do not have investment experience with crowdfunding [3]. Crowdfunding systems appeal to a wide range of audiences, as they are open to anyone with and without investment experience. This facilitates fund raising.

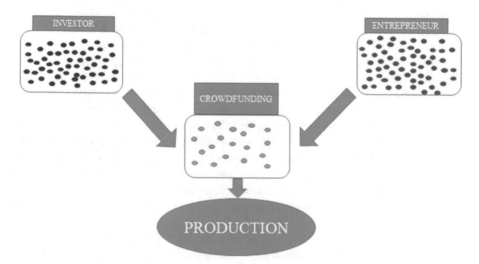

**Fig. 1.** Crowdfunding system basic components.

Figure 1 shows the main elements of crowdfunding systems. Crowdfunding sys-tems fall into categories such as reward-oriented, donation-oriented, and capital-oriented. Kiva are donation-oriented, Indeegogo reward-oriented, Microventures capi-tal-focused plat-forms. Kickstarter is an award-based platform. Crowdfunding systems became especially popular in developed countries after the 2008 economic crisis affected small businesses and entrepreneurs did not find financing with traditional methods [3].

Nowadays, it is easier for donors and individual investors to raise money through the internet [5]. In crowdfunding systems, entrepreneurs can call on large audiences.

When making this call, entrepreneurs can publish their own projects and visual materials such as pictures and videos over the internet with the help of a website [6]. In addition to collecting funds on these platforms, entrepreneurs also pursue objectives such as reducing entrepreneurial risk, seeing the salability of the project, analyzing the share and analysis of projects in the market, and getting feedback from users about their projects [1, 6–8].

Crowdfunding platforms are organizations that seek profit maximization. They are considered successful if a project reaches the targeted fund amount during the time it is published on crowdfunding platforms. Platforms receive commissions from these projects that are considered successful. This commission is between 7–15% [1]. Crowdfunding systems are categorized for different purposes in the literature. In these, the most basic category form is divided into categories such as donation, sponsorship, reward, lending, inkind fund, investment model, presale, income sharing and mixed models [9]. The most known crowdfunding system worldwide is the Kick-starter platform [10]. For this reason, the crowdfunding system discussed in this study was Kickstarter. Apart from this platform, there are many platforms such as Indie-gogo, CrowdFunder, GoFundMe. The basic motivation of those who support the projects in crowdfunding systems has been investigated and the effectiveness of parameters such as reward hunting, solidarity, social responsibility and emotional has been examined. In another definition, crowdfunding can be expressed as the entre-preneurs supplying their needs with the contribution of a large audience other than themselves [11]. Depending on the geographical location, some entrepreneurs may find it difficult to find investments. Entrepreneurs may find it difficult to find funding support due to the political or diplomatic relations of countries, cities, or regions. While such problems prevent entrepreneurs from finding donations and investments, crowdfunding systems eliminate such problems. Time, quality, and cost constitute the basic elements of a project. The most critical level for entrepreneurs is the cost element. If the cost here is considered as economy, financial efficiency plays an important role both in completing the project and turning it into a product, and in completing the prototype and starting mass production. Cost also affects quality and time money meters. Thanks to crowdfunding systems, economic resources are obtained and shortening the conversion times of projects to projects and completing projects with higher quality.

In Turkey in a study evaluating the massive funding system, the average of a supporttiles for the project will be completed in Turkey it was noted that 215 Turkish Liras support and successful projects in average emphasized that 11463 per fund fell [1]. Thies et al. focused on whether successful crowdfunding campaigns are more likely to receive venture capital than failed campaigns [12]. Jensen et al. focused on what could have caused the crowdfunding campaign to fail. In his study, he analyzed 144 technology-focused campaigns successfully funded on Kickstarter. These campaigns are technology-oriented and are physical equipment pre-ordered by their supporters [13]. Buttice studied the completion and productization of projects after crowdfunding system and how it affects product quality. One of the results of his study stated that there is a positive correlation from the productization of first-time entrepreneurs in crowdfunding systems. It was stated that those who supported the projects contributed to the design of the products, especially those who collected funds for the first time provided an advantage when converting their projects into products [14]. Zhang researched the motivation of the supporters who supported the projects of entrepreneurs in crowdfunding systems. It has been demonstrated that the supporters' own feelings of individual satisfaction are more effective than the feeling of helping a project [15]. In his study, Li researched 1,058 successful Kickstarter pro-jects and emphasized that the number of Facebook shares increased when the project approached the financing threshold, and that Facebook shares increased significantly in the last days of the project [16]. The Colombo study revealed that the information provided in crowdfunding systems affected the possibility of receiving capital after the project's deadlines and what information professional investors focused on crowd funding [17].

## 2   Material and Method

The data set was taken from the Github platform [18]. The data set contains 174297 project data belonging to different categories. There are 18 different attributes in the data initially (Table 1).

**Table 1.**  Features in the dataset

| Feature name | Data type | Description |
| --- | --- | --- |
| id | Numeric | The unique number of a Project |
| parent_category | Nominal | The main category of a Project |
| funding_duration_days | Numeric | Project funding period |
| pre_funding_duration_days | Numeric | Pre-fund duration of the Project |
| launch_month | Numeric | Upload date of the Project |
| deadline_month | Numeric | Project deadline |
| Country | Nominal | Country |
| location.type | Nominal | Location type |
| staff_pick | Nominal | Personnel preference |
| is_starrable | Nominal | Is it stable |
| creator_has_slug | Nominal | Project person information |
| blurb_length | Numeric | White paper length |
| blurb_word_count | Numeric | Number of promotional articles |
| name_length | Numeric | Project name length |
| name_word_count | Numeric | Project name word count |
| usd_goal | Numeric | Intended money amount |
| usd_pledged | Numeric | The amount of money promised |
| backers_count | Numeric | Number of supporters |
| state | Label | Status (Class label) |

The feature named id has been removed from the data set since it has no effect on the results. Thus, the number of attributes was 17. Then backer_count was removed with usd_pledge. Also, the newly created attribute named average has been added. Thus, there were 16 different attributes in the data set. In order to determine the effective factor in the data preprocessing stage, a new attribute named average has been created. This average attribute is obtained by dividing the usd_pledge attribute into the backer_count. This refers to the average donation of a donor with the new attribute. In addition, 80% of the data in the data set is reserved for training and 20% for testing.

The methods used in this study are determined as rule-based decision tree algorithms. These methods are Hoeffding Tree, J48, RandomTree, Decision Stump and OneR algorithms.

The Hoeffding Tree algorithm, which works by reading and processing the data in the data set at most once, is an effective method in big data [19]. The algorithm uses a statistical value called Hoeffding limit when branching. This value is shown in Eq. (1) [20]. In this way, it is also determined how to distinguish the nodes while creating the decision tree.

$$\epsilon = \sqrt{\frac{R^2 \ln\frac{1}{\delta}}{2n}} \tag{1}$$

R: Range of real-valued random variable

n: number of independent observations of the variable

J48 decision tree algorithm is one of the most frequently used methods in the literature. It is a method that can work with both continuous and discrete data. It is a matter that makes the method successful when it is chosen according to the qualities with high knowledge gain while branching [21].

As can be understood from the name, the RandomTree algorithm is an algorithm that creates a tree by selecting it from a random tree set. It is one of the RandomTree supervised algorithms. This algorithm can be used for classification and regression. RandomTree selects a random dataset to create a decision tree. All nodes in a normal tree must be arranged to indicate the best divider between variables. Each tree makes a classification and a tree stands out for this class. In this way, all trees in the dataset are called forests and the most prominent tree determines the classification [22].

A single node tree is created with the Decision Stump algorithm. In this way, only one attribute is considered in the classification process. It is an easy method to understand and interpret as the tree formed with Decision Stump has a single knot and leaf knots directly attached to it [23].

The OneR algorithm is a classification algorithm based on creating a rule tree through an attribute. A rule is created by selecting the data group containing the largest number of independent data, and this rule is expanded, and operations are performed [24].

## 3 Application

The classification process is the determination of the class of the data whose class label is unknown by creating a feature set with the help of the features in the data. In this study,

J48, HoeffdingTree and RandomTree methods were used for classification procedures. In addition, Decision Stump and OneR algorithms, one of the rule-based algorithms, were used to determine the most important attribute.

When the Average attribute was removed from the data set, the accuracy value decreased in the HoeffdingTree, J48 and RandomTree algorithms. Removing the average attribute caused the accuracy values to get the lowest value. In this respect, this attribute, which is not initially in the dataset, but which is produced later, is very important for the classification process.

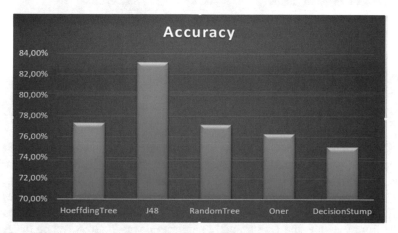

**Fig. 2.** Classification was performed using HoeffdingTree, J48 and RandomTree, Oner, Decision Stump and accuracy values were obtained.

Figure 2, HoeffdingTree, J48 and RandomTree, Oner, Decision Stump algorithms are classified with 17 attributes and their accuracy values are shown. In this study, effective factor was determined, and the classification success was increased. There are 16 different attributes in the data set.

**Table 2.** Accuracy values after feature extraction

| Feature name | HoeffdingTree | J48 | RandomTree |
|---|---|---|---|
| parent_category | 82.02 | 83.06 | 83.78 |
| funding_duration_days | 78.27 | 83.23 | 77.15 |
| pre_funding_duration_days | 78.86 | 83.31 | 76.20 |
| launch_month | 77.54 | 83.27 | 77.40 |
| deadline_month | 77.58 | 83.26 | 76.93 |
| Country | 77.72 | 82.96 | 76.42 |
| location.type | 77.41 | 83.27 | 77.41 |

(continued)

**Table 2.** (*continued*)

| Feature name | HoeffdingTree | J48 | RandomTree |
|---|---|---|---|
| staff_pick | 76.76 | 81.86 | 74.83 |
| is_starrable | 77.36 | 83.19 | 77.28 |
| creator_has_slug | 77.3 | 83.37 | 76.47 |
| blurb_length | 79.66 | 83.28 | 77.62 |
| blurb_word_count | 76.98 | 83.32 | 76.40 |
| name_length | 77.36 | 83.33 | 76.92 |
| name_word_count | 77.54 | 83.29 | 76.94 |
| usd_goal numeric | 78.42 | 78.52 | 72.19 |
| average | 70.29 | 71.06 | 65.50 |
| state | Class label | - | - |

As shown in Table 2, when the average attribute was removed from the data set, the accuracy values decreased in the HoeffdingTree, J48 and RandomTree algorithms. Removal of the average attribute caused the accuracy values to get the lowest. In this respect, it is very important for the classification process with this attribute, which is not initially in the dataset, but which is produced later. When the Average attribute was removed from the data, the HoeffdingTree algorithm was 70.26%. When the Average attribute was removed from the data, the J48 algorithm was 71.06%. When the Average attribute was removed from the data, the HoeffdingTree algorithm was 65.50% (Fig. 3).

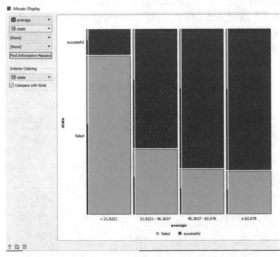

**Fig. 3.** The distribution of the values of the average attribute in the data is shown. Red regions represent successful projects and blue regions represent projects that failed [26].

For the study, classification process was also done with Decision Stump and OneR algorithm. Accuracy value was determined as 75.04% with Decision Stump. It is determined as 76.29% with OneR algorithm. These two algorithms are based on classi-fication with a single attribute [25]. These algorithms selected the newly created aver-age attribute from the 16 attributes to classify. It is important here that the newly added attribute be present as an active attribute. It is noteworthy that the RandomTree algorithm uses the average attribute as the top node in the decision tree drawing (Figs. 4 and 5).

```
=== Classifier model (full training set) ===

Decision Stump

Classifications

average <= 15.5 : failed
average > 15.5 : successful
average is missing : successful

Class distributions

average <= 15.5
failed   successful
0.9311351199663253        0.06886488003367476
average > 15.5
failed   successful
0.2959714990408331        0.7040285009591669
average is missing
failed   successful
0.4258306224433008        0.5741693775566992
```

**Fig. 4.** The average attribute came to the fore with rule-based algorithms. The Decision Stump algorithm states that 15 is critical for the average [20].

```
=== Classifier model (full training set) ===

average:
        < 18.5  -> failed
        < 19.5  -> successful
        < 20.5  -> failed
        < 24.5  -> successful
        < 25.5  -> failed
        < 99.5  -> successful
        < 100.5 -> failed
        < 267.5 -> successful
        < 268.5 -> failed
        < 324.5 -> successful
        < 325.5 -> failed
        < 339.5 -> successful
        < 340.5 -> failed
        < 376.5 -> successful
        < 377.5 -> failed
        < 382.5 -> successful
        < 383.5 -> failed
        < 397.5 -> successful
        < 399.5 -> failed
        < 402.5 -> successful
```

**Fig. 5.** The average attribute came to the fore with rule-based algorithms. The OneR algorithm revealed critical values for its average attribute [20].

Average is also the first node of the tree in RandomTree and J48 methods, among other decision tree algorithms. In this respect, the average attribute stands out as the most basic part of the classification.

## 4 Conclusion and Suggestions

With this study, it is aimed to contribute to the development of production by helping entrepreneurs to explain their projects better in the entrepreneurship ecosystem. With crowdfunding systems, money can be collected from all over the world. This will also contribute to the gross national products of the countries of the project developers. With crowdfunding systems, the number of entrepreneurs is increasing. Depending on the increasing number of entrepreneurs, the number of innovative projects will also increase. This will lead to the realization of many projects that will improve the life quality of people and other living things. With crowdfunding systems, the need for material resources, although not sufficient, will lead to the realization of high-quality ideas. In this study, the classification of successful and unsuccessful projects in crowdfunding systems has been investigated. In this study, classification was made using different machine learning algorithms. The classification has been made by producing a new attribute from the existing data. In this respect, the average value stands out for the project owners looking for funds in crowdfunding systems. With the average value, it can be envisaged that the project will obtain the targeted fund. While one of the important results of this study is determining reward mechanisms, awards can be given by referring to the average number of donations recommended in this study. For example, the project team can send a prototype of the product to someone who donates an average of $65, send a tshirt, or try one of the other types of rewards. For future work, different feature selection operations can be applied for this study. In addition, classification success rates can be increased with different classification algorithms.

## References

1. Çubukçu, C.: Kitlesel fonlama: Türkiye'deki kitlesel fonlama platformları üzerinden bir değerlendirme. Girişimcilik ve İnovasyon Yönetimi Dergisi **6**(2), 155–172 (2017)
2. Sakarya, Ş, Bezirgan, E.: Kitlesel Fonlama Platformları: Türkiye Ve Yurtdışı Karşılaştırması. Düzce Üniversitesi Sosyal Bilimler Enstitüsü Dergisi **8**(2), 18–33 (2018)
3. Çağlar, M.T.: Yeni Nesil Alternatif Finansman Yöntemi Olarak Kitlesel Fonlama: Dünya Ve Türkiye'deki Uygulamalarının Etkinliği. Uygulamalı Sosyal Bilimler Dergisi **3**(2), 18–34
4. Fisk, R.P., Patrício, L., Ordanini, A., Miceli, L., Pizzetti, M., Parasuraman, A.: Crowd-funding: transforming customers into investors through innovative service platforms. J. Serv. Manage. (2011)
5. Best, J., Neiss, S., Swart, R., Lambkin, A.: Scaling innovation: Crowdfunding's potential for the developing world. Information for Development Program (infoDev), The World Bank (2013)
6. Atsan, N., Erdoğan, E.O.: Girişimciler için Alternatif bir Finansman Yöntemi: Kitlesel Fonlama Crowdfunding. Eskişehir Osmangazi Üniversitesi İktisadi ve İdari Bilimler Dergisi **10**(1), 297–320 (2015)

7. Gerber, E.M., Hui, J.S., Kuo, P.Y.: Crowdfunding: why people are motivated to post and fund projects on crowdfunding platforms. In: Proceedings of the İnternational Workshop on Design, İnfluence, and Social Technologies: Techniques, İmpacts And Ethics, vol. 2, no. 11, p. 10. Northwestern University Evanston, IL, February 2012
8. Gulati, S.: Crowdfunding: A Kick Starter for Startups. TD Economics (2014)
9. De Buysere, K., Gajda, O., Kleverlaan, R., Marom, D., Klaes, M.: A framework for European crowdfunding (2012)
10. Ercan, S.: Türkiye'de Kitlesel Fonlama: Destekçilerin Motivasyonu. Ekonomi, İşletme ve Yönetim Dergisi **1**(1), 1–9 (2017)
11. Howe, J.: The rise of crowdsourcing. Wired Mag. **14**(6), 1–4 (2006)
12. Thies, F., Huber, A., Bock, C., Benlian, A., Kraus, S.: Following the crowd—does crowdfunding affect venture capitalists' selection of entrepreneurial ventures? J. Small Bus. Manage. **57**(4), 1378–1398 (2019)
13. Jensen, L.S., Özkil, A.G.: Identifying challenges in crowdfunded product development: a review of Kickstarter projects. Des. Sci. **4** (2018)
14. Butticè, V., Noonan, D.: Active backers, product commercialisation and product quality after a crowdfunding campaign: a comparison between first-time and repeated entrepreneurs. Int. Small Bus. J. **38**(2), 111–134 (2020)
15. Zhang, H., Chen, W.: Backer motivation in crowdfunding new product ideas: Is it about you or is it about me? J. Prod. Innov. Manag. **36**(2), 241–262 (2019)
16. Li, G., Wang, J.: Threshold effects on backer motivations in reward-based crowdfunding. J. Manage. Inf. Syst. **36**(2), 546–573 (2019)
17. Colombo, M.G., Shafi, K.: Receiving external equity following successfully crowdfunded technological projects: an informational mechanism. Small Bus. Econ. **56**(4), 1507–1529 (2019). https://doi.org/10.1007/s11187-019-00259-1
18. Github Kickstarter Project. https://github.com/rajatj9/Kickstarter-projects. Accessed 05 May 2020
19. Domingos, P., Hulten, G.: Mining high-speed data streams. In: Proceedings of the sixth ACM SIGKDD International Conference on Knowledge Discovery and Data Mining, pp. 71–80 (2000)
20. Waikato Environment for Knowledge Analysis Version 3.8.4, 1999-2019, The University of Waikato, Hamilton, New Zealand
21. Quinlan, J.R.: C4.5: Programs for Machine Learning, Morgan Kauffman, San Francisco, CA, USA (1993)
22. Yıldırım, P.: Pattern classification with imbalanced and multiclass data for the pre-diction of albendazole adverse event outcomes. Procedia Comput. Sci. **83**, 1013–1018 (2016)
23. Onan, A.: Şirket İflaslarının Tahminlenmesinde Karar Ağacı Algoritmalarının Karşılaştırmalı Başarım Analizi. Bilişim Teknolojileri Dergisi **8**(1) (2015)
24. Atagün, E., Timuçin, T., Bayburtlu, Ö., Çelen, S., Yücedağ, İ.: Data Mining Based Decision Support System for Users via Computer Hardware Components (2018)
25. Wu, X., et al.: Top 10 algorithms in data mining. Knowl. Inf. Syst. **14**(1), 1–37 (2008)
26. Data Mining. https://orangedatamining.com/. Accessed 05 May 2020

# Predictive Analysis of the Cryptocurrencies' Movement Direction Using Machine Learning Methods

Tunahan Timuçin[✉], Hacer Bayiroğlu, Hakan Gündüz, Tuba Karagül Yildiz, and Ercan Atagün

Department of Computer Engineering, Duzce University, Duzce, Turkey
{tunahantimucin,hacerbayiroglu,hakangunduz,tubakaragul, ercanatagun}@duzce.edu.tr

**Abstract.** Cryptocurrencies are among the most interesting financial instruments of recent years. Unlike the classical understanding that money exists as a means of change from one hand to another, this digital economy has begun to attract people's attention. The most popular currency emerging from this concept of cryptocurrency is "Bitcoin". As the popularity of Bitcoin started to increase since 2016, the number of academic studies on cryptocurrencies has increased in parallel. In light of these developments, our study proposes predictive models of price change directions of high market value cryptocurrencies. Bitcoin, Ethereum and Litecoin were selected as cryptocurrencies and daily opening, closing, high and low prices of these currencies were collected from financial websites. Preprocessing was performed on the collected data to create input vectors. These vectors were given regression algorithms which are Multiple Linear, Polynomial, Support Vector, Decision Tree and Random Forest Regression. As evaluation metrics, R-square Method ($R^2$) and Root Mean Square Error (RMSE) were chosen. After doing experiments with different parameter settings, it was found out that the chosen machine learning models showed satisfactory performances in predicting the directions of then mentioned cryptocurrencies.

**Keywords:** Cryptocurrency · Regression · Data mining · Prediction

## 1 Introduction

One of the best and most popular technologies of recent years is the blockchain, the platform on which cryptocurrencies are located. Bitcoin, the first cryptocurrency in 2009, emerged with blockchain technology and is still a popular cryptocurrency. Understanding cryptocurrency is possible with the answer to the question of what is cryptology. In addition to cryptology being an encryption science, it can also be ex-plained as the process of sending the encrypted data according to a particular system to the receiver with the help of secure environments and re-decrypting this data in the receiver. The definition of the cryptocurrency can be made as a blockchain-based, digitally executed

© The Author(s), under exclusive license to Springer Nature Switzerland AG 2021
J. Hemanth et al. (Eds.): ICAIAME 2020, LNDECT 76, pp. 256–264, 2021.
https://doi.org/10.1007/978-3-030-79357-9_26

virtual currency that can be used in exchange processes known as cryptology science to ensure its security.

After the technology boomed in the 90s, there have been many entrepreneurs and software developers who want to release their digital currency. Systems such as Flooz, Beenz and DigiCash, which are among the developed systems, fell into internal con-flicts during these attempts and failed when economic difficulties were added to this situation.

Although these projects were unsuccessful, they led to the emergence of a system development idea that was not controlled by a particular center. Bitcoin is the first digital coin to come out of the current cryptocurrencies. When 2009 began, it emerged with a study published by Satoshi Nakamoto. It is not clear whether Nakatomo is a single person or a group of workers. However, thanks to this open source software, the world has moved towards a new order. This network, created with the help of blockchain, is similar to platforms used to share files that are encrypted end-to-end.

However, while applying these systems, this problem in the payment system must be solved, since the existing money is likely to be spent twice. In order to solve this problem, the traditionally practiced and best known method is the service of a bank-like central tool in which transactions and the entire database are recorded. Thanks to this intermediary system, all payment transactions are recorded. However, the method in which this central intermediary service exists also causes reliability problems, as it will keep the entire economic balance under control.

In the light of this information, considering this equilibrium, transactions must be performed by a system that includes distributed systems and servers, such as the bitcoin network.

These transactions can now be done via blockchain. Blockchain network is a reli-able system since it is a decentralized system that needs to be approved separately on each server. Each transaction on the blockchain network is public and everyone can see it. All transfers are made by specifying the wallet address of the sender open to the public and the wallet address of the recipient and the amount to be sent. First of all, this transfer is verified by the sender and ends with the system confirming the accuracy of the transaction. All these verification operations are done by cryptocurrency miners, known as the decoders of passwords. Once a transaction is approved by miners, it is added to the network and is irreversible. The most important risk for cryptocurrencies is a 51% attack on the blockchain network. 51% attack is the risk that more than half of the servers in the blockchain network can be intercepted by cyber-attack. Because the blockchain network needs to exceed half of the servers for verification.

The production of cryptocurrencies takes place through a distributed system, such as a blockchain network. In other words, it is a user-based system. However, each cryp-tocurrency has a production capacity. For this reason, whenever the production of crypto money increases, the ciphers that need to be deciphered while producing a new one of this crypto money become difficult.

Cryptocurrencies also have the problem of being legal, varying from country to country. In some countries, the state of legality remains uncertain, while in some countries cryptocurrency use and even trade and permits are allowed. In some coun-tries, the use of cryptocurrencies is completely prohibited and restricted.

Advantages of Cryptocurrencies;

- There is no time and space limitation.
- Transfers are very fast.
- It is very easy to store and carry.
- It can be processed in all regions of the world.
- As in printed money, there is no expense and expense.
- It is not affected by the economic or sociopolitical situation of the countries.
- There is no lower limit for payments.
- It eliminates the transaction costs of the companies.
- It is not possible to freeze or confiscate the account.

The concept of crypto money, which we have come across in recent years, is a new alternative exchange tool for everyone. Unlike the classical understanding that money continues to exist as a tool for hand-to-hand exchange, this digital economy has started to attract people's attention rapidly. Although this new structure has many risks, the high returns that cryptocurrencies have recently made lead investors seeking new investment products to cryptocurrencies, thereby pushing their goals to achieve a higher return. Countries have not yet made an effective regulation on these digital currencies. However, in some Asian countries (South Korea, Thailand, Vietnam and the Philippines), four major stock market companies were opened [1]. Interest in this is increasing not only in these countries but also in the world. Considering that many transactions in the world are realized over the dollar, the main purpose has gained even more interest with the pursuit of a safer transaction and profit.

As a digital currency, forecasting price changes of cryptocurrencies, which are traded not only for online payments but also as an investment tool in the market, will facilitate future investment and payment decisions. For these reasons, studies on the prediction of future prices of cryptocurrencies are increasing. In this section, the re-sults obtained from the literature review on the price prediction of crypto money are shared. In these studies, there are also researchers working on single cryptocurrencies. One of them, Gencer, has developed artificial intelligence algorithms to estimate bitcoin price by taking into account the monthly average prices of the last three years [2]. Chen et al. argued that Ethereum, which has the second highest market value, supports much more functionality than Bitcoin. In their study, they made an estimate of the ether price, arguing that Ether's price predictability could cause it to differ significantly from Bitcoin [3]. One of the studies focused on predicting the future price of big cryptocurrencies is the work of Prashanth and Das. Six major currencies with larger market value were selected for price estimation, and the future price was estimated with the LSTM model by collecting open, high, low and recent price historical data of currencies [4]. Indicating that the ability to predict price volatility will facilitate future investment and payment decisions, Wang and Chen stated, however, that cryptocurrencies are extremely difficult to predict in the price movement. For this purpose, they have created a model with various machine learning and deep learning algorithms based on the transaction data of three different markets and the number and content of user comments and responses from online forums [5]. In addition to making direct price estimates, there are also researchers working on predicting whether the next move of selected cryptocurrencies, such as Bai et al., will be up or down [6]. Another field of study is to analyze the comments on social media and try to make predictions. Lamon et al. analyzed the ability of news and social media data to predict price fluctuations for three cryptocurrencies (bitcoin, litecoin and ethereum) [7]. Abraham et al. provided a method to estimate changes in Bitcoin and Ethereum prices using Twitter data and Google Trends data [8]. Valencia et al. In their work, they proposed using common machine learning tools and available social media data to estimate the price movement of Bitcoin, Ethereum, Ripple and Litecoin crypto market movements using Twitter and market data [9]. Another work that contributed to the guesswork on social media comments belongs to Bremmer. In his study, Bremmer aimed to investigate the relationship between the comments made on the social media website Reddit and the daily prices or market direction of cryptocurrencies Ethereum, Litecoin and Ripple [10].

Financial price balloons were previously linked to the outbreak-like spread of an investment idea; Phillips and Gorse, who say that such balloons are common in cryptocurrency prices, have sought to predict such balloons for a number of cryptocur-rencies, using a hidden Markov model previously used to detect flu outbreaks, based on the behavior of new online social media indicators [11]. In addition to all these areas, there are also studies indicating that methods such as artificial intelligence, machine learning, and time series should be used to estimate crypto money prices more accurately. Desev et al. Analyzed the effectiveness of crypto money markets by applying econometric models to different short-term investment horizons [12]. As an example of machine learning, Akyildirim et al. analyzed the predictability of the twelve cryptocurrencies used at frequencies at the daily and minute level using machine learning classification algorithms. While making this estimate; support vector machines, including logistic regression, arterial neural networks, and random forests. They stated that the average classification accuracy of the four algorithms is consistently above the 50% threshold for all cryptocurrencies and all time periods indicating that there is a certain degree of predictability in the cryptocurrency markets [13]. František has written a master's thesis on researching the efficiency and implementation of machine learning algorithms to estimate the future price of a cryptocurrency [14]. Yao et al. Contributed to this field, suggesting a new method to estimate the price of the cryptocurrency, taking into account various factors such as market value, volume, circulating supply and maximum supply based on deep learning techniques [15]. Möller's thesis work is to create a technique that includes Neural Networks to try to predict the trends of cryptocurrencies like Bitcoin and Ethereum, and in this context, this includes building, designing and training a deep learning Neural Network based on historical trade data for cryptocurrencies [16]. Explaining the creation of a short-term forecasting model using the machine learning approach of cryptocurrency prices, Derbentsev et al. Adapted the modified model of the Binary Auto Regression Tree (BART) from the standard models of the regression trees and data from the time series [17]. Spilak, who wanted to offer a Neural Network framework to provide a deep machine learning solution to the price prediction problem, has contributed to this area [18].

## 2 Material and Method

Regression is an analysis method used to measure the relationship between a dependent variable and one or more independent variables. It allows making predictions about unknown future events from known findings [19]. The dependent variable is the predicted variable in the regression model. The independent variable is the descriptive variable in the regression model; used to estimate the value of the dependent variable. There may be a linear or nonlinear relationship between variables. In this section, the definitions of the regression algorithms used in the study are given.

## 2.1 Linear Regression

Simple Linear Regression tries to model the relationship between two continuous variables by developing a linear equation to estimate the value of a dependent variable based on the value of an independent variable. If there is more than one independent variable, it is named Multiple Linear Regression. So model is built on a linear relationship regardless of data size and sometimes linearity acceptance is an error. Simple Linear Regression Equality:

$$y = c + b * x + e \tag{1}$$

Multiple Linear Regression Equality:

$$y = c + b_1 * x_1 + b_2 * x_2 + \cdots \ldots \ldots + b_i * x_i + e \tag{2}$$

## 2.2 Polynomial Regression

It is the analysis method used when the connection between the variables is not linear but in the form of a parabolic curve. In this model polynomial degree is very important for success. Multiple decision trees are created and merged into those trees to be a consistent structure.

$$y = c + b_1 * x_1 + b_2 * x_2^2 + \cdots \ldots \ldots \ldots + b_i * x_i^i + e \tag{3}$$

## 2.3 Support Vector Regression

Support Vector algorithm is used for classification at first although it is used in regression. The purpose of the support vector regression is to ensure that the range we will draw is within the maximum point. The points cut by these maximum gaps are called support points [20]. In this model data must scale and choosing a right kernel is very important.

## 2.4 Decision Tree Regression

It divides the independent variables according to the information gain [21]. In this model data does not need to scale and it can be use both linear and nonlinear prob-lems. When asked for a value from this range during estimation, the answer tells the mean in this range which learned during training process. Therefore, it produces the same results for the desired predictions in a certain range and in small data sets are highly likely to be overfitting.

## 2.5 Random Forest Regression

The Random Forest Regression method divides the data into small pieces and applies different decision trees to all of them, predict with the average of the results obtained from them. In this model data does not need to scale and it can be use both linear and nonlinear problems like Decision Tree Regression. And problems such as overfit-ting and achieving fixed results are low.

# 3 Statistical Evaluation Criteria of Fitting Model

## 3.1 $R^2$ Method

This method explains that how much of the variation in y can be explained by the regression curve. So we can understand what percentage of the total change in the dependent variable is explained by the changes in the independent variables. $R^2$ coefficient is a number ranging from 0 to 1. 1 means that the result of perfect algorithm.

$$R^2 = 1 - \frac{\sum (y_i - \hat{y}_i)^2}{\sum (y_i - \bar{y}_i)^2} \tag{4}$$

## 3.2 Root Mean Square Error (RMSE)

Root Mean Square Error (RMSE) measures how much error there is between predicted values and observed. Its value ranges from 0 to $\infty$. The smaller an RMSE value means that the closer predicted and observed values are we have.

Therefore, it produces the same results for the desired predictions in a certain range and in small data sets are highly likely to be overfitting.

$$R = \sqrt{\frac{1}{N} \sum_{i=1}^{N} (y_i - \hat{y}_i)^2} \tag{5}$$

## 3.3 Data Set

The data set used in this study was taken from the website https://www.cryptodatado wnload.com/. Bitcoin, Ethereum and Litecoin, which are the 3 highest currencies traded in the highest density, have been obtained hourly since 2017, with a data set of more than 23 thousand data each. This data set has been made suitable for use in prediction algorithms based on machine learning (Fig. 1).

**Fig. 1.** The cryptocurrencies discussed in this study are show [22].

## 4    Results and Discussion

In this study, we use regression algorithms which are Decision Tree, Polynomial, Multiple Linear, Support Vector, and Random Forest Regression, to forecast 3 cryptocurrencies. For each cryptocurrency we have more than 23 thousands data obtained per hour so it is like a time series. In our study first we apply standard scaler to the data then as a preprocessing we used a window to organize the data. This window is containing 5 h of data, and by sliding the window one hour every time, we estimate the data at the 6th hour. Thus, each time 5 values are our property and 6th value becomes the data that should be estimated. Then we split data set %70 training and %30 test sets. We compare each models' successes with two statistical methods, R-square Method ($R^2$) and Root Mean Square Error (RMSE). We measure each models training success with $R^2$ and test success with RMSE. We wanted to see how it would get results on the data set it had never seen. For all cryptocurrencies after train-ing phase, $R^2$ results tell us Decision Tree Regression Model is the best with highest score. But after test phase RMSE results tell us Linear Regression and Decision Tree Regression Model can fit our problem with a low score. RMSE and $R^2$ scores are given in tables for all models and for all cryptocurrencies.

**Table 1.** Methods and values

| Methods | $R^2$ values | RMSE values |
|---|---|---|
| Linear Regression | 0.99891 | 0.03075 |
| Polynomial Regression | 0.99895 | 0.03201 |
| Support Vector Regression | 0.99806 | 0.04263 |
| Decision Tree Regression | 0.99997 | 0.03075 |
| Random Forest Regression | 0.99972 | 0.03418 |

Table 1 shows the values obtained as a result of the regression analysis for Litecoin data.

**Table 2.** Methods and values

| Methods | $R^2$ values | RMSE values |
|---|---|---|
| Linear Regression | 0.99924 | 0.02594 |
| Polynomial Regression | 0.99926 | 0.02640 |
| Support Vector Regression | 0.99848 | 0.03619 |
| Decision Tree Regression | 0.99999 | 0.02594 |
| Random Forest Regression | 0.99982 | 0.03037 |

Table 2 shows the values obtained as a result of the regression analysis for Ethereum data.

**Table 3.** Methods and values

| Methods | $R^2$ values | RMSE values |
|---|---|---|
| Linear Regression | 0.99875 | 0.03380 |
| Polynomial Regression | 0.99877 | 0.03392 |
| Support Vector Regression | 0.99849 | 0.03850 |
| Decision Tree Regression | 0.99999 | 0.03380 |
| Random Forest Regression | 0.99972 | 0.03783 |

Table 3 shows the values obtained as a result of the regression analysis for Bitcoin data.

## 5 Conclusions and Suggestions

In this study, we use regression algorithms which are Decision Tree, Polynomial, Multiple Linear, Support Vector, and Random Forest Regression, to forecast 3 cryptocurrencies. For each cryptocurrency we have more than 23 thousands data obtained per hour so it is like a time series. In our study first we apply standard scaler to the data then as a preprocessing we used a window to organize the data. This window is containing 5 h of data, and by sliding the window one hour every time, we estimate the data at the 6th hour. Thus, each time 5 values are our property and 6th value becomes the data that should be estimated. Then we split data set %70 training and %30 test sets. We compare each models' successes with two statistical methods, R-square Method ($R^2$) and Root Mean Square Error (RMSE). We measure each mod-els training success with $R^2$ and test success with RMSE. We wanted to see how it would get results on the data set it had never seen. For all cryptocurrencies after train-ing phase, $R^2$ results tell us Decision Tree Regression Model is the best with highest score. But after test phase RMSE results tell us Linear Regression and Decision Tree Regression Model can fit our problem with a low score. RMSE and $R^2$ scores are given in tables for all models and for all cryptocurrencies.

# References

1. Kesebir, M., Günceler, B.: Kripto Para Birimlerinin Parlak Geleceği. Igdir University J.Soc. Sci. (17) (2019)
2. Çelik, U.: Veri işleme grup yöntemi türünde sinir ağları algoritması ile bıtcoın fiyat tahmini. Scientific Committee, 1322 (2019)
3. Chen, M., Narwal, N., Schultz, M.: Predicting price changes in Ethereum. Int. J. Comput. Sci. Eng. (IJCSE) (2019). ISSN 0975-3397
4. Prashanth, J., Das, V.S.: Cryptocurrency Price Prediction Using Long-Short Term Memory Model (2018)
5. Wang, Y., Chen, R.: Cryptocurrency price prediction based on multiple market sentiment. In: Proceedings of the 53rd Hawaii International Conference on System Sciences (2020)
6. Bai, C., White, T., Xiao, L., Subrahmanian, V.S., Zhou, Z.: C2P2: a collective cryptocurrency up/down price prediction engine. In: 2019 IEEE International Conference on Blockchain (Blockchain), pp. 425–430. IEEE (2019)
7. Lamon, C., Nielsen, E., Redondo, E.: Cryptocurrency price prediction using news and social media sentiment. SMU Data Sci. Rev 1(3), 1–22 (2017)
8. Abraham, J., Higdon, D., Nelson, J., Ibarra, J.: Cryptocurrency price prediction using tweet volumes and sentiment analysis. SMU Data Sci. Rev. 1(3), 1 (2018)
9. Valencia, F., Gómez-Espinosa, A., Valdés-Aguirre, B.: Price movement prediction of cryptocurrencies using sentiment analysis and machine learning. Entropy 21(6), 589 (2019)
10. Bremmer, G.: Predicting tomorrow's cryptocurrency price using a LSTM model, historical prices and Reddit comments (2018)
11. Phillips, R.C., Gorse, D.: Predicting cryptocurrency price bubbles using social media data and epidemic modelling. In 2017 IEEE Symposium Series on Computational Intelligence (SSCI), pp. 1–7. IEEE, November 2017
12. Desev, K., Kabaivanov, S., Desevn, D.: (2019). Forecasting cryptocurrency markets through the use of time series models. Bus. Econ. Horizons (BEH), 15(1232-2020-345), 242–253 (2019)
13. Akyildirim, E., Goncu, A., Sensoy, A.: Prediction of cryptocurrency returns using machine learning. Ann. Oper. Res. 297(1–2), 3–36 (2020). https://doi.org/10.1007/s10479-020-035 75-y
14. František, G.: Predikce vývoje ceny vybraných kryptoměn pomocí strojového učení (Bachelor's thesis, České vysoké učení technické v Praze. Vypočetní a informační centrum.) (2019)
15. Yao, Y., Yi, J., Zhai, S., Lin, Y., Kim, T., Zhang, G., Lee, L.Y.: Predictive analysis of cryptocurrency price using deep learning. Int. J. Eng. Technol. 7(3.27), 258–264 (2018)
16. Möller, J.: Studying and Forecasting Trends for Cryptocurrencies Using a Machine Learning Approach. Bachelor's Theses in Mathematical Sciences (2018)
17. Derbentsev, V., Datsenko, N., Stepanenko, O., ezkorovainyi, V.: Forecasting cryptocurrency prices time series using machine learning approach. In: SHS Web of Conferences, vol. 65, p. 02001. EDP Sciences (2019)
18. Spilak, B.: Deep neural networks for cryptocurrencies price prediction (Master's thesis, Humboldt-Universität zu Berlin) (2018)
19. Akgül, A., Çevik, O.: İstatistiksel Analiz Teknikleri "SPSS' de İşletme Yönetimi Uygulamaları". Emek Ofset Ltd., Şti, Ankara (2003)
20. GITHUB, Support Vector Regression, https://yavuz.github.io/destek-vektor-regresyonuve-makineleri/, last accessed 2020/05/05.
21. VERIBILIMI, Decision Tree Regression, https://www.veribilimiokulu.com/karar-agaciile-regresyon-decision-tree-regression-python-ornek-uygulama/, last accessed 2020/05/05.
22. Cryptocurrency Analysis, with Python—Buy and Hold, https://towardsdatascience.com/cryptocurrency-analysis-with-python-buy-and-hold-c3b0bc164ffa, last accessed 2021/06/06

# Sofware Quality Prediction: An Investigation Based on Artificial Intelligence Techniques for Object-Oriented Applications

Özcan İlhan and Tülin Erçelebi Ayyıldız[(✉)]

Department of Computer Engineering, Başkent University, Ankara, Turkey
21810032@mail.baskent.edu.tr, ercelebi@baskent.edu.tr

**Abstract.** Exploring the relationship between Object Oriented software metrics and predicting the accuracy rate in defect prediction using the Chidamber & Kemerer (CK) metrics suite are the main purposes of this study. For these purposes, eleven machine learning (ML) techniques were analyzed to predict models and estimate reliability for eight open-source projects. Therefore, evaluating the relation between CK metrics and defect proneness was determined. Moreover, the performances of eleven machine learning techniques were compared to find the best technique for determining defect prone classes in Object Oriented software. The techniques were analyzed using Weka and RapidMiner tools and were validated using a tenfold cross-validation technique with different kernel types and iterations. Also, Bayesian belief's network forms show which metrics are being the primary estimators for reliability. Due to the imbalanced nature of the dataset, receiver operating characteristic (ROC) analysis was used. The accuracy of the projects was evaluated in terms of precision, recall, accuracy, area under the curve (AUC) and mean absolute error (MAE). Our analyses have shown that Random Forest, Bagging and Nearest Neighbors techniques are good prediction models with AUC values. The least effective model is Support Vector Machines (SVM).

**Keywords:** Artificial intelligence techniques · Software quality metrics · Software reliability · Defect prediction · Machine learning · Object Oriented metrics

## 1 Introduction

Software reliability is an important issue for customer and product satisfaction. It is usually measured by the number of defects found in the developed software and defects can be associated with software complexity. Maintenance is a challenging task and software failure rate related to the complexity level of software. So, defect-free software is nearly impossible. Because when the line of code size increasing, the prediction of faulty class more difficult and need a higher software testing effort. But minimization of software defect is very important. Therefore, in the software development lifecycle, detecting defects as early as possible is significant for minimization of defects and improving quality. In this research, reliability for the detection of software quality is

J. Hemanth et al. (Eds.): ICAIAME 2020, LNDECT 76, pp. 265–279, 2021.
https://doi.org/10.1007/978-3-030-79357-9_27

considered. Defect prediction, software reliability and reducing the cost of testing are challenging tasks. Reliability is usually measured by the number of defects found in the developed software and defects counts vary from different parameters such as complexity, time, cost constraints. So, the important issue is how can be used these constraints efficiently. Software metrics especially Object Oriented useful for determining the quality and performance of any software system. To detect and predict software reliability, ML techniques are useful technologies. Different ML techniques in literature have been used with different software metrics suite and limited small dataset for software reliability. All these ML techniques performance results are different because of the dataset. The important constraints are dataset size and its characteristic. In this research, comparative performance analysis between eleven ML techniques (Decision Trees, Bayesian Network, Rule-Based Classification, K-Nearest Neighbors (KNN), Naïve Bayes, Support Vector Machine (SVM), Logistic Regression (LR), Random Forest (RF), Bagging, Boosting and deep learning technique which is called Artificial Neural Network (ANN)) on eight open source project to improve software reliability and determine of the best ML techniques with performance parameters. The rest of the paper is organized as follows. Section 2 summarizes the related works. Section 3 introduces a summary of software metrics and dataset. In Sect. 4, the applied ML techniques are considered. In Sect. 5, our findings are discussed in the result and performance evaluation results are given. As a result, in Sect. 6, the conclusion and future works are discussed.

## 2 Related Works

There are various research and models that have been proposed about software metrics to be a good indicator and predict software reliability in the past. This section presents the related works and performance of ML techniques on the use of various software metrics.

Chidamber & Kemerer Object Oriented (OO) design metrics aim to predict the defects efficiently and it can be indicator of quality. These six metrics suite such as Coupling Between Object Class (CBO), Number of Children (NOC), Weighted Methods Per Class (WMC), Depth of Inheritance Tree (DIT), Lack of Cohesion (LCOM), Response for a Class (RFC) of methods measure the maintainability and reusability of software and focus on the class and class hierarchy [37].

Basili et al. studied the impact of OO metrics such as CK metrics suite on fault prediction and in development life cycle fault prediction should be in the early stages. OO metrics good predictor of reliability rather than the traditional metrics. Multivariate logistic regression prediction techniques were used in this research [43].

Li & Henry have recommended several OO software metrics such as Message Passing Coupling (MPC), Number of Methods (NOM), Data Abstracting Coupling (DAC), Number of Methods (NOM), Size of Procedures or Functions (SIZE1), Properties Size defined in a class (SIZE2) evaluated the correlation of these metrics and the maintenance effort. These metrics measure the amount of change, coupling, size and level of complexity in the code with the use of various OO metrics [44].

Tang et al. research the dependency between CK metrics suite and the OO system faults. The result of the research showed that WMC can be a major predictor for faulty classes [30].

Briand et al. predicting fault proneness of class, the relationship between fault and proposed extracted 49 metrics using multivariate logistic regression models. All metrics except NOC to be a significant predictor of fault prone classes [24].

Quah & Thwin used General Regression Neural Network (GRNN) and Ward Neural Network for estimating reliability and maintainability of software quality using OO software metrics. General Regression Neural Network model and Ward Neural Networks compared with each other. GRNN network model is the best indicator of reliability and maintainability [31].

Gyimothy et al. considered the relationship between CK Object Oriented metrics and fault prediction. The research compared the efficiency of sets of CK metrics using the Mozilla dataset. Different ML techniques are used such as Decision Tree, Neural Network and Logistic Regression (LR). Coupling between objects (CBO) metric is the best indicator for fault detection. Number of Children (NOC) has a low impact [40].

Zimmermann et al. research on predicting faulty and non-faulty classes of Java source code of Eclipse bug dataset with 14 different size metrics. A combination of complexity metrics can predict faults efficiently [42].

Singh et al. evaluate the KC1 NASA dataset and analyzed Decision Tree, Neural Network using Receiver Operating Characteristic (ROC) analysis for fault proneness to find which technique performs better and validate CK metrics. Based on the performance results, the machine learning method results better than the logistic regression method [46]. Other research showed the effectiveness of OO metrics and predictors of fault-prone systems [33].

Pai and Dugan's research evaluated the NASA KC1 dataset using Bayesian methods and WMC, CBO, RFC, Source Lines of Code (SLOC) are effective for software fault proneness prediction [34].

Elish et al. used the Support Vector Machine (SVM) method to predict the defect-prone software and evaluated the performance of SVM. Eight machine learning techniques compared with the NASA dataset. Results show that the SVM is better than other compared models [19]. Different alternatives of machine learning techniques and models analyzed for identifying defects [16, 17]. Another study is Arisholm et al. used Neural Networks, Logistic Regression, Support Vector Machine techniques and compared various variants of Decision Tree techniques with Java telecom systems. AdaBoost combined with C4.5 gave the best results for the prediction of fault proneness [3].

Malhotra, Shukla and Sawhney compared the performances of different 17 ML techniques on Xerces OO software to find defect prone classes. The bagging technique is the best performance machine learning for defect prediction models in this research [32].

Okutan & Yildiz (2017) evaluated the Bayesian Networks method to explore the relation between software metrics and defect proneness. Research defined two different metrics related to the quality of source code (LOCQ) and based on the number of developers (NOD). Results indicate that LOCQ, LOC and RFC are more effective to predict software quality [2].

Reddivari and Raman compared the performance of eight machine learning techniques for reliability and maintainability with using PROMISE data repository and

QUES systems. Random Forest provided the best predictor for reliability and maintenance. Also, WMC, the Coupling Between Methods (CBM), DIT and MPC are good contributors to quality [38].

In our research not only compare performance of machine learning techniques with default parameters but also optimize the parameters to find the best performance results and avoid overfitting. Further, different ML tools were applied for performance comparisons such as Weka and RapidMiner. More machine learning methods and extended dataset were used for the precision of prediction models.

## 3 Software Metrics and Dataset

### 3.1 Data Collection

Dataset was obtained from eight open source real-world software projects (3189 instances, 1236 defects) in our experiments. The public datasets (software defect prediction) were obtained from PROMISE data repository [10]. Project names of the dataset are Tomcat, Velocity, Ivy, Jedit, Workflow, Poi, Forrest, Ant. The dataset contains OO software metrics. Details of the dataset are given in Table 1. In this study, the project name and version were removed from the dataset for Weka. In RapidMiner, name and version were not selected as attributes. Because these attributes did not affect the determination of the target class for classification. In our dataset, defect labels contain number of defects in the module. For our research, these defect numbers transform into a binary classification such as 0 is non-defective, 1 is a defective module.

**Table 1.** Details of dataset.

| Dataset | Version | No.of instances | % defective |
|---------|---------|-----------------|-------------|
| Tomcat | 6.0 | 858 | 9 |
| Ant | 1.7 | 745 | 22 |
| Forrest | 0.8 | 32 | 18 |
| Jedit | 4.3 | 492 | 2 |
| Workflow | 1 | 39 | 51 |
| Poi | 3.0 | 442 | 63 |
| Ivy | 2.0 | 352 | 11 |
| Velocity | 1.6 | 229 | 34 |

### 3.2 Software Metrics

Software metrics have been used to measure different parameters of the software code and measure software complexity level, characteristics. Software metrics suite intent to solve different problems such as reliability and maintainability of a software product. A lot of OO software metrics have been proposed in the past. Software traditional metrics and mostly used metrics suite are McCabe's cyclomatic complexity [41], MOOD Metrics [9], Li and Henry [44], Halstead [27], Lorenz and Kidd [29], Robert Martin [35], Tegarden et al. [8], Bansiya and Davis [4], Melo and Abreu [45], Briand et al. [24], Etzkorn et al. [26], Sharble and Cohen [39], Kim and Ching [22], Harrison et al. [14], Genero et al. [12], Tang, Kao and Chen [30] and CK metrics suite [37] are used for evaluating software maintainability and reliability. In this research, CK metrics suite was selected because CK metrics are one of the most popular OO metrics. The important point is understanding of the relation between OO metrics and reliability. CK metrics have also been found to be effective in predicting faults in a software system. A lot of OO metrics have been evaluated and proposed but CK metrics are widely used for external quality attributes such as reusability and maintainability. It can be measured OO software internal characteristics such as inheritance, coupling, cohesion, encapsulation, class complexity and polymorphism. Eight software project contains 20 software attributes. Line of Code (LOC) is a McCabe's metric and it is a count of source code line with non-commented mean does not include blank lines and comments. Average Method Complexity (AMC) using McCabe's Cyclomatic Complexity method. In our dataset included some of Martin's Metrics such as Afferent Couplings (CA) and Efferent Couplings (CE). CA is the number of classes that calling specific or given class. CE is the number of classes called by a specific or given class. Weighted Methods Per Class (WMC) is CK metric. It is measure and sum of the complexities of the methods defined in a class. The high value of WMC shows that more complex class. Maintainability of the class is more difficult if complexities increase [31, 37]. Another CK metrics are Lack of Cohesion Among Methods of a Class (LCOM), Number of Children (NOC), Coupling Between Objects (CBO), Response for a Class (RFC), Depth of Inheritance Tree (DIT). LCOM is a count of methods pairs which not having any common attributes. Low cohesion increase complexity and high LCOM illustrates poor encapsulation. Value range between 0 and 1. CBO is number of classes to couple with or given class. A high number of couplings cause difficulty in maintainability. DIT indicates the maximum path from the class to the root of the tree [37]. More inherited methods mean high number of steps from the root to the leaf node increase complexity and reusability. NOC measures the number of direct children or subclasses of a given class. It is related to inheritance. RFC is the set of all methods that can be executed when the response to a message received by the object of some method or class [37]. External and internal methods in a class. Lack of Cohesion Among Methods of a Class 3 (LCOM3) is minor improvements related to calculation of cohesion level of the LCOM by Henderson-Sellars. Another metric in our dataset is Coupling between Methods (CBM). CBM measures the total number of redefined or new methods to which the inherited methods are coupled [31]. The Number of Public Methods (NPM) is the number of accessible public methods in a given class. Cohesion Among Methods of a Class (CAM) is a similarity method based on prototypes, signatures. Data Access Metric (DAM) is a measure of the ratio of all protected (non-public) and private attributes to all attributes of the class. It is related to encapsulation. Measure of Functional Abstraction (MFA) is the ratio in the number of methods inherited by a class to the total number of defined methods available. [43]. It is related to inheritance. Measure of Aggression (MOA) is the percentage of user defined data declarations in a class. It is related to composition. Inheritance Coupling (IC) is the number of parent classes to which are coupled with a specific or given class. In the same class, the maximum values of methods are MAX_CC and the mean values of methods are AVG_CC. Defect Count is shown as presences or absence of the defects. It is a binary classification.

# 4  Machine Learning Techniques and Experiments

Machine learning techniques are proven to be effective in terms of software reliability and enable retrieve useful information for software reliability. Various ML techniques are studied. Researchers proposed different ML techniques for reliability. The most popular and widely used ML algorithms were selected in this research. Default settings were used in Weka tool for each technique in Table 2. Improved performance results of Weka and RapidMiner [36] results are shown in Table 3. Followings are the brief description of selected algorithms.

## 4.1  Decision Trees

Decision tree is a predictive and classification model that tree classifies the examples by sorting them down the tree in which each node represents all possible decisions with edges. Decision making used attributes which information gain is high. C4.5 uses information gain and decision tree is implemented as J48 in Weka [28]. Default value of confidence factor is 0.25 in Weka. For this reason, the confidence factor of RapidMiner was specified 0.25 in Table 2 and Table 3. There were not any additional filters in both tools. In Table 3, reduced error pruning option [47] was enabled in both tools and minimal leaf size was specified 6 in RapidMiner to improve the results. LCOM metric was found to be good predictors in pruned tree followed by RFC and DIT.

## 4.2  Random Forest

Random Forest (RF) is ensemble learning techniques for supervised classification and operates by constructing multiple decision trees. RF consists of bagging of individual decision trees. The selection of features at each split randomize. Predictions depend on the individual predictor strength [23]. The maximum depth was specified 0 for both tools in Table 2 and Table 3. There were not any additional filters in both tools. In Table 3, criterion was specified to information gain in RapidMiner and the number of iterations was specified to 150 in both tools to improve the results. LCOM and WMC metrics are important contributors to fault prediction.

## 4.3  Bayesian Classification

Bayesian classifier is a simple probabilistic classifier. It is based on Bayes theorem with independence assumptions between the predictors and features. Bayesian Network represents a graphical description of attributes and their conditional dependencies. It uses K2 as a search algorithm [18, 20]. Naïve Bayes is a supervising prediction model used to make classifications of dataset based on a series of independent feature "naive" assumptions. Default values were used in both tools for Naïve Bayes and Bayes Network in Table 2. Naïve Bayes (NB) is a widely used model because of its effectiveness and robustness. Software metrics in our dataset have non-normal distribution. Problem of the non-normal distribution of NB can be solved with kernel density estimation [13]. In Table 3, Kernel Estimator were used for Naïve Bayes in Weka to help improve prediction accuracy. For Bayesian network, simple estimator and K2 greedy search algorithm were used in Weka in Table 2 and Table 3.

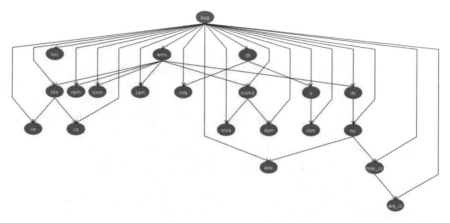

**Fig. 1.** Bayesian Network indicates the relationships among all software metrics in dataset. Maximum parents were specified 2 in the K2 search algorithm.

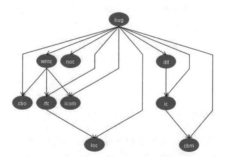

**Fig. 2.** Bayesian Network formed for most effective defect prediction metrics.

Figure 1 and Fig. 2 shows that WMC, NOC and DIT directly effective on reliability followed by LCOM, RFC, IC, CBO, NPM, MFA, CAM. According to Fig. 1 and Fig. 2, AMC, AVG _CC and MAX_CC metrics are not effective for determining the defect-prone class.

**4.4    Rule-Based Classification**

Rule-Based classification use of IF-THEN rules for classifying records for class prediction. Rule-based classifiers are classification rules extracts from decision trees. In this research, PART rule-based was used in Weka [28]. In Table 2, default values were used such as confidence factor. It is 0.25 in Weka. In Table 3, if confidence factor is specified to 0.01, accuracy rate of Rule-Based classification increases.

**4.5    SVM – SMO, LibSVM, LibLINEAR**

SVM is a supervised learning model that can be used for classification or regression problems. Support Vector Machine algorithm is available in Weka as Sequential Minimal Optimization (SMO) and it is a binary classifier [28]. Polynomial kernel function is used for non-linear mapping of data and find an optimal decision boundary. SVM attempts to find a plane that has the maximum margin or optimal decision boundary in the mapped dimension in other words try to find an optimal boundary between the possible outputs. LibSVM and LibLINEAR are widely used outside of Weka. SMO and LibSVM are non-linear SVMs and can use the L1 norm. LibLINEAR is faster than SMO and LibSVM [15]. SMO default kernel type is polynomial in Weka. For this reason, default kernel type was specified to polynomial in RapidMiner for comparison with each other. In Table 2, there were not use any additional filters for SMO, LibSVM and LibLINEAR. In Table 3, the Pearson VII kernel function (PUK) kernel type was used in Weka and radial kernel type was used in RapidMiner to the improvement of results of SMO. Default values provide the best results for LibLINEAR and LibSVM for both tools in Table 3.

**4.6    Logistic Regression**

Logistic Regression (LR) is based on a statistical model and the type of predictive regression analysis. It is a form of binary regression and used for categorical target value and method for binary classification problems. In Table 2 and Table 3, there were not use any additional filters for LR and default values provide the best results of LR in both tools.

**4.7    Bagging and Boosting**

Bagging and Boosting is an ensemble method is that a group of weak learners combined to create a strong learner and try to reduce bias and variance of weak learners for better performance. For this reason, the accuracy of these models high and increases the robustness, improves the stability and helps to avoid overfitting [25]. In Table 2, the default bagging classifier is REPTree also called Regression Tree and Boosting algorithms default classifier is Decision Stump in Weka. For this reason, Decision Stump were selected for the Boosting algorithm and REPTree were selected for the Bagging algorithm in RapidMiner tool. The iterations are same in both tools for comparison with each other. In Table 3, Random Forest were used in Weka and RapidMiner tools to improve prediction accuracy for Bagging and Boosting algorithms.

## 4.8   Artifical Neural Networks

Artificial Neural Network (ANN) is a learning algorithm and known as a Multilayer Perceptron (MLP) in Weka [11, 28]. It is inspired by neural networks of the human brain. It is a feed-forward Artificial Neural Network model. ANNs are composed of multiple nodes (processing units) and links (connection between nodes). ANN has input, output and hidden layers [1]. Input layers take the CK metrics values as input at the input layer nodes. Weights are assigned on the links which connect the nodes and weights are used for computation of the output. Output is the fault prediction accuracy rate. These rules are used for specifying the number of neurons in the hidden nodes such as the number of hidden layer neurons should be 70% to 90% of the size of the input layer, the hidden layer neurons count should be less than twice of the neurons count in input layer [5–7]. Besides, the activation function complexity is important for determining hidden layers and if hidden layers up to three, the complexity of neural network is increased [21]. Less hidden layers count will result in decreased accuracy. A large number of hidden layers may result in overfitting.

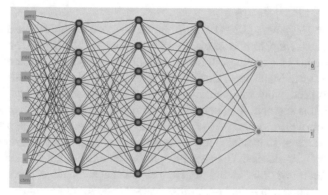

**Fig. 3.** The graphical user interface of the Neural Network for most effective defect prediction software metrics.

Neural network formed has 9 input and 2 output nodes. Input nodes are software metrics in Fig. 3.

Also, all metrics which is in dataset were represented in graphical user interface with 20 input nodes in Fig. 4. In Table 2, MLP using 0.3 learning rate, 0.2 momentum with 500 epochs are default values in Weka. Same values were specified in RapidMiner for comparing with each other. In Table 3, training cycle was specified 200 in RapidMiner and learning rate were set to 0.1 in Weka tool to improve our results.

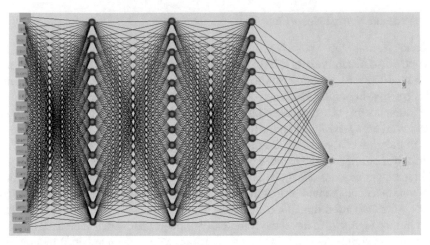

**Fig. 4.** The graphical user interface of the Neural Network for all software metrics in dataset.

### 4.9 Nearest Neighbors

The K-Nearest Neighbors (KNN) algorithm solves both classification and regression problems. Nearest Neighbors is implemented in Weka as Ibk [28] and an example of a non-parametric lazy learner algorithm. Commonly used distance measure is Euclidean distance. Others are Manhattan, Cosine, Minkowski and Hamming distance methods. KNN model is used in our study. Finding the best value for k is not easy. If the value of k is too small, it is susceptible to noise in the data. Too large value of k made it computationally expensive, less sensitive to noise and it considers more neighbors. Commonly choosing the value of k calculated as the square root of N which is the number of samples in your training dataset. The algorithm finds the closest data points, neighbors in the training dataset in order to classify the new instance. In Table 2, according to our experience, the number of neighbor K value was set to 60 for more stable predictions and Euclidean distance function was used for both Weka and RapidMiner to compare with each other. In Table 3, Manhattan distance function improves of Weka and RapidMiner's accuracy results.

## 5 Result and Evaluation

### 5.1 Performance Evaluation Results

Details of used performance metrics for evaluation of results listed below. True positive (TP) represents the correctly classified positive class. True negative (TN) represents the correctly classified non-fault classes. False positive (FP) represents the incorrectly classified positive class. False negative (FN) represents the incorrectly classified negative class.

Accuracy is proportion of the total number of correct classifies amongst the total number of predictions. Accuracy is shown by the Eq. 1 expression.

$$\text{Accuracy} = \frac{\text{TP} + \text{TN}}{(\text{TP} + \text{FP} + \text{TN} + \text{FN})} \tag{1}$$

**Table 2.** Confusion matrix for faulty and not faulty class.

|  |  | Actual class | |
| --- | --- | --- | --- |
|  |  | True | False |
| Predicted class | Positives | True positives (TP) | False positives (FP) |
|  | Negatives | False negatives (FN) | True negatives (TN) |

Precision is ratio of the correctly predicted positive instance to all the positive instances. Precision is calculated in Eq. 2

$$\text{Precision} = \frac{TP}{(TP + FP)} \tag{2}$$

Recall (Sensitivity) is proportion of correctly classify fault-prone classes to all fault prone classes. Recall is calculated in Eq. 3

$$\text{Recall} = \frac{TP}{(TP + FN)} \tag{3}$$

Area under the curve (AUC) is receiver operating characteristic curve (ROC). It is a two-dimensional graphical representation where TP values on the y-axis and FP values on the x-axis [26]. AUC method can deal with noisy or unbalanced datasets and it is the most effective performance evaluation technique.

Mean Absolute Error (MAE) measures the average of all absolute errors, differences between the original and predicted values of the data. The lower the value, the better is the model's performance.

Root Mean Square Error (RMSE) is one of the performance indicators. It compares values predicted by a model and the observed value. The smaller an RMSE value, the closer predicted values are (Table 4).

The DIT, WMC and NOC metrics were found to be significant predictors. The results of this study could be effective for predicting reliability in software systems. In this way, making software systems less defect prone and more reliable.

The results were evaluated by comparing the average AUC values and AUC values were obtained by all the ML techniques. AUC evaluation metrics were used while comparing our performance measurement of models in Weka and RapidMiner. The evaluation results indicate in Table 3. According to Weka, Bagging and Random Forest has the highest AUC values with accuracy rate followed by KNN and Bayesian Network. There are not any significant differences between KNN and Bayesian Network in terms of AUC values. Part and Logistic Regression AUC values are same but the accuracy rate of Part is higher than Logistic Regression. Performance of LibSVM is good over SMO in Weka and RapidMiner implementation over the same dataset in Table 2. After improvement of models, SMO is more reliable than LibSVM. Nevertheless, SVMs are the least effective model for defect prediction. In RapidMiner, Bagging and Random Forest methods have the highest AUC values like Weka. Besides, ANN method has high AUC values in RapidMiner result. Figure 1 and Fig. 2 shows that DIT, WMC and NOC are important factors to the prediction of software defect.

**Table 3.**  Results for reliability prediction.

| Technique | Accuracy weka | AUC weka | AUC rapidMiner | Precision weka | Recall weka | MAE weka |
|---|---|---|---|---|---|---|
| J48 | 81.500% | 0.720 | 0.697 | 0.802 | 0.815 | 0.227 |
| Random forest | 84.258% | 0.823 | 0.791 | 0.832 | 0.843 | 0.234 |
| Naïve Bayes | 76.763% | 0.734 | 0.736 | 0.735 | 0.768 | 0.234 |
| Bayesian network | 81.000% | 0.785 | – | 0.802 | 0.811 | 0.227 |
| Part | 81.812% | 0.749 | – | 0.797 | 0.818 | 0.242 |
| KNN | 82.439% | 0.772 | 0.749 | 0.810 | 0.824 | 0.256 |
| SMO | 78.833% | 0.500 | 0.696 | 0.788 | 0.788 | 0.211 |
| LibSVM | 80.652% | 0.545 | 0.698 | 0.832 | 0.807 | 0.193 |
| LibLinear | 74.098% | 0.582 | 0.696 | 0.726 | 0.741 | 0.259 |
| ANN | 82.596% | 0.751 | 0.773 | 0.808 | 0.826 | 0.243 |
| Bagging | 83.819% | 0.806 | 0.801 | 0.826 | 0.838 | 0.237 |
| AdaBoost | 80.620% | 0.769 | 0.499 | 0.793 | 0.806 | 0.277 |
| Logistic regression | 79.178% | 0.761 | 0.765 | 0.750 | 0.792 | 0.282 |

**Table 4.**  Improvement of results for reliability prediction.

| Technique | Accuracy weka | AUC weka | AUC rapidMiner | Precision weka | Recall weka | MAE weka |
|---|---|---|---|---|---|---|
| J48 | 82.376% | 0.748 | 0.758 | 0.806 | 0.824 | 0.243 |
| Random Forest | 84.289% | 0.824 | 0.831 | 0.832 | 0.843 | 0.234 |
| Naïve Bayes | 79.397% | 0.768 | 0.769 | 0.778 | 0.794 | 0.210 |
| Bayesian Network | 81.000% | 0.785 | – | 0.802 | 0.811 | 0.227 |
| Part | 82.815% | 0.761 | – | 0.814 | 0.828 | 0.238 |
| KNN | 82.659% | 0.780 | 0.764 | 0.813 | 0.827 | 0.248 |
| SMO | 83.537% | 0.637 | 0.773 | 0.831 | 0.835 | 0.164 |
| LibSVM | 80.652% | 0.545 | 0.698 | 0.832 | 0.807 | 0.193 |
| LibLinear | 74.098% | 0.582 | 0.696 | 0.726 | 0.741 | 0.259 |
| ANN | 83.537% | 0.773 | 0.786 | 0.822 | 0.835 | 0.239 |
| Bagging | 84.227% | 0.828 | 0.831 | 0.833 | 0.842 | 0.237 |
| AdaBoost | 84.352% | 0.783 | 0.608 | 0.833 | 0.844 | 0.159 |
| Logistic Regression | 79.178% | 0.761 | 0.765 | 0.750 | 0.792 | 0.282 |

# 6 Conclusion and Future Work

The aim of this paper, comparing the performance of eleven different ML techniques with used tenfold cross-validation for prediction of software reliability and metrics were obtained from PROMISE open source data repository for reliability. In this stage, the important findings of the works are summarized. The results show that DIT, WMC, NOC metrics are more useful and significant predictors in determining the quality. This is followed by LCOM, RFC, IC, CBO, NPM, MFA and CAM. The evaluation results indicate that Bagging and Random Forest are the best techniques for reliability and SMO, LibSVM, LibLINEAR SVMs are the least effective model for defect prediction. As future work, this research can further be conducted by increasing the number of Object Oriented software projects and applied other ML techniques.

# References

1. Jain, A.K., Mao, J., Mohiuddin, K.M.: Artificial neural networks: a tutorial. IEEE Comput., 31–44 (1996)
2. Okutan, A., Yildiz, O.: Software defect prediction using bayesian networks. Empirical Softw. Eng., 1–28 (2012)
3. Arisholm, E., Briand, L.C., Johannessen, E.B.: A systematic and comprehensive investigation of methods to build and evaluate fault prediction models. J. Syst. Softw. **83**(1), 2–17 (2010)
4. Bansiya, J., Davis, C.: A hierarchical model for quality assessment of object-oriented designs. IEEE Trans. Software Eng. **28**(1), 4–17 (2002)
5. 5.Berry, M.J.A., Linoff, G.: Data Mining Techniques. Wiley, New York (1997)
6. 6. Blum, A.: Neural Networks in C++. Wiley, New York (1992)
7. Boger, Z., Guterman, H.: Knowledge extraction from artificial neural network models. In: IEEE Systems, Man and Cybernetics Conference, Orlando, FL, USA (1997)
8. Tegarden, D.P., Sheetz, S.D., Monarchi, D.E.: A software complexity model of object-oriented systems. Decis. Support Syst. **13**(3–4), 241–262 (1995)
9. Abreu, F.B.: Design metrics for object-oriented software systems. In: ECOOP 1995 Quantitative Methods Workshop, Aarhus (1995)
10. Boetticher, G., Menzies, T., Ostrand, T.J.: Promise repository of empirical software engineering data (2007). http://promisedata.org/repository
11. Holmes, G., Donkin, A., Witten, I.: Weka: a machine learning workbench. In: Proceedings 2nd Australian New Zealand Conference on Intelligent Information Systems, pp. 1269–1277 (1994)
12. Genero, M.: Defining and validating metrics for conceptual models. Ph.D. thesis, University of Castilla-La Mancha (2002)
13. Haijin, J.I., Huang, S., Lv, X., Wu, Y., Nonmembers, Feng, Y.: Empirical studies of a kernel density estimation based Naive Bayes method for software defect prediction. IEICE Trans. **102**-D(1), 75–84 (2019)
14. Harrison, R., Counsell, S., Nithi, R.: Coupling metrics for object-oriented design. In: 5th International Software Metrics Symposium Metrics, pp. 150–156 (1998)
15. Hsu, C.W., Chan, C.C., Lin, C.J.: A practical guide to support vector classification (2010). http://www.csie.ntu.edu.tw/~cjlin/papers/guide/guide.pdf
16. Gondra, I.: Applying machine learning to software fault proneness prediction. J. Syst. Softw. **81**(2), 186–195 (2008)

17. Laradji, I.H., Alshayeb, M., Ghouti, L.: Software defect prediction using ensemble learning on selected features. Inf. Softw. Technol. **58**, 388–402 (2015)
18. Huang, K.: Discriminative Naive Bayesian classifiers. Department of Computer Science and Engineering, the Chinese University of Hong Kong (2003)
19. Elish, K.O., Elish, M.O.: Predicting defect-prone software modules using support vector machines. J. Syst. Softw. **81**(5), 649–660 (2008)
20. Murphy, K.P.: Naive Bayes Classifiers, Technical report, October 2006
21. Karsoliya, S.: Approximating number of hidden layer neurons in multiple hidden layer BPNN architecture. Int. J. Eng. Trends Technol. (2012)
22. Kim, E.M., Chang, O.B., Kusumoto, S., Kikuno, T.: Analysis of metrics for object-oriented program complexity. In: Computer Software and Applications Conference, IEEE Computer, pp. 201–207 (1994)
23. Breiman, L.: Random forests. Mach. Learn. **45**(1), 5–32 (2001)
24. Briand, L., Daly, J., Porter, V., Wust, J.: Exploring the relationships between design measures and software quality. J. Syst. Softw. **5**, 245–273 (2000)
25. Brieman, L.: Bagging predictors. Mach. Learn. **24**, 123–140 (1996)
26. Etzkorn, L., Bansiya, J., Davis, C.: Design and code complexity metrics for OO classes. J. Object-Oriented Program. **12**(1), 35–40 (1999)
27. Halstead, M.H.: Elements of Software Science. Elsevier Science, New York (1977)
28. Hal, M., et al.: The WEKA data mining software: an update. SIGKDD Explor. **11**(1), 10–18 (2009). Cover, T.M., Hart, P.E.: Nearest neighbour pattern classification. IEEE Trans. Inf. Theory **IT-13**(1), 21–27 (1967)
29. Lorenz, M., Kidd, J.: Object-Oriented Software Metrics. Prentice Hall, Englewood (1994)
30. Tang, M.-H., Kao, M.-H., Chen, M.-H.: Empirical study on object-oriented metrics. In: Proceedings of the 6th International Software Metrics Symposium, pp. 242–249, November 1999
31. Thwin, M.M.T., Quah, T.-S.: Application of neural networks for software quality prediction using object-oriented metrics. In: Proceeding IEEE International Conference Software Maintenance (ICSM) (2003)
32. Malhotra, R., Shukla, S., Sawhney, G.: Assessment of defect prediction models using machine learning techniques for object-oriented systems. In: 5th International Conference on Reliability, Infocom Technologies and Optimization (ICRITO) (Trends and Future Directions) (2016)
33. Yu, P., Systa, T., Muller, H.: Predicting fault-proneness using OO metrics: an industrial case study. In: Proceedings Sixth European Conference Software Maintenance and Reengineering, (CSMR 2002), pp. 99–107, March 2002
34. Pai, G.J., Dugan, J.B.: Empirical analysis of software fault content and fault proneness using Bayesian methods. IEEE Trans. Softw. Eng. **33**(10), 675–686 (2007)
35. Martin, R.: OO design quality metrics—an analysis of dependencies. In: Proceedings of the Workshop Pragmatic and Theoretical Directions in Object-Oriented Software Metrics (OOPSLA 1994) (1994)
36. RapidMiner Open Source Predictive Analytics Platform Available. https://rapidminer.com/get-started/
37. Chidamber, S.R., Kemerer, C.F.: A Metrics suite for object oriented design. IEEE Trans. Softw. Eng. **20**(6), 476–493 (1994)
38. Reddivari, S., Raman, J.: Software quality prediction: an investigation based on machine learning. In: IEEE 20th International Conference on Information Reuse and Integration for Data Science (2019)
39. SharbleR, C., Cohen, S.S.: The object oriented brewery: a comparison of two object oriented development methods. ACM SIGSOFT Softw. Eng. Notes **18**(2), 60–73 (1993)

40. Gyimothy, T., Ferenc, R., Siket, I.: Empirical validation of object-oriented metrics on open source software for fault prediction. IEEE Trans. Softw. Eng. **31**(10), 897–910, October 2005
41. McCabe, T.J.: A complexity measure. IEEE Trans. Software Eng. **2**(4), 308–320 (1976)
42. Zimmermann, T., Premraj, R., Zeller, A.: Predicting defects for eclipse. In: International Workshop on Predictor Models in Software Engineering, PROMISE 2007, ICSE Workshops, p. 9, May 2007
43. Basili, V., Briand, L., Melo, W.L.: A validation of object oriented design metrics as quality indicators. IEEE Trans. Software Eng. (1996)
44. Li, W., Henry, S.: Object-oriented metrics that predict maintainability. J. Syst. Softw. **23**(2), 111–122 (1993)
45. Melo, W., Abreu, F.B.E.: Evaluating the impact of object oriented design on software quality. In: Proceedings of the 3rd International Software Metrics Symposium, Berlin, Germany, pp. 90–99, March 1996
46. Singh, Y., Kaur, A., Malhotra, R.: Empirical validation of object-oriented metrics for predicting fault proneness models. Software Qual. J. **18**(1), 3–35 (2010)
47. Aluvalu, R.: A reduced error pruning technique for improving accuracy of decision tree learning. Int. J. Eng. Adv. Technol. (IJEAT) **3**(5) (2014). ISSN: 2249–8958

# A Multi Source Graph-Based Hybrid Recommendation Algorithm

Zühal Kurt[1]([⊠]), Ömer Nezih Gerek[2], Alper Bilge[3], and Kemal Özkan[4]

[1] Department of Computer Engineering, Atılım University, Ankara, Turkey
zuhal.kurt@atilim.edu.tr
[2] Department of Electrical and Electronics Engineering, Eskişehir Technical University,
Eskişehir, Turkey
[3] Department of Computer Engineering, Akdeniz University, Antalya, Turkey
[4] Department of Computer Engineering, Eskişehir Osmangazi University, Eskişehir, Turkey

**Abstract.** Images that widely exist on e-commerce sites, social networks, and many other applications are one of the most important information resources integrated into the recently deployed image-based recommender systems. In the latest studies, researchers have jointly considered ratings and images to generate recommendations, many of which are still restricted to limited information sources, sources namely, ratings with another input data, or which require the pre-existence of domain knowledge to generate recommendations. In this paper, a new graph-based hybrid framework is introduced to generate recommendations and overcome these challenges. Firstly, a simple overview of the framework is provided and, then, two different information sources (visual images and numerical ratings) are utilized to describe how the proposed framework can be developed in practice. Furthermore, the users' visual preferences are determined based on which item they have already purchased. Then, each user is represented as a visual feature vector. Finally, the similarity factors between the users or items are evaluated from the user visual-feature or item visual-feature matrices, to be included the proposed algorithm for more efficiency. The proposed hybrid recommendation method depends on a link prediction approach and reveals the significant potential for performance improvement in top-N recommendation tasks. The experimental results demonstrate the superior performance of the proposed appraoch using three quality measurements - hit-ratio, recall, and precision - on the three subsets of the Amazon dataset, as well as its flexibility to incorporate different information sources. Finally, it is concluded that hybrid recommendation algorithms that use the integration of multiple types of input data perform better than previous recommendation algorithms that only utilize one type of input data.

**Keywords:** Graph · Visual-feature · Link prediction

## 1 Introduction

In the current information age, the amount of data accessible online has ex-panded exponentially. It is often hard for users to cope with this information overload and obtain

© The Author(s), under exclusive license to Springer Nature Switzerland AG 2021
J. Hemanth et al. (Eds.): ICAIAME 2020, LNDECT 76, pp. 280–291, 2021.
https://doi.org/10.1007/978-3-030-79357-9_28

the data they need without the help of data mining systems and artificial intelligence. Recommender systems operate as information filtering systems to recommend or help the user find an item among a collection of items that will most probably be liked by the user. Currently, recommender systems are utilized by many different applications [1, 4–6], for instance, recommendations made for movies, TV programs, clothing, or music that can be found enjoyable by the user.

Previous studies indicate that incorporating richer input data beyond numerical ratings can improve the recommendation performance of the recommender systems. This observation has led to development of hybrid recommender systems, which have been getting more attention in recent years [1, 2]. In the beginning, these hybrid systems commonly integrated CBF and CF [2, 3] techniques for recommendation generation. Then, the integration can be in different forms; for example, switching, weighting, mixing, etc., based on the reachable content profiles and meta-recommendation algorithms. Currently, many e-commerce applications aggregate a rich source of information, namely text, rating, image, etc., which express different aspects of user preferences. Hence, the concept of hybrid recommendation is extended with the integration of various information sources. With new developments, integrating heterogeneous information to generate recommendations became possible by translating the different information sources into a unified representation space [4, 5]. For example, a visual representation space is generated to map certain items by using their visual images based on similar features, and also by utilizing the information concerning the users' preferences [6]. Moreover, a visual-based Bayesian personalized ranking approach that applies a top-N recommendation task, has been proposed as an image-based recommendation system to extract co-purchase recommendations [6].

In the last few years, there has been growing interest in image-based recom-mender systems since e-commerce sites, social networks, and other applications continue to accumulate a rich source of visual information. For example, a visually explainable recommendation approach that utilizes deep learning-based models to obtain visual explanations is introduced in [5]. In the samework, a deep neural network is developed to generate image regional highlights as explanations for target users. Quite recently, considerable attention has been paid to combine ratings and images to generate recommendations [6–8]. However, even though the performance of recommendation algorithms has been improved in recent years, these approaches continue to utilize only limited information sources, or need the pre-existence of domain knowledge to generate a recommendation. Nonetheless, it is possible to further improve the performance of recommender systems by either hybridizing the algorithms, or doing so with heterogeneous information sources [4, 13].

With this goal in perspective, this paper presents a new graph-based hybrid framework for recommendation generation. Initially, we give a simple overview of the proposed graph-based model, and then apply this model by using two different information sources to explain how the model can be used in practice. The remainder of the paper is organized as follows: The detailed representation of the proposed recommendation algorithm is explained in Sect. 2. The evaluation measurements that are used in this study are given in Sect. 3. The application of the experiments in three real-world datasets and the discussion of the experimental results are included in Sect. 4. Finally, the results and future research directions are summarized.

## 2  Motivation

It is commonly agreed by experts incorporating different information sources beyond numerical ratings can increase performance of recommendation systems. As a result, studies are currently focused on hybrid recommendation algorithms [6, 7]. Formerly, hybrid recommendation techniques where commonly designed by integrating content-based [14] and CF-based [15] algorithms. For example, a recommendation algorithm depends on a link prediction approach with the weights in the graph expressed by complex numbers, as proposed in [11]. This algorithm depends on a link prediction approach with the weights in the graph represented by complex numbers that can accurately differentiate the 'similarity' between two users and two items, and identify the 'like' from a user for an item. However the above-stated study, mostly ignores the similarity relationship between users and items. Then, in [10], the extension of this method is proposed as the similarity-inclusive link prediction method (SIMLP), which also utilizes the similarity between two users (or two items). Accordingly, user-user and item-item cosine similarity values are applied by carefully incorporating the relational dualities to enhance the accuracy of the recommendation algorithm. Furthermore, a large amount of different/various information sources about users, items, and their interactions are gathered from many e-commerce applications that help to develop the new concept of hybrid recommendation algorithm as the combination of various information sources [4, 5].

## 3  Multi-source Hybrid Recommendation Algorithm

The similarity-inclusive link prediction method (SIMLP) for hybrid recommendation systems (Hybrid-SIMLP) differs slightly from the proposed algorithm described in [10]. The modeling process of the adjacency matrix for the Hybrid-SIMLP algorithm is different from the SIMLP algorithm, while the processes of computing the powers of the adjacency matrix and providing the final recommendation are the same procedure and based on the top-N recommendation task.

For the present work, first a user-item rating/preference matrix is generated from the datasets, as described in [9]. Then, the ratings in this matrix are converted to $j$ or $-j$ based on whether the rating is greater than or equal to 3. If the rating is less than 3, it is replaced with $-j$, which means that the user states '*dislike*' for the item; when the rating is greater than or equal to 3, it is replaced with $j$ to represent '*like*'. Moreover, if the $(u, i)$ pair is not included in the training set, the corresponding component of the user-item preference matrix becomes zero. After the process of rating conversion, the generated matrix is denoted as $A_{UI}$. Furthermore, the conjugate transpose of $A_{IU}$ can be represented as $A_{IU} = -A_{UI}^T$, and these preference matrices $A_{UI}$ and $A_{IU}$ are denoted as complex matrices [10] (see Fig. 5.1.(c)). An example of the user-item rating matrix and the bipartite graph model of this matrix is illustrated in Fig. 1 (a) and (b). The rating conversion of the user-item preference matrix is given in Fig. 1 (c). The user-item interaction (bipartite signed) graph after the rating conversion of the user-item preference

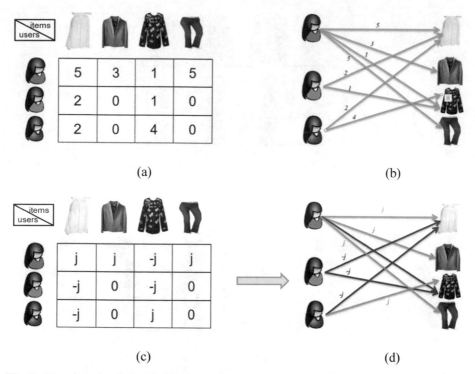

**Fig. 1.** User–item graph model. (Fig. 1.a: User-item rating matrix, Fig. 1.b: Bipartite graph model, Fig. 1.c: Rating conversion of rating matrix, Fig. 1.d: Bipartite signed graph).

matrix is drawn in Fig. 1 (d). Green links are represented as the '*like*' edges and denoted as $j$, red links are represented as the '*dislike*' edges and denoted as $-j$ in this bipartite signed graph (see Fig. 1 (d)).

Secondly, the item-feature matrix is generated by using all items of datasets, which is introduced in [9]. These item features, represented as AlexNet, are extracted in the same way that feature extraction takes place, as described in [4]. The users' visual preference towards different items' images is identified to generate each users' visual features. In the first step, all the items that users have rated or bought before are detected. In the second step, the AlexNet feature vectors of these items are computed. Then, these feature vectors, that belong to items previously rated by a given user, are summed up. Finally, the mean of this summation is calculated with the number of items that users' rated before. Hence, each user can be denoted as a 4096-dimensional visual-feature vector. An example of this procedure is depicted in Fig. 2. An illustration of an item-feature matrix is given in Fig. 2 (a). The process of generating a user feature vector is illustrated in Fig. 2. (b), and the generated users' feature matrix is given in Fig. 2. (c).

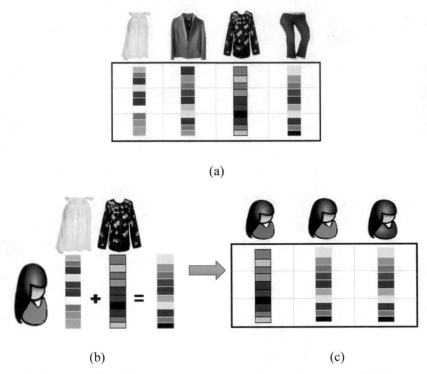

(a)

(b)                                        (c)

**Fig. 2.** User and item feature matrices. (Fig. 2.a: Item feature matrix, Fig. 2.b: The generation model of a user feature vector, Fig. 2.c: User feature matrix).

The process of generating the main adjacency matrix for the proposed hybrid recommender system is somewhat varied from the generation of the adjacency matrix, which is described in [10]. These variations depend on generating user-user and item-item similarity matrices. The user-user similarity matrix is developed from the generated user-feature matrix, and the item-item similarity matrix is computed based on the generated item-feature matrix by utilizing the cosine similarity measurement. Finally, the main adjacency matrix is generated with the combination of user-item preference matrix $A_{UI}$, the conjugate transpose of this matrix $A_{IU}$, and the item-item and user-user similarity matrices $A_{II}, A_{UU}$ as in [10]. The main adjacency matrix, which is given in Eq. (1), is reformulated for the proposed hybrid recommendation algorithm. This reformulation can be represented as:

$$A = \begin{pmatrix} u_{11} & \cdots & u_{1n} & r_{11} & \cdots & r_{1n} \\ \vdots & \ddots & \vdots & \vdots & \ddots & \vdots \\ u_{m1} & \cdots & u_{mn} & r_{m1} & \cdots & r_{mn} \\ -r_{11} & \cdots & -r_{1n} & k_{11} & \cdots & k_{1n} \\ \vdots & \ddots & \vdots & \vdots & \ddots & \vdots \\ -r_{m1} & \cdots & -r_{mn} & k_{m1} & \cdots & k_{mn} \end{pmatrix}, \tag{1}$$

where $u_{ij}$ denotes the cosine similarity value between the $i^{th}$ and $j^{th}$ users' visual feature vectors, $k_{ij}$ denotes the cosine similarity value between the $i^{th}$ and the $j^{th}$ items visual feature vectors, $r_{ij}$ indicates the like/dislike relationship between the $i^{th}$ user and the $j^{th}$ item, and $-r_{ij}$ expresses the like/dislike relationship between the $i^{th}$ user and the $j^{th}$ item in Eq. (1).

Furthermore, the hyperbolic sine of this adjacency matrix is implemented as a link prediction function for the proposed hybrid system, as in [10, 11]. The hyperbolic sine of the adjacency matrix can be formulated as in Eq. (2).

$$sinh(A) = U \cdot sinh(\Lambda) \cdot U^T \tag{2}$$

However, the hyperbolic sine of the biggest eigenvalue cannot be a number. Then, we need to normalize $\Lambda$ by dividing it by the biggest eigenvalue. After the normalization of $\Lambda$, Eq. (2) is reformulated as

$$sinh(A) = U \cdot sinh(\frac{\Lambda}{the\ biggest\ \Lambda}) \cdot U^T \tag{3}$$

In another approach, the eigenvalue vector $\Lambda$ can be normalized as in the range of [0, 1]. However, the eigenvalue vector $\Lambda$ includes negative values, and when $\Lambda$ normalizes as in the range of [0, 1], the negative information will be discarded. Hence, $\Lambda$ is normalized by dividing it by the biggest eigenvalue in order not to lose any information, (see in Eq. (3)).

## 4   Evaluation Measures

The following three typical top-N recommendation measures are used for evalua-tion. These are hit-ratio, precision, and recall measurements, as in [4, 5].

### 4.1   Hit-Ratio

The hit-ratio is evaluated by counting the number of hits; i.e., the number of items in the test set that are also presented in the top-N recommendation item list for each user. Then, the hit-ratio of the recommender system can be represented as:

$$Hit - Ratio(N) = \frac{\#hits}{n} \tag{4}$$

where $n$ represents the total number of users and $\#hits$ is the overall number of hits by recommendation system.

### 4.2   Recall

The percentage of items in the test set is also presented in the users' top-N recommendation list. The recall of the recommender system can be represented as:

$$Recall(N) = \frac{\#hits}{|T|} \tag{5}$$

where $|T|$ is the number of test ratings.

## 4.3  Precision

The percentage of correctly recommended items in the test set is also presented in the users' top-N recommendation list. The precision of the recommended system can be written as:

$$\text{Pre}cision(N) = \frac{\#hits}{|T| \cdot N} \tag{6}$$

where $|T|$ is the number of test ratings and $N$ is the length of the recommendation list.

## 5  Experimental Results and Datasets

The proposed hybrid recommendation algorithm and other comparison methods are implemented in a real-world Amazon review dataset [9], which comprises user interactions (ratings, reviews, helpfulness votes, etc.) on items, along with the item metadata descriptions, price, brand, image features, etc. concerning 24 product categories in the period of May 1996 – July 2014. Also, each product category covers a sub-dataset. Three product categories that have various sizes and density levels are selected for evaluation, namely: Beauty, Cell Phone, and Clothing. The standard five-core datasets (subsets) of these three categories are utilized for the experiments, as in [4]. The five-core data is a subset of the data in which all users and items have at least 5 reviews and ratings. The statistics of these three datasets are demonstrated in Table 1.

**Table 1.**  Statistics of the 5-core datasets.

| Datasets | #Users | #Items | #Interactions | Density |
|----------|--------|--------|---------------|---------|
| Clothing | 39387 | 23033 | 278677 | 0.0307% |
| Cell phones | 27879 | 10429 | 194493 | 0.0669% |
| Beauty | 22363 | 12101 | 198502 | 0.0734% |

In Table 1, the total number of ratings in each dataset is indicated as the number of interactions. Each product/item is described by an image, which has already been processed into a 4096-dimensional AlexNet feature vector in these datasets. Then, these vectors are utilized to generate item representation and user representation, as explained in the generation process of item visual-feature and user visual-feature vectors. These visual feature vectors are extracted by using Alex Net, as in [6].

The testing methodology for the proposed hybrid recommendation algorithm is identical to the previous studies [4, 5]. The ratings are divided into two subsets, training and test sets, for each dataset. The test set contains only the 5-star ratings and items that are relevant to the corresponding users. The generation procedure of the training set and the test set can be described as follows: In the first step, 30% of items rated by each user is randomly selected to generate a temporary test set, while the temporary training set contains other ratings as well. In the second step, the 5-star ratings in the temporary test set are selected to generate the final test set, and the remaining ratings in the temporary test set are merged into the temporary training set to construct the final training set. Then, the training set is used to predict ratings or relational dualities (like or dislike) for each item-user pair. Furthermore, the 5-core datasets used for the experiments have at least 5 interactions for each user. Thus, at least 3 interactions are applied per user for training, and at least 2 interactions are applied per user for testing. All test items are selected as the same way as [4].

**Table 2.** The performance comparison of the proposed algorithm and previous methods for top-10 recommendation.

| Datasets | Beauty | | | Clothing | | | Cell Phones | | |
|---|---|---|---|---|---|---|---|---|---|
| Measures (%) | Recall | Hit ratio | Precision | Recall | Hit ratio | Precision | Recall | Hit Ratio | Precision |
| CFKG [5] | 10,34 | 17,13 | 1,96 | 5,47 | 7,97 | 0,76 | 9,50 | 13,46 | 1,33 |
| JRL [4] | 6,95 | 12,78 | 1,55 | 2,99 | 4,64 | 0,44 | 7,51 | 10,94 | 1,1 |
| CORLP [11] | 4,76 | 4,23 | 0,43 | 5,35 | 6,03 | 0,54 | 2,75 | 4,03 | 0,28 |
| Hybrid-SIMLP | **29,1** | **49,15** | **2,91** | **28,59** | **25,63** | **2,56** | **32,45** | **46,52** | **3,25** |

Table 2 provides a performance comparison of the proposed hybrid (Hybrid-SIMLP) recommendation algorithm with previous methods, namely Complex Representation-based Link Prediction (CORLP), Collaborative Filtering with Knowledge Graph (CFKG), and Joint Representation Learning (JRL), framework for top-10 recommendations. The results of the CFKG and JRL methods are taken from [4, 5]. The implementation of these methods appears in [12]. When considering the advantages of a graph-based hybrid recommendation our system (Hybrid-SIMLP) algorithm, the results indicate that the performance of the proposed hybrid recommendation method is better than CFKG and JRL due to using three quality measurements: hits ratio, recall, and precision in the Beauty, Clothing, and Cell Phone datasets. As for the CORLP method, the experimental results are not satisfactory. Hence, it can be observed that hybrid recommendation algorithms that use the integration of multiple types of input data perform better than previous recommendation algorithms that only utilize the user-item interaction matrix.

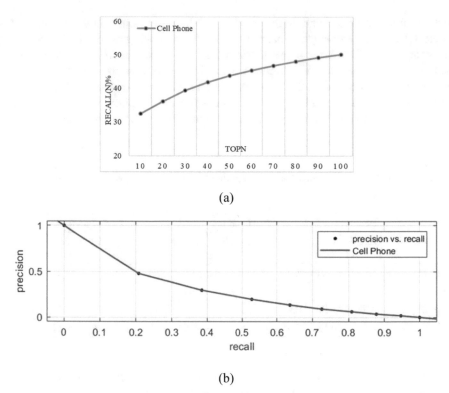

**Fig. 3.** The results based on Cell Phone dataset. (Fig. 3.a: Recall-at-N, and Fig. 3.b: Precision-versus-recall on all items).

The recall-at-N results of the proposed hybrid recommendation algorithm on Beauty, Cell Phone, and Clothing datasets appear in Fig. 3 (a), Fig. 4 (a), and Fig. 5(a), respectively. The recall-at-N and precision-at-N results of the proposed hybrid recommendation algorithm are indicated by utilizing the top-N recommendation task, where N is ranged from 10 to 100. Moreover, the precision-versus-recall comparison of the results is shown in Fig. 3 (b), Fig. 4 (b) and Fig. 5 (b).

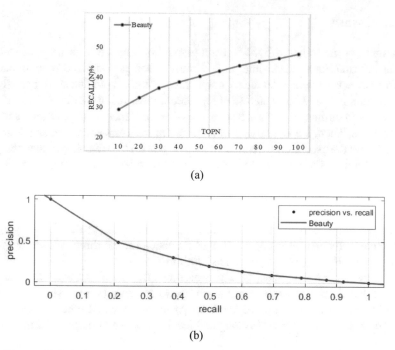

(a)

(b)

**Fig. 4.** The results based on Beauty dataset, (Fig. 4a: Recall-at-N, and Fig. 4b: Precision-versus-recall on all items).

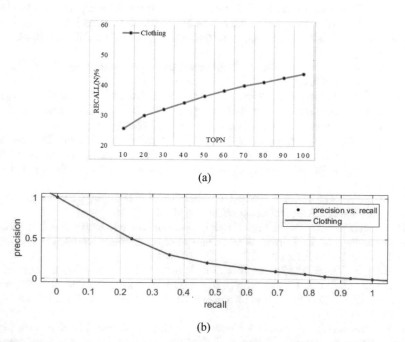

(a)

(b)

**Fig. 5.** The results based on Clothing dataset. (Fig. 5a: Recall-at-N, and Fig. 5b: Precision-versus-recall on all items).

## 6   Conclusion

In this study, a graph-based hybrid recommendation algorithm is proposed utilizing different information sources (images and ratings) for personalized recommendation. Furthermore, a user-item interaction graph incorporates both the user behaviors for each user-item pair and the visual item representations. Additionally, the users' visual preference is identified towards different items' visual images to generate each users' visual features. Then, the similarity factors between the users and between the items are evaluated from the user-visual feature and item-visual feature matrices. Later, the impact of the inclusion of similarity factors on the recommendation performance is explored. The experimental results indicate that the proposed Hybrid-SIMLP method, using visual similarity factors, provides a more efficient recommendation algorithm for the top-N recommendation task. Futhermore, the performance of the proposed recommendation algorithm is better than CORLP [11], CFKG [4], and JRL [5] due to applying three quality measurements: hit-ratio, recall, and precision on the three subsets of Amazon dataset, namely Beauty, Cell-Phone, and Clothing.

The proposed method successfully overcomes the deficiencies of the previously introduced hybrid recommendation algorithms, apart from being more practical with real-world datasets when compared to others. As for future work, other number systems, such as the quaternions, can be investigated for the proposed link prediction method.

## References

1. Wu, C.Y., Diao, Q., Qiu, M., Jiang, J., Wang, C.: Jointly modeling aspects, ratings and sentiments for movie recommendation. In: Proceedings of the 20th ACM SIGKDD International Conference on Knowledge Discovery and Data Mining, New York, USA, pp. 193–202. ACM (2014)
2. Burke, R.: Hybrid recommender systems: survey and experiments. User Model. User-Adap. Inter. **12**(4), 331–370 (2002)
3. Melville, P., Mooney, R.J., Nagarajan, R.: Content-boosted collaborative filtering for improved recommendations. In: Proceedings of 18th National Conference on Artificial Intelligence, Edmonton, Alberta, Canada, pp. 187–192 (2002)
4. Zhang, Y., Ai, Q., Chen, X., Croft, W.B.: Joint representation learning for top-n recommendation with heterogeneous information sources. In: Proceedings of the 2017 ACM on Conference on Information and Knowledge Management, New York, USA, pp. 1449–1458. ACM (2017)
5. Zhang, Y., Ai, Q., Chen, X., Wang, P.: Learning over knowledge-base embeddings for recommendation. arXiv preprint arXiv:1803.06540 (2018)
6. He, R., McAuley, J.: VBPR: Visual bayesian personalized ranking from implicit feedback. In: Proceedings of the Thirtieth AAAI Conference on Artificial Intelligence (AAAI 2016, pp. 144–150. AAAI Press (2016)
7. McAuley, J., Targett, C., Shi, Q., Van Den Hengel, A.: Image-based recommendations on styles and substitutes. In: Proceedings of the 38th International ACM SIGIR Conference on Research and Development in Information Retrieval, New York, USA, pp. 43–52. ACM (2015)
8. He, R., McAuley, J.: Ups and downs: modeling the visual evolution of fashion trends with one-class collaborative filtering. In: Proceedings of the 25th International Conference on World Wide Web, Québec, Canada, pp. 507–517 (2016)

9. Amazon review data (2016). http://jmcauley.ucsd.edu/data/amazon/
10. Kurt, Z., Ozkan, K., Bilge, A., Gerek, O.N.: A similarity-inclusive link prediction based recommender system approach. Elektronika IR Elektrotechnika **25**(6), 62–69 (2019)
11. Xie, F., Chen, Z., Shang, J., Feng, X., Li, J.: A link prediction approach for item recommendation with complex number. Knowl. Based Syst. **81**, 148–158 (2015)
12. Joint Representation Learning Model (JRLM) (2017). https://github.com/QingyaoAi/Joint-Representation-Learning-for-Top-N-Recommendation
13. Zhang, Y., Chen, X.: Explainable recommendation: a survey and new perspectives. arXiv preprint arXiv:1804.11192 (2018)
14. Schafer, J.B., Frankowski, D., Herlocker, J., Sen, S.: Collaborative filtering recommender systems. In: Brusilovsky, P., Kobsa, A., Nejdl, W. (eds.) The Adaptive Web. LNCS, vol. 4321, pp. 291–324. Springer, Heidelberg (2007). https://doi.org/10.1007/978-3-540-72079-9_9
15. Ekstrand, M.D., Riedl, J.T., Konstan, J.A.: Collaborative filtering recommender systems. Found. Trends Hum. Comput. Interact. **4**(2), 81–173 (2011)

# Development of an Artificial Intelligence Based Computerized Adaptive Scale and Applicability Test

Mustafa Yagci[✉]

Faculty of Engineering and Architecture, Ahi Evran University, 40100 Kirsehir, Turkey

**Abstract.** The number of questions in the scales used to determine the attitudes of individuals is one of the biggest problems with this type of scale. Because the number of questions is high, individuals may be distracted, and their motivation may decrease. This has a negative effect on the reliability and validity of the scale. In this study, a decision tree algorithm with an artificial neural networks (ANN) model was developed to determine the learning approaches of students. Thus, the number of questions students will have to answer will be minimized. The research was conducted in two phases. In the first phase, the decision tree model was developed with the ANN J48 algorithm for the data obtained from the original form of the scale in a previous study by the researcher. This phase was carried out with 186 students. In the second phase, an artificial intelligence-based computerized adaptive scale (CAS) was designed by using the decision tree model developed. Then, CAS and the original scale were applied to a different group of students (113 students) from the first phase, and the relationship between these two was investigated. The results of the research are very promising. Students' learning approach preferences were estimated with very high reliability dand accuracy with only a few questions. 95.7% accuracy was achieved with very few questions. According to this result, it can be said that the validity and reliability of the scale developed by the decision tree algorithm is quite high. In addition, it was observed that all participants filled in the scale with great care as it required very few items. This is another factor that increases the reliability of the scale.

**Keywords:** Computerized adaptive scale · Artificial intelligence · Learning approaches · Artificial neural networks · Decision trees

## 1 Introduction

For the design of learning environments, a user model needed to be created; this means collecting information about users and transferring this information to the model [1]. User characteristics are often determined through surveys or scales designed to learn individual differences. Based on the fact that individuals learn in different ways, various techniques have been developed to classify individual differences. In the design of the learning environment, the most important of the individual characteristics frequently used for the purpose of adaptation is the students' learning approach.

© The Author(s), under exclusive license to Springer Nature Switzerland AG 2021
J. Hemanth et al. (Eds.): ICAIAME 2020, LNDECT 76, pp. 292–303, 2021.
https://doi.org/10.1007/978-3-030-79357-9_29

Students acquire and process information in different ways depending on their learning style. A learning environment should ensure that all students learn despite their different learning styles [2]. An e-learning environment should provide the most appropriate information according to individual differences and needs of the students [3]. To achieve this goal, it is necessary to determine how students learn.

The decision tree method and classification method are widely used in the literature in the industrial and engineering fields. However, in the social sciences and educational sciences, very few are used in the analysis of individual evaluation forms such as scales and scale. [1] used decision trees to identify individuals' learning approaches for the design of learning management systems. In the study, the artificial neural networks (ANN) model used the C4.5 algorithm, and it was able to determine the students' learning style preferences with high accuracy with only a few questions. [4] developed a model based on the ID3 algorithm to predict users' response based on online survey data. As a result, the decision tree was found to have higher response rates based on the prediction model. [5] tried to determine the student profile for online courses through data mining. The researchers concluded that determining a student profile with data mining methods had a positive effect on success and motivation. [6] found that data mining methods provide very effective results in determining the factors that affect students' academic achievement. [7] also used decision tree techniques to classify students' internet addiction levels. Similarly, [8] examined students' school performances using educational data mining techniques. [3] classified students according to behavioral factors using the NBTree classification algorithms for an e-learning environment. [2] created an application that determined the learning styles of the students through Bayesian networks in web-based courses. In recent years, the success of different algorithms and artificial intelligence techniques has been examined with a computerized adaptive test (CAS) [9–11].

When the literature is examined, it is seen that the data obtained are classified by ANN methods such as decision trees or data mining, and it is also seen that the number of studies in which [2, 5] e-learning environments are designed according to the results obtained is limited. In this study, the learning approach preferences of the students are classified with the J48 algorithm of the ANN model. Then, an artificial intelligence-based computerized adaptive scale was developed in accordance with the decision tree model created by the J48 algorithm. With the developed scale, students' learning approach preferences were determined. The statistical relationship between the results obtained from the original scale form consisting of 20 questions was examined. Thus, the validity and reliability test of the developed artificial intelligence application was performed. This study is very important in terms of showing that artificial intelligence applications can be developed for scales and scale.

It can sometimes take time for students to answer questions on the scale of learning approaches, and this can be quite tedious. In such scales, the goal is to determine the tendency of the learning approach according to the student's score, and practicality is important in the scales. If a high level of accuracy can be achieved with a small number of questions, it can be said that the validity of the scale is high, and it appears to be a valid and reliable result.

In this study, an approach was proposed to minimize the number of questions needed to determine students' learning approach tendency. In this context, the data collected with the original form of the learning approach scale consisting of 20 questions were classified with the ANN model, and a decision tree model was developed. With the developed decision tree, an artificial intelligence-based CAS was designed.

Artificial intelligence-based computerized adaptive scales are computer software that asks questions to respondents based on their response to previous questions. Since the development of these software requires expertise in the field of IT, very few studies have been conducted [10].

The aim of this study is to develop a decision tree algorithm that can determine the tendency of students' learning approach with a minimum number of questions and by asking different questions to different students. In conclusion, the main problem that this study addresses is how to develop and evaluate a decision tree algorithm with the ANN model. For this purpose, the answers to the following sub-problems were sought:

1. What is the accuracy rate of the decision tree classification algorithm developed by the ANN model of the learning approaches scale?
2. Is there a significant relationship between the classification obtained with the application of the artificial intelligence-based computerized adaptive scale and the classification obtained with the application of the original scale?

## 2   Methodology

In this study, a decision tree algorithm was developed in order to classify students' learning approach tendency with as few questions as possible. In this algorithm, the next question is chosen by taking into consideration the answers given by the student to the previous questions (Fig. 1). In this study, an algorithm was developed to select the most appropriate items from the original learning approaches scale in order to determine the learning approach of each student.

Classification is one of the main objectives of the data mining techniques [12]. In general, these techniques (pre-classified) learn a classification model from the processing of samples. Once the model is learned, it can be used to classify new samples of unknown class [1]. This study explains how to use classification techniques that determine which questions should be asked to each student in order to reduce the number of answers required to classify learning approaches.

This study was carried out in two phases. In the first phase, in a previous study by the researcher, the data collected by the learning approaches scale developed by [13] were classified with ANN. ANN, computer programs that mimic the neural networks of the human brain, can be thought of as a parallel information processing system. ANN make various generalizations by training existing data so it has the ability to learn, and thus, solutions to the events or situations to be produced later are produced. In this context, it is thought that the tendency of individuals' learning approaches can be determined by ANN in a very short time with many fewer questions and with a much higher reliability coefficient.

The data set obtained in this study was analyzed with the J48 algorithm in the Weka Workbench program used in data mining. Each instance in the data set has a set of attributes, one of which is class information. The classification is based on a learning algorithm. The data set knows to which class the student belongs (educational set) is used for educational purposes, and a model is created [12]. The model is tested with the data set (experiment set) that is not included in the learning set, and its success is measured. Using this model, it is possible to determine to which class the samples belong [14, 15]. Classification algorithms learn a model based on instances of the data set in which each instance is defined as a collection of attributes [1]. Thus, an event or an example is formed by a specific student's answer to each question of the scale.

The J48 algorithm was used to determine the most relevant questions in the scale of the learning approaches. The J48 algorithm is an advanced version of the C4.5 decision tree algorithm; J48 forms decision trees from a set of training data [1]. One of the main characteristics of decision trees is the flexibility they provide; for example, a decision tree has the ability to use strong decision rules to classify subsets with different characteristics and different instances [16].

Considering the purpose of this study, decision trees are the most appropriate tools. Decision trees are the most commonly used machine learning algorithms, especially for small data sets where other techniques, such as Bayesian networks, cannot be applied. Decision trees are more advantageous than some other machine learning techniques because the results are more transparent and can be interpreted by field experts without knowledge of computers or artificial intelligence [17]. Nodes in a decision tree include testing a particular property of the sample to be classified. Depending on the attribute value, the most appropriate landing branch is followed. This procedure is repeated repeatedly until a leaf is reached. Generally, each leaf has a label with a class to be assigned to the samples reaching that leaf. Different characteristics can be used from each sample to be classified along the path from root to leaf [1].

In the second phase of the study, an artificial intelligence-based computerized adaptive scale was designed and implemented. At this phase, first, an artificial intelligence-based computerized adaptive scale was developed according to the decision tree model. MySQL, an open source database management system, was used to develop the scale. The interfaces were designed with PHP, HTML and javascript languages. PHP is a server-side, HTML-embedded scripting language for creating platform-independent, dynamic web pages and intranet applications.

Then, the developed scale was applied to 113 students except the first phase students. New data were obtained from the answers of each student to different number of different questions. After a three-week interval, the original form of the scale consisting of 20 questions was applied to the same group of students, and the correlation between these two applications was calculated. This phase of the study was carried out in the factorial design of the experimental design types according to the experimental conditions. The factorial design within the groups allows the main effects and common effects of at least two independent variables on the dependent variable to be tested [18]. At this phase of the study, the relationship between the classification using the original form of the scale and the classification using the artificial intelligence based computerized adaptive scale form using ANN model was examined.

For this general purpose, an artificial intelligence-based computerized adaptive scale based on learning approaches scale items was developed. This test asks as few questions as possible (one, two, three or up to four questions) to determine the general preference of each student regarding the learnisng approach. The resulting model represents students' learning approach preferences but is not a definite result. On the other hand, the ANN msodels [1, 5], which are formed by the data mining method on the data obtained at different times from the related scale, show that this model is sufficient for determining the students' learning style preferences.

### 2.1 Workgroup

The first phase, in which the original form of the scale was used, was conducted with 186 pre-service teachers studying in the department of Computer Education and Instructional Technology at a public university of the 2014–2015 academic year. The second phase in which the accuracy of the developed ANN model was tested was carried out with 113 students studying in the department of Computer Engineering Department a public university of the 2019–2020 academic year.

### 2.2 Analysis and Interpretation of Data

In the first phase of the study conducted in two phases, the decision tree algorithm of the data set obtained from the original scale was developed. In the second phase, the artificial intelligence based computerized adaptive scale was designed by using the developed decision tree algorithm. Figure 1 shows a section of the program codes. Figure 2 shows an image of the developed interfaces. The CAS and the original form of the 20-item scale was applied to a group of students other than the first phase students. The relationship between the classification results obtained from the two applications was calculated by Pearson Correlation analysis. The significance level of the differences and relationships was examined at the $p < .01$ level. Correlation analysis is a statistical method used to test the linear relationship between two variables and to measure the degree of this relationship, if any. In correlation analysis whether there is a linear relationship and the degree of this relationship, if any, is calculated with the correlation coefficient which is between $-1$ and $+1$. The fact that the correlation coefficient is close to $+1$ indicates that there is a positive relationship between the variables, and if it is close to $-1$, there is

```
if (responses$Radio1==1) ogr_yak1=2
    else if (responses$Radio1==3||responses$Radio1==5) ogr_yak1=1
    else if (responses$Radio1==2){
        i=19
        responses<-artificial.template.show(12,params=list(id=i,question=items[i,"question"],a=it
ems[i,"a"],b=items[i,"b"],c=items[i,"c"],d=items[i,"d"],e=items[i,"opt_e"]))
        if (responses$Radio1==1) ogr_yak1=1
            else if (responses$Radio1==3||responses$Radio1==4||responses$Radio1==5) ogr_yak1=2
            else if (responses$Radio1==2){
                i=4
                responses<-artificial.template.show(12,params=list(id=i,question=items[i,"question"],a=it
ems[i,"a"],b=items[i,"b"],c=items[i,"c"],d=items[i,"d"],e=items[i,"opt_e"]))
                if (responses$Radio1==1||responses$Radio1==2) ogr_yak1=1
                    else if (responses$Radio1==5) ogr_yak1=2
                    else if (responses$Radio1==3){
                        i=13
                        responses<-artificial.template.show(12,params=list(id=i,question=items[i,"question"],
a=items[i,"a"],b=items[i,"b"],c=items[i,"c"],d=items[i,"d"],e=items[i,"opt_e"]))
                        if (responses$Radio1==1||responses$Radio1==3||responses$Radio1==4||responses$Radio1==5) o
gr_yak1=1
                        else if (responses$Radio1==2) ogr_yak1=2
    }
                    else if (responses$Radio1==4){
                        i=18
```

**Fig. 1.** A section of artificial intelligence-based CAS program codes

an excellent negative relationship. The fact that the correlation coefficient is 0 indicates that there is no linear relationship between the variables [19].

| Soru no: 6 | Yeni konuların pek çoğunu ilginç bulurum ve konulara ilgili daha fazla bilgi bulabilmek için sık sık ekstra zaman harcarım. |
| --- | --- |
| ⦿ | Hiç |
| ○ | Bazen |
| ○ | Yarı Yarıya |
| ○ | Sık Sık |
| ○ | Her Zaman |
| | **Gönder** |

**Fig. 2.** Artificial intelligence-based CAS interface

# 3 Findings

1. What is the accuracy rate of the developed decision tree classification algorithm?

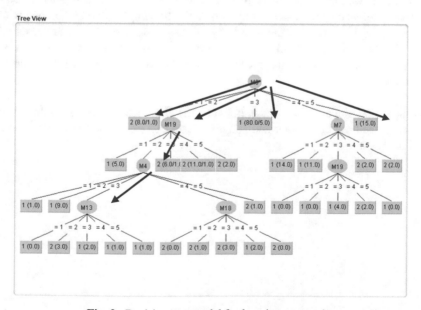

**Fig. 3.** Decision tree model for learning approaches

Figure 3 shows the classification tree obtained by the J48 algorithm developed for the scale of learning approaches. The circles represent the scale item while the rectangles represent the student's learning approach. It is seen that the decision tree starts with the sixth question and the classification is made with at least one question and a maximum four questions. Figure 4 shows the classification model of the decision tree obtained from the sample.

The questions are represented by the circles (Mxx). Here xx is the question number of the relevant item on the scale of learning approaches. The answer given to each question is between one and five. The student is asked a new question from the classification tree depending on the answer (one to five). This process continues until it reaches the leaves.

In the classification tree, the leaves are represented by squares. If the value of the squares is one, the student has a deep learning approach and, if it is two, has a superficial learning approach. According to the classification tree, the sixth item of the scale represents the root of the tree. If the student responds to the sixth item of the scale by selecting the first, third, or fifth options, the learning approach tendency is determined without the student having to answer any other questions. For example, the student who answers the sixth item by selecting the first option has a superficial approach tendency, and the student who answers the third or fifth options has adopted a deep learning approach. The three roads indicated by red arrows in Fig. 1 are the shortest roads (with one item). The path indicated by the purple arrow is the longest (6th, 19th, 4th and 13th) roads. In the longest path, the student has to answer articles 6, 19, 4 and 13 (Fig. 1, purple arrows).

Table 1 provides a summary of the decision tree. Of the total 186 sample data, 178 were classified as desired. Accurate classification success was calculated as 95.7%. The Kappa coefficient was found to be 0.877. Accordingly, the success of the classification agreement was very high. This value is a measure of reliability, indicating that success does not depend on chance.

Table 2 shows the decision tree matrix table. From the table, it is understood that 178 of 186 sample data are correctly classified. Of the sample data, 143 belong to class a (deep learning approach), while 43 belong to class b (surface learning approach). 140 out of 143 data belonging to class a and 38 out of 43 data belonging to class b were correctly classified. That is, only three data from class a and only five from class b are misclassified.

2. Is there a significant relationship between the classification based on the artificial intelligence based computerized adaptive scale and the classification based on the original scale?

```
=== Classifier model (full training set) ===

J48 pruned tree
------------------

M6 = 1: 2 (8.0/1.0)
M6 = 2
|   M19 = 1: 1 (5.0)
|   M19 = 2
|   |   M4 = 1: 1 (1.0)
|   |   M4 = 2: 1 (9.0)
|   |   M4 = 3
|   |   |   M13 = 1: 1 (0.0)
|   |   |   M13 = 2: 2 (3.0)
|   |   |   M13 = 3: 1 (2.0)
|   |   |   M13 = 4: 1 (1.0)
|   |   |   M13 = 5: 1 (1.0)
|   |   M4 = 4
|   |   |   M18 = 1: 2 (0.0)
|   |   |   M18 = 2: 2 (1.0)
|   |   |   M18 = 3: 2 (3.0)
|   |   |   M18 = 4: 1 (2.0)
|   |   |   M18 = 5: 2 (0.0)
|   |   M4 = 5: 2 (1.0)
|   M19 = 3: 2 (6.0/1.0)
|   M19 = 4: 2 (11.0/1.0)
|   M19 = 5: 2 (2.0)
M6 = 3: 1 (80.0/5.0)
M6 = 4
|   M7 = 1: 1 (14.0)
|   M7 = 2: 1 (11.0)
|   M7 = 3
|   |   M19 = 1: 1 (0.0)
|   |   M19 = 2: 1 (0.0)
|   |   M19 = 3: 1 (4.0)
|   |   M19 = 4: 2 (2.0)
|   |   M19 = 5: 1 (0.0)
|   M7 = 4: 2 (2.0)
|   M7 = 5: 2 (2.0)
M6 = 5: 1 (15.0)

Number of Leaves  :      29

Size of the tree :       36
```

**Fig. 4.** Classifier model (full training set)

Table 3 summarizes the results of the correlation analysis of the relationship between two separate classifications.

When Table 3 is examined, it is seen that there is a high level, positive, and significant relationship between an artificial intelligence-based computerized adaptive scale and classification according to the original scale ($r = .729$, $p < .01$).

**Table 1.** Summary of decision tree

| Correctly classified instances | 178 (% 95.7) |
|---|---|
| Kappa statistic | 0.877 |
| Mean absolute error | 0.0785 |
| Root mean squared error | 0.1982 |
| Relative absolute error | 22.0014% |
| Root relative squared error | 47.0059% |
| Total number of instances | 186 |

**Table 2.** Confusion matrix

| a | b | ←classified as |
|---|---|---|
| 140 | 3 | a = 1 |
| 5 | 38 | b = 2 |

**Table 3.** The relationship between classification based on artificial **intelli**gence-based CAS and classification based on the original scale

| Artificial intelligence based computerized adaptive scale | Grand Average | |
|---|---|---|
| Original version of the scale | r | .729** |
| | p | .000 |
| | N | 113 |

**Correlation is significant at the 0.01 level (2 tailed)

## 4   Discussion and Conclusion

In this study, an approach was proposed to minimize the number of questions needed to determine students' learning approach tendencies. In this context, the data obtained from the learning approaches scale were classified with ANN, and a decision tree model was developed. With the developed decision tree model, an artificial intelligence-based computerized adaptive test was designed and evaluated.

As a result of the application of the J48 decision tree algorithm, one of the ANN models, classification can be made with at least one and a maximum four items. With this decision tree, students' learning approaches were determined with only a few items, and a very high level of accuracy was achieved. Practicality is important in scales. It can be said that the validity of the scale is high if a high level of accuracy can be achieved with few questions. The results of this study show that the J48 decision tree algorithm provides a high level of accuracy for the learning approaches scale; therefore, the validity of the scale is high. So, the J48 decision tree algorithm gave a valid and reliable result for the scale of learning approaches. In addition, due to the small number of questions to be answered, it was observed that all participants filled out the scale with great care. This positively affected the motivation and reliability of the scale.

However, the small number of samples used in the classification caused both larger errors and a requirement for longer paths to the roots [1]. The classification tree created with 186 samples determined for this study provided access to the leaf using very short paths with a very high level of accuracy. Accordingly, it can be said that the number of samples determined is sufficient. In addition, the sample size and the application results in two phases show that the classification tree produced with different sample groups will not be very different.

[1] also developed a decision tree with 97% accurate classification which required an average of four or five questions to classify students' learning styles. Using the NBTree classification algorithm, [3] the students were classified according to their interests. As a result, they found that the results obtained from the decision tree algorithm and the results obtained through the traditional survey were consistent with each other. [5] tried to determine the student profile through data mining for use in online courses. As a result, based on the individual characteristics of the students, six decision trees were developed to classify the participation rates in online courses. To determine students' learning styles, [2] also collected data with the Bayesian network in an online course. From the literature, it is understood that ANN are very successful in determining student profiles.

In this study, a computerized adaptive scale based on artificial intelligence was developed. Students' learning approaches could be classified with only a few items (an average of 2.5 items) using this scale. It was concluded that there was a high, positive, and significant relationship with the classification results obtained from the original scale. Accordingly, it can be said that the classification using the decision tree algorithm and the classification using the original scale yield the same result. This proves that the developed decision tree algorithm works correctly. According to this result, it can be said that the algorithm developed in this study can be used to determine students' learning approaches. When the literature is reviewed, it is seen that various scale data are classified with ANN. However, a study that developed an application using the developed algorithm and compared the results could not be found; thus, this aspect can be said to be an original study.

Students' responses to scale items vary according to their individual differences, so, accordingly, the individual differences of students can be determined. The scales used to determine individual differences consisted of 30 items, on average. In fact, the number of items in some scales used by researchers can reach 50, 60, or even 70. In this case, individuals can become bored while answering the scale items and may even randomly mark the items. Therefore, the reliability and accuracy level of the research results decreases. In this context, the artificial intelligence based computerized adaptive scale developed in this study is thought to prevent such problems.

# 5  Suggestions

Artificial intelligence-based version of the type of scales frequently used by researchers can be designed. Thus, more reliable and valid results can be obtained. It is also proposed that an internet-based platform for the design of such artificial intelligence-based computerized adaptive tests be designed.

# References

1. Ortigosa, A., Paredes, P., Rodriguez, P.: AH-questionnaire: an adaptive hierarchical questionnaire for learning styles. Comput. Educ. **54**, 999–1005 (2010). https://doi.org/10.1016/j.compedu.2009.10.003
2. García, P., Schiaffino, S., Amandi, A.: An enhanced Bayesian model to detect students' learning styles in Web-based courses. J. Comput. Assist. Learn. **24**, 305–315 (2008). https://doi.org/10.1111/j.1365-2729.2007.00262.x
3. Özpolat, E., Akar, G.B.: Automatic detection of learning styles for an e-learning system. Comput. Educ. **53**(2), 355–367 (2009). https://doi.org/10.1016/j.compedu.2009.02.018
4. Luo, N., Wu, S., Zou, G., Shuai, X.: Online survey prediction model for high response rate via decision tree. In: 12th UIC-ATC-ScalCom, pp. 1792–1797, August 2015
5. Topîrceanu, A., Grosseck, G.: Decision tree learning used for the classification of student archetypes in online courses. Procedia Comput. Sci. **112**, 51–60 (2017). https://doi.org/10.1016/j.procs.2017.08.021
6. Kurt, E., Erdem, A.O.: Discovering the factors effect student success via data mining techniques. J. Polytech. **15**(2), 111–116 (2012)
7. Kayri, M., Günüç, S.: Turkey's secondary students' internet addiction affecting the level decision trees with some of the factors investigation methods. Educ. Admin. Theor. Pract. **10**(4), 2465–2500 (2010)
8. Bilen, Ö., Hotaman, D., Aşkın, Ö.E., Büyüklü, A.H.: Analyzing the school performances in terms of LYS successes through using educational data mining techniques: İstanbul sample 2011. Educ. Sci. **39**(172), 78–94 (2014)
9. Sabbaghan, S., Gardner, L.A., Chua, C.: A Q-sorting methodology for computer-adaptive surveys - style "research". In: ECIS-2016, pp. 1–14 (2016)
10. Sabbaghan, S., Gardner, L., Eng Huang Chua, C.: A threshold for a Q-sorting methodology for computer-adaptive surveys. In: ECIS-2017, vol. Spring, pp. 2896–2906 (2017). http://aisel.aisnet.org/ecis2017_rip/39
11. Sabbaghan, S., Gardner, L., Eng Huang Chua, C.: Computer-adaptive surveys ( CAS ) as a means of answering questions of why. In: PACIS, vol. 57, pp. 1–15 (2017)
12. Witten, I., Frank, E.: Data Mining–Practical Machine Learning Tools and Techniques. Elsevier, Amsterdam (2005)
13. Biggs, J., Kember, D., Leung, D.Y.P.: The revised two-factor study process questionnaire: R-SPQ-2F. Br. J. Educ. Psychol. **71**(1), 133–149 (2001). https://doi.org/10.1348/000709901158433
14. Quinlan, J.R.: Induction of decision trees. Res. Dev. Expert Syst. XV **1**(1), 81–106 (1986). https://doi.org/10.1023/A:1022643204877
15. Cihan, P., Kalıpsız, O.: Determining the most successful classification algorithm for the student project questionnaire. TBV J. Comput. Sci. Eng. **8**(1), 41–48 (2015)
16. Safavian, S.R., Landgrebe, D.: A survey of decision tree classifier methodology. IEEE Trans. Syst. Man Cybern. **21**(3), 660–674 (1991)

17. Sánchez-Maroño, N., Alonso-Betanzos, A., Fontenla-Romero, O., Polhill, J.G., Craig, T.: Empirically-derived behavioral rules in agent-based models using decision trees learned from questionnaire data. In: Agent-Based Modeling of Sustainable Behaviors, pp. 53–76. Springer, Cham (2017)
18. Büyüköztürk, Ş: Experimental Patterns. Pegem Publishing, Ankara (2001)
19. Kalaycl, Ş: SPSS Applied Multivariate Statistical Techniques. Asil Publishing, Ankara (2009)

# Reduced Differential Transform Approach Using Fixed Grid Size for Solving Newell–Whitehead–Segel (NWS) Equation

Sema Servı[1][(✉)] and Galip Oturanç[2]

[1] Department of Computer Engineering, Selcuk University, Konya, Turkey
semaservi@selcuk.edu.tr
[2] Department of Mathematics, Karamanoglu Mehmetbey University, Karaman, Turkey
goturanc@kmu.edu.tr

**Abstract.** In this study, the Equation of Newell-Whitehead Segel (NWS) is solved with a new method named as *Reduced Differential Transform by using Fixed Grid Size* (RDTM with FGS) [1–3, 12]. The method is quite useful for linear and nonlinear differential equation solutions. The simplicity efficiency, success and trustworthiness of the mentioned method, two equations were solved in the implementation part and results were compared through the *Method of Variational Iteration* (VIM) [5–18].

**Keywords:** Reduced differential transform method (RDTM) · VIM · Newell–Whitehead–Segel equation

## 1 Introduction

The problems involving partial differential equations have emerged not only in mathematics but in many Applied Sciences. Partial differential equations are very difficult and time consuming to solve. Therefore, approximate solution methods have gained importance such as *Variational Iteration* [7, 26], *Homotopy Analysis* [8], *Homotopy Perturbation* [9], *Differential Transform* (DTM) [10], *Adomian's Decomposition* (ADM) [7–11, 26], *Reduced Differential Transform* (RDTM) [1–4, 6, 12, 14–17], *Exponential Finite Difference* [23, 24], *Mahgoub Adomian Decomposition* [25], *Exponential B-Spline Collocation* [27] methods and *Modified Variational Iteration Algorithm*-II [28], etc.

RDTM is a method for the development of DTM and has been applied to many problems [1, 6, 14–17]. With this method, the results were obtained by simple operations by converting partial differentiable equations into algebraic equations [4, 6, 14, 16]. In this study, we developed RDTM with fixed grid size algorithm by dividing the intervals given in the equations into equal parts. Fixed grid size algorithms were earlier applied in the literature [3, 12] for achieving the approximate solution of initial value problems with DTM, either linear or non-linear. By using these, positive results were obtained. We also applied RDTM on the NWS equation, which is an amplitude equation known as *Generalized Fisher* and NWS equations in the literature. The equation is an amplitude

© The Author(s), under exclusive license to Springer Nature Switzerland AG 2021
J. Hemanth et al. (Eds.): ICAIAME 2020, LNDECT 76, pp. 304–314, 2021.
https://doi.org/10.1007/978-3-030-79357-9_30

equation and describes strip design in two-dimensional systems. Also, models the interplay of diffusion term impact with the non-linear impact of the reaction term [5–20]. The NWS is given in Eq. (1.1)

$$u_t(x, t) = ku_{xx} + au(x, t) - b^q u(x, t) \tag{1.1}$$

where a, b, k: real numbers, f(x): an analytic function, q and k: positive integers. The initial condition for Eq. (1.1),

$$u(x, 0) = f(x) \tag{1.2}$$

## 2   RDTM with Grid Size Solution

The definitions given in the literatüre for RDTM can be seen in several references [1–5]:

**Definition 2.1.** If the $u(x, t)$ function is analytic and constantly differentiated regarding time t and space x in the related field, t-dimensional $U_k(x)$ transformed function is as in Eq. (2.1):

$$U_k(x) = \frac{1}{k!} \left[ \frac{\partial^k}{\partial t^k} u(x, t) \right]_{t=0} \tag{2.1}$$

In this study, while the original function is represented with the lowercase (i.e. $u(x, t)$). The transformed function is represented with uppercase (i.e. $U_k(x)$).

The definition of the differential inverse transform for $U_k(x)$ is given below:

$$u(x, t) = \sum_{k=0}^{\infty} U_k(x) t^k \tag{2.2}$$

When Eqs. (2.1) and (2.2) are combined, this equation below can be written:

$$u(x, t) = \sum_{k=0}^{\infty} \frac{1}{k!} \left[ \frac{\partial^k}{\partial t^k} u(x, t) \right]_{t=0} \tag{2.3}$$

Considering the definitions above, it can be said that the power series expansion leads us to the notion of the RDTM.

From the Table 1, we can write the following iteration formula:

**Table 1.** Reduced differential transformation

| Functional form | Transformed form |
|---|---|
| $u(x, t)$ | $U_k(x) = \frac{1}{k!} \left[ \frac{\partial^k}{\partial t^k} u(x, t) \right]_{t=0}$ |
| $w(x, t) = u(x, t) \pm v(x, t)$ | $W_k(x) = U_k(x) \pm V_k(x)$ |
| $w(x, t) = \alpha u(x, t)$ | $W_k(x) = \alpha U_k(x)$ ($\alpha$ is a constant) |
| $w(x, y) = x^m t^n$ | $W_k(x) = x^m \delta(k - n)$ |
| $w(x, y) = x^m t^n u(x, t)$ | $W_k(x) = x^m U(k - n)$ |
| $w(x, t) = u(x, t) v(x, t)$ | $W_k(x) = \sum_{r=0}^{k} V_r(x) U_{k-r}(x) = \sum_{r=0}^{k} U_r(x) V_{k-r}(x)$ |
| $w(x, t) = \frac{\partial^r}{\partial t^r} u(x, t)$ | $W_k(x) = (k + 1) \dots (k + r) U_{k+1}(x) = \frac{(k+r)!}{k!} U_{k+r}(x)$ |
| $w(x, t) = \frac{\partial}{\partial x} u(x, t)$ | $W_k(x) = \frac{\partial}{\partial x} U_k(x)$ |

Our purpose here is to investigate the solution of (1.2) the interval and the interval is divided into subintervals which are equally interspaced by the grid points as indicated in Fig. 1

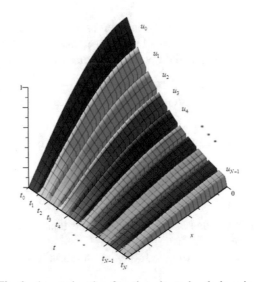

**Fig. 1.** Approximation functions in each sub domain.

Here $t_0 = 0$, $t_N = T$ as in (2.4) for each i = 0, 1, 2, ..., N

$$t_i = t_0 + ih, \ h = \frac{T}{N} \tag{2.4}$$

Here approximate solution function $u(x, t)$ can be indicated as $u^0(x, t)$ at the first interval and after RDTM procedures are applied, Taylor polynomials from nth order can be obtained at the point $t_0 = 0$ as

$$u^0(x, t) = U_0^0(x, t_0) + U_1^0(x, t_0)(t - t_0) + U_2^0(x, t_0)(t - t_0)^2 + \dots + U_n^0(x, t_0)(t - t_0)^n$$

(2.5)

Remember that the equation for initial condition (1.2) was,

$$u(x, t_0) = U_0^0(x, t_0)$$

(2.6)

So the approximate value of $u(x, t)$ at $t_1$ can be evaluated by (2.5) as

$$u(x, t_1) \approx u^0(x, t_1)$$
$$= U_0^0(x, t_0) + U_1^0(x, t_0)(t_1 - t_0) + U_2^0(x, t_0)(t_1 - t_0)^2 +$$
$$\dots + U_n^0(x, t_0)(t_1 - t_0)^n$$
$$= U_0^0 + U_1^0 h + U_2^0 h^2 + \dots + U_n^0 h^n$$
$$= \sum_{j=0}^{n} U_j^0 h^j$$

(2.7)

The value of $u^0(x, t)$ is initial value of the second sub-domain $u^0(x, t)$ at $t_1$ so formula (2.8) can be written,

$$u^1(x, t) = U_0^1(x, t_1) + U_1^1(x, t_1)(t - t_1) + U_2^1(x, t_1)(t - t_1)^2 +$$
$$\dots + U_n^1(x, t_1)(t - t_1)^n$$
$$u^1(x, t_1) = u^0(x, t_1) = U_0^1(x, t_1)$$

(2.8)

In a similar way, $u^2(x, t_2)$ can be calculated as

$$u(x, t_2) \approx u^1(x, t_2)$$
$$= U_0^1(x, t_1) + U_1^1(x, t_1)(t_2 - t_1) + U_2^1(x, t_1)(t_2 - t_1)^2 +$$
$$\dots + U_n^1(x, t_1)(t_2 - t_1)^n$$
$$= U_0^1 + U_1^1 h + U_2^1 h^2 + \dots + U_n^1 h^n$$
$$= \sum_{j=0}^{n} U_j^1 h^j$$

(2.9)

With the same process, the solution of $u^i(x, t)$ at $t_{i+1}$ grid point can be found.

$$u(x, t_{i+1}) \approx u^i(x, t_{i+1})$$
$$= U_0^i(x, t_i) + U_1^i(x, t_i)(t_{i+1} - t_i) + U_2^i(x, t_i)(t_{i+1} - t_i)^2 +$$
$$\dots + U_n^i(x, t_i)(t_{i+1} - t_i)^n$$
$$= U_0^i + U_1^i h + U_2^i h^2 + \dots + U_n^i h^n$$

$$= \sum_{j=0}^{n} U_j^i h^j \tag{2.10}$$

From here, if the values of $u^i(x, t)$ are calculated, analytical solution of $u(x, t)$ is obtained as

$$u(x, t) = \begin{cases} u^0(x, t), & 0 \le x \le 1, 0 < t \le t_1 \\ u^1(x, t), & 0 \le x \le 1, t_1 \le t \le t_2 \\ \vdots \\ u^{N-1}(x, t), & 0 \le x \le 1, t_{N-1} \le t \le t_N \end{cases}$$

We get more sensitive results when we increase the value of $N$.

## 3  Application

The numerical results obtained by the proposed approach are encouraging. The following examples are considerable about the method.

**Example 3.1**  The nonlinear NWS Equation, is given by the following equation [5, 19]:

$$u_t = u_{xx} + 2u - 3u^2 \tag{3.1}$$

The initial condition is given by,

$$u(x, 0) = \lambda \tag{3.2}$$

and exact solution for the (3.1) equation is,

$$u(x, t) = \frac{-\frac{2}{3}\lambda e^{2t}}{-\frac{2}{3} + \lambda - \lambda e^{2t}}$$

By choosing $N = 5$ we get t values $\{t_0, t_1, t_2, t_3, t_4, t_5\}$ as $\{0, \frac{1}{5}, \frac{2}{5}, \frac{3}{5}, \frac{4}{5}, 1\}$ respectively.

In accordance with the process of the solution, firstly RDTM of (3.1) and (3.2) can be reached through Table 1 and the Eq. (3.3) is obtained.

$$(k + 1)U_{k+1}^0(x) = \frac{\partial^2}{\partial x^2} U_k^0(x) + 2U_k^0(x) - 3 \sum_{r=0}^{k} U_r(x)U_{k-r}(x)$$
$$u(x, t_0) = U_0^0(x, t_0) = \lambda \tag{3.3}$$

Then, following graphics show the behavior of the results RDTM with FGS and VIM [5]. The our method of solution is convergent to the exact solution. So it's obvious that it gives results better than VIM (Fig. 2 and Table 2).

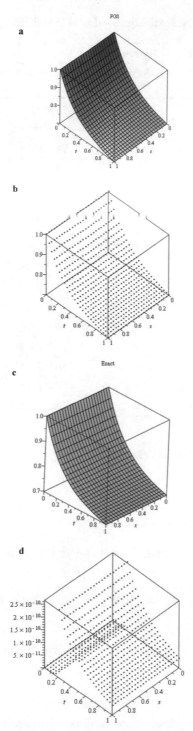

**Fig. 2.** For $\lambda = 1$, In this graphics, for $N = 5$, **a** The area indicates the FGS solution, **b** The blue spots indicate the result we obtained by VIM, **c** The green area indicates the exact solution, **d**: The surfaces show the absolute error of RDTM with FGS

**Table 2.** Numerical results of RDTM with FGS solution.

| | X | For N = 5 | |
|---|---|---|---|
| | | Absolute error (RDTM) | Absolute error (VIM) |
| $t = 0.2$ | 0.2 | $2 * 10^{-9}$ | $73 * 10^{-8}$ |
| | 0.4 | $2 * 10^{-9}$ | $73 * 10^{-8}$ |
| | 0.6 | $2 * 10^{-9}$ | $73 * 10^{-8}$ |
| | 0.8 | $2 * 10^{-9}$ | $73 * 10^{-8}$ |
| $t = 0.4$ | 0.2 | $3 * 10^{-9}$ | $509 * 10^{-7}$ |
| | 0.4 | $3 * 10^{-9}$ | $509 * 10^{-7}$ |
| | 0.6 | $3 * 10^{-9}$ | $509 * 10^{-7}$ |
| | 0.8 | $3 * 10^{-9}$ | $509 * 10^{-7}$ |
| $t = 0.6$ | 0.2 | $8 * 10^{-10}$ | $10751 * 10^{-10}$ |
| | 0.4 | $8 * 10^{-10}$ | $10751 * 10^{-10}$ |
| | 0.6 | $8 * 10^{-10}$ | $10751 * 10^{-10}$ |
| | 0.8 | $8 * 10^{-10}$ | $10751 * 10^{-10}$ |
| $t = 0.8$ | 0.2 | $4.1 * 10^{-9}$ | $16477 * 10^{-10}$ |
| | 0.4 | $4.1 * 10^{-9}$ | $16477 * 10^{-10}$ |
| | 0.6 | $4.1 * 10^{-9}$ | $16477 * 10^{-10}$ |
| | 0.8 | $4.1 * 10^{-9}$ | $16477 * 10^{-10}$ |

**Example 3.2** The linear NWS *equation* is

$$u_t = u_{xx} - 2u \tag{3.4}$$

when the initial condition is [5, 19],

$$u(x, 0) = e^x,$$

The analytical solution is given in the literature [14] is as

$$u(x, t) = e^{x-t} \tag{3.5}$$

According to procedure, first the RDTM is got via Table 1 and then the Eq. (3.6) is obtained,

$$(k + 1)U_{k+1}^0(x) = \frac{\partial^2}{\partial x^2} U_k^0(x) - 2U_k^0(x) \tag{3.6}$$

and initial conditions,

$$u(x, t_0) = U_0^0(x, t_0) = e^x$$

Following graphics show the results RDTM with FGS-Exact and VIM-Exact [5] (Fig. 3 and Table 3).

**a**                              FGS-Exact

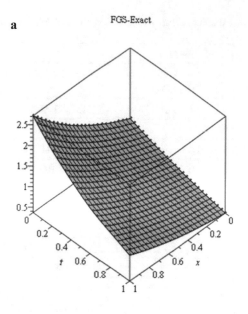

**b**                              VIM-Exact

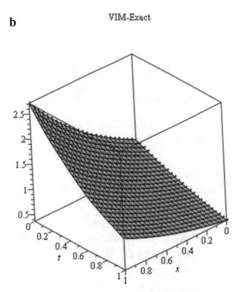

**Fig. 3.** In this graphics, for N = 5, **a** the green area indicates the exact solution and the red spots indicate the result we obtained by RDTM with FGS, **b** the green area indicates the exact solution and the blue spots indicate the result we obtained by VIM.

**Table 3.** Absolute errors of RDTM with FGS solution and VIM

|  | X | for N = 5 | |
|---|---|---|---|
|  |  | Absolute error (RDTM) | Absolute error (VIM) |
| $t = 0.2$ | 0.2 | $2 * 10^{-20}$ | $2.243711172*10^{-9}$ |
|  | 0.4 | 0 | $2.7404750148*10^{-9}$ |
|  | 0.6 | 0 | $3.3472237417*10^{-9}$ |
|  | 0.8 | 0 | $4.088308310*10^{-9}$ |
| $t = 0.4$ | 0.2 | $1 * 10^{-20}$ | $2.0820070810267*10^{-7}$ |
|  | 0.4 | $1 * 10^{-20}$ | $2.542969191275*10^{-7}$ |
|  | 0.6 | 0 | $3.105989584140*10^{-7}$ |
|  | 0.8 | $1 * 10^{-19}$ | $3.79366424488*10^{-7}$ |
| $t = 0.6$ | 0.2 | $2*10^{-20}$ | $2.58491491795128*10^{-6}$ |
|  | 0.4 | $3 * 10^{-20}$ | $3.15722221039506*10^{-6}$ |
|  | 0.6 | 0 | $3.8562399159011*10^{-6}$ |
|  | 0.8 | 0 | $4.7100220694089*10^{-6}$ |
| $t = 0.8$ | 0.2 | $1 * 10^{-20}$ | $1.410612033284603*10^{-5}$ |
|  | 0.4 | $1 * 10^{-20}$ | $1.722925428147740*10^{-5}$ |
|  | 0.6 | $1 * 10^{-20}$ | $2.104385870043942*10^{-5}$ |
|  | 0.8 | 0 | $2.57030270590496*10^{-5}$ |

## 4 Results and Discussion

In this paper, two examples are given in order to demonstrate that the reduced differentials transform method with fixed grid size is an efficient, good-performing method to solve this partial nonlinear differential equation. Examples were solved with RDTM with FGS and VIM, and the results were exhibited with tables and graphs. Both examples show that RDTM with FGS is a strong approximate method for solving the NWS Equation and other linear or nonlinear differential equations.

## 5 Conclusions

The method RDTM with FGS has been applied to the *Newell - Whitehead – Segel* partial differential equation, which is nonlinear, and the method is described on two examples. The method can be programmed quite easily, and is extremely efficient and effective because it can be simply adapted to algebraic equation. It is more successful than the VIM method seen in the results obtained. The proposed approach in this study gave promising results in solving nonlinear partial differential equations of diverse type. The exact solution in a fast convergent sequence can be calculated by an analytic approach and this can be presented as the main advantage of the method. In addition, the calculations made in this study were done with the Maple program.

**Acknowledgements.** The authors would like to acknowledge the anonymous referee for his/her careful reading and precious comments.

# References

1. Servi, S.: Ph. D. Thesis, Selcuk University, İmproved reduced differential transform method for solution of linear and nonlinear partial differential (in Turkish), Konya (2016)
2. Keskin, Y., Sema, S., Galip, O.: Reduced differential transform method for solving Klein Gordon equations. In: Proceedings of the World Congress on Engineering, vol. 1 (2011)
3. Jang, M.-J., Chen, C.-L., Liy, Y.-C.: On solving the initial-value problems using the differential transformation method. Appl. Math. Comput. **115**(2), 145–160 (2000)
4. Keskin, Y., Oturanc, G.: Reduced differential transform method for partial differential equations. Int. J. Nonlinear Sci. Numer. Simul. **10**(6), 741–750 (2009)
5. Prakash, A., Kumar, M.: He's variational iteration method for the solution of nonlinear Newell–Whitehead–Segel equation. J. Appl. Anal. Comput. **6**(3), 738–748 (2016)
6. Keskin, Y., Oturanç, G.: Reduced differential transform method for generalized KdV equations. Math. Comput. Appl. **15**(3), 382–393 (2010)
7. Bildik, N., Konuralp, A.: The use of variational iteration method, differential transform method and Adomian decomposition method for solving different types of nonlinear partial differential equations. Int. J. Nonlinear Sci. Numer. Simul. **7**(1), 65–70 (2006)
8. Liao, S.: Notes on the homotopy analysis method: some definitions and theorems. Commun. Nonlinear Sci. Numer. Simul. **14**(4), 983–997 (2009)
9. Mohyud-Din, S.T., Noor, M.A.: Homotopy perturbation method for solving partial differential equations. Zeitschriftfür Naturforschung-A **64**(3), 157 (2009)
10. Kurnaz, A., Oturanç, G., Kiris, M.E.: n-Dimensional differential transformation method for solving PDEs. Int. J. Comput. Math. **82**(3), 369–380 (2005)
11. Adomian, G.: Solutions of nonlinear PDE. Appl. Math. Lett. **11**(3), 121–123 (1998)
12. Kurnaz, A., Oturanç, G.: The differential transform approximation for the system of ordinary differential equations. Int. J. Comput. Math. **82**(6), 709–719 (2005)
13. Wazwaz, A.M.: Partial Differential Equations and Solitary Waves Theory. NPS, Springer, Heidelberg (2009). https://doi.org/10.1007/978-3-642-00251-9
14. Keskin, Y., Oturanç, G.: Application of reduced differential transformation method for solving gas dynamic equation. Int. J. Contemp. Math. Sci. **22**(22), 1091–1096 (2010)
15. Keskin, Y., Sema, S., Galip O.: Reduced differential transform method for solving Klein Gordon equations. In: Proceedings of the World Congress on Engineering, vol. 1 (2011)
16. Keskin, Y., Oturanc, G.: Numerical solution of regularized long wave equation by reduced differential transform method. Appl. Math. Sci. **4**(25), 1221–1231 (2010)
17. Servi, S., Yildiray, K., Galip, O.: Reduced differential transform method for improved Boussinesq equation. In: Proceedings of the İnternatıonal Conference on Numerical Analysis and Applied Mathematics 2014 (ICNAAM-2014), vol. 1648, no. 1. AIP Publishing (2015)
18. Saravanan, A., Magesh, N.: A comparison between the reduced differential transform method and the Adomian decomposition method for the Newell–Whitehead–Segel equation. J. Egyptian Math. Soc. **21**(3), 259–265 (2013)
19. Nourazar, S.S., Mohsen, S., Nazari-Golshan, A.: On the exact solution of Newell-Whitehead-Segel equation using the homotopy perturbation method. arXiv preprint arXiv:1502.08016 (2015).

20. Debnath, L.: Nonlinear Partial Differential Equations for Scientists and Engineers. Springer, Cham (2011). https://doi.org/10.1007/978-0-8176-8265-1
21. Drazin, P.G., Johnson, R.S.: Solitons: an Introduction, vol. 2. Cambridge University Press, Cambridge (1989)
22. Ablowitz, M.J., Clarkson, P.A.: Solitons, Nonlinear Evolution Equations and Inverse Scattering, vol. 149. Cambridge University Press, Cambridge (1991)
23. Hilal, N., Injrou, S., Karroum, R.: Exponential finite difference methods for solving Newell–Whitehead–Segel equation. Arab. J. Math. **9**(2), 367–379 (2020). https://doi.org/10.1007/s40 065-020-00280-3
24. Baleanu, D., et al.: Analytical and numerical solutions of mathematical biology models: the Newell-Whitehead-Segel and Allen-Cahn equations. Math. Methods Appl. Sci. **43**(5), 2588–2600 (2020)
25. Almousa, M., et al.: Mahgoub adomian decomposition method for solving Newell-Whitehead-Segel equation (2020)
26. Gupta, S., Goyal, M., Prakash, A.: Numerical treatment of Newell-Whitehead-Segel equation. TWMS J. App. Eng. Math. **10**(2), 312–320 (2020)
27. Wasima, I., et al.: Exponential B-spline collocation method for solving the gen-eralized Newell-Whitehead-Segel equation
28. Ahmad, H., et al.: Analytic approximate solutions for some nonlinear parabolic dynamical wave equations. J. Taibah Univ. Sci. **14**(1), 346–358 (2020)

# A New Ensemble Prediction Approach to Predict Burdur House Prices

Mehmet Bilen[1](✉), Ali H. Işık[2], and Tuncay Yiğit[3]

[1] Golhisar School of Applied Sciences, Mehmet Akif Ersoy University, Burdur, Turkey
mbilen@mehmetakif.edu.tr
[2] Department of Computer Engineering, Mehmet Akif Ersoy University, Burdur, Turkey
ahakan@mehmetakif.edu.tr
[3] Department of Computer Engineering, Suleyman Demirel University, Isparta, Turkey
tuncayyigit@mehmetakif.edu.tr

**Abstract.** Looking at the hierarchy of needs, although housing is among the most basic physical needs of an individual, houses have also been used for a long time to generate income. However, determining the price of the house in the purchase and sale transactions is a challenging study since many independent variables must be examined. In order to overcome this difficulty, an ensemble estimation approach was developed in which many different prediction models were brought together in this study. In order to test the developed approach, a new data set was created by compiling the house sales advertisements in the province of Burdur. The data set samples created contain basic characteristics such as the square meter size of the houses, the number of rooms, and the age of the building. In addition to these qualities, the distances of the residence from educational institutions, hospitals, city center and other public institutions were also calculated and included in the data set in order to make the estimation process better. The data set obtained was used to train the developed approach, compare and test the predictions. The findings obtained as a result of the training and testing processes were shared.

**Keywords:** House price prediction · Ensemble algorithm · Regression

## 1 Introduction

The determination of house prices is an important issue that concerns many people and the field of work, such as those who are considering buying a house, real estate agents, investors, tax calculation experts, insurers [1]. Considering that only in the second quarter of 2020, 283.731 houses were sold with an increase of 14% compared to the previous year, the importance of this transaction is seen once more [2]. However, many independent variables included in the calculation make it difficult to accurately determine house prices. Hedonic method, which is based on the consumer theory of Lancaster [3], has been used frequently in the literature for many years to estimate house prices by evaluating many parameters such as the needs of the consumers and the characteristics of the house. Nevertheless, different methods are also encountered in the literature due to the difficulty in determining the independent variables [4].

J. Hemanth et al. (Eds.): ICAIAME 2020, LNDECT 76, pp. 315–324, 2021.
https://doi.org/10.1007/978-3-030-79357-9_31

In this study, a new approach is proposed in which the heuristic estimation algorithms used as an alternative to the hedonic method work in an ensemble structure and produce a new prediction. The proposed approach includes powerful prediction algorithms such as Deep Learning (DL), Decision Trees (DT), Gradient Boosting Trees (GBT), Generalized Linear Model (GLM), Random Forest (RF). It is aimed to obtain better prediction results by optimizing the prediction accuracy of individual algorithms with a collaborative structure.

This study has been prepared under four main headings: Introduction, Materials and Methods, Research Findings and Results. In the introduction sentence, the necessity of house prices estimation, which constitutes the motivation of the study, and the difficulties that arise while making this estimate are mentioned and the studies in the literature are mentioned. In the Material and Method section, the data set created, the algorithms used in the study and the details of the developed approach are given. The performance values obtained after the analyzes performed on the data set created in the research findings, different algorithms and the estimations of the developed approach on this data set are given. Finally, it was concluded by giving all the results within the scope of the work carried out.

## 2    Material and Methods

In this section, initially the details of the data set created are given. Then the algorithms and ensemble approach used in the study were mentioned.

### 2.1    Burdur House Prices Dataset

In order to obtain house prices in the province of Burdur, the advertisements posted on Sahibinden.com website were examined and the houses with the most information were selected. Net and gross square meter sizes of the selected residences, number of rooms, age of the building, floor, total number of floors (duplex, triplex, etc. no), whether it is on the site, basic information such as dues have been compiled. In addition, new attributes were created by calculating the distances to the city square, the nearest public school, university, hospital, city hall, governorship building and city bus station using the location information of the relevant residences. As a result, a data set with a total of 120 samples with 22 attributes and price information was created.

### 2.2    Ensemble Prediction Approach

Ensemble approaches are defined in the literature as an interim process of multiple algorithms for the solution of a common problem [5]. The flow diagram of the training process of the approach proposed in this context is given in Fig. 1.

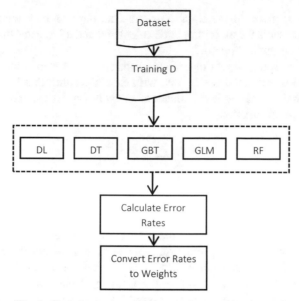

**Fig. 1.** Training process of proposed ensemble approach

Training subset is used for training operations instead of the entire data set as can be seen in the given figure, a. In this data set used in the training step, there are 80 randomly selected samples from 120 samples belonging to the original data set. The remaining 40 samples are reserved for use in the testing process.

Each sample in the training dataset is first trained with each of the algorithms Deep Learning, Decision Trees, Gradient Incrementing Tree, Generalized Linear Model, Random Forest algorithms to obtain predictive values. The error values of these algorithms are calculated with the method given in Eq. 1 by using the difference between the estimated values and the expected values.

$$H = \frac{1}{n} \sum_{i=1}^{n} \frac{\left| Y_i - \widehat{Y}_i \right|}{Y_i} \tag{1}$$

While the symbol H given in the equation represents the error value, $n$ represents the total number of samples in the training set, $Y_i$ is used for the actual price information and $\widehat{Y}_i$ indicates the estimated price information realized by the relevant algorithm.

These error values obtained are used in calculating the weight value of each algorithm. The algorithm with the lowest error should have more influence within the ensemble algorithm, and the algorithm with the highest error rate should have the least. For this reason, the obtained error rates must be weighted inversely and each algorithm must affect the result by the amount of their weights. The calculation given in Eq. 2 is used to convert the obtained error values into weights.

$$W_X = \frac{\sum_{i=1}^{a} H_i}{H_X} \tag{2}$$

The $W_X$ value given in the equation represents the weight symbol, while X the weight value represents the algorithm to be calculated and a indicates the number of the algorithm in the ensemble approach.

After calculating the weight of each algorithm, the test step is starts. In this step, each test sample is estimated one by one with each algorithm that it is in the training step, then the final estimation is calculated as given in Eq. 3 by using these estimation results and weight information.

$$\ddot{Y}_k = \sum_{i=1}^{a} \widehat{Y}_{ki} W_i \tag{3}$$

## 3    Findings

The performance values obtained by the algorithms as a result of the train and test processes carried out with the prediction algorithms and the developed approach are given in Table 1. Generalized Linear Model, which has the lowest Relative Error and Mean Square Error (MSE) value during the training phase, was the most successful algorithm. When the training times are compared, it is seen that the Decision Trees completed the training process first.

**Table 1.** Performance values obtained after train and test phases

|  |  | DL | DT | GBT | GLM | RF | Ensemble |
|---|---|---|---|---|---|---|---|
| Train | Relative error | 0,1426 | 0,1577 | 0,1447 | **0,1207** | 0,1673 | – |
|  | MSE | 69672 | 59296 | 84237 | **40069** | 99355 | – |
|  | Time (ms) | 259704 | **6560** | 478494 | 107347 | 22943 | – |
| Test | Relative error | 0,1167 | 0,1422 | 0,1181 | 0,1059 | 0,1588 | **0,1004** |
|  | MSE | 43959 | 53545 | 45791 | **33591** | 60137 | 36203 |

It is seen that the proposed approach has the lowest error in the test phase according the test results in the given table. It has been more successful by falling below the error rate that the algorithms that created it individually.

The representation of real results on the graph with the prediction operations performed with the algorithms of DL, DT, GBT, GLM, RF and the proposed ensemble approach are given in Fig. 2, 3, 4, 5, 6 and 7, respectively.

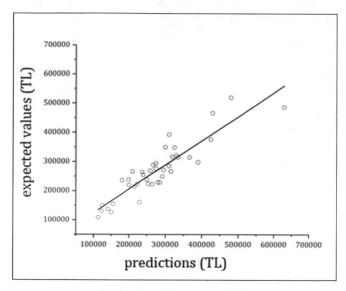

**Fig. 2.** Prediction results (DL)

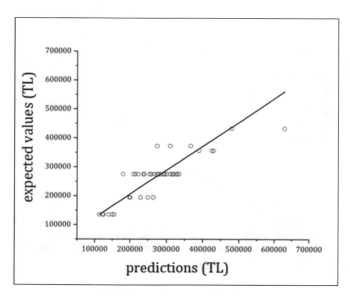

**Fig. 3.** Prediction results (DT)

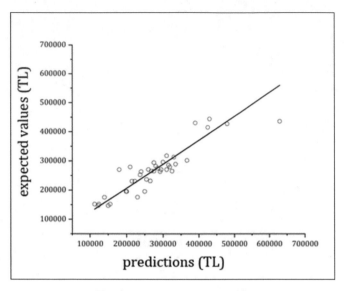

**Fig. 4.** Prediction results (GBT)

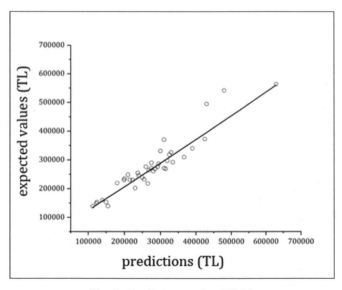

**Fig. 5.** Prediction resultes (GLM)

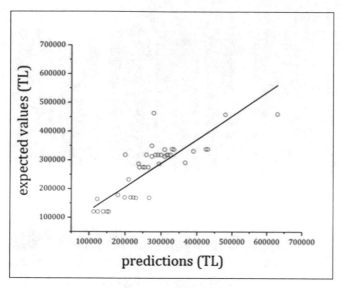

**Fig. 6.** Prediction results (RF)

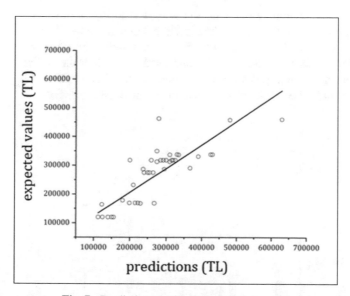

**Fig. 7.** Prediction results (Proposed Approach)

When the graphs given are examined, it is seen that the proposed approach realizes the closest predictions to the real values compared to other algorithms. The average success rates of algorithms can also be seen on the chart given in Fig. 8.

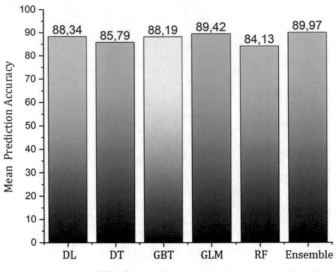

**Fig. 8.** Average success rate

Box graphics containing many details are given in Fig. 9 in order to examine the estimation processes performed by the algorithms in more detail. When the given chart is examined, it is seen that although the error value is higher than the proposed ensemble approach, the Generalized Linear Model successfully predicts the outliers and achieves a balanced error rate in its predictions. It is seen that the Random Forest algorithm is the worst algorithm in the prediction of House Prices of Burdur province.

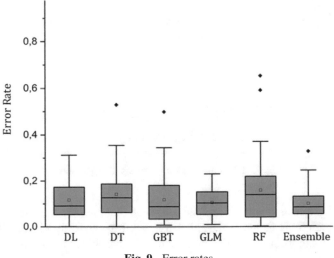

**Fig. 9.** Error rates

The weights assigned by each algorithm to the attributes during the training phase in order to estimate the House Price are given in Table 2. When these weights are examined, it is seen that the most important factor in determining the price of a house is the m2 width of the house. However, each algorithm performed prediction using different attributes. Even if an important feature is not selected by an algorithm, it is one of the biggest advantages of the ensemble approaches that it is selected by another algorithm to obtain better predictive values.

**Table 2.** Selected attributes and its weights

| DL | | DT | | GBT | | GLM | | RF | |
|---|---|---|---|---|---|---|---|---|---|
| Attr | W. | Attr. | W. | Attr. | W.. | Attr. | W. | Attr. | W. |
| m2 (gross) | 0,98 | m2 (net) | 0,72 | m2 (net) | 0,76 | m2 (gross) | 0,99 | Room Count | 0,59 |
| Location (X) | 0,08 | Room count | 0,07 | m2 (gross) | 0,22 | Floor Count | 0,14 | Age of Build | 0,14 |
| Bath. Count | 0,08 | Distance(7) | 0,06 | Location (Y) | 0,16 | Room Count | 0,11 | – | – |
| Age of Build | 0,04 | Status | 0,05 | Due | 0,08 | Bath. Count | 0,03 | – | – |
| Distance(5) | 0,02 | Location (X) | 0,04 | Distance(4) | 0,07 | - | – | – | – |
| – | – | Due | 0,04 | Room Count | 0,01 | - | – | – | – |
| – | – | Floor Count | 0,01 | Bath. Count | 0,01 | - | – | – | – |

## 4  Conclusion

In this study a data set that can be used in the training of prediction algorithms with the house prices obtained from the sales advertisements of the province of Burdur was created. The obtained dataset has been used in the training of Deep Learning, Decision Trees, Gradient Boosting Tree, Generalized Linear Model, Random Forest algorithms. In addition, a new ensemble prediction approach is also proposed. When the success achieved by the proposed approach is compared with the individual success rates achieved by the algorithms that create it, it is seen that the developed approach produces results closer to real values with a lower error rate.

In future studies, it is aimed to obtain a richer data set for training prediction algorithms by expanding the data set to include different provinces. However, by including different prediction algorithms in the ensemble approach, it is aimed to reduce the resulting error rate even more.

# References

1. Frew, J., Jud, G.D.: Estimating the value of apartment buildings. J. Real Estate Res. **25**, 77–86 (2003)
2. Tang, J., Rangayyan, R.M., Xu, J., El Naqa, I., Yang, G.: GÖSTERGE - Türkiye Gayrimenkul Sektörü, 2. Çeyrek Raporu (2020)
3. Lancaster, K.J.: A new approach to consumer theory. J. Pol. Econ. **74**, 132–157 (1966)
4. Selim, S., Demirbilek, A.: Türkiye'deki Konutların Kira Değerinin Analizi: Hedonik Model ve Yapay Sinir Ağları Yaklaşımı. Aksaray Üniversitesi İktisadi ve İdari Bilimler Fakültesi Dergisi **1**(1), 73–90 (2009)
5. Moreira, J.M., Jorge, A.M., Soares, C., Sousa, J.F.: Ensemble approaches for regression: a survey. ACM Comput. Surv. **45**(1), 1–10 (2012)

# The Evidence of the "No Free Lunch" Theorems and the Theory of Complexity in Business Artificial Intelligence

Samia Chehbi Gamoura[1]([✉]), Halil İbrahim Koruca[2], Esra Gülmez[2],
Emine Rümeysa Kocaer[2], and Imane Khelil[3]

[1] EM Strasbourg Business School, Strasbourg University, HuManiS, UR 7308 Strasbourg,
France
samia.gamoura@em-strasbourg.eu
[2] Süleyman Demirel Üniversitesi, Endüstri Mühendisliği Bölümü, Isparta, Türkiye
{halilkoruca,eminekocaer}@sdu.edu.tr
[3] Research Center of Legal Studies, Ibn Khaldun University, Tiaret, Algeria

**Abstract.** For more than decades, the central drivers of business growth have been technological tools. The most important of these is Artificial Intelligence (AI). We name the paradigm "Business Artificial Intelligence (BAI)" the category of AI approaches that can create profitable new business models, earnest business-values, and more competitiveness, particularly Machine Learning (ML). However, ML algorithms are plentiful, and most of them necessitate specific Data structures, particular applicability features, and in-depth analysis of the business context. Because of these considerations, an increasing number of industrial ML-based applications fail. Therefore, it is natural to ask the question of applicability limits and ML algorithms selection, giving the business context. Accordingly, in this paper, we propose the evidence of the "No-Free-Lunch" (NFL) theorems to understand ML use's applicability in business organizations.

**Keywords:** Business Artificial Intelligence · Machine learning · Theory of complexity · No-Free-Lunch-Theorem · ML applicability

## 1 Introduction

In their ultimate objective of looking for competitive power through Artificial Intelligence (AI), companies start by examining how AI algorithms can add business values [1]. A broad range of those algorithms and heuristics are already used in predictive analytics [2], automated processes [3], and additive manufacturing [4]. We name this sub-field of AI approaches specifically applicable to business issues, the "Business Artificial Intelligence" (BAI), where the sub-class of "Machine Learning" (ML) algorithms take a large place in research.

Because of their ability to learn from the experience and determine the hidden patterns, ML algorithms can serve companies in finding worthy business values [5]. Their main motive is to enhance the predictions and accuracy in operations [6]. For instance,

J. Hemanth et al. (Eds.): ICAIAME 2020, LNDECT 76, pp. 325–343, 2021.
https://doi.org/10.1007/978-3-030-79357-9_32

today, we observe BAI-based digital marketing's rapid progress, making customers' experience much more manageable than ever [7].

However, it is not always feasible to apply ML techniques to business problems [8]. According to [9], one of the most misunderstandings associated with ML integration in business systems is the wide variety of algorithms and their technical characteristics and complexities. In the business field, selecting the right ML approach is one of the central dares that industrials and researchers face nowadays [10].

The objective of this paper is twofold: First, it aims to examine the limits of ML approaches applicability in BAI. Second, it tries to find evidence of the No Free Lunch Theorems of Wolpert [11], which is part of the Theory of Complexity (TOC).

The paper's structure is as follows: Sect. 2 presents the background overview and the related works, including ML characteristics. Section 3 situates the problem and the research gap. Then, Sect. 4 outlines the proposed approach and discussions about the results. Finally, Sect. 5 completes this paper with a conclusion and some open views.

## 2    Background and Related Works

### 2.1    Business Artificial Intelligence (BAI)

By the paradigm "Business Artificial Intelligence" (BAI), we intend the branch of Artificial Intelligence (AI) approaches that can be integrated into business systems to generate additional business values, particularly the Machine Learning (ML) algorithms. We are not the first to use this paradigm; as far as we know, the paradigm "Business Artificial Intelligence" has been used the first time in an academic context [12].

ML approaches belong to the sub-class of computational methods that automate the acquisition of knowledge from the environment and experiences [13]. Technically, an ML algorithm builds a model automatically (learns) that maps a relationship between the Dataset of the observed features and the output patterns [14]. Theoretically, the learning process involves a classifier's approximation to find the correct mapping between the input and the output patterns [1]. The generic model of ML integration in business commonly involves five phases: Data collection, features' selection, algorithm selection, training, and performance evaluation. The output patterns come from this induced model's processing and not from the algorithm's processing in ML (Fig. 1).

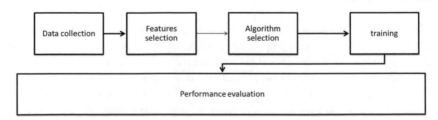

**Fig. 1.** Generic model of ML algorithm development

Within the recent few years, the ML algorithms have become more effective and widely available in enterprises and management researches, even for managers and academics without a technical background [15]. Indeed, the examination of the academic bibliography in BAI shows an extensive list of algorithms, methods, heuristics, and techniques in different activities; in stock prediction [16], in marketing analytics [17], in accounting liability [18], and so forth. More than thousands of algorithms exist, and about tens of variants are created every day in publications [14]. That point raises the problem of selection; because; there is no known method to apply in defining or finding the 'best' algorithm that can fit the business issue [9]. Besides, the challenge of optimization of algorithms is related to the theorems of the No Free Lunch (NFL): "for a problem; no search algorithm is better than another when its performance is averaged over all possible discrete functions" [11].

Briefly, the NFL are theorems in Computational Complexity and Optimization (CCO). They are part of the Theory of Complexity (TOC) [19] as TOC discourses the algorithms' costs of time and space, in addition to the class of NP-complete problems [9].

## 2.2   Theory of Complexity and No Free Lunch Theorems in BAI

The NFL theorems' foundations state that all the optimization approaches (methods) of a given problem; perform similarly well when averaged over all the possible issues [20]. NFL theorems are part of the Theory of Complexity (TOC) that, tacitly, denotes an algorithm's complexity for unraveling a problem and quantifying performance [21].

Technically, the NFL theorems are mathematical folkloric statements of optimization that prove the equality of algorithms when their performance is averaged through all possible problems [11]. These theoretical concepts gave birth to an entire research field through which significant investigations reveal the main issues, restrictions, and circumstances, or even the inability to apply algorithms [22]. Analyzing the complete NFL theorems is beyond this paper's scope. For further information, we orient to the most relevant reference of Wolpert [11].

The success of ML at intelligent tasks is mainly due to its ability to discover a complex structure that is not specified in advance [23]. Today, ML algorithms are easy to find and apply: everyone can download the suitable package (R, Python, NumPy, Scikit-learn, Matplotlib, etc.) that can fit with the appropriate problem (for instance: Artificial Neural Networks [18], K-Means clustering [24], multi-agent systems [16], etc.). However, this also increases the risk of mistakes and misinterpreted outputs [22].

The model complexity and the Dataset structure are the imperative bases for knowing the correctness in applying ML algorithms. Nonetheless, their theoretical and research implications are often non-existent among publications related to the business field [25]. Moreover, in selecting an ML algorithm, there is no predefined method of choosing the target type (the Output Business Values - TOV); are they discrete or continuous? Are they labels or quantified? Are they probabilities-based?

# 3   Problematic and Research Gap

## 3.1   Problematic of the Study

In BAI, the hurdle of the diversity in the ML classifiers becomes obvious. Indeed, there is no single classifier of induction (model) [8]. Furthermore, some authors believe that

ML is considered an ill-posed problem because of two main reasons: firstly, the business output value – TOV) depends strongly on the algorithm's choice and the input Dataset used in training. Secondly, in most business cases, the suitable business value is not guaranteed [12]. Other scientists argue through the challenge of 'curse of dimensionality' [26], where the robustness of the algorithm is extremely sensitive to the dimension of the Dataset, that is in turn highly scalable in business, unlike the other fields.

The review of many publications attests that most industrial executives often have concerns in identifying where applying ML to business issues [8]. The black box nature and the mathematical-statistical origins of ML approaches make this mission complex for managers and business practitioners [27]. Therefore, finding the optimal ML algorithm to solve an issue is sometimes particularly problematic and challenging task. Given the earlier argumentation, the question is that: "Given the performance of a learning algorithm on the training Dataset, is it possible to acquire the best business value for patterns outside the Dataset?" (Fig. 2).

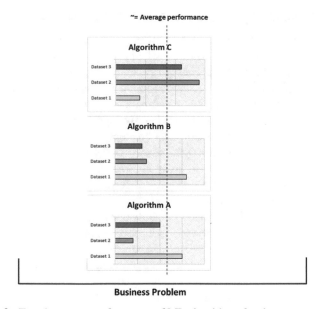

**Fig. 2.** Equal average performance of ML algorithms for the same problem

## 3.2 Scope of Related Research

The existing research works present a short range of publications on applicability and ML algorithm optimization and selection in management and business. We conducted a review to try elucidating these publications' scope through *Dimensions.ai®* tool [28]. To proceed, we queried out all academic publications by using the query's parameters in Table 1. The screenshot of the analytic overview is illustrated in Fig. 3.

As clarified in Fig. 3, a few numbers of papers discussed the topics of Machine learning applicability, algorithms selection, algorithms optimization, or the no free lunch

**Table 1.** The input query in Dimensions.ai® for the study in the paper

| Parameter in Dimension® query | Values | |
|---|---|---|
| Export date | Nov 2019 | |
| Criteria | Keywords | Machine learning, applicability, algorithm selection, review, algorithm optimization, no free lunch theorem |
| | Fields of research | Economics, Applied Economics, Commerce, Management, Tourism and Services, Information systems, Business and management, marketing, econometrics |
| | Sub-fields | Business and Management, Banking, Finance and Investment, Information and Computing Sciences, Economics, Information Systems, Applied Economics, Marketing, Studies in Human Society |
| | Years | 2010 to 2019 |
| Publication | Articles in journals | |
| Aggregation factor | Means | |

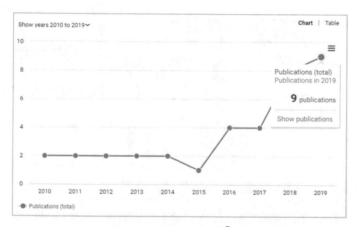

**Fig. 3.** Analytics overview in Dimension.ai® of the query of Table 1.

theorem in a review or original research. The maximum number of publications in all the sub-fields of business and management does not exceed 9 (achieved in 2019).

# 4  Proposed Approach and Findings

## 4.1  Proposed Classifications

### 4.1.1  Business Applications Classification

To understand the applicability of algorithms and their limits in business organizations, building a classification system that separates the different business functions is necessary for a preliminary phase [20]. Therefore, we opted for the well-known Pyramid Multi-Levels (PML) classification of Ivanov [29] (Fig. 4). Four different levels build the PMC classification:

1) Operational-Level Systems (OLS): include the transactional operations support by monitoring the daily tasks and business trades,
2) Management-Level Systems (MLS): gather all activities of decision making, supervising, controlling, and administrative tasks,
3) Knowledge-Level Systems (KLS): capitalize all knowledge, information, and meta-information about the flows of designing products and improvement of activities,
4) Strategic-Level Systems (SLS): comprise the planning and the activities related to the organization's macro-management.

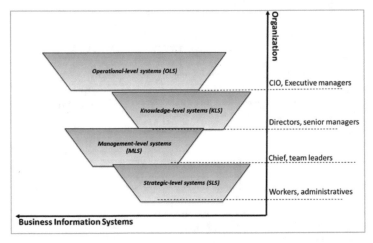

**Fig. 4.** Pyramid Multi-Levels (PML) classification of Ivanov (Ivanov 2010)

Our review of literature lists a large number of ML algorithms in publications related to business applications. We classified them in a matrix and reported the most relevant top 3 in each level in Table 2.

**Table 2.** Some relevant researches of ML use following PML classification

| Business classification | ML application in business | Reference |
|---|---|---|
| Operational-Level Systems (OLS) | Sales forecasting | [30] |
| | Production planning | [31] |
| | Transport planning and scheduling | [32] |
| Knowledge-Level Systems (KLS) | Market analysis | [7] |
| | Designing, analyzing, engineering | [13] |
| | Quality ruling | [2] |
| Management-Level Systems (MLS) | Stores sales management | [33] |
| | Product process management | [34] |
| | Stock handling | [15] |
| Strategic-Level Systems (SLS) | Orders processing, reporting | [1] |
| | Payment of creditors and cash | [35] |
| | Installation, maintenance | [36] |

### 4.1.2  Machine Learning Algorithms Classification

The most renowned classification of ML algorithms is based on four classes:

1) Supervised Learning (SL): the ML algorithm automatically defines the assignment of Data into a set of known patterns (labeled classes) [37]. Artificial Neural Networks (ANN) [38] and Support Vector Machines (SVM) [15] are well-known SL approaches.
2) Unsupervised Learning (UL): the ML algorithm is self-organized to determine the unknown patterns in Datasets without labeling [39]. UL's prominent families are clustering and association rules [40].
3) Reinforcement Learning (RL): algorithms enable capitalizing knowledge from the feedback received through interactions with the external environment [12]. The theoretical model is based on the Markov Decision Process (MDP), and Dynamic Programming (DP) approaches [41],
4) Ensemble Learning (EL): Multiple ML techniques exist as hybrid approaches but after dividing the main problem, considered as problematic, into a set of different sub-problems with different predictors [4]. Table 3 below provides the most relevant algorithms in publications from our literature review. The table classifies algorithms following the given classification: supervised, unsupervised, reinforcement, and ensemble learning.

In this paper, we name this classification "SURE classification"; as "Supervised-Unsupervised-Reinforcement-Ensemble" classification.

**Table 3.** Some relevant researches of ML use following SURE classification

| Approaches | Types | Some relevant references |
|---|---|---|
| Artificial Neural Network (ANN) | Supervised | [30, 42–44] |
| Fisher Linear Discriminant (FLD) | | [34, 45] |
| Logistic Regression (LRE) | | [46, 47] |
| Support Vector Machine (SVM) | | [15, 41] |
| K-Nearest Neighbour (KNN) | | [10, 33] |
| Case-Bases Reasoning (CBR) | | [38, 48] |
| Hidden Markov Models (HMM) | | [49–51] |
| Naive Bayes Classifier (NBC) | | [52, 53] |
| Bayesian Networks (BNE) | | [54, 55] |
| Decision Trees (DTR) | | [29, 56] |
| Conditional Random Fields (CRF) | | [57] |
| Linear Regression (LRE) | | [18] |
| Non-Linear Regression (NLR) | | [58] |
| Polynomial regression (PRE) | | [59] |
| Maximum Entropy (MEN) | | [60–62] |
| Fuzzy Set Inference (FSI) | | [1, 32, 63, 64] |
| Online Mirror Descent learning (OMD) | | [49] |
| Recurrent Neural Networks (RNN) | | [65] |
| K-Means (KME) | Unsupervised | [24] |
| Mixture Models (MMO) | | [40], |
| Hierarchical Cluster Analysis (HCA) | | [66, 67] |
| Hebbian Learning (HLE) | | [4] |
| Generative Adversarial Networks (GAN) | | [68] |
| Frequent Pattern Growth (FRG) | | [69] |
| Apriori Algorithm (AAL) | | [70] |
| Eclat Algorithm (EAL) | | [71] |
| Sarsa (SAR) | Reinforcement | [72, 73] |
| Q-Learning (QLE) | | [25, 74–76] |
| Divergence de Kullback-Leibler (DKL) | | [77] |
| Random Decision Forest (RDF) | Ensemble | [78–80] |
| AdaBoost Algorithm (ABA) | | [81] |
| LogitBoost Algorithm (LBA) | | [82] |

## 4.2 Discussion and Results

We classified 917 publications from our comprehensive review and then cleansed them to keep only the peer-reviewed papers in the category of ML's original researches (without review papers). We then determined, for each publication: which field, with which ML algorithm, and which business context for each publication. Based on our review, ML algorithms' use in the business problems following Ivanov's PML classification [29] shows different applications' trends and intensity. We synthesized and ranked them in illustrations in the following four positions.

**Position 1: Knowledge-Level Systems (KLS) (34.00% Applications of ML)**
We found that KLS are the most popular systems integrating ML with 31,77%. We analyze in Fig. 5 the distribution of algorithms and found the majority are Non-Linear Regression (NLR) approaches with 25.36%. Then come the Decision Trees (DTR) (~=11.66%) and Logistic Regression (LRE) techniques (~=8.75%). The three approaches belong to supervised learning, but they are different in their modes of infer-ring. The NLR and DTR are information-based, where LRE is error-based learning. Likewise, we found several algorithms that are absent in KLS (0.00%), such as Fuzzy Classifiers (FUC) and the Opus algorithm (OPA).

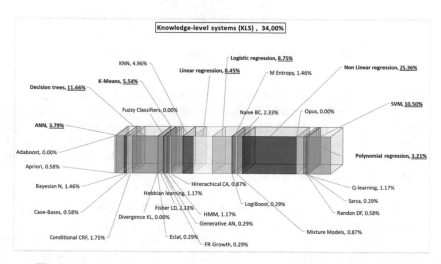

**Fig. 5.** Distribution of ML algorithms use in knowledge-level systems (KLS)

At the KLS level, the following sub-fields are ranked in the decreasing order of ML intensity of use (Fig. 6): 'Knowledge Processing Systems (KPS)' with 52,90%, 'Product and Facilities (PAF)' with 32,37%, and 'Process Automation (PAU)' with 14,73%. At this level, we found that ML algorithms occur in the 'Accuracy in Assessment (AIA)' with more than the half rate of use (52.51%), mostly higher than their use in the 'Ontology Management (ONM)' (19.63%), the 'Natural Language Processing (NLP)' (13.70%), and the 'Content Discovery Platforms (CDP)' (14.16%). We think the ML advantage in knowledge-based systems is mainly due to their capacity to infer to discover the hidden

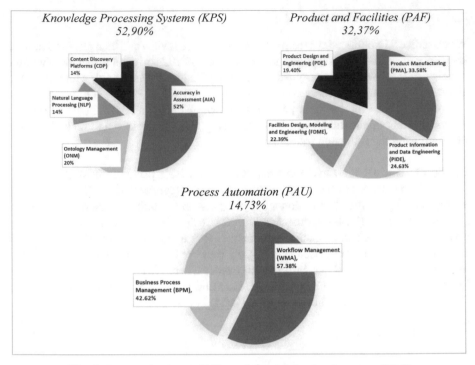

**Fig. 6.** Panoramic views of ML use in knowledge-level systems (KLS)

knowledge that is related to the business values the companies are considering. Besides, we think that ML techniques naturally enhance the business values related to design and intellectual production of goods and services [83], in addition to industrial creation [46], development of goods and services [15], engineering fulfillment, and data processing [84]. Publications related to Business Process Management (BPM) [23] and workflow automation [3] are perfect examples.

**Position 2: Management-Level Systems (MLS) (31,77% Applications of ML)**

MLS are secondary in the top used level-based systems, with a rate of 31,77%. As illustrated in Fig. 7, applications in the MLS level mainly apply Regression algorithms: the Non-Linear Regression (NLR) with 19.19% and the Linear Regression (REG) with 18.18%. However, we observe that both algorithms belong to supervised learning. Our impression is that MLS publications use linear approach first and then try obtaining more adequate results through using the non-linear form of regression. About the null rate of 0.00%, we see that no publication incorporates Generative Adversarial Networks (GAN), Maximum Entropy (MEN) algorithm, and Hierarchical Clustering.

As illustrated in the pie charts of Fig. 8, publications in MLS that cover ML integration use ML in the 'Customer-oriented analytics' with more than half publications rate (52,14%) and the 'Decision making and decision-aid' systems (47,86%). With respectively 23.51%, 18.35%, and 13.44%, the 'Opinion Mining & Sentiment Analysis (OMSA)', 'Recommender Systems (RSY),' and the 'Predictive Aid-Decision (PAD)'

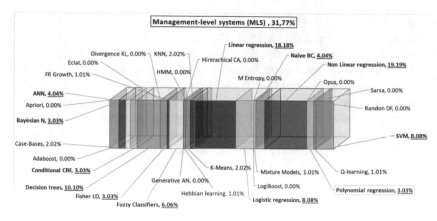

**Fig. 7.** Distribution of ML algorithms use in management-level systems (MLS)

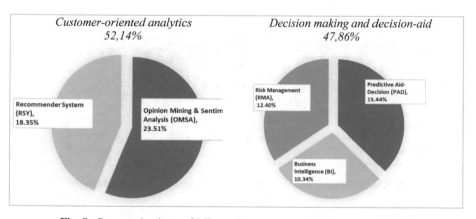

**Fig. 8.** Panoramic views of ML use in management-level systems (MLS)

are the top three business systems that are using learning algorithms in MLS. This ranking is aligned and rational with our hypothesis in using ML mostly for recommending systems and social media analysis through opinion and sentiment mining. In contrast, the ML-based approaches in Business Intelligence (BI) and Risk Management (RMA) are respectively (10.34%) and (12.40%). We explain these figures by the fact ML algorithms have been used to reinforce the decisional systems, including risk management since the first publications as BI appeared in the years before 2000.

## Position 3: Strategic-Level Systems (SLS) (18,06% Applications of ML)

The SLS are in the third position, with a global rate of use equal to 18,06% (Fig. 9). The Non-Linear Regression (NLR) algorithms are the top most used learning approaches (23.97%), followed by the Linear Regression (REG) approaches with a less rate of 10.62% and Logistic Regression (LOR) with 8.56%. K-Nearest Neighbor (KNN), Support Vector Machine (SVM), Artificial Neural Network (ANN) are also applied with low percentages less than 10.00%. We find this distribution disparate as it merges the top three learning modes: information-based learning (NLR, REG), similarity-based learning (KNN), and error-based learning (SVM, ANN).

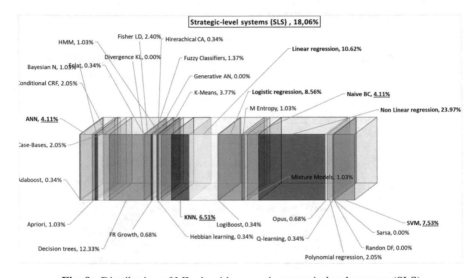

**Fig. 9.** Distribution of ML algorithms use in strategic-level systems (SLS)

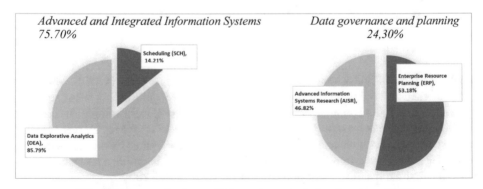

**Fig. 10.** Panoramic views of ML use in strategic-level systems (SLS)

At this strategic level, we distinguish between two categories (Fig. 10): (1) the class of the 'Advanced and Integrated Information Systems' with (75.70%); divided into a bulk of

ML applications in 'Data Explorative Analytics' (DEA) (85.79%) and Scheduling (SCH) (4.21%). (2)The 'Data governance and planning' with (24,30%) have the uppermost interest of academics with half-rate of papers (about 53.18%) in the 'Enterprise Resource Planning (ERP),' and 46.82% of use in the 'Advanced Information Systems Research (AISR).' Our pre-hypothesis about ML use in planning is confirmed in these figures, but the presence of the ad-hoc learning approaches in advanced information systems at this high rate is an unexpected new finding for us.

### Position 4: Operational-Level Systems (OLS) (16,17% Applications of ML)

With 16.17%, the OLS take the fourth and last position in ML business applications (Fig. 11). At this level, the high rate of using Non-Linear Regression (NLR) algorithms is still on the top with 21.93%. The figure shows that the OLS is head-led relatively by Regression algorithms and Decision Trees (DTR). These two classes of algorithms occur in different business values related to accuracies such as risk assessment [4], business Data accuracy [15], and price prediction [63]. The rest of the publications are scattered between Support Vector Machines (8.56%), and Linear Regression (LRE) (6.95%), Artificial Neural Networks (5.88%), and Fisher Linear Discriminant algorithm (4.28%). Besides, the few ~9% of publications, in unsupervised learning for accuracy and assessment, are exclusively using Clustering algorithms (K-means, Mixture Learning Models). However, surprisingly, the hypothesis we had in mind about supervised learning only for accuracy business values is reversed in this finding. We note here that most algorithms are based on information-based algorithms in the predictive tasks because of the operational systems' transactional nature. Many other ML algorithms exist with low percentages of less than 3%, such as Bayesian Networks (2.67%), Apriori algorithm (0.53%), Q-Learning (1.60%)), and so forth. Some others do not exist (0.00%) like Hebbian learning, Opus, and Eclat algorithms.

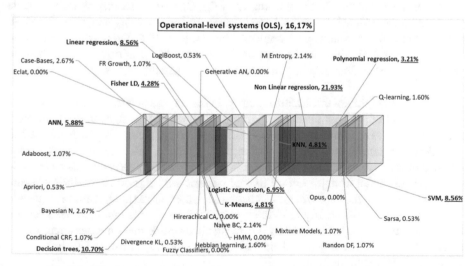

**Fig. 11.** Distribution of ML algorithms use in operational-level systems (OLS)

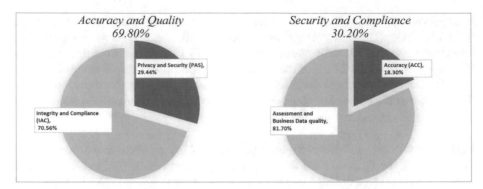

**Fig. 12.** Panoramic views of ML use in operational-level systems (OLS)

More than 69% of publications wrap 'Accuracy and Quality' systems (Fig. 12). In this field, come 'Privacy and Security (PAS)' with 29.44% and 'Integrity & Compliance (IAC)' with 70.56%. This distribution argues that managers require ML to look for more accurate information and prioritize tasks and activities in assessment, particularly in finances [85] and pricing [33]. Also, managers prefer predictive analytics, such as statistical methods [42].

## 5   Conclusion and Future Works

In this paper, the research questions deal with the applicability of Machine Learning (ML) in Business Artificial Intelligence (BAI). Through the analysis of an extensive review and classifications, we tried to provide the evidence of the "No-Free-Lunch" (NFL) theorems, which is connected to the Theory of Complexity (TOC). Our approach proposes a double classification for both ML techniques and levels-based business organizations. Through this classification, we examined a comprehensive review with different analytics figures. We tried to reveal at which level, why, and how these ML algorithms are embedded (prediction, optimization, accuracy, enhancement, sustainability, competitiveness, and so forth).

The findings highlight three critical points; 1) ML approaches do not follow any pre-designed model of BAI applications, 2) a large number of ML algorithms in different modes can resolve the same business issue in the same challenge, 3) intensities of ML use follow disparate trends in each business level. These outcomes clearly state the evidence of the "No-Free Lunch Theorem" in business. Our perspective is to propose an assessment model of applicability to evaluate any ML algorithm feasibility before application. Indeed, the deployment and development of any ML algorithm is time and budget consuming, particularly in training phases of algorithms that may need several error-try attempts. In addition, having a best-practices guide and an assessment model can benefit all the stakeholders and encourages more intelligent use of ML in business for now and the upcoming future.

# References

1. Adam, S.P., Alexandropoulos, S.A.N., Pardalos, P.M., Vrahatis, M.N.: No free lunch theorem: a review. In: Demetriou, I.C., Pardalos, P.M. (eds.) Approximation and Optimization. SOIA, vol. 145, pp. 57–82. Springer, Cham (2019). https://doi.org/10.1007/978-3-030-12767-1_5
2. Akter, S., Bandara, R., Hani, U., Wamba, S.F., Foropon, C., Papadopoulos, T.: Analytics-based decision-making for service systems: a qualitative study and agenda for future research. Int. J. Inf. Manage. **48**, 85–95 (2019)
3. Al-Dmour, A.H., Al-Dmour, R.H.: Applying multiple linear regression and neural network to predict business performance using the reliability of accounting information system. Int. J. Corp. Finan. Account. (IJCFA) **5**(2), 12–26 (2018)
4. Alzubi, J., Nayyar, A., Kumar, A.: Machine learning from theory to algorithms: an overview. J. Phys. Conf. Ser. **1142**(1), 012012 (2018)
5. Antonescu, D.: Are business leaders prepared to handle the upcoming revolution in business artificial intelligence? Quality-Access to Success **19**(166) (2018)
6. Aswani, R., Kar, A.K., Ilavarasan, P.V., Dwivedi, Y.K.: Search engine marketing is not all gold: insights from Twitter and SEOClerks. Int. J. Inf. Manage. **38**(1), 107–116 (2018)
7. Basani, Y., Sibuea, H.V., Sianipar, S.I.P., Samosir, J.P.: Application of sentiment analysis on product review e-commerce. J. Phys. **1175**(1), 012103 (2019)
8. Bousoño-Calzón, C., Bustarviejo-Muñoz, J., Aceituno-Aceituno, P., Escudero-Garzás, J.J.: On the economic significance of stock market prediction and the no free lunch theorem. IEEE Access **7**, 75177–75188 (2019)
9. Buckel, J., Parker, A., Oehlert, C., Shipley, S.: Estimating persistence in employee business expense correspondence examinations using hidden Markov and Semi-Markov models. Stat. J. IAOS **34**(1), 63–75 (2018)
10. Chan, S.L., Lu, Y., Wang, Y.: Data-driven cost estimation for additive manufacturing in cyber manufacturing. J. Manuf. Syst. **46**, 115–126 (2018)
11. Chang, P.C., Lin, J.J., Dzan, W.Y.: Forecasting of manufacturing cost in mobile phone products by case-based reasoning and artificial neural network models. J. Intell. Manuf. **23**(3), 517–531 (2012)
12. Chehbi Gamoura, S., Koruca, H.I., Köker, U.: Challenge of supply chain resilience in big data era: the butterfly effect. Int. J. Appl. Math. Electron. Comput. **10**(4) (2018)
13. Chehbi-Gamoura, S.: Smart workload automation by swarm intelligence within the wide cloud computing. Int. J. Mech. Eng. Autom. **3**, 2163–2405 (2016)
14. Chehbi-Gamoura, S., Derrouiche, R., Damand, D., Barth, M.: Insights from big data analytics in supply chain management: an all-inclusive literature review using the SCOR model. Prod. Plann. Control **31**, 355–382 (2020)
15. Chen, W., Liu, H., Xu, D.: Dynamic pricing strategies for perishable product in a competitive multi-agent retailers market. J. Artif. Soc. Soc. Simul. **21**(2) (2018)
16. Chen, Z.Y., Fan, Z.P., Sun, M.: A hierarchical multiple kernel support vector machine for customer churn prediction using longitudinal behavioral data. Eur. J. Oper. Res. **223**(2), 461–472 (2012)
17. Chong, F.S., et al.: Understanding consumer liking of beef using hierarchical cluster analysis and external preference mapping. J. Sci. Food Agric. **100**(1), 245–257 (2020)
18. Das, R.D., Winter, S.: A fuzzy logic based transport mode detection framework in urban environment. J. Intell. Transp. Syst. **22**(6), 478–489 (2018)
19. De Caigny, A., Coussement, K., De Bock, K.W., De Caigny, A., Coussement, K., De Bock, K.W.: A new hybrid classification algorithm for customer churn prediction based on logistic regression and decision trees. Eur. J. Oper. Res. **269**(2), 760–772 (2018)

20. Devarajan, Y.: A study of robotic process automation use cases today for tomorrow's business. Int. J. Comput. Tech. **5**(6), 12–18 (2018)
21. Dezi, L., Santoro, G., Gabteni, H., Pellicelli, A.C.: The role of big data in shaping ambidextrous business process management. Bus. Process Manage. J. (2018)
22. Di Persio, L., Honchar, O.: Artificial neural networks architectures for stock price prediction: comparisons and applications. Int. J. Circ. Syst. Sig. Process. **10**, 403–413 (2016)
23. Diaz-Rozo, J., Bielza, C., Larrañaga, P.: Clustering of data streams with dynamic gaussian mixture models: an IoT application in industrial processes. IEEE Internet Things J. **5**(5), 3533–3547 (2018)
24. DigitalScience: Dimensions.ai (2020). https://www.dimensions.ai/. Accessed 10 Oct 2020
25. Dixit, V., Chaudhuri, A., Srivastava, R.K.: Assessing value of customer involvement in engineered-to-order shipbuilding projects using fuzzy set and rough set theories. Int. J. Prod. Res. **57**(22), 6943–6962 (2019)
26. Dixon, M.: Sequence classification of the limit order book using recurrent neural networks. J. Comput. Sci. **24**, 277–286 (2018)
27. Farhan, M.S., Abed, A.H., Ellatif, M.A.: A systematic review for the determination and classification of the CRM critical success factors supporting with their metrics. Future Comput. Inf. J. **3**(2), 398–416 (2018)
28. Feng, X., et al.: Selecting multiple biomarker subsets with similarly effective binary classification performances. J. Vis. Exp. **140**, e57738 (2018)
29. Fung, T.C., Badescu, A.L., Lin, X.S.: Multivariate Cox hidden Markov models with an application to operational risk. Scand. Actuar. J. **8**, 686–710 (2019)
30. Ghatasheh, N.: Business analytics using random forest trees for credit risk prediction: a comparison study. Int. J. Adv. Sci. Technol. **72**(2014), 19–30 (2014)
31. Ghetas, M., Yong, C.H.: Resource management framework for multi-tier service using case-based reasoning and optimization algorithm. Arab. J. Sci. Eng. **43**(2), 707–721 (2018)
32. Gomber, P., Kauffman, R.J., Parker, C., Weber, B.W.: On the Fintech revolution: interpreting the forces of innovation, disruption, and transformation in financial services. J. Manage. Inf. Syst. **35**(1), 220–265 (2018)
33. Gómez, D., Rojas, A.: An empirical overview of the no free lunch theorem and its effect on real-world machine learning classification. Neural Comput. **28**(1), 216–228 (2016)
34. Gupta, A., Aggarwal, A., Bareja, M.: A review on classification using machine learning. Int. J. Inf. Syst. Manage. Sci. **1**(1) (2018)
35. Hamadouche, A., Kouadri, A., Bakdi, A.: A modified Kullback divergence for direct fault detection in large scale systems. J. Process Control **59**, 28–36 (2017)
36. Henrique, B.M., Sobreiro, V.A., Kimura, H.: Stock price prediction using support vector regression on daily and up to the minute prices. J. Fin. Data Sci. **4**(3), 183–201 (2018)
37. Höpken, W., Eberle, T., Fuchs, M., Lexhagen, M.: Improving tourist arrival prediction: a big data and artificial neural network approach. J. Travel Res. 0047287520791244 (2020)
38. Huang, J.C., Shao, P.Y., Wu, T.J.: The study of purchase intention for men's facial care products with K-nearest neighbour. Int. J. Comput. Sci. Inf. Technol. (IJCSIT) **1**(10) (2018)
39. Ivanov, D.: An adaptive framework for aligning (re) planning decisions on supply chain strategy, design, tactics, and operations. Int. J. Prod. Res. **48**(13), 3999–4017 (2010)
40. Jahirabadkar, S., Kulkarni, P.: Clustering for high dimensional data: density based subspace clustering algorithms. Int. J. Comput. Appl. **63**(20) (2013)
41. Kartal, A.: Balance scorecard application to predict business success with logistic regression. J. Adv. Econ. Finan. **3**(1), 13 (2018)
42. Khanvilkar, G., Vora, D.: Sentiment analysis for product recommendation using random forest. Int. J. Eng. Technol. **7**(3), 87–89 (2018)

43. Khorsheed, R.M., Beyca, O.F.: An integrated machine learning: Utility theory framework for real-time predictive maintenance in pumping systems. Proc. Inst. Mech. Eng. Part B J. Eng. Manuf. 0954405420970517 (2020)
44. Kim, D.H., et al.: Smart machining process using machine learning: a review and perspective on machining industry. Int. J. Precis. Eng. Manuf. Green Technol. 5(4), 555–568 (2018)
45. Köker, U., Ibrahim Koruca, H., Chehbi-Gamoura, S.: A comparison of current and alternative production characteristics of a flow line: case study in a yarn producer's packaging unit. Int. J. Appl. Math. Comput. Sci. 7, 15–21 (2018)
46. Kramarić, T.P., Bach, M.P., Dumičić, K., Žmuk, B., Žaja, M.M.: Exploratory study of insurance companies in selected post-transition countries: non-hierarchical cluster analysis. CEJOR 26(3), 783–807 (2017). https://doi.org/10.1007/s10100-017-0514-7
47. Law, S.H., Lee, W.C., Singh, N.: Revisiting the finance-innovation nexus: evidence from a non-linear approach. J. Innov. Knowl. 3(3), 143–153 (2018)
48. Lee, B.R., Kim, I.S.: The role and collaboration model of human and artificial intelligence considering human factor in financial security. J. Korea Inst. Inf. Secur. Cryptol. 28(6), 1563–1583 (2018)
49. Lee, J.S., Hwang, S.H., Choi, I.Y., Kim, I.K.: Prediction of track deterioration using maintenance data and machine learning schemes. J. Transp. Eng. Part A Syst. 144(9), 04018045 (2018)
50. Leong, L.Y., Hew, T.S., Ooi, K.B., Wei, J.: Predicting mobile wallet resistance: a two-staged structural equation modeling-artificial neural network approach. Int. J. Inf. Manage. 102047, 51 (2020)
51. Levner, E., Ptuskin, A.: Entropy-based model for the ripple effect: managing environmental risks in supply chains. Int. J. Prod. Res. 56(7), 2539–2551 (2018)
52. Li, Z.Y., Zhang, Y.B., Zhong, J.Y., Yan, X.X., Lv, X.G.: Research on quantitative trading strategy based on neural network algorithm and fisher linear discriminant. Int. J. Econ. Finan. 9(2), 133–141 (2017)
53. Liu, Y., Liu, Y.Z., Zhang, H., Li, T.: An RFID data cleaning strategy based on maximum entropy feature selection. J. Dig. Inf. Manage. 14(2), 86 (2015)
54. Lockamy, A., III: Benchmarking supplier risks using Bayesian networks. Benchmarking Int. J. 18(3), 409–427 (2011)
55. Maroofi, F.: Investigating Q-learning approach by using reinforcement learning to decide dynamic pricing for multiple products. Int. J. Bus. Inf. Syst. 31(1), 86–105 (2019)
56. Massaro, A., Galiano, A., Meuli, G., Massari, S.F.: Overview and application of enabling technologies oriented on energy routing monitoring, on network installation and on predictive maintenance. Int. J. Artif. Intell. Appl. (IJAIA) 9(2) (2018)
57. Moore, J.C., Smith, P.W., Durrant, G.B.: Correlates of record linkage and estimating risks of non-linkage biases in business data sets. J. R. Stat. Soc. A. Stat. Soc. 181(4), 1211–1230 (2018)
58. Mossalam, A., Arafa, M.: Using artificial neural networks (ANN) in projects monitoring dashboards' formulation. HBRC J. 14(3), 385–392 (2018)
59. Ntakaris, A., Magris, M., Kanniainen, J., Gabbouj, M., Iosifidis, A.: Benchmark dataset for mid-price forecasting of limit order book data with machine learning methods. J. Forecast. 37(8), 852–866 (2018)
60. Pourjavad, E., Shahin, A.: The application of Mamdani fuzzy inference system in evaluating green supply chain management performance. Int. J. Fuzzy Syst. 20(3), 901–912 (2017). https://doi.org/10.1007/s40815-017-0378-y
61. Puchkov, E.V., Osadchaya, N.Y.A., Murzin, A.D.: Engineering simulation of market value of construction materials. J. Adv. Res. Law Econ. 9, 615–625 (2018)
62. Ralston, B.: Does payroll tax affect firm behaviour? Econ. Pap. J. Appl. Econ. Policy 39(1), 15–27 (2020)

63. Rathod, A., Dhabariya, M.A., Thacker, C.: A survey on association rule mining for market basket analysis and Apriori algorithm. Int. J. Res. Advent Technol. **2**(3) (2014)

64. González Rodríguez, G., Gonzalez-Cava, J.M., Méndez Pérez, J.A.: An intelligent decision support system for production planning based on machine learning. J. Intell. Manuf. **31**(5), 1257–1273 (2019). https://doi.org/10.1007/s10845-019-01510-y

65. Roig, J.S.P., Gutierrez-Estevez, D.M., Gündüz, D.: management and orchestration of virtual network functions via deep reinforcement learning. IEEE J. Sel. Areas Commun. **38**(2), 304–317 (2019)

66. Sabbeh, S.F. Machine-learning techniques for customer retention: a comparative study. Int. J. Adv. Comput. Sci. Appl. **9**(2) (2018)

67. Salhi, M., Korde, K.A.: Optimal feature selection in order to bank customer credit risk determination. IT Manage. Stud. **6**(22), 129–154 (2018)

68. Scherer, M.: Multi-layer neural networks for sales forecasting. J. Appl. Math. Comput. Mech. **17**(1) (2018)

69. Shahin, A., Kianersi, A., Shali, A.: Prioritizing key supply chain risks using the risk assessment matrix and Shannon fuzzy entropy—with a case study in the home appliance industry. J. Adv. Manuf. Syst. **17**(03), 333–351 (2018)

70. Shrivastava, A.: Usage of machine learning in business industries and its significant impact. Int. J. Sci. Res. Sci. Technol. **4**(8) (2018)

71. Sinha, S., Bhatnagar, V., Bansal, A.: Multi-label Naïve Bayes classifier for identification of top destination and issues to accost by tourism sector. J. Glob. Inf. Manage. (JGIM) **26**(3), 37–53 (2018)

72. Sojan, S., Raphy, S.K., Thomas, P.: Techniques used in decision support system for CRM-a review. Int. J. Inf. Technol. Infrastruct. **3**(1) (2014)

73. Sulova, S.: Association rule mining for improvement of IT project management. TEM J. **7**(4), 717–722 (2018)

74. Tratkowski, G.: Identification of nonlinear determinants of stock indices derived by Random Forest algorithm. Int. J. Manage. Econ. (2020)

75. Varol, S., Marquez, A.: An empirical study on assessing brand loyalty in automobile industry using hidden Markov model. Acad. Market. Stud. J. **1**, 24 (2020)

76. Vidhate, D.A., Kulkarni, P.: Improved decision making in multiagent system for diagnostic application using cooperative learning algorithms. Int. J. Inf. Technol. **10**(2), 201–209 (2017). https://doi.org/10.1007/s41870-017-0079-7

77. Viji, D., Banu, S.K.Z.: An improved credit card fraud detection using k-means clustering algorithm. Int. J. Eng. Sci. Invention (IJESI) (2018)

78. Wahana, A., Maylawati, D.S., Irfan, M., Effendy, H.: Supply chain management using FP-growth algorithm for medicine distribution. J. Phys. Conf. Ser. **1**(978), 012018 (2018)

79. Wang, H., Sarker, B.R., Li, J., Li, J.: Adaptive scheduling for assembly job shop with uncertain assembly times based on dual Q-learning. Int. J. Prod. Res. 1–17 (2020)

80. Wang, Kwok, T.H., Zhou, C., Vader, S.: In-situ droplet inspection and closed-loop control system using machine learning for liquid metal jet printing. J. Manuf. Syst. **47**, 83–92 (2018)

81. Wang, Y.-F.: Adaptive job shop scheduling strategy based on weighted Q-learning algorithm. J. Intell. Manuf. **31**(2), 417–432 (2018). https://doi.org/10.1007/s10845-018-1454-3

82. Whitley, D., Watson, J.P.: Complexity theory and the no free lunch theorem. In: Search Methodologies, pp. 317–339. Springer, Boston (2005). https://doi.org/10.1007/0-387-28356-0_11

83. Wolpert, D.H., Macready, W.G.: No free lunch theorems for optimization. IEEE Trans. Evol. Comput. **1**(1), 67–82 (1997)

84. Zhang, S., Wong, T.N.: Integrated process planning and scheduling: an enhanced ant colony optimization heuristic with parameter tuning. J. Intell. Manuf. **29**(3), 585–601 (2014). https://doi.org/10.1007/s10845-014-1023-3
85. Zhu, Y., Zhou, L., Xie, C., Wang, G.J., Nguyen, T.V.: Forecasting SMEs' credit risk in supply chain finance with an enhanced hybrid ensemble machine learning approach. Int. J. Prod. Econ. **211**, 22–33 (2019)

# Text Mining Based Decision Making Process in Kickstarter Platform

Tuba Karagül Yildiz[✉], Ercan Atagün, Hacer Bayiroğlu, Tunahan Timuçin, and Hakan Gündüz

Department of Computer Engineering, Duzce University, Duzce, Turkey
{tubakaragul,ercanatagun,hacerbayiroglu,tunahantimucin,
hakangunduz}@duzce.edu.tr

**Abstract.** Kickstarter is a platform that supports the transformation of projects into products by embracing entrepreneurial investor interaction. Entrepreneurs who register on the Kickstarter website exhibit presentations and visuals on their projects. People who work or have new ideas in the fields of art, comics, design, technology, film, games, music and publishing are looking for support to improve their ideas. Entrepreneurs display the project details in a textual expression in order to attract investors' interest. Investors make a decision to invest in the project by reading these expressions and consider them logically. Investors need to solve the difficulties they encountered in decision-making through various artificial intelligence or language processing methods. Natural Language Processing (NLP) is one of the preferred methods for such decision-making problems. In this study, NLP techniques and word attributes were obtained from the explanations of the projects in order to provide support to the investors. In addition, temporal attributes were extracted by considering the start and end times of the projects. According to these attributes, whether the projects can reach the targeted budget has been determined by artificial learning methods and predictable effects of some words or phrases in reaching the targeted budget have been highlighted.

**Keywords:** Kickstarter · Natural Language Processing · Text mining

## 1 Introduction

Crowdfunding systems are systems where many investors come together and invest in projects they like. Crowdfunding systems are the most intense platforms for entrepreneur-investor interaction. Kickstarter is one of the most popular crowdfunding systems.

Although crowdfunding systems, venture capital, and angel investment are seen as similar concepts, there are fundamental differences between them. The main purpose of these three financing methods is to create funds to support young entrepreneurs who do not have sufficient resources to receive loans from banks and other credit institutions. Angel investors generally support business ideas that they see fit individually. Venture capital investment partnerships provide institutional funding. Crowdfunding systems,

J. Hemanth et al. (Eds.): ICAIAME 2020, LNDECT 76, pp. 344–349, 2021.
https://doi.org/10.1007/978-3-030-79357-9_33

on the other hand, can obtain funds from both investors and people without investment experience with the help of the internet without a time and place constraint [1].

Crowdfunding systems have been growing in recent years. According to World Bank research, these systems are expected to reach a volume of 96 billion dollars in 2025. This amount represents a growth of 1.8 times the world's risk capital. Besides, it is predicted that a fund resource of 13.8 billion dollars will be created in Europe and Central Asia [2].

Many economic habits have changed with the impact of the 2008 world economic crisis. One of these changing habits is the new financing methods recommended for investment. And one of these new financing methods is crowdfunding systems. The Kickstarter platform, one of the crowdfunding systems, was established in 2009. Kickstarter is a digital platform for providing the financial support needed by an entrepreneur with a project idea.

Electronic environments have caused changes in many areas as well as transforming the field of investment. With electronic crowdfunding systems such as Kickstarter, entrepreneur-investor cooperation that cannot come together due to geographical, political, or other reasons is provided.

Ettert et al. Used time series and tweet data to predict the pass/fail status of Kickstarter projects [3]. Scientists have been investigating the links between language and social behavior for years. They investigate the harmony between the languages and words people use in daily life and social behaviors. On the other hand, Mitra et al. Have tried to determine which linguistic features of entrepreneurs are taken into consideration by investors investing in Kickstarter. They analyzed the phrases and variables commonly found on crowdfunded sites [4]. Buttice et al. Examined the impact of narcissism on the fundraising of projects in Kickstarter and found a negative relationship between crowdfunding success and narcissism [5]. In another study on Kickstarter, fundraising was predicted. Wang et al. Applied deep learning, decision tree, Random Forest, Logistic Regression, SVM, and KNN methods. They predicted the total value that the project could generate with a strong success rate with deep learning [6].

Blaseg et al. researched consumer protection on Kickstarter. Kickstarter advertising campaigns have analyzed and focused on the failure of promised discounts [7]. Shafqat et al., Based on ignoring investors' comments, combined neural networks with language modeling techniques and hidden patterns of debate left in comments. They developed a neural network architecture by modeling language against fraud and fraud with RNN and LSTM [8]. Crowdfunding systems accommodate not only entrepreneurs but also social responsibility project owners, non-governmental organizations' projects, and artists' projects to find funds [9]. In Kickstarter, some studies are specialized and focused on certain categories and even certain projects. Lolli based his work on the project on Kickstarter, which is in the video game category and received the most funds in this field [10]. Ryoba et al. Presented a metaheuristic algorithm to perform a complete search of a subset of features with high success on Kickstarter. They proposed the Whale optimization algorithm with the KNN classifier algorithm [11]. Lee et al. Emphasized that the average project success rate in crowdfunding systems is 41% and that this rate is gradually decreasing. In this respect, they predicted the project's success by determining the important factors of successful projects. They created a data set about campaigns, updates, and comments in crowdfunding. As a method, they tried to predict project success by developing a deep neural network model [12]. Yu and his colleagues developed a model that predicts the success of the crowdfunding project through deep learning. They proposed an MLP model by using the historical information of Kickstarter projects in the data they received over Kaggle. They applied the model they developed in the study to different crowdfunding platforms that were not previously addressed [13]. Alamsyah and Nugroho have used data mining methods to map the success of Kickstarter projects and investor trends. In this context, they used KNN, ANN, and K-means methods [14]. Zhou et al. Linked the Kickstarter project owners' fundraising process from supporters to the project disclosure partial. They generally benefited from the project descriptions and benefited from the past experiences and expertise of the project owners. With the help of this information, they tried to predict whether a project was successful or not [15].

Guo et al. combined data mining and economic models. The study examined trends in distance information of a Kickstarter project during the funding period. They provided a theoretical basis for improving the promotion of crowdfunding projects by conducting online analysis [16].

In this study, the effect of the language used by entrepreneurs who develop projects and introduce their projects to other people on project success is examined.

## 2   Material and Method

Natural Language Processing (NLP) is a subcategory of artificial intelligence. NLP is used to express and analyze human language mathematically [17]. NLP has been used in many areas in recent years. With text mining processes, semantic and syntactic analysis can be done, and large amounts of natural languages can be processed [18]. Text mining is the process of discovering information from large text data. It is the process of determining interesting patterns and relationships from texts [19]. Increasing data on the internet and other electronic media in recent years have increased the usability of text mining. This data set is taken from the web page called web robots [20] (Table 1).

**Table 1.** Features in the dataset

| Feature name | Data type | Description |
| --- | --- | --- |
| Blurb | String | Project Description |
| State | Label | Status (Class label) |

50000 pieces of data are used in this study. This data has the project description 'blurb' and class tag. The relationship between 'blurb' and the class tag has been revealed by text mining. Cross validated while classifying. It was determined as k = 10 for cross-validation.

# 3   Application

In this study, text mining was performed using the Naïve Bayes algorithm (Fig. 1).

**Fig. 1.** This is the word cloud obtained from the project definitions in the data set obtained for the study [21].

Word cloud is the creation of a cloud according to the frequency of occurrence of words. In the word cloud, words are positioned on the cloud according to the numbers they are used in (Fig. 2).

```
Correctly Classified Instances        35372                70.744 %
Kappa statistic                        0.3268
Mean absolute error                    0.3406
Root mean squared error                0.4529
Relative absolute error               73.9004 %
Root relative squared error           94.3539 %
Total Number of Instances        50000
```

**Fig. 2.** This figure contains information on the classification success rate, Kappa statistics, and error types [22].

The success rate in this study is measured as 70.744% (Figs. 3 and 4).

| | TP Rate | FP Rate | Precision | Recall | F-Measure | MCC | ROC Area | PRC Area | Class |
|---|---|---|---|---|---|---|---|---|---|
| | 0,845 | 0,537 | 0,737 | 0,845 | 0,787 | 0,335 | 0,731 | 0,815 | failed |
| | 0,463 | 0,155 | 0,627 | 0,463 | 0,533 | 0,335 | 0,731 | 0,629 | successful |
| Weighted Avg. | 0,707 | 0,399 | 0,697 | 0,707 | 0,695 | 0,335 | 0,731 | 0,748 | |

**Fig. 3.** TP Rate, FP Rate, precision, accuracy, recall, etc. such values are shown. Here, the TP rate is higher than the FP rate [22].

```
=== Confusion Matrix ===


    a      b    <-- classified as
  27031   4962 |    a = failed
   9666   8341 |    b = successful
```

**Fig. 4.** The Confusion Matrix obtained with this study is shown here [22].

The developed model recognized unsuccessful projects more. 35372 projects were correctly estimated. 14628 projects were wrongly estimated.

## 4 Conclusions and Suggestions

With crowdfunding systems such as Kickstarter, project owners can deliver their projects to all people to find funds. In systems such as Kickstarter, project owners introduce their projects using photos, videos and text.

With this study, the text expressions used by Kickstarter project owners for project promotion were discussed. The relationship of textual expressions with the success of the projects was investigated by text mining. Of the information describing the projects, only text expressions were focused. In the study, 70% success was obtained by using Naïve Bayes algorithm. Since the projects in the data set are mostly unsuccessful, the developed model tends to identify unsuccessful projects. The success rate can be increased by using larger data sets in future studies. Lemmatization and Stemming operations can be done to increase the success rate. Customized algorithms can be developed for Lemmatization and Stemming operations. Stopword improvements can be made in data preprocessing steps. Besides, the fact that the number of successful and unsuccessful projects with a class label is more balanced in the data set will affect the success of the model to be developed. Also, different text mining algorithms can be used. Another suggestion for future studies is that a model can be developed with different data structures by combining different elements such as pictures, videos, used in project definitions with the text expression, which is the project description, and thus more descriptive and more successful models can be developed. The success rate can be increased by applying deep learning techniques with the text mining process.

# References

1. Fettahoğlu, S., Khusayan, S.: Yeni finansman olanağı: kitle fonlama. Uşak Üniversitesi Sosyal Bilimler Dergisi **10**(4), 497–521 (2017)
2. Sancak, E.: Applicability and readiness of crowdfunding in Turkey. International Journal of Business and Social Science, vol. 7, no.1 (2016)
3. Etter, V., Grossglauser, M., Thiran, P.: Launch hard or go home! Predicting the success of Kickstarter campaigns. In: Proceedings of the First ACM Conference on Online Social Networks, pp. 177–182, October 2013
4. Mitra, T., Gilbert, E.: The language that gets people to give: phrases that predict success on kickstarter. In: Proceedings of the 17th ACM Conference on Computer Supported Cooperative Work & Social Computing, pp. 49–61, February 2014
5. Butticè, V., Rovelli, P.: Fund me, I am fabulous!" Do narcissistic entrepreneurs succeed or fail in crowdfunding?. Pers. Individ. Differ. **162**, 110037 (2020)
6. Wang, W., Zheng, H., Wu, Y.J.: Prediction of fundraising outcomes for crowdfunding projects based on deep learning: a multimodel comparative study. Soft Computing, pp. 1-19 (2020)
7. Blaseg, D., Schulze, C., Skiera, B.: Consumer protection on Kickstarter. Mark. Sci. **39**(1), 211–233 (2020)
8. Shafqat, W., Byun, Y.C.: Topic predictions and optimized recommendation mechanism based on integrated topic modeling and deep neural networks in crowdfunding platforms. Appl. Sci. **9**(24), 5496 (2019)
9. Tekeoğlu, N.: Kitlesel fonlamaile alternatif film finansmanı oluşturma ve bir film analizi: sıradışı İnsanlar. Int. J. Soc. Sci. **38**, 295–302 (2015)
10. Lolli, D.: 'The fate of Shenmue is in your hands now!': Kickstarter, video games and the financialization of crowdfunding. Convergence **25**(5–6), 985–999 (2019)
11. Ryoba, M.J., Qu, S., Zhou, Y.: Feature subset selection for predicting the success of crowdfunding project campaigns. Electronic Markets, pp. 1–14 (2020).https://doi.org/10.1007/s12 525-020-00398-4
12. Lee, S., Lee, K., Kim, H.C.: Content-based success prediction of crowdfunding campaigns: a deep learning approach. In: Companion of the 2018 ACM Conference on Computer Supported Cooperative Work and Social Computing, pp. 193–196 October 2018
13. Yu, P.F., Huang, F.M., Yang, C., Liu, Y.H., Li, Z.Y., Tsai, C.H.: Prediction of crowdfunding project success with deep learning. In: 2018 IEEE 15th International Conference on e-Business Engineering (ICEBE), pp. 1–8. IEEE October 2018
14. Alamsyah, A., Nugroho, T.B.A.: Predictive modelling for startup and investor relationship based on crowdfunding platform data. In: IOP Conferences Series: Journal of Physics: Conferences Series, vol. 971, March 2018
15. Zhou, M.J., Lu, B., Fan, W.P., Wang, G.A.: Project description and crowdfunding success: an exploratory study. Inf. Syst. Front. **20**(2), 259–274 (2018). https://doi.org/10.1007/s10796-016-9723-1
16. Guo, L., Guo, D., Wang, W., Wang, H., Wu, Y.J.: Distance diffusion of home bias for crowdfunding campaigns between categories: insights from data analytics. Sustainability **10**(4), 1251 (2018)
17. Khurana, D., Koli, A., Khatter, K., Singh, S.: Natural language processing: state of the art, current trends and challenges (2017). arXiv preprint arXiv:1708.05148
18. Ferreira-Mello, R., André, M., Pinheiro, A., Costa, E., Romero, C.: Text mining in education. Wiley Interdisc. Rev. Data Min. Knowl. Discovery **9**(6), e1332 (2019)
19. Hark, C., et al.: Doğal dil İşleme yaklaşimlari ile yapisal olmayan dökümanlarin benzerliği. In: 2017 International Artificial Intelligence and Data Processing Symposium (IDAP), pp. 1–6. IEEE, September 2017
20. Kickstarter Project. https://webrobots.io/kickstarter-datasets/. Accessed 06 June 2020
21. https://orangedatamining.com/
22. https://www.cs.waikato.ac.nz/ml/weka/

# Deep Q-Learning for Stock Future Value Prediction

Uğur Hazir$^{(\boxtimes)}$ and Taner Danisman

Computer Engineering Department, Akdeniz University, Antalya, Turkey
20185156008@ogr.akdeniz.edu.tr, tdanisman@akdeniz.edu.tr

**Abstract.** There are many models at Finance sector to decide which stock to buy or sell and when to act. Financers apply Stochastic oscillator with Bollinger bands, MACD Indicator (Moving Average Convergence Divergence), RSI (Relative Strength Index) Indicator and Chaikin Money Flow and much more techniques and try to find out some patterns to make predictions and decisions. There are hundreds of indicators but there is lack of knowledge and research how successful these techniques are. At this project more realistic financial environment is used to create more efficient method for real life. For every buy and sell action a transaction cost is applied to create a more harsh conditions. The agent starts with very limited money and tries to maximize it. For those conditions it is highly possible to lose all the money that will be called as bankrupt throughout at this paper. Deep Q-Learning is used at this project because it is very complicated problem to predict how a stock will perform and there are many parameters effecting the performance of the stock. As classical financial models do; MACD and Stochastic parameters are given as well as daily parameters of a stock as input for the project. This project focuses on daily predictions and tries to answer which action will be more beneficial and when to act. Also this project tries to answer the questions that if we invest certain amount of money how much profit expected to be earn and what is the possibility of bankruptcy for a given test period time after learning process.

**Keywords:** Deep Q-Learning · Stock · Future value prediction

## 1 Introduction

Technical analysis is focused in this project, because the goal is to perform actions at daily bases in a short period of time rather than long term investment. Also, Stochastic oscillator with Bollinger bands, MACD Indicator (Moving Average Convergence Divergence), RSI (Relative Strength Index) Indicator and Chaikin Money Flow and much more techniques are used for technical analysis. These indicators produce signals that are used by investors to make a decision about when to buy or sell a stock. In this project it is aimed to create an agent which decides its actions itself by using Deep Q-Learning. The agent decides when to buy or sell a stock or does nothing by waiting (or holding a stock).

In the article named "Performance of technical trading rules: evidence from Southeast Asian stock markets" [1] Southeast Asian stock markets and some technical analysis

J. Hemanth et al. (Eds.): ICAIAME 2020, LNDECT 76, pp. 350–359, 2021.
https://doi.org/10.1007/978-3-030-79357-9_34

indicators are compared with simple buy and hold methodology. They found out that "unprofitable strategies like RSI and STOCH still perform worse than a simple Buy-and-Hold strategy (BH) even with optimized parameters except in the Singaporean market, where they manage to beat a BH with the long-only strategies" [1], overall, their results show that "these five Southeast Asian stock markets are, to a varying degree, at least close to weak-form efficient as most popular technical trading strategies could not earn statistically significant returns, particularly after transaction costs. The only exception is the Thai market" [1]. They also found out that "in terms of market timing", [1] "using technical indicators does not help much" [1] and "even profitable strategies such as MACD and STOCH-D could not reliably predict subsequent market directions as their profitable trades are usually less than fifty percent of total number of trades" [1].

Daily trading is a very risky method for investment. According to NASAA "Seventy percent of the accounts lost money and were traded in a manner that realized a 100% RISK OF RUIN (loss of all funds). Only three accounts of the twenty-six evaluated (11.5% of the sample), evidenced the ability to conduct profitable short-term trading" [2]. Also according to [3] "The fact that at least 64% of the day traders in this study lost money suggests that being a profitable day trader is more difficult than the industry maintains."

In this project it is aimed to create an agent which decides its actions itself by using Deep Q-Learning. The agent decides when to buy or sell a stock or does nothing by waiting (or holding a stock). Also this project tries to answer the questions that if we invest certain amount of money how much profit expected to be earn and what is the possibility of bankruptcy for a given test period time after learning process which is a very important concept for investors.

## 2  Dataset

We used 4 BIST-30 (Borsa Istanbul 30 Index) stocks: ARCLK [4], ASELS [5], SAHOL [6] and TUPRS [7]. We gather our data from Mynet website [4–7]. The data is gathered for each stock between 12.01.2000–21.11.2019. The training data cover the dates between 08.03.2000–31.12.2014 (almost 14 years) and the test data cover the dates between 02.01.2015–21.11.2019 (almost 4 years). We used Close, Low, High and Volume parameters as input. We also preprocessed the data to create MACD and Stochastic indexes so that the dates between 12.01.2000–07.03.2000 are used to calculate those indexes.

EMA12, EMA26, MACD and MACDsignal9 indicators are generated using following equations [8]:

$$EMA_n = Closing\ Price_n \frac{2}{Time\ Period + 1} + EMA_{n-1}\left(1 - \frac{2}{Time\ Period + 1}\right) \quad (1)$$

For the Eq. (1) EMA12 and EMA26 indices are generated setting Time Period as 12 and 26 (days) respectively. n stands for today and n−1 represents yesterday.

$$MACD_n = EMA12_n - EMA26_n \quad (2)$$

For a given day MACD is just the subtraction of EMA12 and EMA26 values of the given day (Eq. (2)).

$$signal_n = MACD_n \frac{2}{Time\ Period + 1} + signal_{n-1}\left(1 - \frac{2}{Time\ Period + 1}\right) \quad (3)$$

Using Eq. (3) MACDsignal9 indicator is calculated by setting Time Period to 9 where n stands for today and $n-1$ represents yesterday.

StockhasticMax14, StockhasticMin14, StockhasticK14, StockhasticD14, StockhasticMax5, StockhasticMin5, StockhasticK5 and StockhasticD5 indicators are generated using following equations [9, 10]:

$$\%K = 100 * \frac{C - L_{Time\ Period}}{H_{Time\ Period} - L_{Time\ Period}} \quad (5)$$

$$\%D = \frac{K_1 + K_2 + K_3}{3} \quad (6)$$

StochasticMax14 and StochasticMax5 are two indicators that they are the max value of the High values of last 14 days and 5 days respectively. Likewise, StochasticMin14 and StochasticMin5 are two indicators that they are the min value of the Low values of last 14 days and 5 days respectively. StockhasticK14 and StockhasticK5 are calculated using Eq. (5) by setting Time Period as 14 and 5 respectively. StockhasticD14 and StockhasticD5 are calculated using Eq. (6) as they are the average of last 3 days K% values.

## 3 Assumptions

Using Investment Calculator [11] if we set Starting Amount to 1000$ after 14 years with 5% yearly return rate without any contribution our End Balance will be 1979.93$. Likewise; if we set Starting Amount to 1000$ after 4 years with 5% yearly return rate without any contribution our End Balance will be 1215.51$. So that our first assumption is that we invest 1000 TL and if our end balance is more than 2000 TL it is at profit for the training period. Second assumption is for the test period if we invest 1000 TL to make profit our end balance must be more than 1220 TL. And the third assumption is for the test period if the initial investment is 10000 TL to make profit end more than 12200 TL. The fourth assumption is that if the balance is lower than 250 TL than it is a bankrupt. The fifth assumption is that we buy or sell at Close value for that day. The final assumption for this project is assuming that the only cost is transaction cost. So that for every buy and sell action 1 TL constant cost is applied.

## 4 Method

Deep Q-Learning methodology is used and for DQN (Deep Q Network) we have three action space which are: 0 for wait, 1 for buy and 2 for sell. Close, Low, High, Volume, Year, EMA12, EMA26, MACD, MACDsignal9, StockhasticMax14, StockhasticMin14,

StockhasticK14, StockhasticD14, StockhasticMax5, StockhasticMin5, StockhasticK5, StockhasticD5 are our properties for DQN.

The neural network has 540 neurons at first layer, 180 neurons at second layer, 64 neurons at third layers, 32 neurons at fourth layer, 8 neurons at fifth layer and 3 output neurons that represents wait, buy and sell actions. RELU is used as activation for all layers, but for the output layer linear is used which is equal to the value itself directly means no activation function used. For the neural network 5% dropout is used by assigning gamma to 0.95, the learning rate is set to 0.001 and Adam optimizer is used for backpropagation.

For the DQN transaction cost (reward at the same time) is set to 1 for every buy or sell action, waiting reward is set to −0.2 when holding a stock and doing nothing reward is set to −0.5 to encourage making an action. Epsilon is set to 1 initially for exploration and decay by multiplying epsilonDecay constant 0.995 until it reached below epsilon_min constant which is equal to 0.01 so that epsilon greedy strategy is applied for exploitation. Window size is set to 10 to focus last 10 operation day. For the replay memory, the memory size is set to 1000 and batch size is set to 32 to break the correlation between consecutive samples.

$$q_t(s, a) = E\big[R_{t+1} + \gamma_{a'}^{max} q_t(s'a')\big] \tag{7}$$

Q values are calculated using Eq. (7) where s stands for state, a stands for action, $q_t$ stands for q value for the given time, E comes from experience replay memory, $R_{t+1}$ is the reward if we apply the action, $\gamma$ is set to 0.95 max a' is the max q value we will get from next action s' and a' is the next state and next action respectively.

So that we start as for each episode we reset the environment to the very beginning. After that for each day we take an action while we are considering epsilon greedy strategy the first actions will be random and after epsilon decays it is determined from q values which gives maximum reward. Then we make a step and store our experience in replay memory. Then we assign next state as current state. We apply replay memory method and epsilon greedy method. We finalize the episode after we processed the last day value or if we bankrupt.

We test the DQN by loading the trained models we saved. Epsilon is set to 0.01 and epsilon greedy is not necessary now because it is also the minimum. The remaining settings are same. We made 2 tests that the first test has 1000 TL initial balance and the second has 10000 TL initial balance and each are run for 100 episodes.

## 5   The Combined DQN Method

During the tests we observed that as the DQN learns number of actions decreases. Also, there are long idle periods which agent does nothing. We also considered that there are many buy signals following each other as well as sell signals which are not used. To use them and to make a better mechanism we modified our test application. Rather than run each stock DQN's as a single period, we run them simultaneously. The Combined DQN Method which we invented selects and attaches itself to the best performed DQN for the given time buy buying a stock as if the best performing DQN gives buy signal during that period. The Combined DQN Method sells if the attached DQN give signal to sell or if it is %5 profit or loss.

## 6  Results

We trained each stock for 5000 episodes. Training period took almost 45 days for a single stock.

For the training of ARCLK we have found that moving average for 100 episodes the bankruptcy rate drops from 47% to below 5%, profitability rises 30% to above 90%, number of operations drops 901 to 39 and average end balance rises from 1975 to 5004 (Figs. 1 and 2).

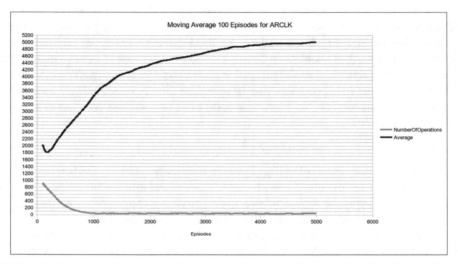

**Fig. 1.** Number of operations and average end balance moving average 100 Episodes for the ARCLK stock

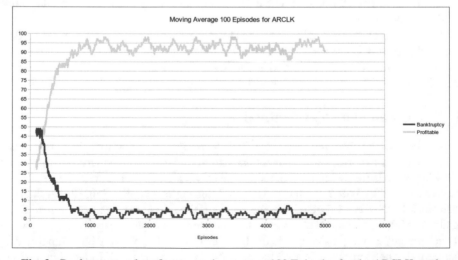

**Fig. 2.** Bankruptcy and profit rate moving average 100 Episodes for the ARCLK stock

For the training of SAHOL we have found that moving average for 100 episodes the bankruptcy rate drops from 58% to below 5%, profitability rises 12% to above or equal to 85%, number of operations drops 818 to 41 and average end balance rises from 724 to 3281 (Figs. 3 and 4).

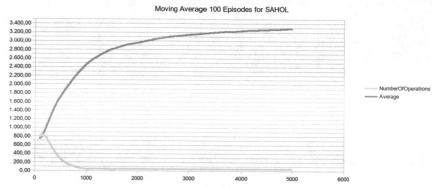

**Fig. 3.** Number of operations and average end balance moving average 100 Episodes for the SAHOL stock

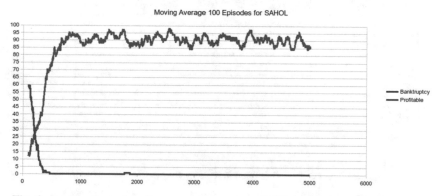

**Fig. 4.** Bankruptcy and profit rate moving average 100 Episodes for the SAHOL stock

For the training of ASELS we have found that moving average for 100 episodes the bankruptcy rate drops from 56% to below 5%, profitability rises 35% to above or equal to 95%, number of operations drops 691 to 39 and average end balance rises from 4090 to 7469 (Figs. 5 and 6).

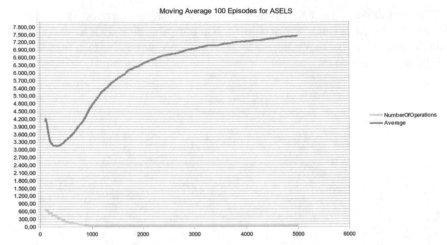

**Fig. 5.** Number of operations and average end balance moving average 100 Episodes for the ASELS stock

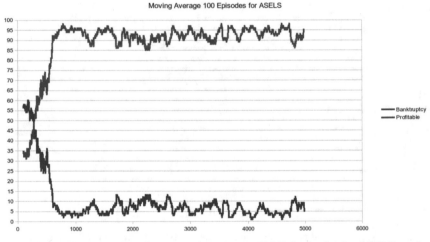

**Fig. 6.** Bankruptcy and profit rate moving average 100 Episodes for the ASELS stock

For the training of TUPRS we have found that moving average for 100 episodes the bankruptcy rate drops from 28% to below 5%, profitability rises 36% to above 95%, number of operations drops 1050 to 40 and average end balance rises from 2546 to 4856 (Figs. 7 and 8).

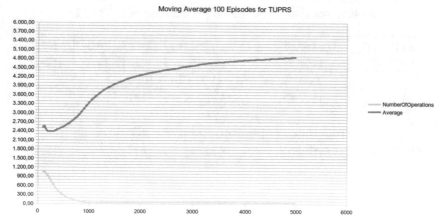

**Fig. 7.** Number of operations and average end balance moving average 100 Episodes for the TUPRS stock

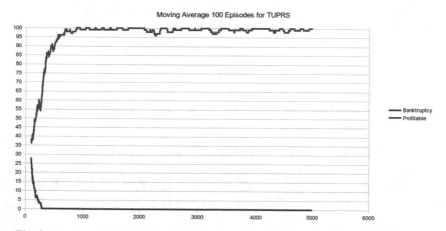

**Fig. 8.** Bankruptcy and profit rate moving average 100 Episodes for the TUPRS stock

These are the test results for 100 episodes if we invest 1000 TL (Table 1):

**Table 1.** Test results for when the initial investment is 1000 TL.

| MaxARCLK | MaxSAHOL | MaxASELS | MaxTUPRS | MaxC |
|---|---|---|---|---|
| 1847.62 | 1157.74 | 5207.16 | 3239.24 | 3584.50 |
| **LossRates (%) --- Lessthan 1000** | | | | |
| LrARCLK | LrSAHOL | LrASELS | LrTUPRS | LrC |
| 23 | 45 | 17 | 14 | 55 |
| **Profit Rates (%) --- Morethanorequalto 1220** | | | | |
| PrARCLK | PrSAHOL | PrASELS | PrTUPRS | PrC |
| 35 | 0 | 28 | 23 | 23 |
| **Avgnumber of Operations** | | | | |
| AvgOp ARCLK | AvgOp SAHOL | AvgOp ASELS | AvgOp TUPRS | AvgOp C |
| 13.83 | 9.68 | 14.04 | 13.39 | 166.31 |
| **Avg Value** | | | | |
| Avg ARCLK | Avg SAHOL | Avg ASELS | Avg TUPRS | Avg C |
| **1146.97** | **1157.74** | 1333.87 | 1221.73 | **1087.95** |

C stands for combined methodology

These are the test results for 100 episodes if we invest 10000 TL (Table 2):

**Table 2.** Test results for when the initial investment is 10000 TL.

| MaxARCLK | MaxSAHOL | MaxASELS | MaxTUPRS | MaxC |
|---|---|---|---|---|
| 17,635.85 | 12214.13 | 41864.28 | 30951.26 | 36375.93 |
| **LossRates (%) --- Lessthan 10000** | | | | |
| LrARCLK | LrSAHOL | LrASELS | LrTUPRS | LrC |
| 20 | 38 | 25 | 13 | 30 |
| **Profit Rates (%) --- Morethanorequalto 12200** | | | | |
| PrARCLK | PrSAHOL | PrASELS | PrTUPRS | PrC |
| 42 | 1 | 39 | 22 | 46 |
| **Avgnumber of Operations** | | | | |
| AvgOp ARCLK | AvgOp SAHOL | AvgOp ASELS | AvgOp TUPRS | AvgOp C |
| 11.65 | 9.11 | 12.34 | 11.61 | 170.91 |
| **Avg Value** | | | | |
| Avg ARCLK | Avg SAHOL | Avg ASELS | Avg TUPRS | Avg C |
| **12062.13** | **10222.14** | 15687.91 | 12257.79 | 13472.07 |

C stands for combined methodology

# 7 Conclusion

We have found out that while the DQN learns the number of operations decreases during the training process and it learns how to not bankrupt. Bankruptcy rate decreases below 5% during the training process and 0% for the test phase. If the initial investment rises costs drop and profit increases. We believe that if we increased the number of stocks the Combined DQN Test Method will be more successful.

# References

1. Tharavanij, P., Siraprapasiri, V., Rajchamaha, K.: Performance of technical trading rules: evidence from Southeast Asian stock markets. Springerplus **4**(1), 1–40 (2015). https://doi.org/10.1186/s40064-015-1334-7
2. NASAA Day Trading Report (1999). https://www.nasaa.org/wp-content/uploads/2011/08/NASAA_Day_Trading_Report.pdf. Accessed 20 Nov 2020
3. Jordan, D.J., Diltz, J.D.: The profitability of day traders. Financ. Anal. J. **59**(6), 85–94 (2003). https://doi.org/10.2469/faj.v59.n6.2578
4. ARCELIK, ARCLK TarihselVeriler | MynetFinans. http://finans.mynet.com/borsa/hisseler/arclk-arcelik/tarihselveriler/. Accessed 23 May 2020
5. ASELSAN, ASELSTarihselVeriler | MynetFinans. http://finans.mynet.com/borsa/hisseler/asels-aselsan/tarihselveriler/. Accessed 23 May 2020
6. SABANCI-HOLDING, SAHOLTarihselVeriler | MynetFinans. http://finans.mynet.com/borsa/hisseler/sahol-sabanci-holding/tarihselveriler/. Accessed 23 May 2020
7. TUPRAS, TUPRSTarihselVeriler | MynetFinans. http://finans.mynet.com/borsa/hisseler/tuprs-tupras/tarihselveriler/. Accessed 23 May 2020
8. How to Calculate MACD in Excel. http://investexcel.net/how-to-calculate-macd-in-excel/. Accessed 23 May 2020
9. Stochastic oscillator – Wikipedia. https://en.wikipedia.org/wiki/Stochastic_oscillator/. Accessed 24 May 2020
10. How to Calculate the Stochastic Oscillator. http://investexcel.net/how-to-calculate-the-stochastic-oscillator/. Accessed 24 May 2020
11. Investment Calculator. https://www.calculator.net/investment-calculator.html/. Accessed 24 May 2020
12. Reinforcement Learning - Goal Oriented Intelligence. https://www.youtube.com/playlist?list=PLZbbT5o_s2xoWNVdDudn51XM8lOuZ_Njv/. Accessed 24 May 2020
13. Reinforcement Learning - Goal Oriented Intelligence – deeplizard. https://deeplizard.com/learn/playlist/PLZbbT5o_s2xoWNVdDudn51XM8lOuZ_Njv/. Accessed 24 May 2020

# Effect of DoS Attacks on MTE/LEACH Routing Protocol-Based Wireless Sensor Networks

Asmaa Alaadin, Erdem Alkım, and Sercan Demirci[✉]

Department of Computer Engineering, Ondokuz Mayıs University, Samsun, Turkey
asmaa.alaadin@bil.omu.edu.tr, {erdem.alkim,sercan.demirci}@omu.edu.tr

**Abstract.** In recent times, Wireless Sensor Networks (WSN) have become used in all fields of life, especially military and commercial ones. These networks are so widespread that it can be used in small enclosed spaces, or even to cover large geographical areas. But as usual, there are some threats that sabotage the functionality of these networks or misuse it. One of the most common of these threats is Denial-of-Service (DoS) attacks. In this article, we will summarize some of the most common techniques used in these attacks and classify it according to the targeted layer. Last, we will show the effect of DoS attacks on the lifetime of WSN's that using MTE and LEACH protocol.

## 1 Introduction

Wireless sensor network (WSN) is a network which is consisted of multiple sensing devices called sensor nodes. The number of these nodes in one WSN may reach thousands. These devices operate according to certain algorithms so that they can work together to accomplish a specific role. These networks can be used for military surveillance, forest fire detection, or building security monitoring [1]. These sensors collect data from the surrounding environment, share it with each other and send it back to base station using wireless communication network. When this base station receives sent data, it analyzes it and takes appropriate action and sends the information to the end-user. The base station is usually connected to the end-user using existing infrastructures of the network such as the Internet or GSM networks [6].

Since the sensors are small in size and their energy resources are usually limited, so their ability to process data is also poor. One of the most challenges facing WSN operators is the limited energy problem. In some environments, connecting nodes to power sources may be very difficult. Therefore, the use of algorithms that help reduce energy consumption is an effective solution in such situations. Some algorithms periodically compress the collected data and send it to other nodes. After that the node goes into hibernation mode so that it saves the energy and increases node's lifetime [9]. The limited power problem in networks can be exploited to attack it. Many DoS attacks increase the consumption of network's energy to keep them out of action. In this paper, we wanted to shed

© The Author(s), under exclusive license to Springer Nature Switzerland AG 2021
J. Hemanth et al. (Eds.): ICAIAME 2020, LNDECT 76, pp. 360–368, 2021.
https://doi.org/10.1007/978-3-030-79357-9_35

light on the energy consumption attacks that is targeting data-link layer. We wanted also to analyze the effects of these attacks on network's lifetime.

### 1.1  Organization

In the remainder of this paper, related works are listed in Sect. 2. we will give an overview of the most common DoS attacks in the third section. In Sect. 4, we will discuss the practical results of testing DoS attacks on a MTE/LEACH protocol-based networks. Last Section will give the conclusion.

## 2  Related Works

Because of the high risk posed by DoS attacks on networks, many researches have been done. Researchers have studied the techniques used in attacks and their impact on the network in general and layers in particular. [11] discusses the differences between DoS attacks and how to avoid them. In [13], a survey was performed to classify DoS attacks as active or passive attacks. Performance of WSN under DoS attacks in physical and network layers was analyzed in [3]. In [4], Depending on the impact of DoS attacks on the network, simulations was performed and the attacks has been classified as below:

- Attacks causing loss in packets.
- Attacks causing interference.
- Attacks causing noise.
- Attacks that change node's software.

In this article, we will summarize DoS attacks and their effects on WSNs. We have simulated the Energy Consumption attack on two different routing protocol (MTE and LEACH) in the data-link layer.

## 3  DoS Attacks

Denial of Service (DoS) attacks is defined as an event that reduces the capacity of a network to perform its expected role [15]. These attacks are not defined only as the attempts that seek to completely stop the functionality of the networks, but also on any attempt to limit it's functions too. Figure 1 shows an example of DoS attack on a part of WSN. According to the interruption of the communication law, two main categories have been defined to classify DoS attacks: passive attacks and active ones [7]. Table 1 shows the classification of DoS attacks according to the targeted network layers.

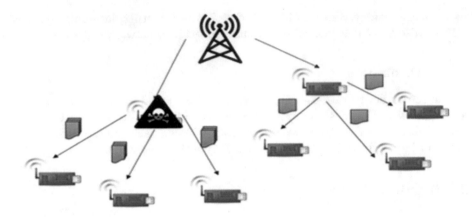

**Fig. 1.** An example of DoS Attack on a Wireless Sensor Network

## 3.1  Physical Layer

Modeling signals and data using frequency determination methods and data modeling, which are highly dependent on the physical infrastructure of the network, are the main tasks of this layer. So, the attacks try to sabotage or even destroy this network's infrastructure.

**Table 1.** DoS attacks classification according to the targeted network layers.

| Layer | Task of layer | DoS Attacks |
|---|---|---|
| Application | Interaction with the end user<br>Node localization<br>Time synchronization | - |
| Transport | Reliable end-to-end data transmission | 1. Flooding [6]<br>2. Desynchronization [6] |
| Network | Orientation<br>Topology management | 1. Selective forwarding [15]<br>2. Sinkhole [15]<br>3. Sybil [15]<br>4. Wormholes [15]<br>5. Hello flood [15] |
| Data-Link | Data flow<br>Creating Data frames<br>Middle access and control errors | 1. Collision [11]<br>2. Consumption [10]<br>3. Unfairness [12] |
| Physical | Frequency determination<br>Signal modulation<br>Transmission and reception<br>Data encryption | 1. Material sabotage [11]<br>2. Interference [11]<br>3. Jamming [16] |

The first example of this layer's attacks is **"Material Sabotage"** where the attacker may gets a physical access to the node. Attacker can changes node's functionality or change the data in node's memory. Some attacks generate network traffic as waves to interfere with network ones. This traffic can be generated continuously or periodically. This is called **"Interference"** attack. Sometimes, in wireless sensor networks, signals may get tangled, which can cause the message to be distorted or even lost as in **"Jamming"** attacks.

## 3.2  Data-Link Layer

The importance of this layer can be summarized in formatting and framing the data before maintaining the properly flowing of it. This layer's attacks target the data flowing by targeting networking paths. In **"Collision"** attacks attacker will make two or more of the network's nodes send data at the same time and using same frequencies. This may lead to loss of packets during transmission process. In other words, the attacker will induct one of the nodes to send data that will prevent other data receiving sent by another nodes. This may similar to **"Consumption"** attacks, where the attacking node tries to block network's channels by sending many sequenced messages or send requests. Sometimes, attackers can exploit their abilities to control the data flow of the network, this network becomes vulnerable to **"Unfairness"** attack so that data is sent continuously to block network's channels.

## 3.3  Network Layer

Since this layer is responsible of routing data and managing network's topology, attacks of this layer will attempt to distort the network structure or topology. This may cause changing in the flowing data and the destinations of this data. As an example, attacking nodes can reject or accept any packet. This can be done by several different criteria like blocking all incoming packets from a specific nodes and allowing the other ones. This type of attacks is called **"Selective Forwarding"**. When the attacking node rejects all packets from the other nodes, these nodes will declare this node as invalid or malicious node and will begin to send data using different paths from the previous ones. Other type of attacks occurs when a node rejects to receive all packets from other nodes, this is look like a data **"Sinkhole"**. Usually, sinkhole node is chosen as the most effective node in the network. Attacking node must have more energy to perform such these actions. Attacker may set up a connection with the base station also and sends referral message to the other node. After that, all nodes will direct their traffic to the malicious node, therefore a sinkhole attack has been launched [2]. When an attacker has the ability to assign more than one ID to one of the nodes in the network, this node becomes a destination for a greater number of data, thus the attacker can obtain all of this data. Of course, the more IDs assigned to a node, the more data is sent to it. This attack is called **"Sybil Attack"**.

Many protocols require a "hello" message to begin communicating with other nodes [14]. In **"Hello flooding"** attack, the attacking node sends fake hello packets so that the other nodes will see attacking node as a neighbor node. Thus the data will start to be sent to this node. There is also another type of attacks **"Wormhole"**, where the attacker creates a tunnel in the network structure, so that the malicious node is one of the ends of this tunnel. The attacker can use any of the pre-mentioned attacks to accomplish his attack.

### 3.4   Transport Layer

Ensuring reliable communication and securing data transmission is the responsibility of this layer. So a request to establish a connection must be sent when we want to establish a connection between two nodes. As the meaning of **"Flooding"** word says, an attacker sends out multiple requests to establish a connection, so that nodes will consume more resources.

**"Synchronization"** attack happens when malicious node sends a request to establish a connection with other nodes. When the connection is established, the synchronization requests will be sent continuously, thus, nodes will consume more power while responding to the malicious node.

**Table 2.** Parameters' values used in DoS attack simulation

| Parameter | Value |
|---|---|
| Area Size | $250 * 250$ m |
| Number of nodes | 30 node |
| BS Coordinates | $(125, 125)$ |
| Initial energy of Nodes | 2 J |
| Data message size | 4000 bit |
| Package header size | 150 bit |
| Communication protocols | MTE [5], LEACH [8] |
| Number of clusters | 5 |
| Energy of the attacking node | 20 J |
| Simulation time | 13 s |

## 4   Experimental Simulations and Results

Due to the limited works and researches on Data-Link layer DoS attacks, we wanted to demonstrate the effect of "Energy Drain" attack on this layer. In this section, we performed a practical simulations to see the effects of DoS attacks on WSNs which uses MTE/LEACH routing protocols. "Energy Drain" DoS attacks will try to disable the WSN by consuming its nodes' energy. As mentioned in the first section, power consumption is one of the most challenging problems which faces the developers and operators of these networks. All simulations and tests was performed using C++ and Python. Other simulation properties and used parameters' values are listed in Table 2.

In Order to make these simulations success, the following attacking scenario was applied:

- Selecting randomly one of our network's nodes.
- Properties of the selected node has been modified by increasing its energy and the size of the node's message.
- The attacking node will try to consume receiver nodes' energy in each transmission of data.

In the second simulation test, The difference between LEACH routing protocol-based network's lifetime before applying DoS attack (Fig. 4) and after applying it (Fig. 5) wasn't too much. This is because of the structure of LEACH protocol. In MTE routing protocol-based networks, the attacks will effect more nodes than the nodes will be effected in LEACH routing protocol-based networks. This depends on the path which the packet will take to reach the destination.

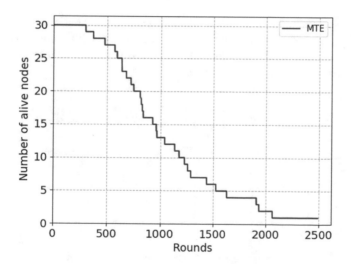

**Fig. 2.** MTE routing protocol-based network's lifetime before attacking

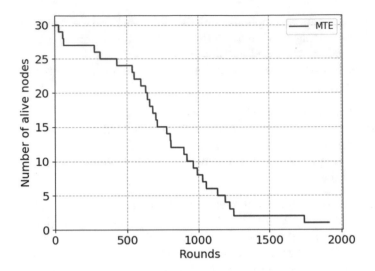

**Fig. 3.** MTE routing protocol-based network's lifetime after attacking

The first performed simulation was on a MTE routing protocol-based network. Figure 2 shows the network's lifetime before attacking the network. As we can see in Fig. 3, network's lifetime has been reduced due to incremental energy consuming applied by attacking nodes.

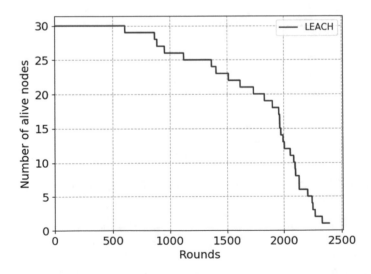

**Fig. 4.** LEACH routing protocol-based network's lifetime before attacking

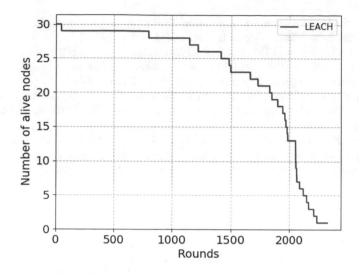

**Fig. 5.** LEACH routing protocol-based network's lifetime after attacking

## 5    Conclusions

With the increasing usage of WSNs, the threats to the security of these networks
have become common. One of the most common these threats is DoS attacks. In
this paper we summarized the most used techniques and types of these attack.
Of course, each type can effect a specific layer. In the practical section, we
performed DoS attacking simulations on MTE/LEACH routing protocol-based
networks. LEACH-based networks showed that it has more capability to resist
these attack than MTE-based networks. In future works, we will try to develop
a defense mechanism in wireless sensor networks against the attacks that we
simulated in this work.

## References

1. Akyildiz, I.F., Su, W., Sankarasubramaniam, Y., Cayirci, E.: A survey on sensor
   networks. Comm. Mag. **40**(8), 102–114 (2002). https://doi.org/10.1109/MCOM.
   2002.1024422
2. Chaudhry, J.A., Tariq, U., Amin, M.A., Rittenhouse, R.G.: Dealing with sinkhole
   attacks in wireless sensor networks. Adv. Sci. Technol. Lett. **29**(2), 7–12 (2013)
3. Das, R., et al.: Performance analysis of various attacks under AODV in WSN
   & MANET using OPNET 14.5. In: 2016 IEEE 7th Annual Ubiquitous Computing, Electronics & Mobile Communication Conference (UEMCON), pp. 1–9. IEEE
   (2016)
4. Diaz, A., Sanchez, P.: Simulation of attacks for security in wireless sensor network.
   Sensors **16**(11), 1932 (2016)
5. Ettus, M.: System capacity, latency, and power consumption in multihop-routed
   SS-CDMA wireless networks. In: Proceedings RAWCON 98. 1998 IEEE Radio and
   Wireless Conference (Cat. No. 98EX194), pp. 55–58. IEEE (1998)

6. Fan, C.S.: High: a hexagon-based intelligent grouping approach in wireless sensor networks. Adv. Electr. Comput. Eng. **16**(1), 41–47 (2016)
7. Lupu, T.G., Rudas, I., Demiralp, M., Mastorakis, N.: Main types of attacks in wireless sensor networks. In: WSEAS International Conference. Proceedings. Recent Advances in Computer Engineering, vol. 9. WSEAS (2009)
8. Ma, X.W., Yu, X.: Improvement on leach protocol of wireless sensor network. In: Applied Mechanics and Materials, vol. 347, pp. 1738–1742. Trans Tech Publications (2013)
9. Oliveira, F., Semente, R., Fernandes, J., Júnior, S., Melo, T., Salazar, A.: Eewes: an energy-efficient wireless sensor network embedded system to be applied on industrial environments. Ingeniería e Investigación **35**(2), 67–73 (2015)
10. Ould Amara, S., Beghdad, R., Oussalah, M.: Securing wireless sensor networks: a survey. EDPACS **47**(2), 6–29 (2013)
11. Raymond, D.R., Midkiff, S.F.: Denial-of-service in wireless sensor networks: attacks and defenses. IEEE Pervasive Comput. **7**(1), 74–81 (2008)
12. Saxena, M.: Security in wireless sensor networks-a layer based classification. Purdue University, Department of Computer Science (2007)
13. Shahzad, F., Pasha, M., Ahmad, A.: A survey of active attacks on wireless sensor networks and their countermeasures. International Journal of Computer Science and Information Security (IJCSIS), vol. 14, no. 12 (2016)
14. Singh, V.P., Jain, S., Singhai, J.: Hello flood attack and its countermeasures in wireless sensor networks. Int. J. Comput. Sci. Issues (IJCSI) **7**(3), 23 (2010)
15. Wood, A.D., Stankovic, J.A.: Denial of service in sensor networks. Computer **35**(10), 54–62 (2002)
16. Xu, W., Ma, K., Trappe, W., Zhang, Y.: Jamming sensor networks: attack and defense strategies. IEEE Network **20**(3), 41–47 (2006)

# On Connectivity-Aware Distributed Mobility Models for Area Coverage in Drone Networks

Mustafa Tosun[2], Umut Can Çabuk[1(✉)], Vahid Akram[1],
and Orhan Dagdeviren[1]

[1] Ege University, Izmir, Turkey
{umut.can.cabuk,vahid.akram,orhan.dagdeviren}@ege.edu.tr
[2] Pamukkale University, Denizli, Turkey
mustafatosun@pau.edu.tr

**Abstract.** Drone networks are becoming increasingly popular in recent years and they are being used in many applications such as area coverage, delivery systems, military operations, etc. Area coverage is a broad family of applications where a group of connected drones collaboratively visit the whole or parts of an area to fulfill a specific objective and is widely being researched. Accordingly, different mobility models have been designed to define the rules of movements of the participating drones. However, most of them do not consider the network connectivity which is crucial, plus many models lack the priorities and optimization strategies that are important for drone networks. Therefore within this study, three known connectivity-aware mobility models have been analyzed comparatively. Two non-connectivity-aware mobility models have further been implemented to catch the placebo effect if any. Per the detailed experiments on the mobility models, coverage rates, connectivity levels, and message traffic have been evaluated. The study shows that the Distributed Pheromone Repel (DPR) model provides a decent coverage performance, while the Connectivity-based model and the Connected Coverage model provide better connectivity and communication quality.

**Keywords:** Area coverage · Distributed algorithms · Mobility models · Connectivity · Drone networks · UAV

## 1 Introduction

Recent advances on the communication technologies and the evolution of small size, low cost and low energy hardware have triggered the boost of a new generation of devices that are going to make a revelation on human life. Unmanned aerial vehicles (UAV), especially the multi-rotor aircrafts that are called drones, are one of the future technologies which are increasingly being used in commercial, military, healthcare, transportation, monitoring, security, entertainment, and surveillance applications. UAV's can also fly autonomously and/or in groups, called swarms, to perform such missions [6].

© The Author(s), under exclusive license to Springer Nature Switzerland AG 2021
J. Hemanth et al. (Eds.): ICAIAME 2020, LNDECT 76, pp. 369–380, 2021.
https://doi.org/10.1007/978-3-030-79357-9_36

Due to the wide range of popular applications and particular challenges, many recent research have focused on UAVs. One obvious drawback of using UAVs is their production/sales costs. Generally, UAVs with wide communication ranges, long flight times, powerful processors and different sensor/actuator equipment are very expensive. The cost constraints, limits the usability of such UAVs, usually to the military applications and make them a less profitable approach for many commercial applications. Using a large swarm of low-cost UAVs, instead of a single (or a few) expensive UAV, is a worthwhile and feasible alternative that has many advantages, such as agility, cost-efficiency and reliability over using a single UAV. By utilizing a swarm, the joint coverage area can easily be extended via adding more UAVs to the network, the cooperative task can be done by collaboration of UAVs and persistent services can be guaranteed by periodically replacing the worn-out UAVs. However, the swarms, have also some challenges, such as communication issues, limited range, mobility control, path planning and dynamic routing.

The maximum radio range of inexpensive UAVs are typically limited, which eventually restricts the allowable distances to the neighboring UAVs. Therefore, each flying UAV should remain in the range of at least one other UAV, which is connected to some other UAVs. In case of disconnections due to limited ranges, it is proved that the restoring the connectivity between the UAVs in a disconnected drone network is an NP-hard problem [2], hence to avoid losing the resources the drones should always be kept connected to the rest of the swarm. To face this problem, many research have proposed different mobility models that describe how the drones should change their position while they communicate with other drones in their radio range. The mobility of flying drones in 3 dimensional space is completely different from terrestrial vehicles, which mostly utilizes the 2 dimensional space. Hence, the available mobility models for traditional mobile ad-hoc networks such as the Random Waypoint (RWP) model [7] are not neatly applicable to UAV swarms. In the case of swarms, mobility models that are specifically designed for flying objects are required.

In this paper, the selected mobility models for flying ad-hoc networks (FANET) have been described and further analyzed to evaluate their performances from various perspectives. Five of the notable mobility models for FANETs were implemented on a discrete event simulator and their performances were compared accordingly, using different metrics including the speed of coverage (convergance rate), the fairness of coverage, the connectivity percentage (convergance degree), the average connected components throughout the mission and the required message traffic volume. Briefly, the main contributions of this work are as follows:

- A review highlighting the advantages and the disadvantages of available mobility models for FANETs consisting of rotating-wing drones has been provided.
- Five notable mobility models have been implemented successfully in Python using the SimPy library.
- The performance of implemented models have been comparatively evaluated using five different metrics (listed above).

The rest of the paper has been structured as follows: Section 2 presents a review about the available mobility models for FANETs. Section 3 describes the details of the five selected mobility models. Section 4 explains the important metrics that can be used to analyze the performance of mobility models. Section 5 provides the simulation results of the implemented mobility models and compares the performances, thereof. Finally, Section 6 draws a conclusion and points out the potential future works.

## 2  Related Works

Various mobility models for UAV swarms have been proposed by different studies, in which each model has focused on specific properties such as application scenarios, area coverage, connectivity, speed and/or particular UAV types. Most papers consider the fixed-wing UAVs, while rotating-wing UAVS (drones) are also researched in some recent works.

The self-organizing aerial ad-hoc network mobility model proposed by Sanchez et al. in [11] focused on disaster recovery applications. In this model, the UAVs keep their connectivity alive by pushing and pulling each other to some extent according to their similarities and a threshold value. This model focuses on victim spotting and saving scenarios, but ignores the area coverage constraints.

Kuiper and Tehrani has proposed a mobility model, named pheromone, for UAV groups in reconnaissance applications [8]. This model only considers the coverage constraint and ignores the connectivity issues. More, the model has only three directions which is only suitable for fixed-wing UAVs. To preserve the stability of the connections in autonomous multi-level UAV swarms performing wide-area applications, a connectivity maintenance strategy, which combines the pheromone-based mobility model and some mobile clustering algorithms has been proposed by Danoy et al. in [5]. Yanmaz has further investigated the coverage and connectivity of autonomous UAVs in monitoring applications and proposed a connectivity-based mobility model to sustain the connectivity between the UAVs and also to the base station [13]. The proposed model focuses on connectivity and tries to maximize the covered area by a swarm of connected drones.

A distributed mobility model for autonomous interconnected UAVs for area exploration has been proposed by Messous et al. in [9]. In this model, the UAVs, explore a target area while they maintain the connectivity with their neighboring UAVs and the base station. This model takes the energy consumption constraint into account to increase the overall network lifetime, which is a noteworthy feature. Another decentralized and localized algorithm for mobility control of UAVs has been proposed by Schleich et al. in [12]. The proposed algorithm is designed to perform surveillance missions while preserving the connectivity. The UAVs maintain the connectivity between themselves by establishing a tree-based overlay network, where the root of the tree is the base station. However, fully ad-hoc networks that do not include a base station would be out of the scope.

A survey about the available mobility models for different applications regarding flying ad-hoc networks has been presented by Bujari et al. in [4]. This survey compares the models only from the application perspective and does not consider the other (i.e. performance) metrics.

# 3    Mobility Models

Within the scope of this study, the following models have been implemented: the random-walk as in [8]), the Distributed Pheromone Repel model [8], the Connectivity-based model [13], the KHOPCA model [5], and the Connected Coverage model [12]. This section briefly explains the details of each model.

## 3.1    Random Mobility Model

The random mobility (also called random-walk) is a very simple approach, in which the UAVs move independently and randomly in the target area. While there are already more than a few random-walk models in the literature, the variations can even be increased by extending them with suitable connectivity maintenance algorithms [1]. In the considered basic model, UAVs fly in the area in a randomly chosen direction for each second (or for another pre-defined time slot). To change the direction in the target area, the authors in [8] have proposed three actions being "go straight", "turn left" and "turn right". UAVs can decide on their actions randomly, based on fixed probabilities. The same strategy has been implemented in this study. The random mobility model is commonly used in many applications such as search and explore missions and monitoring or surveillance operations. The random model is also widely used as a benchmark indicator in many studies.

## 3.2    Distributed Pheromone Repel Mobility Model

In this model the drones may take the same three actions as in the random-walk model which are "go straight", "turn left" and "turn right" [8]. However, instead of fixed probabilities for these decisions, in the pheromone-based model, a so called "pheromone" map is used to guide the UAVs. The pheromone map contains information regarding the visited/explored areas and is shared/synchronised among the UAVs. An aircraft exchanges information about the scanned area, and according to this information, the other UAVs decide to turn left, right or go straight ahead.

## 3.3    Connectivity-Based Mobility Model

The connectivity-based mobility model is a self-organizing model, in which the UAVs use local neighborhood information to sustain the connectivity of the network [13]. The UAVs can enter or leave the network at any time. The target area can also be changed during the algorithm execution. The main goal is to cover the maximum possible area, while preserving the connectivity between all UAVs and the base station. Each UAV explores its neighbors and computes its location periodically. Then it decides if it needs to change its direction. The information that is exchanged between the UAVs contain the current locations and the directions.

In every $t$ seconds, each UAV checks whether it is located within the transmission range of the sink UAV. If it has a live communication link to the sink, it checks if it still would be in the range after $t$ seconds, given its current direction and the speed. If the UAV estimates that it will be in the range of the sink, it keeps its previous flying direction. Otherwise, the UAV changes its direction randomly towards a point in the transmission range of the sink. If the sink is not in the radio range, then the UAV tries to find a multi-hop path to the sink. If it can find such a path, it estimates the next location of itself and its neighbors and determines whether it would still be connected to the sink after $t$ seconds. If the path still exists, it keeps its previous direction. If the UAV estimates that the connection will not exist in the next $t$ seconds, then it changes its direction such that it stays connected to a neighbor UAV that provides a multi-hop path to the sink. If a UAV can not find such a path to the sink, it tries to keep the connection to a neighboring UAV to avoid a complete isolation. Finally, when a UAV becomes isolated, it keeps its previous direction until reaching the boundary of the mission field or until meeting another UAV. At the boundaries of the field, the UAV changes its direction randomly towards a point in he field.

### 3.4 KHOPCA-Based Mobility Model

This model combines the ACO-based mobility model given in [8] with the KHOPCA clustering algorithm presented by Brust et al. in [3] for the UAV swarms to optimize their surveillance capabilities, coverage area and stability of the connectivity. Each UAV locally executes the KHOPCA-based clustering algorithm. The UAVs select their directions using the pheromone-based approach and the KHOPCA-based probability. The cluster-head UAVs or the UAVs with a lower weight than the KHOPCA threshold value, uses the pheromone-based mobility model to select their direction. For the other UAVs, the location of a neighbor with the lowest KHOPCA weight is selected as the destination.

### 3.5 Connected Coverage Mobility Model

This model has three sequential steps as follows: 1- Each UAV selects a preferred neighbor UAV among its one-hop neighbors to stay connected with them. 2- The UAVs compute their current and future positions using the direction and speed values, alternative directions are also computed to stay connected with one of the selected neighbors. 3- A pheromone-based behaviour helps selecting the best direction from the previously computed set. This model maintains a tree structure, where the root is the base station. To compute the alternative directions, each UAV estimates the upcoming position of each neighbor UAV using their current positions and the current destinations of every single neighbor. Yet, the

UAVs check where they would be in this predefined future and decide a correct direction to satisfy the connectivity constraint of the model. In this model, UAVs regularly drop virtual pheromones on their ways and these pheromones gradually disappear after some time. Thus, the UAVs will be "attracted" to the places that are marked by the lowest pheromone concentration.

## 4   Performance Metrics

This section introduces the performance metrics that show the quality or usefulness of the models considering the area coverage features, connectivity maintenance and communication costs. The speed of coverage (convergence rate) and the fairness of coverage have been measured to see the quality of the area coverage capabilities. Also, the ratio of the connected UAVs, the ratio of connectivity to the root and the average connected component number shows how well connected the topologies are, by using each mobility model. Finally, one of the most important metrics, namely the communication costs, is calculated. The total number of the sent messages (by each UAV) and the size of these messages are taken into consideration in that calculation.

*Speed of Coverage*: Also the convergence speed or rate. In most area coverage missions, mobility models need to cover the whole area as fast as possible. This metric shows how much time required to scan the specific percentage of the mission field. The coverage time is getting higher and higher as the desired percentage (threshold) increases. Hence, instead of the whole area, the time passed until covering the 80% and 95% of the field is recorded to find out the coverage speeds.

*Fairness of Coverage*: The fairness of coverage is the metric that shows how balanced the whole area is scanned. To compute the fairness, it is necessary to record for how many seconds (or times) each square meter (or unit area in general) is already scanned by the UAVs. If two or more UAVs scan the same part of the field (i.e. a cell) at the same second, it should be counted as one second. Hence, the coefficient of variations gives how balanced the scanning times are. It's expected that the fairness to be close to 0.

*Connected Topology Percentage*: Connected topology percentage is the ratio of the non-continuous cumulative time period, in which all the UAVs are connected to the network, to the total time period, which the mobility model needs to cover the 95% of the mission field. To compute that percentage, the number of the connected components are gathered at each second of the execution time, then the ratio to the execution time is calculated.

*Connectivity Percentage To Root*: This metric is taken into account in the Connectivity Based and the Connected coverage model. Since the drone network applications use the one UAV as a gateway and main processor, these models tries to make drones are connected to the root during execution. Its computed as the average of the connectivity percents of the drones to the root (single-hop or multi-hop).

*Average Connected Component*: This metric is the mean of the connected component numbers which gathered at each second. It shows how much the topology is dispersed during the execution time.

*Message Count*: Message count is the number of messages sent by UAVs during the execution of the mission. This metric was not considered in the related previous works. It gives an insight regarding the communication costs of the models. As the message counts increase, the energy required to send the messages increases, too. It's expected to be lower as a quality/performance indicator.

*Total Message Size*: The message counts are not merely enough to find out the whole communication costs. Because, the energy spent by UAVs (during the communications) is also related to the size of the sent (and received) messages. The message size is directly proportional to the communication cost. Thus, the message counts should be multiplied by the average message sizes for each model throughout the mission to get comparable values.

## 5 Simulation Results

The mobility models are implemented using Python and the SimPy library [10]. The parameters of the simulations are given in Table 1. The size of the area is defined as 2000 m × 1000 m in a rectangular shape. For the pheromone-based models, the map's cell size is set as 5 m. However, the scan times have been measured for each square meter to get more precise results. The UAVs' radio ranges and operational (e.g. sensor) ranges are fixed to 400 m and 20 m respectively. The programmed autopilot mimics a simple rotating-wing drone behaviors. Accordingly, the speeds of the UAVs are fixed at 5 m/s. The decision intervals are specified differently for each model as specified in their original papers. The communication infrastructure is assumed to be optimal (no loss, no collision, unlimited

**Table 1.** Simulation parameters

| Parameter name | Parameter value |
| --- | --- |
| Simulation area size | 2000 m × 1000 m |
| Map's cell size | 5 m × 5 m |
| Measurement cell size | 1 m × 1 m |
| UAV Speed | 5 m/s |
| Radio range | 400 m |
| Coverage range | 20 m |
| Decision intervals Random | 1 s |
| DPR | 10 s |
| Conn. Based | 2 s |
| KHOPCA Based | 30 s and 0.2 probability |
| Connected Coverage | 30 s |
| Experiments # of UAVs | [4,6,8,10,15,20,30,40,50] |
| # of runs per experiment | 30 |

bandwidth etc.) as in [5,8,12]. The number of UAVs used varies between 4 to 50 to test the models in sparse and dense environments. Each model was simulated 30 times to get statistically consistent results.

Figure 1a and 1b shows the average execution times of the models to cover 80% and 95% of the mission field, respectively. According to the results, the time the algorithms need to cover the 80% of the area is approximately half of the time needed to cover the 95%, which is an interesting outcome. The DPR has performed better than the other models as expected. Especially in sparser networks, the DPR model is roughly two times faster than the others. The models, that consider connectivity, cover the area more slowly. Moreover, the Connectivity-based and the KHOPCA-based models are even slower than the random-walk model. As the density of the UAVs increases, the ratios between the Connectivity-based model and the others decrease. Even more, the Connectivity-based model has covered the area faster than the other models when the topologies include 30, 40 and 50 drones.

The fairness of coverage policy of the models are given in Fig. 1c. As it can be seen that the KHOPCA-based and the Connectivity-based models cover the area less balanced than the others. Surprisingly, the random-walk model has given similar results with the DPR and the Connected Coverage models. As the UAV density increases, the Connectivity-based and the KHOPCA-based models scan the area more balanced. The change of the density doesn't affect the fairness of the random-walk and Connected Coverage models dramatically.

Figure 2a shows that in what percentage of the execution time the UAVs stayed connected within each of the models. Per to the figure, when the number of UAVs is something from 4 to 10, the Connectivity-based model provides much better connectivity rates than the others. However, as the UAV density increases, its results are getting worse, while the Connected Coverage's, the DPR's and the random-walk's performances are getting significantly better. The average connected component numbers are shown in Fig. 2b. The results are similar to the results presented in the connectivity percentage shown in Fig. 2a, as expected. Because, the more dispersed the network means the less connected the network. The KHOPCA-based model has provided a worse connectivity than the others, as also expected, because it aims to perform clustering over the topology.

Another metric about the connectivity is that the connectivity percentage from any UAV to the root UAV, shown in Fig. 2c. While KHOPCA-based model has provided bad connectivity rates to the root, The Connectivity-based model has provided better, as expected. The DPR, the Connected Coverage and the random-walk models have achieved similar connectivity rates. However, the Connected Coverage model has better connectivity to the root than the DPT and the random-walk models. The Connectivity-based model has provided two times better connectivity to the root than the others in sparse topologies, while they are all similar in dense topologies.

The total message counts sent by all UAVs for each model is given in Fig. 3a. The message count is directly related to the decision interval times and the execution times of the models. The shorter interval times or the larger execution

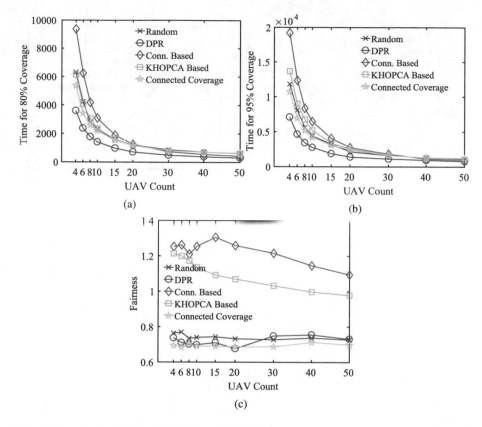

**Fig. 1.** Speed of achieving 80% (a) and 95% (b) coverage versus number of UAVs. (c) Fairness of coverage versus number of UAVs.

times naturally mean more messages to be sent. Since the random-walk model does not need any message to be transmitted, its message number is 0 for all situations. The Connectivity-based model is the one that uses the maximum number of messages due to its short decision interval and long execution time. The Connected Coverage used less messages than the DPR model for 4 to 15 UAVs. For 20 to 50 UAVs, however, they have similar results. The KHOPCA-based model's message count is not affected by the varying UAV numbers, notably. The message counts do not show the precise cost of the communication alone. It is, in fact, required to consider the size of the sent messages, too. Figure 3b shows the total message sizes sent by the models. Although the Connectivity-based model has sent the highest number of messages, its message sizes are the smallest, because the other models periodically send the maps data sized $100 \times 100$. Since the DPR cause the UAVs to send maps more frequently than the others, the DPR has the maximum message size. The Connected Coverage and the KHOPCA-based models have comparable message sizes.

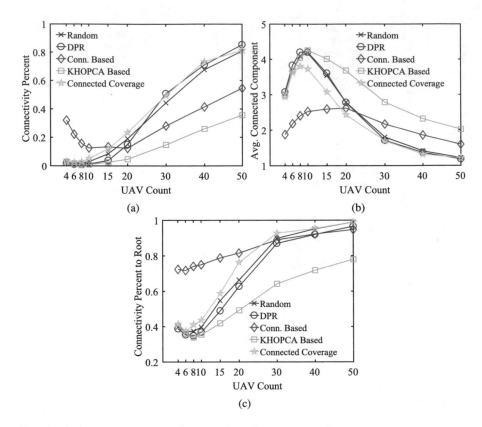

**Fig. 2.** (a) Connectivity percentage versus number of UAVs. (b) Average connected component number versus number of UAVs. (c) Connectivity to the root (sink) versus number of UAVs.

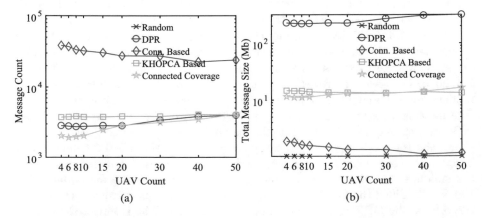

**Fig. 3.** (a) Total message counts versus number of UAVs. (b) Cumulative message sizes versus number of UAVs.

# 6    Conclusion

In this study, an extensive performance evaluation of some noteworthy connectivity-aware distributed mobility models tailored for drone networks is presented considering area coverage applications. In accordance, Connectivity-based, KHOPCA-based, Connected Coverage, random-walk and DPR models are implemented. Through simulations, their coverage performances, connectivity qualities, and communication costs are compared over seven designated performance metrics, which are namely the speed of coverage, the fairness of coverage, the connectivity rate, the average connected component number, the connectivity rate to the root, the total message count and the total message volume generated by all UAVs. From a large number of measurements, it is concluded that there is an obvious trade-off between the connectivity rates and the coverage levels. Besides, as it seems, when the models consider the connectivity, their coverage performances are getting lower. According to the results, the DPR model has the best coverage metrics, while the Connectivity-based model pose the best connectivity metrics. The Connected Coverage model uses the least number of messages during its executions, while it generally provides average results over all other performance metrics. In future, it is planned to design and analyze new mobility models, especially to deal with the $k$-connectivity restoration problem in drone networks for arbitrary $k$ values.

# References

1. Çabuk, U., Dalkılıç, G.: Topology control in mobile wireless sensor networks. International Journal of Applied Mathematics Electronics and Computers, pp. 61–65 (2016)
2. Akram, V.K., Dagdeviren, O.: On hardness of connectivity maintenance problem in drone networks. In: 2018 IEEE International Black Sea Conference on Communications and Networking (BlackSeaCom), pp. 1–5. IEEE (2018)
3. Brust, M.R., Frey, H., Rothkugel, S.: Adaptive multi-hop clustering in mobile networks. In: Proceedings of the 4th International Conference on Mobile Technology, Applications, and Systems and the 1st International Symposium on Computer Human Interaction in Mobile Technology, pp. 132–138. ACM (2007)
4. Bujari, A., Palazzi, C.E., Ronzani, D.: Fanet application scenarios and mobility models. In: Proceedings of the 3rd Workshop on Micro Aerial Vehicle Networks, Systems, and Applications, pp. 43–46. ACM (2017)
5. Danoy, G., Brust, M.R., Bouvry, P.: Connectivity stability in autonomous multi-level UAV swarms for wide area monitoring. In: Proceedings of the 5th ACM Symposium on Development and Analysis of Intelligent Vehicular Networks and Applications, pp. 1–8. ACM (2015)
6. Hayat, S., Yanmaz, E., Muzaffar, R.: Survey on unmanned aerial vehicle networks for civil applications: a communications viewpoint. IEEE Commun. Surv. Tutorials 18(4), 2624–2661 (2016)
7. Hyytiä, E., Virtamo, J.: Random waypoint model in n-dimensional space. Oper. Res. Lett. 33(6), 567–571 (2005)

8. Kuiper, E., Nadjm-Tehrani, S.: Mobility models for UAV group reconnaissance applications. In: 2006 International Conference on Wireless and Mobile Communications (ICWMC 2006), p. 33. IEEE (2006)
9. Messous, M.-A., Senouci, S.-M., Sedjelmaci, H.: Network connectivity and area coverage for UAV fleet mobility model with energy constraint. In: 2016 IEEE Wireless Communications and Networking Conference, pp. 1–6. IEEE (2016)
10. Müller, K., Vignaux, T.: Simpy: Discrete event simulation for python
11. Sanchez-Garcia, J., Garcia-Campos, J., Toral, S., Reina, D., Barrero, F.: A self organising aerial ad hoc network mobility model for disaster scenarios. In: 2015 International Conference on Developments of E-Systems Engineering (DeSE), pp. 35–40. IEEE (2015)
12. Schleich, J., Panchapakesan, A., Danoy, G., Bouvry, P.: Uav fleet area coverage with network connectivity constraint. In: Proceedings of the 11th ACM International Symposium on Mobility Management and Wireless Access, pp. 131–138. ACM (2013)
13. Yanmaz, E.: Connectivity versus area coverage in unmanned aerial vehicle networks. In: 2012 IEEE International Conference on Communications (ICC), pp. 719–723. IEEE (2012)

# Analysis of Movement-Based Connectivity Restoration Problem in Wireless Ad-Hoc and Sensor Networks

Umut Can Cabuk[✉], Vahid Khalilpour Akram, and Orhan Dagdeviren

Ege University, Izmir, Turkey
{umut.can.cabuk,vahid.akram,orhan.dagdeviren}@ege.edu.tr

**Abstract.** Topology control, consisting of construction and maintenance phases, is a crucial conception for wireless ad-hoc networks of any
type, expressly the wireless sensor networks. Topology maintenance, the
latter phase, concerns several problems, such as optimizing the energy
consumption, increasing the data rate, making clusters, and sustain-
ing the connectivity. A disconnected network, among other strategies,
can efficiently be connected again using a Movement-based Connectivity
Restoration (MCR) method, where a commensurate number of nodes
move (or are moved) to the desired positions. However, finding an optimal route for the nodes to be moved can be a formidable problem. As
a matter of fact, this paper presents details regarding a direct proof
of the NP-Completeness of the MCR Problem by a reduction from the
well-studied Steiner Tree Problem with the Minimum number of Steiner
Points and Bounded Edge Length.

**Keywords:** Wireless sensor networks · Mobile ad-hoc networks ·
Connectivity restoration fault tolerance · NP-completeness

## 1 Introduction

As an integral part of the revolutionary Internet of Things (IoT) concept, wireless ad-hoc networks are one of the several technologies those are shaping the
century. Sensors, actuators, and other lightweight electronic devices with wireless
networking capabilities (also called nodes) are used to build smart systems to be
used in industries, commercial services and everyday life [1]. However, due to the
cost-oriented restrictions of these devices, such as limited battery lifetime, scarce
bandwidth, weak processing power, low memory and possibly others, make these
nodes prone to various problems, which may degrade the network's stability and
even can completely interrupt the services they provide.

As a result of these restrictions; say, a battery discharge, software faults,
a hacking/hijacking attack or a physical damage may be occurred, and these
may cause removal, displacement or death of the subject node(s). Otherwise,
external factors like strong wind, rain, floods, ocean currents, stray/wild animals
or curious/malevolent people may cause the same on a larger scale.

© The Author(s), under exclusive license to Springer Nature Switzerland AG 2021
J. Hemanth et al. (Eds.): ICAIAME 2020, LNDECT 76, pp. 381–389, 2021.
https://doi.org/10.1007/978-3-030-79357-9_37

Removal, displacement or scatter of the nodes of a wireless ad-hoc network, may eventually disconnect the whole network, whenever "enough" number of nodes have been placed on places far "enough". The definition of "enough" is totally scenario dependent. Also, death of the nodes may or may not cause a disconnection, depending on the initial placement. The connectivity can be restored essentially in two ways; either new nodes (in sufficient numbers) should be placed in appropriate places to bind (or bridge) the existing nodes, or the existing nodes should be move towards each other and/or to a determined rendezvous-point.

Both strategies may work well, again depending on the scenario. Besides, both are easy to implement using naïve methods, as long as it is possible to add new nodes to the network (for the first solution) or it is possible to move the nodes (for the second solution). But as stated earlier, in such applications, resources are scarce. So, the number of nodes to be added should be minimum (for the first solution), or likewise, the total distance to be travelled by the nodes should be minimum. Please note that the total distance moved is actually a function of the energy consumed and the time spent, which are also scarce resources as well as the node count. In fact, the former strategy is a fundamental variant of the well-known Steiner Tree Problem, and already proven to be NP-Complete [2]. The latter (as we call the Movement-based Connectivity Restoration Problem - MCR), however, was studied less and no assertive claims have been found in the literature regarding its complexity class, which is then proven to be also NP-Complete within this work.

The remaining parts of this paper have been organized as follows; In Sect. 2 we provide a brief survey about the existing works on the connectivity restoration in wireless ad-hoc and drone networks. Section 3 contains the network model description and preliminaries. We have proved NP-Hardness of movement based connectivity restoration problem in Sect. 4. Finally the conclusion has been presented in Sect. 5.

## 2   Related Works

Connectivity restoration is one the important challenges in wireless ad-hoc networks and many researches have focused on this problem from different perspectives. One of the approaches for connectivity restoration is deploying new nodes [3–14]. In these approaches the aim is restoring the network connectivity with placing minimum number of nodes to the optimal locations. These approaches are generally used in static networks where the required locations are always reachable by humans. Some other researches focus on the determining the radio power of nodes to create a connected network [15–18]. The radio hardware limitation is main restriction for these methods because if the distance between disconnected partitions is long, increasing the radio power may be insufficient to restore the connectivity.

Some other researches focus on the preventing the disconnection of the network by finding current $k$ value of a network [19–25]. A network is $k$-connected if it can tolerate at least $k$ nodes without losing the connectivity of active nodes. In 1-connected networks losing a single node can separate the network to disconnected partitions. In a 4-connected network at least 4 nodes must stop working to lose the network connectivity. Finding the $k$ value and preserving it in a safe level reduces the risk of disconnection however, finding or preserving the $k$ value is an energy consuming task which needs a large amount of message passing.

For mobile ad-hoc networks various movement-based connectivity restoration algorithms have been proposed [26–30] which move mobile nodes to their neighbors or some new locations according different principles, however, to the best of our knowledge there is no optimal algorithm form movement-based connectivity restoration. This process is more complicated for the networks with $k > 1$. Despite of proposed different heuristics algorithms in different researches for movement based connectivity restoration, the existing researches neither prove the NP-Hardness nor present an optimal approach. In this paper we prove that the movement-based connectivity restoration in wireless ad-hoc networks is NP-Hard.

## 3   Problem Formulation

We can model any wireless ad-hoc network as an undirected graph $G = (V, E)$, where $V$ is the set of vertices (representing the wireless sensor nodes) and $E$ is the set of edges (representing the wireless communication links), considering the following assumptions:

1. All nodes have access to the medium and have broadcasting capabilities (unicasting is trivial).
2. All wireless links are bidirectional.
3. All nodes have the same communication range $r$.
4. A link is automatically established between any pair of nodes that have an Euclidean distance of $d \leq r$.
5. No link can be removed to form a specific topology, unless the above condition is no longer satisfied. In the case of $d > r$, the link is automatically removed.
6. Operational ranges (i.e. sensing area) are omitted.

Without loss of generality, the given assumptions make it easier to work on the problem. Nevertheless, the provided proof can easily be expanded considering the absence or change of any of the above statements. A not-to-scale drawing representing the nodes (smaller red circles with numbers inside) and the radio coverage (larger blue circles with dashed lines) of an example wireless ad-hoc network is given in Fig. 1.a. The same wireless ad-hoc network is then re-drawn in Fig. 1.b using the graph representation, which will be referred in the rest of the paper.

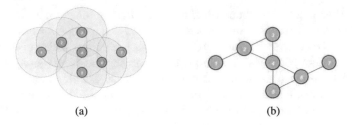

(a)                                                    (b)

**Fig. 1.** a) An example wireless ad-hoc network, showing the nodes and their radio coverage b)Modeling a wireless ad-hoc network as an undirected graph.

Table 1 shows the symbols, which are used in the rest of the paper and for the proof.

**Table 1.** Used symbols in the paper

| Symbol | Meaning |
|---|---|
| $|V|$ | Cardinality of set V |
| $U \backslash V$ | Items in U that are not in V |
| $\mathbb{R}^2$ | Two-dimensional Euclidean plane |
| $\mathbb{R}$ | The set of real numbers |
| $||v, u||$ | Distance between points $u$ and $v$ in $\mathbb{R}^2$ |

## 4  NP-Completeness of Movement Based Connectivity Restoration

In this paper we prove that the movement-based connectivity restoration in Ad-hoc Sensor Networks is an NP-Complete problem. The Steiner Tree Problem with Minimum number of Steiner Points and Bounded Edge Length is a known NP-Complete problem [2] and we show that it can be reduced to the movement based connectivity restoration in polynomial time. The Steiner Tree Problem with Minimum number of Steiner Points and Bounded Edge Length (ST) is formally defined as follow:

**ST Problem:** Given a set $V$ of $n$ points in $\mathbb{R}^2$ and a positive number $r \in \mathbb{N}$, the problem is establishing a spanning tree $T = (V \cup S, E)$ in the graph $G$ with nodes $V$ and minimum number of Steiner points $S$ such that the maximum length of each edge $e \in E$ is $r$.

Figure 2 below shows an example instance of the ST Problem where $V = \{1, 2, 3, 4, 5, 6, 7\}$ and $E = \emptyset$. In general, $E$ does not necessarily be an empty set, as long as $G$ is disconnected. The red circles (in both Fig. 2a and b) represent the initially deployed nodes, while the green circles with dotted circumference (in Fig. 2b) represent the add-on nodes.

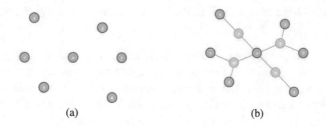

(a)                                    (b)

**Fig. 2.** a) A disconnected network as an example instance of the ST Problem. b) The same network connected by placing new nodes on arbitrary Steiner points.

As given in Fig. 3b; using the pre-calculated Steiner points to place new nodes provide a connected network with the minimum required number of nodes. Although this solution provides means of cost efficiency in restoring the connectivity; depending on the network type and the scenario, there may be cases where adding new nodes to the network is expensive, or impractical, or not even possible. In these cases, if it is possible to move the nodes or if they can move autonomously in an efficient manner, then addressing the MCR Problem instead can help restoring the connectivity of the network better.

**MCR Problem:** Let $V \subset \mathbb{R}^2$ be the set of position of nodes in a wireless ad-hoc network where the communication range of each node is $r$ and $E = \{\{u, v\} \mid u, v \in V$ and $\|u, v\| \leq r\}$ be the set of links between the nodes. Given a moving cost function $c \ V \times \mathbb{R}^2 \to \mathbb{R}$, the problem is finding a mapping function $M \ V \to \mathbb{R}^2$ such that $\sum_{v \in V} c(v, M(v))$ is minimum and $G = (M(V), E)$ is connected.

Figure 3a shows an example instance of the MCR problem where $V = \{1, 2, 3, 4, 5, 6, 7\}$ and $E = \emptyset$ (again not necessarily). Figure 3b–d give representations of hypothetical solutions to the instance given in Fig. 3a. Even though the drawings are not-to-scale, the edge length condition $\|u, v\| \leq r$ is exclusively satisfied. A comprehensive run of a valid algorithm optimally (or sub-optimally) solving the MCR problem may find the real results, which include the optimal (or sub-optimal) route and the final position of each node that can move or be moved. There might also be multiple optimal solutions.

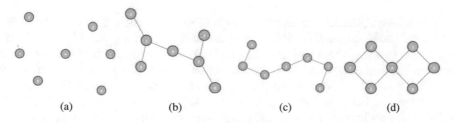

(a)                (b)                (c)                (d)

**Fig. 3.** a) A disconnected network as an example instance of the MCR problem. b) A possible connected network in the form of a balanced tree (iff root is 4). c) A possible connected network in the form of a line. d) A possible connected network in the form of a polygon.

**Theorem 1:** *There is a polynomial time reduction from ST Problem to MCR Problem.*

**Proof:** Let $V$ be the set of $n$ nodes which must be covered by a spanning tree with the minimum number of Steiner points and $r$ be the maximum length of each edge in the spanning tree. If we assume that the set $S$ is an optimum solution for ST problem, then an optimal algorithm for ST Problem should correctly find the set $S = \{s_1, s_2 \ldots s_h\}$ of $h \geq 0$ Steiner points such that $G = (S \cup V, E)$ is connected where $E = \{\{u, v\} \mid u, v \in S \cup V \ and \ \|u, v\| \leq r\}$.

Assume that we have an optimal algorithm $MCR\,(V, r, c)$ for MCR Problem which returns a set $M$ of new locations of nodes in $V$ such that $\sum_{v \in V} c\,(v, M\,(v))$ is minimum and $G = (M(V), E)$ is connected. Let $ST(V, r)$ algorithm returns minimum number of Steiner nodes which are required for creating a connected spanning tree. We must show that a polynomial time algorithm can uses $MCR\,(V, r, c)$ to produce optimal solution for $ST(V, r)$. The resulting optimum spanning tree in ST Problem can be considered as a connected graph. The $MCR$ algorithm accepts the initial nodes set, the $r$ value and a cost function $c$. The $V$ and $r$ are common inputs for both $MCR$ and $ST$ algorithms. We define the cost function $c$ as follow:

$$c\,(v, p) = \begin{cases} 1 & if \ v \in V \ and \ v \neq p \\ 0 & otherwise \end{cases}$$

According the above function the cost of moving node $v \in V$ to a new location $p$ is 1. Note that, the cost of moving each node to its own location (keeping node unmoved) is 0. The cost of moving any node $v \notin V$ to any point is 0. Using this cost function and MCR algorithm, the following algorithm produces an optimal solution for the ST problem.

Algorithm **ST** $(V, \ r)$
   1:$U \leftarrow \emptyset$
   2:$M \leftarrow MCR(V, r, c)$
   3: **while** $\sum_{v \in V} c\,(v, M\,(v)) > 0$ **do**
   4:   $U \leftarrow U \cup t \notin (V \cup U)$ .
   5:   $M \leftarrow MCR(V \cup U, r, c)$
   6:**return** $U$.

The above algorithm repeatedly selects a random node $t \notin (V \cup U)$ and adds it to set $U$ and calls $MCR(V \cup U, r, c)$ to find a mapping for $MCR$ problem. According to the defined cost function if $MCR$ moves any node $v \in V$ to a new location the resulting total cost will be higher than 0. If the nodes in $U$ is not enough for creating a connected graph, the $MCR$ will move some the nodes in $V$ and the total cost will be higher than 0. When $MCR$ moves only the newly added nodes and create a connected graph then the resulting total cost will be 0. In other words. when the nodes in $U$ are enough for creating a connected graph, the total cost of moving will be 0. Formally we can write:

$$\sum_{v \in V} c\left(v, M\left(v\right)\right) = 0 \;\; \rightarrow \;\; \nexists v \in V \; s.t \; v \neq M\left(v\right) \;\; \rightarrow$$

$$\forall \, u \, \in U \;\; u \in M$$

Since we add the nodes to $U$, one by one, upon the condition $\sum_{v \in V} c(v, M(v)) = 0$ becomes true, the set $U$ will have the optimal number of Steiner points. The maximum iteration of while loop is $|U \backslash V|$ or the number of added Steiner nodes. For any $n \geq 0$ and $r > 0$, an $n \times n$ two-dimensional Euclidean plane can be completely covered by at most $((n/r) + 1)^2$ circles with radios $r$ such that the distance between the center of neighboring circles becomes less than or equal to $r$. So $|U \backslash V| \leq ((n/r) + 1)^2$ and the algorithm has polynomial-time complexity.

According to Theorem 1, if we have a polynomial-time algorithm for $MCR$ problem, then we can solve the $ST$ problem in polynomial time which proves that the $MCR$ problem is NP-Hard, because the NP-Hardness of $ST$ problem has already been proved.

# 5   Conclusion

Wireless ad-hoc networks, especially the wireless sensor networks made of constrained devices, are susceptible to displacement and disconnection caused by various internal (i.e. battery outage) and external factors (i.e. storm). The two main approaches used to re-establish the connection are adding new nodes to appropriate positions and moving the existing nodes to appropriate positions (if they are mobile or movable). A hybrid solution may also be possible but neglected in this study.

For the first strategy; yielding an optimal solution that requires the minimum possible number of nodes to reconnect the network is equivalent to the well-studied Steiner Tree Problem and was already proven to be NP-Complete. For the second; finding and optimal solution that requires coverage of the minimum possible distance in total to bring the network together is defined as the MCR Problem. Within the scope of this work; the MCR Problem on a 2-dimensional plane is thoroughly shown to be reducible to the Steiner Tree Problem in polynomial time. Hence, it is proven that the MCR problem is also NP-Complete.

# References

1. Akyildiz, I.F., Su, W., Sankarasubramaniam, Y., Cayirci, E.: A survey on sensor networks. IEEE Commun. Mag. **40**, 102–114 (2002)
2. Guo-Hui, L., Xue, G.: Steiner tree problem with minimum number of Steiner points and bounded edge-length. Inf. Process. Lett. **69**(2), 53–57 (1999)
3. Han, Z., Swindlehurst, A.L., Liu, K.R.: Optimization of MANET connectivity via smart deployment/movement of unmanned air vehicles. IEEE Trans. Veh. Technol. **58**, 3533–3546 (2019)

4. Li, N., Hou, J.C.: FLSS: a fault-tolerant topology control algorithm for wireless networks. In: Proceedings of the MobiCom 2004, pp. 275–286. ACM (2004)
5. Gupta, B., Gupta, A.: On the k-connectivity of ad-hoc wireless networks. In: Proceedings of the 2013 IEEE Seventh International Symposium on Service-Oriented System Engineering, SOSE 2013, pp. 546–550. IEEE Computer Society, Washington (2013)
6. Tian, D., Georganas, N.D.: Connectivity maintenance and coverage preservation in wireless sensor networks. Ad Hoc Netw. **3**, 744–761 (2005)
7. Segal, M., Shpungin, H.: On construction of minimum energy k-fault resistant topologies. Ad Hoc Netw. **7**, 363–373 (2009)
8. Deniz, F., Bagci, H., Korpeoglu, I., Yazici, A.: An adaptive, energy-aware and distributed fault-tolerant topology-control algorithm for heterogeneous wireless sensor networks. Ad Hoc Netw. **44**, 104–117 (2016)
9. Bagci, H., Korpeoglu, I., Yazici, A.: A distributed fault-tolerant topology control algorithm for heterogeneous wireless sensor networks. IEEE Trans. Parallel Distrib. Syst. **26**, 914–923 (2015)
10. Bredin, J.L., Demaine, E.D., Hajiaghayi, M.T., Rus, D.: Deploying sensor networks with guaranteed fault tolerance. IEEE/ACM Trans. Netw. (TON) **18**, 216–228 (2010)
11. Yun, Z., Bai, X., Xuan, D., Lai, T.H., Jia, W.: Optimal deployment patterns for full coverage and k-connectivity ($k \leq 6$) wireless sensor networks. IEEE/ACM Trans. Netw. (TON) **18**, 934–947 (2010)
12. Younis, M., Akkaya, K.: Strategies and techniques for node placement in wireless sensor networks: a survey. Ad Hoc Netw. **6**, 621–655 (2008)
13. Bai, X., Xuan, D., Yun, Z., Lai, T.H., Jia, W.: Complete optimal deployment patterns for full-coverage and k-connectivity ($k \leq 6$) wireless sensor networks. In: Proceedings of the 9th ACM International Symposium on Mobile Ad-hoc Networking and Computing, MobiHoc 2008, pp. 401–410. ACM, New York (2008)
14. Barrera, J., Cancela, H., Moreno, E.: Topological optimization of reliable networks under dependent failures. Oper. Res. Lett. **800**, 132–136 (2015)
15. Nutov, Z.: Approximating minimum-power k-connectivity. Ad Hoc Sensor Wirel. Netw. **9**(1–2), 129–137 (2010)
16. Wan, P.-J., Yi, C.-W.: Asymptotic critical transmission radius and critical neighbor number for k-connectivity in wireless ad-hoc networks. In: Proceedings of the 5th ACM International Symposium on Mobile Ad-hoc Networking and Computing, pp. 1–8. ACM (2004)
17. Jia, X., Kim, D., Makki, S., Wan, P.-J., Yi, C.-W.: Power assignment for k-connectivity in wireless ad-hoc networks. J. Comb. Optim. **9**, 213–222 (2005)
18. Zhang, H., Hou, J.: On the critical total power for asymptotic k-connectivity in wireless networks. In: Proceedings IEEE 24th Annual Joint Conference of the IEEE Computer and Communications Societies, vol. 1, pp. 466–476. IEEE (2005)
19. Jorgic, M., Goel, N., Kalaichelvan, K., Nayak, A., Stojmenovic, I.: Localized detection of k-connectivity in wireless ad-hoc, actuator and sensor networks, pp. 33–38. IEEE, ICCCN (2007)
20. Cornejo, A., Lynch, N.: Fault-tolerance through k-connectivity. In: Workshop on Network Science and Systems Issues in Multi-Robot Autonomy: ICRA 2010, vol. 2 (2010)
21. Cornejo, A., Lynch, N.: Reliably detecting connectivity using local graph traits. In: Proceedings of the 14th International Conference on Principles of Distributed Systems, OPODIS 2010, vol. 825, pp. 87–102. Springer, Berlin (2010)

22. Dagdeviren, O., Akram, V.K.: PACK: path coloring based k-connectivity detection algorithm for wireless sensor networks. Ad Hoc Netw. 1(64), 41–52 (2017)
23. Censor-Hillel, K., Ghafari, M., Kuhn, F.: Distributed connectivity decomposition. In: Proceedings of the 2014 ACM Symposium on Principles of Distributed Computing, pp. 156–165. ACM (2014)
24. Akram, V.K., Dagdeviren, O.: DECK: a distributed, asynchronous and exact k-connectivity detection algorithm for wireless sensor networks. Comput. Commun. **116**, 9–20 (2018)
25. Szczytowski, P., Khelil, A., Suri, N.: DKM: distributed k-connectivity maintenance in wireless sensor networks. In: WONS, pp. 83–90. IEEE (2012)
26. Younis, M., Senturk, I.F., Akkaya, K., Lee, S., Senel, F.: Topology management techniques for tolerating node failures in wireless sensor networks: A survey. Comput. Netw. **58**, 254–283 (2014)
27. Atay, N., Bayazit, O.B.: Mobile wireless sensor network connectivity repair with k-redundancy. In: WAFR, Vol. 57 of Springer Tracts in Advanced Robotics, pp. 35–49. Springer, Berlin (2008)
28. Almasaeid, H.M., Kamal, A.E.: On the minimum k-connectivity repair in wireless sensor networks. In: Proceedings of the IEEE ICC 2009, pp. 195–199. IEEE Press, Piscataway (2009)
29. Abbasi, A.A., Younis, M., Akkaya, K.: Movement-assisted connectivity restoration in wireless sensor and actor networks. IEEE Trans. Parallel Distrib. Syst. **20**, 1366–1379 (2009)
30. Wang, S., Mao, X., Tang, S.-J., Li, X., Zhao, J., Dai, G.: On movement-assisted connectivity restoration in wireless sensor and actor networks. IEEE Trans. Parallel Distrib. Syst. **22**, 687–694 (2011)

# Design of External Rotor Permanent Magnet Synchronous Reluctance Motor (PMSynRM) for Electric Vehicles

Armagan Bozkurt[1](✉), Yusuf Oner[2], A. Fevzi Baba[3], and Metin Ersoz[4]

[1] Department of Electronics and Computer Education, Pamukkale University, Denizli, Turkey
armbozkurt@pau.edu.tr
[2] Department of Electrical and Electronics Engineering, Pamukkale University, Denizli, Turkey
yoner@pau.edu.tr
[3] Department of Electrical and Electronics Engineering, Marmara University, Istanbul, Turkey
fbaba@marmara.edu.tr
[4] Senkron Ar-Ge Engineering A.S., Denizli, Turkey
mersoz@senkronarge.com

**Abstract.** In this study, it is aimed to design an external rotor permanent magnet synchronous reluctance motor (PMSynRM) for an electric vehicle (EV). In recent years, developments in EV technology have increased the need for electric motor design. Electric motors designed to be used in EVs are considered to have high efficiency, high torque and high power density in a wide speed range. Considering these situations, one of the recently designed electric motors is the synchronous reluctance motor (SynRM). However, the disadvantage of a synchronous reluctance motor is the torque ripple. In this study, a 2-kW, three-phase, eight-pole, 24-slots hub motor with an external rotor (PM-SynRM) with a rotational speed of 750 rpm is designed. The design was modeled using the finite element method (FEM) and the necessary analysis was performed. When designing the flux barrier $L_d/L_q$ rate was considered. By using permanent magnets in the design of electric motors, torque and efficiency values have been increased. For this reason, in our PMSynRM design, Neodymium Iron Boron (NdFeB) permanent magnet material is placed in the rotor flux barriers to increase torque. In addition, it is aimed to reduce the torque ripples at low speeds by selecting the distributed winding in the stator windings. Simulation results show that high torque is obtained with low torque ripple. In addition, good results were obtained in terms of efficiency, current, speed and power density.

**Keywords:** Permanent magnet synchronous reluctance motor (PMSynRM) ·
Finite element method (FEM) · Optimization · Electric vehicle (EV)

## 1 Introduction

Recently, the popularity of electric vehicles (EV) has increased for many reasons. One of these reasons is that it aims to reduce greenhouse gas (GHG) emissions [1]. Due to its effect on the ozone layer, it has become imperative to significantly reduce gas emissions [2]. With

J. Hemanth et al. (Eds.): ICAIAME 2020, LNDECT 76, pp. 390–399, 2021.
https://doi.org/10.1007/978-3-030-79357-9_38

the shortage of energy and environmental pollution, the search for new electric vehicles is getting more and more attention from both governments and manufacturers [3].

The power system of an EV consists of only two components: the motor that supplies the power and the controller that controls the application of this power. There are various electric motors in industrial applications. They are used to drive all kinds of industrial devices. They are also used to propel EVs [4]. Many motor types have been used to drive EVs such as DC motor, induction motor, and permanent magnet synchronous motor (PMSM) [3]. For an electric motor to drive EV successfully, some performance indexes should be taken into consideration such as high efficiency, high power density, affordable cost, and having good dynamic properties [4, 5].

Recently, three-phase interior permanent magnet (IPM) motors and synchronous reluctance motors (SynRMs) have been an alternative to induction motors (IMs) because they offer higher torque density and efficiency [6].

The IPM is widely used in automotive applications such as Hybrid Electric Vehicle (HEV) and Electric vehicle (EV). One of the major disadvantages of the IPM is that it requires the use of large quantities of rare earth magnets in machine design [7].

Today, the latest developments in power electronics converters and with modern technologies, electric drives are also offered as an important alternative for synchronous reluctance motor (SynRM) and especially permanent magnet-assisted synchronous reluctance motor (PMaSynRM), interior magnet motor (IPM) [8]. Recently, the outer rotor PMaSynRM has gained popularity, including situations such as high torque density, low torque vibration and low harmonic back electromotive force (EMF) [7].

When the literature is examined; they have designed PMa-SynRMs with five phases, two inner rotors with the same outer dimensions, and two outer rotors, rare earth (neodymium alloy) and rare earth-free (ceramic) magnets [9]. In [10], they have designed an outer rotor synchronous reluctance motor (ORSynRM) and an outer rotor permanent magnet synchronous reluctance motor (ORPMASynRM) and made comparative performance analyzes. They have also proposed to reduce the weight of the rotor, without affecting the motor's electromagnetic performance, to improve the performance of the motor. In [11], they have proposed a new rotor topology for an outer rotor Fa-SynRM to achieve high efficiency and wide constant power speed ratio (CPSR). They have accomplished multi-purpose optimization using a Finite Element Analysis (FEA) based model for 1.2 kW electric/hybrid electric two-wheel application with Genetic Algorithms (GA). In [12], SynRMs with three different slots and two different rotor topologies are designed and comparison is made on the air gap magnetic flux density distribution, the variation of $L_d$ and $L_q$ inductances, and the average torque value and ripple torque value. In [13], they have analyzed the effect of permanent magnet volume on motor performance, although lamination geometry and stack length are kept constant. They have designed the PMASR motor based on finite element analysis (FEA) to operate in a very high field weakening speed range.

SynRM working principle is based on reluctance concept to produce torque, the ratio and difference of direct and quadrature axis inductances should be as high as possible [12]. When SynRM is examined in structure, there are two rotor types, axially laminated anisotropic (ALA), transversely laminated anisotropic (TLA) [14]. Synchronous reluctance motor (SynRM) structure with transverse flux barrier transverse lamination rotor

(TLA) shows very similarity to the asynchronous motor structure. The stator structure and windings of SynRM are the same as the asynchronous motor [15]. In this study, TLA structure is used.

The paper is organized as follows: in Sect. 2 mathematical expressions and design of external rotor PMSynRM, in Sect. 3, magnetic analysis and results of the designed motor are given, and in Sect. 4, the paper concludes with the conclusions.

## 2   Mathematical Expressions and Design of the External Rotor PMSynRM

The designed external rotor PMSynRM motor is shown in Fig. 1. PMSynRM is designed to be mounted in the wheels of the EVs. The design parameters of the designed motor are given in Table 1. The d-axis and q-axis magnetization inductances play an important role in determining the electromagnetic performance of the motor [8] and this is important in defining the rotor geometry parameters of the motor. The rotor structure of the designed PMSynRM has a transversely laminated rotor (TLA) and is very similar to the asynchronous motor structure. PMSynRM's model is created in ANSYS RMxprt program. The number of rotor barriers is selected as 2 and the type of magnet used is NdFe35. The number of rotor pole pairs is 4. Both stator and rotor materials are chosen as M15_29. All motor sizing and design is done in ANSYS RMxprt.

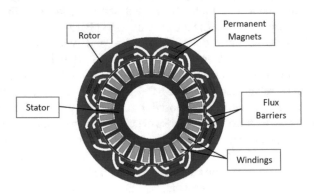

**Fig. 1.** The designed external rotor PMSynRM

**Table 1.** Design parameters of the motor

| | |
|---|---|
| Rated speed | 750 rpm |
| Rated power | 2 kW |
| Rotor outer diameter | 265 mm |
| Slot/pole | 24/8 |

In Fig. 2, the stator structure of designed PMSynRM is shown. The stator design is similar to the asynchronous motor. The stator consists of twenty-four slots.

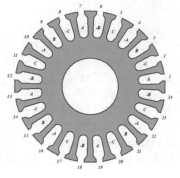

**Fig. 2.** 24-slotted stator structure of the designed PMSynRM

In Fig. 3, the three-phase coil distribution of the designed PMSynRM is shown. The windings of the motor are similar to those of the asynchronous motor. The designed PMSynRM consists of three-phase windings. A distributed winding shape has been chosen to achieve uniform torque. Star is chosen as the connection type of the windings.

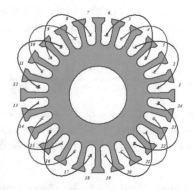

**Fig. 3.** Three phase coil distribution of the designed PMSynRM

High torque must be achieved for the designed PMSynRM to be used in an EV. Torque value produced by the motor is defined as (1).

$$T_{em} = \frac{3}{2}p(\lambda_d i_q - \lambda_q i_d) \tag{1}$$

$$\lambda_d = L_d i_d \tag{2}$$

$$\lambda_q = L_q i_q - \lambda_p \tag{3}$$

If (1) rearranged

$$T_{em} = \frac{3}{2}p(\lambda_p i_d + (L_d - L_q)i_q i_d) \tag{4}$$

is obtained.

Here, p is the number of motor poles, $\lambda_d$ and $\lambda_q$ are d-axis flux linkage and q-axis flux linkage, respectively. $i_d$ and $i_q$ are d- and q-axis magnetization currents of the stator, $L_d$ and $L_q$ are the magnetization inductance components on the d-axis and q-axis, respectively. $\lambda p$ is the permanent magnet flux linkage. The performance of PMSynRM depends on the d-axis and q-axis magnetization inductance values. In the expression given in (4), the first part is related to the flux produced by the permanent magnet, the second part uses the difference of the d-axis and q-axis inductances associated with the rotor structure. Thus, the torque obtained from the motor can be increased in two ways, the first is to use too many permanent magnets, and the second is to increase the barriers on the rotor. The saliency ratio, $\xi = L_d/L_q$, determines the many operating characteristics of the motor, the power factor, sensivity to parameter variation and field weakening performance.

The maximum output power factor of the PMSynRM is defined as

$$PF|_{max} = \cos(\varphi)|_{max} = \frac{\zeta - 1}{\zeta + 1} \tag{5}$$

Using permanent magnet in PMSynRM increases power factor and efficiency. Increasing the power factor reduces the currents of the motor. Reducing currents reduces stator ohmic losses, and thereby the efficiency of the motor increases [16].

The efficiency of the motor can be written as

$$\eta = \frac{P_{out}}{P_{out} + P_{core} + P_{cu} + P_{fw} + P_{str}} \tag{6}$$

Here, $P_{out}$ is the output power of the motor, $P_{core}$ is the core loss of the motor, $P_{cu}$ is copper loss of stator, $P_{fw}$ is friction and ventilation losses and $P_{str}$ is stray load losses. The efficiency of PMSynRM is higher than the efficiency of the asynchronous motor.

## 3    Magnetic Analysis and Results of the Designed Motor

The designed PMSynRM is shown in three dimensions from different angles, in Fig. 4. 3D sizing is done to make magnetic analysis of the motor. Then, performance analysis and verifications have carried out at ANSYS Maxwell. The torque generated by the motor, three-phase currents passing through the stator windings, flux linkages, magnetic flux density analysis, and magnetic field analysis have performed.

**Fig. 4.** 3D model of the designed PMSynRM from different angles

In Fig. 5, magnetic flux density analysis of the designed PMSynRM is shown.

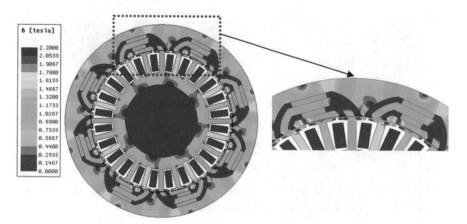

**Fig. 5.** Magnetic flux density of the designed PMSynRM

Magnetic flux lines distribution of the designed PMSynRM is shown in Fig. 6. The flux lines distribution of the external rotor PMSynRM shown in the Fig. 6. can be determined by the flux lines added by the PM along the q-axis.

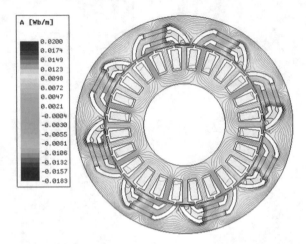

**Fig. 6.** Magnetic flux lines distribution of the designed PMSynRM

The torque-time chart of the designed PMSynRM is shown in Fig. 7. The motor produces an average of 25.52 Nm of torque when loaded.

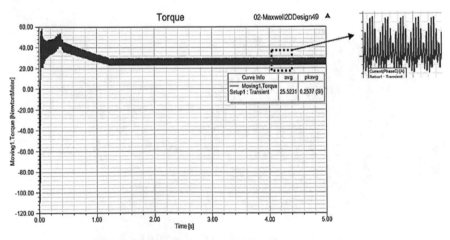

**Fig. 7.** Torque-time chart of the designed PMSynRM

In Fig. 8, the current-time chart of the designed PMSynRM is shown. As seen in the chart, the motor draws an average of 17.97A current when it is loaded.

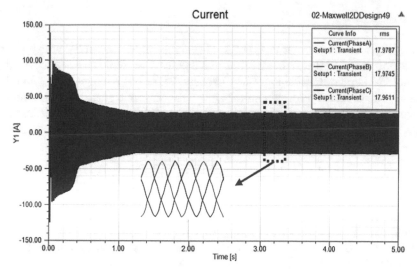

**Fig. 8.** Current-time chart of the designed PMSynRM

In Fig. 9, the flux linkage-time chart of the designed PMSynRM is shown. When PM is placed inside the flux barriers, it produces a flux in the opposite direction of the q-axis.

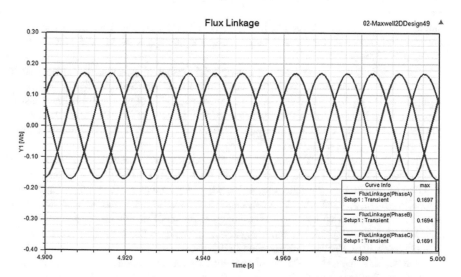

**Fig. 9.** Flux linkage-time chart of the designed PMSynRM

## 4   Conclusions

In this study, an external rotor PMSynRM is designed to be used in EV with a power of 2 kW at 750 rpm. The design has 24 slots and 8 poles. The rotor has two barriers at each pole and PM is placed inside these flux barriers. 25.52 Nm average torque and 92.7% efficiency are obtained when the motor is loaded. Magnetic analysis of the motor are performed and simulation results are shown. As seen from the simulation results, the designed external rotor PMSynRM motor has provided the desired performance.

## References

1. Fuad, U.-N., Sanjeevikumar, P., Lucian, M.-P., Mohammad, N.M., Eklas, H.: A comprehensive study of key electric vehicle (EV) components, technologies, challenges, impacts, and future direction of development. Energies **10**, 1217, 1–82 (2017)
2. Zarma, T.A., Galadima, A.A., Aminu, M.A.: Review of motors for electric vehicles. J. Sci. Res. Rep. **24**(6), 1–6 (2019)
3. Cai, C.-L., Wang, X.-G., Bai, Y.-W., Xia, Y.-C., Liu, K.: Key technologies of EV motor drive system design. Procedia Eng. **16**, 356–362 (2011). Elsevier
4. Xue, X.D., Cheng, K.W.E., Cheung, N.C.: Selection of electric motor drives for electric vehicles. In: 2008 Australasian Universities Power Engineering Conference, 14–17 December 2008
5. Singh, K.V., Bansal, H.O., Singh, D.: A comprehensive review on hybrid electric vehicles: architectures and components. J. Mod. Transp. **27**(2), 77–107 (2019). https://doi.org/10.1007/s40534-019-0184-3
6. Baek, J., Bonthu, S.S.R., Choi, S.: Design of five-phase permanent magnet assisted synchronous reluctance motor for low output torque ripple applications. IET Electric Power Appl. **10**(5), 339–346 (2016)
7. Bonthu, S.S.R., Choi, S., Gorgani, A., Jang, K.: Design of permanent magnet assisted synchronous reluctance motor with external rotor architecture. IEEE (2015)
8. Aghazadeh, H., Afjei, E., Siadatan, A.: Comparative analysis of permanent magnet assisted synchronous reluctance motor with external-rotor. In: 2019 10th International Power Electronics, Drive Systems and Technologies Conference (PEDSTC), 12–14 February, Shiraz University, Iran, pp.22–27 (2019)
9. Bonthu, S.S.R., Islam, M.Z., Choi, S.: Performance review of permanent magnet assisted synchronous reluctance traction motor designs. In: 2018 IEEE Energy Conversion Congress and Exposition (ECCE), Portland, OR, USA, 23–27 September 2018, pp.1682–1687 (2018)
10. Inte, R.A., Jurca, F.N., Martis, C.: Analysis performances of outer rotor synchronous reluctance machine with or without permanent magnets for small electric propulsion application. In: The 11th International Symposium on Advanced Topics in Electrical Engineering, ATEE, 28–30 March 2019, Bucharest, Romania (2019). IEEE
11. Deshpande, Y., Toliyat, H.A.: Design of an outer rotor ferrite assisted synchronous reluctance machine (Fa-SynRM) for electric two wheeler application. In: 2014 IEEE Energy Conversion Congress and Exposition (ECCE), 14–18 September 2014, Pittsburgh, PA, USA, pp. 3147–3154 (2014)
12. Oprea, C., Dziechciarz, A., Martis, C.: Comparative analysis of different synchronous reluctance motor topologies. In: 2015 IEEE 15th International Conference on Environment and Electrical Engineering (EEEIC), 10–13 June 2015, Rome, Italy (2015)

13. Barcaro, M., Bianchi, N., Magnussen, F.: Permanent-magnet optimization in permanent-magnet-assisted synchronous reluctance motor for a wide constant-power speed range. IEEE Trans. Industr. Electron. **59**(6), 2495–2502 (2012)
14. Taghavi, S., Pillay, P.: A Sizing methodology of the synchronous reluctance motor for traction applications. IEEE J. Emerg. Sel. Top. Power Electron. **2**(2), 329–340 (2014)
15. Ersöz, M., Öner, Y., Bingöl, O.: Akı bariyerli TLA tipi senkron relüktans motor tasarımı ve optimizasyonu. J. Fac. Eng. Archit. Gazi Univ. **31**(4), 941–950 (2016)
16. Haataja, J., Pyrhönen, J.: Permanent magnet assisted synchronous reluctance motor: an alternative motor in variable speed drives. In: Parasiliti, F., Bertoldi, P. (eds.) Energy Efficiency in Motor Driven Systems, pp. 101–110. Springer, Heidelberg (2003). https://doi.org/10.1007/978-3-642-55475-9_16

# Support Vector Machines in Determining the Characteristic Impedance of Microstrip Lines

Oluwatayomi Adegboye, Mehmet Aldağ, and Ezgi Deniz Ülker[(⊠)]

Department of Computer Engineering, European University of Lefke, Mersin-10, Lefke, Turkey
eulker@eul.edu.tr

**Abstract.** In this work, the estimation of the characteristic impedance of a microstrip transmission line with the aid of support vector machines was conducted. In the estimation, first data was obtained by using empirical analytical formulas and this data was used for training and testing in support vector machine. Both classification and regression analysis were conducted using different kernel functions. Kernel functions used in this case were linear kernel and radial kernel. In the analysis, different regression techniques as $\nu$-regression and $\varepsilon$-regression were also used. It is concluded that using radial kernels produce more accurate results. Moreover, it can be concluded that support vector machines can be efficiently used in determining the characteristic impedance of microstrip lines.

**Keywords:** Support vector machine · Support vector regression · Microstrip line · Kernels

## 1 Introduction

Support vector machines are supervised learning models with learning algorithms for analyzing the data for classification and regression. Although there are many different implementations for support vector machines, it is still a current research topic for many researchers around the world in terms of exploring different additions to the method [1–10] as well as applying them to different application problems [11–14]. Support vector machines have also found variety of applications to microwave engineering problems. In this work, support vector machines are used for both classification and regression for determining the characteristic impedance of a microstrip transmission line. We had three objectives in this work. The first task was to see if successful classification can be done for this problem, the second was to see what kind of kernel function can yield better result in regression, and the third one was what type of regression technique leads to a more accurate result.

## 2 Support Vector Machine

Support vector machine theory was first suggested by Vapnik in 1968 in his work on the convergence of relative frequencies of events to their probabilities [15]. A set of data

© The Author(s), under exclusive license to Springer Nature Switzerland AG 2021
J. Hemanth et al. (Eds.): ICAIAME 2020, LNDECT 76, pp. 400–408, 2021.
https://doi.org/10.1007/978-3-030-79357-9_39

is grouped together for different categories and these separate categories are divided by a gap. When a new set of data is tested their position in these groups determine which category they belong to. In simple two dimensional plane we can think this separation to be done by a hyperplane. A good separation is done when a hyperplane has the largest distance to the nearest training data. When linear separation is not possible, it was proposed that the original finite dimensional space to be mapped to higher dimensional space to aid the separation easier [16, 17]. Support vector machines can be found in many different classification problems such as face recognition [18], text recognition [19], image segmentation [20], remote sensing [21], classification of cancer issues [22] and classification of web [23].

For regression one of the earlier proposals occurred in 1997 by Vapnik et al. [24]. Similar to support vector machine, in support vector regression a subset of training data produces the model and predictions take place accordingly. Different variants for support vector machine exist such as least square support vector machine (LS-SVM) by Suykens and Vandewalle [25] and adaptive support vector machine by Rosipal and Gorilami [26]. Support vector machine was used as a regression method in many different variety of applications, for example forecasting power output of voltaic systems [27], design of a rectangular patch antenna feed [28], and GDP prediction [29].

## 3   Microstrip Line

Microstrip line is a is a type of a planar transmission line which can be fabricated using the printed circuit board technology and is usually preferred at radio frequencies due to its simplicity. They are very popular because of ease of fabrication, ease of integration with other monolithic devices, good heat sinking and good mechanical support properties. Microstrip line consists of a metallic strip on a grounded dielectric substrate as shown in Fig. 1.

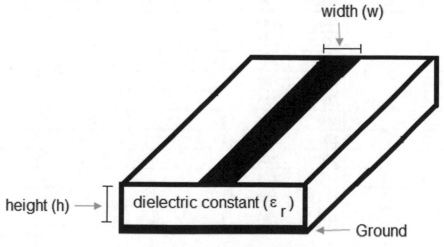

**Fig. 1.**  Simple microstrip line.

In microstrip line there are many parameters which affect the characteristic impedance value of the line. The empirical analytical formulas are derived for microstrip lines by Owens [30]. The dielectric constant ($\varepsilon_r$), width of the line (w), thickness of the substrate (h) are some of these parameters which affect the characteristic impedance value of a microstrip transmission line.

Design of microstrip lines using support vector regression were done in terms of analyzing and predicting impedance values and effective dielectric constant thoroughly by Gunes et al. [31–33].

## 4 Classification for Microstrip Line

Dimensions of the microstrip line play a crucial role in the design of microstrip circuits since they directly affect the characteristic impedance of the line. In the classification problem, a hypothetical problem is created which the line is considered to be usable (e.g. acceptable for fabrication) or not depending on the impedance value obtained from the parameters in question. In this case the dielectric constant of the substrate ($\varepsilon_r$), substrate thickness (h) and width of the line (w) were considered as the input parameters, and decision was set depending on these values. A set of 50 data was obtained via using empirical analytical calculations. In the analysis for classification, two experiments were done. In one of the cases 28 of the data were used in training and 22 in testing the data. In this set, it was obtained that 15 of the test data were classified correctly (about 68% correct). In one other case we used 22 data for training and 28 data for testing. While testing this case we observed that 20 of these test data were classified correctly result yielding 71%. The test data and classifications are displayed in Table 1 and Table 2 below.

**Table 1.** Support vector machine for classification (22 test data).

| Dielectric constant | h (mm) | w (mm) | Acceptable ( −1 or 1) | Classified correctly |
|---|---|---|---|---|
| 2.20 | 2.00 | 21.83 | −1 | Yes |
| 2.20 | 2.00 | 20.54 | −1 | Yes |
| 2.20 | 2.00 | 12.45 | −1 | Yes |
| 2.20 | 2.00 | 8.49 | 1 | Yes |
| 2.20 | 2.00 | 6.16 | 1 | Yes |
| 2.20 | 2.00 | 4.64 | 1 | Yes |
| 2.20 | 2.00 | 3.60 | 1 | Yes |
| 2.20 | 2.00 | 2.82 | 1 | Yes |
| 2.20 | 2.00 | 2.24 | 1 | Yes |
| 2.20 | 2.00 | 1.79 | 1 | Yes |
| 2.20 | 2.00 | 1.44 | 1 | Yes |

*(continued)*

**Table 1.**  (*continued*)

| Dielectric constant | h (mm) | w (mm) | Acceptable ( −1 or 1) | Classified correctly |
|---|---|---|---|---|
| 2.20 | 2.00 | 1.38 | 1 | Yes |
| 2.20 | 2.00 | 1.32 | −1 | No |
| 2.20 | 2.00 | 1.16 | −1 | No |
| 10.8 | 2 | 2.74 | 1 | Yes |
| 10.8 | 2 | 1.77 | 1 | Yes |
| 10.8 | 2 | 1.44 | 1 | Yes |
| 10.8 | 2 | 1.17 | −1 | No |
| 10.8 | 2 | 0.95 | −1 | No |
| 10.8 | 2 | 0.77 | −1 | No |
| 10.8 | 2 | 0.63 | −1 | No |
| 10.8 | 2 | 0.52 | −1 | No |

**Table 2.**  Support vector machine for classification (28 test data).

| Dielectric constant | h (mm) | w (mm) | Acceptable (−1or 1) | Classified correctly |
|---|---|---|---|---|
| 2.2 | 2 | 20.54 | −1 | No |
| 2.2 | 2 | 1.32 | −1 | No |
| 2.2 | 2 | 1.16 | −1 | No |
| 10.8 | 1.5 | 31.07 | −1 | No |
| 10.8 | 1.5 | 1.33 | −1 | Yes |
| 10.8 | 1.5 | 1.08 | −1 | Yes |
| 10.8 | 1.5 | 0.88 | −1 | Yes |
| 10.8 | 2 | 24.55 | −1 | No |
| 10.8 | 2 | 0.77 | −1 | Yes |
| 10.8 | 2 | 0.63 | −1 | Yes |
| 10.8 | 2 | 0.52 | −1 | Yes |
| 2.2 | 1.5 | 3.48 | 1 | Yes |
| 2.2 | 1.5 | 2.7 | 1 | Yes |
| 2.2 | 1.5 | 2.11 | 1 | Yes |
| 2.2 | 1.5 | 1.68 | 1 | Yes |
| 2.2 | 1.5 | 1.37 | 1 | Yes |
| 2.2 | 2 | 12.45 | 1 | Yes |

(*continued*)

**Table 2.** (*continued*)

| Dielectric constant | h (mm) | w (mm) | Acceptable (−1or 1) | Classified correctly |
|---|---|---|---|---|
| 2.2 | 2 | 4.64 | 1 | Yes |
| 2.2 | 2 | 3.6 | 1 | Yes |
| 2.2 | 2 | 2.82 | 1 | Yes |
| 2.2 | 2 | 2.24 | 1 | Yes |
| 2.2 | 2 | 1.79 | 1 | Yes |
| 2.2 | 2 | 1.44 | 1 | Yes |
| 2.2 | 2 | 1.38 | 1 | Yes |
| 10.8 | 1.5 | 18.41 | 1 | Yes |
| 10.8 | 1.5 | 14.23 | 1 | Yes |
| 10.8 | 1.5 | 5.99 | 1 | No |
| 10.8 | 1.5 | 3.33 | 1 | No |

## 5    Regression for Microstrip Line

As mentioned previously, rather than classification, support vector machine can be used as a regression tool inheriting the main features of the algorithm. In this analysis rather than having a classification of data the output is a real number. In regression a tolerance margin is used to set the approximation to the support vector machine.

The dimensions of the microstrip line play a crucial role in the design of microstrip circuits since they directly affect the characteristic impedance of the line. In this case, the data is trained with varying input parameters and the output, which is the characteristic impedance of the line, is determined. In this case, similar to the classification problem, the dielectric constant of the substrate ($\varepsilon_r$), substrate thickness (h) and width of the line (w) are considered as the input parameters. A set of 50 data points are used in the regression. In creating the model we used randomly 45 of this data, and 5 of them for the testing. Moreover we used all the data to see how closely the regression model represents the real impedance values. Two different kernels namely the linear kernel and radial kernel were used. Also analysis is done with two different regression methods: $\nu$-regression and $\varepsilon$-regression. Table 3 shows the analysis for linear kernel and for radial kernel.

**Table 3.** Linear and radial kernel comparison with regression types.

| Kernel | Regression type | Average error (test data) | Average error (all data) |
|---|---|---|---|
| Linear | $\nu$-regression | 10.989 | 12.810 |
| Linear | $\varepsilon$-regression | 10.283 | 12.490 |
| Radial | $\nu$-regression | 6.531 | 9.692 |
| Radial | $\varepsilon$-regression | 6.852 | 9.275 |

Table 3 indicates the average error obtained by applying different kernels and different regression types. As we can see in the above table we deduce that comparatively both of the kernels have similar behavior. On the other hand radial kernel produced relatively better approximation in the model to the real data. Similarly, when average error is considered with the prediction of all the data in terms of regression type $\varepsilon$ regression produced better results than $\upsilon$ regression. The Figs. 2 and 3 show the modelling of all data using linear and radial kernels respectively for better visualization. In both linear kernel and radial kernel, it was noticed that predicting the correct characteristic impedance values at low impedances was relatively better compared with the high impedances.

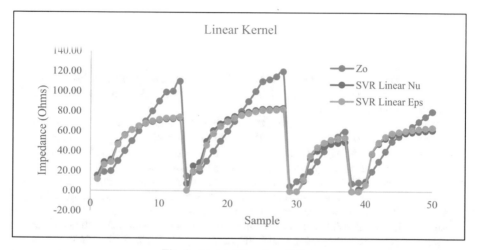

**Fig. 2.** Linear Kernel modeling.

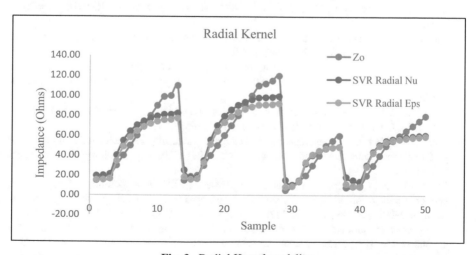

**Fig. 3.** Radial Kernel modeling.

## 6  Conclusions

In this work application of support vector machine for classification and regression is done for microstrip lines. It is observed that classification, though being very difficult because of the three different parameters and small number of data considered for training, 68% and 71% correct decisions were observed. In regression analysis, it was observed that radial kernel worked better than linear kernel. Also, it seemed that in either kernels average error is less when $\varepsilon$-regression is used compared with the $\nu$-regression.

## References

1. Wang, C., Han, D.: Credit card fraud forecasting model based on clustering analysis and integrated support vector machine. Cluster Comput. **22**(6), 13861–13866 (2018). https://doi.org/10.1007/s10586-018-2118-y
2. Tang, J., Chen, X., Hu, Z., Zong, F., Han, C., Li, L.: Traffic flow prediction based on combination of support vector machine and data denoising schemes. Phys. A **534**, 120642 (2019). https://doi.org/10.1016/j.physa.2019.03.007
3. Ahmadi, M.H., Ahmadi, M.A., Nazari, M.A., Mahian, O., Ghasempour, R.: A proposed model to predict thermal conductivity ratio of Al2O3/EG nanofluid by applying least squares support vector machine (LSSVM) and genetic algorithm as a connectionist approach. J. Therm. Anal. Calorim. **135**(1), 271–281 (2018). https://doi.org/10.1007/s10973-018-7035-z
4. Mi, X., Liu, H., Li, Y.: Wind speed prediction model using singular spectrum analysis, empirical mode decomposition and convolutional support vector machine. Energy Convers. Manage. **180**, 196–205 (2019). https://doi.org/10.1016/j.enconman.2018.11.006
5. Liu, W., Ci, L., Liu, L.: A new method for fuzzy support vector machine algorithm for intrusion detection. Appl. Sci. **10**(3), 1065 (2020). https://doi.org/10.3390/app10031065
6. Blanquero, R., Carrizosa, E., Jimenez-Cordero, A., Martin-Barragan, B.: Functional-bandwidth kernel for support vector machine with functional data: an alternating optimization algorithm. Eur. J. Oper. Res. **275**(1), 195–207 (2019). https://doi.org/10.1016/j.ejor.2018.11.024
7. Guo, H., Wang, W.: Granular support vector machine: a review. Artif. Intell. Rev. **51**(1), 19–32 (2017). https://doi.org/10.1007/s10462-017-9555-5
8. Ding, S., Sun, Y., An, Y., Jia, W.: Multiple birth support vector machine based on recurrent neural networks. Appl. Intell. **50**(7), 2280–2292 (2020). https://doi.org/10.1007/s10489-020-01655-x
9. Peng, X.: TSVR: An efficient twin support vector machine for regression. Neural Netw. **23**(3), 365–372 (2010). https://doi.org/10.1016/j.neunet.2009.07.002
10. Rastogi, R., Sharma, S.: Fast laplacian twin support vector machine with active learning for pattern classification. Appl. Soft Comput. **74**, 424–439 (2019). https://doi.org/10.1016/j.asoc.2018.10.042
11. Monteiro, R.P., Cerrada, M., Cabrera, D.R., Sanchez, R.V., Bastos-Filho, C.J.A.: Using a support vector machine based decision stage to improve the fault diagnosis on gearboxes. Comput. Intell. Neurosci. 1383752 (2019). https://doi.org/10.1155/2019/1383752.
12. Sharif, U., Mehmood, Z., Mahmood, T., Javid, M.A., Rehman, A., Saba, T.: Scene analysis and search using local features and support vector machine for effective content-based image retrieval. Artif. Intell. Rev. **52**(2), 901–925 (2018). https://doi.org/10.1007/s10462-018-9636-0

13. Zhou, Y., Chang, F., Chang, L., Kao, I., Wang, Y., Kang, C.: Multi-output support vector machine for regional multi-step-ahead PM2.5 forecasting. Sci. Total Environ. **651**, 230–240 (2019). https://doi.org/10.1016/j.scitotenv.2018.09.111

14. Richhariya, B., Tanveer, M., Rashid, A.H.: Diagnosis of Alzheimer's disease using universum support vector machine based recursive feature elimination (USVM-RFE). Biomed. Signal Process. Control **59**, 101903 (2020). https://doi.org/10.1016/j.bspc.2020.101903

15. Vapnik, V.N., Chervonenkis, A.Ya.: On the uniform convergence of relative frequencies of events to their probabilities. Dokl. Akad. Nauk USSR, **181**(4), 781–787 (1968). (Russian)

16. Cortes, C., Vapnik, V.N.: Support vector networks. Mach. Learn. **20**, 273–297 (1995). https://doi.org/10.1007/BF00994018

17. Schölkopf, B., Burges, C., Vapnik, V.: Incorporating invariances in support vector learning machines. In: von der Malsburg, C., von Seelen, W., Vorbrüggen, J.C., Sendhoff, B. (eds.) ICANN 1996. LNCS, vol. 1112, pp. 47–52. Springer, Heidelberg (1996). https://doi.org/10.1007/3-540-61510-5_12

18. Li, Y., Gong, S., Liddell, H.: Support vector regression and classification based multi-view face detection and recognition. In: Proceedings Fourth IEEE International Conference on Automatic Face and Gesture Recognition (Cat. No. PR00580), Grenoble, France, pp. 300–305 (2000). https://doi.org/10.1109/AFGR.2000.840650

19. Schlapbach, A., Wettstein, F., Bunke, H.: Estimating the readability of handwritten text - a support vector regression based approach. In: 2008 19th International Conference on Pattern Recognition, Tampa, FL, pp. 1–4, (2008). https://doi.org/10.1109/ICPR.2008.4761907

20. Wang, X., Wang, T., Bu, J.: Color image segmentation using pixel wise support vector machine classification. Pattern Recogn. **44**(4), 777–787 (2011). https://doi.org/10.1016/j.patcog.2010.08.008

21. Gualtieri, J.A., Cromp, R.F.: Support vector machines for hyperspectral sensing classification. In: Proceedings of the SPIE 3584 27th AIPR Workshop: Advances in Computer-Assisted Recognition (1999). https://doi.org/10.1117/12.339824

22. Furey, T.S., Cristianini, N., Duffy, N., Bednarski, D.W., Schummer, M., Haussler, D.: Support vector machine classification and validation of cancer tissue samples using microarray expression data. Bioinformatics **16**(10), 906–914 (2000). https://doi.org/10.1093/bioinformatics/16.10.906

23. Sun A., Lim E-P., Ng W-K.: Web classification using support vector machine. In: WIDM 2002 Proceedings of the 4th International Workshop on Web İnformation and Data Management, pp. 96–99 (2002). https://doi.org/10.1145/584931.584952

24. Müller, K.-R., Smola, A.J., Rätsch, G., Schölkopf, B., Kohlmorgen, J., Vapnik, V.: Predicting time series with support vector machines. In: Gerstner, W., Germond, A., Hasler, M., Nicoud, J.-D. (eds.) ICANN 1997. LNCS, vol. 1327, pp. 999–1004. Springer, Heidelberg (1997). https://doi.org/10.1007/BFb0020283

25. Suykens, J.A.K., Vandewalle, J.: Least squares support vector machine classifiers. Neural Process. Lett. **9**, 293–300 (1999). https://doi.org/10.1023/A:1018628609742

26. Rosipal R., Gorilami M.: An adaptive support vector regression filter: a signal detection application. In: 9th International Conference on Artificial Neural Networks: ICANN 1999 (1999). https://doi.org/10.1049/cp:19991176

27. Shi, J., Lee, W., Liu, Y., Yang, Y., Wang, P.: Forecasting power output of photovoltaic systems based on weather classification and support vector machines. IEEE Trans. Ind. Appl. **48**(3), 1064–1069 (2012). https://doi.org/10.1109/TIA.2012.2190816

28. Ülker S.: Support vector regression analysis for the design of feed in a rectangular patch antenna. In: 2019 3rd International Symposium on Multidisciplinary Studies and Innovative Technologies (ISMSIT), Ankara, Turkey, pp. 1–3 (2019). https://doi.org/10.1109/ISMSIT.2019.8932929

29. Ülker E. D., Ülker S.: Unemployment rage and GDP prediction using support vector regression. In: AISS 2019: Proceedings of the International Conference on Advanced Information Science and System, pp. 1–5 (2019). https://doi.org/10.1145/3373477.3373494.Article no. 17
30. Owens, R.P.: Accurate analytical determination of quasi-static microstrip line parameters. Radio Electron. Engineer **46**(7), 360–364 (1976). https://doi.org/10.1049/ree.1976.0058
31. Güneş, F., Tokan, N.T., Gürgen, F.: Support vector design of the microstrip lines. Int. J. RF Microwave Comput. Aided Eng. **18**(4), 326–336 (2008). https://doi.org/10.1002/mmce.20290
32. Tokan, N.T., Gunes, F.: Knowledge-based support vector synthesis of the microstrip lines. Prog. Electromagn. Res. **92**, 65–77 (2009). https://doi.org/10.2528/PIER09022704
33. Tokan, N.T., Gunes, F.: Analysis and synthesis of the microstrip lines based on support vector regression. In: 2008 38th European Microwave Conference, Amsterdam, pp. 1473–1476 (2008). https://doi.org/10.1109/EUMC.2008.4751745

# A New Approach Based on Simulation of Annealing to Solution of Heterogeneous Fleet Vehicle Routing Problem

Metin Bilgin and Nisanur Bulut[✉]

Department of Computer Engineering, Bursa Uludağ University, Bursa, Turkey
metinbilgin@uludag.edu.tr, nisanurbulutnb@gmail.com

**Abstract.** In this study, the problem of vehicle routing, which arises from supplying a heterogeneous fleet of vehicles with different capacities and features, is discussed. First, a two-level method has been developed to address the transport-carrying assignment and then the route approach. In the first stage of the proposed method, the transported persons are matched with the transported vehicles considering their suitability. In the second stage, the routing process is completed by using the simulated annealing algorithm, which is a metaheuristic method. The novelty aspect of this study is the creation of a solution model by simulating the scheduling model of parallel machines. The implementation of the study was carried out on the distribution plans obtained from a large-scale logistics company performing cost analysis over the total distance. The proposed solution model is compared with the solution method used by the firm and it is concluded that the total distance and the resulting costs are reduced.

**Keywords:** Vehicle routing problem · Travelling vendor problem · Heterogeneous fleet · Annealing simulation algorithm · Parallel machine scheduling

## 1 Introduction

Due to today's logistics practices, diversified customer demands and legal obligations determined, vehicle routing constraints have been diversified, while vehicle routing problems have become increasingly complex. Routing that cannot be done effectively is reflected the company owners as an increase in cost, excessive resource consumption, and due to the fuel consumption of vehicles, it returns to nature as an increase in carbon footprint. Vehicle routing problem belongs to NP (Non-deterministic polynomial Time)-difficult problems class [1]. With the increase in vehicle diversity and the increase in the number of customer demands in the vehicle fleet participating in the problem, it becomes difficult to find a solution. In the multi-vehicle routing problem, the level of this difficulty increases one more time and the solution time is extended.

Vehicle routing problem has been brought to the literature by Dantzig and Ramser. This problem is the determination of the route by minimizing the distance travelled and the time travelled from n source points to m destination points [2]. In this problem, there is

© The Author(s), under exclusive license to Springer Nature Switzerland AG 2021
J. Hemanth et al. (Eds.): ICAIAME 2020, LNDECT 76, pp. 409–417, 2021.
https://doi.org/10.1007/978-3-030-79357-9_40

more than one vehicle that is separated from the centre point to the destination points. For this reason, the problem is called multiple traveller seller problems [3]. When vehicles are not identical in terms of capacity and type, the vehicle fleet is named as heterogeneous fleet [4]. It is not possible to deliver with a fleet that is always homogeneous. The use of a heterogeneous fleet is a common situation in practical logistics applications due to the customization of customer demands and the need for special equipment for the transported goods.

While supply customer demands by vehicle types are compared to an assignment problem, finding the route with the minimum distance for the vehicles is compared to the scheduling problem. Scheduling is the process of performing the works expected in a time frame in the proper order and sharing the resources to be used in the best way [5]. In this problem, tools are the resources assigned and customer requests are sequential jobs that need to be done.

In this study, the transportation problem that a company will make from its central warehouse to m distribution centre is discussed. The constraints that must be taken into account when making the distribution are that the transported assets are suitable with the vehicles and the distance travelled during the distribution is minimized. For the solution of this problem, a two-level solution has been proposed based on the scheduling of parallel machines. At the first level, the assets to be moved are assigned to the appropriate vehicles by random selection. At the second level, routing is performed for the vehicles with certain assignments.

## 2 Simulated Annealing Problem

The annealing simulation algorithm proposed by Kirkpatrick et al. is modelled by mimicking the process of slowly cooling the solids after heating. By annealing, the material is softened by heating and cooled gradually, ensuring the molecules of the material become regular. In the optimization application, this behaviour is imitated. The search process starts at a high temperature. Thus, the solution space is kept wide. A certain number of cycles occur at each reduction of the temperature value. Neighbouring solutions are produced in each cycle. Thus, research is made among possible solutions. The annealing simulation algorithm is based on randomness. Thus, the possibility of acceptance in bad solutions that do not contribute to the purpose function occurs. This prevents it from sticking to the local best spots [6]. The pseudocode of the annealing simulation algorithm is seen in Algorithm 1.

**Algorithm 1.** Annealing simulation algorithm pseudo code

| | |
|---|---|
| 1. | Set start solution $C_0$ |
| 2. | Calculate purpose function value $f(C_0)$ |
| 3. | Set temperature for start S>0 |
| 4. | Set Temperature Reduction Rate 0<SAO<1 |
| 5. | Set the number of cycles for each temperature reduction N>0 |
| 6. | Repeat |
| 7. | S>1 |
| 8. | Repeat |
| 9. | N>1 |
| 10. | Generate a neighboring solution $C_1$ |
| 11. | Calculate purpose function value $f(C_1)$ |
| 12. | Calculate energy change $\Delta \leftarrow f(C_1) - f(C_0)$ |
| 13. | İf $\Delta \geq 0$ $f(C_0) \leftarrow f(C_1)$ |
| 14. | Else if r←random(0,1) |
| 15. | İf $r < \exp\left(\frac{-\Delta}{S}\right)$ then $f(C_0) \leftarrow f(C_1)$ |
| 16. | Else $f(C_1) \leftarrow f(C_0)$ |
| 17. | N←N-1 |
| 18. | S←S*SAO |

Annealing simulation is started according to a certain rule or with a randomly generated starting solution. The temperature value is reduced according to a specified rule. Each time the temperature is reduced, a certain number of cycles are made and neighbouring solutions are investigated. The temperature parameter also directly affects the neighbouring solution's acceptance as the current solution. While the temperature parameter is too high, the research space is kept very wide, while keeping the temperature parameter too low reduces the leaps in the solution space, so that possible good solutions can be overlooked. For this reason, the temperature should be reduced gradually starting from a high-temperature value. Rapid reduction of temperature shortens solution time, but it can also prevent sufficient research from being done for the best solution [7].

# 3 Problem Definition

The logistics company, where the application tests are carried out, pays its drivers at the rate of the distance travelled during the distribution process. For this reason, minimizing the distance taken while providing customer demands is a critical value for the company. The basis of the problem is to assign the goods to be transported to suitable vehicles and then to ensure that each vehicle reaches the distribution targets by taking the minimum road distance.

## 3.1 Constraints in the Problem

1. Each transport should start from the central warehouse and end at the central warehouse.
2. There is only one central storage point.
3. Every customer request must be carried by the vehicle that suits it.
4. Each customer request must be made with only one vehicle.

5. The total distance of the route used when deploying should be minimized.
6. There can be no transport without vehicle assignment.

### 3.2  Flexibilities in the Problem

1. Target points for vehicles may be the same.
2. No number of restrictions for vehicles.
3. Each shipping does not have to leave the warehouse at the same time

The test plans used are 14, two of each problem size. Plan naming consists of the number of transports and the total number of destination points included in each transport. For example, Plan1–6 contains 6 transports and 8 destination points for each transport. These test plans were obtained from the application company.

## 4  Application

Customer demands are handled as a distribution plan. To facilitate solution representation, modelling was made on distribution plans. This modelling shown in Fig. 1; It includes the parameters of transportation, transportation, centre point, destination point. From this description, the shipping definition will be used for customer requests.

1. Shipping: Includes all points included in a distribution plan.
2. Transport: It is the transport of goods that takes place in the distribution plan from one point to another, in other words, the point transition.
3. Target Point: These are the points in the distribution plan to be delivered.
4. Central Point (M): It is the starting and ending point of transportation. Refers to the central warehouse.

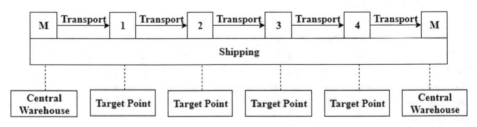

**Fig. 1.** Modelling definitions

The application consists of two stages. In the first step, each transport is assigned to existing vehicles randomly by means of one-to-one matching. If suitable vehicles cannot be found due to the inspection status of the vehicles, malfunctioning or reserving, virtual vehicles are produced. This prevents the application from being affected by the vehicle resource constraint.

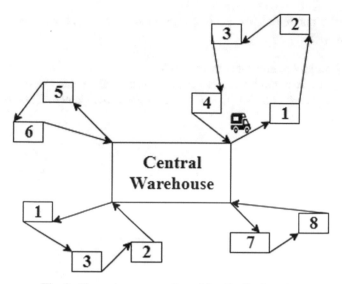

**Fig. 2.** General representation of the distribution plan

As seen in Fig. 2, all vehicles load goods from the central warehouse and then set off towards the destination points in the distribution plan. The arrival time of the vehicle to the destination is not the subject of the problem. For this reason, instantaneous positions of vehicles and their distance to the central warehouse were not taken into account during the solution phase. Parameters of Annealing Simulation Algorithm are shown in Table 1.

## 4.1   Using Annealing Simulation Algorithm in Problem Solution

1. It starts with the initial temperature value.
2. Neighbouring solutions are produced as much as the number of cycles determined. The production of neighbour solutions for each cycle is as follows:

   2.1. Two randomly determined target points are displaced for each transport.
   2.2. The total path distance is calculated using the Euclidean formula for the destination points in the shipping.

3. The total distance of all shippings within the neighbouring solution set is listed. The path distance value, which is the largest value, is accepted as the score value of the solution set.
4. The difference ($\Delta$) between the score value of the current solution set and the score value of the neighbourhood solution set is calculated.

   4.1. If $\Delta \geq 0$, the neighbour solution set is accepted.
   4.2. If $\Delta < 0$, a value of r is produced in the range of $0 < r < 1$, and if $r < \exp((-\Delta) /$ temperature) the neighboring solution set is accepted.

4.3. If the conditions for Δ are not provided, the neighbour solution set is not accepted.

5. The loop control parameter is increased by 1.
6. The temperature control parameter is reduced by multiplying the value of 0.99.
7. If the temperature value is >1, 1, 2, 3, 4, 5, 6 steps are repeated. If not, the algorithm will stop running.

**Table 1.** Annealing Simulation Algorithm Parameter Values

| Temperature values | Temperature reduction rate value | Loop values |
|---|---|---|
| 5, 10, 15, 20, 25, 50, 75 | 0,99 | 250, 500, 750, 1000 |

## 4.2 Purpose Function Evaluation Method

As shown in the first part of Fig. 3, each vehicle is modelled as a parallel machine. The target points in the shipping are modelled as sequential works of the parallel machine. The maximum value of completion of all jobs in parallel machines is the CMAX value [8]. With the solution set approach, the target point change is made in all shippings at once. The total road distance of the shipping, which has the largest total distance in the solution set, is considered as the score value. In the second part of Fig. 3, it is shown that the score value is considered as the total distance of all shippings in the solution set. The score value was determined by the CMAX approach and by calculating the total distance of the road. The score values obtained by both methods were compared with the score value, which is the result of the initial solution of the plan, and how much reduction was found in the initial score value.

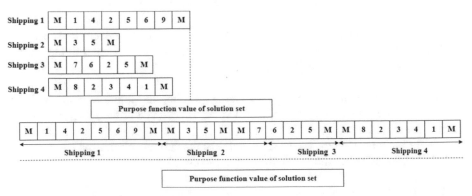

**Fig. 3.** Indication of the score value of the purpose function

# 5   Result

Information on the work performed in this section will be presented. The results of the study are given in Table 2 and Fig. 4. As can be seen in Fig. 4, the initial score value refers to the score value of the plan before routing. As a result of the study conducted with the proposed method, the total kilometre values in the distribution lists were distributed balanced and the starting score value was reduced.

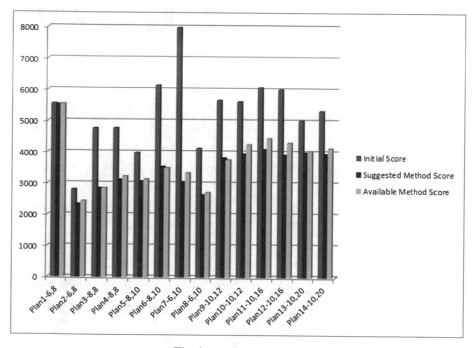

**Fig. 4.** Application results

The data presented in Table 2 includes the parameters of the annealing simulation algorithm used in the solution method for each plan and the best score values compared to the starting score.

Score: Indicates the largest total kilometre value for shipping in the plan.

Initial score: Indicates the score value before optimizing the plan.

Result score: Indicates the score value after optimizing the plan.

Method: Specifies the proposed and used the score calculation method.

When Table 2 is analyzed, it is noticed that the current method provides an improvement on the initial score as the proposed method, but it is seen that the proposed method provides a greater improvement compared to the available method.

**Table 2.** Application results of distribution plans.

| Plan | Temperature | Loop | Initial score | Result score | Method |
|---|---|---|---|---|---|
| P1–6,8 | 5 | 1000 | 5550.063984 | 5523.23516 | Suggested |
| P1–6,8 | 5 | 250 | 5550.063984 | 5539.527066 | Available |
| P2–6,8 | 5 | 250 | 2805.246257 | 2337.382633 | Suggested |
| P2–6,8 | 5 | 250 | 2805.246257 | 2434.264203 | Available |
| P3–8,8 | 5 | 500 | 4769.069649 | 2850.675297 | Suggested |
| P3–8,8 | 10 | 1000 | 4769.069649 | 2850.675297 | Available |
| P4–8,8 | 25 | 1000 | 4774.481587 | 3122.652722 | Suggested |
| P4–8,8 | 50 | 750 | 4774.481587 | 3230.324691 | Available |
| P5–8,10 | 10 | 500 | 3983.391731 | 3070.508009 | Suggested |
| P5–8,10 | 50 | 1500 | 3983.391731 | 3138.640627 | Available |
| P6–8,10 | 15 | 750 | 6138.924608 | 3539.108844 | Suggested |
| P6–8,10 | 10 | 750 | 6138.924608 | 3501.504287 | Available |
| P7–6,10 | 20 | 750 | 7975.010216 | 3050.894745 | Suggested |
| P7–6,10 | 10 | 1000 | 7975.010216 | 3345.792264 | Available |
| P8–6,10 | 5 | 750 | 4122.383091 | 2637.726108 | Suggested |
| P8–6,10 | 10 | 750 | 4122.383091 | 2712.785779 | Available |
| P9–10,12 | 20 | 1000 | 5668.715257 | 3822.948637 | Suggested |
| P9–10,12 | 50 | 500 | 5668.715257 | 3764.590177 | Available |
| P1010,12 | 75 | 750 | 5622.764481 | 3953.770845 | Suggested |
| P10–10,12 | 75 | 750 | 5622.764481 | 4248.893896 | Available |
| P11–10,16 | 75 | 750 | 6070.451051 | 4098.831179 | Suggested |
| P11–10,16 | 50 | 750 | 6070.451051 | 4449.676716 | Available |
| P12–10,16 | 25 | 750 | 6011.785587 | 3919.192682 | Suggested |
| P12–10,16 | 25 | 500 | 6011.785587 | 4311.302766 | Available |
| P13–10,20 | 20 | 1000 | 5028.129453 | 3998.537206 | Suggested |
| P13–10,20 | 20 | 750 | 5028.129453 | 4046.785558 | Available |
| P14–10,20 | 15 | 750 | 5328.986009 | 3952.326316 | Suggested |
| P14–10,20 | 5 | 1000 | 5328.986009 | 4134.83959 | Available |

# References

1. Keskintürk, T., Topuk, N., Özyeşil, O.: Classification of vehicle routing problems and solving techniques and an implementation. İşletme Bilimi Derg. **3**(1), 77–107 (2015)
2. Dantzig, G.B., Ramser, J.H.: The truck dispatching problem. Manage. Sci. **6**(1), 80–91 (1959)
3. Yıldırım, T., Kalaycı, C.B., Mutlu, Ö.: A novel metaheuristic for traveling salesman problem: blind molerat algorithm. Pamukkale Univ Muh Bilim Derg. **22**(1), 64–70 (2016)

4. Keçeci, B., Altınparmak, F., Kara, İ.: Heterogeneous vehicle routing problem with simultaneous pickup and delivery: mathematical formulations and a heuristic algorithm. J. Fac. Eng. Archit. Gazi University **30**(2), 185–195 (2015)
5. Kaya, S., Karaçizmeli, İ.H.: Solving multi-objective identical parallel machine scheduling problems with a common due date and set-up times. Harran University J. Eng. **3**(3), 205–2013 (2018)
6. Ayan, T.Y.: Kaynak Kısıtlı Çoklu Proje Programlama Problemi İçin Tavlama Benzetimi Algoritması. Atatürk Üniversitesi İktisadi ve İdari Bilimler Derg. **23**(2), 101–118 (2009)
7. Kirkpatrick, S., Gelatt, C.D., Vecchi, M.P.: Optimization by simulated annealing. Sci. New Ser. **220**(4598), 671–680 (1983)
8. Eren, T., Güner, E.: Minimization of total completion time and maximum tardiness on parallel machine scheduling. J. Fac. Eng. Archit. Selçuk University **21**(1–2), 21–32 (2006)

# Microgrid Design Optimization and Control with Artificial Intelligence Algorithms for a Public Institution

Furkan Üstünsoy[1(✉)], Serkan Gönen[2], H. Hüseyin Sayan[3], Ercan Nurcan Yılmaz[3], and Gökçe Karacayılmaz[4]

[1] Institute of Natural and Applied Sciences, Gazi University, Ankara, Turkey
[2] Faculty of Engineering and Architecture, Gelisim University, Istanbul, Turkey
sgonen@gelisim.edu.tr
[3] Faculty of Technology, Gazi University, Ankara, Turkey
{hsayan,enyilmaz}@gazi.edu.tr
[4] Institute of Natural and Applied Sciences, Hacettepe University, Ankara, Turkey

**Abstract.** In this study, a microgrid in Matlab/Simulink environment controlled by artificial intelligence algorithm was designed for the campus area of a public institution. In addition, the performance analysis and optimization of this microgrid was realized by using Hybrid Optimization of Multiple Energy Resources (HOMER) software. In both simulation studies, the solar energy potential of the institution and the actual electricity consumption data collected for one year were used. With the HOMER simulation study, feasibility of renewable energy integration for the public institution was demonstrated. In addition, all combinations of microgrids were optimized and the results were evaluated. Simulink analysis was performed using artificial intelligence algorithms for 24 h energy flow analysis. According to the simulation results, it was seen that renewable energy sources can feed the existing loads and even supply energy to the grid when necessary. As a result, it has been proved that integrating the designed microgrid into the campus energy network will be profitable and efficient for the public institution in question.

**Keywords:** Artificial intelligence · Microgrid · Matlab/Simulink · HOMER · Solar energy · System optimization

## 1 Introduction

Scientists have sought different power grids to improve reliability and quality due to increasing social demands, increased energy consumption, and inadequate existing power grids [1, 2]. For these reasons, significant changes occur in the electrical systems of developed countries. These changes increase the use of renewable energy sources. Efficient use of renewable energy sources and utilization of their potentials in the best way in medium and long-term energy plans and targets in electric power generation applications are a critical issue of today's energy sector [3]. In order to achieve these

© The Author(s), under exclusive license to Springer Nature Switzerland AG 2021
J. Hemanth et al. (Eds.): ICAIAME 2020, LNDECT 76, pp. 418–428, 2021.
https://doi.org/10.1007/978-3-030-79357-9_41

goals, each distribution transformer region should be designed as microgrid and renewable energy sources should be integrated to these points. As a matter of fact, microgrids using renewable energy sources are rapidly expanding in the power industry thanks to their ability to reduce global warming and eliminate environmental constraints [4]. The microgrid paradigm is an attractive way to future intelligent distribution networks, thanks to its ability to operate in both grid-connected and islanding mode. Dynamic islanding operations provide greater flexibility in the integration of Distributed Generation (DG) units and also provide a more reliable electrical service [5].

Having a correct feasibility of a microgrid to be designed is very important in terms of installation cost. Even though the transition to renewable energy sources is mandatory in the future, many issues such as whether the system to be implemented in order to implement these plans are profitable, how many years it will pay for the installation cost or the service life are critical. At this point, the management of the microgrid according to the most appropriate scenarios or control techniques is very important in terms of profitability and efficiency. In other words, each microgrid should be optimized, a feasibility study should be implemented and an efficient energy management algorithm should be designed.

## 2   Related Works

In recent years, many researchers have worked on microgrid design and opti-mization and control methods. For example, the League Championship Algorithm, a new method for determining the optimum values of the proportional-integral-derivative (PID) controller's gains used in frequency control in microgrid systems, has been proposed in [6]. In another study, optimization of a microgrid consisting of a wind turbine, solar panel, diesel generator, inverter and loads has been investigated with multi-purpose hybrid metaheuristic algorithms [7]. In a similar study, the design, performance analysis and optimization of a mixed microgrid using HOMER software for the hospital complex on the campus of Eskişehir Osmangazi University (ESOGU) is presented [3]. Optimization is very important in a microgrid design, but the energy management strategy is also critical for efficient operation. As a matter of fact, opti-mum energy management was achieved by using fuzzy logic control in a microgrid to provide a more reasonable and efficient strategy [8]. In another study, a robust optimization method was used to eliminate uncertainties related to real-time market price signals (trading) [9]. In another study in which artificial intelligence algorithms were applied, the operation of the microgrid was controlled depending on certain factors such as SOC level, source current and tariff by using fuzzy logic controller [10]. In another similar study, a 25-rule Fuzzy Logic Controller was designed to control the charge level of the battery in the microgrid and to prevent mains power deviations [11]. In another study, SMES (super conducting magnetic energy storage) with high PV penetration levels and a fuzzy logic approach to improve the reliability of useful microgrids are presented [12]. Another similar, fuzzy logic control system is presented for charging and discharging the battery energy storage system in microgrid applications [13]. It is also possible to design the storage group with different scenarios according to the demands of the designer and the user. As a matter of fact, an isolated dc microgrid was proposed using photovoltaic panels as the primary power source during the day and spare dc generator at other times [14]. In a similar study, an AC-DC hybrid energy router structure that serves as an interface between the power consumer and the distribution network was proposed [15]. In another study implemented from perspective of a smart grid, a two-level hierarchical supervisory control system was proposed for electric vehicles (PEVs) involved in frequency regulation in microgrids with interconnected areas [16].

When the literature is examined, valuable studies related to a microgrid design, optimization and control has been presented. In this study, a microgrid was designed and optimized for a real corporate campus and controlled according to artificial intelligence algorithms. In addition, the feasibility of the system was extracted and the results were revealed.

## 3    Matlab/Simulink with an AC Microgrid Design

Microgrid with voltage level between phases 380 V was modeled in Mat-lab/Simulink environment and the results were presented. The microgrid model consists of the power grid, PV production unit, storage unit and load as shown in Fig. 1. The high voltage level of 154 kV in the power network were reduced to 34.5 kV medium voltage level through the transformer. After approximately three kilometers of transmission line, the 34.5 kV voltage level was reduced to 380 V distribution voltage level and connected to the AC microgrid.

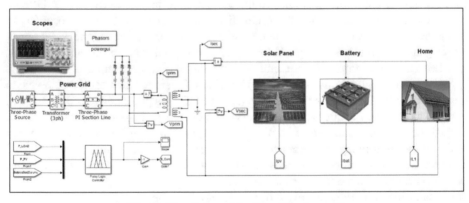

**Fig. 1.** Simulink model of the AC microgrid.

The photovoltaic generation unit is modeled as a controlled current source. It was sized to produce a maximum of 40 kW during 24 h of operation. The battery energy storage unit had a capacity of 50 kWh and was modeled as a controlled current source. Semiconductor tools cannot be used directly during phasor analysis in Simulink program. Instead a mathematical model is subtracted for each block. In Fig. 2, the mathematical model designed for the storage group is given as an example. The models created for PV panel unit and current load were designed in a similar way.

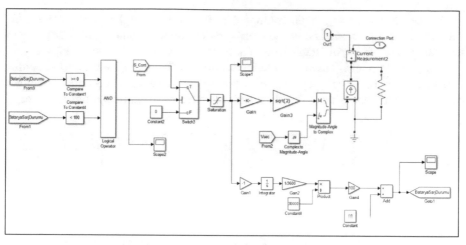

**Fig. 2.** Storage grup simulink model.

## 3.1 Artificial Intelligence Algorithm Designed for a Microgrid Control

Fuzzy logic control is the most suitable artificial intelligence method for energy management of the microgrid. Hence, the designed fuzzy logic controller was implemented using the fuzzy editor in the simulink. Storage group SOC level, instantaneous power generation of PV system and instantaneous power consumption of current load was defined as inputs of controller while battery energy flow amount was defined as output. Input membership functions and output membership functions are given in Fig. 3.

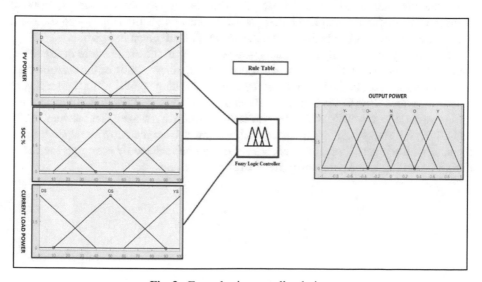

**Fig. 3.** Fuzzy logic controller design.

In the designed fuzzy logic controller, min-max (clipping) of mamdani method was used. Mean-max method was preferred as the defuzzification unit. Surface view according to input-output membership functions and rule table is given in Fig. 4.

The table of rules is based on experience and how the microgrid behaves according to the situation. In the design phase, the most appropriate control scenario was realized by revising both membership functions and rule table by using fuzzy anfis interface.

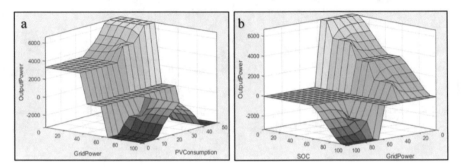

**Fig. 4.** I/O Surface (Fig. 1a: Gridpower-PVPower, Fig. 1b: SOC-Gridpower).

## 3.2 AC Microgrid Simulink Model Simulation Results

The PV production unit produced energy during bright hours. It produced a maximum of 40 KW of energy and firstly supplied the load. Between 12.00–13.00, excess energy was transferred from the PV production unit to the power grid. The power grid, on the other hand, supplied energy to the load during the time intervals when the energy supplied from the PV production unit and battery storage unit was insufficient to feed the load. The measurement results obtained from the simulation are given in Fig. 5.

The battery unit was filled or discharged according to the architecture in arti-ficial intelligence algorithms. Here, the instantaneous power consumption of the current load, the production status of the PV unit and the charge-discharge power of the battery were determined according to the battery's full charge rate. The initial value of the Storage group was selected as 50%. When the results are examined, it is seen that between 00:00 and 24:00, the storage group supported or drew power from the network depending on the state of the inputs. Here, different scenarios can be realized by changing the fuzzy design according to the demands of the designer or users.

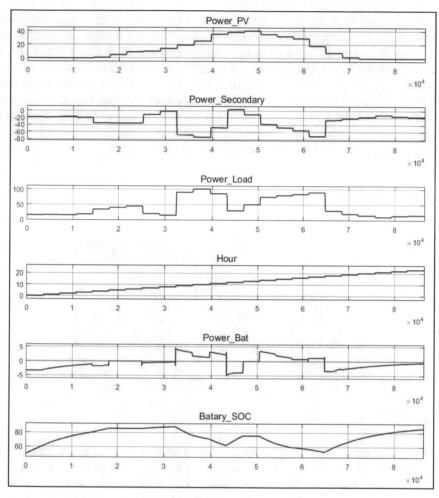

**Fig. 5.** Simulink simulation result.

## 4  Simulation and Optimization of a Microgrid with Homer

At the design stage, first of all, microgrid architecture was created. Later, in the created architecture, each microgrid component is modeled. First, a powergrid is modeled and connected to the AC busbar to feed the priority load. Sales capacity was set to 1600 kW and purchase capacity was set to 600 kW. When the network is being modeled, purchase and sale pricing was loaded as daylight between 06.00–17.00, peak between 17.00–20.00 and night between 20.00–06.00, by setting the "scheduled rates" mode. Photovoltaic production unit was designed and connected to DC bus. The Solar-Max 500RX A with Generic with a nominal capacity of 499.95 kW PV brand PV has been selected through the program. Here, investment, renovation, operation and maintenance cost values for 1 kW PV capacity are determined. Life expectancy was settled as 25 years. According to the latitude and longitude coordinates determined for the PV power system, the monthly average solar radiation, the cloudlessness index for each month, the average daily solar radiation and temperature values were downloaded from the Internet and uploaded to the system. A battery was used as energy storage unit and connected to DC bus. The Generic 100 kWh Li-Ion battery with a nominal capacity of 100 kWh was used from the catalog. Hereby, investment, replacement, operation, maintenance cost and life span values were determined for the battery. Between the AC and DC busbar, a converter with ABB MGS100 60 kW capacity and 95% efficiency was used. Investment, replacement, operation, maintenance cost and life span values determined for 1 kW converter capacity is depicted. The microgrid architecture designed in Fig. 6 is given.

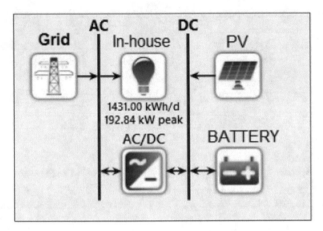

**Fig. 6.** Microgrid architecture in HOMER.

Finally, the coordinates of the public institution were loaded into the program. Thus, solar radiation, cloudlessness index for each month, daily average solar radiation and temperature values could be obtained via Internet. Then the load selection was performed. The load was added to the system as the priority load and the values close to the load profile made in the simulink program were added to the system over 24 h. The load profile in the Homer program is depicted in Fig. 7. Here, the distribution of the load according to the months can be observed with the plot button.

The HOMER program has five control strategies: Cycle Charging, Load Fol-lowing, MATLAB Link, Generator Order and Combined Dispatch. In this study, "Load fol-lowing" strategy was preferred. After determining the control strategy, modeling of economy and constraints was performed. Consequently, simulation and optimization were performed and the results were examined.

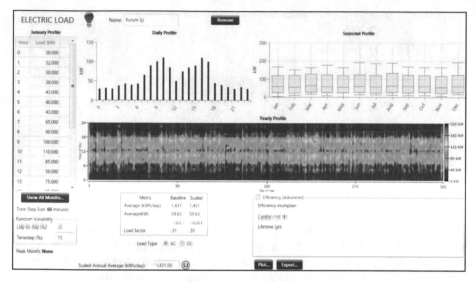

**Fig. 7.** Load profile.

## 4.1 Simulation and Optimization of System Designed in Homer Program

The results obtained after the simulation and optimization of the grid-connected micro-grid are given in Fig. 8. All combinations of the system have been optimized and the results have been presented.

The cost of the system is given in Fig. 9. Net present cost was calculated as 1.428.826,00 TL, energy cost was calculated as 0,09659 TL and working cost was calculated as −38.078,00 TL. Investment, replacement, operation and maintenance, scrap and total costs corresponding to each component are also depicted in Fig. 9.

| FV (kW) | FV-MPPT (kW) | Batarya Birimi | Şebeke (kW) | AC/DC (kW) | Dispatch | NPC (₺) | COE (₺) | Operating cost (₺/yr) | Initial capital (₺) | Ren Frac (%) | Total Fuel (L/yr) | Capital Cost (₺) | Production (kWh/yr) |
|---|---|---|---|---|---|---|---|---|---|---|---|---|---|
| 593 | 500 | | 600 | 475 | LF | ₺1.43M | ₺0.0966 | -38,078 | ₺1.92M | 82.2 | 0 | 1,778,697 | 989,751 |
| 589 | 500 | | 600 | 475 | LF | ₺1.43M | ₺0.0970 | -37,219 | ₺1.91M | 82.1 | 0 | 1,767,625 | 984,531 |
| 593 | 500 | | 600 | 474 | LF | ₺1.43M | ₺0.0966 | -38,055 | ₺1.92M | 82.2 | 0 | 1,778,522 | 989,669 |
| 584 | 500 | | 600 | 475 | LF | ₺1.43M | ₺0.0976 | -35,889 | ₺1.89M | 81.9 | 0 | 1,750,645 | 976,445 |
| 589 | 500 | | 600 | 476 | LF | ₺1.43M | ₺0.0971 | -37,081 | ₺1.91M | 82.0 | 0 | 1,765,817 | 983,674 |
| 604 | 500 | | 600 | 475 | LF | ₺1.43M | ₺0.0955 | -40,581 | ₺1.95M | 82.4 | 0 | 1,811,515 | 1,004,889 |
| 577 | 500 | | 600 | 475 | LF | ₺1.43M | ₺0.0984 | -34,252 | ₺1.87M | 81.7 | 0 | 1,731,140 | 967,050 |
| 599 | 500 | | 600 | 473 | LF | ₺1.43M | ₺0.0960 | -39,443 | ₺1.94M | 82.3 | 0 | 1,797,783 | 998,625 |
| 588 | 500 | | 600 | 471 | LF | ₺1.43M | ₺0.0972 | -36,844 | ₺1.91M | 82.0 | 0 | 1,764,828 | 983,205 |
| 589 | 500 | | 600 | 470 | LF | ₺1.43M | ₺0.0972 | -37,089 | ₺1.91M | 82.0 | 0 | 1,768,428 | 984,912 |
| 597 | 500 | | 600 | 478 | LF | ₺1.43M | ₺0.0963 | -38,937 | ₺1.93M | 82.3 | 0 | 1,790,000 | 995,025 |
| 597 | 500 | | 600 | 471 | LF | ₺1.43M | ₺0.0964 | -38,782 | ₺1.93M | 82.2 | 0 | 1,790,000 | 995,025 |
| 580 | 500 | | 600 | 470 | LF | ₺1.43M | ₺0.0982 | -34,863 | ₺1.88M | 81.8 | 0 | 1,739,656 | 973,858 |
| 582 | 500 | | 600 | 478 | LF | ₺1.43M | ₺0.0979 | -35,453 | ₺1.89M | 81.9 | 0 | 1,745,250 | 973,858 |
| 586 | 500 | | 600 | 469 | LF | ₺1.43M | ₺0.0975 | -36,317 | ₺1.90M | 82.0 | 0 | 1,759,156 | 980,509 |
| 607 | 500 | | 600 | 474 | LF | ₺1.43M | ₺0.0953 | -41,238 | ₺1.96M | 82.5 | 0 | 1,821,389 | 1,009,348 |
| 581 | 500 | | 600 | 478 | LF | ₺1.43M | ₺0.0980 | -35,249 | ₺1.89M | 81.8 | 0 | 1,742,699 | 972,631 |
| 586 | 500 | | 600 | 479 | LF | ₺1.43M | ₺0.0975 | -36,345 | ₺1.90M | 82.0 | 0 | 1,756,660 | 979,319 |
| 589 | 500 | | 600 | 479 | LF | ₺1.43M | ₺0.0971 | -37,143 | ₺1.91M | 82.1 | 0 | 1,766,904 | 984,189 |

**Fig. 8.** Optimization results.

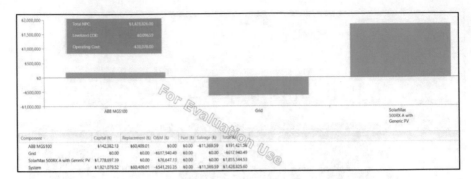

**Fig. 9.** System cost status.

The economic summary of the system is given in Fig. 10. The most optimal system is based on the PV power system-power grid as a result of optimization. The present value is the difference of the net present cost between the base system and the selected system and has been calculated as 794,671 TL. This value shows that the net present cost of the selected system is lower. Annual value is 61,471 TL. The profitability of the investment is 8.2%, the internal profitability ratio is 11.3%, the simple repayment period is 8.37 years and the work controlled repayment period was calculated as 11.86 years.

A summary of the system's electricity usage is given in Fig. 11. 82.9% of electricity generation was purchased from PV power system and 17.1% was purchased from power grid. 45.6% of the electricity consumption was consumed by AC load and 54.4% was sold to the power grid. 82.2% of the renewable energy source was utilized in the production of electricity.

|  | | | | | FV (kW) | FV-MPPT (kW) | Batarya Birimi | Şebeke (kW) | AC/DC (kW) | NPC (₺) | Initial capital (₺) |
|---|---|---|---|---|---|---|---|---|---|---|---|
| | | | | | | | Architecture | | | Cost | |
| Base system | | | | | 89.5 | 500 | 3 | 600 | 242 | ₺2.22M | ₺551,000 |
| Current system | | | | | 593 | 500 | | 600 | 475 | ₺1.43M | ₺1.92M |

| Metric | Value |
|---|---|
| Present worth (₺) | ₺794,671 |
| Annual worth (₺/yr) | ₺61,471 |
| Return on investment (%) | 8.2 |
| Internal rate of return (%) | 11.3 |
| Simple payback (yr) | 8.37 |
| Discounted payback (yr) | 11.86 |

**Fig. 10.** Economic summary of the system.

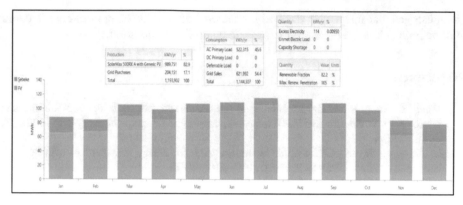

**Fig. 11.** Electricity usage summary.

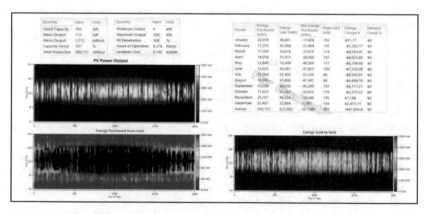

**Fig. 12.** PV power usage summary and grid usage summary.

A summary of the system's PV power system usage and the usage of the energy purchased and sold to the power grid for a year are given in Fig. 12. The power system, with a nominal capacity of 593 kW, produced 989,751 kWh of energy in one year and manufactured for 4374 h.

## 5    Results

As the need for renewable energy sources increases, the concept of microgrid is increasing day by day. These microgrids, capable of maintaining the production-load balance, need energy management with optimum gains. However, microgrid and renewable energy sources should be designed to achieve the right feasibility and minimum installation cost and maximum profit. Therefore, in this study, microgrid was designed and optimized for a real corporate campus and energy management was realized in the most appropriate way. In addition, all design options and cost analyzes were made by optimization with the HOMER program for the designed microgrid. As a result, it has been demonstrated that integrating the designed microgrid into the campus energy network will be profitable and efficient for the public institution in question.

# References

1. Wang, Z., Yang, R., Wang, L.: Intelligent multi-agent control for integrated building and micro-grid systems. In: ISGT 2011, pp.1–7, January 2011. https://doi.org/10.1109/ISGT. 2011.5759134
2. Üstünsoy, F., Sayan, H.H.: Sample laboratory work for energy management with SCADA supported by PLC. J. Polytech. **21**(4), 1007–1014 (2018)
3. Çetinbaş, İ, Tamyürek, B., Demirtaş, M.: Design, analysis and optimization of a hybrid micro-grid system using HOMER software: Eskişehir Osmangazi University example. Int. J. Renew. Energy Dev. **8**(1), 65–79 (2019)
4. Bevrani, H., Habibi, F., Babahajyani, P., Watanabe, M., Mitani, Y.: Intelligent frequency control in an AC microgrid: online PSO-based fuzzy tuning approach. IEEE Trans. Smart Grid **3**, 1935–1944 (2012)
5. Saveen, G., Prudhvi, R.P., Manikanta, D.V., Satya, P.D.: Design and implementation of energy management system with fuzzy control for multiple Microgrid. In: ICISC 2018, June 2018. https://doi.org/10.1109/ICISC.2018.8399003
6. Özdemir, M.T., Yıldırım, B., Gülan, H., Gençoğlu, M.T., Cebeci, M.: İzole Bir AA Mikroşebekede Lig Şampiyonası Algoritması İle Optimum Otomatik Üretim Kontrolü. Fırat Üniversitesi Mühendislik Bilimleri Dergisi **29**(1), 109–120 (2017)
7. Ağır, T.T.: New swarm intelligence based optimization algorithms for the optimization of microgrids. Eur. J. Tech. **8**(2), 196–208 (2018)
8. Lagouir, M., Badri, A., Sayouti, Y.: An optimal energy management system of islanded hybrid AC/DC microgrid. In ICOA 2019. doi: https://doi.org/10.1109/ICOA.2019.8727621.
9. Hussain, A., Bui, V., Kim, H.: Robust Optimal Operation of AC/DC Hybrid Microgrids Under Market Price Uncertainties. IEEE Access **6**, 2654–2667 (2017)
10. Nair, D. R., Devi, S., Nair, M. N., Ilango, K.: Tariff based fuzzy logic controller for active power sharing between microgrid to grid with improved power quality. In: ICEETS 2016, April 2016. https://doi.org/10.1109/ICEETS.2016.7583789.
11. Prathyush, M., Jasmin, E.: A fuzzy logic based energy management system design for AC microgrid. In: ICICCT 2018, April 2018. https://doi.org/10.1109/ICICCT.2018.8473317.
12. Said, S., Aly, M., Hartmann, B., Alharbi, A., Ahmed, E.: SMES-Based Fuzzy Logic Approach for Enhancing the Reliability of Microgrids Equipped With PV Generators. IEEE Access **7**, 92059–92069 (2019)
13. Faisal, M., Hannan, M.A., Ker, P., Uddin, M.: Backtracking Search Algorithm Based Fuzzy Charging-Discharging Controller for Battery Storage System in Microgrid Applications. IEEE Access **7**, 159357–159368 (2019)
14. Kushal, T., Karin, M.: Fuzzy logic controller for lithium-ion battery in standalone DC microgrid. In: WIECON-ECE 2015, December 2015. https://doi.org/10.1109/WIECON-ECE.2015.7444005.
15. Liu, B., et al.: An AC/DC Hybrid Multi-Port Energy Router With Coordinated Control and Energy Management Strategies. IEEE Access **7**, 109069–109082 (2019)
16. Janfeshan, K., Masoum, M.: Department of Electrical and Computer Engineering, Curtin University, Perth, WA, AustraliaHierarchical Supervisory Control System for PEVs Participating in Frequency Regulation of Smart Grids. IEEE Power and Energy Technology Systems Journal **4**(4), 84–93 (2017)

# Determination of Vehicle Type by Image Classification Methods for a Sample Traffic Intersection in Isparta Province

Fatmanur Ateş[1]([✉]), Osamah Salman[2], Ramazan Şenol[1], and Bekir Aksoy[2]

[1] Faculty of Technology, Electrical and Electronics Engineering, Isparta University of Applied Sciences, Isparta 32200, Turkey
{fatmanurates,ramazansenol}@isparta.edu.tr
[2] Faculty of Technology Mechatronics Engineering, Isparta University of Applied Sciences, Isparta 32200, Turkey
y1193065400/@stud.sdu.edu.tr, bekiraksoy@isparta.edu.tr

**Abstract.** Today, technologies related to artificial intelligence are used in almost every area of life. Scientists are working to maximize the predictive accuracy of artificial intelligence. Thus, they aim to reduce human errors. They also aim to make human life more comfortable by using artificial intelligence technology. Artificial intelligence models are tried on various data sets, and various analyzes are carried out to increase accuracy. In this study, a data set of vehicle types was created at an intersection in Isparta city center. Classification studies were performed with Capsule Networks and Convolutional Neural Networks (CNN) on this data set. Besides, some parameters were changed, and the accuracy of the models was observed. The results are given in the form of a table.

**Keywords:** Convolutional neural networks · Capsule networks · Image classification · Vehicle type determination

## 1 Introduction

Deep learning is a machine learning technique that imitates the human brain's abilities using large amounts of data [1]. Deep learning consists of multiple computational layers and has a structure that enables data to be learned. Deep learning is one of the techniques frequently used in solving many interdisciplinary problems [2]. Deep learning is used in many fields, such as face recognition [3, 4], speech recognition [5], biomedical [6], and defense [7, 8]. Among the deep learning methods, Capsule Networks (CapsNet), Deep Neural Networks (DSA), Recursive Neural Networks (RNN), and Convolutional Neural Networks (CNN) are frequently used models. The CNN model is especially a deep learning structure created by modeling the human visual system and has achieved the most successful results in computer vision [9]. However, CNN causes information loss due to pooling and has some built-in limitations, such as not being resistant to affine transformations [10]. To overcome these problems in the CNN model, Hinton et al. Capsule networks architecture has been created by [11, 12].

J. Hemanth et al. (Eds.): ICAIAME 2020, LNDECT 76, pp. 429–438, 2021.
https://doi.org/10.1007/978-3-030-79357-9_42

The classification success on CNN and CapsNet images; It is frequently used in many areas such as medical images, face recognition, and defense field. One of the important usage areas of CNN and CapsNet is traffic systems where the traffic flow is continuously monitored by mobese cameras. These two models are frequently used, especially in traffic flow analysis. When the academic studies in the literature are examined; There are different studies on vehicle type classification. Shavai et al. proposed a CNN-based method of vehicle classification for electronic toll collection. They classified the vehicle with 99.1% accuracy, using a total of 73,638 images [13]. Zhang et al. made a classification study using capsule networks as passenger cars, jeeps, buses, and minibusses. They used 1900 images for training and 470 images for testing [14]. Körez and Barışçı made object classification by using unmanned aerial vehicle (IHA) images. The classification process was made using capsule nets for six objects (human, skateboarder, cyclist, car, bus, and golf cart). Changes in accuracy were observed by changing the data increase, training time, and initial weights of the model [15]. Mekhalfi et al. Classified solar panels and cars taken at different angles by unmanned aerial vehicles with capsule networks. Accuracy changes are given by changing the capsule network [16]. R. Chen and colleagues used CNN and capsule networks to identify the vehicle logo. They achieved the highest classification accuracy of 100% with capsule networks [17]. Y. Chen et al. Classified the vehicles with CNN. The images of the vehicles taken from the rear were trained in 32 × 32. They classified the vehicles into 3 types: automobile, jeep and microbus [18]. Huo et al. made a multi-tag image classification using CNN. They classified the vehicle images by creating a CNN-based model considering both the type (truck, car, van, bus), vehicle angle (front view, rear view, and side view), and the lighting conditions (day or night) [19].

## 2   Material and Method

### 2.1   Material

In the study, the camera images taken from a sample intersection in Isparta's city center were transformed into two-dimensional images, classification with CNN and CapsNet, which are deep learning methods, and changes in accuracy were observed by changing some hyperparameters. The most accurate method was determined considering the changed parameters. Detailed information about the data set and the deep learning methods used are given below.

### 2.1.1   Creating Data Set

In creating the data set, the first three-dimensional images taken from the intersection in the city center of Isparta were obtained as two-dimensional images by using image processing methods. Since there is more than one object in these images, the image was cropped to separate the vehicle images to be classified. A data set was created with images of 60 trucks, 50 motorcycles, 60 buses, and 188 cars. Among the two methods, 80% of the data set is randomly allocated for training and 20% for testing. Also, the images that were trained and tested for the two models are 28 × 28 × 3.

## 2.1.2  CNN Networks

CNN is called the type of multilayer perceptrons. The design of this algorithm was inspired by the visual center of the animals. There are successful studies on image/video processing, biomedical signal/image processing, and other applications [20]. CNN model consists of three basic layers: convolution layer, common layer, and full link layer.

### 2.1.2.1 Convolution Layer

Different filters such as $3 \times 3, 5 \times 5, 7 \times 7$ are applied to the input image in this layer, and low and high-level features are extracted from the image.

The activation function is applied after each convolution layer [21]. Although there are many different functions as the activation function, the Relu function is generally preferred [22].

### 2.1.2.2 Pooling Layer

Pooling is a layer that is added between convolution layers. It is used to reduce the parameters and the number of calculations. Thus, time is saved. The pooling layer causes information loss in CNN architectures.

### 2.1.2.3 Fully Connected Layer

It is the layer where the classification is made according to the features obtained after the feature extraction process. The process of estimating the image is performed by weighing the features extracted in this layer. This situation is similar to the architecture of artificial neural networks.

## 2.1.3  CapsNet Method

The CapsNet method has a very similar structure to the CNN method. However, the difference between CapsNet from CNN is that it performs neuron-based operations with capsules that have a group of neurons. To ensure nonlinearity, squash function (squash) is added to each capsule here. The mathematical expression of the crush function is given in Eq. 1 [23].

$$V_j = \frac{\|sj\|^2}{1 + \|sj\|^2} \frac{sj}{\|sj\|} \tag{1}$$

In the equation, Vj is the crush focussing, and sj is the weighted total prediction vector. In the CNN model, the outputs of the layers are scalar. However, in CapsNet, the outputs of the capsule layers are vectorial. Mathematical expressions of weights and affine transformations are given in Eq. 1, 2 [23].

$$s_j = \sum_i c_{ij}.u_{j|i} \tag{2}$$

$$u_{j\backslash i} = W_{ij}.u_i \tag{3}$$

In the equations, cij is the coefficient of coupling, uj l i is the prediction vector, Wij is the weight matrix, and ui is i. shows the capsule exit in the layer.

The last step in the CapsNet method is to perform the thresholding process with a function called softmax. In Eq. 4, the mathematical expression of the differentiable thresholding (softmax) function is given [23].

$$C_{ij} = \frac{\exp{(bij)}}{\sum_k \exp{(bik)}} \tag{4}$$

In the equation, bij specifies the initial log probability that connects capsule i to capsule j.

### 2.1.4 Some Hyper Parameters Changed to Increase Accuracy in Deep Learning Methods

#### 2.1.4.1 Optimization Algorithms
The weight values found for the training of artificial intelligence models are important. Weight values are updated according to error. Various optimization algorithms are used to minimize the error. The optimization algorithms used in the study are given below.

In the stochastic gradient descent (SGD) optimization algorithm, which is the first of the optimization methods used, the calculation is made over a training sample instead of the whole training data. Briefly, the error is calculated for one sample at a time and the weights are updated [24]. The second optimization method, root mean square prop (RMS-Prob), is one of the algorithms that uses an adaptive learning rate. It uses a different learning rate in each parameter and the learning rate is updated according to the result obtained. Convergence can be fast in this algorithm [24]. In the Adadelta algorithm, which is another opmization algorithm, the parameters have their own learning speed. As the education increases, the learning speed decreases and the learning process does not take place after a while [24]. The fourth optimization algorithm, Adam algorithm, is an adaptive optimization algorithm. The learning rate is updated according to each parameter [24]. Adagrad was used as the fifth optimization method. In this algorithm, the learning speed can be adapted according to the parameters. There are major updates for sparse parameters, while smaller updates are made for frequent parameters. Each parameter has its own learning speed [25, 26]. The last optimization method is Nadam. This algorithm is a combination of two algorithms. Nesterov Accelerated Gradient (NAG) algorithm has been included in the Adam algorithm to make changes in the momentum expression [25].

#### 2.1.4.2 Applied Filter Criteria
Filtering methods are used to extract features from the image. Filters are actually a neuron belonging to artificial neural networks. The weights of the neuron are the values in the filter. As the filter size increases, the output matrix gets smaller, which causes information loss [27].

#### 2.1.4.3 Epoch, Number of Loop and Packet Size
An epoch is when the entire data set passes through the entire network, back and forth. The number of cycles is the data of the specified packet size passing through the network in forward and reverse directions. Packet size is the amount of data received from the

data set for forward and backward propagation. The higher the packet size, the more memory is required [28].

### 2.1.4.4 Batch Size

During the training, the whole data set does not enter the training at the same time, the data set is divided into parts and the divided parts enter the training one by one [29]. The number of pieces the data set will be divided into is determined by the batch size.

### 2.1.5 Performance Evaluation Criteria

The performance of classification models is measured using classification metrics. It is very important to know the concepts of true positive, false positive, correct negative, false negative in order to understand these metrics. Correct positive (TP), correctly predicted positive class; false positive (FP), false predicted positive class; True negative (TN) refers to the correctly predicted negative class and false negative (FN) refers to the incorrectly predicted negative class. Mathematical expressions of sensitivity, specificity, accuracy and F1 score are given in Eqs. 5,6,7and8 [30–34].

$$Sensitivity = \frac{TP}{TP + FN} \tag{5}$$

$$Specificity = \frac{TN}{TN + FP} \tag{6}$$

$$Accuracy = \frac{TP + TN}{TP + FN + TN + FP} \tag{7}$$

$$F1\ Score = \frac{2 * TP}{2 * TP + FN + FP} \tag{8}$$

Another frequently used evaluation metric is the ROC curve. ROC is a probability curve. It is one of the measurements used in cases where there are unbalanced data sets [35].

### 2.2  Method

The scheme of the CNN model used in the study is shown in Fig. 1. The images were first passed through the convolution layer consisting of 32 filters in $24 \times 24 \times 3$ dimensions. This process is followed by the pooling layer, the reconvolution layer and then the pooling layer. The maximum pooling parameter is used in partnership. The information was given to the fully connected layer and the classification process was carried out. With dropout layers, it is aimed to prevent over learning by ensuring that some neurons are forgotten. Dense layer has nodes used.

CapsNet architecture used in the study is given in Fig. 2. Images of $28 \times 28 \times 3$ dimensions were given to the input layer, and the convolution process was applied twice. The transactions up to this section are classical CNN transactions. The convolution layer output is given as the entrance to the primary capsule. The squash function, in which its input and output are vectors, has been implemented in the primery layer as the activation

| Layer (type) | Output Shape | Param # |
|---|---|---|
| conv2d_8 (Conv2D) | (None, 24, 24, 32) | 2432 |
| max_pooling2d_3 (MaxPooling2 | (None, 12, 12, 32) | 0 |
| conv2d_9 (Conv2D) | (None, 8, 8, 32) | 25632 |
| max_pooling2d_4 (MaxPooling2 | (None, 4, 4, 32) | 0 |
| dropout_3 (Dropout) | (None, 4, 4, 32) | 0 |
| flatten_2 (Flatten) | (None, 512) | 0 |
| dense_6 (Dense) | (None, 128) | 65664 |
| activation_3 (Activation) | (None, 128) | 0 |
| dropout_4 (Dropout) | (None, 128) | 0 |
| dense_7 (Dense) | (None, 4) | 516 |
| activation_4 (Activation) | (None, 4) | 0 |

Total params: 94,244
Trainable params: 94,244
Non-trainable params: 0

**Fig. 1.** CNN architecture used in classification

| Layer (type) | Output Shape | Param # | Connected to |
|---|---|---|---|
| input_3 (InputLayer) | (None, 28, 28, 3) | 0 | |
| conv2d_3 (Conv2D) | (None, 24, 24, 128) | 9728 | input_3[0][0] |
| conv2d_4 (Conv2D) | (None, 10, 10, 256) | 819456 | conv2d_3[0][0] |
| reshape_3 (Reshape) | (None, 3200, 8) | 0 | conv2d_4[0][0] |
| primarycap_squash (Lambda) | (None, 3200, 8) | 0 | reshape_3[0][0] |
| densecaps (DenseCapsule) | (None, 4, 16, 1) | 1638400 | primarycap_squash[0][0] |
| input_4 (InputLayer) | (None, 4) | 0 | |
| mask_1 (Mask) | (None, 4) | 0 | densecaps[0][0] input_4[0][0] |
| capsnet (CapsuleLength) | (None, 4) | 0 | densecaps[0][0] |
| decoder_out (Sequential) | (None, 28, 28, 3) | 2969392 | mask_1[0][0] |

Total params: 5,436,976
Trainable params: 5,436,976
Non-trainable params: 0

**Fig. 2.** Capsule Network Architecture Used in Classification

function. Dynamic routing was applied in the dense capsule layer. Then the masking process was applied. The decoder part was passed, the network was restructured, and the classification process was carried out.

The results obtained were evaluated using the accuracy rate, one of the performance evaluation metrics.

## 3  Research Findings

Before the data set was trained, the number of epochs was determined as 250. Initially, the batch size was set at 32. By changing the kernel number and optimization algorithm, the training is repeated continuously. The highest test results of the models are given in Table 1 and Table 2.

**Table 1.** The highest test results of CapsNet model trained with different filter sizes and different optimization algorithms

| CASPNET | | | | | | |
|---|---|---|---|---|---|---|
| | SGD | Adam | Adadelta | Adagrad | Nadam | RMS-prob |
| 3 × 3 | % 70,83 | % 95,83 | % 88,89 | %88,89 | **% 97,22** | **% 97,22** |
| 5 × 5 | % 69,44 | % 94,44 | % 93,06 | % 52,78 | % 90,28 | % 95,83 |
| 7 × 7 | % 68,06 | % 88,89 | % 93,06 | % 52,78 | % 52,78 | % 94,44 |
| 9 × 9 | % 66,67 | % 90,28 | % 91,67 | % 52,78 | % 52,78 | % 87,50 |

**Table 2.** The highest test results of the CNN model trained with different filter sizes and different optimization algorithms

| CNN | | | | | | |
|---|---|---|---|---|---|---|
| | SGD | Adam | Adadelta | Adagrad | Nadam | RMS-prob |
| 3 × 3 | % 52,78 | % 84,72 | % 86,11 | % 84,72 | % 84,72 | % 87,50 |
| 5 × 5 | % 52,78 | **% 90,28** | % 88,89 | % 87,50 | **% 90,28** | % 86,11 |
| 7 × 7 | % 52,78 | % 84,72 | % 86,11 | % 83,33 | % 83,33 | % 87,50 |
| 9 × 9 | % 70,83 | % 83,33 | % 81,94 | % 79,17 | % 79,17 | % 81,94 |

According to the results in the tables, the highest accuracy value in the CapsNet model was found to be 97.22% by using Nadam and RMS-prop algorithms with filter size of 3 × 3. In the model created using CNN, the highest accuracy value was determined as 90.28% by using Nadam and Adam algorithms, where the filter size is 5 × 5. In the next stage, the accuracy rate was tried to be increased by changing the batch size. The batch size has been changed by using the filter measurements and optimization algorithms with the highest accuracy values, and the results obtained are given in Table 3. and Table 4.

**Table 3.** The results of the effect of change of batch size on accuracy using the filter size and optimization algorithms that give the highest accuracy for the CapsNet model

| High accuracy model | Batch size | | | |
|---|---|---|---|---|
| | 16 | 32 | 64 | 128 |
| 3 x 3, Nadam | % **97,22** | % **97,22** | % 95,83 | % 84,72 |
| 3 x 3, RMS-prob | % **97,22** | % **97,22** | % 94,44 | % 91,67 |

**Table 4.** The results of the effect of the change of batch size on accuracy using the filter size and optimization algorithms that give the highest accuracy for the CNN model.

| High accuracy model | Batch size | | | |
|---|---|---|---|---|
| | 16 | 32 | 64 | 128 |
| 5 x 5, Adam | % **91,67** | % 90,28 | % **91,67** | % 84,72 |
| 5 x 5, Nadam | % **91,67** | % 90,28 | % 86,11 | % 88,89 |

According to the accuracy values given in Table 3 and Table 4, it has been determined that changing the batch size does not contribute to increasing the accuracy of the model established with CapsNet. For CNN, it has been found to increase the accuracy of the model from 90.28% to 91.67%.

To briefly summarize the study results, the CapsNet model gave more successful results in classifying the data set compared to the CNN model. Classification accuracy of 97.22% for CapsNet and 91.67% classification accuracy for CNN is considered a successful accuracy value, considering the number of samples in the data set.

## 4   Results

Today, when information technology is developing in a dizzying way, artificial intelligence methods are frequently used in many applications. One of the important usage areas of artificial intelligence methods is to extract information from images. In the study, the vehicle type classification process was carried out using CNN and CapsNet deep learning methods for a sample intersection in Isparta city center. The results of both deep learning models are given below.

- The sample intersection videos in Isparta's central district were transformed into two-dimensional pictures with an application prepared in the Python programming language.
- The data set was created by determining the tools from the images transformed into two dimensions by image cropping.
- First, the classification process was performed with the CNN deep learning method on the obtained data set, and the classification was made with an accuracy of 90.28%.

- In the CapsNet method, another deep learning method, the classification accuracy is 97.22%.
- Among the two methods, the highest classification accuracies were found by using different optimization algorithms and different filter measures.
- Later, batch size values were changed and applied to CNN and CapsNet deep learning methods to increase the accuracy rate. While the accuracy rate in CNN increased from 90.28% to 91.67% with the batch size change, there was no change in CapsNet.

**Acknowledgment.** We would like to thank the Isparta Directorate of Transportation and Traffic Services for providing the data set that constitutes a resource for the study.

# References

1. Kayaalp, K., Süzen, A.A.: Derin Öğrenme ve Türkiye'deki Uygulamaları, p. 1. IKSAD International Publishing House, Basım sayısı, Yayın Yeri (2018)
2. LeCun, Y., Bengio, Y., Hinton, G.: Deep learning. Nature **521**(7553), 436–444 (2015)
3. Prasad, P.S., Pathak, R., Gunjan, V.K., Rao, H.R.: Deep learning based representation for face recognition. In ICCCE 2019, pp. 419–424. Springer, Singapore (2020)
4. Bashbaghi, S., Granger, E., Sabourin, R., Parchami, M.: Deep learning architectures for face recognition in video surveillance. In: Deep Learning in Object Detection and Recognition, pp. 133–154. Springer (2019)
5. Demır, A., Atıla, O., Şengür, A.: Deep learning and audio based emotion recognition. In: 2019 International Artificial Intelligence and Data Processing Symposium (IDAP), pp. 1–6. IEEE, September 2019
6. Toğaçar, M., Ergen, B., Sertkaya, M.E.: Zatürre Hastalığının Derin Öğrenme Modeli ile Tespiti. Firat University J. Eng. **31**(1) (2019)
7. Çalışır, S., Atay, R., Pehlivanoğlu, M.K., Duru, N.: Intrusion detection using machine learning and deep learning techniques. In: 2019 4th International Conference on Computer Science and Engineering (UBMK), pp. 656–660. IEEE, September 2019
8. Koşan, M.A., Benzer, R.: Siber Güvenlik Alanında Derin Öğrenme Yöntemlerinin Kullanımı (2019)
9. Kizrak, M.A., Bolat, B.: Derin öğrenme ile kalabalık analizi üzerine detaylı bir araştırma. Bilişim Teknolojileri Dergisi **11**(3), 263–286 (2018)
10. Kınlı, F., Kıraç, F.: FashionCapsNet: clothing classification with capsule networks. Int. J. Inf. Technol. **13**(1) (2020)
11. Sabour, S., Frosst, N., Hinton, G.E.: Dynamic routing between capsules. In: Advances in Neural İnformation Processing Systems, pp. 3856–3866 (2017)
12. Hinton, G.E., Krizhevsky, A., Wang, S.D.: Transforming autoencoders. In: International Conference on Artificial Neural Networks, pp. 44–51. Springer, Heidelberg, June 2011
13. Shvai, N., Hasnat, A., Meicler, A., Nakib, A.: Accurate classification for automatic vehicle-type recognition based on ensemble classifiers. IEEE Trans. Intell. Transp. Syst. (2019)
14. Zhang, Z., Zhang, D., Wei, H.: Vehicle type recognition using capsule network. In: 2019 Chinese Control And Decision Conference (CCDC), pp. 2944–2948. IEEE, June 2019
15. Körez, A., Barışcı, N.: İnsansız Hava aracı (İHA) Görüntülerindeki Nesnelerin Kapsül Ağları Kullanılarak Sınıflandırılması (2019)
16. Mekhalfi, M.L., Bejiga, M.B., Soresina, D., Melgani, F., Demir, B.: Capsule networks for object detection in UAV imagery. Remote Sens. **11**(14), 1694 (2019)

17. Chen, R., Jalal, M.A., Mihaylova, L., Moore, R.K.: Learning capsules for vehicle logo recognition. In: 2018 21st International Conference on Information Fusion (FUSION), pp. 565–572. IEEE, July 2018

18. Chen, Y., Zhu, W., Yao, D., Zhang, L.: Vehicle type classification based on convolutional neural network. In: 2017 Chinese Automation Congress (CAC), pp. 1898–1901. IEEE, October 2017

19. Huo, Z., Xia, Y., Zhang, B.: Vehicle type classification and attribute prediction using multitask RCNN. In: 2016 9th International Congress on Image and Signal Processing, BioMedical Engineering and Informatics (CISPBMEI), pp. 564–569. IEEE, October 2016

20. Şeker, A., Diri, B., Balık, H.H.: Derin öğrenme yöntemleri ve uygulamaları hakkında bir inceleme. Gazi Mühendislik Bilimleri Dergisi (GMBD) 3(3), 47–64 (2017)

21. Tümen, V., Söylemez, Ö.F., Ergen, B.: Facial emotion recognition on a dataset using convolutional neural network. In: 2017 International Artificial Intelligence and Data Processing Symposium (IDAP), pp. 1–5. IEEE, September 2017

22. Bilgen, I., Saraç, Ö.S.: Gene regulatory network inference from gene expression dataset using autoencoder. In: 2018 26th Signal Processing and Communications Applications Conference (SIU), pp. 1–4. IEEE, May 2018

23. Beşer, F., Kizrak, M.A., Bolat, B., Yildirim, T.: Recognition of sign language using capsule networks. In: 2018 26th Signal Processing and Communications Applications Conference (SIU), pp. 1–4. IEEE, May 2018

24. Gülcü, A., Kuş, Z.: Konvolüsyonel Sinir Ağlarında Hiper-Parametre Optimizasyonu Yöntemlerinin İncelenmesi. Gazi Üniversitesi Fen Bilimleri Dergisi Part C: Tasarım ve Teknoloji 7(2), 503–522 (2019)

25. Ruder, S.: An overview of gradient descent optimization algorithms. arXiv preprint arXiv: 1609.04747(2016)

26. Çarkacı, N.: Derin Öğrenme Uygulamalarında En Sık Kullanılan Hiperparametreler (2018). Erişim Tarihi 29 May 2020. https://medium.com/deeplearning-turkiye/derin-ogrenme-uygulamalarinda-en-sikkullanilan-hiper-parametrele r-ece8e9125c4

27. Kutlu, N.: Biyoistatistik Temelli Bilimsel Araştırmalarda Derin Öğrenme Uygulamaları, Yüksek Lisans Tezi, Lefkoşa (2018)

28. Kurt, F.: Evrişimli Sinir Ağlarında Hiper Parametrelerin Etkisinin İncelenmesi, Yüksek Lisans Tezi (2018)

29. Tan, Z.: Derin Öğrenme Yardımıyla Araç Sınıflandırma, Yüksek Lisans Tezi (2019)

30. Šimundić, A.M.: Measures of diagnostic accuracy: basic definitions. Ejifcc 19(4), 203 (2009)

31. Zhu, W., Zeng, N., Wang, N.: Sensitivity, specificity, accuracy, associated confidence interval and ROC analysis with practical SAS implementations. In: NESUG Proceedings: Health Care and Life Sciences, Baltimore, Maryland, vol. 19, p. 67 (2010)

32. Lalkhen, A.G., McCluskey, A.: Clinical tests: sensitivity and specificity. Continuing Educ. Anaesthesia Critical Care Pain 8(6), 221–223 (2008)

33. Eusebi, P.: Diagnostic accuracy measures. Cerebrovascular Diseases 36(4), 267–272 (2013)

34. Chicco, D., Jurman, G.: The advantages of the Matthews correlation coefficient (MCC) over F1 score and accuracy in binary classification evaluation. BMC Genomics 21(1), 6 (2020)

35. Taş, B.: Roc Eğrisi ve Eğri Altında Kalan Alan (Auc) (2019). Erişim Tarihi 29 May 2020. https://medium.com/@bernatas/roc-e%C4%9Frisi-vee%C4%9Fri-alt%C4%B1nda-kalan-alan-auc-97b058e8e0cf

# Prediction of Heat-Treated Spruce Wood Surface Roughness with Artificial Neural Network and Random Forest Algorithm

Şemsettin Kilinçarslan[1]([✉]), Yasemin Şimşek Türker[1], and Murat İnce[2]

[1] Civil Engineering Department, Engineering Faculty, Suleyman Demirel University, Isparta,
Turkey
semsettinkilincarslan@sdu.edu.tr
[2] Isparta Vocational School of Technical Sciences, Isparta University of Applied Sciences,
Isparta, Turkey
muratince@isparta.edu.tr

**Abstract.** Wood material has a wide range of uses due to its many positive properties. In addition to its positive features, it also has negative features that limit the usage area of wooden material. One of the commonly used methods to minimize these negative properties is heat treatment application. In the study, the surface roughness values of Spruce (*Picea abies*) samples heat treated with ThermoWood method were investigated. Surface roughness measurements were carried out in the radial and tangential directions with the Mitutoyo SJ-201M tactile surface roughness tester. Then, the contact angle values of the samples in the tangential and radial direction were determined. TS 4084 standard was used to determine the swelling and shrinkage amounts of the samples whose contact angle values were determined. Surface roughness values of the samples were estimated by artificial neural network (ANN) and random forest algorithm. In the estimation of contact angle with random forest algorithm and ANN method, swelling and shrinkage amounts were entered as input. In the study, it has been determined that the predictions made in the radial direction with artificial neural networks give the most accurate results. In predictions made in the radial direction with artificial neural networks, $R^2 = 0.98$ and RMSE = 0.03. In the radial study conducted with Random Forest Algorithm, $R^2 = 0.96$ and RMSE = 0.11. As a result, it has been determined that the surface roughness of a wood material can be estimated by ANN and Random forest algorithm.

**Keywords:** Heat treatment · Spruce · Surface roughness · Artificial neural network · Random forest algorithm

## 1 Introduction

Wood material has been used in many fields since the first years of human beings [1, 2]. Until today, the wood material, which meets the raw material need for fuel, shelter, tool making and gun production, has played an important role in the advances made by

J. Hemanth et al. (Eds.): ICAIAME 2020, LNDECT 76, pp. 439–445, 2021.
https://doi.org/10.1007/978-3-030-79357-9_43

human beings in the technological field [3]. Despite the continuous decrease of forest areas, the increase in per capita consumption makes it necessary for the useful life of the wood to be longer. It is the most common method to cover the material with various layers of protective top surface in order to ensure that the wood material surface will be durable for longer periods against adverse environmental conditions [4].

Since it has a significant effect on the adhesion resistance, surface roughness must be determined [5]. Roughness characterizes minor irregularities occurring on a machined surface under the influence of the production process or material properties [6]. The surface roughness of the wood is one of the most important criteria in determining the quality of the final product. Therefore, it is important to consider the parameters related to processing conditions and wood properties to improve the surface quality of the final product [7].

Exposure of wood material to nitrogen or similar inert gas or normal atmosphere in a certain period of time between 100 °C–250 °C is called heat treatment and is accepted as a wood modification method [8]. The main purpose of the heat treatment is to provide dimensional stabilization in wood material and to increase wood durability without using any protective chemicals. The main effect of heat treatment on the physical properties of wood material is that it reduces the amount of moisture in the balance [9–12]. Jamsa and Vitaniiemi (2001) [13] explain the decrease in the balance of moisture in the heat-treated wood by being able to absorb fewer water molecules by chemical changes that cause a decrease in hydroxyl groups during heat treatment. In the study, it was aimed to estimate the surface roughness values of Spruce (*Picea abies*) samples heat treated with ThermoWood method with artificial neural networks and random forest algorithm with high accuracy.

## 2 Material and Method

### 2.1 Provision of Experiment Samples and Conducting Experiments

Spruce (*Picea abies*) wood obtained from Antalya dealership of Naswood Ltd. Sti., a manufacturer of ThermoWood®, was used as material in the study. According to TS 2471, the samples were kept in the air conditioning cabinet with a temperature of $20 \pm 2$ C and it is ensured that their moisture is approximately $12 \pm 0,5\%$. Mitutoyo SJ-201M tactile surface roughness tester device was used to determine the surface roughness values of the samples (Fig. 1).

Radial and tangent directions of Ra values of all test samples used in the study were determined. Contact angle, expansion and contraction amount values were determined to estimate surface roughness values. Therefore, the contact angles of the samples whose surface roughness values were determined, were determined by performing dynamic wetting test. Samples were left in the air-conditioning cabinet at $20 \pm 2°$ C and $65 \pm 5\%$ relative humidity until they were unchanged. Drops of 5 μl of pure water drawn into a syringe were dropped onto the surfaces of the samples and after the dropping process, the image of the drop on the surface was taken at the end of the 30th second [14, 15]. Experiments were performed on the tangent and radial surfaces of each sample and images were taken. In order to determine the contact angles over the drop images, the "Image J" image analysis program was used.

**Fig. 1.** Surface roughness tester

Experiments for determining the contraction amounts were carried out based on the TS 4083 standard. Until the weight of the test samples have remained unchanged, the samples have been kept in pure water at $20 \pm 5$ °C and the unchanged samples have been measured. The samples, whose measurements are finished, are kept under room conditions until the moisture content is approximately 12–15%. Then, the samples were dried in oven at $103 \pm 2$ °C until they were unchanged and their measurements were made. TS 4084 standard was used to determine the amount of expansion. Samples were first kept at $103 \pm 2$ °C until they reached constant sizes and their measurements were made. Then, the samples were kept in distilled water at $20 \pm 5$ °C until they were unchanged and their measurements were made.

## 2.2 Machine Learning Methods Used for Predictive Purposes

Machine learning is a model that allows predicting new data using mathematical relationships from existing data [16]. Machine learning, which is one of the subfields of artificial intelligence, is divided into three as Supervised, Unsupervised and Reinforcement Learning [17]. Supervised learning; It aims to find correlations by using the signed data input and output values and to generate outputs to new input values that the system has never seen using these relations [18]. Classification and curve fitting-regression algorithms are examples of this learning method. Unsupervised learning; It aims to find unknown structures from unmarked data entries [19]. Clustering is the most known method of unsupervised learning. Reinforced learning is; It is a learning method that includes goal-reward-action stages and is inspired by behavioral psychology and is used in areas such as robot navigation, real-time decision making [20]. In machine learning methods, algorithms such as artificial neural networks, random forest algorithm, decision trees and support vector machine are used [21].

In this study, artificial neural networks and random forest algorithm are used as supervised learning method. Artificial neural networks have been developed inspired by the human nervous system [22]. ANN tries to predict the result of new data that are not shown to system by learning (training) [23]. For this purpose, it uses the neurons and activation function in the layers. In this study, radial and tangent belonging to spruce tree; radial and tangential surface roughness was tried to be estimated by using contact angle, expansion and contraction amounts. For this purpose, two hidden layered artificial neural networks with 6 inputs, 2 outputs and 10 neurons were used. Training rate was taken as 0.5 and sigmoid was used as activation function.

The random forest algorithm consists of many randomly generated decision trees. This algorithm is a method that can perform both regression and classification tasks with the use of multiple decision trees and is generally known as bagging [24]. In the bagging method, it involves educating each decision tree on a different data sample that is made by changing the sampling [25]. The main idea behind this is to combine multiple decision trees in determining the final output rather than relying on individual decision trees. The output of the decision trees is chosen as the output for random forest, for which option is the majority [26]. For this reason, random forest algorithm does not have any problem such as over fitting [27]. Also, as it can work on real numerical values, it does not need normalization as in artificial neural networks. For this reason, it is a fast, simple and flexible algorithm.

### 2.3 Creating the Data Set

Surface roughness is one of the most important factors affecting top surface treatments in wood material. The longevity of the material is an important feature for its sustainability and durability. In the study, the surface roughness of the specimens of the heat treated Spruce (*Picea abies*) type was estimated according to the contact angle, expansion and contraction values. In the study, the radial and tangent of the spruce tree; 30 data were used as contact angle, expansion and contraction amounts. Of these data, 22 were used as training data and 8 as test data. These data are; tested on the same computer with artificial neural networks and random forest algorithm using the same configuration and optimization methods.

### 2.4 Performance Measurement Metrics

The effects and successes of estimation methods need to be tested. Root Mean Square Error (RMSE) is the standard deviation of the prediction errors. The regression line is a measure of how far away from data points and how far it spreads [28]. RMSE is widely used in prediction and regression analysis. It is desired to have small value. R-Square ($R^2$) shows how close the regression line is to real data values [29]. R squared is between 0 and 1, and it is desired to be close to 1. Mean Absolute Error (MAE) is calculated as the average of absolute differences between target values and estimates [30]. MAE is a linear score, that is, all individual differences are on average equal weight. It is the situation that is desired to have small value.

## 3 Results

In the study, the surface roughness of the materials was estimated according to the contact angle, expansion and contraction data of the heat treated spruce samples. Contact

**Table 1.** Heat treated spruce samples test results

|  | Contact angle ($^O$) | Surface roughnes (um) | Swelling (%) | Shrinkage (%) |
|---|---|---|---|---|
| Tangential | 79.38 | 4.55 | 4.43 | 4.43 |
| Radial | 75.92 | 5.54 | 3.42 | 3.50 |

angle, surface roughness, contraction and expansion amount values of heat treated spruce samples are given in Table 1.

In the study, artificial neural networks and random forest algorithm were used to estimate surface roughness from contact angle, contraction and expansion amount values. In Table 2, statistical analysis results of tangential and radial prediction data of heat treated spruce samples are given.

**Table 2.** Statistical analysis of forecast data of heat treated samples

|  | Artificial neural network | | Random forest algorithm | |
|---|---|---|---|---|
|  | Tangential | Radial | Tangential | Radial |
| RMSE | 0.04220632 | 0.03791579 | 0.13027175 | 0.11824636 |
| $R^2$ | 0.97161895 | 0.98679366 | 0.91466367 | 0.96441962 |
| MAE | 0.03576223 | 0.02989962 | 0.1001 | 0.1010125 |

In the surface roughness estimates made in the study, the best estimate is the tangent direction ($R^2 = 0.98$) estimation in the artificial neural network method. In both estimation methods, R2 values were higher and RMSE and MAE values were lower. Therefore, surface roughness values of heat treated wood material can be estimated by both methods.

## 4  Conclusions

In the study, the contact angle, contraction and expansion amount values of spruce wood, which are widely used in the industrial sense, were determined and surface roughness was estimated with artificial neural networks and random forest algorithm. In the heat treated samples, it was determined that the best estimate was made in the radial direction ($R^2 = 0.98$) in the estimations made by the artificial neural networks method, and the best estimate was made in the radial direction ($R^2 = 0.96$) in the estimations made by the random forest algorithm. In both estimation methods, $R^2$ values were higher and RMSE and MAE values were lower. Therefore, surface roughness values of heat treated wood material can be estimated by both methods.

Surface roughness is one of the most important factors affecting top surface treatments in wood material. The longevity of the material is an important feature for its sustainability and durability. Therefore, it is important to estimate surface roughness values with these methods used in the study.

# References

1. Kilincarslan, S., Simsek Turker, Y.: Physical-mechanical properties variationwith strengthening polymers. Acta Phys. Pol., A **137**(4), 566–568 (2020)
2. Kilincarslan, S., Simsek Turker, Y.: Evaluation in terms of sustainability of wood materials reinforced with FRP. J. Techn. Sci. **10**(1), 22–30 (2020)
3. Budakçı M.: Pnomatik Adezyon Deney Cihazı Tasarımı, Üretimi ve Ahşap Verniklerinde Denenmesi, Doktora Tezi, Gazi Üniversitesi Fen Bilimleri Enstitüsü (2003)
4. Highley, T.L., Kicle, T.K.: Biological degradation of wood. Phytopsthology **69**, 1151–1157 (1990)
5. Ratnasingam, J., Scholz, F.: Optimal surface roughness for high-quality finish on rubberwood (Hevea brasiliensis). Holz als Roh- und Werkstoff **64**, 343–345 (2006)
6. Magoss, E.: General regularities of wood surface roughness. Acta Silvatica Lignaria Hungarica **4**, 81–93 (2008)
7. Singer, H., Özşahin, Ş: Employing an analytic hierarchy process to prioritize factors influencing surface roughness of wood and wood-based materials in the sawing process. Turk. J. Agric. For. **42**, 364–371 (2018)
8. Yıldız S.: Isıl İşlem Uygulanan Doğu Kayını ve Doğu Ladini Odunlarının Fiziksel, Mekanik, Teknolojik ve Kimyasal Özellikleri, Doktora Tezi, Karadeniz Teknik Üniversitesi Fen Bilimleri Enstitüsü (2002)
9. Ates, S., Akyildiz, M.H., Özdemir, H.: Effect of heat treatment on calabrian pine (Pinus brutia Ten.) wood. BioResources **4**(3), 1032–1043 (2009)
10. Esteves, B.M., Pereira, H.M.: Wood modification by heat treatment: a review. BioResources **4**(1), 370–404 (2009)
11. Militz H.: Heat treatment of wood: European process and their background. In: International Research Group Wood Protection, Section-4 Processes, No: IRG/WP 02–40241 (2002)
12. Tjeerdsma, B.F., Boonstra, M., Pizzi, A., Tekely, P., Militz, H.: Characterisation of thermally modified wood: molecular reasons for wood performance improvement. Holz Roh-und Werkstoff **56**, 149–153 (1998)
13. Jamsa, S., Viitaniemi, P.: Heat treatment of wood – better durability without chemicals. In: Review on heat treatments of wood, COST Action E22, EUR 19885, pp. 17–22 (2001)
14. Kocaefe, D., Poncsak, S., Dore, G., Younsi, R.: Effect of heat treatment on the wettability of white ash and soft mapple by water. Holz Roh und Werkstof **66**(5), 355–361 (2008)
15. Kilincarslan, S., Simsek Turker, Y.: Determination of contact angle values of heat-treated spruce (*Picea abies*) Wood with image analysis program. Biomed. J. Sci. Tech. Res. **18**(4), 13750–13751 (2019)
16. Shrestha, D.L., Solomatine, D.P.: Machine learning approaches for estimation of prediction interval for the model output. Neural Netw. **19**(2), 225–235 (2006)
17. Sasakawa, T., Hu, J., Hirasawa, K.: A brainlike learning system with supervised, unsupervised, and reinforcement learning. Electr. Eng. Japan **162**(1), 32–39 (2008)
18. Ghosh-Dastidar, S., Adeli, H.: A new supervised learning algorithm for multiple spiking neural networks with application in epilepsy and seizure detection. Neural Netw. **22**(10), 1419–1431 (2009)
19. Sanger, T.D.: Optimal unsupervised learning in a single-layer linear feedforward neural network. Neural Netw. **2**(6), 459–473 (1989)

20. Singh, S., Jaakkola, T., Littman, M.L., Szepesvári, C.: Convergence results for single-step on-policy reinforcement-learning algorithms. Mach. Learn. **38**(3), 287–308 (2000)
21. Rodriguez-Galiano, V., Sanchez-Castillo, M., Chica-Olmo, M., Chica-Rivas, M.J.O.G.R.: Machine learning predictive models for mineral prospectivity: an evaluation of neural networks, random forest, regression trees and support vector machines. Ore Geol. Rev. **71**, 804–818 (2015)
22. Van Gerven, M., Bohte, S.: Artificial neural networks as models of neural information processing. Front. Comput. Neurosci. **11**, 114 (2017)
23. Zhang, G., Patuwo, B.E., Hu, M.Y.: Forecasting with artificial neural networks: the state of the art. Int. J. Forecast. **14**(1), 35–62 (1998)
24. Lahouar, A., Slama, J.B.H.: Day-ahead load forecast using random forest and expert input selection. Energy Convers. Manage. **103**, 1040–1051 (2015)
25. Vitorino, D., Coelho, S.T., Santos, P., Sheets, S., Jurkovac, B., Amado, C.: A random forest algorithm applied to condition-based wastewater deterioration modeling and forecasting. Procedia Eng. **89**, 401–410 (2014)
26. Belgiu, M., Drăguţ, L.: Random forest in remote sensing: a review of applications and future directions. ISPRS J. Photogramm. Remote. Sens. **114**, 24–31 (2016)
27. Li, C., Sanchez, R.V., Zurita, G., Cerrada, M., Cabrera, D., Vásquez, R.E.: Gearbox fault diagnosis based on deep random forest fusion of acoustic and vibratory signals. Mech. Syst. Signal Process. **76**, 283–293 (2016)
28. Nevitt, J., Hancock, G.R.: Improving the root mean square error of approximation for nonnormal conditions in structural equation modeling. J. Exp. Educ. **68**(3), 251–268 (2000)
29. Recchia, A.: R-squared measures for two-level hierarchical linear models using SAS. J. Stat. Softw. **32**(2), 1–9 (2010)
30. Willmott, C.J., Matsuura, K.: Advantages of the mean absolute error (MAE) OVER the root mean square error (RMSE) in assessing average model performance. Climate Res. **30**(1), 79–82 (2005)

# Design and Implementation of Microcontroller Based Hydrogen and Oxygen Generator Used Electrolysis Method

Mustafa Burunkaya[1] and Sadık Yıldız[2(✉)]

[1] Faculty of Technology, Department of Electrical and Electronics Engineering,
Gazi University, 06500 Ankara, Turkey
bmustafa@gazi.edu.tr
[2] Graduate School of Natural and Applied Sciences, Gazi University, 06500 Ankara, Turkey

**Abstract.** In this study, the parameters of *h*ydrogen and oxygen gas production used electrolysis method were determined and the automatic control of these parameters was performed. For this purpose, the different types of solutions and physical gravities which yield the most efficient product output under certain conditions were investigated. Electrolyte quantity/volume, solution concentration, solution temperature, the current and the applied voltage were determined as the control parameters. The mixture of water which is rich in hydrogen ions and sulphuric acid which increases H+ ions of solution was determined as the most suitable electrolyte. Today, since the most widely used type of energy is electrical energy, electrolysis method was preferred for the production of hydrogen. The determined parameters were controlled by the performed *microcontroller-based* control system.

**Keywords:** Electrolysis · Hydrogen · Oxygen · Microcontroller · Clean energy

## 1 Introduction

There are a lot of energy sources for use in daily life. However, if the energy is obtained from the primary sources being limited, the energy shortage will be in the near future. Fossil-based energy sources are estimated to be exhausted between 35 and 200 years [1, 2]. For these reasons a long-term, production of simple, clean and economical energy sources are needed. Some important alternatives are nuclear, solar, wind and hydrogen energy [3, 4]. Among them, the hydrogen energy is considered as the most important clean energy source in the future [4].

Hydrogen energy, whose input and the most important waste is water ($H_2O$) do not have a negative impact on the human and environmental health [2, 5]. Hydrogen which is abundant in nature is not in free form in large quantities in nature so it is not a primary energy source. For this reason, the conversion process is required to produce hydrogen by spending an amount of energy. Hydrogen could be produced using various power sources so the method to obtain hydrogen has great importance for energy efficiency [4, 6].

© The Author(s), under exclusive license to Springer Nature Switzerland AG 2021
J. Hemanth et al. (Eds.): ICAIAME 2020, LNDECT 76, pp. 446–454, 2021.
https://doi.org/10.1007/978-3-030-79357-9_44

Since today the most widely used type of energy is electrical energy, hydrogen production by electrolysis method has been used in this study. Firstly, the necessary parameters were determined for the most efficient way to perform electrolysis process. In the second phase, the electrolysis system mechanics, in which the obtained parameters could be applied, was designed. After that the production of hydrogen and oxygen gas was performed manually by using this system. Finally, the parameters of this system were controlled by the designed electronic system with feedback. The performance tests were made by comparing the measurement results of manual and automatic control system.

## 2 Electrolysis

The process by which water is decomposed hydrogen and oxygen gases by using direct current is called "electrolysis" [7]. Acid or base substances can be used as an electrolyte. These processes can be performed in ambient conditions. The required minimum voltage is 1.229 Volt (V) (25 °C). Some chemical processes occur during electrolysis of the electrolyte [8–10].

The oxygen gas is collected in tube which is connected to the positive (+) pole of power supply, the hydrogen gas is collected in tube which is connected to the negative (−) pole of power supply because the hydrogen is positive (+) sign and the oxygen is negative (−) sign. The volume of obtained hydrogen gas is twice of the volume of obtained oxygen gas. The good quality water should be used and any metal equipment should not be used to prevent corrosion [7, 9, 11]. If the current flow in the circuit and time increase, the amount of gas increases. If the voltage is constant, the resistances of the electrolyte must be reduced to increase the current.

The resistance of electrolyte depends on the pH value of electrolyte. While the PH value is 7, the resistance has largest value. If the pH value of electrolyte is reduced to 1 or, increased to 14, the resistance reduces. Acids increase the [H+] value of water and the bases increase the [OH−] value of water [11, 12].

Another result of the electrolysis process is the product of oxygen. Oxygen is the most abundant element [4, 11]. The most important compound of it is the water. The oxygen is in gas form in the air and found as the dissolved oxygen in water.

## 3 Hydrogen Energy

Hydrogen (H) is a colorless, odorless, tasteless, and transparent gas. Hydrogen is the lightest element in nature. It burns easily in the air and oxygen. The most abundant element in the universe is hydrogen and it is the 75% of all the materials' mass [13]. Hydrogen production resources are extremely abundant and diverse. Some of the most important of these are natural gas and water. Today, a lot of methods are used for hydrogen production like electrolysis, waste gas purification, radiolysis, biomass and thermo-chemical conversion [6].

Hydrogen is the most clear, high performance and the unlimited fuel. When the hydrogen is burned with pure oxygen, only the heat energy and water vapor occur. Hydrogen has the highest energy content in the known fuel per unit. When hydrogen is converted into liquid, the volume is reduced 700 times. When the hydrogen is burned with oxygen, approximately 2600 °C heat occurs [4, 5, 13, 14].

Hydrogen is not more dangerous than other fuels. The safety coefficient of fossil fuels is between 0.50–0.80. This value is around 1 for hydrogen. Properties of hydrogen, is more relevant to the characteristics of an ideal fuel. The use of hydrogen is easy and widespread. Currently, the storage of the large quantity of energy is not still possible. Most important feature of the hydrogen may be the storage feature of it. Since hydrogen could be produced everywhere, the energy transport losses are less [14, 15]. For these reasons, hydrogen is one of the most importance alternative energy sources in the future [4].

## 4    Designed Hydrogen and Oxygen Generator

The block structures of hydrogen and oxygen gas generator systems, which is used electrolysis method, with auto-controlled is given in Fig. 1.

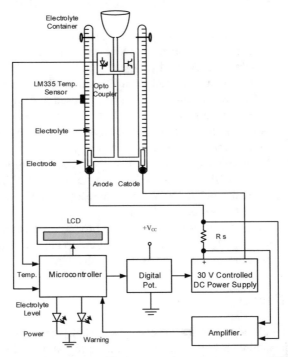

**Fig. 1.** The block structure of hydrogen and oxygen generator with feedback

### 4.1    Determination of Water Electrolysis Parameters with Experimental Analysis Method

Firstly, mechanic system was designed and performed in order that the control parameters of hydrogen and oxygen production to be determined and to be applied. The appropriate solution type and condition were researched for the efficient hydrogen production. Salt, acids and bases were added to the water to increase the conductivity and the electrolysis processes was performed. Below the results from many experimental works are summarized.

When the water was only used as the electrolyte, the hydrogen and oxygen didn't occur. Some salt was added to accelerate reaction. Since the copper electrode entered the chemical reaction with oxygen, oxidation didn't occur, it was covered with yellow layer. For this reason, the oxygen didn't occur. The used energy was calculated as 780 Joule (J). In another experiment, Copper and Chromium-Nickel electrodes were used. Sodium Carbonate (Base, $Na_2CO_3$) was used as electrolyte. In this time, hydrogen was also yielded in addition to oxygen. The used energy was calculated as 216 J.

The designed mechanic system was used for the last experiment which was performed to determine the parameters of electrolysis (Fig. 1). 160 milli liter (ml) water, 3 ml $H_2SO_4$ (sulphuric acid) solution and platinum electrodes were used [9]. Measurement results are given in Table 1.

**Table 1.** Manually electrolysis of solution consisted of water and 3 ml $H_2SO_4$ by using designed mechanic system.

| Water quantity in solution | | 160 ml |
|---|---|---|
| $H_2SO_4$ quantity in solution | | 3 ml |
| Electrode type | | Platinum |
| Temperature | | 19 °C |
| Applied voltage | Current | Time |
| 12 V | 62 mA | 11:10 |
| 12 V | 63 mA | 11:12 |
| 12 V | 61 mA | 11:13 |
| 12 V | 61 mA | 11:15 |
| 12 V | 62 mA | 11:17 |
| 12 V | 59 mA | 11:20 |
| 12 V | 56 mA | 11:25 |
| 12 V | 55 mA | 11:28 |
| 12 V | 53 mA | 11:30 |

Any chemical reaction did not occur between the acid solution and platinum electrodes. For this reason, platinum material was determined as the most suitable type of electrode. At the end of experiment, 18 ml $H_2$ and 9 ml $O_2$ were yielded. 864 J energy and 72 C electric load were used to produce 33.58 cm$^3$ $H_2$. and 16.78 cm$^3$ $O_2$.

In this experiment, sulfuric acid ($H_2SO_4$) was mixed with water then it was electrolyzed. The number of moles of hydrogen is a more then the number of moles of hydrogen as a result of electrolysis [11]. Conductivity was increased with pH value closed the 1, in other words the acid was used [9]. Thus, the more hydrogen production was yielded.

## 4.2 Microcontroller Based Control Unit

According to experimental studies accomplished in the previous section, the input parameters of the control system was determined as the amount of solution level, solution concentration, solution temperature and the feedback current (voltage). The output parameters are the applied voltage (energy) to the electrode which is proportional to measured current and the controlling of the digital potentiometer which controls this energy (Fig. 1). The output parameters of the system are hydrogen and oxygen gas.

The PIC 16F877 microcontroller (MCU) was used for the control and displaying processes of these parameters. Internal 8-channel, 10 Bit A/D converter of the MCU was used for the A/D conversion processes of the measured parameters [16]. Since the MCU was used for control processes, the flexible system controlled by software was obtained. Processes were controlled automatically, rapidly and effectively according to mechanical system is smaller than the average value (58 milli Amper (mA), 4.7 V), applied voltage of electrolyte vas increased. If this value is bigger than the value of 62 mA (5 V), it was decreased. For this purpose, a digital potentiometer (DS1869-10k, 64 steps, wiper position is maintained in the absence of power) was controlled according to comparison result [17]. The serial element (2N30055 BJT) of power supply was controlled with the obtained this proportional voltage. So, when the concentration of the solution reduces, the more voltage was applied to increase the current (Max 30 V) (Fig. 2). After these software processes were performed, delay functions in the value of 500 ms were implemented for the stability of the system. Measurement result of manual control (Table 1) and automatic control (Table 2) processes can be compared. Current fluctuations of automatic control in electrolysis processes were less than those of mechanical system.

## 4.3 Analog Measurement Circuits

Four optical receiver-transmitter units (IR diode and Opto-BJT) were placed on the glass tube being in the middle for measuring the level of solution. Since the first sensor was placed in the bottom of the tube, 10 ml measuring resolution was provided. The concentration of solution wasn't measured. The temperature of the solution was measured by using the LM335 sensor [18]. Sensor was tightly pasted on the glass tube by using silicon material with low thermal resistance.

The current flow from linear power supply changed from 38 mA to 1 A depending on the concentration of solution. The best results were obtained by using 160 ml water, 3 ml $H_2SO_4$ and the platinum electrode. So automatic control processes were carried out according to this experiment results (Table 1). Initially, voltage level of power supply is in the value of 12 V. The current of solutions (53 mA–62 mA) was converted to voltage (53 mV–62 mV) with 1 ohm ($\Omega$) resistor connected to output of power supply (Fig. 1). As the current was so small, it was not a problem that the resistance was greater than the usual. First, this output was filtered. After that, it was applied to a Amplifier, which has a gain of 81, to get the 5 V output at the biggest value of the physical parameter (current)

[19]. It was a float amplifier. Required energy for operation of it was also supplied from a floating power supply. The output voltage between 5.00 V and 4.293 V was applied to A/D input of MCU. If the measured value.

**Table 2.** Automatic control of electrolysis processes of solution used water and $H_2SO_4$

| Water quantity in solution | | 160 ml |
| H$_2$SO$_4$ quantity in solution | | 3 ml |
| Electrode type | | Platinum |
| Temperature | | 19 °C |
| Applied voltage | Current | Time |
| 12 V | 63 mA | 11:55 |
| Auto | 62 mA | 11:57 |
| Auto | 62 mA | 12:00 |
| Auto | 61 mA | 12:11 |
| Auto | Break | |
| Auto | 62 mA | 12:51 |
| Auto | 61 mA | 12:53 |
| Auto | 61 mA | 12:55 |
| Auto | 60 mA | 12:57 |
| Auto | 60 mA | 13:10 |
| Auto | 61 mA | 13:15 |
| Auto | 59 mA | 13:50 |

## 4.4  Software

The CCS C Compiler was used for software development, Microchip MPLAB PM3 programmer was used for programming. A/D interrupts were used to avoid unnecessary delays during the conversion process. Watchdog timer (WDT) was used to prevent the software locks. Flow chart of the software which important steps were shown was given in Fig. 2.

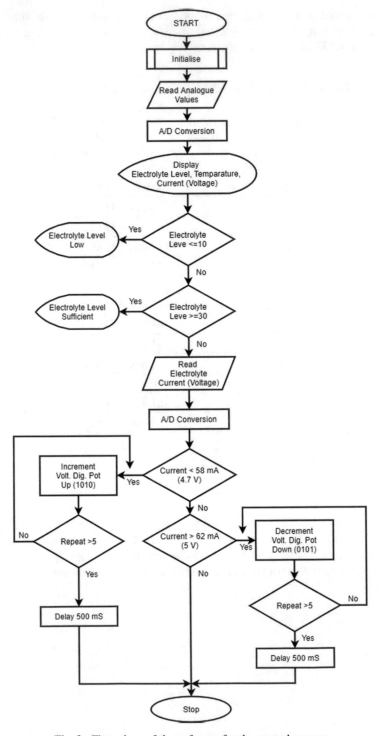

**Fig. 2.** Flow chart of the software for the control process

# 5 Results and Suggestions

Hydrogen energy which is an unlimited, renewable the clean energy source, may be the solution to currently emerging energy needs. Since today the most widely used type of energy is electrical energy, hydrogen was produced by electrolysis method in this study. The parameters which affect the efficiency of hydrogen production were researched and these parameters control was performed by MCU based control system with feedback. So processes were controlled automatically, rapidly and effectively according to mechanical system. The sulfuric acid ($H_2SO_4$) which reduce the electrical resistance of water and increase H+ ions in water was used to produce more hydrogen. As platinum electrode was used, oxidation and corrosion at least were on the electrodes. Since the solution concentration was low, electrolysis current was in the range of 53–62 milli volt (mV). The resulting noise problem was reduced. If the concentration is higher, this problem will not be occurred. In this case, the measuring resistance value should be lower than 1 $\Omega$. If this range is adjusted by employing Zero and Span circuits, the resolution will be higher. Electrolyte level was measured in the value of 10 cm resolution. This value can be increased easily by increasing the number of sensors. If the pH meter is designed for measuring the concentration, the diminishing electrolyte can be put in place by using a mechanical system. If the control functions of the MCU become more complex, its' performance will decrease. This problem should be solved by using dsPIC controller which has signal processing feature.

Electrical energy is used in the electrolysis processes. It is extremely important that which source of energy is used for these purposes. Required energy is available in natural or unnatural sources in large quantities as a waste. Such as solar, wind, automobile etc. These energies can be converted into electrical energy. Then it can be applied to a similar system which is investigated in this study.

# References

1. Sookananta, B., Galloway, S., Burt, G.M., McDonald, J.R.: The placement of facts devices in modern electrical network. In: Proceedings of the 41st International Universities Power Engineering Conference, Newcastle-upon-Tyne, pp. 780–784 (2006). https://doi.org/10.1109/UPEC.2006.367585
2. Ying, L., Yin, C., Yuan, R., Yong, H.: Economic incentive mechanism of renewable energy generation. In: 2008 International Conference on Electrical Machines and Systems, Wuhan, pp. 2689–2694 (2008)
3. Chompoo-inwai, C., Banjongjit, S., Leelajindakrairerk, M., Faungfoo, P., Lee, W.: Design optimization of wind power planning for a country of low-medium wind speed profile. In: 2007 IEEE/IAS Industrial & Commercial Power Systems Technical Conference, Edmonton, Alberta, pp. 1–6 (2007). https://doi.org/10.1109/ICPS.2007.4292093
4. Wyczalek, F.A., Suh, C.M.: The end of petroleum. In: 2002 37th Intersociety Energy Conversion Engineering Conference, IECEC 2002, Washington, DC, USA, pp. 775–781 (2002). https://doi.org/10.1109/IECEC.2002.1392148
5. Lymberopoulos, N.: Hydrogen from renewables. In: Sheffield, J.W., Sheffield, Ç. (eds.) Assessment of Hydrogen Energy for Sustainable Development. NATO Science for Peace and Security Series C: Environmental Security. Springer, Dordrecht (2007). https://doi.org/10.1007/978-1-4020-6442-5_4

6. Uchida, H.-H., Kato, S., Suga, M.: Environmental impact of energy distribution using hydrogen. In: Proceedings Second International Symposium on Environmentally Conscious Design and Inverse Manufacturing, Tokyo, Japan, pp. 1124–1127 (2001). https://doi.org/10.1109/ECODIM.2001.992536

7. Petrucci, H., Harwood, S.W., Herring, F.G., Atasoy, B., et al.: Genel Kimya (Cilt 1). Palme Yayıncılık, Ankara (2002)

8. Teschke, O.: Theory and operation of a steady-state pH differential water electrolysis cell. J. Appl. Electrochem. **12**, 219–223 (1982). https://doi.org/10.1007/BF00616904

9. Kostin, V.I., Fateev, V.N., Bokach, D.A., et al.: Hydrogen and sulfuric acid production by electrolysis with anodic depolarization by sulfurous anhydride. Chem. Petrol. Eng. **44**, 121–127 (2008). https://doi.org/10.1007/s10556-008-9022-x

10. Lee, H.H., Yang, J.W.: A new method to control electrolytes pH by circulation system in electrokinetic soil remediation. J. Hazard Mater. **77**(1–3), 227–240 (2000). https://doi.org/10.1016/s0304-3894(00)00251-x. PMID: 10946130

11. The Editors of Encyclopaedia Britannica: Electrolysis (06 Feb 2020). https://www.britannica.com/science/electrolysis

12. Wolfberg, C.J.: pH meets electrolysis: the test. California State Science Fair 2006 Project Summary. http://csef.usc.edu/History/2006/Projects/J0538.pdf

13. Hydrogen. https://www.eia.gov/kids/energy-sources/hydrogen/

14. Chabak, A.F., Ulyanov, A.I.: Storage and use of hydrogen. Russ. Eng. Res. **27**(4), 202–206 (2007)

15. Kusko, A., Dedad, J.: Stored energy - short-term and long-term energy storage methods. IEEE Ind. Appl. Mag. **13**(4), 66–72 (2007). https://doi.org/10.1109/MIA.2007.4283511

16. Microchip, PIC16F87X Data Sheet 28/40-Pin 8-Bit CMOS FLASH Microcontrollers, pp. 1–3 (2009)

17. Dallas Semiconductor, DS1869, 3V Dallastat TM Electronic, Digital Rheostat, pp. 1–8 (2009)

18. National Semiconductor, LM335 Precision Temperature Sensors, pp. 1–15 (17 December 2008)

19. Linear Technology, LTC1050 Precision Zero-Drift Op-Amp with Internal Capacitor, pp. 1–16 (2009)

# ROSE: A Novel Approach for Protein Secondary Structure Prediction

Yasin Görmez[(⊠)] and Zafer Aydın

Abdullah Gül Üniversitesi, Kayseri, Turkey
{yasin.gormez,zafer.aydin}@agu.edu.tr

**Abstract.** Three-dimensional structure of protein gives important information about protein's function. Since it is time-consuming and costly to find the structure of protein by experimental methods, estimation of three-dimensional structures of proteins through computational methods has been an efficient alternative. One of the most important steps for the 3-D protein structure prediction is protein secondary structure prediction. Proteins which contain different number and sequences of amino acids may have similar structures. Thus, extracting appropriate input features has crucial importance for secondary structure prediction. In this study, a novel model, ROSE, is proposed for secondary structure prediction that obtains probability distributions as a feature vector by using two position specific scoring matrices obtained by PSIBLAST and HHblits. ROSE is a two-stage hybrid classifier that uses a one-dimensional bi-directional recurrent neural network at the first stage and a support vector machine at the second stage. It is also combined with DSPRED method, which employs dynamic Bayesian networks and a support vector machine. ROSE obtained comparable results to DSPRED in cross-validation experiments performed on a difficult benchmark and can be used as an alternative to protein secondary structure prediction.

**Keywords:** Protein secondary structure prediction · Protein structure prediction · Machine learning · Deep learning · Recurrent neural network

## 1 Introduction

The function of a protein depends on the protein tertiary structure in the structural biology. Because determining protein structure with experimental methods is costly, three-dimensional structure estimation from amino acid sequence provides an effective alternative and is one of the crucial aims in theoretical chemistry and bioinformatics. PSP is also used in drug design problems in cases where experimental methods are inadequate to detect structures of proteins to which drug molecules bind. Various structural properties of the target protein, such as secondary structure, torsion angle, solvent accessibility, are estimated before the 3D structure is predicted. Among those prediction tasks, protein secondary structure prediction (PSSP), is one of the essential steps of PSP. In PSSP, the goal is to find the corresponding secondary structure label of each amino acid in the protein sequence.

© The Author(s), under exclusive license to Springer Nature Switzerland AG 2021
J. Hemanth et al. (Eds.): ICAIAME 2020, LNDECT 76, pp. 455–464, 2021.
https://doi.org/10.1007/978-3-030-79357-9_45

For the PSSP problem, machine learning approaches have been used commonly. These methods include, artificial neural networks (ANN) [1, 2], dynamic Bayesian networks (DBN) [3, 4], hidden markov models (HMM) [5], support vector machines (SVM) [6, 7], k-nearest neighbor (knn) [8, 9] and several deep learning approaches [10–12].

The success of structure estimation depends on attributes extracted for training as much as classification algorithms used for training. The propensity of each amino acid to generate beta-strands or helices and rules for the formation of secondary structural elements generated the basis of secondary structure prediction methods that were firstly developed. These approaches obtained 60% accuracy in estimating 3-state secondary structures (loop, helix, strands) of an amino acid sequence of a protein. Thanks to the feature vectors derived using the multiple alignment methods in the form of position specific scoring matrices (PSSM), the prediction success rate reached to 80–82% [3, 13]. Although the PSI-BLAST algorithm is frequently used for multiple alignments, it has been discovered that employing features extracted using the HHblits algorithm also increases the prediction success rate [3, 15]. In addition to deriving input features using sequence alignment algorithms, it is also possible to compute distribution scores that represent the propensities of secondary structure elements for each amino acid of the target and use these as input features in prediction methods. These scores can be in the form of structural profiles derived by aligning the target to template proteins with known structures or they can be obtained as the outputs of prediction methods. When structural profiles were used as an additional feature matrix, the estimation accuracy meaningfully increased to 84–85% [14, 16, 17].

Among the few methods that employ various forms of feature representations DSPRED was developed as a two-stage classifier that includes dynamic Bayesian networks (DBN) as time-series estimators and a support vector machine [3]. In this study, ROSE method is proposed for PSSP, which replaces the dynamic Bayesian networks of DSPRED with one dimensional bidirectional recurrent neural networks (1D-BRNN) [18]. BRNN is also a time-series model that obtained state-of-the-art results in secondary structure prediction. In the second step of ROSE, a SVM is used similar to DSPRED. In addition, a hybrid model CRD is also developed that employs both the 1D-BRNN and DBN models which are followed by an SVM.

## 2    Methods

### 2.1    Problem Definition and Secondary Structure Labels

Once the 3D structures of proteins are obtained by experimental methods, a program called DSSP is used to assign a secondary structure element of each amino acid [19]. This is the labeling process. DSSP uses a total of 8 class labels to characterize secondary structure elements, which are listed in the first column of Table 1. Since there are three helix types (H, G, and I) and two strand types, the 8-state label representation can be mapped to 3-states (Table 1), which can be sufficient for structure prediction tasks. Once the true labels are assigned by DSSP, protein data sets can be formed, which can be used to train and test machine learning prediction methods.

**Table 1.** Mapping the 8-state representation of secondary structure to 3-states

| 8-state labels | 3-state labels |
|---|---|
| "G", "I", "H" | H |
| "B", "E" | E |
| "", "T", "S" | L |

The secondary structure prediction is the identification of the secondary structural elements starting from the sequence information of the proteins. The aim of PSSP is to assign a secondary structural element (i.e. class label) to each amino acid. There are two versions of secondary structure prediction. The first one uses the 8-state label representation and the second one uses the 3-state labels in Table 1. In this study, we develop secondary structure prediction methods for the 3-state representation (Fig. 1). For PSSP, generally supervised machine learning approaches is used, in which observed secondary structure labels are used while training the model. After training step, unknown secondary structure elements are estimated using trained model.

**Fig. 1.** Estimation of three-state secondary structure. Amino acid sequence is indicated in the first line and secondary structure labels are in the second line

## 2.2 Predicting Secondary Structure Using ROSE

The proposed novel PSSP method, ROSE, can be used as an alternative to DSPRED which is proposed by Aydin et al. [3]. The ROSE method is a two-stage classifier that includes one dimensional bidirectional recurrent neural networks (1D-BRNN) and a SVM classifier. A discrete 1D-BRNN is trained for each (PSSM) produced by PSI-BLAST [20] and the first step of HHblits [21]. Then the predictions are combined with a structural profile matrix (SPM) obtained from the second stage of the HHblits method and the combined matrix will be the input of SVM classifier. Figure 2 illustrates the steps of ROSE.

In Fig. 2, Distribution 1 represents the predicted probability distributions for secondary structure labels and is calculated using PSI-BLAST (PSSM), Distribution 2 represents the probability distributions calculated using HHblits PSSM that is produced in the first phase of HHblits. The PSSM matrices are also referred to as sequence profile matrices. SPM is produced in the second phase of HHBlits by aligning the HMM-profile matrix of the target protein to the HMM-profile matrices of template proteins in PDB [22] and normalizing the frequencies of the label information of PDB proteins. In PSSP problem, the dimension of PSSM matrices are L * 20 (L represents the number of amino

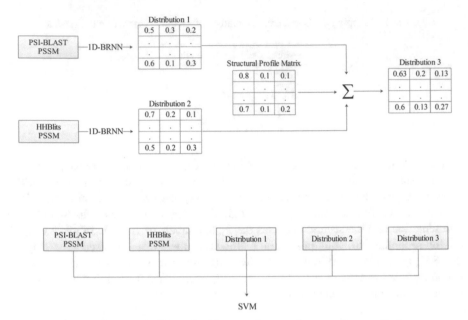

**Fig. 2.** Steps of ROSE method for protein secondary structure prediction

acids in the target protein, and 20 represents the number of different amino acid types) and dimension of distributions and structural profile matrices are L * 3 (3 represents the number of classes). Detailed description for computing the sequence and structural profiles can be found in Aydin et al. [3, 16]. The target amino acid's secondary structure label is correlated to the label of the surrounding amino acids due to chemical interactions. For this purpose, a symmetrical window is used with length 11 to generate the feature vectors for the SVM and central amino acid's secondary structure label is estimated. As a result, 539 features are extracted for each amino acid (11 * 20 = 220 for PSI-BLAST PSSM, 11 * 20 = 220 for HHblits PSSM, 11 * 3 = 33 for Distribution 1, 11 * 3 = 33 for Distribution 2, and 11 * 3 = 33 for Distribution 3).

### 2.3    Combining 1D-BRNN and Dynamic Bayesian Networks

In a second method (called CRD), distributions calculated using 1D-BRNNs are combined with the distributions calculated using dynamic Bayesian networks (DBN), which are sent to SVM as feature vectors. Figure 3 summarizes the steps of the method that combines 1D-BRNNs and DBNs.

In Fig. 2, Distribution 1 represents the predicted probability scores calculated using PSI-BLAST PSSM features and 1D-BRNN, Distribution 2 is predicted using PSI-BLAST PSSM and dynamic Bayesian network (DBN), Distribution 3 is predicted using HHblits PSSM and 1D-BRNN, Distribution 4 is predicted using HHblits PSSM and DBN, Distribution 5 is computed as the average of Distribution 1 and 2, Distribution 6 is the average of Distribution and 4, Distribution 7 is the average of Distributions 5 and 6. Each DBN block implements two DBN models (DBN-past and DBN-future),

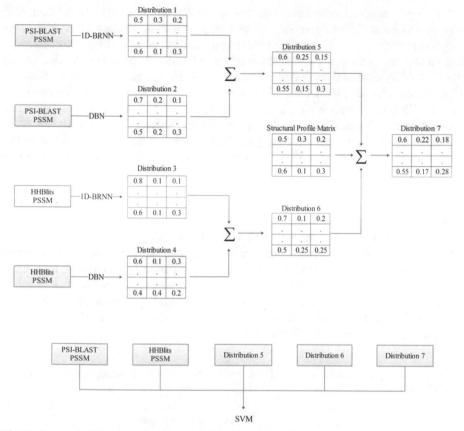

**Fig. 3.** Steps of CRD that combines 1D-BRNNs with DBNs for protein secondary structure prediction

which use a one-sided window for selecting the PSSM columns. The DBN-past uses PSSM columns for the current amino acid position and for the positions in the sequence that precede the target amino acid. The DBN-future uses PSSM columns for the current amino acid and amino acids that come after the current amino acid [3, 16]. For each type of PSSM, the predicted probability distributions generated by taking mean of two distributions produced by DBN-past and DBN-future models to compute the result of the corresponding DBN block in Fig. 2 (i.e. to compute Distribution 2 and Distribution 4). For simplicity, DBN-past and DBN-future models are not shown in this figure. For the SVM model, a symmetrical window of length 11 is used similar to ROSE.

## 3  Application

In this study, CB513 [23] benchmark data set is used for evaluating the accuracy of PSSP methods. CB513 contains 513 proteins and 84119 amino acids. A 7-fold cross-validation experiment is performed on this dataset. For each fold, proteins in the train and test sets are randomly assigned. In the next step, 30% of proteins are randomly selected from each train set to form optimization sets for optimizing the hyper-parameters. Then, 30% of the proteins in each optimization set are selected randomly to generate test set for optimization and the rest of the proteins are used to generate train set for optimization. To compute the probability distributions that are used by SVM, an internal 2-fold cross-validation is performed on each train set of the 7-fold cross-validation using 1D-BRNN and DBN models. Then these models are trained on each train set of 7-fold cross-validation and probability distributions are computed on the corresponding test set. Finally, the SVM model is trained on the train sets (using the distributions predicted by 1D-BRNN and/or DBN models as well as the PSSM features) and predictions are computed on test sets (again separately for each fold). Accuracy, precision, recall, SOV, and Matthew's correlation coefficient [24] are used as the performance measures. The 1D-BRNN models are implemented using the 1D-BRNN software [25] and DBN models using the GMTK software [26]. A one-versus-one support vector machine with RBF kernel is used as the classifier which is implemented by libSVM [27] and all preprocessing is done using Python. For all libSVM experiments, C parameter is configured to 1, and Gamma parameter to 0.00781 which were found as an optimum parameters by Aydin et al. for CB513 [3]. The parameters of 1D-BRNN and DBN models are explained below.

Table 2 shows the accuracy measures of DSPRED for 7-fold cross-validation experiment on CB513 data set. In DBN models, the one-sided PSSM window is set to 5 and the secondary structure label window to 4.

**Table 2.** Accuracy measures of DSPRED evaluated by 7-fold cross validation experiment on CB513 data set

|        | Recall 'L' | Recall 'H' | Recall 'E' | Accuracy |
|--------|------------|------------|------------|----------|
| Fold-1 | 0.837      | 0.671      | 0.860      | 0.804    |
| Fold-2 | 0.818      | 0.711      | 0.850      | 0.805    |
| Fold-3 | 0.824      | 0.698      | 0.871      | 0.814    |
| Fold-4 | 0.828      | 0.710      | 0.853      | 0.812    |
| Fold-5 | 0.812      | 0.741      | 0.826      | 0.802    |
| Fold-6 | 0.821      | 0.720      | 0.853      | 0.814    |
| Fold-7 | 0.853      | 0.719      | 0.855      | 0.828    |

In the second experiment, the performance of the ROSE method is evaluated. For this purpose, 1D-BRNN models are optimized using optimization datasets by performing grid search. A separate optimization is performed for each fold of 7-fold cross-validation. Due to time constraints, only the number of neurons in the hidden layer is optimized by taking values from 30 to 250 with increments of 5. The optimum number of hidden neurons are obtained as 125, 95, 225, 195, 140, 85, and 105 for folds 1–7, respectively. The initial learning rate is set to 0.5, the maximum number of epochs to 10,000, and the mini-batch size to 1. The one-sided window sizes are set to 7 (forward and backward windows including context_fwd, context_bwd, outputs_fwd, outputs_bwd, hidden_fwd, and hidden_bwd) and the remaining hyper-parameters are set to their default values. The 1D-BRNN tool used in this study selects a validation set by randomly sampling from the train set and performs learning rate decay by dividing the learning rate by 2 if the performance on validation set does not improve for 10 epochs. When the learning rate is sufficiently close to zero, the training stops. After optimizing the number of hidden neurons, 1D-BRNN models are trained using train sets and estimates are made using test sets as explained at the beginning of this section. Finally, the estimations are computed by the SVM. The accuracies of different metrics for the ROSE for 7-fold cross-validation on CB513 dataset are shown in Table 3.

**Table 3.** Accuracy measures of rose evaluated by 7-fold cross validation experiment on CB513 data set

|        | Recall 'L' | Recall 'H' | Recall 'E' | Accuracy |
|--------|-----------|-----------|-----------|----------|
| Fold-1 | 0.831 | 0.667 | 0.854 | 0.803 |
| Fold-2 | 0.831 | 0.712 | 0.834 | 0.804 |
| Fold-3 | 0.823 | 0.714 | 0.860 | 0.809 |
| Fold-4 | 0.821 | 0.723 | 0.841 | 0.813 |
| Fold-5 | 0.805 | 0.720 | 0.816 | 0.801 |
| Fold-6 | 0.726 | 0.690 | 0.847 | 0.808 |
| Fold-7 | 0.821 | 0.703 | 0.858 | 0.824 |

In the third experiment, the CRD method is used, which combines 1D-BRNNs, DBNs and an SVM as described in Sect. 2.3. The accuracies of different metric for 7-fold cross-validation on CB513 data set are shown in Table 4.

**Table 4.** Accuracy measures of CRD evaluated by 7-fold cross validation experiment on CB513 data set

|        | Recall 'L' | Recall 'H' | Recall 'E' | Accuracy |
|--------|-----------|-----------|-----------|----------|
| Fold-1 | 0.843 | 0.686 | 0.860 | 0.810 |
| Fold-2 | 0.838 | 0.732 | 0.844 | 0.814 |
| Fold-3 | 0.826 | 0.721 | 0.864 | 0.817 |
| Fold-4 | 0.834 | 0.727 | 0.855 | 0.810 |
| Fold-5 | 0.819 | 0.750 | 0.827 | 0.807 |
| Fold-6 | 0.826 | 0.725 | 0.849 | 0.815 |
| Fold-7 | 0.848 | 0.728 | 0.868 | 0.833 |

Mean segment overlap (SOV) score, mean accuracy (acc), mean Matthew's correlation coefficient (MCC), and standard deviation of difference between accuracy of each fold (STD) are shown in Table 5 for PSSP on CB513 data set obtained by ROSE, DSPRED and CRD methods.

**Table 5.** Mean accuracy measures evaluated by 7-fold cross validation experiment on CB513 data set

| Method | MCC for H | MCC for E | MCC for L | acc | SOV | STD |
|--------|-----------|-----------|-----------|-------|-------|--------|
| DSPRED | 0.73 | 0.67 | 0.80 | 82.78 | 79.79 | 0.0082 |
| ROSE | 0.73 | 0.66 | 0.80 | 82.23 | 79.41 | 0.0075 |
| CRD | 0.72 | 0.66 | 0.79 | 82.52 | 79.50 | 0.0095 |

Because of the random weight initialization in 1D-BRNN, ROSE may obtain different accuracy measures each time the experiments are repeated. For this reason, ROSE was run 4 times and the results were averaged. This produced accuracy values between 80.5%–82.80%. According to Table 5, the accuracy results of the three methods are comparable to each other. If hyper-parameters of the ROSE method are fully optimized including the initial learning rate and mini-batch size, its performance can further improve. In addition, for all methods, standard deviation of the accuracy values obtained for each fold is on acceptable grounds. This shows that the accuracy evaluations are performed by training and testing the models on large enough data sets.

## 4   Conclusions

In this study, ROSE approach is proposed for PSSP and is compared with DSPRED. In addition, distributions obtained using ROSE and DSPRED are combined to form the CRD method. ROSE has many hyper-parameters to optimize. As a future work we will optimize those as well, and train the model on different benchmark datasets with larger number of samples. ROSE contains BRNN models that use deep learning approaches. Therefore, if the sample size is increased, ROSE may perform more accurately. Since ROSE has similar results and is optimizable, it can be used as an alternative to protein secondary structure prediction.

## References

1. Pollastri, G., McLysaght, A.: Porter: a new, accurate server for protein secondary structure prediction. Bioinformatics **21**(8), 1719–1720 (2005)
2. Jones, D.T.: Protein secondary structure prediction based on position-specific scoring matrices. J. Mol. Biol. **292**(2), 195–202 (1999)
3. Aydin, Z., Singh, A., Bilmes, J., Noble, W.S.: Learning sparse models for a dynamic Bayesian network classifier of protein secondary structure. BMC Bioinform. **12**, 154 (2011)

4. Yao, X.-Q., Zhu, H., She, Z.-S.: A dynamic Bayesian network approach to protein secondary structure prediction. BMC Bioinform. **9**, 49 (2008)
5. Martin, J., Gibrat, J.-F., Rodolphe, F.: Analysis of an optimal hidden Markov model for secondary structure prediction. BMC Struct. Biol. **6**, 25 (2006)
6. Yang, B., Wu, Q., Ying, Z., Sui, H.: Predicting protein secondary structure using a mixed-modal SVM method in a compound pyramid model. Knowl.-Based Syst. **24**(2), 304–313 (2011)
7. Zangooei, M.H., Jalili, S.: Protein secondary structure prediction using DWKF based on SVR-NSGAII. Neurocomputing **94**, 87–101 (2012)
8. Yang, W., Wang, K., Zuo, W.: A fast and efficient nearest neighbor method for protein secondary structure prediction. In: 2011 3rd International Conference on Advanced Computer Control, pp. 224–227 (2011)
9. Salamov, A.A., Solovyev, V.V.: Prediction of protein secondary structure by combining nearest-neighbor algorithms and multiple sequence alignments. J. Mol. Biol. **247**(1), 11–15 (1995)
10. Johansen, A.R., Sønderby, C.K., Sønderby, S.K., Winther, O.: Deep recurrent conditional random field network for protein secondary prediction. In: Proceedings of the 8th ACM International Conference on Bioinformatics, Computational Biology, and Health Informatics, New York, NY, USA, pp. 73–78 (2017)
11. Pan, X., Rijnbeek, P., Yan, J., Shen, H.-B.: Prediction of RNA-protein sequence and structure binding preferences using deep convolutional and recurrent neural networks. bioRxiv, p. 146175 (June 2017)
12. Zheng, L., Li, H., Wu, N., Ao, L.: Protein secondary structure prediction based on deep learning. DEStech Trans. Eng. Technol. Res. (ismii) (2017)
13. Cheng, J., Tegge, A.N., Baldi, P.: Machine learning methods for protein structure prediction. IEEE Rev. Biomed. Eng. **1**, 41–49 (2008)
14. Li, D., Li, T., Cong, P., Xiong, W., Sun, J.: A novel structural position-specific scoring matrix for the prediction of protein secondary structures. Bioinformatics **28**(1), 32–39 (2012)
15. Aydin, Z., Baker, D., Noble, W.S.: Constructing structural profiles for protein torsion angle prediction. Presented at the 6th International Conference on Bioinformatics Models, Methods and Algorithms, BIOINFORMATICS (2015)
16. Aydin, Z., Azginoglu, N., Bilgin, H.I., Celik, M.: Developing structural profile matrices for protein secondary structure and solvent accessibility prediction. Bioinformatics **35**(20), 4004–4010 (2019)
17. Pollastri, G., Martin, A.J., Mooney, C., Vullo, A.: Accurate prediction of protein secondary structure and solvent accessibility by consensus combiners of sequence and structure information. BMC Bioinform. **8**, 201 (2007)
18. Magnan, C.N., Baldi, P.: SSpro/ACCpro 5: almost perfect prediction of protein secondary structure and relative solvent accessibility using profiles, machine learning and structural similarity. Bioinformatics **30**(18), 2592–2597 (2014)
19. Kabsch, W., Sander, C.: Dictionary of protein secondary structure: pattern recognition of hydrogen-bonded and geometrical features. Biopolymers **22**(12), 2577–2637 (1983)
20. Altschul, S.F., et al.: Gapped BLAST and PSI-BLAST: a new generation of protein database search programs. Nucleic Acids Res. **25**(17), 3389–3402 (1997)
21. Remmert, M., Biegert, A., Hauser, A., Söding, J.: HHblits: lightning-fast iterative protein sequence searching by HMM-HMM alignment. Nat. Methods **9**(2), 173 (2012)
22. RCSB Protein Data Bank - RCSB PDB (2017). https://www.rcsb.org/pdb/home/home.do
23. Cuff, J.A., Barton, G.J.: Evaluation and improvement of multiple sequence methods for protein secondary structure prediction. Proteins Struct. Funct. Bioinform. **34**(4), 508–519 (1999)
24. Precision and recall (2017). https://en.wikipedia.org/wiki/Precision_and_recall

25. 1D-BRNN. http://download.igb.uci.edu
26. GMTK. https://melodi.ee.washington.edu/gmtk/
27. LIBSVM – A Library for Support Vector Machines. https://www.csie.ntu.edu.tw/~cjlin/lib
svm/

# A Deep Learning-Based IoT Implementation for Detection of Patients' Falls in Hospitals

Hilal Koçak and Gürcan Çetin$^{(\boxtimes)}$

Department of Information Systems Engineering, Muğla Sıtkı Koçman University, Muğla,
Turkey
gcetin@mu.edu.tr

**Abstract.** Falls in hospitalized patients are a major problem for patient safety. Accidental falls are one of the most common incidents reported in hospitals. Thanks to the advances in technology, smart solutions can be developed for hospital environments as well as in all areas of life. Wearable devices, context-aware or computer vision-based systems can be designed to detect patients who fall in hospital. Internet of Things (IoT) can also be placed on wearable health products, and gathered sensors data is processed and analyzed with Machine Learning (ML) and Deep Learning (DL) algorithms. Furthermore, some DL algorithms such as LSTM are also applied to the analysis of time-series data. In this study, to minimize damage caused by falls, we've proposed a model that can achieve real-time fall detection by applying LSTM based deep learning technique on IoT sensor data. In result of the study, falling detection has been realized with 98% F1-score. Moreover, a mobile application has been successfully developed to inform caregivers about patients' fall.

**Keywords:** Deep learning · LSTM · Patients' fall · IoT

## 1 Introduction

Falling is a big problem for people with various diseases, the elderly, disabled people and even healthy people. It has been stated that the injuries caused by falls have increased by 130% recently [1]. Falling can cause loss of function, injury and psychological disturbances in people. People who experience a fall are afraid of falling again, they avoid activities they may perform, and their lives are negatively affected [2]. In the same way, falls are reported as a major patient safety problem in healthcare institutions too. Falling risk is very high for patients who are not assisted in a sickbed and who need to walk for their various needs at night.

Patients' falls may cause injuries, loss of function, prolonged hospital stay, and extra treatment costs. What's more, when they are not noticed, falls can even result in deaths. If the fall occurs in a patient room and the patient is conscious, she/he can use the nurse alert button. Otherwise, the case may not be noticed for a long time. Therefore, fall detection and notification systems are important for the health of patients.

J. Hemanth et al. (Eds.): ICAIAME 2020, LNDECT 76, pp. 465–483, 2021.
https://doi.org/10.1007/978-3-030-79357-9_46

With the development of the Internet of Things (IoT) and data science, the use of these systems become prevalent in smart home/city applications. IoT systems act as a platform to collect data from an environment using sensors, or to access and control devices from remote locations at any time. Due to low cost, its ease of use, compatibility and QoS, IoT techniques are applied in many different applications; healthcare monitoring [3], industrial [4] and home automation applications [5], smart city application [6], environmental monitoring [7] and agriculture [8]. Today, using IoT sensors placed on wearable health products, data such as falling positions and health conditions of falling patients can be obtained in real time. In this study, data from the individuals' head, chest, waist, right thigh, right wrist, right ankle and sensors, each containing a 3-axis accelerometer, gyroscope and magnetometer, were used. However, the large and complex data obtained in motion detection systems made it inevitable that data science methods such as machine learning would be used to obtain meaningful information from these data.

Machine learning is an interdisciplinary subject based on artificial intelligence, involving the subjects of statistics and computer science. Most methods of machine learning deal with a single-layer structure with at most one hidden layer and relatively less data [9]. On the other hand, deep learning, a new field of machine learning, has five, six or more hidden layers of structure and its efficient analysis of large-scale and associated datasets. For this reason, it has come to great attention from researchers in recent years [10, 11]. There are many commonly used deep learning models, such as Convolutional Neural Network (CNN) and Recurrent Neural Network (RNN). With the continued development of deep learning technology, some deep learning models are also applied to the analysis of time series data. RNN is a network model used to process sequential data. However, RNN suffers from gradient disappearance, gradient explosion and insufficient long-term memory problems. LSTM has made RNN more useful by the concept of memory. The memory concept in LSTM supports to capture the underlying data pattern and long dependency among data during the training [12].

This paper proposes a prediction algorithm based on deep LSTM architecture for patients' falling prediction. For this purpose, the patient's fall status was simulated using the 'Simulated Falls and Daily Living Activities' data [13] created by Ozdemir and Barshan [14] with the falls and daily activities of 17 male/female volunteers. However, Ozdemir's another study [15] in 2016 found that using only a waist sensor in the data set was sufficient to detect falls through machine learning methods. In the simulation, motion data taken from IoT devices based on time information was trained and falling detection was performed with 98% F1-score. When a falling event has been determined in testing phase, patients' personal and room information has been successfully reported to the hospital staffs' mobile devices. The following sections of the study are organized as follows; related works are discussed in Sect. 1.1. Material and method information is presented in Sect. 2, and detailed information about the dataset, ANN and LSTM is provided. While the developed model was explained in Sect. 3, the test results for the study were evaluated in Sect. 4.

## 1.1  Related Works

Fall detection systems utilize different tools to determine the fall. In this context, Singh et al. [16], in their study in 2019, classified fall detection systems in three categories using wearable devices, context-aware and vision-based. Depending on this classification, the data required for fall detection can be obtained from camera-based systems, sensors placed in the environment, or wearable sensors. Similarly, Villaseñor et al. [17] created a comprehensive data set by making use of ambient sensors, wearable sensors and vision devices.

Based on the principle that the vibration produced by the ground is different during the fall, the data obtained from the sensors added on the ground can be classified by machine learning methods and fall detection can be made. Rimminen et al. [18] offered a fall detection system using the Bayesian classification method from the data obtained using a floor sensor positioned on the ground and a near field image sensor. In another study, Alwan et al. [19] detected the falling by using threshold based classification via piezoelectric sensor placed on the ground.

Computer vision approaches have been applied to identify the falling events. Miguel et al. [20], in their study in 2017, using a camera and an embedded computer, they detected falling at an accuracy rate of more than 96% through computer vision algorithms such as background subtraction, Kalman filtering and optical flow. On the other hand, Rougier et al. [21] achieved 98% success by using gaussian mixture model classification on the images obtained by using four cameras.

Although the detectors placed in the environment and the studies made with vision-based systems give high success in the detection of falls, the disadvantage that it may be disturbing in terms of privacy of individuals should be taken into consideration. Therefore; the use of wearable sensors in fall detection has been a solution, due to their advantages such as being cheaper and not creating problems for people's privacy.

Since people's smartphone usage rates are very high, falling detection with data obtained from mobile devices is also considered among wearable sensors [22]. Yuwono et al. [23], determined the falling with an accuracy rate of 90% by using Discrete Wavelet Transform, Regrouping Particle Swarm Optimization, Multilayer Perceptron and Augmented Radial Basis Function Neural Networks methods. In this study, the data was obtained from the accelerometer fitted to the waist area. Leu et al. [24] detected the daily activities with 98.46% accuracy and the falling with 96.57% accuracy by using the Decision Tree algorithm with the data received through the application on the mobile phone. Studies on the use of wireless network technologies such as RFID, WSN or Wi-Fi in fall detection are also included in the literature. Chen et al. [25] proposed an intelligent RF-IDH system to detect falls caused by complications in hemodialysis patients using RF signal in their study. In the developed system, after collecting the RSSI signal information, the fall status of the patient after signal preprocessing and residual feature extraction procedures was determined by K-nearest neighbor classification, logistic regression classification, random forest classification and support vector classification algorithm. As a result of the study, it has been shown that the RF-IDH achieved an F1 score of more than 0.99 in both the cross validation phase and the post-test data evaluation phase.

In another study, Hashim et al. [26] designed a wearable fall detection system for Parkinson's disease based on a low power ZigBee Wireless Sensor Network (WSN). Patient falls were accurately detected based on the data event algorithm results of two wireless sensor nodes, an accelerometer and myoware mounted on the patient's body. The falling direction of the patient was correctly determined based on a directional drop event algorithm at the receiving node. The experimental results show that the developed model achieves 100% accuracy and sensitivity in detecting the patient's fall. On the other hand, signal interference seems to be a big problem in systems based on the signal strength received in connection with wireless devices.

The methods vary according to the characteristics of the data used in motion detection. In the study in which the efficiency of machine learning algorithms in detecting the fall situation was tested, [14] applied the k-Nearest Neighbour (k-NN), Least Squares Method (LSM), Support Vector Machines (SVM), Bayesian Decision making (BDM), Dynamic Time Warping (DTW), and Artificial Neural Networks (ANNs) algorithms to the data set obtained from the sensors worn on various parts of people's bodies. Among all tested algorithms, 99% performance was achieved in k-NN and LSM algorithms. At the another study, Micucci et al. [27] used k-NN, SVM, ANN and Random Forest (RF) algorithms in motion and fall detection. The highest success rate was achieved with SVM with an accuracy rate of 97.57%.

Unlike machine learning, Deep Neural Networks (DNN) builds a decision-making architecture based on a variety of weighted combinations of features and not on features we have identified. The use of conventional neural networks does not yield results, especially for sensor data measured over a period of time when the data cannot be labelled line by line. Deep learning methods such as RNN can be used in problems where data varies depending on time. Mutegeki and Han [28] used the CNN - LSTM method to conduct human activity detection/classification in their study in 2020. With the model developed, they achieved 99% accuracy on the iSPL dataset and 92% accuracy on the UCI HAR dataset. In another study, Alarifi and Alwadain [29] performed the fall detection with the intelligence AlexNet convolution network, and modelled system recognizes fall activities with an accuracy as high as 99.45%.

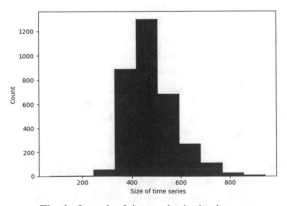

**Fig. 1.** Length of time series in the dataset.

## 2   Materials and Methods

### 2.1   Dataset

The data set used is the "Simulated Falls and Daily Living Activities" dataset, which is created by taking data from the falls and daily activities of 17 male/female volunteers through sensors [13]. In this data set, there are 36 different movements, including 20 falls and 16 daily living activities. Table 1 shows the distinct activity types in the data set [14]. The data set was created by making measurements by repeating each movement five times. The movements of the volunteers were measured with sensors that they wore on the head, chest, waist, right thigh, right wrist, right ankles, each containing 3-axis accelerometer, gyroscope and magnetometer.

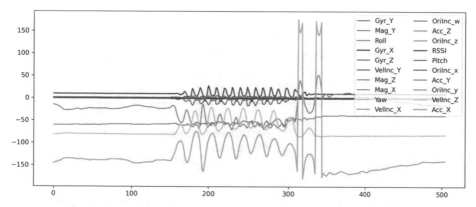

**Fig. 2.** A sample of the daily activity time series change in all columns

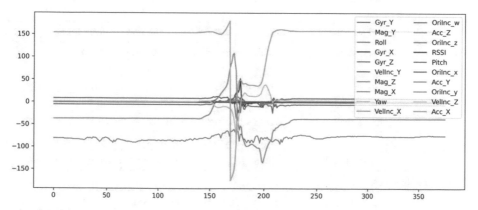

**Fig. 3.** A sample of the fall motion change time series change in all columns.

When the data set is visualized, the number of time steps of each motion measurement in the data set, the change of all columns over time for the daily activity time series and the change in all columns of the time series for the sample falling motion are given

**Table 1.** Fall and non-fall actions (ADLs) considered in this study.

**Fall actions**

| # | Label | Description |
|---|-------|-------------|
| 1 | front-lying | From vertical falling forward to the floor |
| 2 | front-protecting-lying | From vertical falling forward to the floor with arm protection |
| 3 | front-knees | From vertical falling down on the knees |
| 4 | front-knees-lying | From vertical falling down on the knees and then lying on the floor |
| 5 | front-right | From vertical falling down on the floor, ending in right lateral position |
| 6 | front-left | From vertical falling down on the floor, ending in left lateral position |
| 7 | front-quick-recovery | From vertical falling on the floor and quick recovery |
| 8 | front-slow-recovery | From vertical falling on the floor and slow recovery |
| 9 | back-sitting | From vertical falling on the floor, ending sitting |
| 10 | back-lying | From vertical falling on the floor, ending lying |
| 11 | back-right | From vertical falling on the floor, ending lying in right lateral position |
| 12 | back-left | From vertical falling on the floor, ending lying in left lateral position |
| 13 | right-sideway | From vertical falling on the floor, ending lying |
| 14 | right-recovery | From vertical falling on the floor with subsequent recovery |
| 15 | left-sideway | From vertical falling on the floor, ending lying |
| 16 | left-recovery | From vertical falling on the floor with subsequent recovery |
| 17 | syncope | From standing falling on the floor following a vertical trajectory |
| 18 | syncope-wall | From standing falling down slowly slipping on a wall |
| 19 | podium | From vertical standing on a podium going on the floor |
| 20 | rolling-out-bed | From lying, rolling out of bed and going on the floor |

**Non-fall actions (ADLs)**

| # | Label | Description |
|---|-------|-------------|
| 21 | lying-bed | From vertical lying on the bed |
| 22 | rising-bed | From lying to sitting |
| 23 | sit-bed | From vertical to sitting with a certain acceleration onto a bed (soft surface) |

*(continued)*

**Table 1.** (*continued*)

| Non-fall actions (ADLs) | | |
|---|---|---|
| 24 | sit-chair | From vertical to sitting with a certain acceleration onto a chair (hard surface) |
| 25 | sit-sofa | From vertical to sitting with a certain acceleration onto a sofa (soft surface) |
| 26 | sit-air | From vertical to sitting in the air exploiting the muscles of legs |
| 27 | walking-fw | Walking forward |
| 28 | jogging | Running |
| 29 | walking-bw | Walking backward |
| 30 | bending | Bending about 90° |
| 31 | bending-pick-up | Bending to pick up an object on the floor |
| 32 | stumble | Stumbling with recovery |
| 33 | limp | Walking with a limp |
| 34 | squatting-down | Squatting, then standing up |
| 35 | trip-over | Bending while walking and then continuing walking |
| 36 | coughing-sneezing | Coughing or sneezing |

in Fig. 1, Fig. 2 and Fig. 3 respectively. Data set consist of 23 meaningful features such as 'Counter', 'Temperature', 'Pressure', 'VelInc-X', 'VelInc-Y', 'VelInc-Z', 'Acc-X', 'Acc-Y', 'Acc-Z', 'OriInc-x', 'OriInc-y', 'OriInc-z', 'OriInc-w', 'Gyr-X', 'Gyr-Y', 'Gyr-Z', 'Mag-X', 'Mag-Y', 'Mag-Z', 'Roll', 'RSSI', 'Pitch', 'Yaw'.

The graph seen in Fig. 1 shows how many time steps each movement data from the sensors consists of. Given all of the data, the majority of the movements consist of 300 and 700 time intervals. This statistic is required for incomplete data filling or truncate operations since they must all have the same length before delivering the data to the LSTM neural network. The size of the longest time series measured is 945. All data are filled with the last time step each, so that their length is 945.

## 2.2 Artificial Neural Networks

Artificial Neural Networks (ANN) are computer programs that aim to create a humanoid decision system by modeling the nervous system and neural connections of humans. Its ability to solve linear/nonlinear problems is successfully used in many areas due to its success in prediction and classification problems. The information received at the input layer is weighted and transferred to the hidden layer by applying a differentiable activation function. Thanks to hidden layers, the neural network can also solve nonlinear problems. This process is called feed forward. If the predicted value is not correct as a result of this process, or if it is necessary to reduce the cost, feedback should be made and the weights should be updated.

Figure 4 shows the model of an ANN having 3 input units, 3 hidden layers and 4 output layer [30]. In the model, inputs are donated by the circles labeled "x". The circles labeled "+1" are called bias units, and correspond to the intercept term. These inputs is multiplied by their weight (w). Results are aggregated with threshold value and processed by "Activation Function" to find cell output. Parameters are used to adjust the effect of inputs on outputs.

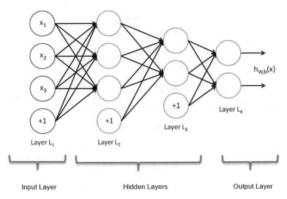

**Fig. 4.** A simple ANN model [30]

Weights are multiplied by input values and transmitted. The most critical point in a network is to calculate the optimum weight values by spreading the error according to the training set of given weights.

## 2.3 Recurrent Neural Networks

Recurrent Neural Network is an ANN model with recurrent cycles that allow keeping past information [31]. It is used in data where sequencing is important. Time series are data sets formed by grouping and expressing parameters whose boundaries change in a given time interval. Data that comes before and after each other in time series feed off each other. In order to identify and analyse the data, the information before and after itself must be kept in memory. There is no memory of ANNs inspired by the working and decision-making mechanism of the human brain. So as they begin to learn the second time series after learning a time series, they forget the first one completely and begin to adapt to the second. They are therefore incapable of learning the relationship between time series.

Nodes in recursive networks, a type of supervised learning, have connections that feed themselves from the past [32]. Thus, the layers transfer the information received from the previous layers to the next layer, and they feed themselves. And so, it is possible for the model to infer meaning from sequential data. Remembering information takes place in hidden layers. At each time step, input is received, the hidden status is updated and a prediction is made [33]. A traditional RNN model is given in Fig. 5. According to the Fig. 5, $t$ is every time step, $a<t>$ expressed by Eq. (1) is the activation, and $y<t>$

given in Eq. (2) is the output. Here; $g1$, $g2$ are activation functions.

$$a^{\langle t \rangle} = g_1(W_{aa}a^{\langle t-1 \rangle} + W_{ax}x^{\langle t \rangle} + b_a \tag{1}$$

$$y^{\langle t \rangle} = g_2\left(W_{ya}a^{\langle t \rangle} + b_y\right) \tag{2}$$

In RNNs, the loss in each time step is taken as the basis, and then the total loss in all time steps is calculated according to Eq. (3).

$$L(\hat{y}, y) = \sum_{t=1}^{T_y} L(\hat{y}^{\langle t \rangle}, y^{\langle t \rangle}) \tag{3}$$

In RNN, backpropagation through time method is used instead of backpropagation method in feed forward neural networks. Weights are updated according to the lost function according to the Eq. (4).

$$\frac{\partial L^{(T)}}{\partial W} = \sum_{t=1}^{T} \frac{\partial L^{(T)}}{\partial W}\bigg| \tag{4}$$

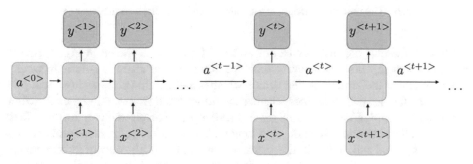

**Fig. 5.** Classical RNN architecture [34]

## 2.4 Long Short Term Memory (LSTM)

LSTM is a type of RNN with a memory cell in its structure [35]. In the recurrent neural networks, close time intervals are taken into account to predict the next step [36]. In some cases, the information that will allow the prediction may have been given in a very old time frame.

In RNN, outputs contain features from inputs that are in very close time steps to them. Therefore, the RNN is inadequate in evaluating knowledge from distant time steps [37]. The reason for this deficiency is the 'Vanishing Gradient' problem encountered in recursive networks. After the neural network is fed forward, the weights must be updated with derivative operations while the back propagation is made. To train neural network, the derivative of the activation function must be different from 0. Otherwise, the weights are not updated, which means learning process is not progress. That's why information come from the distant time steps is lost and what is meant by information here is the gradient. The another problem in RNN is 'Exploding Gradient' which is the problem of never reaching a global solution as a result of overgrowth of gradients [35]. Because of these two problems, RNN is inappropriate for problems requiring long-term memory [38]. As seen in Fig. 6, the basic RNN is a simple structure with a single tanh layer. For this reason, long-term dependencies cannot be learned. LSTM neural network with gate functions was developed to solve these problems [35].

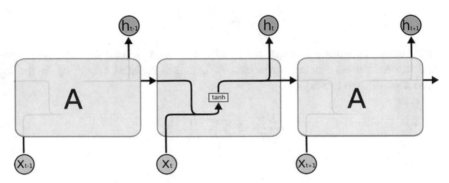

**Fig. 6.** The repeating module in a standard RNN contains a single layer [39]

Figure 7 shows the memory cell of an LSTM model. As shown in Fig. 7, LSTM has a different repetitive structure, although it has a similar chain structure. In where; the x symbol is the structures that decide whether the incoming connections will join or not, and the + symbol decides whether the information from x will be collected or not. Also, the Sigmoid layer returns values like 0 or 1 as a result of incoming data. Tanh function is the activation function used to prevent the vanishing gradient problem. By means of sigmoid and multiplication gates, which take 0 or 1, the flow of information to the neurons is controlled and the properties extracted from the previous time steps are preserved.

In the LSTM model, it is decided which information to delete by using $X_t$ and $h_t$ as input first. These operations are done by using Eq. (5) in the forget layer ($f_t$) and sigmoid is used as the activation function [40].

$$f^t = \sigma(W_f.[C_{t-1}, h_{t-1}, x_t] + b_f \tag{5}$$

Then, the input layer where new information will be determined is activated and the input gate ($i_t$) is updated with the sigmoid function using Eq. (6). Then, with Eq. (7), candidate information that will form the new information is determined by the tanh function. New information is generated using Eq. (8). Finally, in the output layer, the output data is obtained using Eq. (9)[40]

$$i_t = \sigma(W_i.[C_{t-1}, h_{t-1}, x_t] + b_i \tag{6}$$

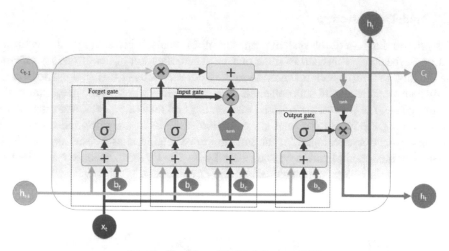

**Fig. 7.** Traditional LSTM diagram [40]

$$c_t = f_t c_{t-1} + i_t \tanh(W_{xc} x_t + W_{hc} h_{t-1} + b_c) \tag{7}$$

$$o_t = \sigma(W_o.[C_t, h_{t-1}, x_t] + b_o \tag{8}$$

$$h_t = o_t \tanh(C_t) \tag{9}$$

In the model, the weight parameters (W) and bias parameters (b) are learned by the model in a way that minimizes the difference between the actual training values and the LSTM output values.

## 3  The LSTM Model for Fall Detection

### 3.1  Data Preprocessing

The data is labeled 1/0 so that each time series represents fall and normal activity. Each move in the data set took place in a time interval of approximately 15 s. The time series forming the movements are of different lengths. Therefore, the data set is arranged so that it is the longest time series size. The missing data is filled with the last line in the time series, that is, the sensor data that arrived at the last moment. Files without sensor data have been deleted. The "Counter" column that may cause an overfitting problem in the data set, and the 'Temperature' and 'Pressure' columns that do not have a separator effect, have been deleted. Afterwards, the data set is divided as 70% train, 15% test and 15% validation. Statistical values for each time series after the data preprocessing stage are given in Table 2.

## 3.2  Model Architecture

The general diagram of the system proposed in this study is given in Fig. 8. According to Fig. 8, the study consisted of three phases. The collection of data from sensors and the creation of data, which is the first stage, is not the focus of this study. Therefore, a ready data set created with data collected from IoT devices was selected in the study. The second phase, the processing of data, visualizing and analyzing data, used the LSTM deep learning model. In the final stage, patient caregivers are informed through the mobile application developed for the fall situations identified as a result of the LSTM model. For this, an application was developed with The Firebird database using the Java programming language in the Android Studio environment.

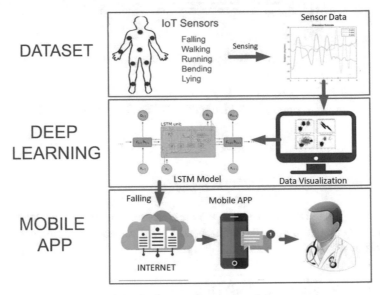

**Fig. 8.** Proposed system model.

## 3.3  The Proposed LSTM Model

The proposed study was carried out in the notebook environment of Google Collabratory running in the cloud and using the Nvidia Tesla k80 GPU. In the study, Tensorflow, an open source deep learning library developed by Google, was used for the model design required for fall detection. Tensorflow is software that enables machine learning in big data and uses data flow graphs for numerical calculations. Tensorflow library, built using [41], supports programming languages such as Python, C++, Java, C#, Javascript and R. Python programming language was used in the development of this project.

In the study, a six-layer DNN was designed using LSTM including Dense, LSTM, Dropout, and Batch Normalization layers for fall estimation. The parameters and flow process of the system to which the LSTM model is applied are shown in Fig. 9.

Performance of the LSTM algorithm, as in other machine learning algorithms, largely depends on the optimum values of the parameters. The optimum values of the parameters are obtained by adjusting different values that the algorithm provides the highest performance value. Hyperparameter refers to all parameters of a model that are not updated during the learning phase, but used to obtain the best accuracy [40].

Since the performance and loss value of the model did not change to a considerable extent after 100 epoch, the epoch 100 was selected by the trial and error method. Training parameters have been modified by manuel and updated to achieve the best results. Dropout regularization method was used in the ratio of (0,1) to prevent the model from overfitting. Droput ensures that the learning synaptic connections are forgotten at a given rate, thus preventing the memorization of the model [42]. After the layers were created, the model was trained using the Adam optimization algorithm with a learning rate of 0.001. Binary-crossentropy was chosen as the objective function, since there will be dual classification as daily life activity and falling. Moreover, the sigmoid activation function is used in the output layer. Table 3 gives the hyper parameters used for the proposed LSTM model.

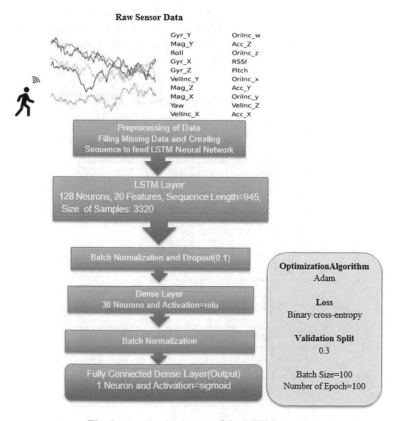

**Fig. 9.** Application steps of the LSTM model.

**Table 2.** Statistical values for each time series in dataset

| Count | 3335.000000 |
|-------|-------------|
| Mean  | 480.199100  |
| Std   | 97.116044   |
| min   | 68.000000   |
| 25%   | 412.000000  |
| 75%   | 9531.000000 |
| max   | 945.000000  |

## 3.4 Result and Discussion

Change of accuracy and loss values while training the model is given in Fig. 9 and Fig. 10. Figure 10 shows the success of training and testing, while Fig. 11 shows the average change in cost for training and testing. Since there was no significant reduction in cost and no increase in success after 100 epochs, the training was restricted to 100 epochs. The reduction of training loss throughout training indicates that the model fits firmly into the data set. Training accuracy and test accuracy values increase almost simultaneously, training loss and test loss values decrease and continue at the same time, indicates that there is no overfitting problem in the model.

**Table 3.** Hyperparameters of the LSTM model

| Sequence size | 3320 |
|---------------|------|
| Epoch | 100 |
| Batch size | 100 |
| Loss function | Binary crossentropy |
| Activation functions | relu, sigmoid |
| Optimization algorithm | adam |
| Learning rate | lr = 0.001 |
| Dropout | 0.1 |

**Table 4.** Confusion matrix of results.

|           | Predicted negatives | Predicted positives |
|-----------|---------------------|---------------------|
| Negatives | 238 true negative   | 304 false positive  |
| Positives | 4 false negative    | 7 true positive     |
| Total     | 242                 | 311                 |

In the study, to verify the performance of our LSTM model confusion matrix of the results is given in Table 4. According to the Table 4, 238 of 242 ADLs were classified as

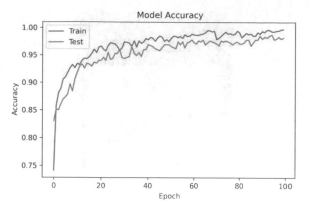

**Fig. 10.** The accuracy performance of the LSTM model.

correct and 4 drops as incorrect. In the fall data, 304 of 311 fall are classified correctly. 7 falls are classified incorrectly. That is, 10 normal life activities were classified as falling. The success rate of the model is 98%.

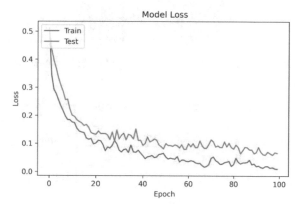

**Fig. 11.** The loss performance of the LSTM model.

The performance of the applied LSTM model was evaluated according to its Precision, Recall, and F1-score values. Precision shows how many of the values predicted as positive are actually positive. The Recall of a model expresses how much of the data to be predicted as positive is predicted as positive. The F1-score value shows the harmonic mean of the precision and Recall values. The main reason for using the F1-score value instead of Accuracy is not to make an incorrect model selection in uneven data sets. The values of Precision, Recall and F1-Score values obtained from Confusion Matrix are given in Table 5 with classification report. According to the Table 5, it is measured that Precision, Recall and the F1-Score values of the ADL and falling datas are 0.98. This shows that the model has a very high level of success.

The next phase of the study is to inform the caregivers of the movements identified as falls through the LSTM model via the mobile app. At this point, an alarm was raised over the application for movements identified as falls during the test phase scripted

**Fig. 12.** A screenshot of the developed mobile application.

**Table 5.** Classification report of the proposed model.

|          | Precision | Recall | F1-Score | Support |
|----------|-----------|--------|----------|---------|
| 0 (ADL)  | 0.98      | 0.98   | 0.98     | 242     |
| 1(FALL)  | 0.98      | 0.98   | 0.98     | 311     |

patient information including patient's room number, patient identification information and time of fall was successfully sent to the authorities. A sample screenshot of the mobile application is given in Fig. 12.

However, it should be stated here that the study is a proof of concept study. The data are processed real-time over a data set, and the patient information that is scripted into the mobile device according to the obtained information is successfully sent. The proposed model proves the usability of IoT devices and Deep learning algorithms for monitoring patients' daily activities and falls in hospitals.

## 4  Conclusion

In this study, a six-layer LSTM model, a deep learning approach, has been proposed to predict the falls of patients staying in hospital. The 'Simulated Falls and Daily Living

Activities' data set created with data collected from sensors including a 3-axis accelerometer, gyroscope and magnetometer has been used to train the proposed model and perform performance evaluation. after the data set has been preprocessed, it is divided into 3 groups: 70% train, 15% test and 15% validation. In the proposed LSTM model, training parameters have been modified by trial and error method and updated to achieve the best scores. Moreover, dropout regularization and bach normalization method have been used to made the network fast and robust in terms of speed and performance accuracy. At the result of the study, the F1-score of the model is 98%. Moreover a developed mobil application has been used to inform caregivers successfully.

# References

1. Dionyssiotis, Y.: Analyzing the problem of falls among older people. Int. J. Gen. Med. **5**, 805–813 (2012)
2. Deshpande, N., Metter, E., Lauretani, F., Bandinelli, S., Guralnik, J., Ferrucci, L.: Activity restriction induced by fear of falling and objective and subjective measures. J. Am. Geriatr. Soc. **56**(4), 615–620 (2008)
3. Akka, M., Sokullu, R., Çetin, H.: Healthcare and patient monitoring using IoT. Internet Things **11**, 1–12 (2020)
4. Mistry, I., Tanwar, S., Tyagi, S., Kumar, N.: Blockchain for 5G-enabled IoT for industrial automation: a systematic review, solutions, and challenges. Mech. Syst. Signal Process. **135**, 106382 (2020)
5. Uma, S., Eswari, R., Bhuvanya, R., Kumar, G.: IoT based voice/text controlled home appliances. Procedia Comput. Sci. **165**(232), 238 (2019)
6. Sharma, A., Singh, P., Kumar, Y.: An integrated fire detection system using IoT and image processing technique for smart cities. Sustain. Cities Soc. **61**, 102332 (2020)
7. Pasika, S., Gandla, S.T.: Smart water quality monitoring system with cost-effective using IoT. Heliyon **6**(7), 1–9 (2020)
8. Terence, S., Purushothaman, P.: Systematic review of Internet of Things in smart farming. Trans. Emerg. Tel. Tech. **31**(6), 1–34 (2020)
9. Wang, P., Fan, E., Wang, P.: Comparative analysis of image classification algorithms based on traditional machine learning and deep learning. Pattern Recognit. Lett. **141**, 61–67 (2020)
10. Islam, M., Huang, S., Ajwad, R., Chi, C., Wang, Y., Hu, P.: An integrative deep learning framework for classifying molecular subtypes of breast cancer. Comput. Struct. Biotechnol. J. **18**, 2185–2219 (2020)
11. Jauro, F., Chiroma, H., Gital, A., Almutairi, M., Abdulhamid, S., Abawajy, J.: Deep learning architectures in emerging cloud computing architectures: recent development, challenges and next research trend. Appl. Soft Comput. **96**, 106582 (2020)
12. Chen, Y.: Voltages prediction algorithm based on LSTM recurrent neural network. Optik **220**, 164869 (2020)
13. UCI Machine Learning Repository: Simulated Falls and Daily Living Activities Data Set Data Set (2018). http://archive.ics.uci.edu/ml/datasets/Simulated+Falls+and+Daily+Living+Activities+Data+Set. Accessed 12 Mar 2020
14. Ozdemir, A., Barshan, B.: Detecting falls with wearable sensors using machine learning techniques. Sensors **14**, 10691–10708 (2014)
15. Ozdemir, A.: An analysis on sensor locations of the human body for wearable fall detection devices: principles and practice. Sensors **16**(8), 10691–10708 (2016)
16. Singh, K., Rajput, A., Sharma, S.: Human fall detection using machine learning methods: a survey. Int. J. Math. Eng. Manag. Sci. **5**, 161–180 (2019)

17. Martínez-Villaseñor, L., Ponce, H., Brieva, J., Moya-Albor, E., Núñez-Martínez, J., Peñafort-Asturiano, C.: Up-fall detection dataset: a multimodal approach. Sensors **19**(9), 1–28 (2019)
18. Rimminen, H., Lindström, J., Linnavuo, M., Sepponen, R.: Detection of falls among the elderly by a floor sensor using the electric near field. IEEE Trans. Inf Technol. Biomed. **14**, 1475–1476 (2010)
19. Rougier, C., Auvinet, E., Rousseau, J., Mignotte, M., Meunier, J.: A smart and passive floor-vibration based fall detector for elderly. Toward Useful Services for Elderly and People with Disabilities, pp. 121–128 (2011)
20. De Miguel, K., Brunete, A., Hernando, M., Gambao, E.: Home camera-based fall detection system for the elderly. Sensors **17**, 1–21 (2017)
21. Alwan, M., Rajendran, P., Kell, S., et al.: fall detection from depth map video sequences. In: 2nd International Conference on Information and Communication Technologies 2006, vol. 1, pp. 1003–1007 (2006)
22. Luque, R., Casilari, E., Moron, M., Redondo, G.: Comparison and characterization of android-based fall detection systems. Sensors **14**, 18543–18574 (2014)
23. Yuwono, M., Moulton, B.D., Su, S.W., Celler, B.G., Nguyen, H.T.: Unsupervised machine-learning method for improving the performance of ambulatory fall-detection systems. Biomed. Eng. Online **11**, 1–11 (2012)
24. Fang-Yie, L., Chia-Yin, K., Yi-Chen, L., Susanto, H., Francis, Y.: Fall detection and motion classification by using decision tree on mobile phone (2017). https://doi.org/10.1016/B978-0-12-809859-2.00013-9
25. Chen, Y., Xiao, F., Huang, H., Sun, L.: RF-IDH: an intelligent fall detection system for hemodialysis patients via COTS RFID. Futur. Gener. Comput. Syst. **113**, 13–24 (2020)
26. Huda, A., Saleem, L., Gharghan, S.: Accurate fall detection for patients with Parkinson's disease based on a data event algorithm and wireless sensor nodes. Measurement **156**, 107573 (2020)
27. Micucci, D., Mobilio, M., Napoletano, P.: UniMiB SHAR: a new dataset for human activity recognition using acceleration data from smartphones. Appl. Sci. **7**, 1101 (2020)
28. Mutegeki, R., Dong, H.: A CNN-LSTM approach to human activity recognition, pp. 362–366 (2020). https://doi.org/10.1109/ICAIIC48513.2020.9065078
29. Alarifi, A., Alwadain, A.: Killer heuristic optimized convolution neural network-based fall detection with wearable IoT sensor devices. Measurement **167**, 108258 (2021)
30. Unsupervised Feature Learning and Deep Learning Tutorial. http://deeplearning.stanford.edu/tutorial/supervised/MultiLayerNeuralNetworks/. Accessed 12 Mar 2020
31. Bengio, Y., Simard, P., Frasconi, P.: Learning long-term dependencies with gradient descent is dicult. IEEE Trans. Neural Networks **5**(2), 157–166 (1994)
32. Haykin, S.: Neural Networks: A Comprehensive Foundation. Prentice Hall PTR, Hoboken (1994)
33. Sutskever, I., Martens, J., Hinton, G.: Generating text with recurrent neural networks. In: Proceedings of the 28th International Conference on Machine Learning, pp. 1017–1024 (2011)
34. Recurrent Neural Networks cheatsheet. https://stanford.edu/~shervine/teaching/cs-230/cheatsheet-recurrent-neural-networks. Accessed 12 Mar 2020
35. Schmidhuber, J., Hochreiter, S.: Long short-term memory. Neural Comput. **9**, 1735–1780 (1997)
36. Sutskever, I., Martens, J., Hinton, G.: Generating text with recurrent neural networks. In: Proceedings of the 28th International Conference on Machine Learning (2011)
37. Hochreiter, S., Bengio, Y., Frasconi, P., Schmidhuber, J.: Gradient flow in recurrent nets: the diculty of learning long-term dependencies. A Field Guide to Dynamical Recurrent Neural Networks (2001)
38. Bengio, Y., Frasconi, P., Simard, P.: Problem of learning long-term dependencies in recurrent networks. In: IEEE International Conference on Neural Networks (1993)

39. OlahLSTM - Neural Network Tutorial-15. https://web.stanford.edu/class/cs379c/archive/2018/class_messages_listing/content/Artificial_Neural_Network_Technology_Tutorials/OlahLSTM-NEURAL-NETWORK-TUTORIAL-15.Pdf. Accessed 30 Apr 2020
40. Rahman, M., Islam, D., Mukti, R.J., Saha, I.: A deep learning approach based on convolutional LSTM for detecting diabetes. Comput. Biol. Chem. **88**, 107329 (2020). https://doi.org/10.1016/j.compbiolchem.107329
41. Sang, C., Di Pierro, M.: Improving trading technical analysis with tensorflow long short-term memory (LSTM) neural network. J. Financ. Data Sci. **5**(1), 1–11 (2019)
42. Srivastava, N., Hinton, G., Krizhevsky, A., Sutskever, I., Salakhutdinov, R.: Dropout: a simple way to prevent neural networks from overfitting. J. Mach. Learn. Res. **15**(56), 1929–1958 (2014)

# Recognition of Vehicle Warning Indicators

Ali Uçar$^{(\boxtimes)}$ and Süleyman Eken

Information Systems Engineering Department, Kocaeli University, 41001 Izmit, Turkey
suleyman.eken@kocaeli.edu.tr

**Abstract.** Testing of a vehicle instrument cluster for design validation is important especially for autonomous vehicles. As autonomous vehicles require dedicated test scenarios, logging warning instrument cluster's indicators/lights and testing their functionality needs to be validated. In addition, testing instrument cluster with different vehicle ECUs, detecting delays and sensitivity are another needs. In this work, we present a camera-based system to detect lights of vehicle instrument cluster. Our goal is to detect lights of instrument cluster through camera-based system in a video-stream. Instead of manual, inspection which is carried out by human, a machine vision-based system is developed for automation of this validation task. We used a cloud system, namely Microsoft Azure Cloud, for object detection. It is used to recognize lights of the cluster. We created our own dataset and it contains 18 kinds of warning indicators and 360 images in which different combinations of indicator lights exist. We tested our model and calculated the precision, recall, and mAPas 97.3%, 82.2%, and 91.4%, respectively. Even under the limited examples of labeled data and unbalanced data set conditions, the results are promising to recognize vehicle warning indicators.

**Keywords:** Vehicle indicator lights · Vehicle instrument cluster · Warning lights · Azure · Cloud · Deep learning · Object detection

## 1 Introduction

Vehicle manufacturers place a set of lights and warning symbols on the vehicle dashboard to inform drivers. In this panel, there are many different symbols such as temperature gauge, oil, battery and brake warning lamp. The logos and figures used for vehicle indicators are universal. Warning lights and signs are designed to give the driver general information about the fault/warning.

Vehicle indicators mean information or warning. Dipped beam, main beam and fog lamps; ESP, ABS and air bag systems; warning symbols such as vehicle lights and door warning lights are for informational purposes only. Oil, battery, brake and engine fault lights indicate a malfunction occurred in the in-vehicle system area, such as engine, ignition or mechanical parts [1].

© The Author(s), under exclusive license to Springer Nature Switzerland AG 2021
J. Hemanth et al. (Eds.): ICAIAME 2020, LNDECT 76, pp. 484–490, 2021.
https://doi.org/10.1007/978-3-030-79357-9_47

In many cases it is not possible for the warning indicators to identify the problem in the vehicle in a specific way. Only the message is given that there is a problem with the functions of certain parts and that the situation poses a risk to driving safety. Solving the problem in the warning lamps is extremely important both for the vehicle and for other people inside the vehicle and in traffic [2, 3].

To the best of our knowledge, there is no study using deep learning or other conventional machine learning techniques to recognize vehicle warning indicators in the literature. However, many works have been done on Intelligent Transportation Systems (ITS) [4, 5] and Advanced Driver Assistant Systems (ADAS) [6, 7]. More specifically, vehicle detection and classification is one of the most frequently studied topics [8–11].

In this paper, vehicle warning indicator recognition, more precisely - the training of neural network, the most computing-intensive part of the software, is achieved based on the Windows Azure cloud technology. The contributions of this paper are as following:

- Building vehicle warning indicator dataset
- Training and applying the state-of-the-art, object detection system Windows Azures cloud technology for recognition

The rest of this paper is organized as follows. "Vehicle Warning Indicators" section presents the most common indicators and their meanings. Next section gives information about Windows Azure cloud platform. In "Experimental setup, results, and analysis" section, the performance results and their analyses are given. The last section is the conclusion part that brings everything together.

## 2 Vehicle Warning Light Symbols and Indicators

The following explains vehicle **warning indicators lights and symbols** on dashboard or cluster. These are almost same for most makes and models of cars, trucks, buses and motorbikes. These indicators can be broadly classified into five different symbols: warning, safety, lighting, common, and advanced feature symbols. A few examples from each class are tabulated in Table 1.

**Table 1.** Vehicle warning light symbols/indicators and their meanings [12]

| Logo | Meaning |
|---|---|
| | Engine Temperature Warning Light means that your engine is running at a temperature is higher than normal limits or overheating. |
| | Oil Pressure Warning Light means that your car oil level is too low or loss of oil pressure. |
| | Traction Control Off means that the vehicles TCS (traction control system) has been disabled. |
| | Seat Belt Indicator means that a seat belt has not been fastened. |
| | Parking Brake Light means park brake is on. |
| | Icy Road Warning Light turns on when the outside air starts to reach freezing temperatures approximately below 4°C. |
| | Distance Warning appears if a vehicle (obstacle) in front is closer than calculated braking safety distance while moving. |
| | Lamp Out means that the connection of the lights needs to be fixed. |
| | Side Light Indicator appears when the standard headlights are on. |
| | Low Fuel Level means that the tank has some fuel but not much and warns drivers to find a gas station. |

*(continued)*

**Table 1.** (*continued*)

| | |
|---|---|
| | Lane Assist light appears when the lane assist system is switched on. |
| | Cruise Control light means the cruise control system is activated. |

# 3   Windows Azure Cognitive Services

Azure is a comprehensive family of AI services and cognitive APIs to develop intelligent applications. There are different Azure Cognitive Services related to decision making (anomaly detector, personalizer), language (translator, text analytics), speech (speech translation, speaker recognition), vision (form recognizer, video indexer), and web search (Bing image search, spell check, entity search). One of the vision services is custom vision that is an AI service and end-to-end platform for applying computer vision to the specific scenario. It supports to build model using the simple interface and run Custom Vision in the cloud or on the edge in containers in enterprise-grade security level. Azure Custom AI is used. First the problem is defined as object detection. Frames are uploaded to Azure training resource. Figure 1 shows our project life cycle.

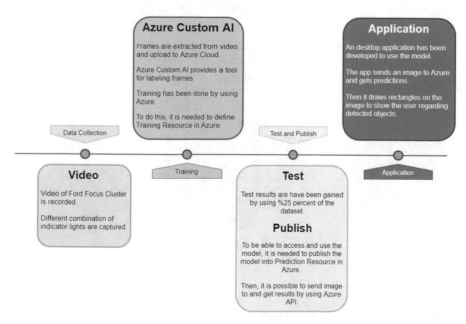

**Fig. 1.**  Project life cycle

# 4    Experimental Analysis

## 4.1    Building Dataset

Ford Focus 2017 is used for collecting data. Dataset is limited to one vehicle model. Video of the vehicle cluster is recorded. The camera is not pinned anywhere. Then, frames are labeled/tagged on Azure as shown in Fig. 2. Dataset is unbalanced as seen Table 2.

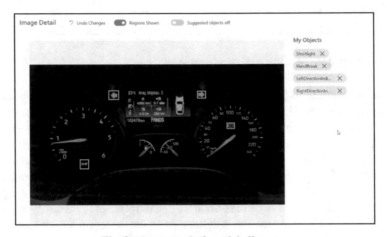

**Fig. 2.** An example from labeling step

**Table 2.** Distribution of used indicators

| Type | # |
|---|---|
| OilLowPressure | 50 |
| LowWasherFluid | 20 |
| LongLight | 35 |
| RightDirectionIndicator | 55 |
| RearFogLight | 62 |
| Airbag | 124 |
| LeftDirectionIndicator | 53 |
| FrontFogLight | 66 |
| EngineEmission | 18 |
| SafetyBelt | 80 |
| Shortlight | 358 |

*(continued)*

**Table 2.** (*continued*)

| Type | # |
|------|---|
| HandBreak | 277 |
| ABS | 58 |
| Battery | 29 |
| LowFuel | 59 |
| ESPoff | 36 |
| ESPon | 37 |
| Warning | 36 |

## 4.2 Experimental Setup and Performance Metrics

In our experiment, a pc with Intel Core i5-5200U 2.2 GHz processor, 8 GB RAM and Windows 10 OS is preferred. We test the system on our own dataset. Recognition result is evaluated using recall, precision, and mAP metrics as the performance measures.

Separating dataset into training, validation and test is done by Azure. Which CNN Architecture is used is selecting by Azure. It is possible to select a general model or custom models. It is possible to select advance training for hyper parameter tuning. Quick training has been selected for this project because of lack of budget. 25% of the dataset are separated from dataset by Azure to assess the performance of the model. We tested our model and calculated the precision, recall, and mAP which are 97.3%, 82.2%, and 91.4%, respectively. The model is published into our Azure Prediction Resource. Then it is possible to use the model from the application by using Azure API. Figure 3 shows the application interface.

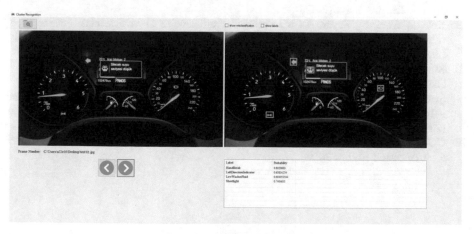

**Fig. 3.** Application interface

## 5   Conclusion

In this study, we firstly created own vehicle warning indicator dataset. Then, Azure Custom AI is used and tested on dataset in terms of recall, precision, mAP metrics. Acquired dataset is unbalanced. This generally causes overfitting, which may affect the final results.

In the future, in order to obtain a generalized model, the dataset should be extended by adding different kind of vehicle cluster. The model can be downloadable, and it is possible to work in local by using docker. (Not allowed test user).

## References

1. Mullakkal-Babu, F.A., Wang, M., Farah, H., van Arem, B., Happee, R.: Comparative assessment of safety indicators for vehicle trajectories on highways. Transp. Res. Rec. **2659**(1), 127–136 (2017)
2. Cronmiller, J.J., Vukosic, S.T., Mack, A., Richardson, J.D., McCann, D.T.: U.S. Patent No. 9,889,795. U.S. Patent and Trademark Office, Washington, DC (2018)
3. Alanazi, M.A.: U.S. Patent No. 9,704,402. U.S. Patent and Trademark Office, Washington, DC (2017)
4. Neilson, A., Daniel, B., Tjandra, S.: Systematic review of the literature on big data in the transportation domain: concepts and applications. Big Data Res. **17**, 35–44 (2019)
5. Zhou, Y., Wang, J., Yang, H.: Resilience of transportation systems: concepts and comprehensive review. IEEE Trans. Intell. Transp. Syst. **20**, 4262–4276 (2019)
6. Stylianou, K., Dimitriou, L., Abdel-Aty, M.: Big data and road safety: a comprehensive review. Mobil. Patterns Big Data Transp. Anal. 297–343 (2019)
7. Borrego-Carazo, J., Castells-Rufas, D., Biempica, E., Carrabina, J.: Resource-constrained machine learning for ADAS: a systematic review. IEEE Access **8**, 40573–40598 (2020)
8. Kul, S., Eken, S., Sayar, A.: A concise review on vehicle detection and classification. In: Proceedings of the Third International Workshop on Data Analytics and Emerging Services, pp. 1–4. IEEE, Antalya (2017)
9. Kul, S., Eken, S., Sayar, A.: Measuring the efficiencies of vehicle classification algorithms on traffic surveillance video. In: Proceedings of International Conference on Artificial Intelligence and Data Processing, pp. 1–6. IEEE, Malatya (2016)
10. Kul, S., Eken, S., Sayar, A.: Distributed and collaborative real-time vehicle detection and classification over the video streams. Int. J. Adv. Rob. Syst. **14**, 1–12 (2017)
11. Şentaş, A., et al.: Performance evaluation of support vector machine and convolutional neural network algorithms in real-time vehicle type and color classification. Evol. Intel. **13**(1), 83–91 (2018). https://doi.org/10.1007/s12065-018-0167-z
12. GOFAR.   https://www.gofar.co/car-warning-lights/car-warning-light-symbols-and-indicators/. Accessed 10 June 2020

# Time Series Analysis on EEG Data with LSTM

Utku Köse, Mevlüt Ersoy, and Ayşen Özün Türkçetin[✉]

Department of Computer Engineering, Suleyman Demirel University, Isparta, Turkey
{utkukose,mevlutersoy}@sdu.edu.tr, aysenozun@gmail.com

**Abstract.** Time series analysis is widely used in the field of health. Data obtained from patients vary according to time. Diagnosis, classification and follow-up of the disease are made with the help of sensors placed on the patients. One of the sensors placed in patients is EEG (Electroencephalogram). EEG is used to receive signals from the brain. These signals are transferred to the computer and the patient's condition is examined. In this study, time series analysis was performed with EEG data set obtained from Kaggle. In the work, LSTM (Long Short Term Memory) repetitive neural network model was used which is one of the deep learning models due to the large data set. In the study, two LSTM based algorithm structures were used, the first algorithm was designed as 8 layers, the second algorithm was added between these 8 layers were increased as a loop through activation and the layers. When the two algorithm models were trained, the loss rates in 16384 sequence time series data were found below 0.01 when calculated with mse and mae metrics.

**Keywords:** Time series analysis · Kaggle · EEG signal · Deep learning · LSTM

## 1 Introduction

Technological advances are increasing day by day in the field of computer and artificial intelligence. Computer interactive systems are developed, compared to these systems, health, military, industry developments are accelerate. In the field of artificial intelligence studies in the field of health, it is developing from disease prediction, disease diagnosis, patient follow-up to robots that perform surgery. Data from patients in the field of health contain very large data information. A lot of time is spent on the classification or analysis of these big data, and the solution to this situation is found in deep learning models, the sub-branch of artificial intelligence.

In order to process the data obtained with embedded systems and computers to learn that data, deep learning methods show a great improvement. The data obtained from the sensors are based on time, temperature and pressure [1]. Analyzes made by time are called time series analysis. Time series analyzes are also frequently used in medicine. As a matter of fact, with the help of the sensors placed in the patients, the condition of the disease is examined the patient's condition and the day. One of the sensors placed on the patients is EEG. EEG is the process of transferring the signals generated by the brain to the computer interface with electrodes connected to the skull [2].

The biggest problem with EEG signals is the noise generated during the information extraction phase. The reason for this noise is due to involuntary eye and muscle movements in the patient when receiving EEG data [3]. In addition, EEG signals have a low

J. Hemanth et al. (Eds.): ICAIAME 2020, LNDECT 76, pp. 491–501, 2021.
https://doi.org/10.1007/978-3-030-79357-9_48

amplitude at the stage of their formation [4]. The process of measuring and interpreting these low amplitude signals in computer environment is called the Brain Computer Interface (BCI) [2]. These data sets that are transferred to the computer are used both in the classification and analysis of EEG signals. EEG signals are kept in very long tables, as they contain time-dependent instant data. It is very difficult to analyze this data. For this, deep learning methods, which is one of the sub-branches of artificial intelligence, are very easy method in the analysis of EEG data. The general naming of these analyzes is referred to as time series analysis.

In deep learning models, time series are used with certain algorithms. One of them is LSTM. The LSTM algorithm gives very good results on speech/text processing. These are a special type of RNN (Recurrent Neural Network) that can learn long-term dependencies.

In this study, LSTM model in RNN network was used to model the EEG data set created with Kaggle to measure the effect of genetic predisposition to alcohol with Artificial Neural Networks. Time series analysis of sensor data obtained via EEG with LSTM was performed.

## 2 Related Work

In this study, Abbasoğlu examined the classification of EEG data using the Attribute Selection (DEOS) method of the Differential Convolution Algorithm, one of the optimization algorithms [2].

In this study, Şentürk developed a mathematical model that will lead the pre-seizure detection of epilepsy by using EEG signals of patients who have undergone epilepsy [3].

Demirel et al. in this study, he used the LSTM network, which is a deep learning model in determining the musical features in any music listened, by analyzing the EEG signals obtained from the Fp1 region, which is the frontal region of the brain. It has achieved high performance as a result of this modeling [5].

In this study, İpek classified the EEG data used in the diagnosis of epilepsy and the Support Vector Machines (DVM), which is one of the deep learning models. As a result of the study, he achieved high performance [6].

In this study, İmak proposed a method to reduce the cost of processing EEG data in the detection of epileptic seizures. In his study, he classified the machine learning algorithms with DVM, Over Learning Machines (OLM), Near K-Neighbor (KNN) and Decision Trees (DT) [7].

In this study for predicting sleep quality, some deep learning methods (CNN (Convolutional Neural Network), LSTM, etc.) were compared with the performance of logistic regression used in machine learning [8].

In this study, a deep learning model using CNN and LSTM algorithms, which learn to classify human activities without prior knowledge, has been successfully applied [9].

LSTM was used to detect offline handwriting [10].

Ay et al. In this study, he made EGG analysis for the diagnosis of depression. In his study, he proposed a hybrid model using CNN architecture. The accuracy rates of LSTM and CNN algorithms used in the study were found to respectively be 97.64 and 99.12 [11].

Chambon et al. In this study, he proposed a deep learning method for the detection of EEG signals during sleep time. It builds on raw EGG signals and relies on the neural network. Various algorithms have been developed to investigate some types of micro and macro events that occur during sleep [12].

Jana et al. In this study, he analyzed the EGG signal received during an epileptic seizure on a prediction of seizure. Converts the raw EEG signal to the EEG signal for automatic extraction and classification of features using the DenseNet neural network model. In the classification, up to 94% verification is obtained by using 5-fold cross-verification [13].

In his Dutta study, he used the LSTM and GRUs (Gated Recurrent Units) algorithms for multi-class time series classification of electrical signals from different regions of the brain [14].

## 3  Material and Method

### 3.1  Dataset

In this study, EGG-Alcohol dataset obtained from Kaggle was used. This data set contains EEG signals of genetic predisposition to alcohol dependence.

The data set consists of EEG signals from 64 electrodes placed on the patient's scalp. There were two groups of subjects: alcoholic and control. A person's EEG data set consists of 16384 line information. Dataset row, column, and data type information is given in Fig. 1 (Table 1).

**Table 1.** EEG dataset information row column table

| # | Column | Non-Null Count | Dtype |
|---|--------|----------------|-------|
| --- | ------ | --------------- | ----- |
| 0 | trial number | 16384 non-null | int64 |
| 1 | sensor position | 16384 non-null | object |
| 2 | sample num | 16384 non-null | int64 |
| 3 | sensor value | 16384 non-null | float64 |
| 4 | subject identifier | 16384 non-null | object |
| 5 | matching condition | 16384 non-null | object |
| 6 | channel | 16384 non-null | int64 |
| 7 | name | 16384 non-null | object |
| 8 | time | 16384 non-null | float64 |

Figure 1 is a graph given person's EEG time-dependent graph of the original data set. The changes in time as shown in the graph shows a great deal of complexity.

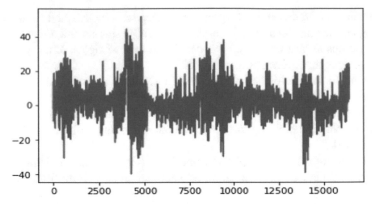

**Fig. 1.** EEG graph of time-dependent data sets

## 3.2 LSTM (Long-Short Term Memory)

LSTM Long Short-Term Memory Network is one of the artificial neural network [15]. LSTM information is transferred to the next time from the previous model. The desired state for a time-dependent variable is the time series to predict the future by looking at their previous values.

LSTM has been proposed by scientists in 1997 [16]. Recurrent called neural network model ensures the continuance of this information is available on the public network model in layers. Each neural network in itself constitutes cycle. Because of this structure ensures the continuance of information and creates a repeating structure [15].

During each iteration, the same function and the same weights are applied to the network every time. LSTM taking the simplest type of neural network input function as that tanh assuming that wxh weight of WHH and input neurons by weight of recurrent neurons in LSTM model for time t case Eq. 1 and is shown in Fig. 2.

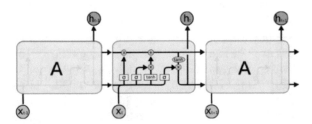

**Fig. 2.** The display of the tanh function of LSTM model [15]

$$h_t = \tan h(W_{hh}h_{t-1} + W_{xh}X_t) \tag{1}$$

Neurons by Eq. 1 takes into account the immediately preceding state. After the last state calculation may continue to produce the output when the current state output function calculated in Eq. 2.

$$y_t = W_{hy}h_t \tag{2}$$

The LSTM function contains 3 sigmoid and 1 tanh function, and these work sequentially in repetitive LSTM. This indicates a non-linear of the LSTM algorithm.

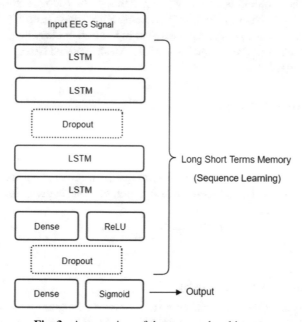

**Fig. 3.** An overview of the proposed architecture

In Fig. 3 a total of 523,171 parameters were used in the LSTM model layer summary. As seen in Fig. 4, 0.2 Dropout layers were added between the layers. The Dropout layer is a layer used to prevent overfitting the system from over-memorizing training data [17].

In addition to the architecture shown in Fig. 3, 15 dense and activation layers were added after the 2nd LSTM layer. In the newly created model, the number of parameters increased to 528,763. The sigmoid function is used as the activation function commonly used in the model. The sigmoid function is a function that maps the logistic regression result in probabilities and returns a value between 0–1. The sigmoid function is shown in Eq. 3:

$$y = \frac{1}{1 + e^{-\sigma}} \tag{3}$$

In logistic regression problems, simplify Eq. 4 is shown as follows:

$$\sigma = b + w_1 x_1 + w_2 x_2 + \ldots + w_n x_n \tag{4}$$

The sigmoid function converts $\sigma$ 'to a value between 0–1'.

## 3.3 Performance Metric

Every artificial intelligence problem has a different purpose. As a result of these purposes, the solution of the problem is made by using different data sets. In solving a problem, it is necessary to understand well what the metric selection is used for. In the problem described, it is necessary to understand whether regression or classification will be done. Because of the study time series analysis, MSE (Mean Squared Error) and MAE (Mean Absolute Error) metrics are used.

### 3.3.1 Mean Squared Error (MSE)

Of substantially MSE measures the mean square error estimation. Each point is the average of squares difference between the predicted and target are calculated resulting values [18]. The higher this value is, the worse the forecast is. The reason for this is that before taking the square sum of the error terms of the individual estimates this value is zero for perfect results. The mathematical function of MSE is given in Eq. (5).

$$MSE = \frac{1}{N} \sum_{i=1}^{N} (y_i - \hat{y}_i)^2 \tag{5}$$

### 3.3.2 Mean Absolute Error (MAE)

MAE is basically calculated as the average of absolute differences between target values and estimates. The mathematical function of the MAE is given in Eq. (6).

$$MAE = \frac{1}{N} \sum_{i=1}^{N} |y_i - \hat{y}_i| \tag{6}$$

MAE metric is not sensitive to outliers. Therefore, it eliminates the major errors in the outcome produced by MSE [18].

## 4   Results

In this study, the EEG data set obtained from Kaggle deep learning model applied time series analysis was conducted. Deep learning algorithm as the LSTM repeated neural network model with 8 layers and datasets are trained to step 100 epoch. To prevent the dataset from over-learning, dropout layers of 0.2 have been added into the LSTM network. The study used laptop during 1050ti GTX video cards and 16385 lines of data, the training period lasted 1 h 5 min. It made this training results in loss rates of EEG time series data sets that have been observed down the MSE and MAE value of 0.01. Alcohol group Fig. 4 and Fig. 5 for the first LSTM model show the loss ratio used in the MSE and MAE calculations.

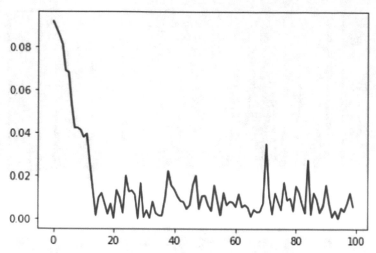

**Fig. 4.** MAE metric value graph of first LSTM model applied to EEG alcohol group data set

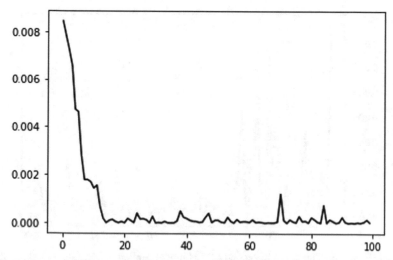

**Fig. 5.** MSE metric value graph of first LSTM model applied to EEG alcohol group data set

Alcohol group Fig. 6 and Fig. 7 for the second LSTM model show the loss ratio used in the MSE and MAE calculations.

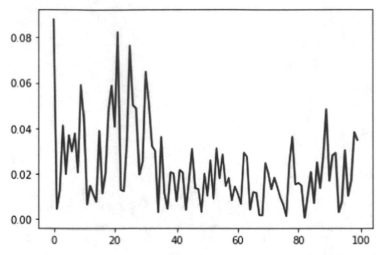

**Fig. 6.** MAE metric value graph of second LSTM model applied to EEG alcohol group data set

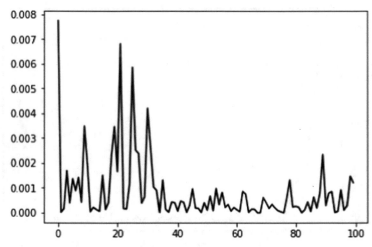

**Fig. 7.** MSE metric value graph of second LSTM model applied to EEG alcohol group data set

Control group Fig. 8 and Fig. 9 for the first LSTM model show the loss ratio used in the MSE and MAE calculations.

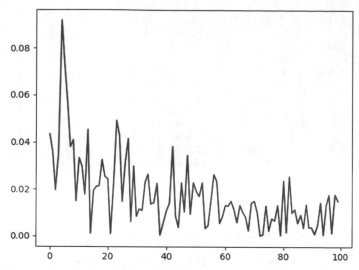

**Fig. 8.** MAE metric value graph of first LSTM model applied to EEG control group data set

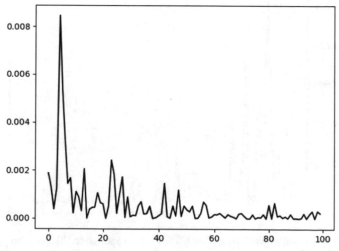

**Fig. 9.** MSE metric value graph of first LSTM model applied to EEG control group data set

Control group Fig. 10 and Fig. 11 for the second LSTM model show the loss ratio used in the MSE and MAE calculations.

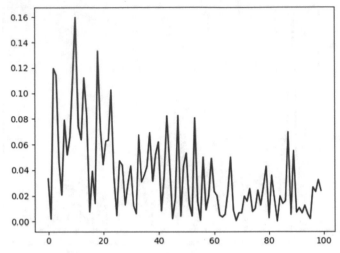

**Fig. 10.** MAE metric value graph of second LSTM model applied to EEG control group data set

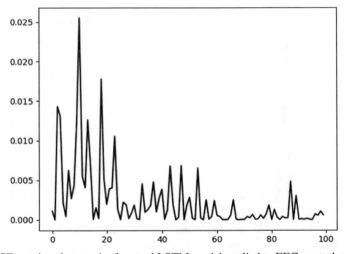

**Fig. 11.** MSE metric value graph of second LSTM model applied to EEG control group data set

## 5    Conclusion

When the first and second proposed LSTM models were compared, it was observed that the first model gave better results and in the second proposed model, the activation and dense layer addition of layers of LSTM layers complicated the training. In the model to be developed, training can be done with both models created using different data sets. LSTM network is recommended when using deep learning-based models especially in time series analysis. As a matter of fact, as seen in the study, it decreases the rate of loss of the model.

# References

1. Coşkun, E., Alma, B.: Büyük Ölçekli Güvenligin Öncelikli Oldugu Sistemlerde Gömülü Zeki Gerçek-Zamanlı Sistemlerin Rolü ve Karmasıklık Ölçümü, Sakarya Üniversitesi (2017)
2. Abbasoğlu, F.: Öznitelik çıkarım ve evrimsel öznitelik seçim metotlarının eeg sinyallerinin sınıflandırma başarısına etkileri (Master's thesis, Fatih Sultan Mehmet Vakıf Üniversitesi, Lisansüstü Eğitim Enstitüsü) (2019)
3. Şentürk, T.: EEG sinyallerinin epilepsinin nöbet öncesi tespitine yönelik analizi. Yüksek Lisans tezi, Erciyes Üniversitesi, Fen Bilimleri Enstitüsü (2019)
4. Olgun, N., Türkoğlu, İ.: Brain–Computer Interface Systems (2017)
5. Demirel, Ç., Akkaya, U.C., Yalçin, M., İnce, G.: Estimation of musical features using EEG signals. In: 2018 26th Signal Processing and Communications Applications Conference (SIU), pp. 1–4. IEEE (May 2018)
6. Mutlu İpek, B.Ü.Ş.R.A.: EEG sinyallerinin epileptik rahatsızlıkların teşhisi için konvolüsyonel sinir ağları ve destek vektör makineleri ile tasnif edilmesi/Classification of EEG signals with convolutional neural networks and support vector machines for diagnosis of epileptic disorders (Doctoral dissertation) (2018)
7. İmak, A., Alçin, Ö.F., İnce, M.C.: Tekrarlılık Ölçüm Analizi ve Aşırı Öğrenme Makinesi Tabanlı Epileptik EEG Sinyallerinin Sınıflandırılması (2019)
8. Sathyanarayana, A., et al.: Sleep quality prediction from wearable data using deep learning. JMIR mHealth uHealth **4**(4), e125 (2016)
9. Baccouche, M., Mamalet, F., Wolf, C., Garcia, C., Baskurt, A.: Sequential Deep Learning for Human Action Recognition. Springer, Heidelberg, pp. 29–39 (2011)
10. Graves, A., Jaitly, N.: Towards end-to-end speech recognition with recurrent neural networks. In: International Conference on Machine Learning, pp. 1764–1772 (2014)
11. Ay, B., et al.: Automated depression detection using deep representation and sequence learning with EEG signals. J. Med. Syst. **43**(7), 205 (2019)
12. Chambon, S., Thorey, V., Arnal, P.J., Mignot, E., Gramfort, A.: A deep learning architecture to detect events in EEG signals during sleep. In: 2018 IEEE 28th International Workshop on Machine Learning for Signal Processing (MLSP), pp. 1–6. IEEE (September 2018)
13. Jana, R., Bhattacharyya, S., Das, S.: Epileptic seizure prediction from EEG signals using DenseNet. In: 2019 IEEE Symposium Series on Computational Intelligence (SSCI), pp. 604–609. IEEE (December 2019)
14. Dutta, K.K.: Multi-class time series classification of EEG signals with recurrent neural networks. In: 2019 9th International Conference on Cloud Computing, Data Science & Engineering (Confluence), pp. 337–341. IEEE (January 2019)
15. Olah, C., Understanding LSTM Networks, ErişimAdresi. https://colah.github.io/posts/2015-08-Understanding-LSTMs/. Erişim Tarihi. Accessed 28 May 2019
16. Hochreiter, S., Schmidhuber, J.: Long short-term memory. Neural Comput. **9**(8), 1735–1780 (1997). https://doi.org/10.1162/neco.1997.9.8.1735
17. Srivastava, N., Hinton, G., Krizhevsky, A., Sutskever, I., Salakhutdinov, R.: Dropout: a simple way to prevent neural networks from overfitting. J. Mach. Learn. Res. **15**(1), 1929–1958 (2014)
18. Drakos, G.: How to select the right evaluation metric for machine learning models: part 1 regression metrics, Erişim adresi. (2018). https://medium.com/@george.drakos62/how-to-select-the-right-evaluation-metric-for-machine-learning-models-part-1-regrression-metrics-3606e25beae0? Erişim Tarihi. Accessed 08 Jan 2020

# Gaussian Mixture Model-Based Clustering of Multivariate Data Using Soft Computing Hybrid Algorithm

Maruf Gögebakan[✉]

Maritime Business Administration Department, Maritime Faculty, Bandirma Onyedi Eylul University, Bandirma/Balikesir, Turkey
mgogebakan@bandirma.edu.tr

**Abstract.** In this study, a new hybrid-clustering algorithm is proposed for model-based clustering of multivariate data. In order to dimension reduction with variable selection in multivariate data, heterogeneous (mixture) structures in the variables are determined according to the univariate Gaussian mixture models (GMM). Component numbers of variables in heterogeneous data are determined comparatively based on univariate Gaussian mixture distribution and k-means algorithms. Based on the number of components (subgroup) in the variables, the number, orientations and structure of the clusters are determined by the mixture models created. The number of candidate mixture models that can be determined by the soft calculation method developed for model-based determination of cluster structures consisting of mixture distributions is determined. Information criteria are used to select the best model that fits the data among candidate mixture models. In selecting the best model, information criteria based on likelihood function, such as AIC (Akaike Information Criteria) and BIC (Bayesian Information Criteria) are used.

**Keywords:** Hybrid clustering algorithm · Gaussian mixture models · Model based clustering · Information criteria

## 1 Introduction

With the advancement of technology and more data, the usage areas of Gaussian mixture models (GMM) are increasing day by day. GMM is used because it provides high success in clustering in many fields such as engineering, health sciences, computer science, economics, behavioral sciences and social sciences (McLachlan and Rathnayake 2014; Bozdoğan 1994; Scrucca and Raftery 2018). Determining the number of components of the variables in Gaussian mixture models is an important and difficult problem. Multivariate techniques are used to reveal components in variables. Determining the number, location and size of the clusters determined based on the number of components of the variables using normal mixture models increases the cluster success in the data (McNicholas 2016; Celeux et al. 2018). Clusters in multivariate normal mixture distributions match the components of variables in heterogeneous data (McLachlan and Chang 2004).

© The Author(s), under exclusive license to Springer Nature Switzerland AG 2021
J. Hemanth et al. (Eds.): ICAIAME 2020, LNDECT 76, pp. 502–513, 2021.
https://doi.org/10.1007/978-3-030-79357-9_49

The number of components in the variables directly affects the number of clusters that will occur in the data. Determining the number and location of clusters that will occur in the data set is a difficult problem in finite mixture distributions (McLachlan and Peel 2000). Finite mixtures of normal distributions are an effective clustering method used to define the group/cluster structure in the data. It is difficult to make statistical inference from real and false roots that can occur in model-based determination of normal mixture distributions in uncovering the cluster in the data set. For this reason, roots in mixture models correspond to different clusters and locations (Seo and Kim 2012). In normal mixture models, clusters are determined by the parameters obtained from the components of each variable. Expectation and Maximization (EM) algorithm is one of the most used methods for parameter estimation in model-based clustering (Dempster et al. 1977). The cluster structure and number depends on the correct estimation of the parameters of the variables and the probability density function of the distributions. Model selection criteria determined based on log-likelihood estimation also use parameters in the same components (Fraley and Raftery 1998). The covariance matrix determines the shape of the clusters in the data and is an important parameter in model-based log-likelihood estimation. Browne and McNicholas used orthogonal steifel manifold optimization to estimate the covariance parameter in mixture models (Browne and McNicholas 2014). It is observed that general covariance matrix structures, where the shape, volume and direction parameters of normal mixture distributions are all different, are used in model-based clustering (Celeux and Govaert 1993).

In this study, the clustering algorithm method based on determining the number and location of the cluster from the number of components in the variables is proposed. It has been investigated that the observations pertaining to the number of components in the variables determine the covariance parameter and thus affect the location and shape of the clusters. For more accurate parameter estimation, K-means algorithm and Gauss mixture models (GMM) were compared in determining the number of observations per component in the general covariance matrix structure. The number of possible mixture models depending on the number of components was determined by the computational mixture models method and mixture models suitable for the model-based clustering method were determined. Parameters in the data were estimated by using EM algorithm, probability density functions and information criteria based on likelihood functions were calculated. Among the appropriate wedge models, the best mixture model matching the data set was determined with the help of information criteria.

## 2   Method and Materials

In multivariate data, the steps of the proposed clustering method on the synthetic dataset are described for clustering of variables based on components.

The synthetic data set was used in the study to compare the k-means and GMM clustering methods to reveal the covariance matrix structures that will occur by assigning variable components and observations that fall on the components. Synthetic dataset was produced by simulation with two components in each variable with two variables and 3 clusters.

## 2.1    Determining the Number of Components of Variables in Multivariate Data

In the data with variable numbers $p \geq 2$, univariate normal mixture models are used to determine the number of components of the variables. Cluster structures in normal mixture models are obtained by converting them to mixture models based on variables. The mixture model of normal distributions is expressed as follows,

$$f(x_j; \theta) = \sum_{i=1}^{g} \pi_i f_i(x_j; \psi_i) \tag{1}$$

where, the probability weights of mixture distributions are derived from the number of observations per component, with the sum totaling $\sum_{i=1}^{g} \pi_i = 1$ and each probability in the range $0 < \pi_i < 1$. Unknown parameters vector consists of $\psi_i = (\mu_i, \Sigma_i)$, $\mu_i$ mean vector and $\Sigma_i$ covariance matrix. EM algorithm is used for parameter estimation for components in univariate normal mixture models. The number of components in the variables is determined from log-likelihood, AIC and BIC values using normal mixture distributions.

## 2.2    Assignment of Expected Observations of Components by K-Means Algorithm

The components in the variables are determined by univariate normal mixture distributions and mixture models with grid structure are created. In mixture models, the number of clusters is obtained by the mixture probability weights, mean vectors and covariance matrix parameters of normal distributions. Here, mixture probability weights ($\pi_i$) are the parameters that determine the size of the set, the mean vector ($\mu_i$) of the set, and the covariance matrix ($\Sigma_i$) shape of the set. With the K-means algorithm, the observations falling on the components are determined according to the distance vector between the observations in the variable. While determining the observation values falling on the components, the Euclid distance is calculated as follows,

$$argmin_s \sum_{i=1}^{k} \sum_{j \in S_i} \|x_j - \mu_i\|^2 \tag{2}$$

where, the distance vector between each observation in the variable and the average of the data group determines the clustering.

## 2.3    Comparison of K-Means and Gaussian Mixture Models in Determining Observations on Components

The number of observations in the components determines the probability weights and size of the mixture model. Components in different volumes create more consistent centers. Clusters of components, in which the number of observations are distributed evenly, form more homogeneous and slightly differentiated models. Decreasing the differences between the models causes an increase in errors in determining the number and location of the cluster. Since the model selection determined with the help of information criteria is prone to biased selection, the differentiation of the clusters that make up the models determines the correct creation of the model. Normal mixture models assign equal numbers of observations while the K-means algorithm assigns a different number of eyes to the components. Although both assignment algorithms have strengths and weaknesses, the k-means algorithm creates more consistent models in the model-based clustering of grid-mixture models.

## 2.4   Determination of Number and Location of Cluster Centers in Normal Mixture Models

The number of components in the variables determines the number of clusters in the mixture model. The minimum and maximum number of clusters that can occur in mixture models of different number of components are obtained from the following equations.

$$C_{max} = \prod_{s=1}^{p} k_s \ \ ve \ \ C_{min} = max\{k_s\} \tag{3}$$

where $s = 1, \ldots, p$ shows the number of variables, $k_s$ shows the number of components in the variables.

**Definition 1:** The function $f : D(f) \rightarrow R(f)$ is defined with the number of variable components $k_s$, such as $\forall k_s \in \{C_{min}, C_{max}\}$ Where, $D(f) = \left[ max\{k_s\}, \prod_{s=1}^{p} k_s \right]$ is the Domain, $R(f)$ shows the image set consisting of suitable candidate mixture models.

## 2.5   Numbers and Structure of Normal Mixture Models

The number of mixture models in grid structure based on variable components is calculated as follows based on the number of cluster centers.

$$M_{Total} = 2^{C_{Max}} - 1 \tag{4}$$

The null model without a set that does not correspond to the components in the variables is subtracted from the calculation.

The multivariate probability density function of mixture models from normal distributions is shown as follows,

$$f_i(x_j; \mu_i, \Sigma_i) = \frac{1}{(2\pi)^{\frac{p}{2}}|\Sigma_i|^{\frac{1}{2}}} \exp\left\{ -\frac{1}{2}(x_j - \mu_i)^T \Sigma_i^{-1}(x_j - \mu_i) \right\} \tag{5}$$

where, $\Omega$ consists of parameters $\theta = \{\pi_i, \psi_i\}$, which is the parameter vector of the multivariate normal mixture distribution space.

## 2.6   Numbers of Grid Structured Normal Mixture Models and Genetic Algorithms

For the density based model based clustering of models obtained from normal mixture distributions with grid structure, vectors obtained from variable components are created for each suitable model.

The vectors consist of sub vectors representing the number, location and size of the mixture model obtained. The structure blocks of vectors in genetic algorithms represent the size and location of the clusters in the $\sigma(H)$ vector, the size of the appropriate mixture model represents the number of clusters in the $o(H)$ model. The size of the vector is determined according to the candidate models so that the number of clusters that can occur in the vector "001001... 0100" is in the range of $1 \leq k \leq C_{max}$. The mixture model calculation method and combinatorics calculation method developed are used to determine the number of valid mixture models. Computational method is preferred as the combinatorial calculation method becomes more difficult and the information complexity increases as the size and number of variables increase in the data set. The number of combinatorial and computational mixture models according to the number of two different variable components is shown in Table 1 and Table 2.

**Table 1.** Two-component and two-variable mixture model combinatorics calculations method.

| # Cluster | Distribution of clusters to components | Combinatorial correlation for mixture model | Current model number |
|---|---|---|---|
| $k = 2$ | 1-1 distributed | $\binom{2}{1}\binom{1}{1} = 2! = 2$ | 2 |
| $k = 3$ | 2-1 distributed | $\binom{2}{1}2! = 2.2 = 4$ | 4 |
| $k = 4$ | 2-2 distributed | $\binom{4}{4} = 1$ | 1 |
| # total models | | $2^{2.2} - 1 = 15$ | |
| # current models | | 7 | |

**Table 2.** Three-component and two-variable mixture model combinatorics calculations method.

| # Cluster | Distribution of clusters to components | Combinatorial correlation for mixture model | Current model number |
|---|---|---|---|
| $k = 3$ | 1 1 1 distributed | $\binom{3}{1}\binom{2}{1}\binom{1}{1} = 3! = 6$ | 6 |
| $k = 4$ | 2 1 1 distributed | $\binom{3}{2}[1 + 1 + 3]\frac{3!}{2!} = 3.5.3 = 45$ | 45 |

(*continued*)

**Table 2.** (*continued*)

| # Cluster | Distribution of clusters to components | Combinatorial correlation for mixture model | Current model number |
|---|---|---|---|
| $k = 5$ | 3 1 1 distributed | $\binom{3}{3}\binom{3}{1}\binom{3}{1}\frac{3!}{2!} =$ <br> $1.3.3.3 = 27$ | $63 + 27 = 90$ |
| | 2 2 1 distributed | $\binom{3}{2}[1 + 3 + 3]\frac{3!}{2!} =$ <br> $3.7.3 = 63$ | |
| $k = 6$ | 3 2 1 distributed | $\binom{3}{3}\binom{3}{2}\binom{3}{1}3! =$ <br> $1.3.3.6 = 54$ | $24 + 54 = 78$ |
| | 2 2 2 distributed | $\binom{3}{2}[2 + 3 + 3]\frac{3!}{3!} =$ <br> $3.8.1 = 24$ | |
| $k = 7$ | 3 3 1 distributed | $\binom{3}{3}\binom{3}{3}\binom{3}{1}\frac{3!}{2!} =$ <br> $1.1.3.3 = 9$ | $27 + 9 = 36$ |
| | 3 2 2 distributed | $\binom{3}{3}\binom{3}{2}\binom{3}{2}\frac{3!}{2!} =$ <br> $1.3.3.3 = 27$ | |
| $k = 8$ | 3 3 2 distributed | $\binom{3}{3}\binom{3}{3}\binom{3}{2}\frac{3!}{2!} =$ <br> $1.1.3.3 = 9$ | 9 |
| $k = 9$ | 3 3 3 distributed | $\binom{3}{3}\binom{3}{3}\binom{3}{3}\frac{3!}{3!} =$ <br> $1.1.1.1 = 1$ | 1 |
| # total models | | $2^{3.3} - 1 = 511$ | |
| # current models | | $0 + 0 + 6 + 45 + 90 + 78 + 36 + 9 + 1 = 265$ | |

The number of current mixture models based on components in the variables is obtained by the soft computing method as follows.

$$f(x; \theta_g) = \sum_{i,j=0}^{n,m} (-1)^{i+j} \binom{n}{i} \binom{m}{j} \binom{(n-i)(m-j)}{k} \tag{6}$$

Where, $g = 1, \ldots, k$ shows the number of clusters in the model, the parameters that distribute the components in variables $n$ *and* $m$ according to indices $i$ *and* $j$ (Cheballah et al. 2013) (Tables 3 and 4).

**Table 3.** Two component bivariate models with current mixture model calculation method.

| # Clusters | # Total models | # current models |
|---|---|---|
| k = 1 | 4 | 0 |
| k = 2 | 6 | 2 |
| k = 3 | 4 | 4 |
| k = 4 | 1 | 1 |
| Total | 15 | 7 |

**Table 4.** Two component and three variable models with current mixture model calculation method.

| # Clusters | # Total models | # current models |
|---|---|---|
| k = 1 | 9 | 0 |
| k = 2 | 36 | 0 |
| k = 3 | 84 | 6 |
| k = 4 | 126 | 45 |
| k = 5 | 126 | 90 |
| k = 6 | 84 | 78 |
| k = 7 | 36 | 36 |
| k = 8 | 9 | 9 |
| k = 9 | 1 | 1 |
| Total | 511 | 265 |

## 2.7  Determination of the Best Model According to Information Criteria Among Grid Structured Mixture Models

In the heterogeneous multivariate data, the parameters of the mixture models in the grid structure created based on the components of the variables are obtained from the sample. Probability weights, mean vectors and covariance matrices based on Bayesian classification estimation are used to reveal the clustering of mixture models. In the proposed algorithm, probability weights determine the weight of the models due to the different number of observations in the variable components. While the locations of the cluster centers to be formed in the mixture model are determined according to the average vector, the resistance matrices are used to determine the shape and structure of the clusters in the model. Since the matrices consisting of different observation numbers are not in the square matrix structure, the reversible matrix is obtained by the method of assigning the average to the variables.

Log-likelihood function of multivariate normal wedge distributions,

$$\log L(\Psi) = \sum_{j=1}^{n} \log\left(\sum_{i=1}^{g} \pi_i f_i(x_j; \theta_i)\right) \qquad (7)$$

It is obtained from the probability density function of multivariate normal distributions. According to likelihood approach, information criteria are obtained in model selection as follows.

$$\log L(\Psi) + dc_n \qquad (8)$$

Where $d = (k - 1) + (kp) + kp\frac{(p+1)}{2}$ is obtained from the number of components and variables in the model to show the number of independent parameters in the model. the cluster penalty term used to determine the number of clusters in the model,

It is selected as $c_n = 2d$ for AIC (Akaike 1974) and $c_n = \log n$ for BIC (Schwarz 1978).

## 3   Results

### 3.1   Application of Proposed Hybrid Clustering Algorithm on Synthetic Data Set

The data set consisting of two variables consisting of normal mixture distributions and two components in each variable was formed as three clusters by simulation.

The best modeling tables and graphs with the help of the number of components on the synthetic data set of the proposed hybrid clustering algorithm and genetic algorithms are given below (Table 5).

**Table 5.** Number of components in the variables in synthetic data with two variables

| # Variables | # Components | AIC | BIC |
|---|---|---|---|
| $X_1$ | k = 1 | 2658 | 2666 |
|  | *k = 2 | **2646** | **2664** |
|  | k = 3 | 2650 | 2679 |
| $X_2$ | k = 1 | 2735 | 2743 |
|  | *k = 2 | **2604** | **2622** |
|  | k = 3 | 2610 | 2639 |

The observations assigned to the components of the variables determined by univariate normal mixture models using k-means algorithms are shown in Table 6 below.

**Table 6.** Components in variables and number of observations assigned to components

| Components | $X_{11}$ | $X_{12}$ | $X_{21}$ | $X_{22}$ |
|---|---|---|---|---|
| # Observations | 89 | 211 | 74 | 226 |

In mixture models in grid structure, the minimum number of clusters is $C_{min} = max\{k_s\} = max\{2, 2\} = 2$, the maximum number of clusters is $C_{max} = \prod k_s = 2.2 = 4$, and the number of variables is obtained according to the number of components. The components of the variables and the cluster centers that can occur in the grid model corresponding to the components are shown in Fig. 1.

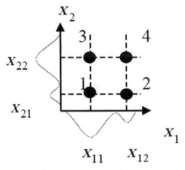

**Fig. 1.** Components in the variable and cluster centers in grid structure corresponding to each component.

Variable numbers of mixture models in the grid structure and parameters of mixture density functions to be obtained from clusters of components in variables are obtained based on the sample. Probability weights for parameters obtained from each center and the components that make up the centers; $\pi_i^k = \frac{n_{i,j}}{\sum_{i,j}^k n_k}$, mean vector; $\mu_i^k = \begin{bmatrix} \mu_{1,i} \\ \mu_{2,j} \end{bmatrix}$ and

covariance matrix; $\Sigma_i^k = \begin{bmatrix} \sigma_{1i}^2 & \rho_{1i,2j}\sigma_{1i}\sigma_{2j} \\ \rho_{2j,1i}\sigma_{2j}\sigma_{1i} & \sigma_{2i}^2 \end{bmatrix}$ to be obtained.

Where, $i, j = 1, 2$ represent variable component numbers and $k = 1, \ldots 4$ represent center numbers. The total number of mixture models in the grid structure is calculated as $M_{Total} = 2^{2.2} - 1 = 15$. In mixture models, the number of valid models corresponding to a set for each component and the vectors obtained from the structure blocks of genetic algorithms (GA) corresponding to mixture models are shown in Table 7.

**Table 7.** The number of current mixture models and GA structure blocks obtained from the distributions in clusters and grid structure corresponding to the components in the variables.

| # Clusters | # Models | # Current models | GA structure blocks $\sigma(H)$ |
|---|---|---|---|
| 1 | 4 | – | – |
| 2 | 6 | 2 | 1001 |
|   |   |   | 0110 |
| 3 | 4 | 4 | 1110 |
|   |   |   | 1101 |
|   |   |   | 1011 |
|   |   |   | 0111 |
| 4 | 1 | 1 | 1111 |

The information criteria obtained based on the structure blocks according to the parameters of the grid mixture normal mixture models and the number of models that may occur are shown in Table 8.

**Table 8.** Information criteria of models obtained based on density functions and structure blocks of mixture models with grid structure.

| Models | AIC | BIC | GA structure blocks $\sigma(H)$ |
|---|---|---|---|
| 1 | 7372 | 7350 | 1001 |
| 2 | 6942 | 6920 | 0110 |
| 3 | 6326 | 6292 | 1110 |
| 4 | 7568 | 7534 | 1101 |
| **5** | **5317** | **5283** | 1011 |
| 6 | 6142 | 6108 | 0111 |
| 7 | 5456 | 5410 | 1111 |

According to the obtained parameter values, the best mixture model matching the data structure of the mixture models was obtained with the help of hybrid clustering algorithm as the fifth model in the table of the vectors obtained from three cluster and genetic algorithms. The surface model of the best model is shown in Fig. 2.

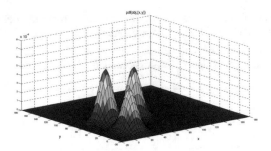

**Fig. 2.** The surface graph of the best mixture model with grid structure based on the components in bivariate data.

## 4   Conclusion and Discussion

In this study, hybrid-clustering algorithm is described for model-based clustering of multivariate data using normal mixture models. The components of the variables in the data have been determined, Genetic Algorithms and soft calculations method have determined the number of mixture models that will consist of the number, location and shape of the clusters to be composed of these components. In order to reveal the best cluster structure in the data, probability density functions of normal mixture distributions have been calculated, and model selection has been made with the information criteria based on the best likelihood estimation with Bayes approach.

As the number of variables increases, the problem of determining the number of clusters in the model is minimized by the inductive approach. The best mixture model among all the possibilities that can occur is provided with the vector representation obtained from genetic algorithms without error.

## References

McLachlan, G.J., Rathnayake, S.: On the number of components in a Gaussian mixture model. Wiley Interdisc. Rev. Data Min. Knowl. Discov. **4**(5), 341–355 (2014)

Bozdogan, H.: Choosing the number of clusters, subset selection of variables, and outlier detection in the standard mixture model cluster analysis. In: Diday, E., et al. (eds.) Invited paper in New Approaches in Classification and Data Analysis, pp. 169–177. Springer, New York (1994). https://doi.org/10.1007/978-3-642-51175-2_19

Scrucca, L., Raftery, A.E.: clustvarsel: a package implementing variable selection for Gaussian model-based clustering in R. J. Stat. Softw. 84 (2018)

McNicholas, P.D.: Model-based clustering. J. Classif. **33**(3), 331–373 (2016)

Celeux, G., Frühwirth-Schnatter, S., Robert, C.P.: Model selection for mixture models perspectives and strategies. In: Handbook of Mixture Analysis, pp. 121–160 (2018)

McLachlan, G.J., Chang, S.U.: Mixture modelling for cluster analysis. Stat. Methods Med. Res. **13**, 347–361 (2004)

McLachlan, G.J., Peel, D.: Finite Mixture Models. Wiley, Hoboken (2000)

Seo, B., Kim, D.: Root selection in normal mixture models. Comput. Stat. Data Anal. **56**(8), 2454–2470 (2012)

Dempster, A.P., Laird, N.M., Rubin, D.B.: Maximum likelihood from incomplete data via the EM algorithm. J. Roy. Stat. Soc. Ser. B (Methodol.) **39**(1), 1–22 (1977)

Fraley, C., Raftery, A.E.: How many clusters? Which clustering method? Answers via model-based cluster analysis. Comput. J. **41**, 578–588 (1998)

Browne, R.P., McNicholas, P.D.: Orthogonal Stiefel manifold optimization for eigen-decomposed covariance parameter estimation in mixture models. Stat. Comput. **24**(2), 203–210 (2014). https://doi.org/10.1007/s11222-012-9364-2

Celeux, G., Govaert, G.: Gaussian parsimonious clustering models (1993)

Cheballah, H., Giraudo, S., Maurice, R.: Hopf algebra structure on packed square matrices. J. Comb. Theory Ser. A **133**, 139–182 (2013)

Akaike, H.: A new look at the statistical model identification. IEEE Trans. Autom. Control **19**(6), 716–723 (1974)

Schwarz, G.: Estimating the dimension of a model. Ann. Statist. **6**, 461–464 (1978)

# Design Optimization and Comparison of Brushless Direct Current Motor for High Efficiency Fan, Pump and Compressor Applications

Burak Yenipinar[1(✉)], Cemal Yilmaz[2], Yusuf Sönmez[3], and Cemil Ocak[3]

[1] OSTİM Technical University, OSTİM, Yenimahalle, Ankara, Turkey
`burak.yenipinar@ostimteknik.edu.tr`
[2] Department of Electrical and Electronics Engineering, Gazi University Technology Faculty, Ankara, Turkey
`cemal@gazi.edu.tr`
[3] Department of Electrical and Energy Engineering, Gazi University Technical Sciences Vocational School, Ankara, Turkey
`{ysonmez,cemilocak}@gazi.edu.tr`

**Abstract.** In this study, design optimization of a 1.5 kW Brushless Direct Current Motor (FDAM) has been carried out for fan and compressor applications. The initial motor design which is obtained by taking the IC 410 cooling method defined in the IEC 60034-6 standard into account, has been optimized in order to offer a high efficiency and compact structure by considering the geometrical dimensions of motor via Genetic Algorithm method. In the design process where the minimum cost/maximum efficiency criterion was taken into consideration, the performance of the optimized motor has been introduced by performing Finite Element Analysis subsequent to analytical calculations. The performance comparison of the optimized BLDC motor and a standard asynchronous motor is also given in the study.

**Keywords:** Brussless direct current motor design · Fan · Compressor · Multiobjective optimization

## 1 Introduction

In recent years, use of Brushless Direct Current Motors (BLDC) has become widespread due to advantages such as high efficiency, high torque to volume ratio, low-noise operation and low-maintenance [1, 2]. Moreover, BLDCs with radial flux surface-mounted magnet structure is quite popular topology which is especially preferred in high torque and power density applications [3]. BLDC motors are structurally similar to the Permanent Magnet Synchronous Motor (PMSM) but evolved from the brushed DC motors. Alternatively, commutation process in BLDC motors is provided electronically and differs from the Brushed DC Motors [4]. Providing the commutation process electronically ensures the motor to be more efficient and low-maintenance due to the losses arising from the brush and collector structure are eliminated.

The highest rate in consumption of electrical energy in the world belongs to electric motors. Energy consumption rate of electric motors vary between 30–80% depending on the development level of the countries [5]. These rates vary depending on the use of high efficiency electric motors and the level of industrial development.

Electric motors are widely used in applications such as compacting, cutting, grinding, mixing, pumps, materials conveying, refrigeration, and air compressors in the industry. Increasing the efficiency of electric motors from 92% to 94% means reducing motor losses approximately by 25%. Generally, to reduce the losses in permanent magnet motors it is required to use low W/kg loss core materials, reducing winding losses and using optimized magnets which have a high BHmax product. Due to unstable price of the rare-earth materials such as dysprosium used in magnets is a risk factor for permanent magnet motors manufacturers in terms of cost estimation. The use of low-cost magnets is an important consideration for permanent magnet motor manufacturers. However, it is inevitable to use magnets containing rare elements such as NdFeB for applications that require of high performance and compact motor structure. So that, manufacturers concentrate their effort on new designs which allows lowest magnet material usage to achieve lowest motor cost without compromising the motor performance As a result, it is necessary to obtain minimum magnet cost by optimizing the magnet geometry, especially in motors where high-cost magnets containing rare elements are used. In this study, deep learning approaches used in cancer diagnosis and treatment are examined. The aim of the study is to show with the support of the literature the extent to which effective results are obtained with a deep learning approach, which is one of the artificial intelligence techniques of a disease such as cancer, which methods and techniques are used, and how these methods are used. In the second part of the study, artificial intelligence and deep learning technique are explained; in the third part, deep learning approaches in cancer diagnosis and treatment processes and examples in the literature; in the fourth part, the results obtained from the study and the suggestions about the study are included.

Low-cost materials usage and high motor efficiency are the two main expectations that cause nonlinear multipurpose problems in design optimization of electric motors. Because of that, the optimum design requires a nonlinear programming technique that meets performance constraints [6, 7]. Using optimization algorithms in this process is a powerful tool for the motor designer [8]. In the literature there are many studies that focused on design optimizations based on motor performance and costs [7–12].

Kondapalli and Azrul have optimized a BLDC motor using GA and Simulated Annealing (SA) techniques and presented the comparison in their study. Both GA and SA optimization techniques have proven to be powerful tools to achieve optimal solutions of the BLDC motor compared to traditional design procedures. It is stated that GA is a more efficient method than SA when compared [7].

Markovic et al. have optimized a slotless surface-mount BLDC motor with classical (gradient-based), direct-search and GA. Total 7 variables have been considered which are based on iron axial length, magnet thickness, winding thickness, stator and rotor yoke thickness, mechanical air gap, ratio: pole arc top ole pitch [8]. Umadevi et al. have performed multiobjective optimization study of a 120 W BLDC motor in MATLAB environment with Enhanced Particle Swarm Optimization (PSO) algorithm. In study, it was tried to obtain maximizing the average torque and minimizing the cogging torque. Obtained results are presented in comparison with conventional PSO and GA [9]. Pichot et al. have proceeded multi-objective optimization with GA to get a high efficiency and low cost BLDC motor rated 30 W. In study, Ferrite magnet SPM motor model and NdFeB magnet IPM motor model were optimized. It has been concluded that the IPM motor is more efficient than the same cost SPM motor [10]. Rahideh et al. have performed an optimization study using the GA algorithm to obtain minimum loss, volume and cost in a slotless surface-mounted BLDC motor [11]. Abbaszadeh et al. have optimized the stator slot opening in a 6-pole surface mounted BLDC motor using Response Surface Methodology (RSM), the GA and The Particle Swarm Optimization (PSO) methods. All three optimum solutions obtained and verified with the FEA method and given comparatively. It was stated that cogging torque decreased by approximately 77% [12].

In this study, cost and efficiency-based design optimization of a high efficiency BLDC motor for powerful fan, pump and compressor applications has been conducted. Fan, pump and compressor applications correspond to approximately 70% of the industrial electric motor energy consumption. Fan, pump and compressors are the applications where maximum efficiency desired. Electric motors used in these applications are required to have the highest efficiency possible.

In this study, it is aimed to decrease motor cost while increasing motor efficiency thanks to the optimization studies. For this reason, in the optimization stage in the study, the purpose function aims to improve both cost and efficiency of the motor. Low-cost high-performance ratio is taken into consideration to achieve optimal BLDC motor design within optimization studies.

Primarily, a 12 Slot 8 Pole, 1500 W and 3000 rpm BLDC motor was designed with analytical methods and it was accepted as the initial model in the study. Stator slot geometry, magnet geometry, conductor per slot, wire cross-section area and core dimensions of the initial model were determined as variable in the optimization process. Selected variables were optimized to achieve maximum efficiency/minimum cost using GA. In addition to optimization studies, a standard induction motor which meets the minimum efficiency class defined by IEC and the optimized BLDC motor given in the study were compared in terms of energy savings.

## 2    Design Optimization of BLDC

### 2.1  Artificial Intelligence

ANSYS RMxprt which makes fast and low-error margin analytical calculations using equivalent circuit parameters was used in the optimization process. Also, GA which is a stochastic optimization method internally available in the software was used. The 3D model of the designed BLDC motor is presented in Fig. 1.

Genetic Algorithm, first formulated by Holland [13], is a probabilistic global search and optimization method that mimics the natural biological evolutionary process. The study of GA can be explained mainly as the selection of random individuals from the existing population, its development towards an optimal population for successive generations using cross and mutation operators [7].

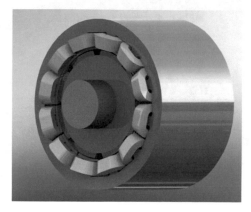

**Fig. 1.** 3D model of designed BLDC

Unlike traditional numerical techniques used in electric motor design, it is much faster and more efficient to reach the global optimum level by the nature of the scholastic search with GA. For this reason, GA is widely used in design optimization process of electric motors.

Geometric variables determined in the optimization process are given in Fig. 2. These variables are based on stator and magnet geometry. Explanations of the variables shown in Fig. 2 are given in Table 1.

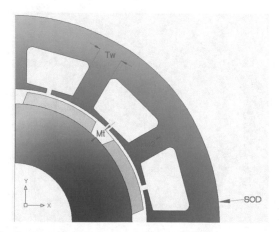

**Fig. 2.** Stator and magnet variables based on motor geometry

The change of geometric parameters in the stator affects both core volume and copper usage in the stator windings. Slot fill factor is determined to be maximum 55% for each design in the optimization process. Therefore, number of turns and wire cross-section in the windings are not defined as a separate variable. Axial core length (L) is also selected as a variable and affects the cost, efficiency and thermal load of the motor.

**Table 1.** Expression of variables

| Variables | Expression |
|-----------|-----------------------|
| SOD | Stator Outer Diameter |
| Hs2 | Stator Slot High |
| Tw | Slot Tooth With |
| Mt | Magnet Thickness |
| L | Iron Axial Length |

Air gap length of the motor is determined as 1 mm and considered as constant for each design. In permanent magnet electric machines, magnet arc to pole arc ratio can be expressed as embrace. Change of embrace value is schematically given in Fig. 3. The embrace value is a function of the magnet volume and affects the total magnet weight. Embrace value directly affects many parameters such as output coefficient, power factor, cogging torque and core flux densities [14]. Also, magnets can be accepted to have the highest unit price within the materials used in the BLDC motor. One of the most critical details in the optimization study is to obtain the optimum embrace value to achieve minimum cost.

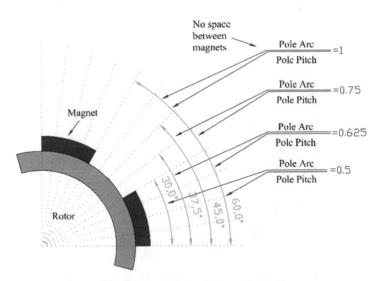

**Fig. 3.** Change of embrace value [15]

Objective function used in the multi-objective optimization study is given in Eq. 1.

$$F(x) = (W_1 \times F_1(x) + W_2 \times F_2(x)) \tag{1}$$

Where;

$F_1(x)$ = Minimization of loss.
$W_1 = 0,6$
$F_2(x)$ = Minimization of material weight.
$W_2 = 0,4$

W1 and W2 are weight factors for minimization of loss and minimization of material weight respectively. Initial, minimum and maximum values of the variables used in the optimization process are given in Table 2. Also, optimum design values obtained as a result of optimization study are given.

**Table 2.** Initial, minimum, maximum and optimum values of the variables.

| Variables | Initial value | Minimum value | Maximum value | Optimum value |
|---|---|---|---|---|
| Emb | 0.75 | 0.7 | 1 | 0.71 |
| Hs2 | 10 | 4.835 | 10 | 7.12 |
| L | 90 | 40 | 120 | 60.67 |
| Mt | 4 | 2 | 6 | 2.26 |
| Sod | 140 | 130 | 160 | 148 |
| Tw | 8 | 4.88 | 14.64 | 10.83 |
| Conductor Per Slot | 48 | Free | Free | 70 |

## 3 Optimization Results

The optimization results show that a high efficiency BLDC motor with optimum cost can be achieved within the specified limits. Material list, weights and costs obtained in the initial and optimum designs are given and compared in Table 3. Prices of 2018 are taken into account to calculate total cost [10]. Performance data of the initial and optimum BLDC motor designs are comparatively presented in Table 4.

Among the materials used in electric motors, neodymium magnets are the material which has the highest unit price in general. Another disadvantage of neodymium magnets can be expressed as a price instability. Obtaining minimum motor cost by the help of optimization study is also minimize the risk of magnet price instability.

It is seen that there is a 7% increase in motor efficiency (mechanical losses are not taken into account) in optimum design while total motor cost is reduced to almost half of the initial design. In addition, stator teeth flux density, specific electric loading parameters are also taken into consideration while selecting the global optimum point from multiple optimum results.

**Table 3.** Comparison of inital and optimal design based on cost and weight

| Parameters | Initial weight [kg] | Optimum weight [kg] | $/kg [13] | Initial cost [$] | Optimum cost [$] |
|---|---|---|---|---|---|
| Stator-Rotor Core | 6.48 | 5.8 | 1.12 | 7.26 | 6.50 |
| Copper | 1.18 | 0.57 | 7.05 | 8.32 | 4.02 |
| NdFe35 | 0.55 | 0.20 | 68.75 | 37.81 | 13.75 |
| Total price | | 53.39 | | 24.27 | |

**Table 4.** Comparison of performance parameters

| Parameters | Unit | Initial design | Optimum design |
|---|---|---|---|
| Core Loss [W] | W | 310.79 | 112.8 |
| Copper Loss [W] | W | 16.11 | 38.1 |
| Output Power [W] | W | 1740.2 | 1586.6 |
| Efficiency | – | %84.18 | %91.31 |
| Speed [rpm] | Rpm | 3487 | 3175 |
| Total Net Weight [kg] | Kg | 8.22 | 6.58 |

## 4   FEA Verification of Optimal Design

FEA is a practical and reliable method used to calculate the electromagnetic performance of the motor in the electrical machine design process. After the optimum solution was obtained with analytical calculations, 2D transient FEA analysis was performed with Ansys Maxwell and the optimum BLDC design was verified.

Magnetic flux distributions in critical parts of the motor such as stator and rotor slots were also observed in the analyses. Two-dimensional mesh structure of BLDC motor is given in Fig. 4-a. As seen in Fig. 4-b, motor is loaded to its nominal power to obtain proper flux density by transient analysis. The rated power of the motor and its performance are summarized in Table 4. The flux densities given in Table 5 are under the critical limits and meets the target values.

The flux density distributions obtained by 2D FEA transient analysis at nominal load are shown in Fig. 4. When the magnetic flux density distributions are examined, it is seen that results are consistent with the desired values.

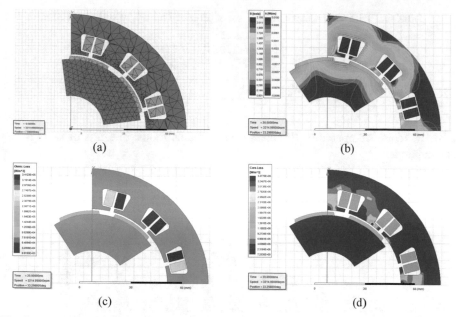

(a)                                                              (b)

(c)                                                              (d)

**Fig. 4.** (a) distribution of mesh, (b) distribution of magnetic flux and magnetic flux density at 20 ms, (c) copper loss at 20 ms, (d) core loss at 20 ms

**Table 5.** Magnetic flux densities of optimum BLDC motor

| Parameters | Unit | Optimum design |
|---|---|---|
| Stator Teeth Flux Density | T | 1.57 |
| Stator Yoke Flux Density | T | 0.78 |
| Air-Gap Flux Density | T | 0.73 |

## 5  FEA Verification of Optimal Design

Minimum efficiency values for different efficiency class are determined by the standards published via International Electrotechnical Commission (IEC) for line-fed asynchronous motors. The minimum efficiency values for 1500 rpm IE2 and IE3 efficiency class asynchronous motors are given in Table 6 and compared to efficiency of proposed BLDC.

It is seen that there is almost %210 difference in loss values between optimum BLDC motor loss and the average loss of IE2 and IE3 class asynchronous motor. The difference is much more with IE2 class as expected but loss of IE3 class considered for saving calculations. The potential energy savings, bill savings and payback time of the optimum BLDC motor are the result of high efficiency. Figure 5 shows the advantages

**Table 6.** Efficiency and loss comparison of proposed BLDC vs IE2 and IE3 IM

| Parameters | Optimum BLDC | IE2 IM | IE3 IM |
|---|---|---|---|
| Efficiency | %91.31 | %81.3 | %84.2 |
| Total Loss [W] | 150.93 | 345.01 | 281.47 |

of BLDC motor compared to IE3 class asynchronous motor in terms of savings and consumptions.

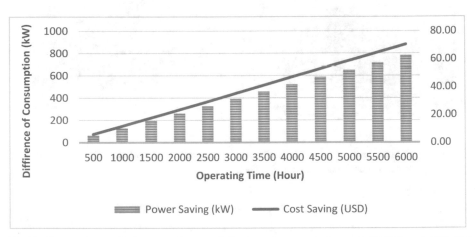

**Fig. 5.** Comparison of Consumptions and Savings between proposed BLDC and IE3 IM (22.05.2020-dated CBRT and EMRA rates taken into consideration for calculation)

## 6   Conclusion

In this study, a high efficiency and low-cost 1.5kW 3000 rpm BLDC motor has been designed and optimized for fan, pump and compressor applications. ANSYS RMxprt was used for analytical calculations and its GA tool was used for optimization studies to obtain minimum cost/maximum efficiency criterion. The FEA of the optimum BLDC motor was carried out via ANSYS Maxwell 2D and verified. As a result of the optimization study, efficiency of the initial design was increased from 84.16% to 91.31%, and the total cost was reduced from $ 53.60 to $ 24.27 by obtaining an optimum design.

In addition, the energy consumption of the optimized BLDC motor has been compared to 3-phase IE2 and IE3 class asynchronous motors whose efficiency values are determined in the IEC 60034-30 standard. It has been observed that proposed BLDC motor can save up to 650 kWh and $ 58.74 after 5000 h of use.

Prices of BLDC motors are generally high due to the magnets in their rotors compared to IMs. In addition, controller requirement for BLDC motors is also increases the initial investment. Nevertheless, the high efficiency of BLDC motors ensures the additional investment cost to be recovered in a short period.

# References

1. Kim, H.C., Jung, T.U.: Analysis of rotor overhang effect considering load torque variance in automobile BLDC fan motor. In 2012 IEEE Vehicle Power and Propulsion Conference, pp. 68–71. IEEE, October 2012
2. Suganthi, P., Nagapavithra, S., Umamaheswari, S.: Modeling and simulation of closed loop speed control for BLDC motor. In: 2017 Conference on Emerging Devices and Smart Systems (ICEDSS), pp. 229–233. IEEE, March 2017
3. Kumar, B.R., Kumar, K.S.: Design of a new dual rotor radial flux BLDC motor with Halbach array magnets for an electric vehicle. In: 2016 IEEE International Conference on Power Electronics, Drives and Energy Systems (PEDES), pp. 1–5. IEEE, December 2016
4. Leitner, S., Gruebler, H., Muetze, A.: Innovative low-cost sub-fractional hp BLDC claw-pole machine design for fan applications. IEEE Trans. Ind. Appl. **55**(3), 2558–2568 (2019)
5. Hasanuzzaman, M., Rahim, N.A., Saidur, R.: Analysis of energy savings for rewinding and replacement of industrial motor. In: 2010 IEEE International Conference on Power and Energy, pp. 212–217. IEEE, November 2010
6. Lee, H.J., Kim, J.H., Lee, J., Joo, S.J., Kim, C.E.: Fluid analysis of hollow shafted BLDC motor for fan blower application. In: 2018 21st International Conference on Electrical Machines and Systems (ICEMS), pp. 329–331. IEEE, October 2018
7. Rao, K.S.R., Othman, A.H.B.: Design optimization of a bldc motor by genetic algorithm and simulated annealing. In: 2007 International Conference on Intelligent and Advanced Systems, pp. 854–858. IEEE, November 2007
8. Markovic, M., Ragot, P., Perriard, Y.: Design optimization of a BLDC motor: a comparative analysis. In: 2007 IEEE International Electric Machines & Drives Conference, vol. 2, pp. 1520–1523. IEEE, May 2007
9. Umadevi, N., Balaji, M., Kamaraj, V.: Design optimization of brushless DC motor using particle swarm optimization. In: 2014 IEEE 2nd International Conference on Electrical Energy Systems (ICEES), pp. 122–125. IEEE, January 2014
10. Pichot, R., Schmerber, L., Paire, D., Miraoui, A.: Robust BLDC motor design optimization including raw material cost variations. In: 2018 XIII International Conference on Electrical Machines (ICEM), pp. 961–967. IEEE, September 2018
11. Rahideh, A., Korakianitis, T., Ruiz, P., Keeble, T., Rothman, M.T.: Optimal brushless DC motor design using genetic algorithms. J. Magn. Magn. Mater. **322**(22), 3680–3687 (2010)
12. Abbaszadeh, K., Alam, F.R., Teshnehlab, M.: Slot opening optimization of surface mounted permanent magnet motor for cogging torque reduction. Energy Convers. Manag. **55**, 108–115 (2012)
13. Holland, J.H.: Adaptation in Natural and Artificial Systems: An Introductory Analysis with Applications to Biology, Control, and Artificial Intelligence. MIT Press, Cambridge (1992)
14. Yılmaz, C., Yenipınar, B., Sönmez, Y., Ocak, C.: Optimization of PMSM design parameters using update meta-heuristic algorithms. In: Hemanth, D.J., Kose, U. (eds.) ICAIAME 2019. LNDECT, vol. 43, pp. 914–934. Springer, Cham (2020). https://doi.org/10.1007/978-3-030-36178-5_81
15. Ocak, C., et al.: Investigation effects of narrowing rotor pole embrace to efficiency and cogging torque at PM BLDC motor. TEM J. **5**(1), 25 (2016)

# Stability of a Nonautonomous Recurrent Neural Network Model with Piecewise Constant Argument of Generalized Type

Duygu Aruğaslan Çinçin[1] and Nur Cengiz[2(✉)]

[1] Süleyman Demirel University, Department of Mathematics, 32260 Isparta, Turkey
`duyguarugaslan@sdu.edu.tr`
[2] Süleyman Demirel University, Graduate School of Natural and Applied Sciences, 32260 Isparta, Turkey

**Abstract.** In this work, a nonautonomous model of recurrent neural networks is developed by generalized piecewise constant argument as discontinuous argument. By this deviating argument, more realistic model is taken into consideration and so more realistic theoretical results are reached. Then, sufficient conditions guaranteeing the uniform asymptotic stability of trivial solution of the related recurrent neural network model with generalized type piecewise constant argument is examined mathematically. For this aim, both Lyapunov-Razumikhin and Lyapunov-Krasovskii methods are considered. The findings are supported by examples and simulations with the help of MATLAB package program.

**Keywords:** Piecewise constant argument of generalized type ·
Equilibrium · Uniform asymptotic stability · Lyapunov-Razumikhin
method · Lyapunov-Krasovskii method · Recurrent neural networks ·
Simulation

## 1  Introduction and Preliminaries

Studies on modeling and developing artificial neural networks (ANN) indicate that it is an area that attracts the attention of many scientists. The first model of artificial neurons was introduced in 1943 by McCulloch and Pitts [1]. Since then, the dynamics of ANN has been one of the most remarkable research areas [2–7]. By the models proposed for ANN, it is aimed to represent the structure of the nervous system appropriately. Accordingly, these equations have been examined in many fields of the science and their various types such as recurrent, cellular, Cohen-Grossberg neural networks have been described. Qualitative properties of ANN have been deeply investigated in [4,8–14,33]. While investigating these properties, it is important to develop models taking the natural paradigms into consideration. For example, state at a past moment of a system can seriously affect the state at the current moment. From a mathematical perspective, this effect can be modeled in accordance with the concept of differential equations with deviating arguments [4,13,15–19].

J. Hemanth et al. (Eds.): ICAIAME 2020, LNDECT 76, pp. 524–539, 2021.
https://doi.org/10.1007/978-3-030-79357-9_51

In our paper, a recurrent neural network model is developed by a deviating argument which is of delayed type. This argument was introduced and called as piecewise constant argument of generalized type (PCAG) [20,21]. Differential equations with PCAG (EPCAG) have been studied in [22–24] and have attracted the attention of many scientists [17–19,22,25,26]. Any piecewise constant function (i.e. PCAG) can be taken as a deviating argument in EPCAG. The length of the intervals determined by the shifting moments of the argument may be variable for EPCAG, and thus PCAG is a more general argument than the greatest integer function [27]. So, the real life situations can be modeled mathematically in a more general form by PCAG. Moreover, the theory of EPCAG is useful while investigating dynamic behavior of the real life problems. For example, many theoretical results in [22–24,28] were presented by combining the concept of EPCAG with the theory in the papers [29–31] and more.

In our paper, the main goal is to examine the stability of the nonautonomous nonlinear recurrent neural network

$$x_i'(t) = -a_i(t)f_i(t, x_i(t)) + \sum_{j=1}^{n} b_{ij}(t)h_j(t, x_j(t)) + \sum_{j=1}^{n} c_{ij}(t)g_j(t, x_j(\beta(t)))$$

$$+ \sum_{j=1}^{n} d_{ij}(t)k_j(t, , x_j(t), x_j(\beta(t))) + e_i(t), \ i = 1, ..., n, \tag{1}$$

with PCAG using Lyapunov-Razumikhin method [22] and Lyapunov-Krasovskii method [29]. Here, we aim to give a different perspective by considering a nonautonomous network system. Neural network systems were rarely studied by considering nonautonomous functions in the literature [8,19,32]. Let $\mathbb{N}$, $\mathbb{R}$ and $\mathbb{R}^+$ describe the sets of all non-negative integers, real numbers and non-negative real numbers, respectively, i.e., $\mathbb{N} = \{0, 1, 2, ...\}$, $\mathbb{R} = (-\infty, \infty)$ and $\mathbb{R}^+ = [0, \infty)$. Denote by $\mathbb{R}^n$, $n \in \mathbb{N}$, the real space of dimension $n$ and consider a norm on $\mathbb{R}^n$ by $\|.\|$ such that $\|u\| = \sum_{i=1}^{n} |u_i|$. In (1), $x(t) = (x_1(t), ..., x_n(t))^T \in \mathbb{R}^n$ is the neuron state vector while $n$ denotes the number of units existing in the network. Here, $f_i \in C(\mathbb{R}^+ \times \mathbb{R}, \mathbb{R})$ is a nonlinear nonautonomous function for each $i = 1, ..., n$. Moreover, nonautonomous and nonlinear activation functions $h_j, g_j \in C(\mathbb{R}^+ \times \mathbb{R}, \mathbb{R})$, $k_j \in C(\mathbb{R}^+ \times \mathbb{R} \times \mathbb{R}, \mathbb{R})$, $j = 1, ..., n$, of neurons denote the measures of activation to its incoming potentials of the $j$th unit, and the nonautonomous inputs $e_i(t) \in C(\mathbb{R}^+, \mathbb{R})$, $i = 1, ..., n$, are external bias on the unit $i$ of the system at $t$. $a_i(t) \in C(\mathbb{R}^+, \mathbb{R}^+)$, $i = 1, ..., n$, correspond to the functions that represent the rate with which the unit $i$ resets its potential to the resting state in isolation when it is not connected to the network and external inputs. Moreover, $b_{ij}(t)$, $c_{ij}(t)$, $d_{ij}(t) \in C(\mathbb{R}^+, \mathbb{R})$, $i, j = 1, ..., n$, are time variable connection weights of the unit $j$ of the system on the $i$th unit at time $t$. We take a real valued sequence $\theta = \{\theta_l\}$, $l \in \mathbb{N}$ with $0 \leq \theta_l < \theta_{l+1}$ for all $l \in \mathbb{N}$ and $\theta_l \to \infty$ as $l \to \infty$. In (1), $\beta(t) = \theta_l$ if $t \in [\theta_l, \theta_{l+1})$, $l \in \mathbb{N}$. Let us presume without loss of generality that the initial time $t_0 \in (\theta_l, \theta_{l+1}]$ for some $l \in \mathbb{N}$.

**Lemma 1.** *Fix a real number* $t_0$. $x(t) = x(t, t_0, x^0) = (x_1(t), ..., x_n(t))^T$ *is a solution of system (1) on* $\mathbb{R}^+$ *if and only if it is a solution of*

$$x_i(t) = x_i^0 + \int_{t_0}^t \left( -a_i(s) f_i(s, x_i(s)) + \sum_{j=1}^n b_{ij}(s) h_j(s, x_j(s)) \right. \tag{2}$$

$$\left. + \sum_{j=1}^n c_{ij}(s) g_j(s, x_j(\beta(s))) + \sum_{j=1}^n d_{ij}(s) k_j(s, x_j(s), x_j(\beta(s))) + e_i(s) \right) ds$$

*on* $\mathbb{R}^+$, *where* $i = 1, ..., n$.

We will use the assumptions given below:

(C1) There exist Lipschitz constants $\ell_i^f$, $\ell_j^h$, $\ell_j^g$, $\ell_j^k > 0$ satisfying the following inequalities

$$|f_i(t, z_1) - f_i(t, z_2)| \le \ell_i^f |z_1 - z_2|,$$

$$|h_j(t, z_1) - h_j(t, z_2)| \le \ell_j^h |z_1 - z_2|,$$

$$|g_j(t, z_1) - g_j(t, z_2)| \le \ell_j^g |z_1 - z_2|,$$

$$|k_j(t, z_1, z_2) - k_j(t, z_3, z_4)| \le \ell_j^k (|z_1 - z_3| + |z_2 - z_4|)$$

for all $t \in \mathbb{R}^+$, $z_m \in \mathbb{R}$, $m = 1, 2, 3, 4$, $i, j = 1, ..., n$;

(C2) The positive numbers $\underline{\theta}, \overline{\theta}$ exist such that the inequality $\underline{\theta} \le \theta_{l+1} - \theta_l \le \overline{\theta}$, $l \in \mathbb{N}$ is true.

For simplicity, let us define the following notations:

$$\overline{a_i} = \sup_{t \in \mathbb{R}^+} (a_i(t)), \ \overline{b_{ij}} = \sup_{t \in \mathbb{R}^+} (|b_{ij}(t)|), \ \overline{c_{ij}} = \sup_{t \in \mathbb{R}^+} (|c_{ij}(t)|), \ \overline{d_{ij}} = \sup_{t \in \mathbb{R}^+} (|d_{ij}(t)|),$$

$$\overline{e_i} = \sup_{t \in \mathbb{R}^+} (|e_i(t)|), \ \sum_{i=1}^n \overline{e_i} := \overline{e}, \ \max_{1 \le i \le n} \left( \sum_{j=1}^n \overline{d_{ji}} \ell_i^k \right) := \overline{d}_k,$$

$$\max_{1 \le i \le n} \overline{a_i} \ell_i^f := \overline{a}_f, \ \max_{1 \le i \le n} \left( \sum_{j=1}^n \overline{b_{ji}} \ell_i^h \right) := \overline{b}_h, \ \max_{1 \le i \le n} \left( \sum_{j=1}^n \overline{c_{ji}} \ell_i^g \right) := \overline{c}_g,$$

$$\sup_{t \in \mathbb{R}^+} |f_i(t, 0)| = M_{f_i}, \ \sup_{t \in \mathbb{R}^+} |h_j(t, 0)| = M_{h_j}, \ \sup_{t \in \mathbb{R}^+} |g_j(t, 0)| = M_{g_j}, \ \sup_{t \in \mathbb{R}^+} |k_j(t, 0, 0)| = M_{k_j},$$

$$\sum_{i=1}^n \overline{a_i} M_{f_i} := M_f^a, \ \sum_{i=1}^n \sum_{j=1}^n \overline{b_{ij}} M_{h_j} := M_h^b, \ \sum_{i=1}^n \sum_{j=1}^n \overline{c_{ij}} M_{g_j} := M_g^c, \ \sum_{i=1}^n \sum_{j=1}^n \overline{d_{ij}} M_{k_j} := M_k^d,$$

$$\rho_* = \sum_{j=1}^n \left( \overline{c_{ji}} \ell_i^g + \overline{d_{ji}} \ell_i^k \right), \ \rho^* = \overline{a_i} \ell_i^f + \sum_{j=1}^n \left( \overline{b_{ji}} \ell_i^h + \overline{d_{ji}} \ell_i^k \right)$$

for $i, j = 1, ..., n$. Here, $\rho_*$, $\rho^*$ are the constants.

Functions $f_i(t, u_i)$ can be written as $f_i(t, u_i) = {}_*f_i(t)u_i + f_i^*(t, u_i)$, where ${}_*f_i(t)$, $f_i^*(t, u_i)$ represent the functions. Suppose that $0 < {}_*\underline{f_i} \leq {}_*f_i(t) \leq \overline{{}_*f_i}$, where ${}_*\underline{f_i}$ and $\overline{{}_*f_i}$ are constants. Then, it can be seen that $\ell_i^f = \ell_i^{f^*} + \overline{{}_*f_i}$ if Lipschitz constants $\ell_i^{f^*} > 0$ for each $i = 1, ..., n$ exist satisfying $|f_i^*(t, u_i) - f_i^*(t, v_i)| \leq \ell_i^{f^*}|u_i - v_i|$, for all $t \in \mathbb{R}^+$, $u_i$, $v_i \in \mathbb{R}$.

Using the notations defined above, we assume that the followings are true:

(C3) $\ell_i^{f^*}\underline{a_i} + \sum_{j=1}^{n}\left(\overline{b_{ji}}\ell_i^h + \overline{c_{ji}}\ell_i^g + 2d_{ji}\ell_i^k\right) < \underline{a_i}\,{}_*\underline{f_i}$ for all $i = 1, ..., n$;

(C4) $\left(\overline{a_f} + \overline{b_h} + 2\overline{c_g} + 3\overline{d_k}\right)\overline{\theta}e^{\left(\overline{a_f} + \overline{b_h} + \overline{d_k}\right)\overline{\theta}} < 1$.

## 1.1 Equilibrium Point and Existence-Uniqueness of Solutions

Denote the constant vector by $x^* = (x_1^*, ..., x_n^*)^T \in \mathbb{R}^n$, where the components are conducted by the following algebraic system

$$0 = -a_i(t)f_i(t, x_i^*) + \sum_{j=1}^{n}b_{ij}(t)h_j(t, x_j^*) + \sum_{j=1}^{n}c_{ij}(t)g_j(t, x_j^*)$$

$$+ \sum_{j=1}^{n}d_{ij}(t)k_j^*(t, x_j^*) + e_i(t), \tag{3}$$

where $k_j^*(t, x_j^*) = k_j(t, x_j^*, x_j^*)$. It is seen that $\left|k_j^*(t, u_j) - k_j^*(t, v_j)\right| \leq 2\ell_j^k|u_j - v_j|$.

**Lemma 2.** *Suppose that the functions $f_i^*$, $i = 1, ..., n$, are globally Lipschitz continuous such that $|f_i^*(t, u_i) - f_i^*(t, v_i)| \leq \ell_i^{f^*}|u_i - v_i|$ for all $t \in \mathbb{R}^+$, $u_i$, $v_i \in \mathbb{R}$, where $\ell_i^{f^*}$ are positive constants. If the conditions (C1) and (C3) hold then there is a unique constant vector $x^* = (x_1^*, ..., x_n^*)$ which satisfies (3).*

*Proof.* Now, we will show that the following system

$$a_i(t)_*f_i(t)x_i^* = -a_i(t)f_i^*(t, x_i^*) + \sum_{j=1}^{n}b_{ij}(t)h_j(t, x_j^*) + \sum_{j=1}^{n}c_{ij}(t)g_j(t, x_j^*)$$

$$+ \sum_{j=1}^{n}d_{ij}(t)k_j^*(t, x_j^*) + e_i(t), \ i = 1, ..., n, \tag{4}$$

has a unique solution based on the contraction mapping principle [33]. Define on $\mathbb{R}^n$ the operator $\Pi$ such that

$$\Pi(z_1, z_2, ..., z_n) = \begin{cases} \Pi_1(z_1, z_2, ..., z_n) \\ \vdots \\ \Pi_n(z_1, z_2, ..., z_n), \end{cases}$$

where

$$\Pi_i(z) = -\frac{f_i^*(t, z_i)}{{}_*f_i(t)} + \frac{1}{a_i(t)_*f_i(t)} \sum_{j=1}^{n} (b_{ij}(t)h_j(t, z_j) + c_{ij}(t)g_j(t, z_j))$$

$$+ \frac{1}{a_i(t)_*f_i(t)} \sum_{j=1}^{n} d_{ij}(t)k_j^*(t, z_j) + \frac{e_i(t)}{a_i(t)_*f_i(t)}, \quad i = 1, ..., n. \tag{5}$$

Let us show that the operator $\Pi : \mathbb{R}^n \to \mathbb{R}^n$ is contractive. So, for $u, v \in \mathbb{R}^n$, it follows from

$$\|\Pi(u) - \Pi(v)\| \leq \max_{1 \leq i \leq n} \left( \frac{\ell_i^{f^*}}{{}_*f_i} + \frac{1}{\underline{a_i}\,{}_*f_i} \sum_{j=1}^{n} (\overline{b_{ji}}\ell_i^h + \overline{c_{ji}}\ell_i^g + 2\overline{d_{ji}}\ell_i^k) \right) \|u - v\|$$

and so from the condition (C3) that the mapping $\Pi : \mathbb{R}^n \to \mathbb{R}^n$ is a global contraction. Therefore, the mapping $\Pi$ has a unique fixed point based on contraction mapping principle. □

Based on Lemma 2, the following result follows.

**Theorem 1.** *If the conditions (C1) and (C3) hold true, then system (1) possesses a unique equilibrium point $x^*$.*

Now, for arbitrary initial moment $\xi$, sufficient conditions relating to the existence-uniqueness of the solution of (1) on $[\theta_l, \theta_{l+1}]$ can be found in Lemma 3.

**Lemma 3.** *Fix $l \in \mathbb{N}$. Suppose that the assumptions (C1)–(C2), (C4) are satisfied. Then, for every $(\xi, x^0) \in [\theta_l, \theta_{l+1}] \times \mathbb{R}^n$, there exists a unique solution $x(t) = x(t, \xi, x^0) = (x_1(t), ..., x_n(t))^T$ of (1) on $[\theta_l, \theta_{l+1}]$ with $x(\xi) = x^0$.*

*Proof.* Existence. Without loss of generality, we take $\theta_l \leq \xi \leq \theta_{l+1}$. For convenience, let $z(t) = x(t, \xi, x^0) = (z_1(t), ..., z_n(t))^T$. Describe a norm $\|z(t)\|_0 = \max_{[\theta_l, \xi]} \|z(t)\|$. By Lemma 1, consider a solution of (1) given by

$$z_i(t) = x_i^0 + \int_{\xi}^{t} \left( -a_i(s)f_i(s, z_i(s)) + \sum_{j=1}^{n} b_{ij}(s)h_j(s, z_j(s)) + \sum_{j=1}^{n} c_{ij}(s)g_j(s, z_j(\theta_l)) \right.$$

$$\left. + \sum_{j=1}^{n} d_{ij}(s)k_j(s, z_j(s), z_j(\theta_l)) + e_i(s) \right), \quad i = 1, ..., n. \tag{6}$$

Construct the sequences $z_i^m(t)$, $z_i^0(t) \equiv x_i^0$, $i = 1, ..., n$, as follows

$$z_i^{m+1}(t) = x_i^0 + \int_\xi^t \left( -a_i(s)f_i(s, z_i^m(s)) + \sum_{j=1}^n b_{ij}(s)h_j(s, z_j^m(s)) \right.$$

$$+ \sum_{j=1}^n c_{ij}(s)g_j(s, z_j^m(\theta_l))$$

$$\left. + \sum_{j=1}^n d_{ij}(s)k_j(s, z_j^m(s), z_j^m(\theta_l)) + e_i(s) \right), \quad i = 1, ..., n, \ m \geq 0 .$$

Using the last equation, the following conclusion can be reached:

$$\left\| z^{m+1}(t) - z^m(t) \right\|_0 \leq \left( (\bar{a}_f + \bar{b}_h + \bar{c}_g + 2\bar{d}_k) \bar{\theta} \right)^m \Omega, \tag{7}$$

where $\Omega = (\bar{a}_f + \bar{b}_h + \bar{c}_g + 2\bar{d}_k) \bar{\theta} \|x^0\| + \left( \bar{e} + M_f^a + M_h^b + M_g^c + M_k^d \right) \bar{\theta}$. The condition (C4) implies that the sequences $z_i^m(t)$ are convergent and their limits satisfy the integral equation (6) on $[\theta_l, \xi]$. Besides, it can be seen by similar computations that the solution can be continued to $[\xi, \theta_{l+1}]$ and this shows the existence of the solution on $[\theta_l, \theta_{l+1}]$.

*Uniqueness.* Denote the solutions of (1) by $x^r(t) = x(t, \xi, x^r)$, where $\theta_l \leq \xi \leq \theta_{l+1}$ $x^r = (x_1^r, ..., x_n^r)^T$ $r = 1, 2$. Now, let us show that $x^1(t) \neq x^2(t)$ for each $t \in [\theta_l, \theta_{l+1}]$ and for $x^1 \neq x^2$. Then, it can be seen that

$$\|x^1(t) - x^2(t)\| \leq e^{(\bar{a}_f + \bar{b}_h + \bar{d}_k)\bar{\theta}} \left( \|x^1 - x^2\| + \bar{\theta} \left( \bar{c}_g + \bar{d}_k \right) \|x^1(\theta_l) - x^2(\theta_l)\| \right) \tag{8}$$

by the Gronwall-Bellman Lemma and by the conditions (C1)–(C2), (C4). While $t = \theta_l$, it is obvious that (8) takes the following form

$$\|x^1(\theta_l) - x^2(\theta_l)\| \leq \left( \|x^1 - x^2\| + \bar{\theta} \left( \bar{c}_g + \bar{d}_k \right) \|x^1(\theta_l) - x^2(\theta_l)\| \right) e^{(\bar{a}_f + \bar{b}_h + \bar{d}_k)\bar{\theta}}. \tag{9}$$

So, (8) and (9) give the following inequality

$$\|x^1(t) - x^2(t)\| \leq \left( 1 - \bar{\theta} \left( \bar{c}_g + \bar{d}_k \right) e^{(\bar{a}_f + \bar{b}_h + \bar{d}_k)\bar{\theta}} \right)^{-1} e^{(\bar{a}_f + \bar{b}_h + \bar{d}_k)\bar{\theta}} \|x^1 - x^2\|. \tag{10}$$

Assume contrarily that there exists $t \in [\theta_l, \theta_{l+1}]$ satisfying $x^1(t) = x^2(t)$. Then,

$$\|x^1 - x^2\| \leq \int_\xi^t \left( \bar{a}_f + \bar{b}_h + \bar{d}_k \right) \|x^1(s) - x^2(s)\| \, ds + \bar{\theta} \left( \bar{c}_g + \bar{d}_k \right) \|x^1(\theta_l) - x^2(\theta_l)\|.$$

Moreover, by (10), the last inequality yields that

$$\|x^1 - x^2\| \leq \frac{(\bar{a}_f + \bar{b}_h + \bar{c}_g + 2\bar{d}_k) \bar{\theta}}{1 - \bar{\theta}(\bar{c}_g + \bar{d}_k)e^{(\bar{a}_f + \bar{b}_h + \bar{d}_k)\bar{\theta}}} e^{(\bar{a}_f + \bar{b}_h + \bar{d}_k)\bar{\theta}} \|x^1 - x^2\|.$$

However, this is a contradiction to (C4). So, uniqueness and thus the lemma are proved.
$\qquad\qquad\qquad\qquad\qquad\qquad\qquad\qquad\qquad\qquad\qquad\qquad\qquad\qquad\qquad\qquad\quad$ $\square$

As a consequence of Lemma 3, the next theorem assures the existence and uniqueness of the solution of system (1) on $\mathbb{R}^+$, which can be proved by the mathematical induction.

**Theorem 2.** *Assume that the conditions (C1)–(C4) are satisfied. Then, for every* $(t_0, x^0) \in \mathbb{R}^+ \times \mathbb{R}^n$, *there exists a unique solution* $x(t) = x(t, t_0, x^0) = (x_1(t), ..., x_n(t))^T$ *of (1) on* $\mathbb{R}^+$ *with* $x(t_0) = x^0$.

*Proof.* $l \in \mathbb{N}$ exists such that $t_0 \in [\theta_l, \theta_{l+1})$. In view of Lemma 3, for $\xi = t_0$, a unique solution $x(t) = x(t, t_0, x^0)$ of (1) exists on the interval $[\xi, \theta_{l+1}]$. Then, existence of a unique solution on the intervals $[\theta_{l+1}, \theta_{l+2}]$, $[\theta_{l+2}, \theta_{l+3}]$, ... can be seen by the mathematical induction with the help of Lemma 3. The proof is completed since the continuation of the unique solution to $\mathbb{R}^+$ is clear.

## 2    Uniform Asymptotic Stability of the Equilibrium Point

Our main goal is to examine the uniform asymptotic stability of the equilibrium of system (1). An approach used in the investigation of EPCAG is remarkable and is based on creating an equivalent integral equation [20, 21, 23]. It follows from the related approach that definitions for the initial value problem for EPCAG and classical ordinary differential equations (ODEs) have similarity. Stability analysis can be performed by choosing any real number as an initial moment. Consequently, definitions of stability for EPCAG coincide with the ones given for ODEs [23]. While doing the stability examination, we take Lyapunov-Razumikhin [22] and Lyapunov-Krasovskii [29] methods into account. For this aim, firstly, we perform a linear transformation. Using (1), we find out that $x_i(t) - x_i^* = y_i(t)$ satisfies the neural network system

$$y_i'(t) = -a_i(t)F_i(t, y_i(t)) + \sum_{j=1}^{n} b_{ij}(t)H_j(t, y_j(t)) + \sum_{j=1}^{n} c_{ij}(t)G_j(t, y_j(\beta(t)))$$

$$+ \sum_{j=1}^{n} d_{ij}(t)K_j(t, y_j(t), y_j(\beta(t))), \tag{11}$$

where $y = (y_1, ..., y_n)^T \in \mathbb{R}^n$, $F_i(t, y_i(t)) = f_i(t, y_i(t) + x_i^*) - f_i(t, x_i^*)$, $H_j(t, y_j(t)) = h_j(t, y_j(t) + x_j^*) - h_j(t, x_j^*)$, $G_j(t, y_j(t)) = g_j(t, y_j(t) + x_j^*) - g_j(t, x_j^*)$ and $K_j(t, y_j(t), y_j(\beta(t))) = k_j(t, y_j(t) + x_j^*, y_j(\beta(t)) + x_j^*) - k_j(t, x_j^*, x_j^*)$, $i = 1, ..., n$. It follows from (C1) that $|F_i(t, z)| \leq \ell_i^f |z|$, $|H_j(t, z)| \leq \ell_j^h |z|$, $|G_j(t, z)| \leq \ell_j^g |z|$ and $|K_j(t, z_1, z_2)| \leq \ell_j^k (|z_1| + |z_2|)$ for all $t \in \mathbb{R}^+$, $z$, $z_1$, $z_2 \in \mathbb{R}$, $i, j = 1, ..., n$. Let the following additional assumptions be satisfied:

(C5) $\overline{\theta} \max_{1 \leq i \leq n} \left\{ \rho_* + \rho^* \left(1 + \overline{\theta}\rho_*\right) e^{\overline{\theta}\rho^*} \right\} < 1$;

(C6) $\delta_{F_i} \leq \dfrac{F_i(t, u)}{u}$, $u \neq 0$, $u \in \mathbb{R}$, where $\delta_{F_i}$ is a positive constant for each $i = 1, ..., n$.

For convenience, we define two notations and assume that the following conditions hold:

$$\rho_1 = \min_{1 \leq i \leq n} \left( \underline{a_i} \delta_{F_i} - \frac{1}{2} \sum_{j=1}^{n} \left( \overline{b_{ji}} \ell_i^h + \overline{b_{ij}} \ell_j^h + \overline{c_{ij}} \ell_j^g + \overline{d_{ji}} \ell_i^k + 2\overline{d_{ij}} \ell_j^k \right) \right),$$

$$\rho_2 = \frac{1}{2} \max_{1 \leq i \leq n} \left( \sum_{j=1}^{n} \left( \overline{c_{ji}} \ell_i^g + \overline{d_{ji}} \ell_i^k \right) \right).$$

(C7) $\rho_2 < \rho_1$;

(C8) $\rho_1 - \rho_2 \kappa^2 - \overline{\theta}\overline{\alpha} > 0$, where $\overline{\alpha} = \max_{1 \leq i \leq n} \left\{ \sum_{j=1}^{n} \alpha_{ji} \right\}$ and $\alpha_{ij}, i = 1, 2, ..., j = 1, 2, ...,$ are arbitrary constants.

The following lemma provides us a notable information on the relation between the values of the solution at any time $t$ and at the deviating argument $\beta(t)$. This relation will be very useful for the proofs of stability through both Lyapunov-Razumikhin and Lyapunov-Krasovskii methods.

**Lemma 4.** *Let the conditions (C1)–(C5) be satisfied and* $y(t) = (y_1(t), ..., y_n(t))^T$ *denote a solution of system (11). Then*

$$\| y(\beta(t)) \| \leq \kappa \| y(t) \|,$$

*is true for all* $t \in \mathbb{R}^+$. *Here* $\kappa = \left( 1 - \overline{\theta} \max_{1 \leq i \leq n} \left\{ \rho_* + \rho^* \left( 1 + \overline{\theta}\rho_* \right) e^{\overline{\theta}\rho^*} \right\} \right)^{-1}$.

*Proof.* Fix $t \in \mathbb{R}^+$ and $l \in \mathbb{N}$. For $t \in [\theta_l, \theta_{l+1})$, one can write

$$y_i(t) = y_i(\theta_l) + \int_{\theta_l}^{t} \left( -a_i(s)F_i(s, y_i(s)) + \sum_{j=1}^{n} b_{ij}(s)H_j(s, y_j(s)) \right.$$

$$\left. + \sum_{j=1}^{n} c_{ij}(s)g_j(s, y_j(\theta_l)) + \sum_{j=1}^{n} d_{ij}(s)K_j(s, y_j(s), y_j(\theta_l)) \right) ds. \quad (12)$$

Therefore,

$$|y_i(t)| \leq |y_i(\theta_l)| \left( 1 + \rho_*\overline{\theta} \right) + \int_{\theta_l}^{t} \rho^* |y_i(s)| \, ds \quad (13)$$

and $\| y(t) \| \leq \| y(\theta_l) \| \max_{1 \leq i \leq n} \left( 1 + \rho_*\overline{\theta} \right) + \int_{\theta_l}^{t} \max_{1 \leq i \leq n} \{ \rho^* \} \| y(s) \| \, ds$ are obtained. In the sequel, based on the Gronwall-Bellman Lemma,

$$\| y(t) \| \leq \| y(\theta_l) \| \max_{1 \leq i \leq n} \left\{ \left( 1 + \rho_*\overline{\theta} \right) e^{\rho^*\overline{\theta}} \right\} \quad (14)$$

is reached.

Moreover, for $t \in [\theta_l, \theta_{l+1})$,

$$|y_i(\theta_l)| \le |y_i(t)| + |y_i(\theta_l)| \, \rho_* \overline{\theta} + \int_{\theta_l}^t \rho^* \, |y_i(s)| \, ds \tag{15}$$

and

$$\|y(\theta_l)\| \le \|y(t)\| + \|y(\theta_l)\| \max_{1 \le i \le n} \{\rho_*\} \overline{\theta} + \int_{\theta_l}^t \max_{1 \le i \le n} \{\rho^*\} \|y(s)\| \, ds \tag{16}$$

are obtained by the integral equation (12). By using (14) in the last inequality,

$$\|y(\theta_l)\| \le \|y(t)\| + \|y(\theta_l)\| \overline{\theta} \max_{1 \le i \le n} \left( \rho_* + \rho^* \left( 1 + \overline{\theta} \rho_* \right) e^{\overline{\theta} \rho^*} \right)$$

and so

$$\|y(\theta_l)\| \le \kappa \, \|y(t)\| \, , \ t \in [\theta_l, \theta_{l+1}) \tag{17}$$

are obtained in the light of the condition (C5). Therefore, it can be seen that the last inequality holds true for all $t \in \mathbb{R}^+$. So, the lemma is proved. □

## 2.1  Lyapunov-Razumikhin Method

Now, we shall consider the Lyapunov-Razumikhin method handled in [22] for EPCAG. Let us describe the following notations:

$$\mathcal{A} = \left\{ u \in C \left( \mathbb{R}^+, \mathbb{R}^+ \right) : u \text{ is strictly increasing, } u(0) = 0 \right\}$$

and

$$\mathcal{B} = \left\{ v \in C \left( \mathbb{R}^+, \mathbb{R}^+ \right) : v(0) = 0, v(s) > 0 \text{ for } s > 0 \right\}.$$

**Lemma 5.** *Suppose that $y(t) : \mathbb{R}^+ \to \mathbb{R}^n$ is a solution of (11) and assumptions (C1)–(C7) hold true. Then the following assumptions are satisfied for $V(y(t)) = \sum_{i=1}^n \dfrac{y_i^2(t)}{2}$:*

*(a)* $\dfrac{\|y(t)\|^2}{2n} \le V(y(t)) \le \dfrac{\|y(t)\|^2}{2}$;

*(b)* $V'(y(t), y(\beta(t))) \le -(\rho_1 - \rho_2 \sigma) \|y(t)\|^2$ *for all $t \ne \theta_l$ in $\mathbb{R}^+$ such that $V(y(\beta(t))) < \sigma V(y(t))$ with $1 < \sigma < \dfrac{\rho_1}{\rho_2}$.*

*Proof.* Consider the function

$$V(y(t)) = \sum_{i=1}^n \frac{y_i^2(t)}{2}. \tag{18}$$

It is clear that the item $(a)$ is true.

Assume that $\psi(s) = \sigma s$ is a continuous nondecreasing function for $s > 0$ and $w(u)$ is a function given by $w(u) = (\rho_1 - \rho_2\sigma)u^2 \in \mathcal{B}$. Let us differentiate the Lyapunov function (18) for $t \neq \theta_l$, $l \in \mathbb{N}$:

$$
\begin{aligned}
V'_{(11)}\left(y, y(\beta(t))\right) = \sum_{i=1}^{n} y_i(t) &\left( -a_i(t)F_i(t, y_i(t)) + \sum_{j=1}^{n} b_{ij}(t)H_j(t, y_j(t)) \right. \\
&\left. + \sum_{j=1}^{n} c_{ij}(t)G_j(t, y_j(\beta(t))) + \sum_{j=1}^{n} d_{ij}(t)K_j(t, y_j(t), y_j(\beta(t))) \right) \\
\leq \sum_{i=1}^{n} &\left( -\underline{a_i}\delta_{F_i} y_i^2(t) + \sum_{j=1}^{n} \overline{b_{ij}}\ell_j^h \, |y_j(t)| \, |y_i(t)| \right. \\
&\left. + \sum_{j=1}^{n} \overline{c_{ij}}\ell_i^g \, |y_j(\beta(t))| \, |y_i(t)| + \sum_{j=1}^{n} \overline{d_{ij}}\ell_i^k \left( |y_j(t)| + |y_j(\beta(t))| \right) |y_i(t)| \right).
\end{aligned}
$$

By the inequality $2\,|u|\,|v| \leq u^2 + v^2$, we obtain

$$
V'_{(11)}\left(y, y(\beta(t))\right) \leq -\rho_1 \sum_{i=1}^{n} y_i^2(t) + \rho_2 \sum_{i=1}^{n} y_i^2(\beta(t)).
$$

Then, we reach $V'_{(11)}\left(y, y(\beta(t))\right) \leq -(\rho_1 - \rho_2\sigma)\|y(t)\|^2$ provided that $V(y(\beta(t))) \leq \psi\left(V(y(t))\right)$ according to the Theorem 3.2.3 in [22].   □

So, it follows from Lemma 5 that the Lyapunov function (18) is suitable for (11) and the next theorem guarantees the uniform asymptotic stability of the trivial (zero) solution of (11) in view of the Lyapunov-Razumikhin method.

**Theorem 3.** *Suppose that (C1)–(C7) hold true. Then, the trivial solution of (11) is uniformly asymptotically stable.*

Since we use a linear transformation, we can say that $x^*$, equilibrium point, of (1) is also uniformly asymptotically stable according to Theorem 3.

## 2.2   Lyapunov-Krasovskii Method

Now, we shall consider the Lyapunov-Krasovskii method [29] and investigate the uniform asymptotic stability of (11).

**Lemma 6.** *Suppose that (C1)–(C6), (C8) are satisfied, and $y(t) : \mathbb{R}^+ \to \mathbb{R}^n$ is a solution of (11). Then items (a) and (b) given below are true for*

$$
V(y(t), y_t) = \frac{1}{2}\sum_{i=1}^{n} y_i^2(t) + \sum_{i=1}^{n}\sum_{j=1}^{n} \alpha_{ij} \int_{\beta(t)-t}^{0} \int_{t+s}^{t} y_j^2(u)\,du\,ds,
$$

*where $\alpha_{ij}, i, j = 1, 2, ..., n$, are arbitrary positive constants:*

(a) $\frac{\|y(t)\|^2}{2n} \leq V(y(t), y_t) \leq \frac{\|y(t)\|^2}{2} \left\{ 1 + \max_{1 \leq i \leq n} \sum_{j=1}^{n} \frac{\alpha_{ji}\left(1+\overline{\theta}\rho_*\right)^2 e^{2\overline{\theta}\rho^*}\overline{\theta}^2}{\left(1-\overline{\theta}\rho_*-(1+\overline{\theta}\rho_*)e^{\overline{\theta}\rho^*}\overline{\theta}\rho^*\right)^2} \right\};$

(b) $V'(y(t), y(\beta(t))) \leq -(\rho_1 - \rho_2\kappa^2 - \overline{\theta}\overline{\alpha}) \|y(t)\|^2$ *for all* $t \neq \theta_l$ *that belongs to* $\mathbb{R}^+$ *with* $\rho_1 - \rho_2\kappa^2 - \overline{\theta}\overline{\alpha} > 0$, $\overline{\alpha} = \max_{1 \leq i \leq n} \left\{ \sum_{j=1}^{n} \alpha_{ji} \right\}.$

*Proof.* Consider

$$V(y(t), y_t) = \frac{1}{2} \sum_{i=1}^{n} y_i^2(t) + \sum_{i=1}^{n} \sum_{j=1}^{n} \alpha_{ij} \int_{\beta(t)-t}^{0} \int_{t+s}^{t} y_j^2(u) du ds. \tag{19}$$

For the item $(a)$, it is obvious that $\frac{\|y(t)\|^2}{2n} \leq V(y(t), y_t)$. However, for the other side, two inequalities for $|y_i(t)|$ and $|y_i(\theta_l)|$ are needed. First, by Gronwall-Bellman Lemma, the inequality (13) gives

$$|y_i(t)| \leq |y_i(\theta_l)| \left(1 + \rho_*\overline{\theta}\right) e^{\rho^*\overline{\theta}}. \tag{20}$$

Then, using (20) for the integral in (15), we obtain

$$|y_i(\theta_l)| \leq |y_i(t)| \left\{ 1 - \overline{\theta} \left( \rho_* + \rho^* \left(1 + \overline{\theta}\rho_*\right) e^{\overline{\theta}\rho^*} \right) \right\}^{-1}. \tag{21}$$

So, substituting (20) into (19), we get that

$$V(y(t), y_t) \leq \frac{1}{2} \sum_{i=1}^{n} y_i^2(t) + \sum_{i=1}^{n} \sum_{j=1}^{n} \alpha_{ij} \int_{\beta(t)-t}^{0} \int_{t+s}^{t} y_j^2(\beta(u)) \left(1 + {}_*\rho\overline{\theta}\right)^2 e^{2({}^*\rho)\overline{\theta}} du ds$$

$$\leq \frac{1}{2} \sum_{i=1}^{n} y_i^2(t) + \sum_{i=1}^{n} \sum_{j=1}^{n} \alpha_{ji} y_i^2(\beta(u)) \left(1 + \rho_*\overline{\theta}\right)^2 e^{2\rho^*\overline{\theta}} \frac{\overline{\theta}^2}{2}.$$

Here, ${}_*\rho = \sum_{i=1}^{n} \left(\overline{c}_{ij}\ell_j^g + \overline{d}_{ij}\ell_j^k\right)$, ${}^*\rho = \overline{a}_j\ell_j^f + \sum_{i=1}^{n} \left(\overline{b}_{ij}\ell_j^h + \overline{d}_{ij}\ell_j^k\right)$. Later, together with (21), the last inequality results in

$$V(y(t), y_t) \leq \frac{1}{2} \sum_{i=1}^{n} y_i^2(t)$$

$$+ \sum_{i=1}^{n} \sum_{j=1}^{n} \alpha_{ji} y_i^2(t) \left(1 + \rho_*\overline{\theta}\right)^2 e^{2\rho^*\overline{\theta}} \frac{\overline{\theta}^2}{2} \left\{ 1 - \overline{\theta} \left( \rho_* + \rho^* \left(1 + \overline{\theta}\rho_*\right) e^{\overline{\theta}\rho^*} \right) \right\}^{-2}$$

$$\leq \frac{1}{2} \left\{ 1 + \max_{1 \leq i \leq n} \sum_{j=1}^{n} \frac{\alpha_{ji} \left(1 + \overline{\theta}\rho_*\right)^2 e^{2\overline{\theta}\rho^*}\overline{\theta}^2}{\left(1 - \overline{\theta}\rho_* - (1 + \overline{\theta}\rho_*) e^{\overline{\theta}\rho^*}\overline{\theta}\rho^*\right)^2} \right\} \|y(t)\|^2.$$

In order to see the item $(b)$, let us evaluate the derivative of the Lyapunov functional (19) with respect to $t$, for $t \neq \theta_l$, $l \in \mathbb{N}$:

$$V'_{(11)}\left(y, y(\beta(t))\right) = \sum_{i=1}^{n} y_i(t) \left[-a_i(t)F_i(t, y_i(t)) + \sum_{j=1}^{n} b_{ij}(t)H_j(t, y_j(t)) \right.$$

$$+ \sum_{j=1}^{n} c_{ij}(t)G_j(t, y_j(\beta(t))) + \left. \sum_{j=1}^{n} d_{ij}(t)K_j(t, y_j(t), y_j(\beta(t))) \right]$$

$$+ \sum_{i=1}^{n}\sum_{j=1}^{n} \frac{d}{dt}\left(\alpha_{ij} \int_{\beta(t)-t}^{0} \int_{t+s}^{t} y_j^2(u)duds\right)$$

$$\leq \sum_{i=1}^{n}\left[-\underline{a_i}\delta_{F_i}y_i^2(t) + \sum_{j=1}^{n} \overline{b_{ij}}\ell_j^h\,|y_j(t)|\,|y_i(t)|\right.$$

$$+ \sum_{j=1}^{n} \overline{c_{ij}}\ell_i^g\,|y_j(\beta(t))|\,|y_i(t)| + \left.\sum_{j=1}^{n} \overline{d_{ij}}\ell_i^k\left(|y_j(t)| + |y_j(\beta(t))|\right)|y_i(t)|\right]$$

$$+ \sum_{i=1}^{n}\sum_{j=1}^{n} \alpha_{ij}\overline{\theta}y_j^2(t).$$

By the fact $2\,|u|\,|v| \leq u^2 + v^2$, we obtain

$$V'_{(11)}\left(y, y(\beta(t))\right) \leq -\rho_1 \sum_{i=1}^{n} y_i^2(t) + \rho_2 \sum_{i=1}^{n} y_i^2(\beta(t)) + \overline{\alpha}\overline{\theta}\sum_{i=1}^{n} y_i^2(t)$$

$$\leq -\left(\rho_1 - \rho_2\kappa^2 - \overline{\alpha}\overline{\theta}\right)\|y(t)\|^2.$$

So, Lemma 6 proves that (18) can be chosen as a Lyapunov functional for the neural network system given by (11). Then the next theorem gives sufficient conditions that guarantee uniform asymptotic stability of the zero solution of (11) depending on the Lyapunov-Krasovskii method.

**Theorem 4.** *If conditions (C1)–(C6), (C8) are satisfied, then the trivial solution of (11) is uniformly asymptotically stable.*

Since a linear transformation is used, it is obvious by Theorem 4 that $x^*$, equilibrium point, of (1) is uniformly asymptotically stable as well.

## 3   Example and Simulation

In the present section, taking the obtained theoretical results into account, we give an example together with simulation obtained by MATLAB package program.

*Example 1.* Take into account the following nonautonomous system

$$x_i'(t) = -a_i(t)f_i(t, x_i(t)) + \sum_{j=1}^{2} b_{ij}(t)h_j(t, x_j(t)) + \sum_{j=1}^{2} c_{ij}(t)g_j(t, x_j(\beta(t)))$$

$$+ \sum_{j=1}^{2} d_{ij}(t)k_j(t,, x_j(t), x_j(\beta(t))) + e_i(t), \ i = 1, 2, \tag{22}$$

where $f_i(t, x_i(t)) = \left(1 + e^{-it}\right) x_i(t)$, $h_j(t, x_j(t)) = \tanh\left(jx_j(t)\right)$, $g_j(t, x_j(\beta(t))) = \tanh\left(\dfrac{x_j(\beta(t))}{2j}\right)$, $k_j(t, x_j(t), x_j(\beta(t))) = 0$, $a_i(t) = 4 + \dfrac{2i}{i + e^{-it}}$, $b_{ij}(t) = c_{ij}(t) = 0.001\sin(ijt)$, $d_{ij}(t) = 0$, $e_i(t) = \dfrac{0.2}{1 + e^{2t}} + 10i$. Let $\theta_l = \dfrac{1}{80}\left(l + \dfrac{(-1)^l}{3}\right)$, $l \in \mathbb{N}$.

Since $\ell_i^f = 2$, $\ell_1^h = 1$, $\ell_2^h = 2$, $\ell_1^g = 0.5$, $\ell_2^g = 0.25$, $\ell_i^k = 0$, $\overline{\theta} = \dfrac{1}{48}$, $\underline{\theta} = \dfrac{1}{240}$, it is obvious that (C1)–(C2) hold true. Moreover, it can be seen that the assumptions (C3)–(C8) are satisfied, respectively:

- $\underline{a_i} \ {}_*\underline{f_i} = 4$ and $\ell_i^{f^*}\,\overline{a_i} + \sum_{j=1}^{n}\left(\overline{b_{ji}}\ell_i^h + \overline{c_{ji}}\ell_i^g + 2\overline{d_{ji}}\ell_i^k\right) = \begin{cases} 0.003 & \text{for } i = 1, \\ 0.0045 & \text{for } i = 2; \end{cases}$
- $\left(\overline{a}_f + \overline{b}_h + 2\overline{c}_g + 3\overline{d}_k\right)\overline{\theta}e^{\left(\overline{a}_f + \overline{b}_h + \overline{d}_k\right)\overline{\theta}} = 0.3212 < 1$;
- $\overline{\theta}\max_{1 \leq i \leq n}\left\{\rho_* + \rho^*\left(1 + \overline{\theta}\rho_*\right)e^{\overline{\theta}\rho^*}\right\} = 0.3212 < 1$ and so $\kappa = 1.4732$;
- $\delta_{F_i} = 1 > 0$ for $i = 1, 2$;
- $\rho_1 = 3.9961$, $\rho_2 = 0.5$;
- $\rho_1 - \rho_2\kappa^2 - \overline{\theta}\overline{\alpha} = 2.8693 > 0$ with $\overline{\alpha} = 1$.

Therefore, the conditions of Theorem 2 are fulfilled and so (22) has a unique solution on $\mathbb{R}^+$. Accordingly, it follows from Theorem 3 or Theorem 4 that the equilibrium solution of (22) is uniformly asymptotically stable since the conditions (C1)–(C7) or the conditions (C1)–(C6), (C8) hold true based on Lyapunov-Razumikhin or Lyapunov-Krasovskii methods, respectively. Simulation results of (22) with the above parameters are shown by Fig. 1.

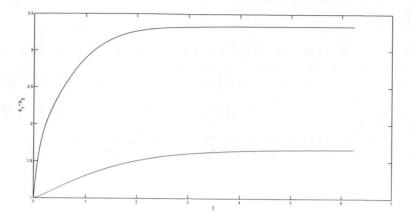

**Fig. 1.** Time responses of variables $x_1(t)$ (red one) and $x_2(t)$ (black one) of (22) for $t \in [0, 6.254166]$ with $\theta_l = \dfrac{1}{80}\left(l + \dfrac{(-1)^l}{3}\right)$, $l \in \mathbb{N}$ and $x_i^0 = 1$, $i = 1, 2$ (Color figure online)

## 4     Conclusion

In the present paper, a nonautonomous recurrent neural network system is considered by taking generalized piecewise constant arguments. It is first in the literature that this kind of system is studied qualitatively. This neural network system is modeled mathematically in a more general form by PCAG. As far as we know, this paper is the first one that investigates existence and stability of equilibrium for this kind of a neural network system. During the stability analysis, we consider Lyapunov-Razumikhin method and Lyapunov-Krasovskii method. Therefore, both a Lyapunov function and a Lyapunov functional providing certain features are taken into consideration. Of course, they can be constructed differently from we consider. For example, the choices for the Lyapunov function/functional could depend on $t$. However, this would require more conditions on the parameters of the system. Even, some conditions concerning the derivative of the parameters of the system could be needed. In terms of applications, taking a time-varying Lyapunov function or functional makes it difficult to choose the parameters of the system. We achieved our goal more easily with the proposed function or functional which are independent of $t$. As a result, it is remarkable that we have achieved simpler and more applicable conditions. Moreover, it can be said that Lyapunov-Razumikhin method seems more convenient since it enables one to reach the results in fewer steps.

**Acknowledgements.** This work is supported by TÜBİTAK (The Scientific and Technological Research Council of Turkey) under project no 118F185.

# References

1. McCulloch, W., Pitts, W.: A logical calculus of the ideas immanent in nervous activity. Bull. Math. Biophys. **5**, 115–133 (1943)
2. Block, H.: The perceptron: a model for brain functioning. Rev. Mod. Phys. **34**, 123–135 (1962)
3. Rojas, R.: Neural Networks: A Systematic Introduction. Springer-Verlag, Heidelberg (1996)
4. Akhmet, M.U., Yılmaz, E.: Neural Networks with Discontinuous/Impact Activations. NSC, vol. 9. Springer, New York (2014). https://doi.org/10.1007/978-1-4614-8566-7_8
5. Hopfield, J.J.: Neurons with graded response have collective computational properties like those of two-state neurons. Proc. Natl. Acad. Sci. **81**, 30–88 (1984)
6. Kosko, B.: Adaptive bidirectional associative memories. Appl. Opt. **26**(23), 4947–4960 (1987)
7. Cohen, M.A., Grossberg, S.: Absolute stability of global pattern formation and parallel memory storage by competitive neural networks. IEEE Trans. Syst. Man Cybern. Syst. **SMC–13**, 815–826 (1983)
8. Yang, Z.: Global existence and exponential stability of periodic solutions of recurrent neural networks with functional delay. Math. Meth. Appl. Sci. **30**, 1775–1790 (2007)
9. Gui, Z., Yang, X.S., Ge, W.: Existence and global exponential stability of periodic solutions of recurrent cellular neural networks with impulses and delays. Math. Comput. Simul. **79**, 14–29 (2008)
10. Akhmet, M., Fen, M.O.: Generation of cyclic/toroidal chaos by Hopfield neural networks. Neurocomputing **145**, 230–239 (2014)
11. Akhmet, M.U., Fen, M.O.: Shunting inhibitory cellular neural networks with chaotic external inputs. Chaos **23**(2), 023112 (2013)
12. Michel, A.N., Farrell, J.A., Porod, W.: Qualitative analysis of neural networks. IEEE Trans. Circuits Syst. **36**(2), 229–243 (1989)
13. Campbell, S.A., Ruan, S., Wei, J.: Qualitative analysis of a neural network model with multiple time delays. Int. J. Bifurcat. Chaos. **9**(08), 1585–1595 (1999)
14. Sudharsanan, S.I., Sundareshan, M.K.: Qualitative analysis of equilibrium confinement and exponential stability of a class of dynamical neural networks. In: International Neural Network Conference, pp. 964–968. Springer, Dordrecht (1990)
15. Qualitative Analysis and Control of Complex Neural Networks with Delays. SSDC, vol. 34. Springer, Heidelberg (2016). https://doi.org/10.1007/978-3-662-47484-6_9
16. Babakhanbak, S., Kavousi, K., Farokhi, F.: Application of a time delay neural network for predicting positive and negative links in social networks. Turk. J. Elec. Eng. Comp. Sci. **24**, 2825–2837 (2016)
17. Akhmet, M.U., Aruğaslan, D., Yılmaz, E.: Stability in cellular neural networks with a piecewise constant argument. J. Comput. Appl. Math. **233**, 2365–2373 (2010)
18. Xi, Q.: Global exponential stability of Cohen-Grossberg neural networks with piecewise constant argument of generalized type and impulses. Neural Comput. **28**(1), 229–255 (2016)
19. Akhmet, M.U., Aruğaslan, D., Cengiz, N.: Exponential stability of periodic solutions of recurrent neural networks with functional dependence on piecewise constant argument. Turk. J. Math. **42**, 272–292 (2018)

20. Akhmet, M.U.: On the integral manifolds of the differential equations with piecewise constant argument of generalized type. In: Agarval, R.P., Perer, K. (eds.) Proceedings of the Conference on Differential and Difference Equations and Applications, pp. 11–20. Hindawi Publishing Corporation, Melbourne (2005)

21. Akhmet, M.U.: On the reduction principle for differential equations with piecewise constant argument of generalized type. J. Math. Anal. Appl. **336**, 646–663 (2007)

22. Akhmet, M.U., Aruğaslan, D.: Lyapunov-Razumikhin method for differential equations with piecewise constant argument. Discrete Contin. Dyn. Syst. Ser. A **25**(2), 457–466 (2009)

23. Akhmet, M.U.: Stability of differential equations with piecewise constant arguments of generalized type. Nonlinear Anal. Theory Methods Appl. **68**, 794–803 (2008)

24. Akhmet, M.U.: Nonlinear Hybrid Continuous Discrete-Time Models. Atlantis Press, Amsterdam, Paris (2011)

25. Chiu, K.S.: Asymptotic equivalence of alternately advanced and delayed differential systems with piecewise constant generalized arguments. Acta Math. Scientia **38**(1), 220–236 (2018)

26. Coronel, A., Maulen, C., Pinto, M., Sepulveda, D.: Dichotomies and asymptotic equivalence in alternately advanced and delayed differential systems. J. Math. Anal. Appl. **450**(2), 1434–1458 (2017)

27. Cooke, K.L., Wiener, J.: Retarded differential equations with piecewise constant delays. J. Math. Anal. Appl. **99**, 265–297 (1984)

28. Akhmet, M.U.: Principles of Discontinuous Dynamical Systems. Springer, New York (2010)

29. Hale, J.: Functional Differential Equations. Springer, New-York (1971)

30. Krasovskii, N.N.: Stability of Motion: Application of Lyapunov's Second Method to Differential Systems and Equations with Time-Delay. SU Press, Stanford (1963)

31. Razumikhin, B.S.: On the stability of systems with delay. Prikl. Mat. Mekh **20**, 500–512 (1956)

32. Oliveira, J.: Global exponential stability of nonautonomous neural network models with unbounded delays. Neural Netw. **96**, 71–79 (2017)

33. Gopalsamy, K.: Stability of artificial neural networks with impulses. Appl. Math. Comput. **154**, 783–813 (2004)

# Dynamics of a Recurrent Neural Network with Impulsive Effects and Piecewise Constant Argument

Marat Akhmet[1], Duygu Aruğaslan Çinçin[2(✉)], and Nur Cengiz[3]

[1] Department of Mathematics, Middle East Technical University, Ankara, Turkey
`marat@metu.edu.tr`
[2] Department of Mathematics, Süleyman Demirel University, 32260 Isparta, Turkey
`duyguarugaslan@sdu.edu.tr`
[3] Graduate School of Natural and Applied Sciences, Süleyman Demirel University, 32260 Isparta, Turkey

**Abstract.** In the present work, a nonautonomous recurrent neural network model with impulsive effects and generalized piecewise constant argument is addressed. Global exponential stability examination for the equilibrium of the considered model is performed mathematically. While investigating the dynamical characteristics of the model, we take the results given in [1,2] into account. The reached theoretical results are exemplified and supported by a simulation with the help of MATLAB package program.

**Keywords:** Recurrent neural networks · Generalized piecewise constant argument · Impulsive differential equations · Equilibrium · Global exponential stability · Simulation

## 1 Introduction

While modeling real life problems mathematically, it is very important not to neglect discontinuous effects. This is because discontinuous arguments allow to think of problems with a more natural concept. Accordingly, differential equations with piecewise constant argument of generalized type (EPCAG) and differential equations including impulsive effects are of great importance [1–6]. Moreover, the qualitative theory of these equations allows to examine the dynamics of problems with these effects. In our study, we develop a recurrent neural networks system with discontinuous arguments by using the principles of these two types of differential equations. In EPCAG introduced by Akhmet [3,4], an argument function of both advanced type and delayed type can be taken as the deviating argument. In order to develop our model, we use the argument function $\beta(t)$ which is of delayed type only. In addition, the impulsive operator in our model corresponds to a nonlinear function. Unlike the differential equations proposed previously for the neural network model, our model includes nonautonomous and

J. Hemanth et al. (Eds.): ICAIAME 2020, LNDECT 76, pp. 540–552, 2021.
https://doi.org/10.1007/978-3-030-79357-9_52

nonlinear functions. For the first time in the literature, this type of recurrent neural networks system is handled within the scope of this paper. Thus, we have analyzed a neural networks model in a more general form when compared with the literature. Hence, the model we propose is expected to have a considerable importance among other studies performed on the neural networks [2,7–15]. In the historical process, after the first model of artificial neurons was given by McCulloch and Pitts [10,16], the models representing the nervous system in terms of mathematical, electronical, biological paradigms have been proposed by many scientists [17–20]. So, the dynamics of artificial neural networks has been a very popular research topic [21,22]. In particular, stability phenomenon of neural network systems has been researched widely [7–9,13–15,23–27]. In present paper, global exponential stability of a nonautonomous recurrent neural networks model has been examined with the help of the theories in [1,2]. For this aim, our paper is arranged as follows. Section 2 defines the parameters of the impulsive neural networks system with piecewise constant argument of generalized type (PCAG) and gives assumptions with some notations. In Subsect. 2.1, we investigate existence of a unique equilibrium point. In Subsect. 2.2, we present conditions guaranteeing the existence-uniqueness of the solutions of the issue model. In Subsect. 2.3, we obtain an auxiliary result and perform the stability investigation using this result. In Sect. 3, we give an example and a simulation. Then, in the final section, we present conclusions.

## 2   Preliminaries

Let $\mathbb{N}$, $\mathbb{R}$ and $\mathbb{R}^+$ describe the sets of all non-negative integers, real numbers and non-negative real numbers, respectively, i.e., $\mathbb{N} = \{0, 1, 2, ...\}$, $\mathbb{R} = (-\infty, \infty)$ and $\mathbb{R}^+ = [0, \infty)$. Denote by $\mathbb{R}^n$, $n \in \mathbb{N}$, the real space of dimension $n$, then consider a norm on $\mathbb{R}^n$ by $\|.\|$ such that $\|u\| = \sum_{i=1}^{n} |u_i|$.

We take into consideration a nonautonomous nonlinear recurrent neural networks model

$$x_i'(t) = -c_i(t)h_i(t, x_i(t)) + \sum_{j=1}^{n} a_{ij}(t)f_j(t, x_j(t))$$

$$+ \sum_{j=1}^{n} b_{ij}(t)g_j(t, x_j(\beta(t))) + d_i(t), \ t \neq \tau_k, \tag{1}$$

$$\Delta x_i|_{t=\tau_k} = I_{ik}\left(x_i(\tau_k^-)\right), \ i = 1, ..., n, \ k \in \mathbb{N},$$

with impact activations and PCAG. In (1), $x(t) = (x_1(t), ..., x_n(t))^T \in \mathbb{R}^n$ is the neuron state vector while $n$ denotes how many units exist in the network. Here, $h_i \in C(\mathbb{R}^+ \times \mathbb{R}, \mathbb{R})$ is a nonlinear nonautonomous function for each $i = 1, ..., n$. Moreover, nonautonomous and nonlinear activation functions $f_j, g_j \in C(\mathbb{R}^+ \times \mathbb{R}, \mathbb{R})$, $j = 1, .., n$, of neurons denote the measures of activation to its incoming potentials of the unit $j$, and the nonautonomous inputs $d_i(t) \in C(\mathbb{R}^+, \mathbb{R})$, are external bias on the unit $i$ of system at $t$ for $i = 1, .., n$.

$c_i(t) \in C(\mathbb{R}^+, \mathbb{R}^+)$, $i = 1, .., n$, corresponds to the functions that represent at which rate unit $i$ resets its potential to the resting position in isolation when it is not connected to the network and external inputs. Additionally, the nonautonomous weight functions $a_{ij}(t)$, $b_{ij}(t) \in C(\mathbb{R}^+, \mathbb{R})$ correspond to the synaptic connection weight of the $j$th unit on the unit $i$ at $t$. $I_{ik} : \mathbb{R} \to \mathbb{R}$, $i = 1, ..., n$, $k \in \mathbb{N}$, represent the impulsive operators. $\tau = \{\tau_k\}$ is a fixed real valued time sequence. When $t = \tau_k$, $k \in \mathbb{N}$, the point is exposed to a change given by $\Delta x_i|_{t=\tau_k} = x_i(\tau_k) - x_i(\tau_k^-)$, where $x_i(\tau_k^-) = \lim_{h \to 0^-} x_i(\tau_k + h)$. It is assumed that the sequence $\{\tau_k\}$ is a nonempty finite set or an infinite set with $|\tau_k| \to \infty$ while $|k| \to \infty$ [1]. Moreover, consider a real valued sequence $\theta = \{\theta_l\}$, $l \in \mathbb{N}$ satisfying $0 \leq \theta_l < \theta_{l+1}$ for all $l \in \mathbb{N}$ and $\theta_l \to \infty$ as $l \to \infty$. Let us assume that $\tau \cap \theta = \varnothing$. In (1), $\beta(t) = \theta_l$ if $t \in [\theta_l, \theta_{l+1})$, $l \in \mathbb{N}$. We presume without loss of generality that the initial time $t_0 \in (\theta_l, \theta_{l+1}]$ for some $l \in \mathbb{N}$.

**Lemma 1.** *Fix a real number $t_0$. $x(t) = x(t, t_0, x^0) = (x_1(t), ..., x_n(t))^T$ is a solution of system (1) on $\mathbb{R}^+$ if and only if it is a solution of*

$$x_i(t) = x_i^0 + \int_{t_0}^t \left( -c_i(s)h_i(s, x_i(s)) + \sum_{j=1}^n a_{ij}(s)f_j(s, x_j(s)) \right.$$

$$\left. + \sum_{j=1}^n b_{ij}(s)g_j(s, x_j(\beta(s))) + d_i(s) \right) + \sum_{t_0 \leq \tau_k < t} I_{ik}(x_i(\tau_k^-)) \quad (2)$$

*on $\mathbb{R}^+$, where $i = 1, ..., n$.*

Suppose that the assumptions given below hold true:

(C1) There exist Lipschitz constants $\ell_i^h$, $\ell_j^f$, $\ell_j^g > 0$ satisfying the inequalities $|h_i(t, u) - h_i(t, v)| \leq \ell_i^h |u - v|$, $|f_j(t, u) - f_j(t, v)| \leq \ell_j^f |u - v|$, $|g_j(t, u) - g_j(t, v)| \leq \ell_j^g |u - v|$ for all $t \in \mathbb{R}^+$, $u$, $v \in \mathbb{R}$, $i, j = 1, ..., n$;
(C2) The impulsive operator $I_{ik} : \mathbb{R} \to \mathbb{R}$ satisfies $|I_{ik}(u) - I_{ik}(v)| \leq \ell_i^I |u - v|$ for all $u$, $v \in \mathbb{R}$, $i = 1, ..., n$, $k \in \mathbb{N}$, where $\ell_i^I > 0$ are Lipschitz constants;
(C3) Positive numbers $\overline{\theta}$, $\underline{\tau}$ exist such that $\theta_{l+1} - \theta_l \leq \overline{\theta}$, $\underline{\tau} \leq \tau_{k+1} - \tau_k$, $k$, $l \in \mathbb{N}$;
(C4) $p = \max_{k \in \mathbb{N}} p_k < \infty$, where $p_k$ is the number of points $\tau_k$ in the interval $(\theta_l, \theta_{l+1})$, $l \in \mathbb{N}$.

For the convenience, let us define some notations:

$$\underline{c_i} = \inf_{t \in \mathbb{R}^+} (c_i(t)), \ \overline{c_i} = \sup_{t \in \mathbb{R}^+} (c_i(t)), \ \overline{a_{ij}} = \sup_{t \in \mathbb{R}^+} (|a_{ij}(t)|), \ \overline{b_{ij}} = \sup_{t \in \mathbb{R}^+} (|b_{ij}(t)|),$$

$$\overline{d_i} = \sup_{t \in \mathbb{R}^+} (|d_i(t)|), \ \sum_{i=1}^n \overline{d_i} := \overline{d}, \ \max_{1 \leq i \leq n} \ell_i^I := \ell^I,$$

$$\max_{1 \leq i \leq n} \overline{c_i} \ell_i^h := \overline{c}_h, \ \max_{1 \leq i \leq n} \left( \sum_{j=1}^n \overline{a_{ji}} \ell_i^f \right) := \overline{a}_f, \ \max_{1 \leq i \leq n} \left( \sum_{j=1}^n \overline{b_{ji}} \ell_i^g \right) := \overline{b}_g,$$

$$\sup_{t \in \mathbb{R}^+} |h_i(t, 0)| = M_{h_i}, \ \sup_{t \in \mathbb{R}^+} |f_j(t, 0)| = M_{f_j}, \ \sup_{t \in \mathbb{R}^+} |g_j(t, 0)| = M_{g_j}, \ \sup_{t \in \mathbb{R}^+} |I_{ik}(0)| = M_{I_i},$$

$$\sum_{i=1}^n M_{I_i} := M_I, \ \sum_{i=1}^n \overline{c_i} M_{h_i} := M_h^c, \ \sum_{i=1}^n \sum_{j=1}^n \overline{a_{ij}} M_{f_j} := M_f^a, \ \sum_{i=1}^n \sum_{j=1}^n \overline{b_{ij}} M_{g_j} := M_g^b$$

for $i, j = 1, ..., n$.

Moreover, let the following conditions be satisfied.

(C5) $\left( \left( \overline{c}_h + \overline{a}_f + 2\overline{b}_g \right) \overline{\theta} + p\ell^I \right) (1 + \ell^I)^p \, e^{(\overline{c}_h + \overline{a}_f)\overline{\theta}} < 1;$

(C6) $\overline{b}_g \overline{\theta} + \left( \overline{\theta} \left( \overline{c}_h + \overline{a}_f \right) + p\ell^I \right) (1 + \overline{b}_g \overline{\theta}) (1 + \ell^I)^p \, e^{(\overline{c}_h + \overline{a}_f)\overline{\theta}} < 1.$

These conditions will be used respectively to show an auxiliary result that gives a relation between the values of the solution of (1) at any time $t$ and at the deviating argument $\beta(t)$ and to prove the existence-uniqueness theorem for (1).

*Remark:* Note that we can write $h_i(t, u_i) = {}_*h_i(t)u_i + h_i^*(t, u_i)$, where ${}_*h_i(t)$ and $h_i^*(t, u_i)$ represent functions. Here, suppose that $0 < \inf_{t \in \mathbb{R}^+} ({}_*h_i(t))$. Then, it can be seen that $\ell_i^h = \ell_i^{h^*} + \overline{{}_*h_i}$ if there exist Lipschitz constants $\ell_i^{h^*} > 0$, $i = 1, ..., n$, such that $\left| h_i^*(t, u_i) - h_i^*(t, v_i) \right| \leq \ell_i^{h^*} |u_i - v_i|$, for all $t \in \mathbb{R}^+$, $u_i$, $v_i \in \mathbb{R}$, and if $0 < \underline{{}_*h_i} \leq {}_*h_i(t) \leq \overline{{}_*h_i}$ is satisfied. Here, $\ell_i^{h^*}$, $\overline{{}_*h_i}$, $\underline{{}_*h_i}$ are constants.

## 2.1 Equilibrium Point

Denote by $x^* = \left( x_1^*, ..., x_n^* \right)^T \in \mathbb{R}^n$, a constant vector, whose components are determined according to the algebraic equation

$$0 = -c_i(t)h_i(t, x_i^*) + \sum_{j=1}^n a_{ij}(t)f_j(t, x_j^*) + \sum_{j=1}^n b_{ij}(t)g_j(t, x_j^*) + d_i(t). \tag{3}$$

In the sequel, we shall present sufficient conditions to have a unique equilibrium point of the differential part of (1) without impact activations. While reaching these conditions, the contraction mapping principle is considered based on Lemma 2.1 given in [7].

**Lemma 2.** *Suppose that the condition (C1) holds and the functions $h_i^*$, $i = 1, ..., n$, are globally Lipschitz continuous such that $\left| h_i^*(t, u_i) - h_i^*(t, v_i) \right| \leq \ell_i^{h^*} |u_i - v_i|$, for all $t \in \mathbb{R}^+$, $u_i$, $v_i \in \mathbb{R}$, where $\ell_i^{h^*}$ are positive constants. If the following inequality*

$$\max_{1 \leq i \leq n} \left( \frac{\ell_i^{h^*}}{{}_*h_i} + \sum_{j=1}^n \left( \frac{\overline{a_{ji}} \ell_i^f + \overline{b_{ji}} \ell_i^g}{c_j \, {}_*h_j} \right) \right) < 1$$

*holds then there is a unique constant vector $x^* = \left( x_1^*, ..., x_n^* \right)$ satisfying (3).*

*Proof.* Now, we will show that the following system

$$c_i(t)_* h_i(t) x_i^* = -c_i(t) h_i^*(t, x_i^*) + \sum_{j=1}^{n} a_{ij}(t) f_j(t, x_j^*) + \sum_{j=1}^{n} b_{ij}(t) g_j(t, x_j^*)$$

$$+ d_i(t), \ i = 1, ..., n, \tag{4}$$

possesses a unique solution based on the contraction mapping principle. Define on $\mathbb{R}^n$ the operator $\Pi$ such that

$$\Pi(z_1, z_2, ..., z_n) = \begin{cases} \Pi_1(z_1, z_2, ..., z_n) \\ \vdots \\ \Pi_n(z_1, z_2, ..., z_n), \end{cases}$$

where

$$\Pi_i(z) = -\frac{h_i^*(t, z_i)}{_* h_i(t)} + \sum_{j=1}^{n} \frac{a_{ij}(t) f_j(t, z_j) + b_{ij}(t) g_j(t, z_j)}{c_i(t)_* h_i(t)}$$

$$+ \frac{d_i(t)}{c_i(t)_* h_i(t)}, \ i = 1, ..., n. \tag{5}$$

Now, we will show that the mapping $\Pi : \mathbb{R}^n \to \mathbb{R}^n$ is a contractive operator. So, for $u, \ v \in \mathbb{R}^n$, it follows from

$$\|\Pi(u) - \Pi(v)\| \le \max_{1 \le i \le n} \left( \frac{\ell_i^{h^*}}{_* h_i} + \sum_{j=1}^{n} \left( \frac{\overline{a_{ji}} \ell_i^f + \overline{b_{ji}} \ell_i^g}{c_j \, _* h_j} \right) \right) \|u - v\|$$

and so from the condition $\max_{1 \le i \le n} \left( \frac{\ell_i^{h^*}}{_* h_i} + \sum_{j=1}^{n} \left( \frac{\overline{a_{ji}} \ell_i^f + \overline{b_{ji}} \ell_i^g}{c_j \, _* h_j} \right) \right) < 1$ that the mapping $\Pi : \mathbb{R}^n \to \mathbb{R}^n$ is a global contraction. Therefore, there is a unique fixed point of the mapping based on the contraction mapping principle. $\square$

It is obvious that $x^* = (x_1^*, ..., x_n^*)^T$ is a unique equilibrium of the differential part of (1). Moreover, let all zeros of the functions that represent impulsive effects be given by the set

$$E^* = \{x_i \in \mathbb{R} | I_{ik}(x_i) = 0, \ k \in \mathbb{N}, \ i = 1, ..., n\}.$$

Assume that the condition given below is valid.

(E) $x_i^* \in E^*$ for each $i = 1, .., n$.

Then, we can state the next theorem.

**Theorem 1.** *Suppose that the assumption (E) is fulfilled in addition to the assumptions of Lemma 2. Then $x^*$ is the uniqe equilibrium point of the system (1).*

## 2.2 Existence and Uniqueness

Results that concern existence and uniqueness of the solutions of (1) on $[\theta_l, \theta_{l+1}]$ including the initial value $t_0$ can be seen in the following lemma.

**Lemma 3.** *Fix $k$, $l \in \mathbb{N}$. Suppose that (C1)–(C5) are satisfied. Then, for every $(t_0, x^0) \in [\theta_l, \theta_{l+1}] \times \mathbb{R}^n$, there exists a unique solution $x(t) = x(t, t_0, x^0) = (x_1(t), ..., x_n(t))^T$ of (1) on $[\theta_l, \theta_{l+1}]$ with $x(t_0) = x^0$.*

*Proof. Existence.* For convenience, denote $z(t) = x(t, t_0, x^0) = (z_1(t), ..., z_n(t))^T$. Let us define a norm $\|z(t)\|_0$ by $\|z(t)\|_0 = \max_{[\theta_l, \theta_{l+1}]} \|z(t)\|$. Based on Lemma 1, consider integral equation

$$z_i(t) = x_i^0 + \int_{t_0}^t \left( -c_i(s)h_i(s, z_i(s)) + \sum_{j=1}^n a_{ij}(s)f_j(s, z_j(s)) \right.$$

$$\left. + \sum_{j=1}^n b_{ij}(s)g_j(s, z_j(\theta_l)) + d_i(s) \right) \tag{6}$$

$$+ \sum_{t_0 \le \tau_k < t} I_{ik}(z_i(\tau_k^-)), \quad i = 1, ..., n, \ t \ge t_0.$$

Construct a sequence $z_i^m(t)$, $z_i^0(t) \equiv x_i^0$, $i = 1, ..., n$, by

$$z_i^{m+1}(t) = x_i^0 + \int_{t_0}^t \left( -c_i(s)h_i(s, z_i^m(s)) + \sum_{j=1}^n a_{ij}(s)f_j(s, z_j^m(s)) \right.$$

$$\left. + \sum_{j=1}^n b_{ij}(s)g_j(s, z_j^m(\theta_l)) + d_i(s) \right)$$

$$+ \sum_{t_0 \le \tau_k < t} I_{ik}(z_i^m(\tau_k^-)), \quad m \ge 0, \ t \ge t_0.$$

Then the following conclusion can be reached:

$$\|z^{m+1}(t) - z^m(t)\|_0 \le \left( \left( \bar{c}_h + \bar{a}_f + \bar{b}_g \right) \bar{\theta} + p\ell^l \right)^m \Omega, \tag{7}$$

where $\Omega = \left( \left( \bar{c}_h + \bar{a}_f + \bar{b}_g \right) \bar{\theta} + p\ell^l \right) \|x^0\| + \left( \bar{d} + M_h^c + M_f^a + M_g^b \right) \bar{\theta} + pM_I$. The condition (C5) implies that the sequences $z_i^m(t)$ are convergent and their limits satisfy the integral Eq. (6) on $[\theta_l, \theta_{l+1}]$, which shows the existence.

*Uniqueness.* Indicate the solutions of (1) by $x^r(t) = x(t, t_0, x^r)$, where $\theta_l < t_0 \le \theta_{l+1}$ $x^r = (x_1^r, ..., x_n^r)^T$ $r = 1, 2$. Now, let us show that $x^1(t) \ne x^2(t)$ for each $t \in [\theta_l, \theta_{l+1}]$ and for $x^1 \ne x^2$. Then, it can be seen by the Gronwall-Bellman Lemma in [1,5] and by the conditions (C1)–(C4) that

$$\|x^1(t) - x^2(t)\| \le \left( \|x^1 - x^2\| + \overline{\theta b_g} \|x^1(\theta_l) - x^2(\theta_l)\| \right) \left( 1 + \ell^l \right)^p e^{(\bar{c}_h + \bar{a}_f)\bar{\theta}}. \tag{8}$$

When $t = \theta_l$, it is obvious that (8) takes the following form

$$\left\|x^1(\theta_l) - x^2(\theta_l)\right\| \leq \left(\left\|x^1 - x^2\right\| + \overline{\theta}\overline{b}_g \left\|x^1(\theta_l) - x^2(\theta_l)\right\|\right) \left(1 + \ell^I\right)^P e^{(\overline{c}_h + \overline{a}_f)\overline{\theta}}. \quad (9)$$

So, (8) and (9) give the following inequality

$$\left\|x^1(t) - x^2(t)\right\| \leq \frac{\left(1 + \ell^I\right)^P e^{(\overline{c}_h + \overline{a}_f)\overline{\theta}}}{\left(1 - \overline{\theta}\overline{b}_g \left(1 + \ell^I\right)^P e^{(\overline{c}_h + \overline{a}_f)\overline{\theta}}\right)} \left\|x^1 - x^2\right\|. \quad (10)$$

Accordingly, for $t = \tau_k^-$, it is seen by (10) that

$$\left\|x^1(\tau_k^-) - x^2(\tau_k^-)\right\| \leq \frac{\left(1 + \ell^I\right)^P e^{(\overline{c}_h + \overline{a}_f)\overline{\theta}}}{\left(1 - \overline{\theta}\overline{b}_g \left(1 + \ell^I\right)^P e^{(\overline{c}_h + \overline{a}_f)\overline{\theta}}\right)} \left\|x^1 - x^2\right\|. \quad (11)$$

Suppose contrarily that there exists $t \in [\theta_l, \theta_{l+1}]$ satisfying the equality $x^1(t) = x^2(t)$. In this case, we have

$$\left\|x^1 - x^2\right\| \leq \int_{t_0}^{t} \left(\overline{c}_h + \overline{a}_f\right) \left\|x^1(s) - x^2(s)\right\| ds + \overline{\theta}\overline{b}_g \left\|x^1(\theta_l) - x^2(\theta_l)\right\|$$
$$+ \ell^I \sum_{t_0 \leq \tau_k < t} \left\|x^1(\tau_k^-) - x^2(\tau_k^-)\right\|. \quad (12)$$

Moreover, by (10), we have

$$\left\|x^1 - x^2\right\| \leq \frac{\left(\overline{c}_h + \overline{a}_f + \overline{b}_g\right)\overline{\theta} + p\ell^I}{1 - \overline{\theta}\overline{b}_g \left(1 + \ell^I\right)^P e^{(\overline{c}_h + \overline{a}_f)\overline{\theta}}} \left(1 + \ell^I\right)^P e^{(\overline{c}_h + \overline{a}_f)\overline{\theta}} \left\|x^1 - x^2\right\|,$$

which is a contradiction to the condition (C6). So, the uniqueness and in turn the lemma are proved.    □

Then, Theorem 2 guarantees the existence-uniqueness of the solutions of (1) on $\mathbb{R}^+$. It can be proved by the mathematical induction using Lemma 3.

**Theorem 2.** *Assume that the conditions (C1)–(C5) are satisfied. Then, for every $(t_0, x^0) \in \mathbb{R}^+ \times \mathbb{R}^n$, there exists a unique solution $x(t) = x(t, t_0, x^0) = (x_1(t), ..., x_n(t))^T$ of (1) on $\mathbb{R}^+$ with $x(t_0) = x^0$.*

### 2.3    Global Exponential Stability of the Equilibrium Point

Our main goal is to examine the global exponential stability of the equilibrium point $x^*$ of the neural network system (1) with PCAG and impact activation. While doing this analysis, we make a linear transformation in order to use the theory in [1,2]. It follows from (1) that $x_i(t) - x_i^* = y_i(t)$ satisfies the network system

$$y_i'(t) = -c_i(t)H_i(t, y_i(t)) + \sum_{j=1}^{n} a_{ij}(t)F_j(t, y_j(t)) + \sum_{j=1}^{n} b_{ij}(t)G_j(t, y_j(\beta(t))), \ t \neq \tau_k$$

$$\Delta y_i|_{t=\tau_k} = W_{ik}\left(y_i(\tau_k^-)\right), \ i = 1, ..., n, \ k \in \mathbb{N}. \tag{13}$$

Here, $y = (y_1, ..., y_n)^T \in \mathbb{R}^n$, $H_i(t, y_i(t)) = h_i\left(t, y_i(t) + x_i^*\right) - h_i\left(t, x_i^*\right)$, $F_j(t, y_j(t)) = f_j\left(t, y_j(t) + x_j^*\right) - f_j\left(t, x_j^*\right)$, $G_j(t, y_j(t)) = g_j\left(t, y_j(t) + x_j^*\right) - g_j\left(t, x_j^*\right)$, $W_{ik}\left(y_i(\tau_k^-)\right) = I_{ik}\left(y_i(\tau_k^-) + x_i^*\right) - I_{ik}\left(x_i^*\right)$, $i = 1, ..., n, \ k \in \mathbb{N}$. It follows from (C1) and (C2) that $|H_i(t, z)| \leq \ell_i^h |z|$, $|F_j(t, z)| \leq \ell_j^f |z|$, $|G_j(t, z)| \leq \ell_j^g |z|$ and $|W_{ik}(z)| \leq \ell_i^I |z|$ for all $t \in \mathbb{R}^+$, $z \in \mathbb{R}$, $i, j = 1, ..., n, \ k \in \mathbb{N}$.

A notable information about the values of the solution at any time $t$ and deviating argument $\beta(t)$ is given by Lemma 4 which has a useful result in the stability proof.

**Lemma 4.** *Let $y(t) = (y_1(t), ..., y_n(t))^T$ denote a solution of (13). Suppose that the assumptions (E), (C1)–(C4) and (C6) hold true. Then, for all $t \in \mathbb{R}^+$, the inequality given by*

$$\|y(\beta(t))\| \leq \lambda \|y(t)\|$$

*is true, where* $\lambda = \left(1 - \left(\overline{b}_g\overline{\theta} + \left(\overline{\theta}\left(\overline{c}_h + \overline{a}_f\right) + p\ell^I\right)\left(1 + \overline{b}_g\overline{\theta}\right)(1 + \ell^I)^p e^{(\overline{c}_h + \overline{a}_f)\overline{\theta}}\right)\right)^{-1}.$

*Proof.* Fix $t \in \mathbb{R}^+$ and $l \in \mathbb{N}$. For the system (13), one can write

$$y_i(t) = y_i(\theta_l) + \int_{\theta_l}^{t} \left(-c_i(s)H_i(s, y_i(s)) + \sum_{j=1}^{n} a_{ij}(s)F_j(s, y_j(s))\right.$$

$$\left. + \sum_{j=1}^{n} b_{ij}(s)G_j(s, y_j(\theta_l))\right) + \sum_{t_0 \leq \tau_k < t} W_{ik}(y_i(\tau_k^-))$$

for $t \in [\theta_l, \theta_{l+1})$. Then,

$$\|y(t)\| \leq \|y(\theta_l)\|\left(1 + \overline{b}_g\overline{\theta}\right) + \int_{\theta_l}^{t} \left(\overline{c}_h + \overline{a}_f\right)\|y(s)\| ds + \ell^I \sum_{t_0 \leq \tau_k < t} \|y(\tau_k^-)\| \tag{14}$$

is obtained for $t \in [\theta_l, \theta_{l+1})$. In the sequel, based on the Gronwall-Bellman Lemma in [1,5],

$$\|y(t)\| \leq \|y(\theta_l)\|\left(1 + \overline{b}_g\overline{\theta}\right)\left(1 + \ell^I\right)^p e^{(\overline{c}_h + \overline{a}_f)\overline{\theta}} \tag{15}$$

is reached. While $t = \tau_k^-$ in (15), it is seen that

$$\|y(\tau_k^-)\| \leq \|y(\theta_l)\|\left(1 + \overline{b}_g\overline{\theta}\right)\left(1 + \ell^I\right)^p e^{(\overline{c}_h + \overline{a}_f)\overline{\theta}}. \tag{16}$$

Moreover, for $t \in [\theta_l, \theta_{l+1})$,

$$\|y(\theta_l)\| \leq \|y(t)\| + \|y(\theta_l)\| \, \overline{b}_g \overline{\theta} + \int_{\theta_l}^{t} (\overline{c}_h + \overline{a}_f) \|y(s)\| \, ds + \ell^I \sum_{t_0 \leq \tau_k < t} \|y(\tau_k^-)\| \quad (17)$$

is achieved by the integral equation for the solution of (13) in the same way as expressed in (2). By using (15) and (16) in the last inequality,

$$\|y(\theta_l)\| \leq \|y(t)\| + \|y(\theta_l)\| \left( \overline{b}_g \overline{\theta} + \left( \overline{\theta} \, (\overline{c}_h + \overline{a}_f) + p\ell^I \right) \left( 1 + \overline{b}_g \overline{\theta} \right) \left( 1 + \ell^I \right)^p e^{(\overline{c}_h + \overline{a}_f)\overline{\theta}} \right)$$

and so

$$\|y(\theta_l)\| \leq \lambda \|y(t)\|, \; t \in [\theta_l, \theta_{l+1}) \quad (18)$$

are obtained in the light of the condition (C6). Therefore, it can be seen that the last inequality is satisfied for all $t \in \mathbb{R}^+$. So, lemma is proved. $\qquad \square$

Next, with the help of Lemma 4, sufficient conditions that ensure the global exponential stability of the trivial solution of (13) are given by the following theorem based on the theory in [1]. For this aim, it is seen that the functions $H_i(t, y_i(t))$ can be separated as $H_i(t, y_i(t)) = {}_*h_i(t)y_i(t) + H_i^*(t, y_i(t))$, where $H_i^*(t, y_i(t)) = h_i^*(t, y_i(t) + x_i^*) - h_i^*(t, x_i^*)$ by virtue of the remark in the Preliminaries section. Additionally, it is obvious that $H_i^*(t, 0) = 0$. Additionally, we give one more assumption as follows:

(C7) $\; {}_*c^h - \overline{{}_*c}_{\ell h} - \overline{a}_f - \overline{b}_g \lambda - \dfrac{\ln \left( 1 + \ell^I \right)}{\tau} > 0$, where the constants ${}_*c^h$ and $\overline{{}_*c}_{\ell h}$ are

defined by ${}_*c^h = \min_{1 \leq i \leq n} \left( \underline{c_i} \, {}_*h_i \right)$ and $\overline{{}_*c}_{\ell h} = \max_{1 \leq i \leq n} \left( \overline{c_i} \ell_i^{h^*} \right)$.

Moreover, the following lemma states the integral equation for the solution of (13) in addition to Lemma 1.

**Lemma 5.** *Fix a real number $t_0$. $y(t) = y(t, t_0, y^0) = (y_1(t), ..., y_n(t))^T$ is a solution of system (13) on $\mathbb{R}^+$ if and only if it is a solution of*

$$y_i(t) = y_i^0 e^{\int_{t_0}^{t} -c_i(s){}_*h_i(s)ds} + \int_{t_0}^{t} e^{\int_{s}^{t} -c_i(u){}_*h_i(u)du} \times \Bigg( -c_i(s)H_i^*(s, y_i(s))$$

$$+ \sum_{j=1}^{n} a_{ij}(s)F_j(s, y_j(s)) + \sum_{j=1}^{n} b_{ij}(s)G_j(s, y_j(\beta(s))) \Bigg) ds \quad (19)$$

$$+ \sum_{t_0 \leq \tau_k < t} e^{\int_{\tau_k}^{t} -c_i(s){}_*h_i(s)ds} W_{ik}(y_i(\tau_k^-))$$

*on $\mathbb{R}^+$, where $i = 1, ..., n$.*

**Theorem 3.** *Let the conditions (E), (C1)–(C7) be satisfied. Then, the trivial solution of (13) is globally exponentially stable.*

*Proof.* Let $y(t) = (y_1(t), ..., y_n(t))^T$ denote a solution of (13). Based on Lemma 5,

$$\|y(t)\| \leq \|y^0\| e^{-\underline{*}c^h(t-t_0)} + \int_{t_0}^t e^{-\underline{*}c^h(t-s)} \left(\overline{*c}_{\ell h} + \overline{a}_f + \overline{b}_g \lambda\right) \|y(s)\| \, ds$$
$$+ \ell^I \sum_{t_0 \leq \tau_k < t} e^{-\underline{*}c^h(t-\tau_k)} \|y(\tau_k^-)\|$$

is obtained. The last inequality can be written as follow

$$\|y(t)\| e^{\underline{*}c^h(t-t_0)} \leq \|y^0\| + \int_{t_0}^t e^{\underline{*}c^h(s-t_0)} \left(\overline{*c}_{\ell h} + \overline{a}_f + \overline{b}_g \lambda\right) \|y(s)\| \, ds$$
$$+ \ell^I \sum_{t_0 \leq \tau_k < t} e^{\underline{*}c^h(\tau_k - t_0)} \|y(\tau_k^-)\|.$$

In the sequel, based on the Gronwall-Bellman Lemma in [1,5], $\|y(t)\| e^{\underline{*}c^h(t-t_0)} \leq \|y^0\| (1 + \ell^I)^{i(t_0,t)} e^{\left(\overline{*c}_{\ell h} + \overline{a}_f + \overline{b}_g \lambda\right)(t-t_0)}$ is reached and so

$$\|y(t)\| \leq \|y^0\| e^{-\left(\underline{*}c^h - \overline{*c}_{\ell h} - \overline{a}_f - \overline{b}_g \lambda - \frac{\ln\left(1 + \ell^I\right)}{\underline{\tau}}\right)(t-t_0)}$$

is found. Condition (C7) leads to the global exponential stability of the trivial solution of (13). □

Since a linear transformation is used, it can be said that $x^*$, the equilibrium point, of (1) is globally exponentially stable as well.

## 3    An Example and a Simulation

This section presents an illustrative example and a simulation using MATLAB package program for the obtained theoretical results.

*Example 1.* Take the neural networks system (1) by the parameters $h_i(t, x_i(t)) = \left(e^{-it} + 1\right) x_i(t) + 0.001 e^{-it} \frac{x_i^2}{1 + x_i^2}$, $f_j(t, x_j(t)) = 0.01 \tanh(j x_j(t))$, $g_j(t, x_j(\beta(t))) = 0.01 \tanh(x_j(\beta(t))/2j)$, $c_i(t) = 0.02 \left(0.1i + \frac{i}{e^{-it} + i}\right)$, $a_{ij}(t) = 0.01 \sin(ijt)$, $b_{ij}(t) = 0.01 \sin(ijt)$, $d_i(t) = 0.0002 \frac{1}{e^{2t} + 1} + i 10^{-3i}$, $I_i\left(x(\tau_k^-)\right) = 10^{-5} \tanh\left(x_i(\tau_k^-)\right) - \frac{i}{k} + 1.2i$, $i, j = 1, 2, 3$, $k \in \mathbb{N}$. Let $\theta_l = \frac{1}{4}\left(l + \frac{(-1)^l}{3}\right)$, $\tau_k = \frac{1}{4}\left(k + \frac{(-1)^k}{3}\right)$, $l, k \in \mathbb{N}$. It can be seen that the assumptions (C1)–(C7) are satisfied by the following calculations: $\ell_i^h = 2.001$, $\ell_i^f = 0.01$, $\ell_i^g = 0.01$, $\ell_i^I = 0.00001$, $\overline{\theta} = \frac{5}{12}$, $\underline{\tau} = \frac{1}{12}$, $p = 2$,

$$\left(\left(\overline{c}_h + \overline{a}_f + 2\overline{b}_g\right)\overline{\theta} + p\ell^I\right)\left(1 + \ell^I\right)^p e^{(\overline{c}_h + \overline{a}_f)\overline{\theta}} = 0.02217434995 < 1,$$

$$\overline{b}_g\overline{\theta} + \left(\overline{\theta}\left(\overline{c}_h + \overline{a}_f\right) + p\ell^I\right)\left(1 + \overline{b}_g\overline{\theta}\right)\left(1 + \ell^I\right)^P e^{(\overline{c}_h + \overline{a}_f)\overline{\theta}} = 0.02226152429 < 1,$$

$$\underline{*}c^h - \overline{*c}_{\ell h} - \overline{a}_f - \overline{b}_g\lambda - \frac{\ln\left(1 + \ell^I\right)}{\underline{\tau}} = 0.00165172376 > 0.$$

Therefore, the conditions of Theorem 2 are satisfied and so (1) possesses a unique solution on $\mathbb{R}^+$. Accordingly, Theorem 3 shows that the equilibrium solution of (1) is globally exponentially stable. Time responses of $x_1(t)$, $x_2(t)$ and $x_3(t)$ in (1) together with parameters above are shown by Fig. 1.

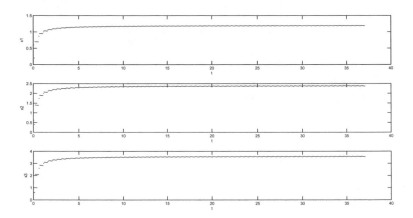

**Fig. 1.** Time responses of state variables of (1) for $t \in [0, 37.5833]$ and $x_i^0 = 0.6$, $i = 1, 2, 3$

## 4    Conclusion

In the present paper, a nonautonomous recurrent neural network system is considered by taking activation functions depending on time. Existence of deviating argument, impulsive effects and nonautonomous functions considered in this paper makes the neural network system more realistic and more complex to analyze. Since it is important for a system to represent a real life problem closely, the related neural network system is aimed to develop further in this study. Actually, it can be seen that nonautonomous activation functions were rarely considered in the literature. So, it is noteworthy to deal with the neural networks model as given in the present paper. It is the first time that the existence-uniqueness and the global exponential stability results for the equilibrium point of such a neural network system are investigated. The given example and the corresponding simulation concerning our findings have an importance in the sense of a verification for the reached theoretical results.

**Acknowledgements.** This work is supported by TÜBİTAK (The Scientific and Technological Research Council of Turkey) under project no 118F185.

# References

1. Akhmet, M.U.: Principles of Discontinuous Dynamical Systems. Springer, New York, Dordrecht, Heidelberg, London (2010). https://doi.org/10.1007/978-1-4419-6581-3
2. Akhmet, M.U., Yılmaz, E.: Neural Networks with Discontinuous/Impact Activations. Springer, New York (2013). https://doi.org/10.1007/978-1-4614-8566-7
3. Akhmet, M.U.: On the integral manifolds of the differential equations with piecewise constant argument of generalized type. In: Agarval, R.P., Perera, K. (eds.) Proceedings of the Conference on Differential and Difference Equations and Applications, pp. 11–20. Hindawi Publishing Corporation, Melbourne (2005)
4. Akhmet, M.U.: On the reduction principle for differential equations with piecewise constant argument of generalized type. J. Math. Anal. Appl. **336**, 646–663 (2007)
5. Lakshmikantham, V., Bainov, D.D., Simeonov, P.S.: Theory of Impulsive Differential Equations. vol. 6. World Scientific, Singapore (1989). Series in modern applied mathematics
6. Akhmet, M.U.: Nonlinear Hybrid Continuous Discrete-Time Models. Atlantis Press, Amsterdam, Paris (2011)
7. Gopalsamy, K.: Stability of artificial neural networks with impulses. Appl. Math. Comput. **154**, 783–813 (2004)
8. Akhmet, M.U., Yılmaz, E.: Impulsive Hopfield-type neural network system with piecewise constant argument. Nonlinear Anal. Real World Appl. **11**, 2584–2593 (2010)
9. Akhmet, M.U., Karacaören, M.: Stability of Hopfield neural networks with delay and piecewise constant argument. Discontinuity, Nonlinearity, Complex **3**(1), 1–6 (2014)
10. Rojas, R.: Neural Networks: A Systematic Introduction. Springer-Verlag, Berlin (1996)
11. Akhmet, M., Fen, M.O.: Generation of cyclic/toroidal chaos by Hopfield neural networks. Neurocomputing **145**, 230–239 (2014)
12. Akhmet, M.U., Fen, M.O.: Shunting inhibitory cellular neural networks with chaotic external inputs. Chaos **23**(2), 023112 (2013)
13. Yang, Z.: Global existence and exponential stability of periodic solutions of recurrent neural networks with functional delay. Math. Meth. Appl. Sci. **30**, 1775–1790 (2007)
14. Gui, Z., Yang, X.S., Ge, W.: Existence and global exponential stability of periodic solutions of recurrent cellular neural networks with impulses and delays. Math. Comput. Simul. **79**, 14–29 (2008)
15. Oliveira, J.: Global exponential stability of nonautonomous neural network models with unbounded delays. Neural Netw. **96**, 71–79 (2017)
16. McCulloch, W., Pitts, W.: A logical calculus of the ideas immanent in nervous activity. Bull. Math. Biophys. **5**, 115–133 (1943)
17. Block, H.: The perceptron: a model for brain functioning. Rev. Mod. Phys. **34**, 123–135 (1962)
18. Hopfield, J.J.: Neurons with graded response have collective computational properties like those of two-state neurons. Proc. Natl. Acad. Sci. **81**, 30–88 (1984)
19. Kosko, B.: Adaptive bidirectional associative memories. Appl. Opt. **26**(23), 4947–4960 (1987)
20. Cohen, M.A., Grossberg, S.: Absolute stability of global pattern formation and parallel memory storage by competitive neural networks. IEEE Trans. Syst. Man Cybern. Syst. **SMC-13**, 815–826 (1983)

21. Stamova, I., Stamov, G.: Applied Impulsive Mathematical Models. Springer, New York (2016). https://doi.org/10.1007/978-3-319-28061-5
22. Campbell, S.A., Ruan, S., Wei, J.: Qualitative analysis of a neural network model with multiple time delays. Int. J. Bifurcat. Chaos **9**(8), 1585–1595 (1999)
23. Akhmet, M.U., Aruğaslan, D., Yılmaz, E.: Stability in cellular neural networks with a piecewise constant argument. J. Comput. Appl. Math. **233**, 2365–2373 (2010)
24. Akhmet, M.U., Aruğaslan, D., Yılmaz, E.: Stability analysis of recurrent neural networks with piecewise constant argument of generalized type. Neural Netw. **23**, 805–811 (2010)
25. Xi, Q.: Global exponential stability of Cohen-Grossberg neural networks with piecewise constant argument of generalized type and impulses. Neural Comput. **28**(1), 229–255 (2016)
26. Akhmet, M.U., Aruğaslan, D., Cengiz, N.: Exponential stability of periodic solutions of recurrent neural networks with functional dependence on piecewise constant argument. Turk. J. Math. **42**, 272–292 (2018)
27. Akhmet, M.U., Yılmaz, E.: Global exponential stability of neural networks with non-smooth and impact activations. Neural Netw. **34**, 18–27 (2012)

# A Pure Genetic Energy-Efficient Backbone Formation Algorithm for Wireless Sensor Networks in Industrial Internet of Things

Zuleyha Akusta Dagdeviren[✉]

International Computer Institute, Ege University, Bornova, Izmir, Turkey

**Abstract.** Industry 4.0 envisions the utilization of Industrial Internet of Things (IIoT) to increase the efficiency and productivity in industrial automation and manufacturing processes. To interconnect various objects, wireless sensor networks (WSNs) are indispensable technologies located in the communication layer of IIoTs. Backbone formation is an important approach to achieve primitive operations such as data aggregation, routing and scheduling in energy-efficient WSNs. One of the fundamental backbone structures is the connected dominating set (CDS) which has been extensively studied by researchers. Although CDS provides a suitable infrastructure for relaying the sensed data packets, node energies are not primary concern during CDS construction phase. To achieve this, nodes are assigned weight values which are taken as a function of their resident energies. Weighted CDS (WCDS) problem is to find a set of nodes with minimum total weight such that every node in network is in the dominating set or a neighbor of at least one element of the dominating set. WCDS provides an energy-efficient backbone which leads to prolong the network lifetime. WCDS problem is in NP-Hard complexity class, so designing polynomial time heuristics is of utmost importance. To this aim, a pure genetic algorithm is proposed for the minimum WCDS problem on WSNs modeled as undirected graph in this paper. Each member of the population used in the algorithm is represented by a chromosome of bits where each bit indicates the corresponding node is in WCDS or not. Fitness value of a member is calculated by considering its total weight as well as the connectivity status of dominators. We analyze the time complexity of the proposed algorithm and show that our algorithm performs well in terms of total weight of the dominators, count of the dominators and the iteration count with respect to graph size, connection probability and maximum energy.

## 1 Introduction

Industry 4.0, fourth industrial revolution, aims to improve the productivity and efficiency in industrial manufacturing and automation by applying information technologies [32]. It has four design principles [14]: interconnection which is the

© The Author(s), under exclusive license to Springer Nature Switzerland AG 2021
J. Hemanth et al. (Eds.): ICAIAME 2020, LNDECT 76, pp. 553–566, 2021.
https://doi.org/10.1007/978-3-030-79357-9_53

ability of sensors, people and devices to connect and communicate with each other, information transparency that provides huge amounts of information to make decisions, technical assistance for aggregating and visualizing information extensively, decentralized decisions achieved by autonomous task executions of components forming the system. Internet of Things (IoT) is a worldwide connected network constructed by billions of objects, whose adoption is envisioned by Industry 4.0 for manufacturing. Industrial IoT (IIoT) brings efficiency, safety and intelligence for industrial automation and production processes. An IIoT system architecture can be divided into three layers: physical, communication and application sorted from bottom to top. The bottom layer includes wide range of physical devices such as computing/data centers, terminals and personal digital assistants. The top layer composes of various applications as smart factories, smart plants and smart supply chain. The communication layer has variety of networking techniques one of which is wireless sensor networks (WSNs) [1,2,33] for providing interconnectivity between objects.

WSNs are used widely due to the features such as low-power consumption, low cost, small size and etc. Since, generally obstacles and environmental challenges are present in industrial environments, multi-hop communication is essential for many application scenarios. To relay the sensed data packets to the sink node, backbone formation is of paramount importance. A backbone is a virtual path of selected nodes in which ends at the sink node [9,18,27,28,30,31,35]. By applying backbone formation, an energy and time-efficient topology is constructed which makes routing significantly easier.

A WSN can be represented with an undirected graph $G(V, E)$ in which $V$ and $E$ are the set of vertices (nodes) and the set of edges (links). Unit disk graph (UDG) is another popular model which can be a suitable model when transmission range is assumed as a circle. On the other side, this assumption fails when the network area includes obstacles [21]. In this situation, undirected graph becomes a better model. Graph theoretic backbone formation algorithms assume that the underlying network is modeled as a graph and provide backbone formation using this property.

Sensor nodes are battery powered, thus minimizing the energy consumption is crucial. Radio component is the dominant energy consumer part on a sensor node; hence packet transmissions between nodes should be optimized. Since backbone nodes are responsible to relay sensed data packets, their energy can be consumed fast. A weighted connected dominating set (WCDS) backbone can handle this problem by assigning node weights to the energies [39]. Finding minimum WCDS is a problem in NP-Hard complexity class, hence approximation algorithms and intelligent heuristics may lead to efficient results in polynomial time. Genetic algorithm is such a technique that is inspired from biological evolution. Although genetic algorithms were proposed to construct dominating sets (DSs) for WSNs such as [17,19,20,24], to the extent of our knowledge a pure genetic algorithm for the minimum WCDS construction on undirected graphs is not reported.

In this study, we propose a pure genetic algorithm for minimum WCDS based backbone formation in WSNs modeled by node weighted undirected graph. We explain the design considerations of the algorithm and analyze its time complexity. We implement the proposed algorithm in Java and compare the algorithm with the common greedy approach. We measure the total weight of the dominating set, dominator count and iteration count produced by the proposed algorithm against varying node counts, connection probabilities, maximum energy, crossover methods and selection methods. The rest of the paper continues as following. The description of DS and its variants are given in Sect. 2. Algorithms present in the literature are discussed in Sect. 3. In Sect. 4, the detailed steps of the proposed algorithm is given. Various measurements against different parameters are discussed in Sect. 5. The paper ends with the concluding remarks and future works in Sect. 6.

## 2   Background

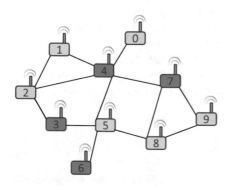

**Fig. 1.** Example DS

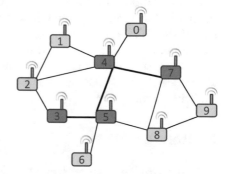

**Fig. 2.** Example CDS

A set of nodes $S$ is dominating if each node is in $S$ or has at least a neighbor in $S$. Finding minimum DS (MDS) is minimizing the cardinality of set $S$. An example DS S={3, 4, 6, 7} is given in Fig. 1. Each node is numbered with its node ID and node 0 is the sink node. $S$ is a connected dominating set (CDS) if $S$ is dominating and all nodes in $S$ are connected. An example CDS S = {3,4,5,7} is given in Fig. 2. Minimum WCDS is a subset of CDS nodes with minimum total weight. An example WCDS is given in Fig. 3. The numbers near to each node are its energy in joule and its weight which is calculated as 1/energy.

Genetic algorithm is a heuristic optimization algorithm. It operates on a population of solution where each solution represented as a chromosome. Initially, a random population is generated, and population is evolved with time. Each chromosome is scored with its fitness value. Chromosomes with higher fitness values are preferred over the other ones. Besides chromosomes with worst fitness values may be removed from the population. To diversify the population,

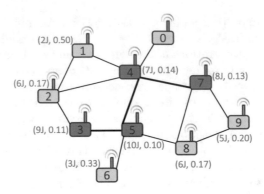

**Fig. 3.** Example WCDS

offspring are produced from parent chromosomes. Elitism method provides the selection of parents with the highest fitness values. In roulette wheel method, the chromosomes that have high fitness values will have better chance for parent selection. Child chromosome can be produced via single point crossover operator in which a random point $p$ in the chromosome data $[0,n]$ is chosen, the first part of the child chromosome $[0,p]$ is produced by copying the $[0,p]$ of first parent chromosome and the second part of the child chromosome $(p,n]$ is produced by copying the $(p,n]$ of second parent chromosome. After a child chromosome is produced, a mutation can be applied. Algorithm continues until termination condition is met.

## 3    Related Work

DS based problems are generally popular and various algorithms have been studied in this manner. Because of DS problem and its most variants are NP-hard class, many researchers focus on solving these problems within a reasonable time. A memetic algorithm to construct the minimum weighted edge DS is proposed by Abdel-Aziz [13], An edge DS is a set of edges which dominates all the other edges in the graph G. In this algorithm, three fitness functions are used and a search method is developed. In our study, we aim to construct node WCDS instead of edge weighted DS since the main goal is to select backbone nodes with higher energy values.

Erlebach et al. study on an approximation algorithm intended to construct weighted DS (WDS) on UDG [10]. Jovanovic et al. proposed an ant colony optimization applied to the minimum WDS problem [16]. An approximation algorithm for the same problem on UDGs is proposed by Zhu et al. [38]. Evolutionary algorithms for minimum WDS problem on undirected graphs are introduced in [26]. An ABC approach is used to solve minimum WDS problem on undirected graphs in [25]. Bouamama and Blum propose a randomized population-based iterated greedy algorithm for the minimum WDS problem [8]. For minimum WDS problem, Albuquerque and Vidal introduce a tabu search based hybrid metaheuristic [4]. In [29], a fast local search algorithm that aims to obtain a good solution within a short time on massive graphs is studied. Another metaheuristic approach for the same purpose is presented in [23]. The main difference of these algorithms from our proposed algorithm is they do not provide a CDS, thus cannot be used for backbone formation.

In the study [17], a genetic algorithm is proposed for minimum CDS problem regarding transmission power in wireless networks modeled as UDGs. Their algorithm tries to find a CDS having at most $k$ nodes. Our proposed algorithm in this study is targeted to run on undirected graphs and it is not limited by the size rule of Kamali. A simple heuristic algorithm on UDG is proposed by Balaji and Revathi to find a MCDS depend on a parameter support [7]. Jovanovic and Tuba study on an ant colony algorithm based on greedy heuristic that is intended to generate CDS [15]. A genetic algorithm is presented to construct a CDS in WANET [24]. Their algorithm works on UDG and does not take into consider node weights same as the previous study. In [11], a heuristic method is proposed to construct a CDS on undirected graphs. Zang et al. propose an approximation algorithm to solve CDS problem [36]. Khalil and Ozdemir study on the construction of CDS by genetic algorithm for UDGs [19,20]. In [3], to construct CDS, three algorithms are proposed in which the first of the algorithms has a performance factor of 5 from the optimal solution and the other two algorithms are variation of the first one. For other unweighted algorithms please refer to [5,12,22,34].

Ambulh et al. present an approximation algorithm to solve the minimum WCDS problem on UDGs [6]. Zhang et al. propose an approximation algorithm to construct WCDS [37]. These algorithms solve the WCDS problem, but they run on UDG model that is not suitable for real world applications.

## 4 Proposed Method

We proposed a genetic algorithm to find a WCDS on a given WSN. The flowchart of the algorithm is shown in Fig. 4. Initially, the algorithm constructs a random graph. The connectivity of the graph depends on the probability parameter which is given as $P$ in the flowchart. If the probability is 1, the graph will be fully connected.

Each node in the graph network has an energy $E$ and weight $w$. Weight is evaluated as in Eq. 1.

$$w_i = \frac{1}{E_i} \tag{1}$$

Total weight of the graph $W_T$ is calculated by the Eq. 2.

$$W_T = \sum_{i=1}^{n} w_i \tag{2}$$

After constructing the graph, population is generated randomly. Each chromosome has the same size with the graph network. While generating chromosomes, 20% of the gens are generated as 1s and the rest bits are as 0s. For each

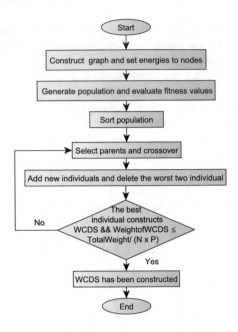

**Fig. 4.** Flowchart of the proposed algorithm

node in the graph, fitness values are evaluated by the fitness function, $ff$ in Eq. 3. In this equation, $N_{covered}$ is the number of the covered nodes and $W_{WCDS}$ is total weight of dominators. *Award* is given for chromosomes which are constructing a DS as adding the number of covered nodes to $ff$. If DS is connected, the number of covered nodes is also added to $ff$ as *Award*. If the DS is not connected, the maximum number of connected dominators is added as *Award*.

$$ff = N_{covered} - W_{WCDS} + Award \tag{3}$$

Then, the population is sorted in descending order according to the fitness values. The first two individuals are selected as parents and mated by single point crossover. The children are replaced with the worst two individuals and the population is sorted again.

$$W_{WCDS} \le \frac{W_T}{N_G \times P_C} \tag{4}$$

These steps are executed until the termination condition is provided. The criteria controls whether the best chromosome constructs a CDS or not. If this chromosome constructs a CDS, an extra control to find minimum WDS is needed. The second termination criterion controls the total weight of the dominators that is shown in Eq. 4 where $N_G$ is graph size and $P_C$ is probability of connectivity. If the best chromosome satisfies these conditions, the minimum WCDS is determined. The detailed steps of the proposed pure genetic algorithm is given below.

---

**Proposed Pure Genetic Algorithm**$(P_C, Max\_Energy, M_P, P_S, N_G, Method_c, Method_s)$

1: **Begin**
2:     Construct a random graph $G(V,E)$ according to connectivity probability $P_C$
3:     Set each $node_i$'s energy $E_i$ to random value between $(0, Max\_Energy)$
4:     Set each $node_i$'s weight $w_i$ to $1/E_i$
5:     Generate random population $P$ having size $P_S$
6:     Evaluate fitness values of each individual as $ff = N_{covered} - W_{WCDS} + Award$
7:     **while** the best individual is not WCDS **or** $W_{WCDS} \geq W_T/(N_G \times P_C)$ **do**
8:         Select parents $p_i$ and $p_j$ according to $Method_s$
9:         Crossover $p_i$ and $p_j$ according to $Method_c$ and obtain new offspring $n_i$, $n_j$
10:        Apply mutation operation to $n_i$, $n_j$ according to mutation probability $M_P$.
11:        Add new offsprings $n_i$, $n_j$ to population $P$
12:        Remove the worst two individuals in $P$
13:    **end while**
14:    Return the best individual
15: **End.**

---

**Theorem 1.** *The time complexity of the proposed pure genetic algorithm is* $O(n^3)$ *where $n$ is the number of nodes, $P_S \in O(n^2)$ and $nM_P \in O(n)$ .*

*Proof.* Constructing a random graph in Line 2 can be achieved in $O(n^2)$ time. Assigning energies and weights of nodes take $O(n)$ time in Lines 3 and 4. Line 5 which generates population of $P_S$ solutions can be executed in $O(nP_S)$ time. Finding the fitness values of each node in Line 6 can be achieved in $O(n^2)$ time. Line 7 includes the population sorting operation as it can be executed in $O(P_S log(P_S))$ time. The while loop condition in Line 8 takes $O(n^2)$ time since the best individual is checked for WCDS formation condition. Line 9 is the parent selection operation and can be achieved in constant complexity. Crossover operation in Line 10 can be achieved in $O(n)$ time. Line 11 is the mutation operation and takes $O(nM_P)$ time. Adding new individuals and removing the worst two individuals operations terminate in constant time complexity, so Line 12 and Line 13 take $O(1)$ time. Similary, Line 15 which is the best individual return operation has constant time complexity. Finally, the total time complexity of the proposed algorithm is $O(n^3)$ regarding Line 8 and Line 10.                                    □

## 5     Results

We implemented the proposed algorithm in Java and measure its performance against various parameters. The algorithms run on networks consisting of 20 to 100 nodes. These networks are randomly connected with the probabilities 0.2 to 1. Besides, the energies of the nodes are randomly chosen from the interval (0, Max_Energy) where Max_Energy is varied from 10 to 50 J. Unless otherwise stated, the probability of connectivity is 0.6, the maximum energy is 30 J, the mutation probability is 0.1, the population size is 30, the node count is 60, the crossover method is single point and the selection method is elitism.

To compare our proposed genetic algorithm, we implement the greedy algorithm where the node having the smallest weight is chosen at each iteration. To show the effectiveness of the algorithms, we run them on three different topologies: grid, random and wheel. Grid topology is frequently used in smart agriculture, random topology is common in industrial applications deployed in harsh environments and wheel topology is an extended star topology which is also applicable in various industrial scenarios. WCDS structures produced by greedy and genetic algorithms for grid topology are given in Figs. 5 and 6, respectively. For all of these figures, red nodes indicate nodes in WCDS and gray nodes are dominatees. Genetic algorithm performs 1.2 times better than the greedy algorithm for this example since the total weight of WCDS produced by our algorithm is 5 whereas greedy algorithm has 6 total weight. WCDS backbones constructed by the algorithms on a random topology are depicted in Figs. 7 and 8. For this example, the backbone of the greedy algorithm has 90 cost whereas the cost of our genetic algorithm is only 30. Lastly, the algorithms are tested on a wheel topology which are given in Figs. 9 and 10. For this network which has 26 nodes, the performance of our algorithm is 11.5 times better than the greedy algorithm.

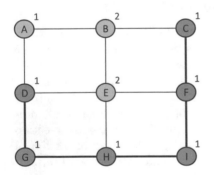

**Fig. 5.** WCDS Backbone in Grid topology produced by Greedy algorithm

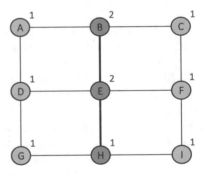

**Fig. 6.** WCDS Backbone in Grid topology produced by Greedy algorithm

Table 1 shows the total weight of WCDS, dominator count and iteration count against varying node count values. As the node count increases linearly, the total weight of WCDS and the dominator count increase by a small factor. The iteration count values fluctuate between 124 and 93 for node count values in 20 to 60 interval. Although a sharp increase is observed for 80 nodes, iteration count values are generally acceptable.

Table 2 shows total weight of WCDS, dominator count and iteration count against connection probability. As connection probability increases, our proposed algorithm produces lower cost WCDS with less number of dominators. This shows us that our algorithm may benefit from connectivity and may produce better WCDSs for densely connected graphs. Besides, iteration count values are

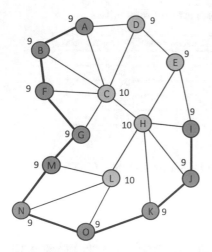

**Fig. 7.** WCDS Backbone in Grid topology produced by Greedy algorithm

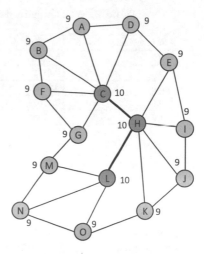

**Fig. 8.** WCDS Backbone in Grid topology produced by Greedy algorithm

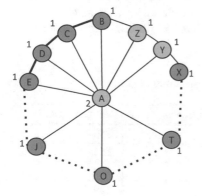

**Fig. 9.** WCDS Backbone in Grid topology produced by Greedy algorithm

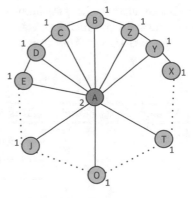

**Fig. 10.** WCDS Backbone in Wheel topology produced by Genetic algorithm

**Table 1.** WCDSs with graph sizes

| Node count | Total weight of WCDS | Dominator count | Iteration count |
| --- | --- | --- | --- |
| 20 | 0.1223 | 3 | 93 |
| 40 | 0.1366 | 3 | 93 |
| 60 | 0.1470 | 4 | 124 |
| 80 | 0.2054 | 5 | 509 |
| 100 | 0.2115 | 5 | 622 |

**Table 2.** WCDSs with different connection probabilities

| Connection probability | Total weight of WCDS | Dominator count | Iteration count |
|---|---|---|---|
| 0.2 | 0.5908 | 10 | 593 |
| 0.4 | 0.3873 | 7 | 130 |
| 0.6 | 0.1738 | 4 | 104 |
| 0.8 | 0.1152 | 3 | 103 |
| 1.0 | 0.1143 | 3 | 101 |

significantly small for dense graphs. Table 3 shows total weight of WCDS, dominator count and iteration count against maximum energy. As maximum energy increases, although dominator count values are stable, total weight of WCDS values decreases. Similarly, with Tables 2, 3 shows that the proposed algorithm chooses backbone nodes with higher energy when maximum energy of nodes increases. Moreover, iteration values are smaller for graphs having nodes with higher energy. Table 4 shows the iteration count for different selection methods and crossover methods. For parent selection methods, elitism performs better than the roulette wheel. For crossover methods, uniform crossover performs worse than the single point crossover and two-point cross over. The best method is to choose elitism with two-point crossover. This shows us that selecting best chromosomes for producing offspring and choosing two points for crossover performs better than the other approaches.

**Table 3.** WCDSs with different maximum energies

| Maximum energy | Total weight of WCDS | Dominator count | Iteration count |
|---|---|---|---|
| 10 | 0.3885 | 4 | 2750 |
| 20 | 0.2252 | 4 | 158 |
| 30 | 0.1652 | 4 | 144 |
| 40 | 0.1152 | 4 | 129 |
| 50 | 0.1004 | 4 | 207 |

**Table 4.** Iteration count for different GA methods

| | | Selection methods | |
|---|---|---|---|
| | | Elitism | Roulette wheel |
| Crossover methods | Single point | 252 | 419 |
| | Two points | 137 | 522 |
| | Uniform | 311 | 853 |

# 6    Conclusion

Industry 4.0 leverages IIoT to increase the efficiency in industrial manufacturing and automation. WSNs provide a critical communication infrastructure for IIoT applications. Utilizing WCDS structure for energy-efficient backbone formation is of paramount importance in terms of prolonging the network lifetime of WSNs. In this study, we propose a genetic algorithm for WCDS based backbone formation in WSNs modeled with undirected graphs. We show the design of the chromosomes and formulated the fitness function. We apply various crossover and parent selection techniques, and showed their performance. We theoretically analyze the proposed algorithm in terms of time complexity. We measure the total weight of WCDSs, dominator count and iteration count against total node count, connection probability and maximum energy. We test and compare our proposed algorithm along with the greedy algorithm on various topologies such as grid, random and wheel. We show that our algorithm produces WCDS with low cost and it is favorable in terms of total weight and dominator count. Our future plan is to employ deep learning techniques to further reduce the weight of produced WCDSs. In this manner, we will first study to model our WSNs with convolutional graph networks then we plan to design, implement and analyze algorithms for this model.

# References

1. Akyildiz, I., Su, W., Sankarasubramaniam, Y., Cayirci, E.: Wireless sensor networks: a survey. Comput. Netw. **38**(4), 393–422 (2002)
2. Akyildiz, I., Vuran, M.C.: Wireless Sensor Networks. John Wiley and Sons Inc., USA (2010)
3. Al-Nabhan, N., Zhang, B., Cheng, X., Al-Rodhaan, M., Al-Dhelaan, A.: Three connected dominating set algorithms for wireless sensor networks. Int. J. Sens. Netw. **21**(1), 53–66 (2016)
4. Albuquerque, M., Vidal, T.: An efficient matheuristic for the minimum-weight dominating set problem. Appl. Soft Comput. **72**, 527–538 (2018). https://doi.org/10.1016/j.asoc.2018.06.052. http://www.sciencedirect.com/science/article/pii/S1568494618303922
5. Alkhalifah, Y., Wainwright, R.L.: A genetic algorithm applied to graph problems involving subsets of vertices. In: Proceedings of the 2004 Congress on Evolutionary Computation (IEEE Cat. No.04TH8753), vol. 1, pp. 303–308 (2004)
6. Ambühl, C., Erlebach, T., Mihalák, M., Nunkesser, M.: Constant-factor approximation for minimum-weight (connected) dominating sets in unit disk graphs. In: Díaz, J., Jansen, K., Rolim, J.D.P., Zwick, U. (eds.) Approximation, Randomization, and Combinatorial Optimization. Algorithms and Techniques, pp. 3–14. Springer Berlin Heidelberg, Berlin, Heidelberg (2006)
7. Balaji, S., Revathi, N.: An efficient heuristic for the minimum connected dominating set problem on ad hoc wireless networks. Int. J. Math. Comput. Phys. Electr. Comput. Eng. **6**(8), 1045–1051 (2012)
8. Bouamama, S., Blum, C.: A randomized population-based iterated greedy algorithm for the minimum weight dominating set problem. In: 2015 6th International Conference on Information and Communication Systems (ICICS), pp. 7–12 (2015)

9. Chen, Y., Liestman, A.: Approximating minimum size weakly-connected dominating sets for clustering mobile ad hoc networks. p. 165 (2002). https://doi.org/10.1145/513819.513821

10. Erlebach, T., Mihalák, M.: A $(4 + \epsilon)$-approximation for the minimum-weight dominating set problem in unit disk graphs. In: Bampis, E., Jansen, K. (eds.) Approximation and Online Algorithms, pp. 135–146. Springer, Berlin, Heidelberg (2010)

11. Fu, D., Han, L., Liu, L., Gao, Q., Feng, Z.: An efficient centralized algorithm for connected dominating set on wireless networks. Proc. Comput. Sci. **56**, 162 – 167 (2015). https://doi.org/10.1016/j.procs.2015.07.190. http://www.sciencedirect.com/science/article/pii/S1877050915016713. The 10th International Conference on Future Networks and Communications (FNC 2015)/The 12th International Conference on Mobile Systems and Pervasive Computing (MobiSPC 2015) Affiliated Workshops

12. He, J.S., Cai, Z., Ji, S., Beyah, R., Pan, Y.: A genetic algorithm for constructing a reliable MCDS in probabilistic wireless networks. In: Cheng, Y., Eun, D.Y., Qin, Z., Song, M., Xing, K. (eds.) Wireless Algorithms, Systems, and Applications, pp. 96–107. Springer, Berlin Heidelberg (2011)

13. Hedar, A.R., Abdel Aziz, N., Sewisy, A.: Memetic algorithm with filtering scheme for the minimum weighted edge dominating set problem. Int. J. Adv. Res. Artif. Intell. **2**(8) (2013). https://doi.org/10.14569/IJARAI.2013.020808

14. Hermann, M., Pentek, T., Otto, B.: Design principles for industrie 4.0 scenarios. In: 2016 49th Hawaii International Conference on System Sciences (HICSS), pp. 3928–3937 (2016)

15. Jovanovic, R., Tuba, M.: Ant colony optimization algorithm with pheromone correction strategy for the minimum connected dominating set problem. Comput. Sci. Inf. Syst. **10**, 133–149 (2013). https://doi.org/10.2298/CSIS110927038J

16. Jovanovic, R., Tuba, M., Simian, D.: Ant colony optimization applied to minimum weight dominating set problem. In: Proceedings of the 12th WSEAS International Conference on Automatic Control, Modelling and Simulation, ACMOS 2010, pp. 322–326. World Scientific and Engineering Academy and Society (WSEAS), Stevens Point, Wisconsin, USA (2010)

17. Kamali, S., Safarnourollah, V.: A genetic algorithm for power aware minimum connected dominating set problem in wireless ad-hoc networks (2006)

18. Karl, H., Willig, A.: Protocols and Architectures for Wireless Sensor Networks. John Wiley and Sons Inc., Hoboken, USA (2005)

19. Khalil, E., Ozdemir, S.: CDs based reliable topology control in WSNS. pp. 1–5 (2015). https://doi.org/10.1109/ISNCC.2015.7238569

20. Khalil, E., Ozdemir, S.: Prolonging stability period of CDs based WSNS. pp. 776–781 (2015). https://doi.org/10.1109/IWCMC.2015.7289181

21. Kuhn, F., Wattenhofer, R., Zollinger, A.: Ad-hoc networks beyond unit disk graphs. In: Proceedings of the 2003 Joint Workshop on Foundations of Mobile Computing, DIALM-POMC 2003, pp. 69–78. Association for Computing Machinery, New York, NY, USA (2003). https://doi.org/10.1145/941079.941089

22. Li, R., Hu, S., Liu, H., Li, R., Ouyang, D., Yin, M.: Multi-start local search algorithm for the minimum connected dominating set problems. Mathematics **7**(12), 1173 (2019). https://doi.org/10.3390/math7121173

23. Lin, G.: A hybrid self-adaptive evolutionary algorithm for the minimum weight dominating set problem. Int. J. Wire. Mob. Comput. **11**(1), 54–61 (2016). https://doi.org/10.1504/IJWMC.2016.079466

24. More, V.S., Mangalwede, S.R.: A genetic algorithm for solving connected domi-
    nating set problem in wireless adhoc network. Int. J. Comput. Commun. Eng. Res.
    **1**(2), 45–48 (2013)
25. Nitash, C.G., Singh, A.: An artificial bee colony algorithm for minimum weight
    dominating set. In: 2014 IEEE Symposium on Swarm Intelligence, pp. 1–7 (2014)
26. Potluri, A., Singh, A.: Hybrid metaheuristic algorithms for minimum weight dom-
    inating set. Appl. Soft Comput. **13**(1), 76–88 (2013). http://www.sciencedirect.
    com/science/article/pii/S1568494612003092
27. Ramalakshmi, R., Radhakrishnan, S.: Weighted dominating set based routing for
    Ad Hoc communications in emergency and rescue scenarios. Wireless Netw. **21**,
    499–512 (2015). https://doi.org/10.1007/s11276-014-0800-4
28. Torkestani], J.A., Meybodi, M.R.: An intelligent backbone formation algorithm
    for wireless ad hoc networks based on distributed learning automata. Com-
    put. Netw. **54**(5), 826–843 (2010). https://doi.org/10.1016/j.comnet.2009.10.007,
    http://www.sciencedirect.com/science/article/pii/S1389128609003260
29. Wang, Y., Cai, S., Chen, J., Yin, M.: A fast local search algorithm for minimum
    weight dominating set problem on massive graphs. In: Proceedings of the 27th
    International Joint Conference on Artificial Intelligence, IJCAI 2018, pp. 1514–
    1522. AAAI Press (2018)
30. Wang, Y., Wang, W., Li, X.Y.: Distributed low-cost backbone formation for wire-
    less ad hoc networks. In: Proceedings of the 6th ACM International Symposium on
    Mobile Ad Hoc Networking and Computing, MobiHoc 2005, pp. 2–13. Association
    for Computing Machinery, New York, NY, USA (2005). https://doi.org/10.1145/
    1062689.1062692
31. Wu, J., Li, H.: On calculating connected dominating set for efficient routing in
    ad hoc wireless networks. In: Proceedings of the 3rd International Workshop on
    Discrete Algorithms and Methods for Mobile Computing and Communications,
    DIALM 1999, pp. 7–14. Association for Computing Machinery, New York, NY,
    USA (1999). https://doi.org/10.1145/313239.313261
32. Xu, H., Yu, W., Griffith, D., Golmie, N.: A survey on industrial internet of things: a
    cyber-physical systems perspective. IEEE Access (2018). https://doi.org/10.1109/
    ACCESS.2018.2884906
33. Yick, J., Mukherjee, B., Ghosal, D.: Wireless sensor network survey. Comput.
    Netw. **52**(12), 2292–2330 (2008). https://doi.org/10.1016/j.comnet.2008.04.002.
    http://www.sciencedirect.com/science/article/pii/S1389128608001254
34. Yu, J., Li, W., Feng, L.: Connected dominating set construction in cognitive radio
    networks. In: 2015 International Conference on Identification, Information, and
    Knowledge in the Internet of Things (IIKI), pp. 276–279. IEEE Computer Society,
    Los Alamitos, CA, USA (2015). https://doi.org/10.1109/IIKI.2015.66
35. Yu, J., Wang, N., Wang, G., Yu, D.: Review: connected dominating sets in wireless
    ad hoc and sensor networks - a comprehensive survey. Comput. Commun. **36**(2),
    121–134 (2013). https://doi.org/10.1016/j.comcom.2012.10.005
36. Zhang, J., Zhou, S.M., Xu, L., wu, W., ye, X.: An efficient connected dominating
    set algorithm in WSNS based on the induced tree of the crossed cube. Int. J. Appl.
    Math. Comput. Sci. **25**, 295–309 (2015). https://doi.org/10.1515/amcs-2015-0023
37. Zhang, Z., Zhou, J., Ko, K.I., Du, D.z.: Approximation algorithm for minimum
    weight connected m-fold dominating set. arXiv:1510.05886 (2017)

38. Zhu, X., Wang, W., Shan, S., Wang, Z., Wu, W.: A PTAS for the minimum weighted dominating set problem with smooth weights on unit disk graphs. J. Comb. Optim.-JCO **23**, 443–450 (2012). https://doi.org/10.1007/s10878-010-9357-z

39. Zou, F., Wang, Y., Xu, X.H., Li, X., Du, H., Wan, P., Wu, W.: New approximations for minimum-weighted dominating sets and minimum-weighted connected dominating sets on unit disk graphs. Theor. Comput. Sci. **412**(3), 198–208 (2011). https://doi.org/10.1016/j.tcs.2009.06.022, http://www.sciencedirect.com/science/article/pii/S0304397509004162. Combinatorial Optimization and Applications

# Fault Analysis in the Field of Fused Deposition Modelling (FDM) 3D Printing Using Artificial Intelligence

Koray Özsoy[1](✉) and Helin Diyar Halis[2]

[1] Department of Electric and Energy, Senirkent Vocational School,
Isparta University of Applied Sciences, Isparta, Turkey
`korayozsoy@isparta.edu.tr`
[2] Faculty of Technology Mechatronics Engineering, Isparta University of Applied Sciences,
Isparta, Turkey

**Abstract.** It is used for 3-dimensional (3D) production, which is called additive manufacturing, which emerged In the 1980s, with the inadequacy of traditional methods. Today, with the development of this technology, all kinds of geometry are produced rapidly. However, with 3D printing technology development, the problems are known as complexity have increased significantly. This situation made the fault analysis during the study a matter to be focused. User intervention is generally required in case of faulty in the parameters during printing using 3D printing technologies. The user must first observe whether there is any faulty and decide what to do. The study focused on automatic diagnostics to help the user diagnose the potential faulty and solve the problem. The data from the sensors belonging to the electronic card designed in the study were taken. The analog values were then read and analyzed, and the faulty conditions were checked according to the analysis. In the field of 3D printing technologies, a new approach has been presented by analyzing the data obtained in faulty analysis with the artificial intelligence method.

**Keywords:** Artificial intelligence · 3D printer technologies · Data analysis

## 1 Introduction

Today, the development of technology has led to the development of existing methods and an increase in their needs in this direction. The term additive manufacturing, which has been in our lives since the 1980s, has been one of the methods developed with the increasing needs and the inadequacy of traditional methods. While forming operations can be performed only on the existing part with traditional methods, parts with complex geometry can be easily manufactured using three-dimensional model data thanks to 3D printing technologies. 3D printing technologies are defined as the manufacturing process where an object is created due to adding the material layer by layer using the model data for manufacturing the three-dimensional model [1]. Thanks to 3D printing technologies,

J. Hemanth et al. (Eds.): ICAIAME 2020, LNDECT 76, pp. 567–577, 2021.
https://doi.org/10.1007/978-3-030-79357-9_54

the part, which is 3D drawn using computer-aided design (BDT), is quickly revealed as a product. This allows freedom in design and then innovation and development [2].

One of the techniques commonly used in 3D printing methods is Fused Deposition Modeling (FDM) technology [3]. FDM is based on layer-by-layer extrusion of filaments, whose raw material is thermoplastic, according to certain coordinates in the nozzle heated to a temperature close to the melting point [4]. The part drawn by using CAD programs is converted to STL format so that the 3D printer can understand the coordinates. The slicing process is performed to determine how the nozzle will move during the extrusion process. Then adjustments such as the fill rate, support angle, layer thickness are made in the software used and sent to the G-code printer. The nozzle moves along the X and Y axes. It leaves the semi-melted thermoplastic material as thin wires on the table. In contrast, the table moves downward when one layer is finished in the Z-axis, and the other layer is formed [5, 6]. This cycle continues until the part is built.

The concept of artificial intelligence was first mentioned at the Dartmouth conference in 1956. However, the studies in this area are ancient [7]. Artificial Intelligence is a set of software and hardware systems that mimic human intelligence, behaving like a human, can reason, develop itself and turn it into action in line with its experiences [8, 9]. Artificial Intelligence analyzes complex data in different fields with various methods and interprets it to be more understandable [10].

Machine learning is considered a branch of artificial intelligence, called predictive modeling [11]. Machine learning makes inferences from existing data and predicts the unknown. In other words, data are analyzed using various algorithms and mathematical and statistical methods to make inferences from input data. Machine learning algorithms are categorized as supervised and unsupervised. In supervised learning, training data is tagged, and the machine's job is to learn how to assign tags to data outside of the training data set. In unsupervised learning, the data in the training data set are unlabeled. The purpose of the machine is to analyze the existing data and determine the relationships. Interpreting data sets and categorizing data in unsupervised learning is performed by the machine learning algorithm [12, 13].

Many different studies have been carried out in 3D printing Technologies using different techniques of artificial intelligence. Among these studies, there is also faulty analysis in three-dimensional printers. When the academic studies on faulty analysis in 3D printers using artificial intelligence are examined;

Bacha et al. (2016) examined the Bayes network use in defect analysis in additive manufacturing. In his studies, current, voltage, limit, and temperature sensors were used by designing a data collection interface. Bayes networks analyzed the data received from the sensors. As a result of the study, it was seen that Bayes networks are used for the classification task in faulty analysis in the industrial field [14].

He et al. (2018) used support vector machines (SVM) for faulty analysis of the three-dimensional printer. An attitude sensor was mounted on the printer moving platform to monitor angular velocity, three axial attitude angles, vibration acceleration, magnetic field density, and data collected for different failure situations. Least-squares support vector machines (LS-SVM) were preferred for the classification of the collected data. Swivel bearing wear was identified as a faulty model, labeling 12 joint bearing failures and normal conditions. A total of 3900 samples were obtained by printing the cylindrical shell model with a 75 mm radius and a height of 0.3 mm for training and diagnostic testing of the model. As a result of the study, an average accuracy of 94.44% was obtained in modeling with LS-SMV [15].

In the first part of the study, 3D printing Technologies, fused deposition modeling (FDM) technology, artificial intelligence, and machine learning were briefly mentioned. Also, related studies on faulty analysis in three-dimensional printers were mentioned. In the second part, the data collection interface used in the study, the method used, XGBoost, and performance evaluation criteria are mentioned. In the third part, the importance of variables for the diagnostic system and performance evaluations in tabular form is shown. As a result, in this study, an approach with the XGBoost algorithm is presented using the data obtained from the faulty analysis interface in 3D printing technology.

## 2    Material and Method

### 2.1    Material

In the study, the variables of temperature, current, and voltage in the x, y, and z axes of the stepper motor driver during printing the part produced by 3D printing technology were examined. A data collection card was created to examine the temperature, current, and voltage variables. A data set was created with the values we obtained through the data collection card. The data set obtained trained using the XGBoost algorithm and K-fold cross-validation. The results obtained were evaluated according to five different performance evaluation criteria: sensitivity, specificity, accuracy, F-score, and ROC curve. Detailed information about the 3D printer, data collection card, models, and performance evaluation criteria used in the study are given below.

#### 2.1.1    3D Printer

The 3D printing process, also known as 3D printing technologies, is a process manufactured by placing parts on a tray in layers in a digital environment. All 3D printing processes require electrical and electronic hardware, software, and materials to work together. The 3D printer used in the study is shown in Fig. 1. The features of the printer are as follows:

1.  The 3D printer is an open-source hypercube model printer with "core XY" mechanism.
2.  Main processor is atmega 2560 MCU.
3.  A4988 stepper motor driver is used.
4.  Drv8825 step motor driver integrated is available.
5.  The dimensions of the printer are 200 × 2000 × 150 mm.

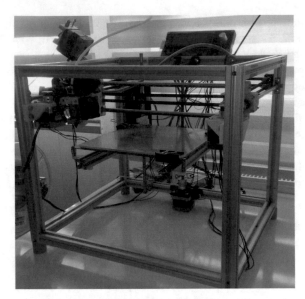

**Fig. 1.** 3D printer

### 2.1.2 Data Acquisition Interface

The data acquisition card shown in Fig. 2 is a microcontroller card that can receive and process data from sensors and transfer data wirelessly by converting from analog to digital. Thanks to this card, it is possible to read the temperature, current, and voltage values during 3D printer printing. LM324, ACS712, DS18B20, HEF4051 are used as hardware in the data acquisition card.

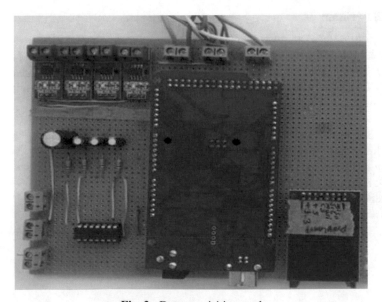

**Fig. 2.** Data acquisition card

### 2.1.2.1 ACS712 Current Sensor

ACS712 current sensor on the data acquisition card is used to provide current control. The ACS712 measures the current through the circuit up to 5 ammeters and generates a signal proportional to the measured current. ACS712 sensor can be used in AC and DC circuits and allows bidirectional current flow. ACS712 current sensor works according to the Hall effect principle [16]. Figure 3 shows the ACS712 current sensor. There are current input (IP+) pin, current output (IP−) pin, Vcc, Out, and Gnd on the sensor. A magnetic field is produced while current flows in the copper strip connect the current input and output pins in the sensor and convert this magnetic field into voltage in the hall effect.

**Fig. 3.** Current sensor (ACS712)

### 2.1.2.2 DS18B20 Temperature Sensor

DS18B20 is a digital temperature sensor. The DS18B20 sensor measures precisely between −10 and +80 °C. The sensor performs reading on 1-Wire, i.e., a single line. As shown in Fig. 4, the DS18B20 temperature sensor has three legs: Vdd, Gnd, and DQ (data). For the sensor to communicate with Arduino, VDD must be connected to +5V, GND to the GND of the Arduino, and the data leg must be connected to +5V with a 4.7 kΩ resistor.

**Fig. 4.** Temperature sensor (DS18B20)

### 2.1.3 XGBoost

XGBoost was proposed by Chen and Guestrin in 2016 [17]. XGBoost is a supervised learning algorithm used in regression and classification problems [18]. The XGBoost algorithm is an optimized high-performance version of the gradient boosting algorithm. XGBoost has become a popular machine learning algorithm in the Kaggle competition due to its high performance and ease of use [19]. The XGBoost algorithm reduces the computation time with its fast operation. This algorithm provides high predictive power and prevents over-learning problems [20, 21].

### 2.1.4 K-Fold Cross Validation

K-Fold Cross Validation is a technique widely used for performance evaluation of machine learning models [22]. The data set is divided into sets as training and testing. The "k" parameter determines how many subsets the data set will be divided into an equal size. While one of the sub-sets as much as the "k" parameter selected is used to verify the test set model, the sub-set as "k-1" is used to create the training set model. The training and testing process is continued for the number of parameters selected [23, 24].

### 2.1.5 Performance Evaluation Criteria

The confusion matrix method and ROC curve, which are frequently used to evaluate the model performance, were preferred. The complexity matrix is shown in Table 1 [25].

**Table 1.** Complexity matrix

| Real value | Predicted value | |
|---|---|---|
| | Positive | Negative |
| Positive | True positive-TP | False positive-FP |
| Negative | False negative-FN | True negative-TN |

Mathematical expressions on how sensitivity, specificity, accuracy, and F-score criteria are calculated using the complexity matrix are given in Eqs. 1, 2, 3, and 4, respectively.

$$Senstivity = \frac{DP}{DP + YP} \tag{1}$$

$$Specificity = \frac{DN}{DN + YN} \tag{2}$$

$$Accuracy = \frac{DP + DN}{DP + DN + YP + YN} \tag{3}$$

$$F - Score = \frac{2DP}{2DP + YP + YN} \tag{4}$$

## 2.2 Method

The workflow diagram of the work performed is given in Fig. 5. In the first stage, the values of temperature, current, and voltage variables in the x, y, and z axes of the stepper motor driver were collected with a data acquisition card while printing on a 3D printer. The data set was created with the collected data. The second stage is the data preprocessing stage. Lost data analysis, labeling, normalization, and feature selection were carried out in the data preprocessing phase. There are nine inputs in the data set and one output value as temperature, current, and voltage. Input variables are determined as T1, T2, T3, I1, I2, I3, V1, V2, V3. It is divided into two classes as it is faulty, and there is no faulty as the output variable. The training data set contains 370 values in total. The training data obtained was skewed using the XGBoost algorithm and k-fold cross-validation in the third stage. In the last stage, the trained model was tested in the evaluation of the model.

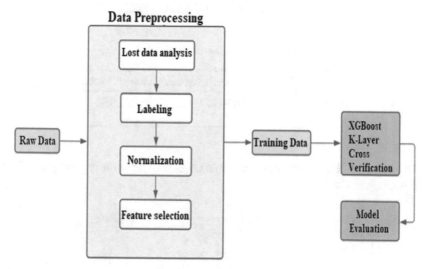

**Fig. 5.** Workflow diagram

# 3  Research Findings

In the study, using the XGBoost machine learning algorithm and k-fold cross-validation, the variables of temperature, current, and voltage in each axis, including the x, y, z axes of the stepper motor driver in a 3D printer, were trained. The following results were obtained by applying the test procedure to the classes. The complexity matrix for the test data set of the XGBoost algorithm is given in Fig. 6. Evaluation criteria according to the complexity matrix are shown in Table 2.

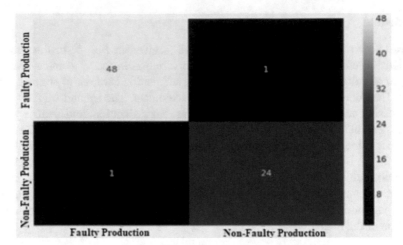

**Fig. 6.** XGBoost karmaşıklık matrisi

**Table 2.** XGBoost evaluation criteria

| Data | Sensitivity | Specificity | Accuracy | F-Score |
|------|-------------|-------------|----------|---------|
| Test data set | 0,96 | 0,97959 | 0,97297 | 0,96 |

ROC curve, which is another performance evaluation criterion of the XGBoost algorithm, is given in Fig. 7.

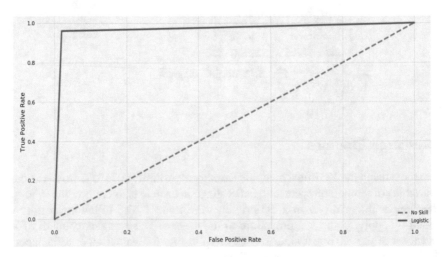

**Fig. 7.** ROC curve

Figure 8 shows the parameters affecting the model and the importance of the parameters.

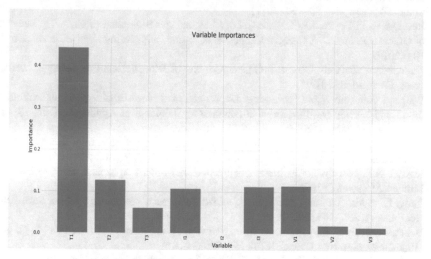

**Fig. 8.** Importance of parameters

## 4 Conclusions

In the study, the XGBoost algorithm was used for faulty analysis in three-dimensional printing technology. In the study, the variation of temperature, voltage, and current variables during printing was investigated. The results obtained were evaluated according to five different performance evaluation criteria.

- The algorithm has been evaluated according to the performance evaluation criterion, complexity matrix, and ROC curve. As a result of the evaluation, it was seen that the algorithm was able to detect faulty in temperature, current, and voltage variables.
- XGBoost algorithm was evaluated according to sensitivity, specificity, accuracy, and F-score performance evaluation criteria. As a result of the evaluation, 96% sensitivity, 97.95% specificity, 97.29% accuracy, and 96% F-Score results were revealed.
- AUC (area under the ROC curve) value of the XGBoost algorithm is calculated. The AUC value of the algorithm used was found to be 0.970.
- The influence of temperature, voltage, and current variables on the model was examined. As a result, it was revealed that the T1 parameter is the most influencing parameter in terms of the model performance.

## References

1. Ngo, T.D., Kashani, A., Imbalzano, G., Nguyen, K.T., Hui, D.: Additive manufacturing (3D printing): a review of materials, methods, applications and challenges. Compos. B Eng. **143**, 172–196 (2018)

2. Herzog, D., Seyda, V., Wycisk, E., Emmelmann, C.: Additive manufacturing of metals. Acta Mater. **117**, 371–392 (2016)
3. Singh, S., Ramakrishna, S., Singh, R.: Material issues in additive manufacturing: a review. J. Manuf. Process. **25**, 185–200 (2017)
4. Rey, D.F.V., St-Pierre, J.P.: Department of Chemical and Biological Engineering, University of Ottawa, Ottawa, ON, Canada. Handbook of Tissue Engineering Scaffolds: Volume One, 109 (2019)
5. Tanzi, M.C., Farè, S., Candiani, G.: Foundations of Biomaterials Engineering. Academic Press, Cambridge (2019)
6. Walker, J.L., Santoro, M.: Processing and production of bioresorbable polymer scaffolds for tissue engineering. In: Bioresorbable Polymers for Biomedical Applications, pp. 181–203. Woodhead Publishing (2017)
7. El Naqa, I., Haider, M.A., Giger, M.L., Ten Haken, R.K.: Artificial intelligence: reshaping the practice of radiological sciences in the 21st century. Br. J. Radiol. **93**(1106), 1–15 (2020)
8. Russell, S.J., Norvig, P.: Artificial Intelligence: A Modern Approach. Pearson Education Limited, Malaysia (2016)
9. Jiang, F., et al.: Artificial Intelligence in healthcare: past, present and future. Stroke Vasc. Neurol. **2**(4), 230–243 (2017)
10. Hosny, A., Parmar, C., Quackenbush, J., Schwartz, L.H., Aerts, H.J.: Artificial Intelligence in radiology. Nat. Rev. Cancer **18**(8), 500–510 (2018)
11. Erickson, B.J., Korfiatis, P., Akkus, Z., Kline, T.L.: Machine learning for medical imaging. Radiographics **37**(2), 505–515 (2017)
12. Shokri, R., Stronati, M., Song, C., Shmatikov, V.: Membership inference attacks against machine learning models. In: 2017 IEEE Symposium on Security and Privacy (SP), pp. 3–18. IEEE, May 2017
13. Biamonte, J., Wittek, P., Pancotti, N., Rebentrost, P., Wiebe, N., Lloyd, S.: Quantum machine learning. Nature **549**(7671), 195–202 (2017)
14. Bacha, A., Benhra, J., Sabry, A.H.: A CNC machine fault diagnosis methodology based on Bayesian networks and data acquisition. Commun. Appl. Electron. **5**, 41–48 (2016)
15. He, K., Yang, Z., Bai, Y., Long, J., Li, C.: Intelligent fault diagnosis of delta 3D printers using attitude sensors based on support vector machines. Sensors **18**(4), 1298 (2018)
16. Shafique, M.T., Kamran, H., Arshad, H., Khattak, H.A.: Home energy monitoring system using wireless sensor network. In: 2018 14th International Conference on Emerging Technologies (ICET), pp. 1–6. IEEE, November 2018
17. Chen, T., Guestrin, C.: Xgboost: a scalable tree boosting system. In: Proceedings of the 22nd ACM SIGKDD International Conference on Knowledge Discovery and Data Mining, pp. 785–794, August 2016
18. Nguyen, H., Bui, X.-N., Bui, H.-B., Cuong, D.T.: Developing an XGBoost model to predict blast-induced peak particle velocity in an open-pit mine: a case study. Acta Geophys. **67**(2), 477–490 (2019). https://doi.org/10.1007/s11600-019-00268-4
19. Zhang, D., Qian, L., Mao, B., Huang, C., Huang, B., Si, Y.: A data-driven design for fault detection of wind turbines using random forests and XGboost. IEEE Access **6**, 21020–21031 (2018)
20. Dhaliwal, S.S., Nahid, A.A., Abbas, R.: Effective intrusion detection system using XGBoost. Information **9**(7), 149 (2018)
21. Mitchell, R., Frank, E.: Accelerating the XGBoost algorithm using GPU computing. PeerJ Comput. Sci. **3**, e127 (2017)
22. Yadav, S., Shukla, S.: Analysis of k-fold cross-validation over hold-out validation on colossal datasets for quality classification. In: 2016 IEEE 6th International Conference on Advanced Computing (IACC), pp. 78–83. IEEE, February 2016

23. Jung, Y.: Multiple predicting K-fold cross-validation for model selection. J. Nonparametric Stat. **30**(1), 197–215 (2018)
24. Wong, T.T., Yeh, P.Y.: Reliable accuracy estimates from k-fold cross validation. IEEE Trans. Knowl. Data Eng. **32**(8), 1586–1594 (2019)
25. Ruuska, S., Hämäläinen, W., Kajava, S., Mughal, M., Matilainen, P., Mononen, J.: Evaluation of the confusion matrix method in the validation of an automated system for measuring feeding behaviour of cattle. Behav. Proc. **148**, 56–62 (2018)

# Real-Time Maintaining of Social Distance in Covid-19 Environment Using Image Processing and Big Data

Sadettin Melenli and Aylin Topkaya[✉]

Istanbul, Turkey
{sadettin.melenli,aylin.topkaya}@itelligence.com.tr

**Abstract.** Recent research in computer vision is increasingly focusing on developing systems to understand people's appearance, movements and activities, provide advanced interfaces for interacting with people, create human models. For any of these systems to work, they need methods to identify people from a particular input image or video. Today, real-time object detection and sizing of objects is an important issue in many areas of the industry. This is a vital issue of computer vision problems. With Covid-19's healing process, it will be very important to maintain social distance. In this research and development, it is aimed to maintain social distance with proposed big data architecture. This article provides an advanced technique to detect objects in video streams in real time and calculate their distance. The system composed research and developments to perform a stream from the camera, such as video stream, distance and object detection model, incoming data stream, data stream collection and report generation. The video stream from the camera is processed with GStream. The frames from the video stream are taken by OpenCV, YOLOV3 is trained by distance and object detection model and developed by Python. Video streaming data trained with Kubeflow is published with Apache Kafka and Apache Spark. It uses HDFS used to store published data. It is used to query and analyze data in Hive, Impala, Hbase HDFS. After that Analytical reports are created. E-mail notifications can be created according to the data in the database using by Apache Oozie. Through the proposed real time big data architecture, people can be safe in closed areas.

**Keywords:** Computer vision · Image processing · Big data · Stream · Detection

## 1 Introduction

Video surveillance is a research topic in employee computer vision that does recognize, track and detect objects on a range of images. Object detection involves finding and detecting objects within a range of video. Each detection method requires an object detection mechanism when the object is first displayed in the video. As the need for powerful computers and the availability of cheap video cameras increase, the interest in video analysis and object detection algorithms has also increased. Image processing refers to processing within the frame of the image or video received as input, and the

J. Hemanth et al. (Eds.): ICAIAME 2020, LNDECT 76, pp. 578–589, 2021.
https://doi.org/10.1007/978-3-030-79357-9_55

resulting processing set may be the image set of the relevant parameters. The purpose of image processing is to observe invisible objects or to detect visible objects.

Analysis of the human movement is one of the newest and most active research topics in image processing. Human movement is an important part of human perception, and motion analysis is to detect human's movements within the range of video. An important research topic for the vision of computers, which has gained great importance in the last few years, is the perception of human activities with a video [1].

In this study, a model has been developed that detects people with image processing and calculates the distance between them. The aim of the study is to ensure that people maintain social distance in Covid-19 coverage. In the second part of the study, the architectural structure and methodology are explained in detail; in the third part, the results obtained from the research and development and suggestions are given.

## 2 Architectural Design and Methodology

### 2.1 Architecture

The video stream from the camera is processed with GStream. The frames from the video stream are taken by OpenCV, YOLOV3 is trained by the distance and object detection model and developed by Python. The objects in the video stream are detected and the distance between them is calculated. Video streaming data trained with Kubeflow is published with Apache Kafka and Apache Spark. It uses HDFS, which is used to store published data. It is used to query and analyze data in Hive, Impala, Hbase HDFS. After that Analytical reports are created. E-mail notifications can be created by Apache Oozie based on data in the database. The system architecture is shown in Fig. 1.

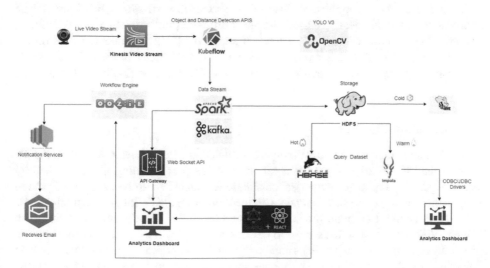

**Fig. 1.** System architecture

## 2.2 Video Stream

Gstreamer is a tool for processing video streams. It is both the software library and the library called from the command line tool. Its purpose in the study is to provide real-time video streaming over the camera IP network. GStreamer is a pipeline-based open source cross-platform multimedia framework that connects a wide variety of media processing systems to complete complex workflows. Usually, the RTSP or HTTP protocol is used by the camera for video streaming [2].

Figure 2 shows the diagram of the Codec system. Video captured from the camera is interfaced to the computer, sent via a wired connection to another computer, called a client, and displayed on the monitor [3].

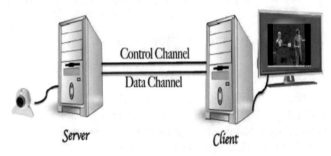

**Fig. 2.** Codec system [3]

GStreamer allows a programmer to create a variety of media-related applications, including simple audio playback, audio and video streaming, playback, recording, and editing. It is used as an application that runs under real-time streaming media and media players. It is software designed to run on many different platforms such as GNU/Linux, Solaris, Mac OS X, Microsoft Windows. It supports all media codecs. Based on the GObject writing system, it can provide all kinds of capabilities that are written in the C programming language, except for a graphical interface for a movie player, for example [3].

## 2.3 KubeFlow Pipelines

The streaming video is trained on the object and detection model developed with OpenCV one of the ML libraries. OpenCV is an open source programming library created by Intel. OpenCV is used for numerical calculations. OpenCV can be used for PC vision applications and an interface is made with the help of OpenCV for AI calculations [4].

ML (Machine Learning) model was created for object and distance detection with OpenCV library. We use KubeFlow to develop and deploy a machine learning system. Kubeflow ML provides a pipelines creation platform. KubeFlow is a workload agnostic pipeline execution framework. This means, every application which can be packaged as a container can be run within KubeFlow. Kubeflow is designed to enable using machine learning pipelines to orchestrate complicated workflows running on Kubernetes. Individual steps of a data processing pipeline are divided into tasks which KubeFlow provisions individually on Kubernetes. This mitigates a common resource assignment problem for high variational resource demands between intermediate pipeline tasks [5]. As seen in Fig. 3, the video stream to Kubeflow is processed using the object and distance detection model. It detects people and calculates the distance between two people.

**Fig. 3.** Object and distance detection

## 2.4 Data Stream

Video streaming data trained with Kubeflow is streamed with Apache Kafka and Apache Spark. Data Stream is data produced continuously by various sources and sent in the simultaneously. It provides never-ending data streams. The data can be analyzed in real time. Stream data includes a wide variety of data. Data such as social media data, banking transactions, location data, mobile and web log files provide this variety [6].

It is very important to collect and analyze big data quickly and accurately. Apache Kafka is a MQ (Message Queuing) system that enables realtime data ingestion and data streaming. RabbitMQ can be characterized as ActiveMQ. Apache Kafka has a queue structure. The data is sent from the source to Kafka, not directly to the analysis tool. In Apache Kafka servers, data is kept distributed. The data to be processed is then withdrawn from Kafka servers. Kafka has a cluster in itself and each node in the Kafka cluster is called a broker. The structure created by these brokers by other services is called Kafka cluster. There are 3 basic building blocks of a Kafka process. These are Producer, Topic and Consumer. Producer is the part that sends the data. Consumer is the part that sends the data to the analysis tool. Topic, the data is kept in the publish subscriber structure. There can be more than one topic here. In a sample system, when data is sent directly to the analysis tool, the analysis tool may cause data loss and the result may be manipulated. Also, when the analysis tool is broken, the data will not be processed and the source will not be able to transfer. This situation is solved when Kafka and Zookeeper are in between. If there is a problem in the analysis machine, the data is kept temporarily in Kafka, thus preventing the problem [7].

MapReduce is a system that enables very large data to be analyzed easily on a distributed architecture. It is inspired by map and reduce functions in functional programming. These two functions are used when processing data.

In the map phase, the master node takes the data and divides it into smaller pieces and distributes it to worker nodes. As worker nodes complete these jobs, they send the result back to the parent node. In the reduction phase, the completed works are combined according to the logic of the work and the result is obtained. To simplify it, the data we want to receive from the data analyzed in the Map stage is drawn, and the analysis we want is performed on this data in the Reduce stage [8] (Fig. 4).

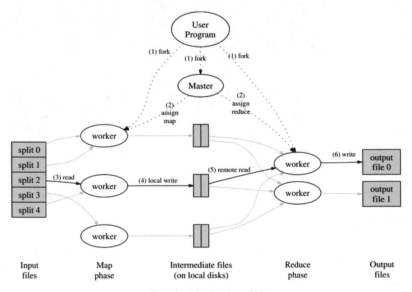

**Fig. 4.** MapReduce [9]

Apache Spark is low volume, high processing technology. The technology it is based on is Apache Mapreduce. Apache Mapreduce has been used for various data processing methods such as interactive querying and flow processing. It can be used in many business methods such as batch applications, interactive algorithms, interactive queries, and streaming. It provides 10 times faster performance while working on Spark disk and 100 times faster in memory. This allows the read/write operations on the disk to be reduced. It also stores intermediate data in memory. Allows API creation with Java, Scala, Python languages. Spark includes 80 top-level operators for interactive querying. Spark supports not only 'Map' and 'Reduce' operations, but also SQL queries, Flow data, ML and graph algorithms. Spark Streaming is the component that uses the fast performance capability of Spark Core for streaming. It holds data in mini-batches and applies RDD (Resilient Distributed Datasets) transforms to the data held in these mini-batches. RDD forms the basis of Spark's data structure. It is to keep these objects irreversibly distributed. Each dataset in RDD is divided into logical partitions so that different nodes in the cluster can handle. RDD can hold Python, Java, Scala objects and user-denifed classes. The purpose of using Spark's RDD is to make faster access and effective MapReduce [10].

There are two models, direct mode and receiver mode, to transfer Kafka data to Spark. Receiver mode, has a receiver. This receiver is used to receive data in the executor. Therefore, it covers the nucleus. In direct mode, the kernel is not used because there is no receiver. In receiver mode, zookeper is maintain the offset of consumer. The direct mod has to protect the offset by itself. It is seen Fig. 5 the working architecture of both models.

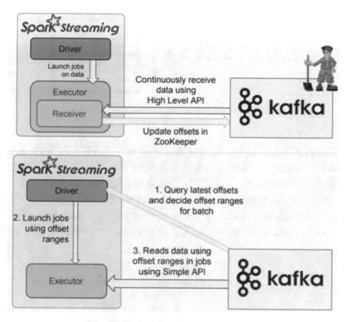

**Fig. 5.** Kafka Spark streaming [11]

## 2.5  Data Storage

HDFS is used to store data streamed using Kafka and Spark. Hadoop provides safe, effective and scalable processing of data in terms of its many features. It works with a backup copy of the data, assuming that the calculated elements and storage may fail [12]. If an error occurs in one node, it will copy a solid copy located in another node. It starts copying again, thus ensuring that data is kept securely. Hadoop paralellime works with the principles, processing the data in parallel, thus protecting its effectiveness. Scalable, petabyte. Allows to process data of size. The lower layer of the Hadoop architecture is the HDFS (Hadoop Distributed File System) [13] file system. The HDFS file system saves data across nodes in the hadoop cluster. From outside HDFS to a client; it looks like a traditional hierarchical file system. CRUD [14] transactions can be made. In other words, files can be created, deleted, and known file operations can be performed. HDFS architecture from private nodes it is formed. These;

1. NameNode: Provides metadata service in HDFS.
2. DataNode: Provides storage blocks for HDFS.

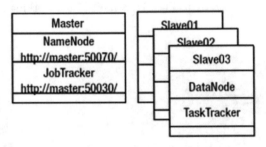

**Fig. 6.** A master 3 slave node model [14].

There is MapReduce engine in the upper layer of HDFS. This engine is from JobTracker and It consists of TaskTracker. It is seen Fig. 6.

Files are divided into blocks in HDFS. These blocks are copied to computers (replicate). Block their size is usually 64 MB. The block size and number of copies of the data can be determined by the user. All file operations are managed with NameNode. All communications in HDFS are with TCP/IP protocol it realized. It is seen Fig. 7.

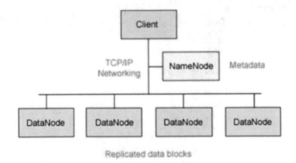

**Fig. 7.** Simplified view of the hadoop cluster [15].

## 2.6 Query Dataset

Hive, Impala and Hbase databases were used to query and analyze data in HDFS. Databases were selected according to the temperature of the data. Hot data are data that are frequently requested to be accessed. Delay rate is too low. The data size is in the MB-GB range. Hot data are data that are less accessible. Latency rate is higher than hot data. The data size is in the GB-TB range. Cold data are rarely requested data. The delay rate is in the minutes-hours range. The data size is in the PB-TB range. With the data temperature and database performance evaluations, hot data in the Hbase database, warm data in the Impala database and cold data in the Hive database are accessed. It is seen Fig. 8 the properties of Hot, Warm and Cold data.

|  | Hot | Warm | Cold |
|---|---|---|---|
| Volume | MB-GB | GB-TB | PB-EB |
| Latency | us,ms | ms-sec | min-hrs |
| Durability | Low-High | High | Very High |
| Request Rate | Very High | High | Low |
| Cost / GB | $$-$ | $-cc | c |

**Fig. 8.** Data temperature

Hive is a platform used to develop SQL type scripts to do MapReduce operations. Facebook promotes the Hive language. Hive because of its SQL like query language is often used as the interface to an Apache Hadoop based data warehouse [16].

Impala is an open-source 'interactive' SQL query engine for Hadoop. The interactive SQL it provides is 4-35 X faster than Hive. It provides a to write SQL queries against your Hadoop data. It is different from Hive in that it uses its own daemons to execute the queries instead of Map-Reduce. These daemons have to be installed alongside data nodes. Impala supports HIVE-QL support (ANSI-92 standard SQL queries with HiveQL) and ODBC drivers [17].

Hadoop works exclusively with batch processing, which means that we can access existing data at regular intervals. For this reason, new technologies have emerged where we can access data randomly. Databases such as Hbase, Cassandra, CouchDB, Dynomo and MongoDB are databases that contain large amounts of data that we can access randomly. Hbase is a column-based distributed database built on HDFS. It is a non-relational database. Its structure is very similar to Google's Big Chart. It allows you to store data with a high level of fault tolerance. They can perform real-time data processing and random read/write operations for very large data sets. The Hbase system is designed to be linear scalable. Many columns and rows contain tables together. Tables should contain items defined as primary keys. It can work with Zookeeper for high performance coordination [18].

## 2.7  Front-End

For real-time reporting, data streamed in Apache Kafka is transferred to React via API Gateway. Real-time reporting is done in React. An API gateway is an API management tool that sits between a client and a collection of backend services. Using the Gateway, RESTful APIs and WebSocket APIs can be created that enable real-time bi-directional communication applications [19].

It is a javascript library produced by Facebook that is used to create user interfaces. React focuses only on the view layer. Other than that, it has nothing to do with any architectural layer. React's only mission is; To do all the operations on the interface in the most logical, easiest, most cost-effective and most efficient way. React is a component based frontend library that uses Virtual DOM architecture [20].

In ReactJS, notification views are available that facilitate the reading and debugging of the code. ReactJS can help the user create some of the easiest or most complex user interfaces and manage their own situation [21]. It is seen Fig. 9 the user interface and dashboards made within the scope of the study.

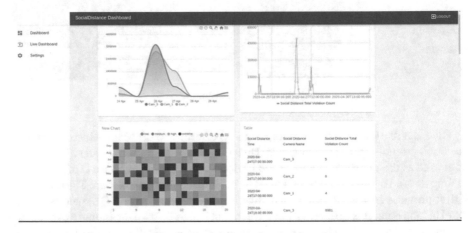

**Fig. 9.** Social distancing dashboard

## 2.8 Notification

E-mail notifications are sent with Oozie workflows. Oozie is a workflow scheduler system for managing Apache Hadoop jobs. Hadoop stack that supports various Hadoop jobs (such as Java map reduction, Flow map reduction, Pig, Hive, Sqoop and Distcp) and system-specific jobs (such as Java programs and shell scripts) out of the box. Oozie is a scalable, reliable and extensible system [22].

Form high level perspective Oozie framework consist of a number of components which are responsible for keeping and managing the batch-based workflows. A single workflow represents a multistage Hadoop job organized into directed acyclic graph which is executed step by step. This concept is similar to other Hadoop processing solutions like Hive or Spark. But if these frameworks work in scope of their own processes, Oozie allows to build cross-process workflows where every step could be executed within different Hadoop component [23] (Fig. 10).

The e-mail action requires SMTP server configuration to be present. SMTP (Simple Mail Transfer Protocol) is a TCP/IP protocol that enables e-mail communication between servers. SMTP service; It works by using a three-tier process model in client, sending server, and receiving server. First, the SMTP protocol is used to send a message from

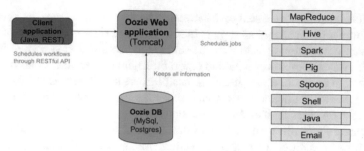

**Fig. 10.** Oozie overall design [23]

Outlook or Webmail and a similar email client to a sending email server. In the second step, the sending e-mail server uses SMTP as a relay service to send the e-mail to the receiving e-mail server. Finally, the receiving server uses an email client (Outlook, Webmail etc.) to download incoming mail via IMAP or POP3.

SMTP can perform operations such as mail forwarding, forwarding to another server name or mail forwarding. IMAP protocol is required to receive and read mail. POP and IMAP are pull protocols, a request is sent to the mail server to access the mailbox. In summary, if we are using a mail client, we need to configure SMTP for sending and IMAP or POP for receiving [24, 25] (Fig. 11).

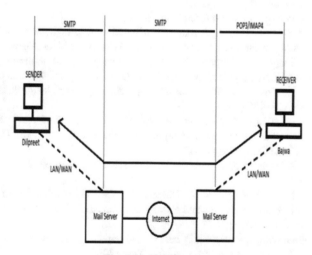

**Fig. 11.** SMTP server

## 3 Conclusion

In this article, the flow data was obtained by trained with the object and distance detection model taken from the camera video. The distance between people with the social distance and object detection model has been found in closed areas. Detected people, distance information, coordinates are sent to the stream. The data coming from the stream to the database are shown in analytical reports in historical and real-time. In cases of extreme social distance, violation densities can be sent via the notification system to inform the user. In this way, the user is provided to report the regions where the violation is density in real-time. In addition, a user interface has been created to enable the user to define cameras, generate notifications, or access real-time dashboards. Thus, it was ensured that the violation cases on the data were followed.

As a future study, different Artificial Intelligence algorithms can be used to improve performance. In addition to person detection, research and developments on mask detection algorithms will continue for analyze whether a person is wearing a mask.

# References

1. Panchal, P., Prajapati, G., Patel, S., Shah, H., Nasriwala, J.: A review on object detection and tracking methods. Int. J. Res. Emerging Sci. Technol. **2**(1), 7–12 (2015)
2. Internet. https://en.wikipedia.org/wiki/GObject#Relation_to_GLib
3. Sundari, G., Bernatin, T., Somani, P.: H. 264 encoder using Gstreamer, pp. 1–4 (2015). https://doi.org/10.1109/ICCPCT.2015.7159511
4. Mittal, N., Vaidya, A., Kapoor, S.: Object detection and classification using Yolo. Int. J. Sci. Res. Eng. Trends **5**, 562–565 (2019)
5. Popp, M.: Comprehensive support of the lifecycle of machine learning models in model management systems. MS thesis (2019)
6. Internet. https://aws.amazon.com/tr/streaming-data/
7. Kreps, J., Narkhede, N., Rao, J.: Kafka: a distributed messaging system for log processing. In: Proceedings of 6th International Workshop on Networking Meets Databases (2011)
8. Capriolo, E., Wampler, D., Rutherglen, J.: https://books.google.com/?hl=en (2012)
9. Dean, J., Ghemawat, S.: MapReduce: simplified data processing on large clusters. In: Proceedings of the 6th Conference on Symposium on Operating Systems Design & Implementation, OSDI 2004, pp. 137–149 (2004)
10. Zaharia, M., Das, T., Li, H., Shenker, S., Stoica, I.: Discretized streams: an efficient and fault-tolerant model for stream processing on large clusters. In: 4th USENIX conference on Hot Topics in Cloud Computing (2012)
11. Internet. https://developpaper.com/optimization-and-comparison-of-reading-kafka-data-by-spark-streaming/
12. Shvachko, K., Hairong, K., Radia, S., Chansle, R.: The Hadoop distributed file system. In: 2010 IEEE 26th Symposium on Mass Storage Systems and Technologies (MSST), Incline Village, NV, USA (2010)
13. Internet. http://en.wikipedia.org/wiki/Create,_read,_update_and_delete (2010)
14. Venner, J.: Pro Hadoop, 1st edn. Apress, New York (2009)
15. Mann, K., Jones, M.T.: Distributed computing with Linux and Hadoop. http://www.ibm.com/developerworks/linux/library/l-hadoop/ (2010)
16. Pol, U.: Big data analysis: comparison of Hadoop MapReduce, pig and hive. Int. J. Innovative Res. Sci. Eng. Technol. **5**, 9687–9693 (2016). https://doi.org/10.15680/IJIRSET.2015.0506026
17. Maposa, T., Sethi, M.: SQL-on-Hadoop: the most probable future in big data analytics (2018)
18. Cattell, R.: Scalable SQL and NoSQL data stores. SIGMOD Rec. **39**(4), 12–27 (2011). https://doi.org/10.1145/1978915.1978919
19. Internet. https://www.nginx.com/learn/api-gateway/

20. Kumar, A., Singh, R.K.: Comparative analysis of AngularJS and ReactJS. Int. J. Latest Trends Eng. Technol. **7**(4), 225–227 (2016)
21. Internet. https://2019-spring-web-dev.readthedocs.io/en/latest/final/taylor/index.html
22. Internet. https://oozie.apache.org/
23. Internet. https://oyermolenko.blog/2017/10/01/scheduling-jobs-in-hadoop-through-oozie/
24. Vijayalakshmi, N., Sivajothi, E., Vivekanandan, P.: efficiency and limitation of secure protocol in email services. Int. J. Eng. Sci. Res. Technol. **1**, 539–544 (2012)
25. Banday, M.T.: Effectiveness and limitations of e-mail security protocols. Int. J. Distrib. Parallel Syst. **2**(3) (2011)
26. Chhabra, G.S., Bajwa, D.S.: Review of e-mail system, security protocols and email forensics. Int. J. Comput. Sci. Commun. Netw. **5**(3), 201–211 (2015)

# Determination of the Ideal Color Temperature for the Most Efficient Photosynthesis of Brachypodium Plant in Different Light Sources by Using Image Processing Techniques

İsmail Serkan Üncü[1], Mehmet Kayakuş[2(✉)], and Bayram Derin[3]

[1] Isparta University of Applied Sciences, 32260 Isparta, Turkey
serkanuncu@isparta.edu.tr
[2] Akdeniz University, 07600 Antalya, Turkey
mehmetkayakus@akdeniz.edu.tr
[3] Süleyman Demirel University, 32260 Isparta, Turkey

**Abstract.** Light is vital for photosynthesis to occur. When plants perform photosynthesis, their give different responses to different light wavelengths. Plants can perform photosynthesis in red and blue light at the most and in green light at the least. In this study, lighting devices consisting of 3 red and blue LEDs were used to measure the photosynthesis efficiency of Brachypodium plant and to find out which color temperature was better. In order to prevent environmental factors from affecting the measurements, a test booth was used. In the system, there are camera for taking images and sensors for measuring temperature, humidity, oxygen, carbon dioxide and FAR. The image of the lighting system (LEDs) is taken with the help of the camera and sensors measure the changing condition as the photosynthesis takes place. An image processing-based software was developed to determine ideal color temperature for photosynthesis. The color temperature of each light design is calculated with this software. As a result of evaluating the information coming from the sensors, it was concluded that the highest efficiency was obtained at 4000 °K, 3000 °K and 2000 °K color temperatures, respectively.

**Keywords:** Photosynthesis · Image processing · Color temperature

## 1 Introduction

Light is one of the most important sources for plant growth and survival [1]. Light affects every aspect of plant growth, starting from seed germination [2]. Numerous studies were conducted with different species of plants to understand the process of plants' adaptation to light intensity [3–6]. The growth and development of plant species are greatly influenced by the process of photosynthesis. Photosynthetic energy provides green plants with almost all of the chemical energy they need and is at the center of their survival and reproductive abilities. Photosynthesis is directly related to with the amount of light that hits the leaves of a plant [6].

J. Hemanth et al. (Eds.): ICAIAME 2020, LNDECT 76, pp. 590–600, 2021.
https://doi.org/10.1007/978-3-030-79357-9_56

Sunlight is one of the important environmental factors affecting photosynthesis and growth. Plants growing in different environments develop with different photosynthetic capacities [7]. It was proven that Solidago Virgaurea plants had different photosynthetic capacities when grown in the sun and in the shade [8]. When plants are grown under certain conditions, they adjust their photosynthetic capacity to suit these conditions [9]. In order to evaluate the different responses of plants to light, it is necessary to have a comprehensive knowledge of light intensity, direction, duration and wavelength [2].

There are many studies on photosynthetic response of plants grown under high and low light levels [10, 11]. Both insufficient light and excessive light energy can prevent photosynthesis efficiency and carbon gain. Plants can adjust their physiological properties and photosynthetic components to growth rays to optimize light absorption and photosynthetic efficiency [12, 13]. In low-light conditions, plants can increase light capture with a variety of approaches to increase leaf area and chlorophyll content and adjust the exposure of leaves to event radiation [6, 14, 15]. Plants grown in high light have smaller leaves with higher nitrogen content, more photosynthetic enzymes, less chlorophyll per unit nitrogen, increased electron carrying capacity per unit chlorophyll and Rubisco, which refers to higher electron carrying capacity in low light [3, 6, 7]. In addition, damage caused by excessive light energy can decrease photosynthetic efficiency [16–18].

Experiments were conducted on different plants to measure the amount of light used by plants in photosynthesis [19, 20]. Plants respond to a wide spectrum of light, from UV-B to light red light [2]. Studies have shown that plants have different photoreceptors that detect UV-B, UV-A, blue, green, red and far red light [21]. The visual range of the human eye is in the range of 400 to 700 mm wavelength. Terrestrial plants also use the light in this range in photosynthesis process [22, 23].

Various scientific study methods are used to measure and analyze photosynthesis in plants. Imaging and image processing methods are among those methods [24]. Image processing methods were used to analyze the effects of water deficiency in leaves on the photosynthesis process [25].

In this study, the most ideal color temperature in different light wavelengths was found for the brachypodium, which was placed inside the test booth. In the study, image processing techniques were used, and environmental conditions were kept constant.

## 2   Materials and Methods

For the study, a software was developed to determine the ideal color temperature required for photosynthesis of the Brachypodium placed in a test booth by image processing methods. The specially designed test booth system ensures that the environmental conditions remain constant while allowing obtaining information about the environment through sensors. The ideal color temperatures were determined by using image processing techniques on the image taken from the source with the reflection system and comparing it with CIE 1931 and CIE 1960 standards.

## 2.1  Photosynthesis from Plants

Plants perform photosynthesis by using two types of chlorophyll. They use chlorophyll A and chlorophyll B and various biological pigments to capture light for photosynthesis [26]. As shown in Fig. 1, Chlorophyll A gives peak response at 430 nm and 680 nm; Chlorophyll B gives peak response at 450 nm and 660 nm [27].

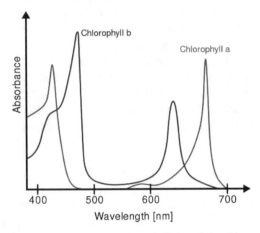

**Fig. 1.**  The absorption spectrum of both the chlorophyll A and the chlorophyll B pigments [28].

The most light absorption and photosynthesis efficiency by chlorophylls take place at the blue and red regions of the electromagnetic spectrum [29]. The absorption percentage of blue or red light by plant leaves is about 90% and the absorption percentage of green light is about 70–80% [30]. While 400 nm (blue light) can activate photosynthesis; plants generally use the range 650–700 nm (red-orange light) [31, 32]. Different light wavelengths play different roles on the development of plants [33]. Red-orange light (600–700 nm wavelength) provides the germination stem to be grow taller and leaf area to increase. Pure red light produces abnormal plants and the optimal plant growing light is blue. Plants react least to green light of approximately 500 nm and reflect this light. That is why the leaves of the plants are green [30, 34]. The photosynthetic response curve given by plants to colored lights is shown in Fig. 2.

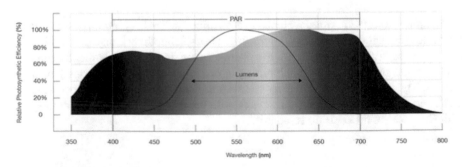

**Fig. 2.**  The photosynthetic response curve given by plants [45].

## 2.2   Brachypodium

Brachypodium is a small, temperate, easily spreading grass with a fast life cycle and a small genome [35]. Brachypodium is an important plant because it is an industrial plant with an economic value as it is used in biodiesel production [36]. In addition, it is used in many researches, most notably in gene mapping studies [37–42]. Because of the importance of the Brachypodium for its fields of use and role in scientific studies, studies are conducted to grow it more efficiently by using effective lighting methods with the greenhouse method. Figure 3 shows the Brachypodium plant.

**Fig. 3.**  Brachypodium plant

## 2.3   Test Booth and Lighting System

In order to prevent the plant from being affected by environmental factors in the experimental environment, a $100 \times 100 \times 30$ mm test booth system that permitted no light and external factors was designed and brachypodium plant was grown in it. Thanks to the electronic devices and sensors placed in the test booth, the humidity and temperature were continuously measured, and the amount of oxygen and carbon dioxide that the plant produced or consumed was recorded by means of the sensors and the electronic mechanism.

The test booth was protected from sudden temperature changes with an integrated resistance and thermostat. If the temperature falls below 25 °C, the thermostat is activated and operates the resistor to prevent the plant from being affected by the change. Since it is a summer plant, we aimed to protect the brachypodium plant from cold. In addition, since the air entrance/exit in and out of the test booth would cause carbon dioxide and oxygen measurements to be made incorrectly, the test booth's cover and cable entries were isolated by use of relevant insulation materials. Figure 4 shows the test booth.

Circuit boards measuring carbon dioxide, oxygen, humidity, temperature and FAR were also integrated into the test booth. Thanks to the electronic mechanisms, the humidity, temperature, carbon dioxide and oxygen in the test booth were measured for seven

**Fig. 4.** Test booth

days with thirty minutes apart for each light ratio. In addition, a FAR sensor was used to determine if there was sufficient light intensity for photosynthesis in the test booth. Apart from these, IP camera was installed in the system and light source circuit was installed in the test booth.

A light source was developed to provide the light needed by the Brachypodium. This light source produced light in different wavelengths, allowing to determine which wavelength was more efficient for brachypodium. Three different wavelengths of light were produced for the red and blue color combinations. 3 lighting systems with light of 50% red – 50% blue, 88% red – 22% blue and 67% red – 33% were designed. For this, 6 red – 3 blue, 7 red – 2 blue and 8 red – 1 blue LED lamps were designed.

## 2.4  Color Temperature of Light Sources

Color temperature is used to describe the color of light produced by a lamp. To determine the color temperature of the light given by a lamp, the lamp light is compared to the Planckian Locus. The chromatic diagram of Planckian Locus 1931 in CIE 1931, which determines the colors and color temperature of light sources, is given in Fig. 5.

The color temperature of the light sources is obtained by using the curves arranged according to CIE 1960 in Fig. 6.

Color change surfaces used for color temperature determination consist of 10 separate regions starting with 1000 °K and separated by transverse lines. These regions reveal the color temperatures arranged according to the light color. The location of the U and V coordinate points is an indicator used to determine the color temperature of the detected light source.

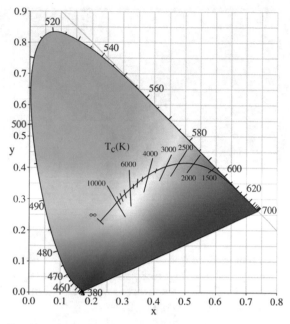

**Fig. 5.** Planckian locus in the CIE 1931 chromaticity diagram [43]

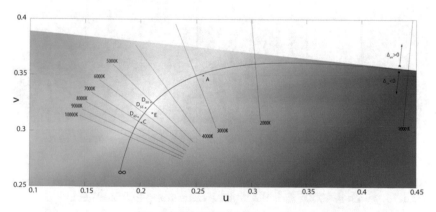

**Fig. 6.** CIE 1960 color space [44]

# 3 Application

The color temperature of the light sources is measured with the software developed. Figure 7 shows the software interface.

Thanks to the developed software, the image taken with the reflection system developed from the light source is calculated by using the image processing techniques according to the Planckian Locus 1931 chromatic diagram in CIE 1931 and CIE 1960 chromaticity space standards, and the light source color temperature is calculated.

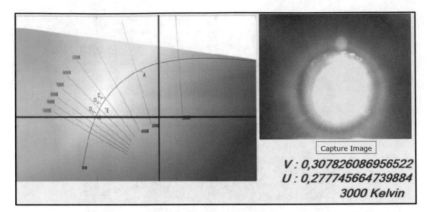

**Fig. 7.** Software interface.

In this study, color temperature was calculated for 6 red – 3 blue, 7 red – 2 blue and 8 red – 1 blue LED lamp systems. Figure 8 shows the color temperature calculations for 6 red – 3 blue colored LEDs, Fig. 9 shows the for 7 red – 2 blue colored LEDs, and Fig. 10 shows the for 8 red – 1 blue colored LEDs.

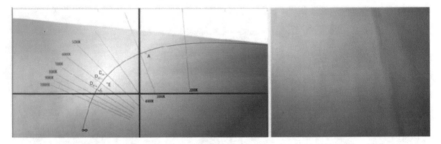

**Fig. 8.** Color temperature calculations for 6 red – 3 blue colored LEDs

The color temperature marked in Fig. 8 is the 6 red – 3 blue ratio. This color temperature, which is 4000 °K, is the highest value compared to others.

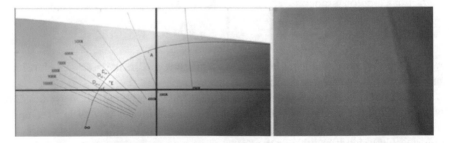

**Fig. 9.** Color temperature calculations for 7 red – 2 blue colored LEDs

The color temperature marked in Fig. 9 is the 7 red – 2 blue ratio. This color temperature, which is 3000 °K, is an average value compared to others.

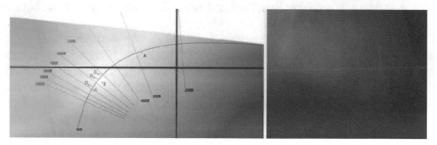

**Fig. 10.** Color temperature calculations for 8 red – 1 blue colored LEDs

The color temperature marked in Fig. 10 is the of 8 red – 1 blue ratio. With 2000°K color temperature, this ratio is the lowest.

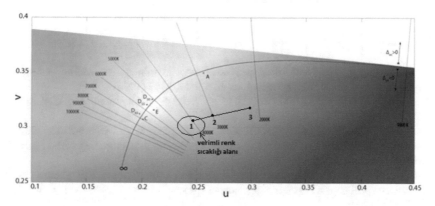

**Fig. 11.** Most efficient color temperature area

Figure 11 shows the most efficient color temperature area for the Brachypodium.

The area indicated by one is the area where the highest efficiency is obtained even though it has a color temperature of 4000 °K. The area indicated by two is the second most efficient area for Brachypodium, although it has a color temperature of 3000 °K. The area marked by three, although it has 2000 °K color temperature, indicates the third most effective area for the brachypodium.

## 4   Conclusion

With this study, color temperature measurement was performed for brachypodium and the most efficient photosynthesis temperature was determined. For this, a test booth was developed to prevent the effects of environmental conditions and to make measurements.

In order to learn the values of the plant and the environment, the sensors which allowed obtaining instantaneous information regarding temperature, humidity, oxygen, carbon dioxide and FAR were used and the data was recorded.

For the system, a combination of red and blue LEDs with which plants have high photosynthesis speed, was designed.

An image processing software was designed for the analysis of the images obtained from the system. Thanks to the developed software, the image taken with the reflection system from the light source was calculated using the image processing techniques according to the CIE 1931 color space chromaticity diagram and CIE 1960 Chromaticity Space standards, and the light source color temperature was calculated. As a result of measurements made with the developed software, 4000 °K, 3000 °K and 2000 °K temperature were obtained for 6 red – 3 blue LEDs, for 7 red – 2 blue LEDs, and for 8 red – 1 blue LEDs, respectively.

The information from the sensors is modeled with expert systems to determine the most efficient light source for the Brachypodium. As a result of evaluating the information obtained from here, the most efficient light source was determined. Using this information, it is concluded that the highest yield for the brachypodium plant is 4000 °K, 3000 °K and 2000 °K, respectively.

# References

1. Whitmore, T.: A review of some aspects of tropical rain forest seedling ecology with suggestions for further enquiry. Man Biosph. Ser. **17**, 3–40 (1996)
2. Fankhauser, C., Chory, J.: Light control of plant development. Annu. Rev. Cell Dev. Biol. **13**(1), 203–229 (1997)
3. Evans, J., Poorter, H.: Photosynthetic acclimation of plants to growth irradiance: the relative importance of specific leaf area and nitrogen partitioning in maximizing carbon gain. Plant Cell Environ. **24**(8), 755–767 (2001)
4. Johnson, D.M., Smith, W.K.: Low clouds and cloud immersion enhance photosynthesis in understory species of a southern Appalachian spruce–fir forest (USA). Am. J. Bot. **93**(11), 1625–1632 (2006)
5. Gorton, H.L., Brodersen, C.R., Williams, W.E., Vogelmann, T.C.: Measurement of the optical properties of leaves under diffuse light. Photochem. Photobiol. **86**(5), 1076–1083 (2010)
6. Givnish, T.J.: Adaptation to sun and shade: a whole-plant perspective. Funct. Plant Biol. **15**(2), 63–92 (1988)
7. Athanasiou, K., Dyson, B.C., Webster, R.E., Johnson, G.N.: Dynamic acclimation of photosynthesis increases plant fitness in changing environments. Plant Physiol. **152**(1), 366–373 (2010)
8. Björkman, O., Holmgren, P.: Adaptability of the photosynthetic apparatus to light intensity in ecotypes from exposed and shaded habitats. Physiol. Plant. **16**(4), 889–914 (1963)
9. Walters, R.G.: Towards an understanding of photosynthetic acclimation. J. Exp. Bot. **56**(411), 435–447 (2005)
10. Boardman, N.T.: Comparative photosynthesis of sun and shade plants. Ann. Rev. Plant Physiol. **28**(1), 355–377 (1977)
11. Berenbaum, M.: Patterns of furanocoumarin distribution and insect herbivory in the Umbelliferae: plant chemistry and community structure. Ecology **62**(5), 1254–1266 (1981)

12. Zhang, W., Huang, W., Zhang, S.-B.: The study of a determinate growth orchid highlights the role of new leaf production in photosynthetic light acclimation. Plant Ecol. **218**(8), 997–1008 (2017). https://doi.org/10.1007/s11258-017-0747-5

13. Rozendaal, D., Hurtado, V., Poorter, L.: Plasticity in leaf traits of 38 tropical tree species in response to light; relationships with light demand and adult stature. Funct. Ecol. **20**(2), 207–216 (2006)

14. Hikosaka, K., Terashima, I.: A model of the acclimation of photosynthesis in the leaves of C3 plants to sun and shade with respect to nitrogen use. Plant Cell Environ. **18**(6), 605–618 (1995)

15. Niinemets, U.: Photosynthesis and resource distribution through plant canopies. Plant Cell Environ. **30**(9), 1052–1071 (2007)

16. Long, S.P., Humphries, S., Falkowski, P.G.: Photoinhibition of photosynthesis in nature. Annu. Rev. Plant Biol. **45**(1), 633–662 (1994)

17. Powles, S.B.: Photoinhibition of photosynthesis induced by visible light. Annu. Rev. Plant Physiol. **35**(1), 15–44 (1984)

18. Kato, M.C., Hikosaka, K., Hirotsu, N., Makino, A., Hirose, T.: The excess light energy that is neither utilized in photosynthesis nor dissipated by photoprotective mechanisms determines the rate of photoinactivation in photosystem II. Plant Cell Physiol. **44**(3), 318–325 (2003)

19. Mohr, H.: The control of plant growth and development by light. Biol. Rev. **39**(1), 87–112 (1964)

20. Von Arnim, A., Deng, X.-W.: Light control of seedling development. Annu. Rev. Plant Biol. **47**(1), 215–243 (1996)

21. Kehoe, D.M., Grossman, A.R.: Similarity of a chromatic adaptation sensor to phytochrome and ethylene receptors. Science **273**(5280), 1409–1412 (1996)

22. Ballantine, J.E.M., Forde, B.: The effect of light intensity and temperature on plant growth and chloroplast ultrastructure in soybean. Am. J. Bot. **57**(10), 1150–1159 (1970)

23. Junge, W., Eckhof, A.: On the orientation of chlorophyll-a I in the functional membrane of photosynthesis. FEBS Lett. **36**(2), 207–212 (1973)

24. Genty, B., Meyer, S.: Quantitative mapping of leaf photosynthesis using chlorophyll fluorescence imaging. Funct. Plant Biol. **22**(2), 277–284 (1995)

25. Omasa, K., Maruyama, S., Matthews, M., Boyer, J.: Image diagnosis of photosynthesis in water-deficit plants. IFAC Proc. Vol. **24**(11), 383–388 (1991)

26. Nishio, J.: Why are higher plants green? Evolution of the higher plant photosynthetic pigment complement. Plant Cell Environ. **23**(6), 539–548 (2000)

27. Lam, E., Ortiz, W., Malkin, R.: Chlorophyll a/b proteins of photosystem I. FEBS Lett. **168**(1), 10–14 (1984)

28. Wikipedia: Chlorophyll b. https://en.wikipedia.org/w/index.php?title=Chlorophyll_b&oldid=932389341 (2020)

29. McCree, K.J.: Test of current definitions of photosynthetically active radiation against leaf photosynthesis data. Agric. Meteorol. **10**, 443–453 (1972)

30. Terashima, I., Fujita, T., Inoue, T., Chow, W.S., Oguchi, R.: Green light drives leaf photosynthesis more efficiently than red light in strong white light: revisiting the enigmatic question of why leaves are green. Plant Cell Physiol. **50**(4), 684–697 (2009)

31. Ohashi-Kaneko, K., Takase, M., Kon, N., Fujiwara, K., Kurata, K.: Effect of light quality on growth and vegetable quality in leaf lettuce, spinach and Komatsuna. Environ. Control. Biol. **45**(3), 189–198 (2007)

32. Kim, H.-H., Goins, G.D., Wheeler, R.M., Sager, J.C.: Green-light supplementation for enhanced lettuce growth under red-and blue-light-emitting diodes. HortScience **39**(7), 1617–1622 (2004)

33. Köksal, N., İncesu, M., Ahmet, T.: LED aydınlatma sisteminin domates bitkisinin gelişimi üzerine etkileri. Int. J. Agric. Nat. Sci. **6**(2), 71–75 (2013)

34. Johkan, M., Shoji, K., Goto, F., Hahida, S.-N., Yoshihara, T.: Effect of green light wavelength and intensity on photomorphogenesis and photosynthesis in Lactuca sativa. Environ. Exp. Bot. **75**, 128–133 (2012)

35. Gomez, L.D., Bristow, J.K., Statham, E.R., McQueen-Mason, S.J.: Analysis of saccharification in Brachypodium distachyon stems under mild conditions of hydrolysis. Biotechnol. Biofuels **1**(1), 15 (2008)

36. Douché, T., et al.: Brachypodium distachyon as a model plant toward improved biofuel crops: search for secreted proteins involved in biogenesis and disassembly of cell wall polymers. Proteomics **13**(16), 2438–2454 (2013)

37. Draper, J., et al.: Brachypodium distachyon. A new model system for functional genomics in grasses. Plant Physiol. **127**(4), 1539–1555 (2001)

38. Catalán, P., Olmstead, R.G.: Phylogenetic reconstruction of the genus Brachypodium P. Beauv. (Poaceae) from combined sequences of chloroplast ndhF gene and nuclear ITS. Plant Syst. Evol. **220**(1–2), 1–19 (2000)

39. Catalán, P., Shi, Y., Armstrong, L., Draper, J., Stace, C.A.: Molecular phylogeny of the grass genus Brachypodium P. Beauv. based on RFLP and RAPD analysis. Bot. J. Linn. Soc. **117**(4), 263–280 (1995)

40. Khan, M.A., Stace, C.A.: Breeding relationships in the genus Brachypodium (Poaceae: Pooideae). Nord. J. Bot. **19**(3), 257–269 (1999)

41. Shi, Y.: Molecular studies of the evolutionary relationships of Brachypodium (Poaceae). University of Leicester (1991)

42. Shi, Y., Draper, J., Stace, C.: Ribosomal DNA variation and its phylogenetic implication in the genusBrachypodium (Poaceae). Plant Syst. Evol. **188**(3–4), 125–138 (1993)

43. Wikipedia Planckian Locus: Planckian locus, May 2020. https://en.wikipedia.org/w/index.php?title=Planckian_locus&oldid=936151098

44. Wikipedia: CIE 1960 color space, May 2020. https://en.wikipedia.org/w/index.php?title=CIE_1960_color_space&oldid=923209050

45. GreenBudGuru: What is PAR light measurement? May 2020. https://www.greenbudguru.com/what-is-par-light-measurement

# Combination of Genetic and Random Restart Hill Climbing Algorithms for Vehicle Routing Problem

E. Bytyçi, E. Rogova$^{(\boxtimes)}$, and E. Beqa

Faculty of Natural and Mathematical Sciences, University of Prishtina "Hasan Prishtina", Prishtina, Kosovo
{eliot.bytyci,ermir.rogova}@uni-pr.edu,
endrit.beqa@studentet.uni-pr.edu

**Abstract.** Vehicle routing problem falls into NP hard problems. Researchers have tried with many techniques to find the most suitable and fastest way for solving the problem. With the usage of divide and conquer approach, the problem is divided into several parts that ensures finding the most suitable algorithms. In this special case, the combination of the genetic algorithm with the random restart hill climbing algorithm is used. The cluster first – route second algorithm provides that the approach will be attributed the necessary speed, from the usage of hill climbing algorithm. Furthermore, usage of genetic algorithm provides for bigger space that can be searched. For the selection step of the genetic algorithms, three types of distributions: uniform, log-normal and exponential were compared with fitness proportionate selection to further speed up the algorithm. The results show that the uniform distribution is the reasonable substitution, while the log-normal and the exponential distribution converge faster with a higher cost penalty.

**Keywords:** Genetic algorithm · Hill climbing algorithm · Vehicle routing problem

## 1 Introduction

The predominant mode of freight transport in Europe and indeed in other parts of the world is road transport. The direct costs associated with this type of transportation have been increasing since year 2000. In addition to this, due to the outbreak of COVID-19, a very high number of companies have expanded their business model to include online shopping and, inevitably, home delivery, resulting in an increase in road transport usage. Furthermore, road transport is essentially related to a good part of indirect costs like blockages, pollution, costs related to safety and security, mobility, delay costs, etc. However, these costs are generally not treated due to the difficulty in determining them [1].

Real-world applications have demonstrated that the use of computerized procedures generates significant savings (generally 5% to 20%) in global transportation costs [2]. Road transport optimization problems are particularly critical in the case of small and

© The Author(s), under exclusive license to Springer Nature Switzerland AG 2021
J. Hemanth et al. (Eds.): ICAIAME 2020, LNDECT 76, pp. 601–612, 2021.
https://doi.org/10.1007/978-3-030-79357-9_57

medium-sized enterprises (SMEs), as they are seldom able to obtain the financial and human resources necessary to implement, maintain and manage effective methods for route optimization.

The vehicle routing problem was first raised by Dantzig and Ramser in 1959 [3]. It is a combinatorial optimization problem that tries to find the optimal routes for fleet vehicles to take in order to visit specific locations. Its goal is to minimize transport costs. Vehicle routing problem falls into NP hard problems. Therefore, it is not practical to find the best solution, due to very large search space. The following example can illustrate the claim easily:

Let us suppose that there are five vehicles in a fleet, that need to visit twenty locations. The number of routes that can be taken by those five vehicles in order to visit all twenty locations and return to the starting point is: $5 * 24! = 3.1 \times 10^{24}$ different routes. Even if, it is presumed that the cost of one route can be calculated by a single CPU cycle, it would take a very fast 3.9 GHz CPU about 9.2 billion days or about 25 million years to complete the calculation. It is therefore absolutely clear that a heuristic algorithm has to be used instead of a brute force approach.

This paper is further organized as followed: Sect. 2 describes the related work on the problem with emphasis on algorithms used. Section 3 described in depth the problem and its constraints, while Sect. 4 describes algorithms used. In Sect. 5 results from experiments are presented and Sect. 6 represents the conclusion and future work.

## 2 Related Work

In their paper [4], authors considered the application of a genetic algorithm (GA) to the basic vehicle routing problem (VRP), in which customers of known demand are supplied from a single depot. Vehicles are subject to a weight limit and, in some cases, to a limit on the distance travelled. Only one vehicle was allowed to supply each customer. According to the authors, this paper's approach could be competitive with other heuristic techniques in terms of solution time and quality.

The paper written by [5], presented a unified heuristic which can solve five different variants of the vehicle routing problem. All problem variants are transformed into a rich pickup and delivery model and solved using the adaptive large neighbourhood search (ALNS) framework. The paper concluded with a computational study, in which the five different variants of the vehicle routing problem were considered on standard benchmark tests from the literature.

Authors in [6], presented an efficient Hybrid Genetic Search with Advanced Diversity Control for a large class of time-constrained VRP, introducing several new features to manage the temporal dimension. New move evaluation techniques were proposed and accounting for penalized infeasible solutions with respect to time-window and duration constraints. Furthermore, geometric and structural problem decompositions were developed to address efficiently large problems.

In their paper [7], authors tackled the generalized vehicle routing problem (GVRP) which is a natural extension of the classical VRP. This paper aimed to present an efficient hybrid heuristic algorithm obtained by combining a GA with a local-global approach to the GVRP and a powerful local search procedure.

However, to the best of our knowledge, there has not been a case that uses the same approach as is done in this paper, to date. Compared to paper [4], our paper has used hybrid approach, which is similar to papers [6] and [7] but in our case un-capacitated VRP was used, with combination of GA to form a single cluster and random restart hill climbing to find the best route for those clusters.

## 3    Formulation of the Problem

Vehicle Routing Problem is not a new problem but a problem that has been around for more than six decades [3]. It is an important problem with both economic and research interest, based on generalization of the Traveling Salesman Problem (TSP) [8].

Traveling Salesman Problem is the problem where a number of cities is given and also the distances between each two cities. The main idea is to find the shortest route, but only by visiting all cities only once and then returning to the starting city. Therefore, two hard constraints of the problem can be formulated as:

1.  visit every city only once, and
2.  return to the starting city.

It should be mentioned that as a variant of the problem, it is considered that the distance from city $i$ to city $j$ is the same as the distance from city $j$ to city $i$. This is known as a symmetric property of the problem.

There exist a number of mathematical formulations on the TSP [9]. One of the first formulations, that of Miller-Tucker-Zemlin given in 1960 [10], ensures a more general formulation for a more general TSP with $n$ nodes, where the salesman has to visit $n$-$1$ nodes exactly once. During his travel, the salesman has to return to home city exactly $t$ times. The main aim is to minimize the cost of the route.

In our specific case, the VRP presented can be viewed as a TSP with multiple vehicles but without a capacity [11]. Therefore, our approach can be categorized as an un-capacitated VRP.

One of the possible visualized solutions of the problem, is presented in Fig. 1.

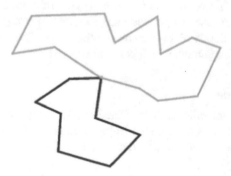

**Fig. 1.** Visual representation of a possible solution of VRP with two vehicles

# 4  Algorithm Parameters Used

This section represents the overall methods used and specific components of the algorithms. It should be noted that the method developed, is the method that provides for better flexibility regarding the algorithms used. This was achieved by using the divide and conquer approach, by splitting the problem in several smaller bits, in order to provide the most suitable algorithm for the specific conditions. Therefore, our approach used genetic algorithm and a special case of the hill climbing algorithm: random restart hill climbing.

## 4.1  Genetic Algorithm

The genetic algorithm (GA) [12] was chosen because of the bigger space that it can control but also for the ability to avoid local maxima. Therefore, in the case of splitting the cities to be visited by each vehicle, GA was used. In order to achieve this, the number of genes and chromosomes that are chosen, are variable. Therefore, a gene consists of the group of the cities, representing the vehicle route, and the chromosome comprises of vehicles.

An example of an individual, containing one chromosome, would look like:

Vehicle 1: City 1, City 3
Vehicle 2: City 2, City 8, City 7, City 10, City 6
Vehicle 3: City 4
Vehicle 4: City 5, City 9

As it is understood that the most important features of the GA are population, fitness, crossover, mutation and selection. Each one of them will be explained in regards to the specifics of the problem.

### 4.1.1  Initial Population

As the first step of the GA is the initialization of the population. It actually represents a subset of solutions. The population is defined as set of chromosomes and usually is created randomly. In our case, the population is formed by taking the available cities and each of the cities is randomly divided on one of the genes. Genes represent the vehicles route. This creates an individual, which is saved in an array.

### 4.1.2  Fitness Function

The fitness function is the other feature of the GA, that evaluates the given solution. In our specific case, as a fitness function the length of the route that an individual pass was used. The fitness is calculated through the function of the TSP and will be discussed further in a later section.

### 4.1.3  Crossover

Crossover is the process of exchanging genes in the chromosomes. It is one of the features of the GA, in which the offspring is created. That offspring, containing genes from both parents, is added to the population. The main problem in our case is that the genes and chromosomes are not linear but are variable. Therefore, a more suitable solution had to be found for crossover.

Since the offspring is created from combining genes of both parents, and in our case the number of genes is not the same in both parents, then the maximum number of genes that the offspring can take, is the same number as of the parent with maximum number of genes. Then, it was checked if there are repeated values (points to visit) in the genes, which are removed, also to add the missing cities to the chosen gene. This process is called the normalization of the offspring. It should be noted that crossover was used only in the case of vehicles but not also values of the genes (points to visit). That could be seen as something that would increase the diversity of the genes, but it does that by breaking up the grouping of cities which is similar to random shuffling.

### 4.1.4  Mutation

Mutation is the process that provides for diversity within population and in order to prevent premature convergence. Premature convergence means that the GA is stuck in local maxima. In our specific case, we have chosen to use the mutation as a change in two genes, by randomly selecting two cities and swapping their positions, otherwise it would end up creating duplicate cities. Another possibility would be to check if both cities are contained in different genes before swapping, ensuring that a change has happened, due to the low probability and increase in computations the check isn't performed.

### 4.1.5  Selection

Selection represents the process of selecting the fittest of individuals and passing their genes to the next generations. In our case, three probability distributions were tested and compared to substitute fitness proportionate selection in order to lower computations needed. The results are discussed in the next section. The substitution of the population is based on the length of the road. That represents the fitness, which is then compared to other individuals, in order to let go some of the individuals and keep the others.

### 4.1.6  End of the Algorithm

Algorithm ends when a change between the first individual and the last one is lower than a predefined number, which is the smallest amount of improvement that a swap between two cities can create. The exact value is not important to be calculated precisely since it can just be underestimated, thereby guarantying that the difference of fitness between the first and last individual will be lower than the smallest improvement achieved by swapping two cities. The algorithm is also terminated when the predefined maximal number of iterations is reached.

## 4.2   Random Restart Hill Climbing

Traveling Salesmen Problem was solved through the usage of Random Restart Hill Climbing. Random Restart Hill Climbing was chosen because of its speed. Traveling Salesman Problem was treated as a problem of ordering the positions in an array, where the first and the last position are selected as the starting point. The algorithm starts with a random order and calculates the distance that should pass, then it appoints a city randomly, that we should move forward or backward. The direction and the number of units that it should move, are also selected randomly. For example, the array of the cities to be visited is:

$$x_1 \ x_2 \ x_3 \ x_4 \ x_5 \ x_6 \ x_1$$

Randomly it was decided that city $x_3$, should move two position ahead and therefore now as a result is:

$$x_1 \ x_2 \ x_5 \ x_4 \ x_3 \ x_6 \ x_1$$

The distance is recalculated again and if the distance was decreased, then the best order was replaced with the actual one and the same city would move in the same direction. If the next move, has returned a longer distance, then previous position would be returned otherwise it would continue ahead. The procedure continues for a certain number of iterations depending on the number cities.

This procedure doesn't eliminate local maxima therefor a shuffle to the current ordering was applied and the hill climbing algorithm was restarted. This explores the search space beyond the local maximum hence increasing the chances of finding the global maximum. The restart procedure is performed several times depending on the number of cities.

## 5   Results from the Experiments

The algorithms used for the problem, where tried in a computer with the I7-5600U processor and 4 Gb of RAM. The data is created by running the algorithm 10 times with the same parameters and taking the average of the 10 runs. The 35 points (cities) were randomly created with the range of the $x$ position being 0–2400 and $y$ position being 0–1400.

The results, partially presented in Table 1, are using Fitness Proportionate Selection with parameters:

- Crossover rate 1.0
- Mutation Rate (0.05–0.003125)
- Population Size (100–500)
- Children per generation (100–500)
- Generation until convergence
- Route length

**Table 1.** Fitness proportionate selection results with different parameters

| Case | Crossover rate | Mutation rate | Population size | Children per generation | Generation until convergence | Route length |
|------|------|------|------|------|------|------|
| 1.00 | 1 | 0.05 | 100 | 100 | 111.3 | 11623.87 |
| 2.00 | 1 | 0.05 | 100 | 200 | 68.1 | 12409.94 |
| 3.00 | 1 | 0.05 | 100 | 300 | 60.7 | 11489.54 |
| 4.00 | 1 | 0.05 | 100 | 400 | 57.6 | 11308.27 |
| 5.00 | 1 | 0.05 | 100 | 500 | 47.5 | 11849.19 |
| 6.00 | 1 | 0.05 | 200 | 100 | 205.8 | 11501.96 |
| 7.00 | 1 | 0.05 | 200 | 200 | 124.7 | 11113.15 |
| 8.00 | 1 | 0.05 | 200 | 300 | 97.2 | 11345.91 |
| 9.00 | 1 | 0.05 | 200 | 400 | 81.7 | 11157.15 |
| 10.00 | 1 | 0.05 | 200 | 500 | 70.8 | 11044.46 |
| ... | ... | ... | ... | ... | ... | ... |
| 116.00 | 1 | 0.00313 | 400 | 100 | 404 | 10800.20 |
| 117.00 | 1 | 0.00313 | 400 | 200 | 212.5 | 11022.10 |
| 118.00 | 1 | 0.00313 | 400 | 300 | 163.4 | 10990.33 |
| 119.00 | 1 | 0.00313 | 400 | 400 | 139.6 | 10817.04 |
| 120.00 | 1 | 0.00313 | 400 | 500 | 124.8 | 10892.62 |
| 121.00 | 1 | 0.00313 | 500 | 100 | 501.6 | 10939.76 |
| 122.00 | 1 | 0.00313 | 500 | 200 | 266.6 | 11053.66 |
| 123.00 | 1 | 0.00313 | 500 | 300 | 212.1 | 10711.85 |
| 124.00 | 1 | 0.00313 | 500 | 400 | 165.1 | 10935.94 |
| 125.00 | 1 | 0.00313 | 500 | 500 | 146.9 | 10822.81 |

## 5.1  Results with Fitness Proportionate Selection

As the results indicate the algorithm performs better with higher crossover rate, while the mutation rate did not affect the results considerably. The number of children created per generation was also considered and the results show that with the same population size a smaller number of children created per generation converges with far less individuals created thus lowering the computations needed. This can be attributed to the increase in average fitness due to the often removal of individuals with a lower fitness score.

As for the population size, it is apparent that with increase of the size the route gets shorten 15% due to the diversity in the population but with a disproportional increase in the number of generations needed for convergence where the increase in generations is 200%–400%. A steady decrease of vehicles used could be seen, as the route length decreased with the shortest routes using three or four vehicles. Figure 2, shows a solution with crossover rate 1.0, mutation rate 0.05, population size 300 and children per generation of 300 using fitness proportionate selection. The route length is 10732 after 160 generations.

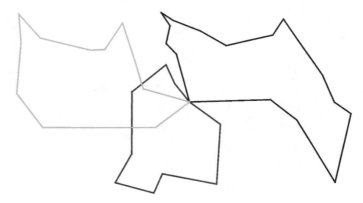

**Fig. 2.** Solution of the problem with crossover rate 1.0

In order to further lower the number of computations needed, also probability distributions to replace the fitness proportionate selection operator, were looked at. The fitness proportionate selection sample is presented in Fig. 3.

Three distributions are analysed: uniform, log-normal and exponential. The selection of these distributions is based on their probability density function (probability mass function for the discrete distributions) since a more aggressive probability mass function like that of the Poisson distribution would be too restrictive to the number of individuals selected. The flatter curve of the probability density function of these distributions help mimic the fitness proportional selection since the algorithm produces a population with difference between the best and the worst score of individuals most of the times under 30%.

## 5.2  Uniform Distribution

From the resulted data, it can be seen that the uniform distribution was the best in terms of route length. In some cases, it found a route that is shorter by a small amount (2%). On average, the uniform distribution found routes 1.77% longer that FPS, with a 13.13% increase in generations needed. Therefore, one can conclude that this distribution is not suited very well for lowering the computational needs, but it can be used as an approach to simplify the algorithm without a large route length penalty. The decrease in route length can be attributed in some cases to the fact that a single bad vehicle route can increase the overall route length so much that it drops an otherwise good individual to the bottom of the population thus potentially losing very good genes. With the uniform distribution every individual has the same probability of being selected so the chances of good genes being lost is far lower, as seen in Fig. 4.

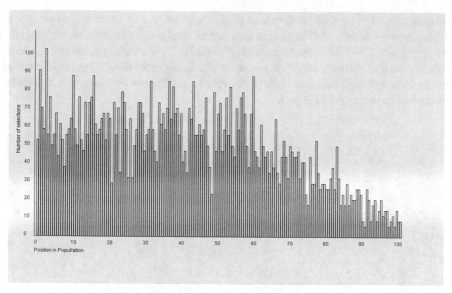

**Fig. 3.** Fitness proportionate selection sample

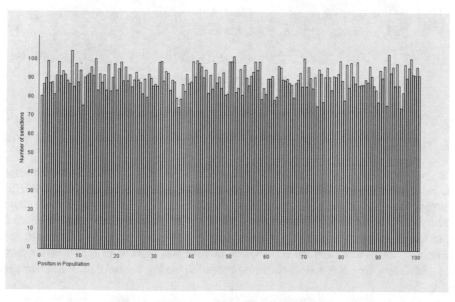

**Fig. 4.** Uniform distribution selection sample

## 5.3 Log-Normal Distribution

The log-normal distribution did not offer a good substitution since it was very sensitive to the parameters used. In the case where a smaller population was used the algorithm would converge to a solution 20%–55% faster with a 5%–10% cost penalty. While in the case where the population size was higher, an increase of route length of 1%–5% with an increase of 5%–25% in generation needed, can be seen. In general, the algorithm converged with 5.74% increase in route length and 12.7% increase in generations needed. With log-normal distribution, individuals at the fifth percentile of the population are heavily favoured, while the rest of the population experiences a very fast drop in selection probability, as presented in Fig. 5.

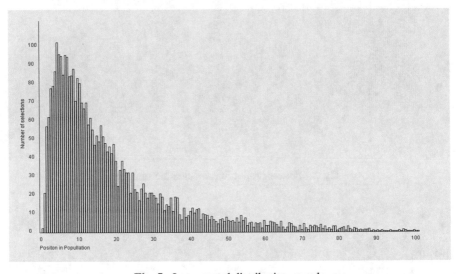

**Fig. 5.** Log-normal distribution sample

## 5.4  Exponential Distribution

The Exponential distribution offered a good compromise between increased route length and faster convergence. On average, the route length increased 10.35% with a 65.75% decrease in generations needed. Since this distribution has a much larger route length penalty it is not suited to be used alone as a selection method. One possibility would be to use its speed as a warm-up phase to increase the populations fitness faster, then switch to a much more refined distribution or selection method to find a better route. The probability drop is much smoother compared to the log normal distribution, as seen in Fig. 6.

In Table 2, results of comparison between distributions and FPS are presented, where clearly is seen that the difference in route favours uniform distribution but the difference in generations favours the exponential distribution.

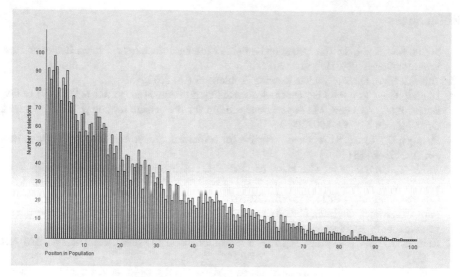

**Fig. 6.** Exponential distribution sample

**Table 2.** Comparison of the results of distributions with FPS

| | Difference in route length compared to FPS | Difference in generations needed until the convergence compared to FPS |
| --- | --- | --- |
| Uniform distribution | +1.77% | +13.13% |
| Log-normal distribution (variance = 1.0) | +5.74% | +12.7% |
| Exponential distribution (mean = 0.5 * population size) | +10.35% | −65.75% |

## 6   Conclusion and Future Work

When performing the experiments, a framework where the compromise between simplicity, speed and flexibility as main priority, was developed. The divide and conquer offers simplicity while the speed is achieved by storing the best solution for genes of an individual and replacing the selection method of the genetic algorithm with a probability distribution. As for flexibility, the individual's normalization process allows for additional constrains, resulting in a slight time penalty.

As future work, the possibility of saving genes which do not fulfil the constrains in a list will be looked at, so that they are removed from the gene pool and don't waste time in computations further increasing speed, similar to a tree pruning process. Another possibility is to use hybrid selection operator, consisting of the exponential distribution – to speed up the process and FPS – to refine the result.

# References

1. Sinha, K.C., Labi, S.: Transportation Decision Making: Principles of Project Evaluation and Programming. Wiley, Hoboken (2011)
2. Toth, P., Vigo, D.: The Vehicle Routing Problem. SIAM (2002)
3. Dantzig, G.B., Ramser, J.H.: The truck dispatching problem. Manage. Sci. **6**(1), 80–91 (1959)
4. Baker, B.M., Ayechew, M.: A genetic algorithm for the vehicle routing problem. Comput. Oper. Res. **30**(5), 787–800 (2003)
5. Pisinger, D., Ropke, S.: A general heuristic for vehicle routing problems. Comput. Oper. Res. **34**(8), 2403–2435 (2007)
6. Vidal, T., Crainic, T.G., Gendreau, M., Prins, C.: A hybrid genetic algorithm with adaptive diversity management for a large class of vehicle routing problems with time-windows. Comput. Oper. Res. **40**(1), 475–489 (2013)
7. Pop, P.C., Matei, O., Sitar, C.P.: An improved hybrid algorithm for solving the generalized vehicle routing problem. Neurocomputing **109**, 76–83 (2013)
8. Reinelt, G.: TSPLIB—a traveling salesman problem library. ORSA J. Comput. **3**(4), 376–384 (1991)
9. Padberg, M., Sung, T.-Y.: An analytical comparison of different formulations of the travelling salesman problem. Math. Program. **52**(1–3), 315–357 (1991)
10. Miller, C.E., Tucker, A.W., Zemlin, R.A.: Integer programming formulation of traveling salesman problems. J. ACM (JACM) **7**(4), 326–329 (1960)
11. Vidal, T., Laporte, G., Matl, P.: A concise guide to existing and emerging vehicle routing problem variants. Eur. J. Oper. Res. **286**(2), 401–416 (2019)
12. Whitley, D.: A genetic algorithm tutorial. Stat. Comput. **4**(2), 65–85 (1994)

# Reliability of Offshore Structures Due to Earthquake Load in Indonesia Water

Yoyok Setyo Hadiwidodo[✉], Daniel Mohammad Rosyid, Imam Rochani, and Rizqullah Yusuf Naufal

Department of Ocean Engineering, Institut Teknologi Sepuluh Nopember (ITS), Surabaya, Indonesia
yoyoksetyo@oe.its.ac.id

**Abstract.** In this study, reliability analysis will be performed on jacket platform to know the probability jacket platform will collapse if an earthquake occurs. Using spectral acceleration at jacket platform's site, earthquake records with similar spectral acceleration can be found. The Peak Ground Acceleration (PGA) from each of the records then used to perform several seismic analysis to obtain base shear values of jacket platform. By performing goodness of fit test with known probability distribution, it can be determined which probability distribution fit the base shear data. With certain base shear value as a limit, the probability of failure, the case when jacket platform will collapse if an earthquake occurs, can be calculated. Base shear limit of jacket platform calculated by performing pushover analysis on nonlinear finite element model of jacket platform. Platform which located at Java Sea will be the subject of this research. Pushover analysis done on Platform shows that base shear values of Platform are 4566.22 kips in X direction and 5195.72 kips in Y direction. The data of base shear values fit with Weibull probability distribution. Using the smaller value between the two base shear result from the pushover analysis as base shear limit, the probability of failure then calculated using Weibull distribution's cumulative density function formula. The result shows that Platform has 5.75% chance to collapse if an earthquake occurs.

**Keywords:** Base shear · Collapse · Earthquake · Jacket platform · Reliability

## 1 Introduction

Indonesia has a vast amount of oil and natural gas reserves. Many of them buried under the sea floor. To be able to extract and process these natural resources, companies installed offshore structures such as jacket platform. Jacket is a type of offshore platform usually installed in shallow water. Most of Indonesia's water can be included in this category. Unfortunately, Indonesia located between four tectonic plates (Eurasian, Indian - Australian, Pacific, and Philippine) and considered as a seismically active region. Pacific Plate contribute to almost 90% of earthquake occurrence and most of it has high magnitude. According to BNPB (Badan Nasional Penanggulangan Bencana), which roughly translate to National Bureau of Disaster Mitigation, in their book Disaster Risk in Indonesia published in 2016, there were 8000 earthquake occurrence in Indonesia with M > 5.0 between

J. Hemanth et al. (Eds.): ICAIAME 2020, LNDECT 76, pp. 613–624, 2021.
https://doi.org/10.1007/978-3-030-79357-9_58

1900–2009 [1, 2]. Seismic activity could cause jacket platform to collapse and result in production shut down, environmental damage, or casualties. These geographic condition make it the utmost priority to make sure jacket platform can withstand earthquake so the effect mentioned before can be avoided.

A number of researchers have performed studies using reliability to determine failure of fixed offshore structures with respect to seismic load uncertainty. Study of the reliability of offshore platform assessed by Hosseini et al. [3] with a case study on location in the Persian gulf with nonlinear time history approach. Meanwhile, Elsayed et al. [4] conducted an investigation of the reliability of the offshore platform in the Gulf of Suez as a case study. Wang et al. [5] presented a response spectrum analysis methodology for evaluating offshore platform reliability and risk under earthquake load. Indonesian waters are prone to earthquakes, there is still no agreement on the earthquake spectrum used to analyze offshore platforms due to earthquake loads. In this study, the earthquake spectrum was selected for the first time from existing earthquake data to suit the location conditions, and used it as a target spectrum. A set of earthquake excitation time histories is generated based on the location spectrum and used as input for seismic analysis. The structural reliability method is then used to estimate the likelihood of platform collapse due to strong earthquake loads. The uncertainty in earthquake loads is taken into account. Finally, a case study is presented to illustrate the application of the proposed spectrum-based seismic reliability approach, for a fixed jacket steel platform located in the Java Sea, Indonesia.

## 2 Objective and Scope of Research

The purpose of this study is to know base shear value when structure collapse, probability distribution that match structure's base shear data as a response when earthquake occurs, and probability structure will collapse if an earthquake occurs.

The scope and limitations of this study are:

- Analysis will be performed on 4 legged-jacket Platform
- Structural analysis will be done using SACS software.
- Structural loads comprised of gravity load and lateral load caused by earthquake.
- For water depth condition, using storm condition with 100-year return period.
- Only performing global analysis.
- Only uses base shear as a variable.

## 3   Results and Discussion

### 3.1   Platform Data

In this study, Platform will be the object of research. Platform is a four-legged steel jacket structure installed in Java Sea, Indonesia Water. Water depth on Platform's location is 49 feet, see Fig. 1. Platform comprised of 3 decks, cellar deck, main deck, dan upper deck. Cellar deck's elevation is at 38 feet, main deck's elevation is at 58 feet 6 inch, and upper deck's elevation is at 78 feet 5 inch. Platform's size and load data can be seen in Tables 1 and 2.

**Fig. 1.** Platform location

**Table 1.** Structural data

| Structure | Description |
|---|---|
| Pile | Outside Diameter = 36 in. Thickness = 2 in. Thickness = 1,5 in. |
| Jacket Legs | Outside Diameter = 40,5 in. Thickness = 1,5 in. |
| Bracing | Outside Diameter = 28 in. Thickness = 0,75 in. Outside Diameter = 26 in. Thickness = 1,25 in. Outside Diameter = 26 in. Thickness = 0,625 in. |

**Table 2.** Structural data

| Deskripsi | Fx (kips) | Fy (kips) | Fz (kips) |
|---|---|---|---|
| Gravity weight | 1.92 | 6.95 | 6338.36 |

## 3.2 Pushover Analysis

Pushover analysis is done using SACS software to get platform capacity. Pushover analysis is a static procedure that uses a simplified linear technique to estimate seismic structural deformations. Structures redesign themselves during earthquakes. An individual component of a structure yield or fail, the dynamic forces on the building are shifted to other components. A pushover analysis simulates this phenomenon by applying loads until the weak link in the structure is found and then revising the model to incorporate the changes in the structure caused by the weak link. A second iteration indicates how the loads are redistributed. The structure is "pushed" again until the second weak link is discovered. This process continues until a yield pattern for the whole structure under seismic loading is identified [6, 7]. In general there are two kind of loads that work on the structure, weight caused by gravity (Table 3) and lateral load caused primarily by earthquake in this study. To put it simply, the lateral load caused by earthquake will be ramped up until structure collapse. Base shear value from 1 step before the structure collapse will be chosen as structure capacity. The structural model can be seen in Fig. 2 and the total gravity weight (comprised of jacket weight, live load, equipments, etc.) can

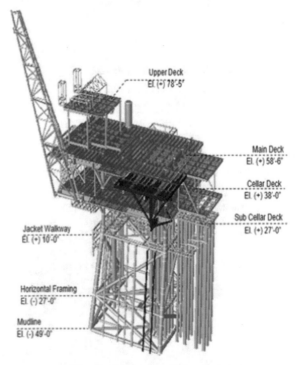

**Fig. 2.** Offshore platform model

be seen in Table 3. Base shear value before structure collapse can be seen in Table 4. Of those two, the lower value will be set as platform capacity.

**Table 3.** Structural data

| Direction | Base shear (Kips) |
|-----------|-------------------|
| X-Dir     | 5195,72           |
| Y-Dir     | 4566,22           |

**Table 4.** Structural loads

| Description | Load (kips) |
|-------------|-------------|
| Sub-Total Bulk Load (Plating, Grating, & Handrail) – Existing | 345.4 |
| Sub-Total Live Load – Existing | 1679.49 |
| Jacket Non Generated Dead Load | 201.75 |
| Wellhead Load – Original | 165 |
| Crane Vertical Load – Original | 231 |
| YY Project | 213 |
| Piping Load | 149.74 |
| Electrical & Instrument Load | 10.36 |
| Total | 2995.74 |

## 3.3 Seismic Analysis

Seismic analysis is a subset of structural analysis and is the calculation of the response of a building (or nonbuilding) structure to earthquakes. One of the important elements in seismic analysis is peak ground acceleration (PGA) input. To obtain PGA, spectral acceleration first generated using ISO 19901-2 as a guide [8, 9]. $S_{a,map}(0,2)$ determined as 0,5 g and $S_{a,map}(1)$ determined as 0,2g. Ca dan Cv obtained from soil testingon site and match it with table Ca dan Cv from ISO 19901-2. The results are as follow, Ca = 1 dan Cv = 0.8. For Abnormal Level Earthquake (ALE), Sa factor is 1.15. ALE chosen because the requirement in ISO states that the structure allowed to exhibit plastic degrading behaviour but catastrophic failures such as global collapse should be avoided. Whereas the in Extreme Level Earthquakes (ELE) requirement, not even little damage to primary structural member is allowed. By inserting T value, spectral acceleration at the site can be obtained (Table 5).

**Table 5.** Spectral acceleration at the site

| T (s) | Sa (g) | Sa ALE (g) |
|---|---|---|
| 0.01 | 0.215 | 0.24725 |
| 0.02 | 0.23 | 0.2645 |
| 0.03 | 0.245 | 0.28175 |
| 0.04 | 0.26 | 0.299 |
| 0.05 | 0.275 | 0.31625 |
| 0.075 | 0.3125 | 0.359375 |
| 0.1 | 0.35 | 0.4025 |
| 0.15 | 0.425 | 0.48875 |
| 0.2 | 0.5 | 0.575 |
| 0.25 | 0.5 | 0.575 |
| 0.3 | 0.5 | 0.575 |
| 0.4 | 0.4 | 0.46 |
| 0.5 | 0.32 | 0.368 |
| 0.75 | 0.21333 | 0.245333 |
| 1 | 0.16 | 0.184 |
| 1.5 | 0.10667 | 0.122667 |
| 2 | 0.08 | 0.092 |
| 3 | 0.05333 | 0.061333 |
| 4 | 0.04 | 0.046 |
| 5 | 0.032 | 0.0368 |

The generated ISO ALE spectral acceleration are then used as 'target spectra' to search a number of time series earthquake acceleration records that have a similar spectral acceleration with the target spectra (Fig. 3). Representative earthquake time series for earthquake acceleration/velocity /displacement are selected from the PEER ground motion database (PGMD) [10–12]. Representative earthquake acceleration time series are selected to provide the best match with the target spectrum, the minimum mean squared error (MSE) between target spectrum and earthquake records is selected. Ten earthquake records with the smallest MSE are selected and can be seen in Table 6.

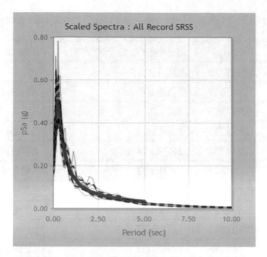

**Fig. 3.** Target spectra compared with earthquake records spectral acceleration

From the selected earthquake records can be obtained PGA value. Each PGA value will be an input for seismic analysis. From seismic analysis, base shear value can be known. Table 6 is a list of each earthquake records PGA value and the resulting base shear from seismic analysis. The higher base shear value of each PGA input (Table 7) will be used as empirical data of a certain probability distribution.

**Table 6.** Earthquake records

| No | Mean square error | Earthquake event | Tahun |
|----|------|------|------|
| 1 | 0.0611 | Chichi | 1999 |
| 2 | 0.0723 | Chuetsuoki | 2007 |
| 3 | 0.0736 | Chichi | 1999 |
| 4 | 0.0799 | Cape Mendocino | 1992 |
| 5 | 0.0828 | Chichi | 1999 |
| 6 | 0.0836 | Duzce | 1999 |
| 7 | 0.0946 | Iwate | 2008 |
| 8 | 0.1064 | Manjil | 1990 |
| 9 | 0.1064 | Chuetsuoki | 2007 |
| 10 | 0.1098 | Chuetsuoki | 2007 |

**Table 7.** PGA and base shear from each earthquake records

| Earthquake event | PGA (g) | Base shear X-Dir (Kips) | Base shear Y-Dir (Kips) |
|---|---|---|---|
| Chichi | 0.033729 | 328 | 281 |
| Chuetsuoki | 0.322435 | 3.20E+03 | 2.74E+03 |
| Chichi | 0.037388 | 367 | 315 |
| Cape Mendocino | 0.265292 | 2.63E+03 | 2.25E+03 |
| Chichi | 0.224338 | 2.22E+03 | 1.90E+03 |
| Duzce | 0.046264 | 4.57E+02 | 3.91E+02 |
| Iwate | 0.160496 | 1.59E+03 | 1.36E+03 |
| Manjil | 0.514564 | 5.10E+03 | 4.37E+03 |
| Chuetsuoki | 0.193342 | 1.92E+03 | 1.64E+03 |
| Chuetsuoki | 0.144486 | 1.64E+03 | 1.22E+03 |

To determine which probability distribution the data follows, The data from Table 8 will be divided into several classes. Class is a section with certain range. If data match the range, data will included in that class.

How to determined class range :

$c = R/k$

where :
Sum of Classes (K)    $= 1 + 3,3 \log n$
$= 4,3$
Range $= 5100 - 328 = 4772$

So,  $c = 4772/4,3$      $= 1109.8$
$c$ = Range of Class (rounded up)
R= Data range
k = Sum of Classes
n = Sum of Data
The result of this distribution can be seen in Figure 4.

**Table 8.** Peak base shear of each PGA input

| Earthquake event | PGA (g) | Peak base shear (Kips) |
|---|---|---|
| Chichi | 0.033729 | 328 |
| Chuetsuoki | 0.322435 | 3.20E+03 |
| Chichi | 0.037388 | 367 |
| Cape Mendocino | 0.265292 | 2.63E+03 |
| Chichi | 0.224338 | 2.22E+03 |
| Duzce | 0.046264 | 4.57E+02 |
| Iwate | 0.160496 | 1.59E+03 |
| Manjil | 0.514564 | 5.10E+03 |
| Chuetsuoki | 0.193342 | 1.92E+03 |
| Chuetsuoki | 0.144486 | 1.64E+03 |

## 3.4  Goodness of Fit

Goodness of Fit is a statistical model that will show how fit the observed or empirical data with theoretical data, where the theoretical data follows known probability distribution shape [13, 14]. Base shear frequency histogram then compared to known probability distribution with the help of Minitab software. The probability distribution that will be used as comparison are normal distribution, lognormal distribution, exponential distribution, and Weibull distribution [15]. The Probability Density Function (PDF) and Cumulative Density Function (CDF) of each probability distribution will be used to help determine which probability distribution match the empirical data (base shear value from seismic analysis). If the distribution is a good fit for the data, the points should fall closely along the fitted distribution line. Departures from the straight line indicate that the fit is unacceptable. The p-value is a probability that measures the evidence against the null hypothesis. If the p-value is less than or equal to the significance level, the decision is to reject the null hypothesis and conclude that your data do not follow the distribution. If the p-value is greater than the significance level, the decision is to fail to reject the null hypothesis [16]. There is not enough evidence to conclude that the data do not follow the distribution. You can assume the data follow the distribution. Significance level ($\alpha$) set as 5% or 0.05. Figures 4, 5, 6 and 7 showing the empirical data compared with 4 kind of probability distribution.

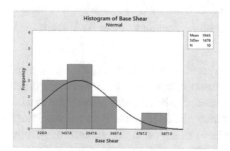 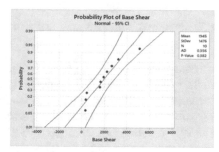

**Fig. 4.**  Base shear histogram compared with probability density function of normal distribution

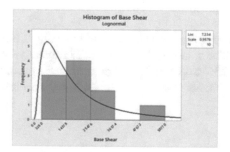 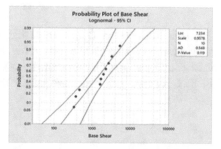

**Fig. 5.**  Base shear histogram compared with probability density function of lognormal distribution

Based on how well the PDF curve follows the histogram, where the empirical data point falls, and p – value, Weibull distribution deemed the most fit. Weibull distribution

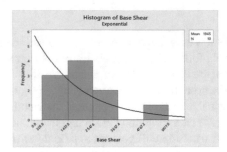 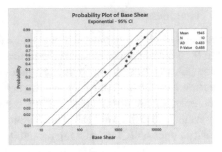

**Fig. 6.** Base shear histogram compared with probability density function of exponential distribution

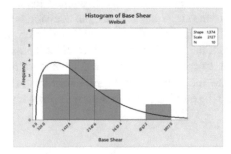 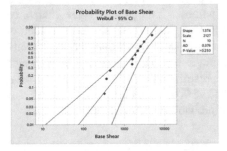

**Fig. 7.** Base shear histogram compared with probability density function of Weibull distribution

has 6 data point fall on linearized CDF line, the most among the other and p-value greater than 0.05 (0.25 > 0.05). If the probability distribution has been determined, the chance of the structure will collapse if an earthquake occurs can be calculated by set certain base shear value as a limit and calculate the area below the PDF curve by applying integral to the PDF function. The limit set as a lower base shear value obtained from pushover analysis which is 4566.22 kips. The integral of Weibull distribution PDF function is as follows:

$$P(x) = 1 - e^{-(\frac{x}{\beta})^{\alpha}}$$

X is the limit obtained from Table 4. $\alpha$ is a shape parameter obtained from Fig. 8. $\beta$ is a scale parameter also obtained from Fig. 8. The value are 1.374 and 2127 respectively. The calculation shows that structure has a 94,25% to not collapse if an earthquake occurs (Fig. 8). Conversely, the structure has a 5,75% chance to collapse if an earthquake occurs.

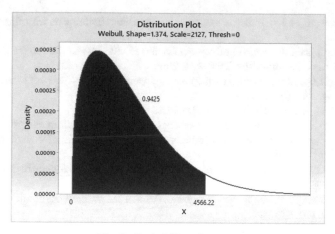

**Fig. 8.** Probability of success

# 4 Conclusion

From the analysis results obtained, the conclusions obtained are as follow:

1. From the pushover analysis can be obtained base shear value 1 step before structure collapses are 4566,22 Kips in Y direction and 5195,72 in X direction
2. Goodness of fit test result shows that base shear data match Weibull distribution the most.
3. Based on the reliability analysis that has been done, structure has a 5.75% chance to collapse when Abnormal Level Earthquake occurs.

# References

1. Chan, E.Y.Y., et al.: Risiko Bencana Indonesia (Disasters Risk of Indonesia). Int. J. Disaster Risk Sci. (2016)
2. Badan Nasional Penanggulangan Bencana: Data Informasi Bencana Indonesia (DIBI), Nopember 2018 (2018)
3. Hosseini, M., Karimiyani, M., Ghafooripour, A., Jabbarzadeh, M.J.: The seismic reliability of offshore structures based on nonlinear time history analyses. In: AIP Conference Proceedings (2008)
4. Elsayed, T., El-Shaib, M., Gbr, K.: Reliability of fixed offshore jacket platform against earthquake collapse. Ships Offshore Struct. (2016)
5. Wang, L., Sari, A., Munro, F.: Structural risk evaluation for offshore platforms under major accidental events. In: Structures Congress 2011 - Proceedings of the 2011 Structures Congress (2011)
6. Khan, M.A.: Seismic design for buildings. In: Earthquake-Resistant Structures (2013)
7. Khan, M.A: Modern earthquake engineering. In: Earthquake-Resistant Structures (2013)
8. ISO 19902: International Standard Petroleum and natural gas industries — Fixed steel offshore structures, 2007-12-01 (2007)

9. BSI: Petroleum and natural gas industries — Specific requirements for offshore structures (2010)
10. Pacific Earthquake Engineering Research Center: PEER Ground Motion Database, Shallow Crustal Earthquakes in Active Tectonic Regimes, NGA-West2 (2017)
11. Christine, A.G., et al.: Pacific earthquake engineering peer NGA-West2 database, Pacific Earthq. Eng. Res. Cent. (2014)
12. Goulet, C.A., et al.: PEER NGA-East Database (2014)
13. Bonett, D.G., Bentler, P.M.: Significance tests and goodness-of-fit in analysis of covariance structures significance tests and goodness of fit in the analysis of covariance structures. Psychol. Bull. (1980)
14. Bray, A., Schoenberg, F.P.: Assessment of point process models for earthquake forecasting. Stat. Sci. (2013)
15. Wu, M.H., Wang, J.P., Ku, K.W.: Earthquake, Poisson and Weibull distributions. Phys. A Stat. Mech. Appl. (2019)
16. Frost, J.: How to interpret regression analysis results: p-values and coefficients, Minitab (2013)

# Unpredictable Oscillations of Impulsive Neural Networks with Hopfield Structure

Marat Akhmet[1](✉), Madina Tleubergenova[2], and Zakhira Nugayeva[2,3]

[1] Department of Mathematics, Middle East Technical University,
06800 Ankara, Turkey
`marat@metu.edu.tr`
[2] Department of Mathematics, Aktobe Regional University,
030000 Aktobe, Kazakhstan
[3] Institute of Information and Computational Technologies,
050010 Almaty, Kazakhstan

**Abstract.** In this study, we consider a new type of oscillations for impulsive models with Hopfield structure. This is the first time, the notions of constant impact changes, impact responses and impact activations are determined. Thus, the electrical sense of the impacts in neural networks is explained, and that has not been done in literature, before. Moreover, we use the constant rates of both signs, positive and negative, which is a novelty for the models. The existence and uniqueness of asymptotically stable unpredictable solutions are proved. Appropriate examples with simulations that support the theoretical results are provided.

**Keywords:** Hopfield-type neural networks · Hopfield-type impulses · Impulsive neural networks · Discontinuous unpredictable oscillations · Asymptotical stability

## 1 Introduction

To show connection of neural networks with physical systems, John J. Hopfield presented a model characterized by differential equations in [1,2]. Today, this model is known as the Hopfield-type neural network, and it is used for image processing, associative memory, pattern recognition, enhancement of X-Ray images and medical image restoration [3–9]. Recently, mathematical models of neural networks with impulses were developed, and used in various fields of science, technology and neurobiology [10–18].

In this paper, we investigate unpredictable solutions of impulsive neural networks of the form

$$x_i'(t) = a_i x_i(t) + \sum_{j=1}^{p} \alpha_{ij} f_j(x_j(t)) + v_i(t), \ t \neq \theta_k,$$

$$\Delta x_i|_{t=\theta_k} = b_i x_i(\theta_k) + \sum_{j=1}^{p} \beta_{ij} g_j(x_j(\theta_k)) + W_{ik}, \tag{1}$$

J. Hemanth et al. (Eds.): ICAIAME 2020, LNDECT 76, pp. 625–642, 2021.
https://doi.org/10.1007/978-3-030-79357-9_59

where $t$, $x_i \in \mathbb{R}$, $x_i(t)$, $i = 1, \ldots p$, corresponds to the membrane potential of the unit $i$, and $p$ is the number of neurons in the network. Moreover, continuous functions $a_i(t), i = 1, \ldots, p$, are rates of self-regulation for the units or reset of potentials, when the units are isolating, $f_i$, $i = 1, 2, \ldots, p$, are activation functions, the constants $\alpha_{ij}, i, j = 1, 2, \ldots, p$, are weights for connection between units $j$ and $i$, $v_i$, $i = 1, 2, \ldots, p$, are input functions.

Similarly, the coefficients $b_i, i = 1, 2, \ldots, p$, in the impulsive equation are constants of self-regulation for the units or reset of potentials, when the units are isolating, $g_i$, $i = 1, 2, \ldots, p$, are activation vectors, the constants $\beta_{ij}, i, j = 1, 2, \ldots, p$, denote the weights for connection between units $j$ and $i$, the sequences $W_{ik}$, $i = 1, 2, \ldots, p, k \in \mathbb{Z}$, are external impulses for the network. We assume that $f_i, g_i : \mathbb{R} \to \mathbb{R}^p$ are continuous functions, the coefficients $a_i$, $b_i$, $\alpha_{ij}$, $\beta_{ij}$, are real numbers, and perturbations $v_i : \mathbb{R} \to \mathbb{R}^p$ are uniformly continuous functions.

In system (1), the impulsive part has the same structure as the differential equation. That is why, it is natural to say that impulses are of the Hopfield-type and the neural network admits the *Hopfield structure*. That is, the model (1) is said to be the *impulsive neural network with Hopfield structure (INNHS)*.

The rates, which are coefficients $a_i(t), i = 1, 2, \ldots, p$, in our paper, are assumed to be positive functions in literature [10,12,14,18]. In the present research the rates can be negative as well as positive and zero issuing from the possibility of negative capacitance in electrical circuits [19,20]. It is the first time that in the jump equation, the *impact response, constant impact change* and *impact activation* are utilized.

Unpredictable oscillations are a completely new type of motion considered in the field of neuroscience [21,22]. Several papers and books have been published confirming the existence and stability of unpredictable and strongly unpredictable oscillations and their applications [23–27]. Using these results, we expanded the theory of unpredictable oscillations for impulsive systems and obtained the new types of discontinuous unpredictable oscillations [28–32].

## 2    Preliminaries

Throughout the work, $\mathbb{N}$, $\mathbb{Z}$ and $\mathbb{R}$ denote the sets of natural numbers, integers and real numbers, respectively. Introduce the norm $\|\psi\| = \max_i |\psi_i|, i = 1, 2, \ldots, p$, where $|\cdot|$ - is the absolute value, $\psi = (\psi_1, \ldots, \psi_p)$ and $\psi_i \in \mathbb{R}, i = 1, 2, \ldots, p$. Consequently, $\|A\| = \max_i \sum_{j=1}^{p} |a_{ij}|, i = 1, 2, \ldots, p$, means the norm for the $p \times p$ matrix $A = \{a_{ij}\}$, $i, j = 1, 2, \ldots, p$.

Consider a set $\mathcal{Q}$ of functions from $\mathbb{R} \to \mathbb{R}^p$ continuous everywhere except of the set of points, countable and unbounded on both sides, left-continuous, where they admit one-sided limits at the points. Two functions $\varphi$, $\phi$ from $\mathcal{Q}$, are said to be $\epsilon$-equivalent on an interval $J \subseteq \mathbb{R}$ if the discontinuity points of the functions can be respectively numerated $\theta_i^\varphi$ and $\theta_i^\phi$, $i = 1, 2, \ldots, m$, such that $|\theta_i^\varphi - \theta_i^\phi| < \epsilon$ for each $i = 1, 2, \ldots, m$, and $\|\varphi(t) - \phi(t)\| < \epsilon$ for each $t \in J$, except those between $\theta_i^\varphi$ and $\theta_i^\phi$ for each $i$. We call the functions are in $\epsilon$-neighborhoods of each other, if $\varphi$ and $\phi$ are $\epsilon$-equivalent on $J$. The topology on the basis of the neighborhoods is said to be $B$-topology [33].

Next, denote by $\widehat{[d_1, d_2]}$, $d_1, d_2 \in \mathbb{R}$, the interval $[d_1, d_2]$, if $d_1 < d_2$ and the interval $[d_2, d_1]$, if $d_2 < d_1$, and fix the set of real numbers $\theta_k$, $k \in \mathbb{Z}$, such that $\underline{\theta} \le \theta_{k+1} - \theta_k \le \overline{\theta}$ for some positive $\underline{\theta}$, $\overline{\theta}$.

**Definition 1** [30]. A function $\omega(t) \in \mathcal{Q}$, $\omega = (\omega_1, \omega_2, ..., \omega_p)$, with the set of discontinuity points $\theta_k$, $k \in \mathbb{Z}$, is said to be discontinuous unpredictable function (d.u.f.) if there exist positive numbers $\epsilon_0$, $\sigma$, sequences $t_n, s_n$ of real numbers and sequences $l_n, m_n$ of integers all of which diverges to infinity such that

(a) $|\theta_{k+l_n} - t_n - \theta_k| \to 0$ as $n \to \infty$ on each bounded interval of integers and $|\theta_{m_n+l_n} - t_n - \theta_{m_n}| \ge \epsilon_0$ for each natural number $n$;
(b) for every positive number $\epsilon$ there exists a positive number $\delta$ such that $\|\omega(t_1) - \omega(t_2)\| < \epsilon$ whenever the points $t_1$ and $t_2$ belong to the same interval of continuity and $|t_1 - t_2| < \delta$;
(c) $\omega(t + t_n) \to \omega(t)$ as $n \to \infty$ in $B$-topology on each bounded interval;
(d) for each natural number $n$ there exists an interval $[s_n - \sigma, s_n + \sigma] \subseteq \widehat{[\theta_{m_n}, \theta_{m_n+l_n}} - t_n]$ which does not contain any point of discontinuity of $\omega(t)$ and $\omega(t + t_n)$, and $\|\omega(t + t_n) - \omega(t)\| \ge \epsilon_0$ for each $t \in [s_n - \sigma, s_n + \sigma]$.

The property (a) is said to be *unpredictability of discrete set* $\theta_k$, $k \in \mathbb{Z}$, the property (b) - *conditional uniform continuity of* $\omega$, the property (c) - *Poisson stability of* $\omega$, and the property (d) - *separation property of* $\omega$.

**Definition 2** [30]. Suppose that $v(t) \in \mathcal{Q}$, $v = (v_1, v_2, ..., v_p)$, is an unpredictable function with the set of discontinuity points $\theta_k$, $k \in \mathbb{Z}$, and $W_k = (W_{1k}, W_{2k}, ..., W_{pk})$, $k \in \mathbb{Z}$, is a bounded sequence in $\mathbb{R}^p$. The couple $(v(t), W_k)$ is called unpredictable if there exist positive numbers $\epsilon_0$, $\sigma$, sequences $t_n, s_n$ of real numbers and sequences $l_n, m_n$ of integers all of which diverge to infinity such that

(a) $|\theta_{k+l_n} - t_n - \theta_k| \to 0$ as $n \to \infty$ on each bounded interval of integers and $|\theta_{m_n+l_n} - t_n - \theta_{m_n}| \ge \epsilon_0$ for each natural number $n$;
(b) for every positive number $\epsilon$ there exists a positive number $\delta$ such that $\|v(t_1) - v(t_2)\| < \epsilon$ whenever the points $t_1$ and $t_2$ belong to the same interval of continuity and $|t_1 - t_2| < \delta$;
(c) $v(t + t_n) \to v(t)$ as $n \to \infty$ in $B$-topology on each bounded interval;
(d) for each natural number $n$ there exists an interval there exists an interval $[s_n - \sigma, s_n + \sigma] \subseteq \widehat{[\theta_{m_n}, \theta_{m_n+l_n}} - t_n]$ which does not contain any point of discontinuity of $v(t)$ and $v(t + t_n)$, and $\|v(t + t_n) - v(t)\| \ge \epsilon_0$ for each $t \in [s_n - \sigma, s_n + \sigma]$;
(e) $\|W_{k+l_n} - W_k\| \to 0$ as $n \to \infty$ for each $k$ in bounded intervals of integers and $\|W_{m_n+l_n} - W_{m_n}\| \ge \epsilon_0$ for each natural number $n$.

We will need the definition of an unpredictable sequence.

**Definition 3** [34]. A bounded sequence $\{\kappa_i\}, i \in \mathbb{Z}$, in $\mathbb{R}^p$ is called unpredictable if there exist a positive number $\varepsilon_0$ and sequences $\{\zeta_n\}, \{\eta_n\}, n \in \mathbb{N}$, of positive integers both of which diverge to infinity such that $\|\kappa_{i+\zeta_n} - \kappa_i\| \to 0$ as $n \to \infty$ for each $i$ in bounded intervals of integers and $\|\kappa_{\zeta_n+\eta_n} - \kappa_{\eta_n}\| \geq \varepsilon_0$ for each $n \in \mathbb{N}$.

The set of discontinuity moments $\theta_k, k \in \mathbb{Z}$, is specified according to the goal of the present research. Let $\gamma_k, k \in \mathbb{Z}$ be an unpredictable sequence. Fix a number $T \geq 4$, such that $\sup_{k \in \mathbb{Z}} |\gamma_k| < \dfrac{T}{h}$, where $h \geq 3$.

Determine the following sequence

$$\theta_k = k\,T + \gamma_k, \ k \in \mathbb{Z}. \tag{2}$$

Since $\gamma_k$ is an unpredictable sequence, there exist a positive number $\epsilon_0$ and sequences $\zeta_n, \eta_n$, both of which diverge to infinity such that $|\gamma_{k+\zeta_n} - \gamma_k| \to 0$ as $n \to \infty$ for each $k$ in bounded intervals of integers and $|\gamma_{\zeta_n+\eta_n} - \gamma_{\eta_n}| \geq \epsilon_0$ for each natural number $n$.

Let us show that $\theta_k$ satisfies the condition (a). We will show that condition (a) specified in Definition 2 is valid for $\theta_k, k \in \mathbb{Z}$, with $t_n = T\zeta_n$, $l_n = \zeta_n$ and $m_n = \eta_n$ for each $n \in \mathbb{N}$. First, we obtain that

$$|\theta_{k+l_n} - t_n - \theta_k| = |(k+\zeta_n)T + \gamma_{k+\zeta_n} - \zeta_n T - kT - \gamma_k| = |\gamma_{k+\zeta_n} - \gamma_k| \to 0,$$

as $n \to \infty$, $k \in \mathbb{Z}$ for each $k$ in bounded intervals of integers. Moreover,

$$|\theta_{m_n+l_n} - t_n - \theta_{m_n}| = |(\eta_n + \zeta_n)T + \gamma_{\eta_n+\zeta_n} - \zeta_n T - \eta_n T - \gamma_{\eta_n}|$$
$$= |\gamma_{\eta_n+\zeta_n} - \gamma_{\eta_n}| \geq \epsilon_0$$

for each natural number $n$. Thus, the set $\theta_k, k \in \mathbb{Z}$, satisfies the property (a).

It can be confirmed that $\theta_k, k \in \mathbb{Z}$, satisfies the inequality $\underline{\theta} \leq \theta_{k+1} - \theta_k \leq \overline{\theta}$ with $\underline{\theta} = T - \dfrac{2T}{h}$ and $\overline{\theta} = T + \dfrac{2T}{h}$.

Let us denote by $x_i(t, s)$ the fundamental solution of the system associated with (1),

$$x_i'(t) = a_i x_i(t), \ t \neq \theta_k,$$
$$\Delta x|_{t=\theta_k} = b_i x_i(\theta_k), \tag{3}$$

where $t \in \mathbb{R}$, constants $a_i, b_i \ i = 1, 2, \ldots, p$, are real valued numbers.

We have that

$$x_i(t, s) = e^{a_i(t-s)} (1 + b_i)^{i([s,t))}, \ t \geq s, \tag{4}$$

where $i([s, t))$ denotes the number of the terms of the sequence $\theta_k$, which belong to the segment $[s, t)$.

The following conditions are needed:

**(C1)** $\lambda = \max_i \left( a_i + \dfrac{1}{T} Ln|1 + b_i| \right) < 0$, for all $i = 1, 2, \cdots, p$ ;

**(C2)** $|f_i(x)| \leq m_f$, and $|g_i(x)| \leq m_g$, where $m_f, m_g$, are positive numbers, for all $i = 1, 2, \ldots, p$, $|x| < H$;

**(C3)** there exist positive numbers $l_f^i$ and $l_g^i$ such that $|f_i(x) - f_i(y)| \leq l_f^i \|x - y\|$, $|g_i(x) - g_i(y)| \leq l_g^i \|x - y\|$, for all $i = 1, 2, \ldots, p$, $|y|, |x| < H$;

**(C4)** there exists a positive number $M$, such that $\sup_{t \in \mathbb{R}} |v_i(t)| + \sup_{k \in \mathbb{Z}} |W_{ik}| = M < \infty$ for all $i = 1, 2, \ldots, p$;

**(C5)** $K\left( \frac{1}{\lambda}\left( \max_i \sum_{j=1}^{p} |\alpha_{ij}| m_f + M \right) + \frac{1}{1 - e^{\lambda \underline{\theta}}} \left( \max_i \sum_{j=1}^{p} |\beta_{ij}| m_g + M \right) \right) < H$;

**(C6)** $K\left( \frac{1}{\lambda} \max_i l_f^i \max_i \sum_{j=1}^{p} |\alpha_{ij}| + \frac{1}{1 - e^{\lambda \underline{\theta}}} \max_i l_g^i \max_i \sum_{j=1}^{p} |\beta_{ij}| \right) < 1$;

**(C7)** $K \max_i l_f^i \max_i \sum_{j=1}^{p} |\alpha_{ij}| + \frac{1}{\underline{\theta}} ln(1 + K \max_i l_g^i \max_i \sum_{j=1}^{p} |\beta_{ij}|) < -\lambda$.

On account of (4), under condition $(C1)$ there exists a number $K \geq 1$ such that

$$|x_i(t, s)| \leq K e^{\lambda(t-s)}, \ t \geq s, \tag{5}$$

for all $i = 1, 2, \ldots, p$, [33].

The following assertion is needed in the proof of the main result of the paper.

**Lemma 1.** *Suppose that the condition $(C1)$ is valid, then for all $i = 1, 2, \ldots, p$, the following inequality holds*

$$|x_i(t + t_n, s + t_n) - x_i(t, s)| \leq \mathcal{K} e^{\lambda(t-s)}, \ t \geq s, \tag{6}$$

*where $\mathcal{K} = K \max(1, |b_i|)$.*

*Proof.* By using (4) and (5), we obtain that

$|x_i(t + t_n, s + t_n) - x_i(t, s)|$

$\leq \left| e^{a_i(t-s)} (1 + b_i)^{i([s+t_n, t+t_n))} - e^{a_i(t-s)} (1 + b_i)^{i([s,t))} \right|$

$\leq \left| e^{a_i(t-s)} (1 + b_i)^{i([s,t))} \right| \left| (1 + b_i)^{|i([s+t_n, t+t_n)) - i([s,t))|} - 1 \right|$

$\leq K \max(1, |b_i|) e^{\lambda(t-s)}$

for all $t \geq s$, $i = 1, 2, \ldots, p$. $\qquad\square$

# 3   Main Result

We will study the problem of the existence and uniqueness of discontinuous unpredictable oscillations for INNHS (1).

**Lemma 2.** *A vector function* $y(t) = (y_1(t), ..., y_p(t))$ *is a bounded solution of system (1), if and only if it is a solution of the following integral equations:*

$$
y_i(t) = \int_{-\infty}^{t} x_i(t, s) \Big[ \sum_{j=1}^{p} \alpha_{ij} f_j(y_j(s)) + v_i(s) \Big] ds
$$

$$
+ \sum_{\theta_k < t} x_i(t, \theta_k+) \Big[ \sum_{j=1}^{p} \beta_{ij} g_j(y_j(\theta_k)) + W_{ik} \Big] \tag{7}
$$

*for all* $i = 1, 2, \ldots, p$, $k \in \mathbb{Z}$.

Consider the set $\mathcal{D} \subset \mathcal{Q}$ of piecewise continuous functions $\psi$ with the set of discontinuity moments of the system (1), applying the norm $\|\psi\|_1 = \sup\limits_{t \in \mathbb{R}} \|\psi(t)\|$, such that

**(K1)** there exists a positive number $H$, which satisfies $\|\psi\|_1 < H$ for all $\psi(t) \in \mathcal{D}$;

**(K2)** for each $\psi(t) \in \mathcal{D}$ it is true that $\psi(t+t_n) \to \psi(t)$ in $B$-topology as $n \to \infty$ on each bounded interval of the real line, where the sequence $t_n$, is the same as for function $v(t)$ in system (1).

Let us introduce the following operator $\Pi\psi(t) = (\Pi_1\psi(t), \Pi_2\psi(t), ..., \Pi_p\psi(t))$ in the space $\mathcal{D}$ such that

$$
\Pi_i\psi(t) = \int_{-\infty}^{t} x_i(t, s) \Big[ \sum_{j=1}^{p} \alpha_{ij} f_j(\psi_j(s)) + v_i(s) \Big] ds
$$

$$
+ \sum_{\theta_k < t} x_i(t, \theta_k+) \Big[ \sum_{j=1}^{p} \beta_{ij} g_j(\psi_j(\theta_k)) + W_{ik} \Big] \tag{8}
$$

*for all* $i = 1, 2, \ldots, p$, $k \in \mathbb{Z}$.

**Lemma 3.** *If* $\psi(t) \in \mathcal{D}$, *then* $\Pi\psi(t) \in \mathcal{D}$.

*Proof.* First, let us prove that the function $\Pi\psi(t)$ satisfies the property $(K1)$. For a function $\psi(t) \in \mathcal{D}$ and all $i = 1, 2, \ldots, p$, we have that

$$
|\Pi_i\psi(t)| = \Big| \int_{-\infty}^{t} x_i(t, s) \Big[ \sum_{j=1}^{p} \alpha_{ij} f_j(\psi_j(s)) + v_i(s) \Big] ds
$$

$$
+ \sum_{\theta_k < t} x_i(t, \theta_k+) \Big[ \sum_{j=1}^{p} \beta_{ij} g_j(\psi_j(\theta_k)) + W_{ik} \Big] \Big|
$$

$$
\leq \int_{-\infty}^{t} |x_i(t, s)| \Big[ \sum_{j=1}^{p} |\alpha_{ij}| |f_j(\psi_j(s))| + |v_i(s)| \Big] ds
$$

$$+ \sum_{\theta_k < t} |x_i(t, \theta_k)| \Big[ \sum_{j=1}^{p} |\beta_{ij}||g_j(\psi_j(\theta_k))| + |W_{ik}| \Big]$$

$$\leq \int_{-\infty}^{t} K e^{\lambda(t-s)} \Big( \max_i \sum_{j=1}^{p} |\alpha_{ij}| m_f + M \Big) ds$$

$$+ \sum_{\theta_k < t} K e^{\lambda(t-\theta_k)} \Big( \max_i \sum_{j=1}^{p} |\beta_{ij}| m_g + M \Big)$$

$$\leq \frac{K}{\lambda} \Big( \max_i \sum_{j=1}^{p} |\alpha_{ij}| m_f + M \Big) + \frac{K}{1 - e^{\lambda \underline{\theta}}} \Big( \max_i \sum_{j=1}^{p} |\beta_{ij}| m_g + M \Big).$$

So, by condition $(C5)$ it is true that $\|\Pi \psi\|_1 < H$.

Let us check that the Poisson stability of $\Pi \psi(t)$, i.e. property $(K2)$, is valid.

Fix a positive number $\epsilon$ and $[a, b]$, $-\infty < a < b < \infty$. Let us prove that $\|\Pi \psi(t + t_n) - \Pi \psi(t)\| < \epsilon$ on $[a, b]$ for sufficiently large $n$. Choose real numbers $c < a$ and $0 < \xi$ satisfying the inequalities

$$\frac{K \Big( \max_i \sum_{j=1}^{p} |\alpha_{ij}| m_f + M \Big) + 2K \Big( \max_i l_f^i \max_i \sum_{j=1}^{p} |\alpha_{ij}| H + M \Big)}{\lambda} e^{\lambda(a-c)} < \frac{\epsilon}{5}, \quad (9)$$

$$\frac{2K \Big( \max_i \sum_{j=1}^{p} |\beta_{ij}| m_g + \max_i l_g^i \max_i \sum_{j=1}^{p} |\beta_{ij}| H + 2M \Big)}{1 - e^{\lambda \underline{\theta}}} e^{\lambda(a-c)} < \frac{\epsilon}{5}, \quad (10)$$

$$\frac{K \Big( \max_i \sum_{j=1}^{p} |\alpha_{ij}| m_f + M \Big) + 2K \Big( \max_i l_f^i \max_i \sum_{j=1}^{p} |\alpha_{ij}| H + M \Big)}{\lambda (1 - e^{\lambda \underline{\theta}})} \Big( e^{\lambda \xi} - 1 \Big) < \frac{\epsilon}{5}, \quad (11)$$

$$\frac{K \xi}{\lambda} \Big( \max_i l_f^i \max_i \sum_{j=1}^{p} |\alpha_{ij}| + 1 \Big) < \frac{\epsilon}{5}, \quad (12)$$

and

$$\frac{K \xi}{1 - e^{\lambda \underline{\theta}}} \Big( \max_i \sum_{j=1}^{p} |\beta_{ij}| (m_g + \max_i l_g^i) + M + 1 \Big) < \frac{\epsilon}{5}. \quad (13)$$

Consider the number $n$ sufficiently large such that $|\theta_{k+l_n} - t_n - \theta_k| < \xi$, $|\psi_i(\theta_{k+l_n}) - \psi_i(\theta_k)| < \xi$, $|W_{i\,k+l_n} - W_{ik}| < \xi$, $|\psi_i(t + t_n) - \psi_i(t)| < \xi$, and $|v_i(t + t_n) - v_i(t)| < \xi$ for all $t \in [c, b]$, $\theta_k \in [c, b]$, $k \in \mathbb{Z}$, $i = 1, 2, \ldots, p$.

In what follows, without loss of generality, assume that

$$\cdots \theta_{m-1} \le c \le \theta_m < \cdots < \theta_q \le t \le \theta_{q+1}$$

where $m, q \in \mathbb{Z}$. If $t \in [a, b], i = 1, 2, \ldots, p$, then we obtain that

$$|\Pi_i \psi(t + t_n) - \Pi_i \psi(t)|$$

$$\le \int_{-\infty}^{c} |x_i(t + t_n, s + t_n) - x_i(t, s)| \left[ \sum_{j=1}^{p} |\alpha_{ij}||f_j(\psi_j(s + t_n))| + |v_i(s + t_n)| \right] ds$$

$$+ \sum_{\theta_k < c} |x_i(t + t_n, \theta_{k+l_n}+) - x_i(t, \theta_k+)| \left[ \sum_{j=1}^{p} |\beta_{ij}||g_j(\psi_j(\theta_{k+l_n}))| + |W_{i\,k+l_n}| \right]$$

$$+ \int_{c}^{t} |x_i(t + t_n, s + t_n) - x_i(t, s)| \left[ \sum_{j=1}^{p} |\alpha_{ij}||f_j(\psi_j(s + t_n))| + |v_i(s + t_n)| \right] ds$$

$$+ \sum_{c \le \theta_k < t} |x_i(t + t_n, \theta_{k+l_n}+) - x_i(t, \theta_k+)| \left[ \sum_{j=1}^{p} |\beta_{ij}||g_j(\psi_j(\theta_{k+l_n}))| + |W_{i\,k+l_n}| \right]$$

$$+ \int_{-\infty}^{c} |x_i(t, s)| \left[ \sum_{j=1}^{p} |\alpha_{ij}||f_j(\psi_j(s + t_n)) - f_j(\psi_j(s))| + |v_i(s + t_n) - v_i(s)| \right] ds$$

$$+ \sum_{\theta_k < c} |x_i(t, \theta_k+)| \left[ \sum_{j=1}^{p} |\beta_{ij}||g_j(\psi_j(\theta_{k+l_n})) - g_j(\psi_j(\theta_k))| + |W_{i\,k+l_n} - W_{ik}| \right]$$

$$+ \int_{c}^{t} |x_i(t, s)| \left[ \sum_{j=1}^{p} |\alpha_{ij}||f_j(\psi_j(s + t_n)) - f_j(\psi_j(s))| + |v_i(s + t_n) - v_i(s)| \right] ds$$

$$+ \sum_{c \le \theta_k < t} |x_i(t, \theta_k+)| \left[ [\sum_{j=1}^{p} |\beta_{ij}||g_j(\psi_j(\theta_{k+l_n})) - g_j(\psi_j(\theta_k))| + |W_{i\,k+l_n} - W_{ik}| \right].$$

Applying Lemma 1, one can verify that

$$\int_{-\infty}^{c} |x_i(t + t_n, s + t_n) - x_i(t, s)| \left[ \sum_{j=1}^{p} |\alpha_{ij}||f_j(\psi_j(s + t_n))| + |v_i(s + t_n)| \right] ds$$

$$+ \sum_{\theta_k < c} |x_i(t + t_n, \theta_{k+l_n}+) - x_i(t, \theta_k+)| \left[ \sum_{j=1}^{p} |\beta_{ij}||g_j(\psi_j(\theta_{k+l_n}))| + |W_{i\,k+l_n}| \right]$$

$$\le \int_{-\infty}^{c} K e^{\lambda(t-s)} \left( \max_i \sum_{j=1}^{p} |\alpha_{ij}| m_f + M \right) ds$$

$$+ \sum_{k=-\infty}^{m-1} 2K e^{\lambda(t-\theta_k)} \left( \max_i \sum_{j=1}^{p} |\beta_{ij}| m_g + M \right)$$

$$< \left( \frac{K}{\lambda} \left( \max_i \sum_{j=1}^{p} |\alpha_{ij}| m_f + M \right) + \frac{2K}{1 - e^{\lambda \underline{\theta}}} \left( \max_i \sum_{j=1}^{p} |\beta_{ij}| m_g + M \right) \right) e^{\lambda(a-c)}$$

for all $i = 1, 2, \ldots, p$. Next, we obtain that

$$\int_c^t |x_i(t + t_n, s + t_n) - x_i(t, s)| \left[ \sum_{j=1}^{p} |\alpha_{ij}| |f_j(\psi_j(s + t_n))| + |v_i(s + t_n)| \right] ds$$

$$+ \sum_{c \leq \theta_k < t} |x_i(t + t_n, \theta_{k+l_n}+) - x_i(t, \theta_k+)| \left[ \sum_{j=1}^{p} |\beta_{ij}| |g_j(\psi_j(\theta_{k+l_n}))| + |W_{i\,k+l_n}| \right]$$

$$\leq \sum_{k=m}^{q} \int_{\theta_k}^{\theta_{k+l_n} - t_n} K e^{\lambda(t-s)} \left( \max_i \sum_{j=1}^{p} |\alpha_{ij}| m_f + M \right) ds$$

$$+ \sum_{k=m}^{q} K e^{\lambda(t-\theta_k)} \xi \left( \max_i \sum_{j=1}^{p} |\beta_{ij}| m_g + M \right)$$

$$< \frac{K(e^{\lambda \xi} - 1)}{\lambda(1 - e^{\lambda \underline{\theta}})} \left( \max_i \sum_{j=1}^{p} |\alpha_{ij}| m_f + M \right) + \frac{K \xi}{1 - e^{\lambda \underline{\theta}}} \left( \max_i \sum_{j=1}^{p} |\beta_{ij}| m_g + M \right)$$

for all $i = 1, 2, \ldots, p$. Moreover, we have

$$\int_{-\infty}^{c} |x_i(t, s)| \left[ \sum_{j=1}^{p} |\alpha_{ij}| |f_j(\psi_j(s + t_n)) - f_j(\psi_j(s))| + |v_i(s + t_n) - v_i(s)| \right] ds$$

$$+ \sum_{\theta_k < c} |x_i(t, \theta_k+)| \left[ \sum_{j=1}^{p} |\beta_{ij}| |g_j(\psi_j(\theta_{k+l_n})) - g_j(\psi_j(\theta_k))| + |W_{i\,k+l_n} - W_{ik}| \right]$$

$$\leq \int_{-\infty}^{c} K e^{\lambda(t-s)} \left( \max_i l_f^i \max_i \sum_{j=1}^{p} |\alpha_{ij}| 2H + 2M \right) ds$$

$$+ \sum_{k=-\infty}^{m-1} K e^{\lambda(t-\theta_k)} \left( \max_i l_g^i \max_i \sum_{j=1}^{p} |\beta_{ij}| 2H + 2M \right)$$

$$< \frac{2K}{\lambda} \left( \max_i l_f^i \max_i \sum_{j=1}^{p} |\alpha_{ij}| H + M \right) e^{\lambda(a-c)}$$

$$+ \frac{2K}{1 - e^{\lambda \underline{\theta}}} \left( \max_i l_g^i \max_i \sum_{j=1}^{p} |\beta_{ij}| H + 2M \right) e^{\lambda(a-c)}$$

for all $i = 1, 2, \ldots, p$. Additionally, we get that

$$\int_c^t |x_i(t, s)| \left[ \sum_{j=1}^{p} |\alpha_{ij}| |f_j(\psi_j(s + t_n)) - f_j(\psi_j(s))| + |v_i(s + t_n) - v_i(s)| \right] ds$$

$$+ \sum_{c \le \theta_k < t} |x_i(t, \theta_k+)| \left[ \sum_{j=1}^{p} |\beta_{ij}| \, |g_j(\psi_j(\theta_{k+l_n})) - g_j(\psi_j(\theta_k))| + |W_{i\,k+l_n} - W_{ik}| \right]$$

$$\le \int_c^t K e^{\lambda(t-s)} \left( \max_i l_f^i \max_i \sum_{j=1}^{p} |\alpha_{ij}| \xi + \xi \right) ds$$

$$+ \sum_{k=m}^{q} \int_{\theta_k}^{\theta_{k+l_n} - t_n} K e^{\lambda(t-s)} \left( \max_i l_f^i \max_i \sum_{j=1}^{p} |\alpha_{ij}| 2H + 2M \right) ds$$

$$+ \sum_{k=m}^{q} K e^{\lambda(t-\theta_k)} \left( \max_i l_g^i \max_i \sum_{j=1}^{p} |\beta_{ij}| \xi + \xi \right)$$

$$< \frac{K\xi}{\lambda} \left( \max_i l_f^i \max_i \sum_{j=1}^{p} |\alpha_{ij}| + 1 \right) + \frac{2\,K(e^{\lambda\xi} - 1)}{\lambda(1 - e^{\lambda\underline{\theta}})} \left( \max_i l_f^i \max_i \sum_{j=1}^{p} |\alpha_{ij}| H + M \right)$$

$$+ \frac{K\xi}{1 - e^{\lambda\underline{\theta}}} \left( \max_i l_g^i \max_i \sum_{j=1}^{p} |\beta_{ij}| + 1 \right)$$

for all $i = 1, 2, \ldots, p$. The inequalities (9)–(13) imply that $\|\Pi\psi(t+t_n) - \Pi\psi(t)\| \le \epsilon$, for $t \in [a, b]$. Therefore, $\Pi\psi(t+t_n) \to \Pi\psi(t)$ uniformly in $B$-topology as $n \to \infty$ on each bounded interval. The function $\Pi\psi(t)$ satisfies the condition $(K2)$, and $\Pi\psi(t) \in \mathcal{D}$. $\qquad\square$

**Lemma 4.** *If conditions (C1)–(C6) are fulfilled, then the operator* $\Pi : \mathcal{D} \to \mathcal{D}$ *is contractive.*

*Proof.* For $\varphi$ and $\psi$ of the set $\mathcal{D}$, we have that

$$|\Pi_i\varphi(t) - \Pi_i\psi(t)| = \int_{-\infty}^{t} |x_i(t, s)| \left[ \sum_{j=1}^{p} |\alpha_{ij}| \, |f_j(\varphi_j(s)) - f_j(\psi_j(s))| \right] ds$$

$$+ \sum_{\theta_k < t} |x_i(t, \theta_k+)| \left[ \sum_{j=1}^{p} |\beta_{ij}| \, |g_j(\varphi_j(\theta_k)) - g_j(\psi_j(\theta_k))| \right]$$

$$\le \int_{-\infty}^{t} K e^{\lambda(t-s)} \max_i \sum_{j=1}^{p} |\alpha_{ij}| \max_i l_f^i |\varphi_j(s) - \psi_j(s)| ds$$

$$+ \sum_{\theta_k < t} K e^{\lambda(t-\theta_k)} \max_i \sum_{j=1}^{p} |\beta_{ij}| \max_i l_g^i |\varphi_j(s) - \psi_j(s)|$$

$$\le \frac{K}{\lambda} \max_i l_f^i \max_i \sum_{j=1}^{p} |\alpha_{ij}| |\varphi_j(t) - \psi_j(t)|$$

$$+ \frac{K}{1 - e^{\lambda\underline{\theta}}} \max_i l_g^i \max_i \sum_{j=1}^{p} |\beta_{ij}| |\varphi_j(t) - \psi_j(t)|$$

$$\leq \left( \frac{K}{\lambda} \max_i l_f^i \max_i \sum_{j=1}^p |\alpha_{ij}| + \frac{K}{1-e^{\lambda \underline{\theta}}} \max_i l_g^i \max_i \sum_{j=1}^p |\beta_{ij}| \right) \|\varphi(t) - \psi(t)\|_1.$$

Therefore, the inequality $\|\Pi\varphi(t) - \Pi\psi(t)\|_1 \leq K \left( \frac{1}{\lambda} \max_i l_f^i \max_i \sum_{j=1}^p |\alpha_{ij}| + \frac{1}{1-e^{\lambda \underline{\theta}}} \max_i l_g^i \max_i \sum_{j=1}^p |\beta_{ij}| \right) \|\varphi(t) - \psi(t)\|_1$. Thus the operator $\Pi$ is contractive by means of condition (C6). $\qquad\square$

**Theorem 1.** *Suppose that the couple $(v(t), W_k)$ in (1) is satisfies Definition 2 and conditions (C1)–(C7) are fulfilled. Then, the INNHS (1) has a unique asymptotically stable discontinuous unpredictable oscillation.*

*Proof.* We first show that $\mathcal{D}$ is complete. Consider a Cauchy sequence $\phi_r(t)$, $r \in \mathbb{N}$, in $\mathcal{D}$, which converges to a limit function $\phi(t)$ on $\mathbb{R}$. It suffices to show that $\phi(t)$ satisfies condition (K2), since the condition (K1) can be easily checked. Fix a closed and bounded interval $I \subset \mathbb{R}$. Denote $\theta_k, k = j, j+1, \cdots, j+m$, the discontinuity points of $\phi(t)$ and $\phi_r(t)$, and $\theta_k^n = \theta_{k+l_n} - t_n, k = j, j+1, \cdots, j+m$, the discontinuity points of $\phi(t+t_n)$ and $\phi_r(t+t_n)$ in the interval $I$, respectively. Let $n$ be a large enough number such that $|\theta_k^n - \theta_k| < \epsilon, k = j, j+1, \cdots, j+m$. Since of the convergence of $\phi_r(t)$ we obtain that $\|\phi(t+t_n) - \phi_r(t+t_n)\| < \frac{\epsilon}{3}$ and $\|\phi_r(t) - \phi(t)\| < \frac{\epsilon}{3}$ if $r$ sufficiently large. Because the sequence $\phi_r(t) \in \mathcal{D}$ and satisfies (K2), we have that for sufficiently large $n$ $\|\phi_r(t+t_n) - \phi_r(t)\| < \frac{\epsilon}{3}$ for $t \notin [\widehat{\theta_k, \theta_k^n}]$, and $|\theta_k^n - \theta_k| < \epsilon, k = j, j+1, \cdots, j+m$. Thus, for sufficiently large $n$ and $r$ it is true that

$$\|\phi(t+t_n) - \phi(t)\| < \|\phi(t+t_n) - \phi_r(t+t_n)\| + \|\phi_r(t+t_n) - \phi_r(t)\|$$
$$+ \|\phi_r(t) - \phi(t)\| < \epsilon \qquad (14)$$

for all $t \notin [\widehat{\theta_k, \theta_k^n}], k = j, j+1, \cdots, j+m$. That is, $\phi(t+t_n) \to \phi(t)$ in $B$-topology as $n \to \infty$ on $I$. The completeness of $\mathcal{D}$ is proved.

Apply the contraction mapping theorem, duo to Lemmas 3 and 4, there exists unique solution $\omega(t) \in \mathcal{D}$ of the system (1).

Next, we will prove that the function $\omega(t)$ satisfies the unpredictability property. Corresponding to the Definition 1 the interval $[s_n - \sigma, s_n + \sigma] \subseteq [\theta_{m_n}, \widehat{\theta_{m_n+l_n}} - t_n]$, does not admit discontinuity points of functions $\omega(t)$, $\omega(t+t_n)$.

Next using the relations

$$\omega_i(t) = \omega_i(s_n) + \int_{s_n}^t a_i \omega_i(s) ds + \int_{s_n}^t \sum_{j=1}^p \alpha_{ij} f_j \omega_j(s)) ds + \int_{s_n}^t v_i(s) ds,$$

and

$$\omega_i(t+t_n) = \omega_i(s_n + t_n) + \int_{s_n}^t a_i \omega_i(s+t_n) ds$$

$$+ \int_{s_n}^{t} \sum_{j=1}^{p} \alpha_{ij} f_j(\omega_j(s+t_n))ds + \int_{s_n}^{t} v_i(s+t_n)ds,$$

we obtain that

$$\omega_i(t+t_n) - \omega_i(t) = \omega_i(s_n+t_n) - \omega_i(s_n) + \int_{s_n}^{t} a_i(\omega_i(s+t_n) - \omega_i(s))ds$$

$$+ \int_{s_n}^{t} \sum_{j=1}^{p} \alpha_{ij} \Big( f_j(\omega_j(s+t_n)) - f_j(\omega_j(s)) \Big) ds$$

$$+ \int_{s_n}^{t} (v_i(s+t_n) - v_i(s))ds. \tag{15}$$

There exist a positive number $\kappa$, natural numbers $l$ and $k$ such that the following inequalities are valid,

$$\kappa < \sigma; \tag{16}$$

$$\kappa \left( \frac{1}{2} - \left( \frac{1}{l} + \frac{2}{k} \right) \right) \left( \max_i a_i + \max_i l_f^i \max_i \sum_{j=1}^{p} |\alpha_{ij}| \right) \geq \frac{4}{3l}; \tag{17}$$

$$||\omega(t+s) - \omega(t)|| < \epsilon_0 \min \left( \frac{1}{k}, \frac{1}{3l} \right), \quad t \in \mathbb{R}, |s| < \kappa. \tag{18}$$

Denote $\Delta = ||\omega(s_n+t_n) - \omega(s_n)||$. Consider two possible cases: (i) $\Delta < \epsilon_0/l$; (ii) $\Delta \geq \epsilon_0/l$ such that the remaining proof falls naturally into two parts.

(i) One can find that from (18) it follows that

$$||\omega(t+t_n) - \omega(t)|| \leq ||\omega(t+t_n) - \omega(s_n+t_n)|| + ||\omega(s_n+t_n) - \omega(s_n)||$$

$$+ ||\omega(s_n) - \omega(t)|| < \frac{\epsilon_0}{k} + \frac{\epsilon_0}{l} + \frac{\epsilon_0}{k} = \epsilon_0 \left( \frac{1}{l} + \frac{2}{k} \right),$$

if $t \in [s_n, s_n + \kappa]$. Therefore, by using relations (15)–(18) we have that

$$|\omega_i(t+t_n) - \omega_i(t)| \geq \int_{s_n}^{t} |v_i(s+t_n) - v_i(s)|ds - \int_{s_n}^{t} a_i|\omega_i(s+t_n) - \omega_i(s)|ds$$

$$- \int_{s_n}^{t} \sum_{j=1}^{p} |\alpha_{ij}| \Big| f_j(\omega_j(s+t_n)) - f_j(\omega_j(s)) \Big| ds$$

$$- |\omega_i(s_n+t_n) - \omega_i(s_n)|$$

$$\geq \frac{\kappa}{2}\epsilon_0 - \left( \max_i a_i + \max_i l_f^i \max_i \sum_{j=1}^{p} |\alpha_{ij}| \right) \epsilon_0 \kappa \left( \frac{1}{l} + \frac{2}{k} \right) - \frac{\epsilon_0}{l}$$

$$= \epsilon_0 \kappa \left( \frac{1}{2} - \left( \frac{1}{l} + \frac{2}{k} \right) \right) \left( \max_i a_i + \max_i l_f^i \max_i \sum_{j=1}^{p} |\alpha_{ij}| \right) - \frac{\epsilon_0}{l}$$

$$\geq \frac{\epsilon_0}{3l}$$

for $t \in [s_n + \frac{\kappa}{2}, s_n + \kappa]$.

ii) We have that

$$\|\omega(t_n + t) - \omega(t)\| \geq \|\omega(t_n + s_n) - \omega(s_n)\| - \|\omega(s_n) - y(t)\|$$
$$- \|\omega(t_n + t) - \omega(t_n + s_n)\| \geq \frac{\epsilon_0}{l} - \frac{\epsilon_0}{3l} - \frac{\epsilon_0}{3l} = \frac{\epsilon_0}{3l},$$

for $t \in [s_n - \kappa, s_n + \kappa]$ and $n \in \mathbb{N}$. That is, $\omega(t)$ satisfies the unpredictability property.

Finally, we will study the asymptotic stability of the oscillation $\omega(t)$. It is true that

$$\omega_i(t) = x_i(t, t_0)\omega(t_0) + \int_{t_0}^t x_i(t, s) \Big[ \sum_{j=1}^p \alpha_{ij} f_j(\omega_j(s)) + v_i(s) \Big] ds$$

$$+ \sum_{t_0 \leq \theta_k < t} x_i(t, \theta_k+) \Big[ \sum_{j=1}^p \beta_{ij} g_j(\omega_j(\theta_k)) + W_{ik} \Big]$$

for all $i = 1, \ldots, p, \; k \in \mathbb{Z}$.

Let $z(t) = (z_1, z_2, \ldots, z_p)$ be another oscillation of system (1). One can write

$$z_i(t) = x_i(t, t_0)z(t_0) + \int_{t_0}^t x_i(t, s) \Big[ \sum_{j=1}^p \alpha_{ij} f_j(z_j(s)) + v_i(s) \Big] ds$$

$$+ \sum_{t_0 \leq \theta_k < t} x_i(t, \theta_k+) \Big[ \sum_{j=1}^p \beta_{ij} g_j(z_j(\theta_k)) + W_{ik} \Big]$$

for all $i = 1, \ldots, p, \; k \in \mathbb{Z}$.

Making use of the relation

$$\omega_i(t) - z_i(t) = x_i(t, t_0)[\omega_i(t_0) - z_i(t_0)]$$

$$+ \int_{t_0}^t x_i(t, s) \sum_{j=1}^p \alpha_{ij} \Big[ f_j(\omega_j(s)) - f_j(z_j(s)) \Big] ds$$

$$+ \sum_{t_0 \leq \theta_k < t} x_i(t, \theta_k+) \sum_{j=1}^p \beta_{ij} \Big[ g_j(\omega_j(\theta_k)) - g_j(z_j(\theta_k)) \Big]$$

for all $i = 1, \ldots, p, \; k \in \mathbb{Z}$, we obtain that

$$|\omega_i(t) - z_i(t)| \leq K e^{\lambda(t-t_0)} |\omega_i(t_0) - z_i(t_0)|$$

$$+ \int_{t_0}^t K e^{\lambda(t-s)} \sum_{j=1}^p |\alpha_{ij}| \big| f_j(\omega_j(s)) - f_j(z_j(s)) \big| ds$$

$$+ \sum_{t_0 \leq \theta_k < t} K e^{\lambda(t-\theta_k)} \sum_{j=1}^p |\beta_{ij}| \big| g_j(\omega_j(\theta_k)) - g_j(z_j(\theta_k)) \big|$$

$$\leq K e^{\lambda(t-t_0)} ||\omega(t_0) - z(t_0)||$$

$$+ K \max_i l_f^i \max_i \sum_{j=1}^p |\alpha_{ij}| \int_{t_0}^t e^{\lambda(t-s)} ||\omega(s) - z(s)|| ds$$

$$+ K \max_i l_g^i \max_i \sum_{j=1}^p |\beta_{ij}| \sum_{t_0 \leq \theta_k < t} K e^{\lambda(t-\theta_k)} ||\omega(\theta_k) - z(\theta_k)||$$

for all $i = 1, 2, \ldots, p,\ k \in \mathbb{Z}$.

Thus, it can be confirmed that

$$||\omega(t) - z(t)|| \leq K e^{\lambda(t-t_0)} ||\omega(t_0) - z(t_0)||$$

$$+ K \max_i l_f^i \max_i \sum_{j=1}^p |\alpha_{ij}| \int_{t_0}^t e^{\lambda(t-s)} ||\omega(s) - z(s)|| ds$$

$$+ K \max_i l_g^i \max_i \sum_{j=1}^p |\beta_{ij}| \sum_{t_0 \leq \theta_k < t} e^{\lambda(t-\theta_k)} ||\omega(\theta_k) - z(\theta_k)||.$$

Now, applying Gronwall-Bellman Lemma [33], one can attain that

$$||\omega(t) - z(t)|| \leq K ||\omega(t_0) - z(t_0)|| e^{(\lambda+D)(t-t_0)} (1 + E)^{k(t_0,t)},$$

where $D = K \max_i l_f^i \max_i \sum_{j=1}^p |\alpha_{ij}|$, and $E = K \max_i l_g^i \max_i \sum_{j=1}^p |\beta_{ij}|$. From the last inequality it follows that

$$||\omega(t) - z(t)|| \leq K ||\omega(t_0) - z(t_0)|| e^{(\lambda+D+\frac{1}{\theta}ln(1+E))(t-t_0)} \tag{19}$$

for $t \geq t_0$.

Consequently, $(C7)$ implies that $\omega(t)$ is the unique asymptotically stable discontinuous unpredictable oscillation of INNHS (1).     □

## 4   Examples

**Example 1.** Consider the logistic map

$$\nu_{k+1} = \mu \nu_k (1 - \nu_k),\ k \in \mathbb{Z} \tag{20}$$

with $\mu = 3.92$, in the interval $[0, 1]$. Then there exists the unpredictable solution $\tau_k, k \in \mathbb{Z}$, [24]. And there exist a positive number $\epsilon_0$ and sequences $\zeta_n, \eta_n$, both of which diverge to infinity such that $|\tau_{k+\zeta_n} - \tau_k| \to 0$ as $n \to \infty$, for each $k$ in bounded intervals of integers and $|\tau_{\eta_n+\zeta_n} - \tau_{\eta_n}| \geq \epsilon_0$ for each $n \in \mathbb{N}$.

Consider the sequence $\theta_k, k \in \mathbb{Z}$, which is defined by relations

$$\theta_k = 5k + \tau_k,\ k \in \mathbb{Z}. \tag{21}$$

Since the sequence (21), is of the form (2) with $T = 5$, it is true that $|\theta_{k+l_n} - t_n - \theta_k| \to 0$ as $n \to \infty$ for each $k$ in bounded intervals of integers and $|\theta_{m_n+l_n} - t_n - \theta_{m_n}| \geq \epsilon_0$ for each natural number $n$, where $t_n = 5\zeta_n$, $l_n = \zeta_n$, and $m_n = \eta_n$. It means that the sequence $\theta_k$ is unpredictable discrete set.

Consider $\Omega(t) = \tau_k$ for $t \in [\theta_k, \theta_{k+1})$, $k \in \mathbb{Z}$. Let us prove that $\Omega(t)$ is a discontinuous unpredictable function.

One can show that $t + t_n \in [\theta_{k+\zeta_n}, \theta_{k+1+\zeta_n})$, $k \in \mathbb{Z}$, if $t \in [\theta_k, \theta_{k+1})$, $k \in \mathbb{Z}$. It can be verified that $\theta'_k \leq t < \theta'_{k+1}$, implies $\theta'_{k+\zeta_n} \leq t+t_n < \theta'_{k+1+\zeta_n}$. That is, the discontinuity points of $\Omega(t + t_n)$ are that ones for $\Omega(t)$. Let us denote them $\theta'_k = \theta_{k+\zeta_n} - t_n$. Accordingly, for all $k \in \mathbb{Z}$ and $n \in \mathbb{N}$, the value of function $\Omega(t+t_n)$ is equal to $\tau_{k+\zeta_n}$. Hence, by using the unpredictability of $\tau_k$, we obtain that $|\Omega(t + t_n) - \Omega(t)| = |\tau_{k+\zeta_n} - \tau_k| \to 0$, as $n \to \infty$ on bounded intervals of time. Moreover, the values $\Omega(t) = \tau_{\eta_n}$ and $\Omega(t+t_n) = \tau_{\eta_n+\zeta_n}$, if $t \in [\theta_{\eta_n}, \theta_{\eta_n+1})$. Consequently, we have that $|\Omega(t + t_n) - \Omega(t)| = |\tau_{\eta_n+\zeta_n} - \tau_{\eta_n}| \geq \epsilon_0$.

Accordingly, $\Omega(t)$ is d.u.f. with $\epsilon_0$, $\sigma = 2$ and sequences $t_n = 5\zeta_n$, $s_n = \frac{\theta_{\eta_n} + \theta_{\eta_n+1}}{2}$, $n \in \mathbb{N}$.

**Example 2.** Let us introduce the following INNHS,

$$x'_i(t) = a_i x_i(t) + \sum_{j=1}^{3} \alpha_{ij} f_j(x_j(t)) + v_i(t),$$

$$\Delta x_i|_{t=\theta_k} = b_i x_i(\theta_k) + \sum_{j=1}^{p} \beta_{ij} g_j(x_j(\theta_k)) + W_{ik}, \qquad (22)$$

where $i = 1, 2, 3$, and $a_1 = -0.2, a_2 = 0.02, a_3 = -0.4$, $b_1 = e^{-6} - 1, b_2 = e^{-2.2} - 1, b_3 = e^{-3.5} - 1$, $f(x(t)) = \frac{2}{25}\sin(x(t))$, $g(x(t)) = \frac{1}{20}arctg(x(t))$,

$$\begin{pmatrix} \alpha_{11} \ \alpha_{12} \ \alpha_{13} \\ \alpha_{21} \ \alpha_{22} \ \alpha_{23} \\ \alpha_{31} \ \alpha_{32} \ \alpha_{33} \end{pmatrix} = \begin{pmatrix} 0.3 \ 0.7 \ 0.4 \\ 0.2 \ 0.5 \ 0.1 \\ 0.6 \ 0.3 \ 0.2 \end{pmatrix}, \quad \begin{pmatrix} v_1(t) \\ v_2(t) \\ v_3(t) \end{pmatrix} = \begin{pmatrix} 0.08\Omega^3(t) \\ 0.09\Omega(t) - 0.5 \\ -0.18\Omega(t)^3 - 0.1 \end{pmatrix},$$

$$\begin{pmatrix} \beta_{11} \ \beta_{12} \ \beta_{13} \\ \beta_{21} \ \beta_{22} \ \beta_{23} \\ \beta_{31} \ \beta_{32} \ \beta_{33} \end{pmatrix} = \begin{pmatrix} 0.6 \ 0.3 \ 0.1 \\ 0.9 \ 0.2 \ 0.3 \\ 0.1 \ 0.5 \ 0.4 \end{pmatrix}, \quad \begin{pmatrix} W_{1k}(t) \\ W_{2k}(t) \\ W_{3k}(t) \end{pmatrix} = \begin{pmatrix} 14\tau_k \\ -6\tau_k \\ 17\tau_k - 0.4 \end{pmatrix},$$

with $\tau_k$, $\theta_k$ are the unpredictable sequences, and $\Omega(t)$ is function from Example 1. The system (22) admits eigenvalues $\lambda_1 = -1.4$, $\lambda_2 = -0.42$, and $\lambda_3 = -1.1$. We have checked that condition ($C1$) is valid for the system with $\lambda = -0.42$ and $K = 1.4$. Moreover, the conditions ($C2$)–($C7$) hold for system (22) with $l_f = 0.08$, $l_g = 0.05$, $m_f = 0.112$ $m_g = 0.109$ and $H = 2.8$. On account of the Theorem 1, there exists the unique discontinuous unpredictable oscillation, $x(t) = (x_1(t), x_2(t), x_3(t))$, of INNHS (22). Moreover, it is asymptotically stable.

The simulation of the unpredictable oscillation $x(t)$ is not possible, because the initial value is not known precisely. For this reason, we will consider another oscillation $\psi(t) = (\psi_1(t), \psi_2(t), \psi_3(t))$, with initial value

$\psi(0) = (0.852, 0.447, 0.965)$. Using (19) one can obtain that $\|\psi(t) - x(t)\| \leq e^{-0.38t}\|\psi(t_0) - x(t_0)\|$, $t \geq 0$. The last inequality shows that the difference $\psi(t) - x(t)$ is decreasing exponentially. Hence, the graph of function $\psi(t)$ approaches to the discontinuous unpredictable oscillation $x(t)$ of the system (22), as $t$ increases. Then, one can consider the graph of $\psi(t)$ instead of the curve of unpredictable oscillation $x(t)$. The coordinates of the oscillation $\psi(t)$ are depicted in Fig. 1. Moreover, Fig. 2 presents the trajectory of this oscillation.

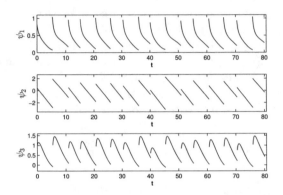

**Fig. 1.** The coordinates of function $\psi(t)$.

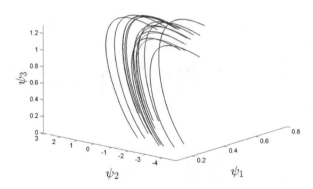

**Fig. 2.** The trajectory of function $\psi(t)$, which approximates the discontinuous unpredictable oscillation $x(t)$.

**Acknowledgements.** M. Akhmet has been supported by 239 2247—National Leading Researchers Program of TÜBITAK, Turkey, N 120C138. Z. Nugayeva has been supported by the Science Committee of the Ministry of Education and Science of the Republic of Kazakhstan (Grant No. AP08856170). M. Tleubergenova has been supported by the Science Committee of the Ministry of Education and Science of the Republic of Kazakhstan (Grants No. AP09258737 and No. AP08955400).

# References

1. Hopfield, J.J.: Neural networks and physical systems with emergent collective computational abilities. Proc. Natl. Acad. Sci. U.S.A. **79**, 2554–2558 (1982)
2. Hopfield, J.J.: Neurons with graded response have collective computational properties like those of two-stage neurons. Proc. Natl. Acad. Sci. U.S.A. **81**, 3088–3092 (1984)
3. Pajares, G.: A Hopfield neural network for image change detection. IEEE Trans. Neural Networks **17**, 1250–1264 (2006)
4. Ramya, C., Kavitha, G., Shreedhara, K.S.: Recalling of images using Hopfield neural network model. In: National Conference on Computers Communication and Controls, vol. 11, pp. 2–7 (2011)
5. Soni, N., Sharma, E.K., Kapoor, A.: Application of Hopeld neural network for facial image recognition. IJRTE **8**, 3101–3105 (2019)
6. Sang, N., Zhang, T.: Segmentation of FLIR images by Hopfield neural network with edge constraint. Pattern Recogn. **34**, 811–821 (2001)
7. Amartur, S.C., Piraino, D., Takefuji, Y.: Optimization neural networks for the segmentation of magnetic resonance images. IEEE Trans. Med. Imaging **11**, 215–220 (1992)
8. Koss, J.E., Newman, F.D., Johnson, T.K., Kirch, D.L.: Abdominal organ segmentation using texture transforms and a Hopfield neural network. IEEE Trans. Med. Imaging **18**, 640–648 (1999)
9. Cheng, K.C., Lin, Z.C., Mao, C.W.: The application of competitive Hopfield neural network to medical image segmentation. IEEE Trans. Med. Imaging **15**, 560–567 (1996)
10. Akca, H., Alassar, R., Covachev, V., Covacheva, Z., Al-Zahrani, E.: Continuous-time additive Hopfield-type neural networks with impulses. J. Math. Anal. Appl. **290**, 436–451 (2004)
11. Akhmet, M.U., Yilmaz, E.: Impulsive Hopfield type neural network systems with piecewise constant argument. Nonlinear Anal. Real World Appl. **11**, 2584–2593 (2010)
12. Li, Y., Lu, L.: Global exponential stability and existence of periodic solution of Hopfield-type neural networks with impulses. Phys. Lett. A **333**, 62–71 (2004)
13. Mohammad, S.: Exponential stability in Hopfield-type neural networks with impulses. Chaos Solitons Fractals **32**, 456–467 (2007)
14. Pinto, M., Robledo, G.: Existence and stability of almost periodic solutions in impulsive neural network models. Appl. Math. Comput. **217**, 4167–4177 (2010)
15. Shi, P., Dong, L.: Existence and exponential stability of anti-periodic solutions of Hopfield neural networks with impulses. Appl. Math. Comput. **216**, 623–630 (2010)
16. Stamov, G.T., Stamova, I.M.: Almost periodic solutions for impulsive neural networks with delay. Appl. Math. Modelling **31**, 1263–1270 (2007)
17. Stamov, G.T.: Almost periodic models of impulsive Hopfield neural networks. J. Math. Kyoto Univ. **49**, 57–67 (2009)
18. Zhang, H., Xia, Y.: Existence and exponential stability of almost periodic solution for Hopfield-type neural networks with impulse. Chaos Solitons Fractals **37**, 1076–1082 (2008)
19. Khan, A., Salahuddin, S.: Negative capacitance in ferroelectric materials and implications for steep transistors. In: 2015 IEEE SOI-3D-Subthreshold Microelectronics Technology Unified Conference (S3S), pp. 1–3 (2015)

20. Khan, A.I., et al.: Negative capacitance in short-channel finfets externally connected to an epitaxial ferroelectric capacitor. IEEE Electron Device Lett. **37**, 111–114 (2016)
21. Akhmet, M., Fen, M.O.: Unpredictable points and chaos. Commun. Nonlinear Sci. Nummer. Simulat. **40**, 1–5 (2016)
22. Akhmet, M., Fen, M.O.: Existence of unpredictable solutions and chaos. Turk. J. Math. **41**, 254–266 (2017)
23. Akhmet, M., Fen, M.O., Tleubergenova, M., Zhamanshin, A.: Unpredictable solutions of linear differential and discrete equations. Turk. J. Math. **43**, 2377–2389 (2019)
24. Akhmet, M.U., Fen, M.O., Alejaily, E.M.: Dynamics with Chaos and Fractals. Springer, Cham (2020). https://doi.org/10.1007/978-3-030-35854-9
25. Akhmet, M.U.: Almost Periodicity, Chaos, and Asymptotic Equivalence. Springer, Cham (2020). https://doi.org/10.1007/978-3-030-20572-0
26. Miller, A.: Unpredictable points and stronger versions of Ruelle-Takens and Auslander-Yorke chaos. Topology Appl. **253**, 7–16 (2019)
27. Thakur, R., Das, R.: Strongly Ruelle-Takens, strongly Auslander-Yorke and Poincaré chaos on semiflows. Commun. Nonlinear Sci. Numer. Simul. **81**, 05018 (2019)
28. Akhmet, M.U., Tleubergenova, M., Nugayeva, Z.: Strongly unpredictable Oscillations of Hopfield-type neural networks. Mathematics **8**, 1791 (2020)
29. Akhmet, M.U., Tleubergenova, M., Zhamanshin, A.: Inertial neural networks with unpredictable oscillations. Mathematics **8**, 1797 (2020)
30. Akhmet, M.U., Tleubergenova, M., Fen, M.O., Nugayeva, Z.: Unpredictable solutions of linear impulsive systems. Mathematics **8**, 1798 (2020)
31. Akhmet, M.U., Tleubergenova, M., Nugayeva, Z.: Unpredictable solutions of impulsive quasi-linear systems. Discontinuity, Nonlinearity and Complexity (accepted)
32. Akhmet, M.U., Arugaslan Cincin, D., Tleubergenova, M., Nugayeva, Z.: Unpredictable oscillations for Hopfield-type neural networks with delay and advanced arguments. Mathematics **9**, 571 (2021)
33. Akhmet, M.U.: Principles of Discontinuous Dynamical Systems. Springer, New York (2010). https://doi.org/10.1007/978-1-4419-6581-3
34. Akhmet, M., Fen, M.O.: Non-autonomous equations with unpredictable solutions. Commun. Nonlinear Sci. Numer. Simul. **59**, 657–670 (2018)

# Performance Analysis of Particle Swarm Optimization and Firefly Algorithms with Benchmark Functions

Mervenur Demirhan[1], Osman Özkaraca[1($\boxtimes$)], and Ercüment Güvenç[2]

[1] Department of Information Systems Engineering, Muğla Sıtkı Koçman University, Mugla, Turkey
osmanozkaraca@mu.edu.tr
[2] Department of Computer Engineering, Süleyman Demirel University, Isparta, Turkey

**Abstract.** In recent years swarm mentality algorithms are usually created with nature inspiration to keep their popularity. One of these optimization techniques is particle swarm optimization, and the other one is firefly algorithm. Firefly algorithm process is working with the lower light intensity directed to higher intensities principle. Particle swarm optimization based on the positions of individuals; swarm keeps following the individual who have great position. This article explains with mathematical testing functions of minimum international points, particle swarm optimization and firefly algorithm. Tried to specify that which function is working better with which algorithm. Also, this research tried to recognize that different parameters are changing the result or not.

**Keywords:** Benchmark function · Firefly algorithm · Particle swarm optimization

## 1 Introduction

Optimization is the process of finding best solution with certain restrictions for purpose or purposes. In the recent years intuitive algorithms are used for the solve problems of separate and constant optimizations. Intuitive algorithms are very useful for the solve optimization mistakes which have not enough mathematical method. A lot of scientific and engineering problems can solve with mathematical methods, and those methods have plenty subjects. Despite these mathematical methods are still not enough to solve optimization problems. So recent years researchers started to find new algorithms which have inspirational by nature. Then a lot of algorithm invented and intuitive algorithms are inspired by nature. In these subjects we can mention about important algorithm in the name of particle swarm optimization (PSO). This algorithm is based on population. According to this algorithm particle number is all about the swarm (population) and those particles are accepted for the solution of this problem. Particle population is moving in the space of the problem, and tries to find the best solution with collective experiences. PSO is a optimization algorithm that changing each particles location in the time. Another algorithm, firefly algorithm (FFA) is inspired by the optical connection of fireflies in

© The Author(s), under exclusive license to Springer Nature Switzerland AG 2021
J. Hemanth et al. (Eds.): ICAIAME 2020, LNDECT 76, pp. 643–653, 2021.
https://doi.org/10.1007/978-3-030-79357-9_60

nature. This algorithm can explain like each individual in swarm moving to the spot where he can improve himself faster to live the best experiences. In the original version of this algorithm's imperfection is improved with firefly swarm strategy in the new versions [1]. These articles mathematical test functions were calculated with international minimum particle swarm optimization and original firefly algorithm. Purpose is finding the true function for true algorithm and it is detected that different parameters are changing the result or not.

## 2  Literature Review

Optimization algorithms are generally created by taking inspiration from nature. The main goal of these algorithms created is to model the solution space using all the real values in the solution set to minimize or maximize a given real function. There are many studies in the literature that have developed solutions to problems in real life using optimization algorithm. Heuristic algorithms are one of them. In computer science, it is an intuitive or heuristic problem solving technique. It does not care whether the accuracy of the result is provable or not, but it usually gets close to good solutions. The PSO and Firefly algorithm, which are heuristic methods, have many examples in the literature. The PSO technique was first proposed by sociologist-psychologist James Kennedy and electrical engineer Russell Eberhart in 1995–1996 to find optimal results for nonlinear numerical problems, inspired by the movements of flocks of birds and fish [2].

These particles try to reach the best value iteratively and in relation to each other (the universal best particle value is reported to all particles). It is possible to find many applications of the PSO algorithm in different fields in the literature. Pluhacek et al. (2015) have worked on the PSO algorithm to show that meta-heuristic performance improves through the application of chaotic sequences. The authors combined the PSO algorithm with chaotic random numbers. The authors have shown that the chaotic PSO method is more successful [3]. Kansal et al. (2013) solved the problem of best placement of different types of distributed generation units such as PV units, fuel cells and synchronous generators using PSO. They used 33-busbar and 69-busbar systems in their work. In addition, the authors have minimized system power losses with this study [4].

Yekrangi (2011) used hybrid neural networks and Particle Swarm Optimization (PSO) to solve the simple pendulum problem known in mechanics [5]. In another study, Raja et al. (2014) solved 2-dimensional Bratu equations using the feed forward ANN model and optimized it with PSO and sequential quadratic programming (SQP) optimization methods [6]. Raja (2014) obtained the solution of the Troesch problem via ANN and optimization method [7]. In his studies, PSO used the active-cluster and optimization approach created by hybridizing these two algorithms.

The PSO algorithm used in many fields in the literature has also been applied in the field of Economics. Alagöz and Kutlu [8] performed portfolio optimization by addressing commodity market products with particle herd optimization technique in their study. Imran, Hashima and Khalidb [9] gave information about the PSO with their work and studied the PSO algorithm. Garlenli, Egrioglu and Chorbaaa [10] worked on portfolio optimization using the particle herd optimization method with stocks in the IMKB 30 index.

The firefly algorithm is another algorithm that used in this study. It is a meta-heuristic algorithm proposed by Xin-She Yang in 2009, inspired by blinking behaviors. In 2010, Xin-She Yang presented a study comparing the Firefly algorithm with other meta-heuristic algorithms [11]. This article shows that the proposed Firefly algorithm is superior to existing meta-heuristic algorithms. In 2013, Shuhao Yu and colleagues proposed the self-adaptive step Firefly algorithm [12]. To avoid falling into the local optimum and reduce the effect of maximum iteration, a self-adaptive step Firefly algorithm is proposed in the article. Gandomi et al. (2011) used ABA, which is an upper heuristic algorithm, for continuous and discrete structural optimization problems. They verified the validity of ABA with the results of studies on six structural optimization problems in the literatüre [13]. Yang and He (2013) focused on the fundamentals of ABA. They conclude that ABA is better than optimum batch search methods compared to discrete search methods. They also made comparisons for high dimensional optimization problems [14].

## 3 Particle Swarm Optimization (PSO)

The PSO technique is a population-based probabilistic optimization method. It is generally used in solutions of multivariate optimization problems. PSO technique has similarities with evolutionary calculation methods. The system is first started by generating random solutions, and generations are updated from these populations, and the optimum solution is sought. Possible solutions, expressed as particles in the PSO method, then act in the problem space, following the optimum particle at that moment. It is a simulation of the movements of the fish and flocks of the PSO method. For example, when birds are flying, searching for food on the ground is compared to seeking a solution for a particular problem. When birds search for food, they follow the other bird that is closest to the food. Each solution, expressed as a particle, is like a bird trying to find the location of the food while flying in the search space. When each of these particles goes into search, it directs its position and coordinates to a function and the suitability value of this particle is calculated. Briefly, its distance from the food is measured.

Representative particles act according to the experience of themselves and other particles, so that the particles exchange information among themselves. The equations of the PSO algorithm are seen in Eqs. 1 and 2. The particles mentioned in the algorithm point to individuals in the swarm [15].

$$v_{i+1} = v_i + c_1 \times rand_1 \times (p_{best} - x) + c_2 \times rand_2 \times (g_{best} - x) \tag{1}$$

$$x_i(k + 1) = x_i(k) + v_i(k + 1) \tag{2}$$

In this equation, "x" particle value, "v" particle's rate of change, $c_1$ and $c_2$: constant values, $rand_1$ and $rand_2$ randomly generated values, the state where the $p_{best}$ particle is closer to solution, $g_{best}$ is the most approximated state among all particles with the following formula calculated. Thanks to this formula, the particle turns to its best solution and to the global best solution. This forces the particle to search for the solution near the best particle and its best state. The pseudo code of the Particle Swarm optimization algorithm is given in Fig. 1.

```
1    Initialize population
2    for t = 1 : maximum generation
3       for i = 1 : population size
4          if f(x_{i,d}(t)) < f(p_i(t))  then  p_i(t) = x_{i,d}(t)
5             f(p_g(t)) = min_t (f(p_i(t)))
6       end
7       for d = 1 : dimension
8          v_{i,d}(t+1) = wv_{i,d}(t) + c_1 r_1 (p_i - x_{i,d}(t)) + c_2 r_2 (p_g - x_{i,d}(t))
9          x_{i,d}(t+1) = x_{i,d}(t) + v_{i,d}(t+1)
10         if  v_{i,d}(t+1) > v_max  then  v_{i,d}(t+1) = v_max
11         else if  v_{i,d}(t+1) < v_min  then  v_{i,d}(t+1) = v_min
12         end
13         if  x_{i,d}(t+1) > x_max  then  x_{i,d}(t+1) = x_max
14         else if  x_{i,d}(t+1) < x_min  then  x_{i,d}(t+1) = x_min
15         end
16      end
17   end
18 end
```

**Fig. 1.** PSO algorithm pseudo code

PSO is a flexible algorithm that can use different values according to different problem types, sizes and variables. Basic parameters are summarized in Table 1 below.

**Table 1.** Description of the parameters of PSO

| Parameters | Description |
|---|---|
| Particle number (herd size) | It generally takes a value between 20 and 40 |
| Particle size | Indicates the number of variables in the problem |
| Learning factors | Express the learning coefficients in the formula. They usually take the '2' value |
| Iteration number | The number of iterations also depends on the problem to achieve a good solution |
| Particle spacing | Particles of different sizes and ranges, which vary depending on the problem we are looking for, can be identified |
| Vmax | It determines the maximum change in the particle and speed, as a result of each iteration |

## 4   Firefly Algorithm (FFA)

Fireflies are known for their flashing lights while flying in spring and summer and there are about two thousand species of insects in nature. Fireflies carry out actions such as breeding, hunting and protection by affecting the opposite sex thanks to their ability to create cold light as a result of some chemical reactions in their bodies, that is, to emit light at short intervals. The firefly optimization algorithm is also the swarm-intelligence approach that optimizes [16]. This algorithm addresses the movement of fireflies in nature to each other or in a random direction, depending on the attractiveness of the light. Three assumptions are accepted to better understand the firefly algorithm. We accept that fireflies got no sex. So one firefly totally can affect the others. Charm is related to the light intensity of the firefly. In this case, from two different light-emitting fireflies, the light intensity moves from the smaller to the brightest. The distance between the fireflies affects the brightness. If the brightness level is equal, the fireflies can move randomly. Light intensity obtained from a light source (Is), r, according to the inverse square law, (I (r)) [17].

$$I(r) = I_S \Big/ r^2 \qquad (3)$$

While light is emitted in an environment, light intensity is absorbed to a certain extent. Therefore, when a constant light absorption coefficient (ɣ) is taken into account, Eq. 4 is obtained. The intensity of the light source when $r = 0$. In dividing the number zero, the distance can be written as a gauss distribution so that Eq. 3 is not in an undefined state.

Thus, the attractiveness of the firefly is calculated by Eq. 4. The charm varies depending on the distance. B0 is the amount of attraction of a firefly when the distance to the other neighboring firefly is $r = 0.B(r)$, on the other hand, is the attractiveness of the firefly with B0 attractiveness at a distance r.

$$B(r) = B_0 e^{-\gamma r^2} \qquad (4)$$

Let $i$ and $j$ be two fireflies and their positions $X_i$ ($x_i$, $y_i$) and $X_j$ ($x_j$, $y_j$) in the two-dimensional plane, respectively. The distance $(r_{i,j})$ between them is calculated by the Euclidean relation, that is Eq. 5.

$$r_{i,j} = \|x_i - x_j\| = \sqrt{\sum_{k=1}^{d}(x_{i,k} - x_{j,k})^2} \qquad (5)$$

Thus, the new position ($X_i$) of a firefly (i) directed towards the more attractive and bright (j) is calculated as in Eq. 6.

$$X_i = X_i + B_0 e^{-\gamma r_{ij}^2}(X_j - X_i) + a\left(rand - \frac{1}{2}\right) \qquad (6)$$

It is the coefficient parameter in the equation that takes a constant value in the $\alpha$ [0,1] range. "*rand*" takes a random value between [0,1]. $B_o$, on the other hand, is the main attraction factor and generally takes value as $B_o = 1$.

## 5   FFA Parameters

The definition of the basic parameters of the firefly algorithm is shown in Table 2 below.

**Table 2.** FFA parameters

| Parameters | Descriptions |
|---|---|
| Particle number (herd size; n) | Usually it takes a value between 20–40 |
| Particle size (d) | Problemdeki değişken sayısını gösterir |
| İteration number | The number of iterations also depends on the problem to achieve a good solution |
| B | Attraction of firefly |
| I | Light intensity |
| γ (gamma) | It is the constant absorption coefficient of light. It usually takes 0.01 and 100 values |
| R | Distance between fireflies |
| α (alpha) | It is a random variable. It usually takes values between 0 and 1 |

The pseudo code of the firefly algorithm is shown in Fig. 2.

Objective function  $f(x)$ ,  $x=(x_1,\ldots\ldots,x_D)^T$
Initialize positions of fireflies  $x_i$  ( $i=1,2,\ldots\ldots,M$ )
Calculate Light intensities  by  $I_i=f(x_i)$
**while** ( $t< $ MaxGeneration $t_{max}$ ) **do**
       **for**  $i = 1$  to M, all M fireflies **do**
          **for**  $j = 1$  to M, all M fireflies **do**
             **if**  $I_j > I_i$  **then**
                 Move firefly  i  toward j by Eq.(8)
             **end if**
          **end for** j
       **end for** i
       Evaluate new solutions  $f(x_i)$
       Rank the fireflies and find the current global
          best g.
**end while**

**Fig. 2.** Firefly algorithm and pseudo code

## 6   Test Functions

This part explains nonlinear benchmark functions which used in the algorithms. Table 3. shows the smallest and largest ranges of functions, formulas, sizes and optimum points of functions.

**Table 3.** Test functions

| Name | Size | Formula | Intervals | Optimum point |
|------|------|---------|-----------|---------------|
| Beale | 2 | $f(x) = (1.5 - x_1 + x_1 x_2)^2$ $+ \left(2.25 - x_1 + x_1 x_2^2\right)^2$ $+ \left(2.625 - x_1 + x_1 x_2^3\right)^2$ | $[-4.5, 4.5]$ | $f(x^*) = 0,$ $x^* = (3,\ 0.5)$ |
| Booth | 2 | $f(x) = (x_1 + 2x_2 - 7)^2$ $+ (2x_1 + x_2 - 5)^2$ | $[-10, 10]$ | $f(x^*) = 0,$ $x^* = (1,\ 3)$ |
| Matyas | 2 | $f(x) = 0.26\left(x_1^2 + x_2^2\right)$ $- 0.48 x_1 x_2$ | $[-10, 10]$ | $f(x^*) = 0,$ $x^* = (0,\ 0)$ |
| Levy no: 13 | 2 | $f(x) = sin^2(3\pi x_1)$ $+ (x_1 - 1)^2\left[1 + sin^2(3\pi x_2)\right]$ $+ (x_2 - 1)^2\left[1 + sin^2(2\pi x_2)\right]$ | $[-10, 10]$ | $f(x^*) = 0,$ $x^* = (1,\ 1)$ |
| Schaffer no: 2 | 2 | $f(x) = 0.5 + \dfrac{sin^2\left(x_1^2 - x_2^2\right) - 0.5}{\left[1 + 0.001\left(x_1^2 + x_2^2\right)\right]^2}$ | $[-100, 100]$ | $f(x^*) = 0,$ $x^* = (0,\ 0)$ |

## 7   Test Results

Original firefly and particle flock optimizations developed with the first randomly generated flock-particles within the maximum and minimum limits specified in the functions were tested with different parameters. If the value of the fitness particle in the population produced in each iteration is better than the previous fitness, its value will be replaced by the new one. This way it will continue until you find the best result. Parameters in Tables 4 and 5. are preferred original parameter values for algorithms. The reason for using different parameters in the study is to see if there are any changes in the success of the algorithms and to see which parameters make the algorithm more successful and what kind of results it gives in which function. The tables below contain the values of the parameters given to the algorithms and the results of the application.

PSO$_{best}$ and FFA$_{best}$ values in Tables 5 and 7. are selected from the values obtained from the solution of the best optimum value to the x and y points. As a result of the experiments, Beale and Matyas functions in the algorithms calculated using the parameters in Table 4. gave better results than FFA since it is closest to the global optimum point in the PSO algorithm. Levy, Schaffer and Booth functions are more successful in the firefly algorithm.

In another experiment, Beale, Matyas, Schaffer and Booth functions in the algorithms calculated using the parameters in Table 6. gave quite successful results in the PSO algorithm. The Levy function is closer to the optimum point in the firefly algorithm.

**Table 4.** Parameters which used on algorithms in the experiment

| Parameter of the FFA algorithm | | Parameter of the PSO algorithm | |
|---|---|---|---|
| Size | 2 | Size | 2 |
| Population size | 20 | Population size | 20 |
| $\alpha$ | 0.5 | c1, c2 learning coefficients | 2 |
| $\beta$ | 0.2 | Weight | 1.2 |
| $\gamma$ | 1.0 | Vmax | 4 |

**Table 5.** Algorithms tested using the parameters in Table 4 according to the functions

| Func. | N | PSO | FFA | PSO Best | ABO Best |
|---|---|---|---|---|---|
| Beale | 1000 | 0.0547762290806 | 0.03096266637283375 | 0.0298903896321 | 0.03096266637283375 |
| | | 0.0298903896321 | 0.024370090104399673 | | |
| | | 0.0427354392981 | 0.024415606399141893 | | |
| | | 0.0419301414284 | 0.023357123763673063 | | |
| | | 0.0428461169585 | 0.02335672840431122 | | |
| Matyas | 800 | 0.000134021607983 | 0.023476665553464615 | 0.000134021607983 | 0.023476665553464615 |
| | | 0.000432696884018 | 0.015256149819020265 | | |
| | | 0.000718020519985 | 0.00044334745613607007 | | |
| | | 0.00106106594181 | 0.00018133024783608797 | | |
| | | 0.00317862315557 | 0.00021127184463437662 | | |
| Levy N.13 | 800 | 0.00153851211201 | 0.000233303468351 | 0.000956471438765 | 0.000233303468351 |
| | | 0.00700051367469 | 0.00688569286481 | | |
| | | 0.0149503281706 | 0.000142801112698 | | |
| | | 0.000956471438765 | 0.000341703889844 | | |
| | | 0.00128009262753 | 0.00678515714984 | | |
| Schaffer | 1000 | 0.417087503337 | 0.00195508438872 | 0.00169556805271 | 0.000131766806753 |
| | | 0.00169556805271 | 0.00874173747945 | | |
| | | 0.0571953085239 | 0.000131766806753 | | |
| | | | 0.00179821097237 | | |
| | | | 0.00488085022578 | | |
| Booth | 1000 | 0.00136440089647 | 0.00029040030386769794 | 0.00129418949453 | 0.00029040030386769794 |
| | | 0.00173328347936 | 0.00032608903205272235 | | |
| | | 0.0017222085764 | 0.00018995780962498227 | | |
| | | 0.00129418949453 | 0.00023558915766899003 | | |
| | | 0.00267355479017 | 0.0039048262879537122 | | |

**Table 6.** Parameters which used on algorithms in the experiment

| Parameter of the FFA algorithm | | Parameter of the PSO algorithm | |
|---|---|---|---|
| Size | 2 | Size | 2 |
| Population size | 20 | Population size | 20 |
| α | 0.2 | c1, c2 learning coefficients | 1.494 |
| β | 0.2 | Weight | 0.729 |

**Table 7.** The results of the algorithms tested using the parameters in Table 6

| Func. | N | PSO | FFA | PSO Best | ABO Best |
|---|---|---|---|---|---|
| Beale | 1000 | 0.0238946561439 | 0.06818679604822045 | 0.0233713328471 | 0.06818679604822045 |
| | | 2.42138608108 | 0.023357746476551082 | | |
| | | 0.0594447938021 | 0.024355761735505407 | | |
| | | 0.0233713328471 | 0.028137422825438213 | | |
| | | 1.09388349531 | 0.02598878297591592 | | |
| Matyas | 800 | 2.07766295113e-06 | 0.000496036227057968 | 0.000373394815637 | 0.000496036227057968 |
| | | 0.00191322236594 | 0.0012830785038233459 | | |
| | | 4.30637724068e-06 | 0.013895147884192355 | | |
| | | 0.00132545591172 | 0.015038431276051867 | | |
| | | 0.000373394815637 | 0.0006877791437066103 | | |
| Levy N.13 | 800 | 0.00161531132093 | 0.000692973488068 | 0.00161531132093 | 0.000104167745552 |
| | | 0.00218892999561 | 0.0531980732794 | | |
| | | 0.221363393009 | 0.000104167745552 | | |
| | | 2.25587152998 | 0.109873685847 | | |
| | | 0.0219945059624 | 0.00115973150719 | | |
| Schaffer | 1000 | 2.63764157797e-05 | 0.000551692229966 | 0.000110895036989 | 0.000262181140018 |
| | | 5.48357779084e-06 | 0.00050810780592 | | |
| | | 0.000110895036989 | 0.000262181140018 | | |
| | | 0.103386607641 | 0.00139369086509 | | |
| | | 0.00912133839777 | 0.00033883179579 | | |
| Booth | 1000 | 0.03698049391263765 | 0.00295485284278537 | 0.00735877871669 | 0.00874642070772932 |
| | | 0.0715936556688 | 0.008247992263569506 | | |
| | | 0.00478233529989 | 0.00874642070772932 | | |
| | | 0.00735877871669 | 0.0006855400135100762 | | |
| | | 0.603035578376 | 0.0003243819036575154 | | |

During the tests carried out, the firefly algorithm had difficulty in finding the optimum point of the matyas function. Schaffer found the optimum point of the function a little earlier than matrix. Table 8 below shows which algorithms the functions perform best.

**Table 8.** Successes in the solution of algorithms according to parameters

|  | The most successful algorithm according to the result of Table 5 | The most successful algorithm according to the result of Table 7 |
|---|---|---|
| Beale | PSO | PSO |
| Matyas | PSO | PSO |
| Booth | FFA | PSO |
| Schaffer | FFA | PSO |
| Levy | FFA | ABO |

## 8 Result and Discussion

Meta-heuristic algorithms, which we have frequently heard of in recent years, are one of the heuristic methods that can yield highly efficient results in the search for solutions for problems. Looking at the literature, it finds hundreds of meta-intuitive algorithms. In this study, particle flock optimization and firefly algorithm were used among meta-heuristic algorithms. The first randomly generated particles/flocks were used on non-linear test functions (Beale, Matyas, Booth, Schaffer, Levy) to find the global optimum point. The goal of any global optimization is to find the best possible solutions. If the search process of the algorithm is poorly designed, it cannot search the function part effectively. This causes an algorithm to get stuck at the local minimum. Depending on the area of the search area, there may be more than one or even an infinite number of optimal solutions. Multimodal functions with many local minima are among the most difficult problem classes for many algorithms. Objective functions can be characterized as continuous, discontinuous, linear, nonlinear, convex, non-external, single mode, multimodal, detachable and non-separable. Functions with flat surfaces make it difficult for the algorithm to find the solution, because the flatness of the function does not provide any information to the algorithm to direct the search process to the minima (Matyas function). The global minimum for problems such as the Schaffer function is very close to the local minima. If functions such as Beale cannot effectively explore the search area, the algorithm fails for such problems, however beale has only a minimum value, which is satisfactorily achieved by all methods. Further iterations will improve their accuracy. The tasks of any global optimization algorithm are to find globally optimal or at least sub-optimal solutions. This article finds and compares global minimum points on the benchmark functions of firefly and particle flock optimization with different parameters. As it can be seen in Table 8. Beale and Matyas function was more successful in PSO algorithm in both different parameters. The parameters in Table 6 had a bad effect on the success of the Booth and Schaffer functions and made the PSO algorithm more successful. Levy function has been more successful in FFA in both different parameters. As a result, the changes in the parameters of the Beale, Matyas and Levy functions did not affect the success. But the Schaffer and Booth function did not achieve this success. In general terms, the particle flock optimization algorithm yielded more successful results than the original firefly algorithm.

# References

1. Merugumalla, M.K., Navuri, P.K.: PSO and Firefly Algorithms based control of BLDC motor drive. In: 2nd International Conference on Inventive Systems and Control (ICISC), Coimbatore, pp. 994–999 (2018). https://doi.org/10.1109/ICISC.2018.8398951
2. Kennedy, J., ve Eberhart, R.: Particle swarm optimization. In: Proceedings of the 1995 IEEE International Conference on Neural Networks, New Jersey, pp. 1942–1948 (1995)
3. luhacek, M., Senkerik, R., Zelinka, I.: PSO algorithm enhanced with Lozi Chaotic Map-Tuning experiment. In: AIP Conference Proceedings, vol. 1648 (2015). https://doi.org/10.1063/1.4912777
4. Kansal, S., Kumar, V., ve Tyagi, B.: Hybrid approach for optimal placement of multiple DGs of multiple types in distribution networks. Int. J. Electr. Power Energy Syst. **75**, 226–235 (2016)
5. Yekrangi, A., et al.: An approximate solution for a simple pendulum beyond the small angles regimes using hybrid artificial neural network and particle swarm optimization algorithm. Procedia Eng. **10**, 3734–3740 (2011)
6. Raja, M.A.Z., Ahmad, S.I., Samar, R.: Solution of the 2-dimensional Bratu problem using neural network swarm intelligence and sequential quadratic programming. Neural Comput. Appl. **25**(7–8), 1723–1739 (2014)
7. Raja, M.A.Z.: Stochastic numerical treatment for solving Troesch's problem. Inf. Sci. **279**, 860–873 (2014)
8. Alagöz, A., Kutlu, M.: Portföy Optimizasyonu Yaklaşımı İle Emtia Piyasasında Portföy Optimizasyonu. Sosyal Ekonomik Araştırmalar Dergisi **12**, 35–50 (2012)
9. Imran, M., Hashima, R., Khalidb, N.E.A.: An overview of particle swarm optimization variants. Procedia Eng. **53**, 491–496 (2013)
10. Çelenli, A.Z., Eğrioğlu, E., Çorba, B.Ş: İMKB 30 İndeksini Oluşturan Hisse Senetleri İçin Parçacık Sürü Optimizasyonu Yöntemlerine Dayalı Portföy Optmizasyonu. Doğuş Üniversitesi Dergisi **16**(1), 25–33 (2015)
11. Yang, X.-S.: Firefly algorithms for multimodal optimization. In: Watanabe, O., Zeugmann, T. (eds.) SAGA 2009. LNCS, vol. 5792, pp. 169–178. Springer, Heidelberg (2009). https://doi.org/10.1007/978-3-642-04944-6_14
12. Yu, S., Yang, S., Su, S.: Self-adaptive step firefly algorithm. J. Appl. Math. (2013)
13. Gandomi, A.H., Yang, X.S., Alavi, A.H.: Mixed variable structural optimization using firefly algorithm. Comput. Struct. **89**(23–24), 2325–2336 (2011)
14. Yang, X.S., ve He, X.: Firefly algorithm: recent advances and applications. arXiv preprint arXiv:1308.3898 (2013)
15. Clerc, M.: From theory to practice in particle swarm optimization. In: Panigrahi, B.K., Shi, Y., Lim, M.H. (eds.) Handbook of Swarm Intelligence. Adaptation Learning and Optimization, vol. 8, pp. 3–36. Springer, Berlin, Heidelberg (2011). https://doi.org/10.1007/978-3-642-173 90-5_1
16. Aydilek, İB.: Değiştirilmiş ateşböceği optimizasyon algoritması ile kural tabanlı çoklu sınıflama yapılması. J. Facul. Eng. Architect. Gazi Univ. **32**(4), 1097–1107 (2017)
17. Yang, X.S.: Firefly algorithm, Levy flights and global optimization. In: Bramer, M., Ellis, R., Petridis, M. (eds.) Research and Development in Intelligent Systems XXVI, pp. 209–218. Springer, London (2010). https://doi.org/10.1007/978-1-84882-983-1_15

# Optimization in Wideband Code-Division Multiple Access Systems with Genetic Algorithm-Based Discrete Frequency Planning

Sencer Aksoy and Osman Özkaraca[✉]

Department of Information Systems Engineering, Muğla Sıtkı Koçman University, Mugla, Turkey
osmanozkaraca@mu.edu.tr

**Abstract.** One of the most serious problems is how the limited number of requencies (carriers) allocated to operators in broadband code division multiple access systems can be distributed to the base station cells in a way that captures the least interference value. This frequency planning is generally used for 2G network. In accordance with its technology, 3.5G systems are defined as interference independent. However, in this study, by applying GA algorithm for frequency planning in Wideband Code Division Multiple Access Systems (WCDMA-3.5G-3.75G) and by trying to obtain the best frequency plan, it was analyzed how the change occurred as a result of this planning. Actual system information was used in the study. Consequently, it was observed that GA decreased the fitness value of 268000 to 53999 levels according to the current plan working in the field with 1% mutation and 80% crossing rate and an improvement of 79% was achieved. Consequently, on the contrary of what is explained theoretically, it is seen that with a better frequency plan, better quality service continuity is provided in cases where interference is reduced.

**Keywords:** Frequency planning · Genetic Algorithm · GSM · UMTS · 3.5G

## 1 Introduction

Mobile technologies have started to increase their impact and their usage areas on the world day by day. Therefore, both GSM operators and product supplier companies realized that efficient networks should be planned as well as efficient design. Cell planning, which is one of the most common problems in mobile networks due to the increasing number of customers and service diversity, has gained importance due to the expanding capacity. The location and capacity of new station to meet expanding and increasing traffic demands emerges as a cell planning problem. Looking at the overall work done, the aim is to minimize the investment cost by optimizing existing base stations instead of making new stations in the first place. In studies, cell planning is carried out to meet the subscriber's coverage and traffic demands [1]. When the literature on this subject is examined, the use of artificial intelligence and similar methods in this field is increasing

J. Hemanth et al. (Eds.): ICAIAME 2020, LNDECT 76, pp. 654–668, 2021.
https://doi.org/10.1007/978-3-030-79357-9_61

day by day. Artificial intelligence and statistical methods such as artificial neural networks, deep learning, fuzzy logic, genetic algorithms, ant colonies, simulated annealing, and intuitional algorithms are the most prominent under the artificial intelligence title used in the solution of difficult problems that could not be put into a complete mathematical model. It is possible to see that existing algorithms have been modified or applied to different areas in order to find a more accurate and faster solution in the studies carried out in this topic [2]. In the literature, it is possible to find many different R&D based studies on this subject. For example, researchers named Vladan Jovanovic and Stefan Scheinert have patented the cell planning method for a special wireless communication system they have developed [3]. In this method, a technique has been developed that determine show to share the load from the cells with higher occupancy to those with more suitable occupancy. Looking at the academic studies, the focus is mainly on the quality of mobile users' base station-mobile mobile phone (downlink) connection. Additionally, in cell planning, mobile phone-base station connection (uplink) service quality of mobile users is also taken into consideration. In their study, Mohamed Kashef et al. suggested a balanced dynamic planning approach that meets both uplink and downlink quality of service (QoS: Quality of service) requirements to indicate the BS switching decision [4]. Contrary to the study carried out by Kashefet al., an algorithm was designed regarding the number of transitions between the two cells in this study. In another study, Afraa Khalifah et al. proposed an algorithm that takes the source block segmentation in hybrid femto cells into account and decides based on link speed and power control to enable user transfer from macro cells to femto cells. With this algorithm proposed in the study, it was reported that the minimum Quality of Service (QoS) requirements of mobile users were met, the network capacity increased, thus the overall performance of dense cellular networks increased [5]. In their study, Shaowei Wang and Chen Ran addressed the problem of cellular network planning in the context of a heterogeneous network, which is considered a cost-effective paradigm to improve cellular system performance. They proposed a dynamic cellular network planning framework that could significantly reduce the system's CAPEX (Initial investment cost) and OPEX (Operational costs) value [6]. Hâkim Ghazzai et al. proposed a new method for the cell planning problem using intuitional algorithm for fourth generation (4G) cellular networks. When looking at the simulation results, a proposed approach was applied for different scenarios with different areas and user distributions and it was observed that the desired network service quality was achieved even for higher capacity problems [7]. Methods that provide intuitive or approach-based solutions ensure near-best results in a reasonable time [8]. Chun-Hung Richard Lin et al. conducted a study on the problem of cell planning to minimize the cost of signaling [9]. The simulation results reveal that the proposed Taboo Search algorithm performs better than previous studies using the traditional Tabu search and genetic algorithms related to the cell planning problem. Dennis M. Rose et al. conducted a study demonstrating the various cases of usage of the SiMoNe simulator used for mobile networks. This study provides an example of network planning and optimization for large-scale urban scenarios and very dense small cell deployments. In the study, five different self-organizing network functions are shown such as MRO, MLB, LTE/GSM DSA, Wi-Fi LTE TS and LTE/V2X TS [10]. In a study conducted by Serkan Kayacan, a software that makes frequency planning using measurement data in mobile communication on 2G technologies was developed. The results of this software were compared with ILSA (Intelligent Local Search Algorithm) in the study. As a result of this comparison, it was seen that a better frequency plan was obtained compared to the ILSA software and all the quality values were improved in the cells in the planned region [11]. In his study, Serdar Biroğul carried out the frequency planning of the GSM network base stations broadcast control channel (BCCH, Broadcast Control Channel) together with the data fusion process using a genetic algorithm. In another study by Serdar Biroğul, he designed epigenetic algorithm (EGA) by adapting epigenetic concepts to classical Genetic Algorithm (GA) structure. The designed epigenetic algorithm has been implemented as a limited optimization probleming BCCH frequency planning of base stations in the GSM network in the study carried out, frequency planning was performed in 3.5G systems, which are said to be independent of frequency, apart from Serdar Biroğul's work in 2G systems [12, 13]. In his study, Charalabos Skianis discussed 3G network planning and optimization methods that deal with the relationship between coverage and capacity [14]. Similarly, Marc St-Hilaire et al. put cost-cutting comments on three problems encountered in Node-B, RNC, MSC topology in 3 Gin their study [15]. The study was conducted to establish an optimum HW system topology. As can be seen from the examples given above, frequency/carrier planning or, in other words, carrier planning problem, improving the services has been a very important issue for GSM operators. The main subject of this study is the cell structure and cell planning in the 3.5G network, the concepts that are important in cell planning and the improvement studies that can be done to meet the increasing number of subscribers rather than the GSM examples currently discussed. In the study which is based on the studies carried out in the literature, a solution was searched for by using Genetic Algorithm (GA), an artificial intelligence component, to minimize the interference problem, which is a very serious problem for mobile base stations.

## 2   Material and Methods

WCDMA technology is a much more complex technology that uses broadband, nonlinear, GSM systems. Since WCDMA systems are code division systems, they are superimposed on the current frequency by multiplying them with a specific code. This code, which is known by the receiver, is reverse processed and the original voice/data is received by the user. The interference problem causes the loss of this code transmitted to the other party by signaling, thus data loss. Just like WCDMA systems, there are source limit in GSM and 3G systems. The band allocated to GSM systems is divided into carriers of 200 kHz each. Similarly, 5 MHz carriers are used as a source in 3G and WCDMA systems. It is evaluated how we can obtain the limited number of frequencies in WCDMA systems to make one another the least interference. Using the GA algorithm, the frequency plan that gives the fastest and least interference value has been investigated.

### 2.1   Frequency Planning

This study was conducted on 30 base stations, using actual network data in one of the mobile operators in Turkey. Since there are an average of 3 cells in each base station and 2-4-6 frequencies can be used in each cell, 30*3*6 = 540540 (but since there are not 6 frequencies in each cell, there may be 2 and 4 frequencies) an average of 400 frequencies are assigned. Due to 30*3*6 = 3G and 3.5G technology, 400/2 approximately 200 frequency pairs are assigned because frequencies are used in pairs. Two issues are important in frequency planning. The first one is Co_channel frequency and the other is Adj_channel frequency. Co_channel frequency is the most effective interference reason for cells that interact with each other. Adj_channel frequency is the successive frequency assignment between cells. Being not as effective as Co_channel, it is the cause of interference due to harmonic effect. The main purpose is to give the same frequencies to the cells with the least interference relationship with a chosen algorithm and to give the frequency of those with interference effect differently. The most important stage after the nominal network is created is the frequency assignment according to this principle. In our study, the handover relations of the cells with each other were used to determine the interference level of the cells with each other. The data in the table below are live system data used in the program, and limited sharing has been made for information security reasons. Column No. 1 Source cells seen in Table 1, column number 2 source cell carrier numbers, column number 3 and 4 target cell and carrier numbers, cell number 5 cell search cell number to the target cell, column 6 all the attempt to the source cell and lastly, column 7 is the percentage of interference obtained according to these number of attempts.

**Table 1.** Actual live system handover data used

| Source | Source carrier num | Target | Target carrier num | Source to target SHO att | Source all SHO Attpt | Interference (%) |
|--------|--------------------|--------|--------------------|--------------------------|----------------------|------------------|
| ASLOJ1 | 4 | ASLOJ1 | 4 | 0 | 182028 | 0,0% |
| ASLOJ1 | 4 | ASLOJ2 | 4 | 2729 | 182028 | 1,5% |
| ASLOJ1 | 4 | ASLOJ3 | 4 | 89688 | 182028 | 49,3% |
| ASLOJ1 | 4 | BAMEV1 | 6 | 3499 | 182028 | 1,9% |
| ASLOJ1 | 4 | MUGDU3 | 4 | 1636 | 182028 | 0,9% |
| ASLOJ1 | 4 | MUGDU4 | 4 | 9311 | 182028 | 5,1% |
| ASLOJ1 | 4 | MUGOR1 | 2 | 0 | 182028 | 0,0% |
| ASLOJ1 | 4 | MUGOR1 | 2 | 0 | 182028 | 0,0% |
| ASLOJ1 | 4 | MUGOR1 | 4 | 60 | 182028 | 0,0% |
| ASLOJ1 | 4 | MUGIN2 | 4 | 1502 | 182028 | 0,8% |
| ASLOJ1 | 4 | MUGOR1 | 4 | 772 | 182028 | 0,4% |
| ASLOJ1 | 4 | MUGOR2 | 4 | 2039 | 182028 | 1,1% |
| ASLOJ1 | 4 | MUGOT1 | 2 | 17919 | 182028 | 9,8% |
| ASLOJ1 | 4 | MUGOT2 | 2 | 10132 | 182028 | 5,6% |

Since WCDMA systems are code division systems, they are superimposed on the current frequency by multiplying them with a specific code. This code, which is known by the receiver, is reverse processed and the original voice/data data is received by the user. The interference problem causes the loss of this code transmitted to the other party by signaling, thus the data loss.

The 400 frequencies were assigned to 30 base stations (approximately 100 cells) using the application and GA. The purpose function used in calculating the suitability value used in the study is given in Eq. 2.1.

$$\sum_{i=0}^{n} \frac{E_i \times T_i \times C_{ci}}{C_i} \tag{2.1}$$

$E_i = i$. Interaction value between source-target in the row
$T_i = i$. Traffic value of source in line
$C_{ci} = i$. Number of overlapping frequencies between the source-target in the row
$C_i = i$. Total frequency number of the source field in the line

## 2.2 Traffic Theorem

The cleaner and quality the frequencies used in a cell, the longer the interview time. It is sized according to the traffic requirement in the cells. Erlang is expressed in telecommunication systems as a unit of users' time to occupy the system. When designing networks in nominal plan, a certain block rate is taken as basis. Block rate; It is expressed as the ratio of requests for access to the network to all access requests and is calculated by Eq. 2.2.

$$B(N) = \frac{\frac{A^N}{N!}}{\sum_{i=0}^{N} \frac{A^i}{i!}} \tag{2.2}$$

## 2.3 Genetic Algorithm

In the study, the parameters chosen first for the optimization cycle are transformed into the appropriate format with the decimal coding method. The sectors in the application can take 3 (0, 1, 2) frequency values. Cycle is used to find the best gene structure in GA, crossing, mutation and replacement of old gene structure with new gene structure. Working principle and sample flow diagram of GA in application are given in Fig. 1. The system initially takes two Excel files as input. One of these files is the file where the traffic information of the sectors is kept, and the other is the Excel file, where the interactions between the sectors are kept.

Possible solutions were randomly sampled from the status space for the creation of the starting population in GA. $Ni_{pop}N_{bit}$ matrices are created during random production of the population. Here, Nipop and Nbit are the total number of start populations, respectively, and the number of bits of the chromosome in N bits, respectively. The population is evaluated by processing this randomly obtained population in the purpose function equation included in the application and by calculating the suitability parameters. In the

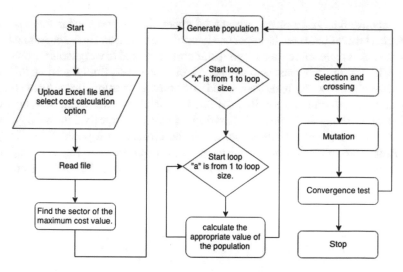

**Fig. 1.** Flow chart of applied GA

study, each population was processed to contain 100 chromosomes, and only one was separated by the selection operator to be the best chromosome. The selection operator selects the chromosomes from the current generation to be used for the new generation. The probability of each chromosome selection is given in Eq. 2.3:

$$Ps(i) = \frac{f(i)}{\sum_{j=1}^{N} f(j)} \tag{2.3}$$

Here $Ps(i)$ and $f(i)$ are the selection and eligibility value for the i chromosome, respectively. Parents are chosen in pairs. After a chromosome is selected, the possibilities are normalized again without the selected chromosome, so that the new parents to be selected are selected from the remaining chromosomes. Weighted random selection method was used in the application. In this method, the probabilities given to the chromosomes in the mating pool are inversely proportional to their cost values. A random number determines which chromosome is selected. This type of weighting is often referred to as roulette wheel weighting. Value weighting method was used in the study. In this method, the probability of selection is calculated from the cost value of the chromosome rather than the predominance in the population. Equation 2.4 is calculated by subtracting the cost values of all chromosomes in the mating pool from the lowest cost value of the chromosomes discarded for each chromosome:

$$C_n = c_n - c_{N_{keep}+1} \tag{2.4}$$

Extraction ensures that all costs are negative. Based on this, the probability of selecting Pn is calculated as in Eq. 2.5.

$$P_n = \left| \frac{C_n}{\sum_{m}^{N_{keep}} C_m} \right| \tag{2.5}$$

This approach tends to further weight the upper chromosome when there is a large spread in cost values between the upper and lower chromosomes. On the other hand, when all chromosomes have approximately the same cost, they tend to weight the chromosomes equally. The possibilities must be recalculated for each generation. Based on this, while selecting the chromosomes to be transferred to the next generation, first elitism is applied and the first two individuals are written to the new generation. After these new individuals are selected, crossing is performed. A ratio is selected primarily for crossing, which ratio determines how much chromosome will be crossed in a population. First, two chromosomes are selected from the pairing pool. As can be seen in Fig. 2, a single cross point "$k$" with a value between 1 and n-1 is randomly selected. The 100 chromosomes that are ranked from the best to the worst according to the purpose function before crossing are grouped as the first 25%, 25–75% and the last 25%, and the groups are subjected to 80% cross between each other. This is the idea of getting even better results than chromosomes with the best genes. As a result of crossing, 2 new genes are formed and transferred directly to the new population.

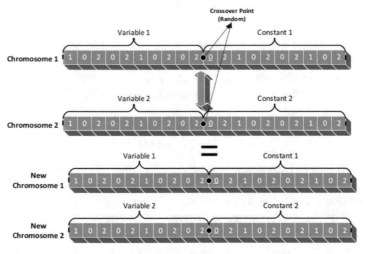

**Fig. 2.** Example of crossing

Individuals in the population can undergo mutation after crossing to increase population diversity and produce new offspring. The number of genes to be mutated in the relevant population ($Gen_{mut}$) according to the percentage of selected mutation is formulated as Eq. 2.6.

$$Gen_{mut} = P \times K \times R \tag{2.6}$$

It is found with. Here; "P" refers to population size, "K" refers to chromosome size, "R" refers to mutation rate. The mutation rate has been made parametrically selectable from the program interface. Trials of 0.1–1.5% were made in the program.

In the number of iterations determined in GA, the program is not terminated because it cannot be known how the program will behave in the next iterations, the defined criteria are observed and terminates according to the amount of the compliance function or penalty value.

In the study, the example of finding the best chromosome given for only 1 iteration is shown in Fig. 3.

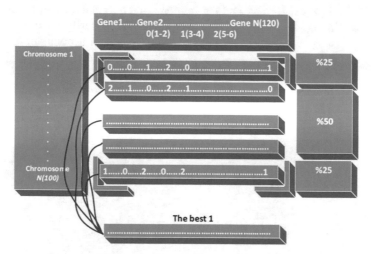

**Fig. 3.** Finding the best chromosome

The subject discussed in this study is 3.5G discrete frequency planning. For 3.5G technology, a maximum of 6 frequencies (carriers) can be used in each cell. Based on the interference table created according to the interaction of the cells with each other, frequency assignments were made to create the least interference.

## 3  Application

Installing, commissioning and operating WCDMA systems is a very difficult and laborious process. WCDMA discrete frequency planning aims to minimize the frequency of interference that will occur between sectors, while "N" is assigning frequency to each sector as much as the number of carriers. Although it can be solved with brute force search method by calculating the total interference values of all permutations and finding the smallest, the number of permutations for the large values of N is N!. This process takes a very long time since it will reach great values. There is a certain resource limit for each station. The limited frequency resource in cellular systems forces operators to optimize to use their limited resources in the best way. Each base station consists of an average of 3 cells. Each cell can be assigned between 1–6 WCDMA frequencies according to user needs. The problem here is to find out how many cells are distributed in such a limited frequency, and with which algorithm the best. In these systems, since the number of cells available is many times more than the number of frequencies, frequency optimization, therefore, re-use of frequencies with a minimum of interference is required.

The main screen of the application developed using the Java programming language can be seen in Fig. 4. Planning was made both for data and voice traffic with GA. Mutation and crossover rates can be selected parametrically from the application screen. The interface that appears when the program first runs is as follows. The program is run with the start button. The results obtained in the realized software are saved in the program folder in excel format, and the purpose function change can be seen on the graphic screen in Fig. 4.

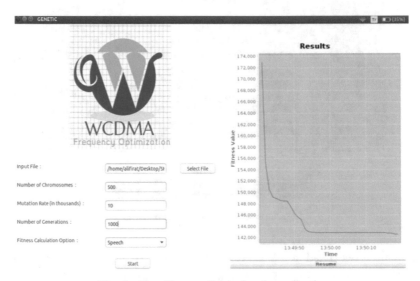

**Fig. 4.** Algorithm result interface in application

When we consider the problem with GA; each chromosome consists of the genes of a total of 400 cells to be assigned to 100 sectors. Each population consists of a total of 100 chromosomes, and the chromosome we are looking for in the solution space must consist of genes that we want to achieve the best in the purpose function. While creating the first population, weighted random selection method is applied. They are subjected to crossing and mutation processes, respectively. The crossover point is selected randomly for crossing the two chromosomes. The number of chromosomes that will be changed as a result of the mathematical operation, the details of which have been given before, is the number of chromosomes to change, and the genes are changed with the randomly generated row column numbers. The chromosomes evaluated in the objective function are listed. It is transferred directly to the semi-new population in the best 2 and good side in the ranking. Elitism is applied by direct selection of the best 2. This approach is a side in our study that we interpret differently than classical GA. Individuals (chromosomes) fight to transfer their genes to the new generation, as in natural heredity in every new population. The best results found iterative in each population are compared to each other, making it the final result with the lowest value in the objective function.

## 4   Experimental Results

With elitism, the best genes are transferred directly to the next generation and the solution is accelerated. The program runs in an infinite loop, and the program outputs up to 2000 iterations are as in Fig. 5 as an example. For 100 populations, 500 populations and 1000 populations, the application was run 20 times under the same conditions. Average values were found using the values obtained at the end of each run. As a result, we were likely to find the best result in 500 populations.

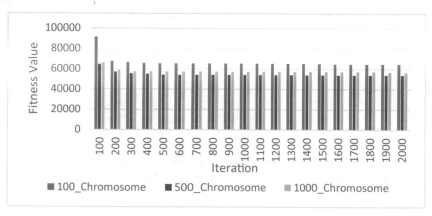

**Fig. 5.** Fitness values obtained in different population sizes

The recovery acceleration captured with GA in 1% mutation values that we obtained the best value is as follows for increasing chromosome values in different populations. The Plan 1 recovery curve obtained in the study performed in the case of 80% crossing, 1% mutation and 100 chromosome length can be seen in Fig. 6. After 400 iterations, the curing acceleration of the program reaches approximately satisfaction.

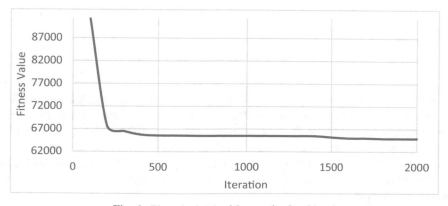

**Fig. 6.** Plan obtained with genetic algorithm 1

The plan 2 recovery curve obtained in the study in the case of 80% crossing, 1% mutation and 500 chromosome length can be seen in Fig. 7. After 1400 iterations, the curing acceleration of the program reaches approximately satisfaction.

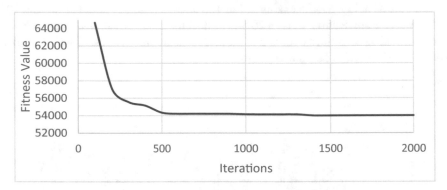

**Fig. 7.** Plan 2 obtained with genetic algorithm

Plan 3 recovery curve obtained in case of 80% crossing, 1% mutation and 1000 chromosome length can be seen in Fig. 8. The best result was observed in 2000 iterations.

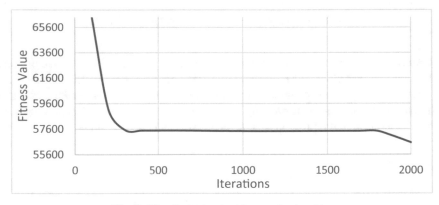

**Fig. 8.** Plan 3 obtained with genetic algorithm

When the results obtained with the GA algorithm are compared, it is ensured that the best frequency plan is applied to 100 cells in the city center. As a result of this application, as seen in Fig. 9, the data download speed of HSDPA (high speed downlink packet access) increased by an average of 2% after the plan was implemented. As can be seen from the figure, after applying the frequency plan to the field on 26.01.2019, the data download speed increased from 2560 kbps to 2630 kbps according to an average weekly observation.

As seen in Fig. 10, the number of interference-induced handover fell from an average of 148 to 143 on 24.01.2019 in the week before the plan was implemented. Thus,

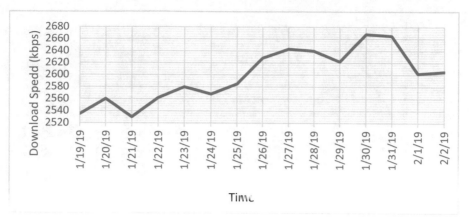

**Fig. 9.** Data download speed change after the applied plan

approximately 3% improvement was observed against the new plan implemented. The quality of communication is improved by decreasing cell-to-cell transitions.

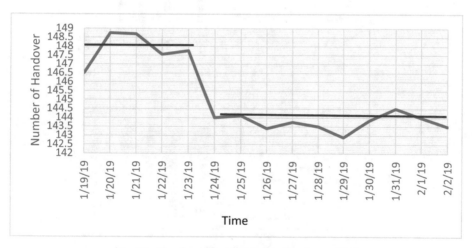

**Fig. 10.** Handover Overhead after the plan applied

As seen in Fig. 11, search drop rate did not change. This means that there is no change in the rate at which interruptions or data rates are interrupted. This shows us that the interference caused by the previous plan is not at a level to create drops.

As can be seen from Fig. 12, in the trials where the mutation rate in GA is variable, the best result was 1% mutation and 80% crossover rate, and the conformity value of 268000 decreased to 53999 levels according to the current plan working in the field and an improvement of 79% was observed. At the same time, in Fig. 12 (b), the same iteration values, 1% mutation rate and the number of chromosomes in the population of 500 were found to be the situation where the best value was achieved in the goal function at 654 iteration and 149050 fitnesss.

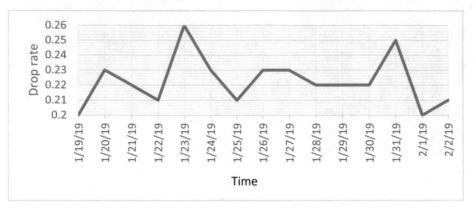

**Fig. 11.** Drop Rate after the applied plan

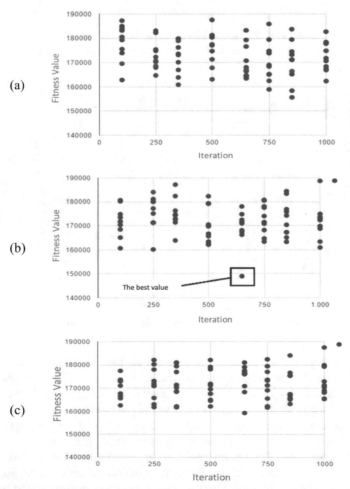

**Fig. 12.** Mutation results in practice (a) 0.1% mutation rate (b) 1% mutation rate (c) 5% mutation rate

# 5  Results and Discussion

Frequency planning is generally used in the literature and practice for the 2G network. Due to its technology, 3.5G systems are defined as interference independent. There are only 6 frequencies in 3.5G. These frequencies must be given to the cells in 2 consecutive sets. In the conducted study, 0–1–2 expresses these frequency sets. It was tried to find the best frequency plan by assigning only 3 frequency sets of 50 fields (50 * 3 = 150cell) in the selected region with an average of 3 cells. In the study carried out, it is observed that 3.5G systems are defined as interference independently as described in the literature, and in case of interference is reduced with a good frequency plan, it is observed that even higher quality service continuity is provided. It was concluded that the service quality, which is a serious problem in the telecom industry, can be improved by using the Genetic Algorithm. In regions where there is no coverage problem, it is seen that instead of making very serious infrastructure modernizations, the existing infrastructure can be made more efficient with such software developments. It is observed that there is a frequency sensitivity in 3.5G systems which is said to be not required for frequency change in practice due to its technology and in the literature. From the results obtained with this study, it was revealed that the frequent interference frequencies changed and there was a significant improvement effect in the network. Accordingly, it has been observed that channel frequencies, overlap, and harmonic effect can be reduced by repeating the frequencies used in 3.5G as a result of capacity and coverage requirements.

# References

1.  Mishra, A.R.: Advanced Cellular Network Planning and Optimisation: 2G/2.5G/3G. Evolution to 4G. John Wiley & Sons, Inc., Hoboken (2007)
2.  Zhu, L., Deng, S., Huang, D.-S.: Editorial: special issue on advanced intelligent computing: theory and applications. Neurocomputing **228**, 1–2 (2017)
3.  Scheinert, S., Jovanovic, V.: Patent No. 9037142 (2015)
4.  Kashef, M., Ismail, M., Serpedin, E., Qaraqe, K.: Balanced dynamic planning in green heterogeneous cellular networks. IEEE J. Sel. Areas Commun. **34**(12), 3299–3212 (2016)
5.  Khalifah, A., Akkari, N., Aldabbagh, G., Dimitriou, N.: Hybrid femto/macro rate-based offloading for high user density network. Comput. Netw. **108**, 371–380 (2016)
6.  Wang, S., Ran, C.: Rethinking cellular network planning and optimization. EEE Wireless Commun. **23**(2), 118–125 (2016)
7.  Ghazzai, H., Yaacoub, E., Alouini, M.-S., Dawy, Z., Abu-Dayya, A.: Optimized LTE cell planning with varying spatial and temporal user densities. IEEE Trans. Veh. Technol. **65**(3), 1575–1589 (2016)
8.  Mutingi, M., Mbohwa, C.: Grouping Genetic Algorithms Advances and Applications. Springer, Cham (2017). https://doi.org/10.1007/978-3-319-44394-2
9.  Lin, C.-H.R., Liao, H.-J., Lin, Y.-C., Liu, J.-S., Huang, Y.-H.: An efficient tabu search for cell planning problem in mobile communication. Wirel. Commun. Mob. Comput. **16**, 486–496 (2016)
10.  Rose, D.M., Hahn, S., Kürner, T.: Evolution from network planning to SON management using the simulator for mobile networks (SiMoNe). In: IEEE 27th Annual International Symposium on Personal, Indoor, and Mobile Radio Communications (PIMRC), Valencia (2016)
11.  Kayacan, S.: GSM'de frekans planlama yöntemleri ve hücrelere frekans ataması yapacak planlama yazılımı gerçekleştirimi. Ege Üniversitesi Fen Bilimleri Enstitüsü, İzmir (2007)

12. Biroğul, S.: GSM Şebekelerinde Frekans Planlamasının Veri Füzyonu ile Gerçekleştirilmesi. Gazi Üniversitesi Fen Bilimleri Enstitüsü, Ankara (2008)
13. Birogul, S.: EpiGenetic algorithm for optimization: application to mobile network frequency planning. Arab. J. Sci. Eng. **41**(3), 883–896 (2015). https://doi.org/10.1007/s13369-015-1869-5
14. Skianis, C.: Introducing automated procedures in 3G network planning and optimization. J. Syst. Softw. **86**(6), 1596–1602 (2013)
15. St-Hilaire, M., Chinneck, J.W., Chamberland, S., Pierre, S.: Efficient solution of the 3G network planning problem. Comput. Ind. Eng. **63**(4), 819–830 (2012)

# Deep Learning Based Classification Method for Sectional MR Brain Medical Image Data

Ali Hakan Işik[2], Mevlüt Ersoy[1], Utku Köse[1], Ayşen Özün Türkçetin[1(✉)], and Recep Çolak[3]

[1] Department of Computer Engineering, Suleyman Demirel University, Isparta, Turkey
{mevlutersoy,utkukose}@sdu.edu.tr, aysenozun@gmail.com
[2] Department of Computer Engineering, Mehmet Akif Ersoy University, Burdur, Turkey
ahakan@mehmetakif.edu.tr
[3] Department of Computer Programming,
Isparta University of Applied Sciences, Isparta, Turkey
recepcolak@isparta.edu.tr

**Abstract.** Recently, deep learning techniques have achieved significant success in medical image analysis. In this article, deep learning methods have been applied to separate brain magnetic resonance (MR) images into different abnormalities and healthy classes. The sectional brain MR of brain images is used as a database from the Open Access Imaging Studies Series (OASIS). Based on Convolutional Neural Network (CNN) method for classification and a thresholding algorithm for image segmentation, the system has been developed. The images that are improved by image processing are transferred to the CNN deep learning model and the classification process is done. Adam algorithm was used as the optimization algorithm for the classification process. As a result of the classification, 80% accuracy rate was obtained. The model's loss rate fell to 0.3 s.

**Keywords:** Deep learning · Convolution Neural Network · MR · Medical imaging · OASIS

## 1 Introduction

Treatment of some diseases has not been found compared to the advancement of technology. One of these diseases is dementia, which causes memory loss. As Alzheimer's disease, an advanced stage of dementia, progresses, the brain shrinks significantly due to a large number of cell deaths. This is reflected in MR and Pet scan images. The images transferred to the computer are examined by doctors and the diagnosis of the disease is revealed as a result of examinations. Many methods are used in computerized diagnosis and treatment in medicine. Utilizing computer science in the medical field has led to the emergence of some terms. One of these is Computer-Assisted Detection (CAD). CAD systems assist in the detection of abnormal structures in medical images by using image processing and pattern recognition techniques. While CAD systems accelerate the decision-making process, they also benefit by reducing the possibility of human error in this process [1].

J. Hemanth et al. (Eds.): ICAIAME 2020, LNDECT 76, pp. 669–679, 2021.
https://doi.org/10.1007/978-3-030-79357-9_62

Computer systems are used in conjunction with medical imaging in many senses such as CAD, detection, and classification of tumors in the body, detection of broken bones, diagnosis and categorization of cancers [2]. Many methods are used in the recognition of images on the computer. Some of these methods are used in different methods such as object recognition, shape detection, and template matching [3]. In medical images, problems such as low resolution, noise, low contrast, geometric deformation may occur. In these images, the development of the CAD system will save doctors from the workload and minimize time loss as well as easier and automatic detection of the disease. In medical imaging, data are obtained using methods such as x-ray radiography, ultrasonography, computed tomography (CT), magnetic resonance (MR), positron emission tomography (PET) [4]. The data obtained by these methods give results as noise due to the limits of the imaging tools, in this case, the readability of the data by the doctors reduces the rate of understanding.

The technology in the development of CAD systems consists of fields such as artificial intelligence, machine learning, and deep learning. Many algorithms developed both processes the image and produce predictions by learning data with certain deep learning methods. Deep learning is in the sub-branch of machine learning and artificial intelligence. As a result of this situation, it relies on artificial neural networks on the basis of deep learning algorithms. Artificial neural networks are a model designed to facilitate scientific studies based on the working structure of the human brain. In the study was used the 10-layer Convolution Neural Network (CNN) model, which is one of the deep learning algorithms.

In this application, a deep learning method was applied to cross-sectional areas from the OASIS open-access study series brain MR image data set. The data set before being trained was subjected to image processing steps. The data set was separated as 80% train, 20% test set, and trained in the deep learning algorithm. It took 6 h and 20 min to train the model. As a result of the training, while the accuracy rate was 80%, the loss rate decreased to 0.2.

## 2 Related Work

Anthimopoulos et al. (2016) proposed the Evolutionary Neural Network (CNN) model to classify different tissue types in lung diseases [5].

Ciompi et al. (2017) proposed a multi-scale Evolutionary Neural Network (CNN) model in the classification of nodules in lung cancer images [6].

The main purpose of the work of Çevik (2011) is to create an application framework that will enable the analysis and measurement of many attributes on medical images by minimizing the effect of user addiction on the results. The designed application allows the application of the medical image processing routines in order; to provide radiologists with a software environment to support neurological degenerative diseases and brain tumors in diagnosis, treatment plan, and treatment verification processes;Thus, it aimed to reduce the variation obtained on the results [7].

A Computer-Aided Detection (CAD) system designed by Dandıl (2017) has been an important method for early diagnosis of lung nodules. One of the most important steps that will affect the success of this system is segmentation. In this study, the segmentation of lung nodules on CT (Computed Tomography) images were performed successfully [8].

Doğan et al. (2016) have developed a CBT system that can reconstruct and classify the interests detected in brain MR images using formal attributes. The developed system consists of four stages: preprocessing, segmentation, area of interest, and tumor detection.

The developed system was evaluated with the REMBRANDT data set consisting of 497 cross-sectional images of 10 patients. In the classification process, the performance of the system has reached 93.36% accuracy with decision trees, 94.89% with artificial neural networks, 96.93% with K-nearest neighbor algorithm, and 96.93% with Meta-Learner algorithm [9].

Fıçıcı (2016) Image processing in brain MR images is of great importance in the clinical analysis of tumors. Tumor volume and location information can be obtained with image processing techniques for successful diagnosis and treatment. In this study, he presented a fully automated method for the detection of brain tumors with high accuracy and volume calculation [9].

Kulkarni and Panditrao (2014) used image processing and Support vector machines and image processing methods in their studies [10].

Turkcetin (2019) thesis studied used image processing in lung cancer PET/CT images. Then applied a deep learning model to lung cancer PET/CT images. She proposed models CNN and DNN deep learning for lung cancer PET/CT images. She found the accuracy rate for CNN model 98.48%, and for DNN model 93.65% [23].

Okyay (2016) studied the use of brain imaging techniques in predicting Alzheimer's disease early diagnosis. In his study, he used magnetic resonance images of 63 samples in three different diseases: 19 Alzheimer's disease, 19 frontotemporal dementia, and 25 vascular dementia. Sliced brain image sets were processed with the Freesurfer brain analysis software tool and formed different feature groups from the statistical information produced after a successful analysis of the program [11].

Ural (2016) has worked on a new system that localizes cancerous areas for stomach images taken using computed tomography (CT) [12].

## 3 Material and Method

### 3.1 Dataset

Brain MR image data set was used with OASIS cross-sectional area. The data set consists of cross-sectional brain images between the ages of 18–96. Of the 100 patients included, 60 were clinically diagnosed with very mild to moderate Alzheimer's disease. The patients consisted of men and women, and 39 patients were asked to use their right hands. Three or four separate T1-weighted magnetic resonance imaging scans obtained in single imaging sessions were included for each patient.

Since the images obtained from the data set are in.gif format, they were first converted to .jpg format. In order to increase the accuracy rate of the deep learning algorithm, one of the image processing methods, the Clahe method has been applied, and visible regions have been made more pronounced in the image. Certain measurement ratios for brain MR images are given in the OASIS data set. One of these is the CDR parameter. According to the given CDR parameter; If CDR = 0, there is no dementia in that person, if CDR = 0.5, it is very mild as dementia, if CDR = 1 is mild dementia and if CDR = 2, it is expressed as a moderate. Although the data set was actually divided into four classes, people with a CDR parameter greater than zero were considered as dementia patients [13].

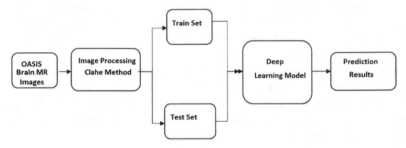

**Fig. 1.** Workflow diagram

The data set is divided into 80% training and 20% testing. After the image processing step, the data set was trained in CNN, the deep learning network. The Python programming language is used for the operations which are rich in libraries. The workflow diagram of the study is shown in Fig. 1.

### 3.2 Image Processing

The Clahe method of the OpenCv library was used to improve the data set. OpenCV is an open-source library developed in C++ language with many computer vision algorithms. It is used in image processing. [14]. First, one 8 * 8 was created Clahe object. Then the

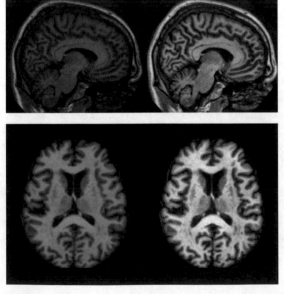

**Fig. 2.** Application of Clahe method on brain MR images

created object was added to the MR image and was improved the image. Figure 2 shows the state of the MR image before and after image processing.

Example Code block;

clahe = cv2.createCLAHE(clipLimit = 2.0, tileGridSize = (8,8)).

cl1 = clahe.apply(img).

### 3.3 Deep Learning Model

CNN and DNN networks are used as a deep learning algorithm. CNN is one of the convolutional network models. Convolutional networks are classification methods that generate a single class tag from the output of image data. In addition, it is necessary to assign a label that indicates that there is a class for each pixel in many visual operations and medical image processing [15]. In the study, the CNN network is composed of 10 layers. The number of trainable parameters was calculated as 575.369. The summary table is given layers of the CNN model in Fig. 3.

| Layer (type) | Output Shape | Param # |
|---|---|---|
| Conv2D | (None, 126, 126, 256) | 7168 |
| MaxPooling2 | (None, 63, 63, 256) | 0 |
| Conv2D | (None, 61, 61, 128) | 295040 |
| Dense | (None, 61, 61, 8) | 1032 |
| Dropout | (None, 61, 61, 8) | 0 |
| Conv2D | (None, 59, 59, 64) | 4672 |
| Conv2D | (None, 57, 57, 64) | 36928 |
| Conv2D) | (None, 55, 55, 64) | 36928 |
| Flatten | (None, 193600) | 0 |
| Dense | (None, 1) | 193601 |

Total params: 575,369
Trainable params: 575,369

**Fig. 3.** Summary table of layers of the model in CNN deep learning

The reason we add the dropout layer between the Convolution layers is to avoid overfitting. Excessive adaptation is the situation where the model memorizes the training sample but gives a poor learning rate on the test sample [16]. Flatten layer was used to reduce MR images to 2 dimensions. The architecture of the proposed network is given in Fig. 4.

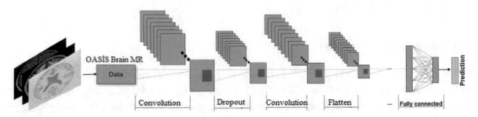

**Fig. 4.** The architecture of the proposed network

The DNN network is the increased number of parameters of the CNN network. In the proposed model, the number of parameters has been increased to 14,770,609. The difference between the DNN deep learning model from the CNN deep learning model is that the hidden layer number is reduced by one after the pooling layer after the first layer and this situation is repeated with a loop. As seen in Fig. 5, DNN architecture consists of 8 layers. Each of the layers contains one convolution, maximum pooling, and a derivative of the normalization steps. But the last three layers are completely interconnected.

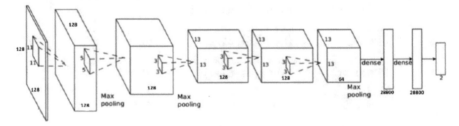

**Fig. 5.** The architecture of the proposed DNN model [22]

### 3.4 Optimizer Parameter

Optimization is a method used to minimize the difference between the output value produced by the network and the actual value [17]. Optimization algorithms are used for nonlinear problems. There are distinct differences in speed and accuracy rates between algorithms. In the studies conducted, it was revealed that the Adam algorithm learned faster and had a more stable structure compared to other parameters [18]. Unlike other algorithms, the Adam algorithm stores the momentum changes as well as the learning rates of each of the parameters [19].

In the study, the Adam method was used in the Keras library of TensorFlow (tf). Keras is a Python library for deep learning calculations. Keras is a neural networks API

(Application Programming Interface) written in Python programming language and can work on structures such as TensorFlow. It was developed to provide fast results [20].

Tensorflow is Google's open-source deep-learning library. It was developed by engineers working on Google Brain to be used in machine learning and deep learning studies. Depending on the graphics card feature of the Tensorflow computer, CPU or GPU can also show operating performance [21].

Setting the learning coefficients in optimization algorithms is very important in the training phase of the model. However, it is not possible to make the best setting for each algorithm. According to the studies, the best learning rate for the Adam parameter was found to be 0.001 [18]. In Fig. 6, the best learning coefficient is given for the Adam algorithm.

Setting the Adam parameter;
tf.keras.optimizers.Adam(learning_rate = 0.001, beta_1 = 0.9, beta_2 = 0.999, epsilon = 1e-07, amsgrad = False, name = "Adam").

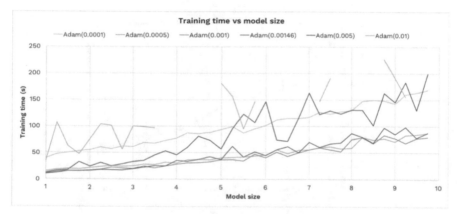

**Fig. 6.** Comparison of the Adam learning parameter according to the training time of the model [18].

According to Fig. 5, the best learning coefficients were found to be 0.0005, 0.001, 0.00146. In this study, the learning coefficient was determined as 0.001.

## 4 Results

In the proposed CNN and DNN deep learning models, firstly image enhancement step was performed for brain MR images. The improved images are presented to the CNN deep learning model in two classes as dementia and not dementia. The model, which is divided into two classes, is divided into 80% education and 20% test sets. A notebook with 1050 Ti GTX graphics card with 7th Generation i5 processor was used to train the model. The training period of the models lasted 6 h 20 min. The data set was retrained on the recorded model. As the highest result, the accuracy rate of the model was found to be 75.55%. As a result of this situation, the rate of loss decreased to 0.3. In OASİS brain MR image data set are given the accuracy rate of Fig. 7 and the loss rate in Fig. 8.

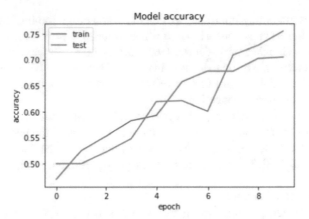

**Fig. 7.** Proposed CNN model accuracy rate

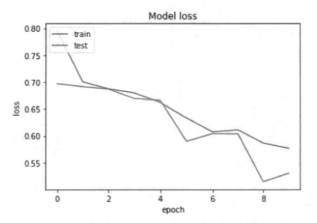

**Fig. 8.** Proposed CNN model loss rate

In the proposed DNN deep learning model, the highest model accuracy result was found to be 78.77%. Loss rates decreased by up to 5%. Figure 9 and Fig. 10 show the accuracy and loss rates of the DNN model.

If the performance rate is low in the proposed deep learning models, this situation is eliminated by increasing the number of layers. In the study, the number of parameters of the DNN deep learning model is around 15 million. Therefore, the success rate of training has increased to 78.77%.

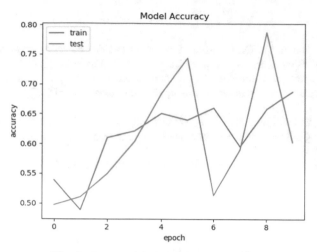

**Fig. 9.** Proposed DNN model accuracy rate

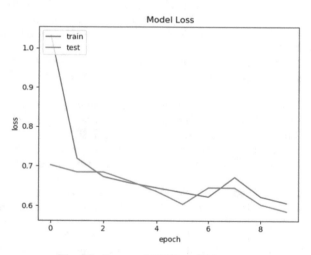

**Fig. 10.** Proposed DNN model loss rate

## 5 Conclusion

The size of the OASIS data set and the reason that the MR images are.gif are scheduled to train the data set by deep learning algorithms. It is for preliminary preparation for the deep learning model to be applied by applying the adjusted image processing. Selecting has reached that the image processing will be more successful before the model is trained for image data sets in the system. After the image processing step, the DNN deep learning model in the sample models was observed according to the better result. The Adam made parameter look better. In future studies, considered that higher accuracy preferred deep learning models can be created in the detection of dementia of OASIS brain MR images.

# References

1. Doğan, B., Demir, Ö., Çalık Kazdal, S.: Computer-aided detection of brain tumors using morphological reconstruction. Uludağ Univ. J. Faculty Eng. **21**(2), 257 (2016)
2. Ekşi, Z., Dandıl, E., Çakıroğlu, M.: Computer aided bone fracture detection. In: 2012 20th Signal Processing and Communications Applications Conference (SIU), pp. 1–4. IEEE, April 2012
3. Ambrosini, R.D., Wang, P., O'dell, W.G.: Computer- aided detection of metastatic brain tumors using automated three-dimensional template matching. J. Magn. Resonan. Imaging **31**(1), 85–93 (2010)
4. Radsite: Consumer Guide to Imaging Modalities Benefits and Risks of Common Medical Imaging Procedures (2011). https://radsitequality.com. Accessed 20 Jan 2019
5. Anthimopoulos, M., Christodoulidis, S., Ebner, L., Christe, A., Mougiakakou, S.: Lung pattern classification for interstitial lung diseases using a deep convolutional neural network. IEEE Trans. Med. Imaging **35**(5), 1207–1216 (2016)
6. Ciompi, F., et al.: Towards automatic pulmonary nodule management in lung cancer screening with deep learning. Sci. Rep. **7**, 46479 (2017)
7. Çevik, A., Eyüpoğlu B.M.: Doku Anomalisi İçeren Beyin MR İmgeleri Üzerinde Mumford -Shah Tabanlı Bölütleme. EMO Bilimsel Dergi, Cilt 1, Sayı 2, Syf 103–107 (2011)
8. Dandıl, E.: Implementation and comparison of image segmentation methods for detection of brain tumors on MR images. In: 2017 International Conference on Computer Science and Engineering (UBMK) (2017)
9. Fıçıcı Ö.C.: MR Görüntülerinde Bulunan Anormal Belirlenmesi Ve Hacimlerinin Hesa-planması. Ankara Üniversitesi Fen Bilimleri Enstitüsü, Yüksek Lisans Tezi.74 s. Ankara (2016)
10. Kulkarni, A., Panditrao, A.: Classification of lung cancer stages on CT scan images using image processing. In: 2014 IEEE International Conference on Advanced Communications, Control and Computing Technologies, pp. 1384–1388. IEEE, May 2014
11. Okyay, S.: Beyin Görüntüleme Tekniklerinde Alzheimer Hastalığı Erken Tanı Tahmininde Kullanılması. Anadolu Üniversitesi Fen Bilimleri Enstitüsü, Yüksek Lisans Tezi. Eskişehir (2016)
12. Ural, A.B.: Görüntü İşleme Metotları İle Girişimsel Olmayan Mide Kanserinin Tespit Edilmesi. Gazi Üniversitesi Fen Bilimleri Enstitüsü, Yüksek Lisans Tezi, Ankara (2016)
13. Marcus, D.S., Wang, T.H., Parker, J., Csernansky, J.G., Morris, J.C., Buckner, R.L.: Open Access Series of Imaging Studies (OASIS): cross-sectional MRI data in young, middle aged, nondemented, and demented older adults. J. Cogn. Neurosci. **19**(9), 1498–1507 (2007)
14. Doxygen. OpenCv Open Source Computer Vision (2015). https://docs.opencv.org/3.1.0/. Accessed 26 Feb 2019
15. Ronneberger, O., Fischer, P., Brox, T.: U-net: convolutional networks for biomedical image segmentation. In: Navab, N., Hornegger, J., Wells, W.M., Frangi, A.F. (eds.) MICCAI 2015. LNCS, vol. 9351, pp. 234–241. Springer, Cham (2015). https://doi.org/10.1007/978-3-319-24574-4_28
16. Genç, Ö.: Keras İle Derin Öğrenmeye Giriş (2016). https://medium.com/turkce/keras-ile-derin-C3%B6%C4%9Frenmeyegiri%C5%9F-40e13c249ea8. Accessed 12 May 2019
17. Gazel, S.E.R., BATİ, C.T.: Derin Sinir Ağları ile En İyi Modelin Belirlenmesi: Mantar Verileri Üzerine Keras Uygulaması. Yüzüncü Yıl Üniversitesi Tarım Bilimleri Dergisi **29**(3), 406–417
18. Mack, D.: How to pick the best learning rate for your machine learning project (2018). https://medium.com/octavian-ai. Accessed 16 July 2020
19. Çarkacı, N.: Derin öğrenme uygulamalarında en sık kullanılan hiper parameteler (2018). https://medium.com/deep-learning-turkiye/derin-ogrenme-uygulamalarinda-en-sik-kullanilanhiper-parametreler-ece8e9125c4. Accessed 16 July 2020

20. Gulli, A., Pal, S.: Deep Learning with Keras. Packt Publishing Ltd, Birmingham (2017)
21. Salomon, J.: Lung Cancer Detection using Deep Learning, Dublin Institute of Technology (2018)
22. Jordan, J.: Common architectures in convolutional neural networks (2018). https://www.jer emyjordan.me/convnet-architectures/. Accessed 30 Sep 2020
23. Turkcetin, A.O., Bayrakci, H.C., Aksoy, B.: Akciger Kanserinin Tespit Edilmesinde Derin Öğrenme Algoritmalarinin Kullanilmasi. Master Degree Thesis, Suleyman Demirel University, Institute of Science (2019)

# Analysis of Public Transportation for Efficiency

Kamer Özgün[1]([✉]), Melih Günay[2], Barış Doruk Başaran[2], Batuhan Bulut[2], Ege Yürüten[1], Fatih Baysan[2], and Melisa Kalemsiz[1]

[1] Antalya Bilim University, Antalya, Turkey
kamer.ozgun@antalya.edu.tr
http://www.antalya.edu.tr
[2] Akdeniz University, Antalya, Turkey
mgunay@akdeniz.edu.tr

**Abstract.** Transit authorities have been searching for the indicators to measure transit service quality and the key factors to attract citizens who do not prefer public transport. The recent advent of data collection technologies such as AVL, APC, GPS and Smart Card (SC) promise opportunities for conducting comprehensive transit system performance measures, improving the quality of service while meeting passenger needs and reducing operation costs.

The aim of this study is to propose metrics to improve transit service. The focus is on bus transportation since it is more flexible compared to rail transportation and widely preferred by the masses in cities. The primary data source of this study comes from the Department of Transportation for the City of Antalya. We load the complete boarding data of December 18, 2019 which is a standard weekday. Most of the analysis is done using Knime software with Python Scripts. As an outcome of this research, the analysis may propose to modify or eliminate inefficient routes, suggest new lines, identify inefficient bus stops, and potentially modify path of a route.

## 1 Introduction

The subject of smart cities is a very new research area in the world. One of the most important issues for smart cities is the design of intelligent transportation systems. Transit authorities have been searching for the indicators to measure transit service quality and the key factors to attract citizens who do not prefer public transport. Various approaches and vast literature are available on evaluation of transit system performance. Quantitative performance indicators or measures are mentioned in two main dimensions: efficiency and effectiveness. Efficiency refers to a system's financial and productivity related performance dimensions; it is concerned with "doing things right". Effectiveness refers to the service's social dimensions; it is concerned with "doing the right things." Early in literature, Fielding et al. [4] proposed various individual performance indicators as efficiency, effectiveness and overall indicators.

J. Hemanth et al. (Eds.): ICAIAME 2020, LNDECT 76, pp. 680–695, 2021.
https://doi.org/10.1007/978-3-030-79357-9_63

In one lane of transit performance literature, afford is on determining significant indicators [12,13]. As a relatively recent study, the TCRP Report 88 provides useful material on more than 400 transit performance measures [14]. In another lane, many researchers have been accepted that efficiency is best captured by using frontier analysis where a transit unit performance is directly compared against the "best practice" of a peer or a combination of peers. Efficiency is achieved by performing at the frontier. Deviations from the frontier are interpreted as inefficiency. Kerstens (1996) [10] efforts in this direction have employed two basic estimation approaches: parametric and nonparametric. Parametric approaches that require a priori information about the trade-offs may be restrictive in most cases [9]. Non parametric approaches does not require an assumption of a functional form relating inputs to outputs and further, inputs and outputs may have very different units. For non-parametric approaches the frontier is determined by 'enveloping' it with piecewise linear functions or hyperplanes [3].

Transit units can be organizations, cities, agencies, transit firms, transit modes, bus lines and so on. For assessing transit agencies, most commonly, labor (total number of employees), capital (total number of buses operated) and energy (amount of fuel consumed) are used as inputs; and vehicle-km, seat-km, or passenger-km are used as outputs. For assessing bus lines, operation time, round-trip distance, number of bus stops are commonly used as inputs and total number of passengers used as outputs [8,9,11,15].

Daraio [2] select and organize input and output variables that are used in the studies they have reviewed and report the list of variables with their occurrence, both in absolute terms and percentage. **Inputs** are 1-Physical measure (such as number of vehicles, seat capacity, drivers and so on); 2-Capital expenses; 3-Operating expenses. **Outputs** are 1-Survice supply (such as vehicles travelled kilometers, vehicles hours of operations, seats offered hours of operations and so on); 2-Service Consumption (such as passengers travelled kilometers, number of passengers, number of trips and so on); 3-Revenue (such as fare revenues and total revenue). Furthermore, **Quality and Characteristics of Service** variables (such as, number of stops, average length of a route, average distance between stops, overlapped route lengths and so on), **Socio/Demographic–Geographic** variables (such as population density, size of the area where the service is accessible and so on) and **Externalities** (i.e. number of accidents and emissions) are reported.

Karlaftis and Tsamboulas (2012a) [9] investigate methods applied in the field of transportation focusing on data analysis applications. They classify methods as 1- Statistics and 2- Computational Intelligence (in particular Neural Networks (NN), as an extremely popular class of CI models). In selecting the appropriate approach – statistical or NN – model simplicity, and model suitability is important as much as model accuracy.

In recent years, the data and information has improved considerably with new technologies. Automated Fare Collection (AFC), Automatic Passenger Counter (APC), Automated Vehicle Location (AVL), and Geographical Positioning

Systems (GPS) have accelerated research on transit systems [7]. It is now possible to consider big data to develop new ways of measuring performance [21,22]. As a particular example, Google Transit can display the location of the activity centers and the mobility between these centers in real-time. Other examples of modern studies on transit performance metrics and visualization of big transit data can be found in report, handbook and manual formats on Transportation Research Board. The report of Valdés et al. (2017) [20], for example is closely related with the aim of our study. Main differences are a- their study is based on a single route hence the approaches are different b- bus stop data in their study is limited due to current APC devices (the stop, boarding, and alighting data were collected manually along a single route). In Turkey, [6] address the problem from monetary perspective (while focusing on pricing policies and automated fare collection systems). To the best of our knowledge, a study to measure bus line performances using big data has not been conducted in Turkey yet.

In this study, the focus is on bus transportation since it is more flexible compared to rail transportation and widely preferred by the masses in cities. The objective of this study is to identify inefficient routes and to propose improvements by a) the evaluation of route efficiencies b) the analysis of bus stop boarding counts and c) the determination of overlapping routes. The primary data source of this study comes from the Department of Transportation for the City of Antalya. The analysis are based on the complete boarding data of December 18, 2019 which is a standard weekday. Most of the analysis is done using Knime software. Novel visualization techniques (i.e. variance-area heatmaps and route similarity dendograms) presented with this study enrich the transit data visualization literature [16] and hopefully provide benefits to transit agencies, policy makers, and the public.

## 2   Methodology

In this study, the focus is on bus transportation since it is more flexible compared to rail transportation and widely preferred by the masses in cities. The primary data source of this study comes from the Department of Transportation for the City of Antalya. We load the complete boarding data of December 18, 2019 which is a standard weekday. The boarding data consists of passenger Id, passenger's boarding stop Id (origin), boarding time, bus Id, route Id (the direction on a particular line). For a particular route, bus stop locations with GPS coordinates required to visualize path on a map, are available from both municipal and commercial websites (such as www.ulasimburada.com).

The data set formed consist of 305 lines and 608 routes. A route consist of a sequential list of bus stops in either forward or backward (return) directions. Each line has opposite two directions except two lines which are omitted in analysis. On December 18, 2019, a total of 7347 trips (single direction services) were made and with these trips a total of 381962 passengers were carried.

Efficiency of a transportation system as a whole requires individual busses which are limited in numbers run as efficient as possible while keeping residents happy through easy access to a bus service, quick access to destination, connectivity to transit network and comfort. Our approach in improving transportation system has 2 stages; 1- improve route efficiency, 2- improve customer satisfaction. However, in this study we solely focus on the route efficiency. In a followup study we aim to include customer satisfaction aspect of the transportation.

Note that most of the analysis is done using Knime [1].

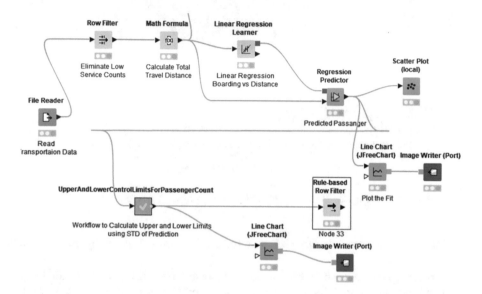

**Fig. 1.** Linear regression and upper and lower control limits for travel distance vs passenger count

## 2.1   Route Efficiency - RE

The simplest way to evaluate route efficiency (RE) is to calculate the number of passengers per unit distance (km) travelled. RE for a route $i$ is given by:

$$RE = \frac{P_{actual}(i)}{D_i} \tag{1}$$

where $P_{actual}(i)$ be the actual number of passengers for Route $i$ and $D(i)$ is the total travel distance in kilometers for this Route.

Currently, there is not a single number on literature that is suggested to be the benchmark for this metric. Because, such metric depends on multiple factors including the dynamics of the population, geography and resources, we calculated RE for all routes and normalize routes with respect to total travel distance via Linear Regression. Figure 1 show the Knime Implementation for this calculation. Using the linear regression, based on total distance travelled we

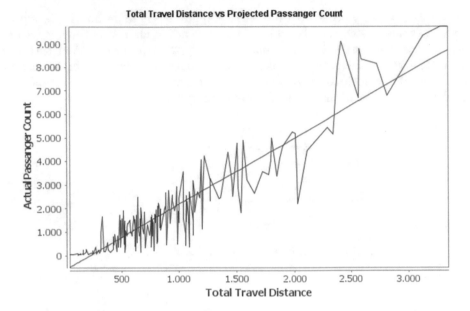

**Fig. 2.** Total travel distance vs passenger count

predicted the expected number of passengers for route i given by $P_{predicted}(i)$ (Fig. 2).

In order to establish outliers per km passenger boarding, the standard deviation of passenger count $(\sigma_r)$ for all routes ($N$ be the total number of routes) is calculated using Eq. 2.

$$\sqrt{\frac{\sum_{i=1}^{N}(P_{actual}(i) - P_{predicted}(i))^2}{N}} \tag{2}$$

Low and high performing routes are determined by identifying routes that are out-of-bounds of upper and lower control limits [18]. They are calculated using Eq. 3:

$$UPC = P_{actual}(i) + \sigma_r$$
$$LCL = P_{actual}(i) - \sigma_r \tag{3}$$

Values beyond these limits may need attention.

### 2.1.1    Traversal Route Evaluation

When lines are designed they often start from an origin and return back to where it start. Each line may have consider to have 2 routes, a forward and backward route. Typically, at the end of the day, when all trips are completed, the number of passengers carried on forward and backward routes are nearly equal. Any significant difference may need attention by the transit authority.

### 2.1.2    Variance-Area Curves for Route Boarding

Simply put, the between-area variance curves are obtained by calculating the variances of gray scale pixel values while varying rectangular unit areas within the 2-D gray scale matrix. This 2-D gray scale matrix is constructed from the heatmap image.

Although there are many ways to choose rectangular unit areas having equal area sizes [5], for the purpose of this study, these unit areas are considered squares.

Let $m$ be the pixel length and $n$ be the pixel width of the entire heatmap matrix. The image is first partitioned into unit areas of size $A_{1x1}$. The between-area variance curve, $CB(A_k)$, is the plot of the coefficients of variation between the varying unit areas $A_{k \times k}$. Mathematically, this relationship is formulated with Eq. 4 [17].

$$CB(A_k) = \frac{100}{\bar{F}_k} \sqrt{\frac{1}{m_k n_k} \sum_{i=1}^{m_k} \sum_{j=1}^{n_k} [F_{i,j} - \bar{F}_k]^2} \tag{4}$$

where  $CB(A_k)$  is the between-area variance among the unit areas of size $A_{k \times k}$,

  $m_k$      is the number of segments in the x direction, that
         is $m_k = m/k$,

  $n_k$      is the number of segments in the y direction, that is $n_k = n/k$,

  $F_{i,j}$     is the value of the property at the cell of at row $i$ and column $j$
         (ex; the gray scale pixel value at the cell),

  $\bar{F}_k$      is the mean value of the property for all unit areas;

$$\bar{F}_k = \frac{\sum_{i=1}^{m_k} \sum_{j=1}^{n_k} F_{i,j}}{m_k n_k}$$

  $m_k n_k$    is the total number of unit areas within a designated heatmap matrix.

## 2.2    Bus Stop Analysis

Analysis of bus stops can be comprehensive. Location of bus stops depends on demand nearby. For the purpose of this study, we do not suggest new bus stops but rather evaluate the boarding demand on existing routes. We identify bus stops with very few boarding as well as stops with high boarding counts. Furthermore, we counted the number of different routes and also the busses pass through a stop during the given day. The heading of bus stop analysis include

– Bus Stop Id/Name
– Boarding Count

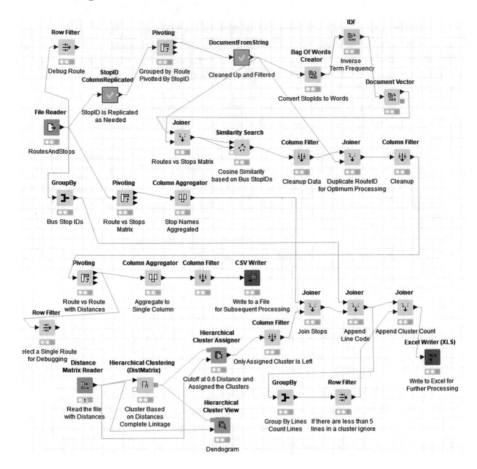

**Fig. 3.** Calculation of route similarity and subsequent hierarchical clustering Knime workflow

- Route Count
- Service Count
- Average Boarding Per Route Arrival at Stop
- Average Boarding Per Bus Service at Stop

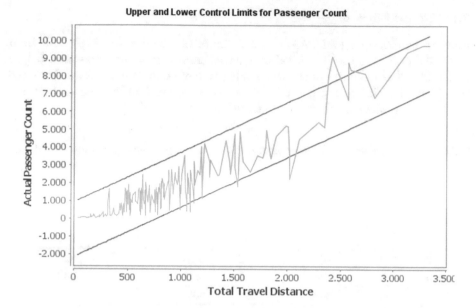

**Fig. 4.** Travel distance vs passenger count upper and lower control limits

## 2.3   Hierarchical Clustering for Bus Stop Similarity

If routes and bus stops had to be eliminated or redesigned. Then it is important to know which routes are close in terms of trip. Bus stops of a route may be used to determine overlapping routes. Several similarity measures are available to determine close routes. Among them cosine similarity with inverse term frequency has advantages over others as it favors less frequent matches over frequent ones [19]. Cosine similarity based distance matrix is formed where Rows and Columns are Routes and each cell in the matrix has a value between 0 and 1 that represents the cosine angle between the two routes. A distance value 0 means perfect match, whereas a distance matrix of 1.0 means routes do not have anything in common. During the hierarchical clustering complete linkage method is used to calculate the cluster distances

Once distance matrix is formed for routes, it has been used to cluster routes with respect to each other via distances calculated in the prior step. Figure 3 show the KNIME workflow for distance matrix calculation and the hierarchical clustering of the routes.

## 3   Results and Discussion

Several metrics are proposed in Methodology section executed on the transportation data obtained from Antalya Municipality. We have discussed Route Efficiency, Traversal Route Evaluation,Variance-Area Curves for Route Boarding, Bus Stop Analysis, and Clustering of Bus Lines with respect to overlapping bus stops.

## 3.1   Route Efficiency - RE

Figure 4 show upper and lower control limits obtained using Eq. 3. As total travel distance increases the total number of passenger counts increases. Upper and lower control limits show one standard deviation (Eq. 2) away from the trajectory line (linear regression). However, there are handful routes that are out of bound, some of which are given in Table 1.

**Table 1.** Selected set of outliers based on total travel distance, passenger count and trip information

| Line code | Route length (km) | Passenger count | Total dist. travelled (km) | Expected passenger count | Route effi- ciency | Bus uti- lization |
|---|---|---|---|---|---|---|
| KC06 | 24 | 9114 | 2412 | 6131 | 3.77 | 93 |
| LF10 | 30 | 8792 | 2570 | 6573 | 3.43 | 103 |
| 511 | 44 | 2209 | 2023 | 5042 | 1.09 | 49 |
| DM85 | 78 | 361 | 1092 | 2437 | 0.33 | 25.7 |
| DM86 | 62 | 428 | 1059 | 2345 | 0.4 | 25.17 |
| CV39 | 10 | 1650 | 329 | 302 | 5.01 | 51.5 |
| 507 | 22.5 | 21 | 113 | −302 | 0.18 | 4.2 |
| DC15A | 40 | 5180 | 2007 | 4998 | 2.58 | 103.6 |
| 531 | 10 | 46 | 128 | −260 | 0.36 | 3.53 |

Table 1 show $KC06$ has 9114 passengers boarded for a total travel distance of 2412 km. Nearly 3.77 passengers per km travelled. There are 100 trips and per trip on average 93 passengers transported throughout the route. Using Table 1, we may conclude that routes $KC06$ and $LF10$ carry a great deal of passengers. On the other hand, routes $DM85$ and $DM86$ under-perform. (Note that expected passenger numbers turn out to be negative for Lines 507 and 531. These two outliers may be omitted while performing regression.)

As an additional metric, we also calculated route efficiency as passengers transported per km distance travelled for all lines before the regression and sort them both ascending and descending. Line $CV39$ has highest passengers per km (5.01) traveled and Line 507 has the lowest passengers per km (0.18) traveled. We also calculated the bus utilization as the number of passengers boarded per bus trip. All routes are sorted for bus utilization in ascending and descending order and top matches are reported in Table 1. Line $DC15A$ had the highest bus utilization as 103 passengers per trip and Line 104 had the lowest 3.53 passengers per trip.

Regression with upper and lower control limits, route efficiency and bus utilization may individually or together be used to evaluate the efficiency of a route.

### 3.1.1   Traversal Route Evaluation

The number of passengers with Forward and Backward trips are expected to be close in count. Any significant difference between line traversals of a trip needs attention. It may be possible identify outliers in order to improve inefficient direction through rerouting. Table 2 show the number of passenger boarding for each direction of highly asymmetric lines. As seen from the table, Line 2 has a huge difference between forward and backward directions. It may be possible to study the path, understand the dynamics and potentially reroute the line for improved service.

**Table 2.** Passenger counts of routes for a line in a round-trip

| Line ID | Forward passenger count | Backward passenger count | Percent difference |
|---------|------------------------|-------------------------|--------------------|
| 2       | 806                    | 65                      | 170                |
| 23      | 1071                   | 144                     | 153                |
| 63      | 4488                   | 1585                    | 96                 |

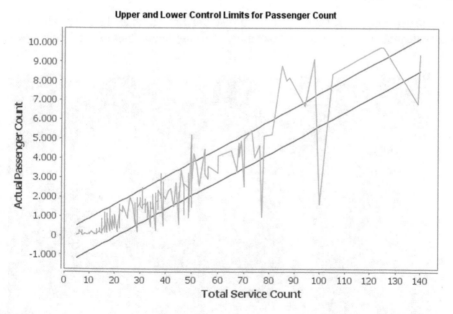

**Fig. 5.** Service count vs passenger count with upper and lower control limits for bus stops

Figure 5 show the total service count versus the number of boardings for a route. As number of busses increase on a route the passenger count increases. However, we identified a few routes which have a high service count, but not enough passengers to support it. Spikes that are below the lower limit shown in the Fig. 5, correspond to the routes VF01-Return and 998-Both directions.

### 3.1.2    Time-Location Variance Analysis of Route Boarding

Time-Location mapping of route boarding provides us to analyze the variation of boarding in both time domain and as well as route bus stop sequences. Mapping involves counting boarding in subsequent bus stops for each trip on a route. Such mapping may be visualized using heatmap plot such as shown in Fig. 6. Figure 6 show high boardings as dark marks distributed in 2-D for select routes. Visually, we can conclude that Route $KC06$ has the lowest variation both in time and location which is also consistent with the route efficiency of this route. Route $LF10$ has medium performance but Route $CV39$ has the most variation both in location and trip.

Variance-Area curve obtained by Eq. 4 allows us to quantify variation in both time and bus stop dimensions. Such numeric evaluation shown in Fig. 7 matches visual experience shown in Fig. 6. $CV39$ has the highest and $KC06$ has the lowest variation.

### 3.2    Bus Stop Analysis

For the selected date, there were 3140 distinct bus stops and among which 2804 has boarding data (at least 1 boarding). Among them infrequent bus stops are

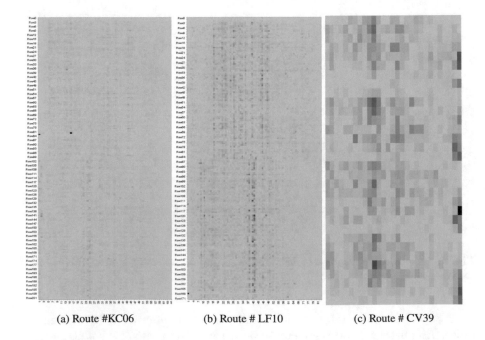

(a) Route #KC06                (b) Route # LF10                (c) Route # CV39

**Fig. 6.** Variance-area analysis of boarding for routes by using heatmap plot

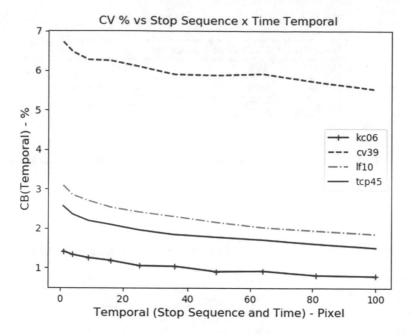

**Fig. 7.** Variance-area curve for boarding on routes

listed in Table 3. One can conclude from the table that there are 567 (% 18) bus stops in Antalya between 1 and 5 passenger boarding. It may be possible to eliminate these bus stops and the stops with no boarding by skipping stops and/or rerouting the bus on a different path to increase route efficiency. Among the 2804 bus stops that have boarding data, only 708 have more than 100 total boardings during a day. The mean boarding count is 136 and the median is 27 boardings. The huge difference between the mean and the median suggest that, few stops with high boarding counts influence the overall distribution, whereas most of the stops generate limited passengers to the network.

**Table 3.** In frequent bus stops

| No boarding | Single boarding | 2 boarding | 3 boarding | 4 boarding | 5 boarding |
|---|---|---|---|---|---|
| 336 | 152 | 131 | 107 | 98 | 79 |

Similarly, bus stops with frequent boarding are monitored in Table 4 According to this table, 8326 passengers appear to board from the same bus stop ( i.e. 10151). Furthermore, as reported in Table 4, total boarding count from 10151, 10366 and 10364 is equal to 18890, that is % 20 of all boarding.

Table 5 show boarding and route count sorted. BusStop #10185 has 35 bus routes passing through. This is the bus stop which is the most connected. However, the number of passengers boarding is only 967, which might be considered

**Table 4.** Top 3 bus stops with frequent boarding

| Bus stop ID | Number of passengers | |
|---|---|---|
| 10151 | 8326 | Vakifbank |
| 10366 | 5492 | MarkAntalya |
| 10364 | 5072 | Vakifbank |

low. The second most connected bus stop is BusStop #10151. Which has a huge number of boarding with 8326 passengers. On the other hand BusStop#15066 has single route with 675 passengers. In order to reduce transfer it is suggested to add new routes to low route count bus stops with high passenger count.

**Table 5.** Number of routes and boarding count at a particular bus stop

| BusStopId | BoardingCount | StopName | RouteCount | AvgBoardingPerRoute |
|---|---|---|---|---|
| 10185 | 967 | Dumplipinar Bulvari | 35 | 27,62 |
| 10151 | 8326 | Vakifbank | 32 | 260,18 |
| 10335 | 1553 | Meltem Blv-5 | 32 | 48,53 |
| 10023 | 3980 | Migros | 31 | 128,38 |
| 15066 | 675 | Uc Kapilar 3 | 1 | 675 |
| 11245 | 505 | 3847.Sk-1 | 1 | 505 |

### 3.3   Hierarchical Clustering of Routes for Bus Stop Similarity

In Antalya, a major reason for low REs, low bus utilization, high boarding variances on bus stops may be resulted from lengthy bus lines with many overlapping sections. This can be investigated by clustering of routes according to their common bus stops. The resulted dendrogram presented in Fig. 8 shows the hierarchical clustering of bus routes. Here x- axis denotes route Ids and y-axis represent similarity distances (y=1 means no similarity whereas y=0 means one hundred percent similarity). For sample shown in Fig. 8, possible interpretations are a) 52111 and 52101 have similar routes as the height of the link that joins them together show how close these 2 routes are; b) Routes are allocated to clusters by drawing a horizontal cut-off line through the dendrogram, By adaptively selecting such cut-off one may decide which routes may be merged in cases where capacity allows.

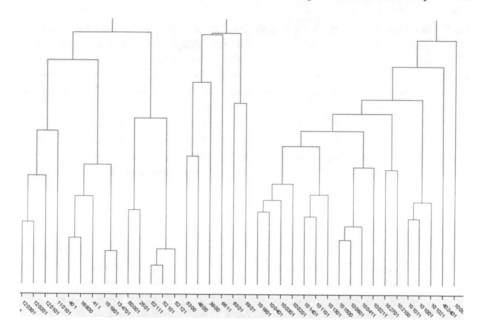

**Fig. 8.** Dendrogram of bus routes

## 4   Conclusion

The aim of this study is to identify inefficient routes and to propose improvements by the evaluation of route efficiencies, the analysis of bus stop boarding counts and the clustering of routes. Several metrics are proposed in Methodology section executed on the transportation data obtained from Antalya Municipality. Results and Discussions are based on the complete boarding data of December 18, 2019 which is a standard weekday. Most of the analysis is done using Knime software.

The simplest way to evaluate route efficiency is to calculate the number of passengers per unit distance (km) travelled. As an additional metric the bus utilization is the number of passengers boarded per bus trip. As total travel distance increases the total number of passenger counts are expected to increase whereas the number of passengers boarded per bus is desired to be nearly constant. Similarly, as number of busses increase on a route the passenger count is expected to increases. In this study, route efficiency and bus utilization for all lines are calculated and regressions with upper and lower control limits are used to determine over-performed and under-performed routes. Traversal analysis is performed to identify routes with significant difference in forward and backward passenger loads.

The boarding demand on existing routes are evaluated by conducting bus stops analysis. A few stops with significantly high boarding counts are noticed whereas most of the stops generate limited passengers to the network. Furthermore, bus stops with high route counts (i.e. most connected bus stops) are determined and boarding counts are compared accordingly. Contrary to expectations, it is observed that there exist scarce boarding counts at the stops with high route counts or larger boarding counts at the stops with few route counts. By bus stop analysis it is possible to suggest new routes to low route count bus stops with high passenger counts.

The variation of boarding demand on both time domain and route bus stop sequences are visualized by using heat maps. Overlapping routes are identified by hierarchical clustering of bus stop similarities. One may decide which routes may be merged in cases where capacity allows by using the dendrogram of bus routes as a result of clustering.

As an outcome of this research, the analysis may propose to modify or eliminate inefficient routes, suggest new lines, identify inefficient bus stops, and potentially modify path of a route.

**Acknowledgement.** The authors would like to thank Ulaşım Planlama ve Raylı Sistem Dairesi Başkanlı ı- Antalya for the data provided in this pilot study.

# References

1. Berthold, M.R., et al.: Knime: The konstanz information miner. In: Studies in Classification, Data Analysis, and Knowledge Organization (2007)
2. Daraio, C., Diana, M., Di Costa, F., Leporelli, C., Matteucci, G., Nastasi, A.: Efficiency and effectiveness in the urban public transport sector: a critical review with directions for future research. Eur. J. Oper. Res. **248**(1), 1–20 (2015). https://doi.org/10.1016/j.ejor.2015.05.059
3. Emrouznejad, A., Yang, G.: A survey and analysis of the first 40 years of scholarly literature in DEA: 1978–2016. Socio-Econ. Plann. Sci. **61**(1), 4–8 (2018). https://doi.org/10.1016/j.seps.2017.01.008
4. Fielding, G.J., Glauthier, R.E., Lave, C.A.: Performance indicators for transit management. Transportation **7**(4), 365–379 (1978)
5. Günay, M., Suh, M., Jasper, W.: Prediction of surface uniformity in woven fabrics through 2-D anisotropy measures, part ii: simulation and verification of the prediction model. J. Text. Inst. **98**, 117–126 (2007). https://doi.org/10.1533/joti.2005.0234
6. Gokasar, I., şimşek, K.: Using big data for analysis and improvement of public transportation systems in Istanbul (2014)
7. Ibarra-Rojas, O., Delgado, F., Giesen, R., Muñoz, J.: Planning, operation, and control of bus transport systems: a literature review. Transp. Res. Part B Methodological **77**, 38–75 (2014). https://doi.org/10.1016/j.trb.2015.03.002
8. Karlaftis, M.: A DEA approach for evaluating the efficiency and effectiveness of urban transit systems. Eur. J. Oper. Res. **152**, 354–364 (2004). https://doi.org/10.1016/S0377-2217(03)00029-8
9. Karlaftis, M.G., Tsamboulas, D.: Efficiency measurement in public transport: are findings specification sensitive? Transp. Res. Part A Policy Pract. **46**(2), 392–402 (2012). https://doi.org/10.1016/j.tra.2011.10.005
10. Kerstens, K.: Technical efficiency measurement and explanation of French urban transit companies. Transp. Res. Part A Policy Pract. **30**(6), 431–452 (1996)

11. Liu, W., Teng, J., Zhang, D.: (2013) Performance evaluation at bus route level: considering carbon emissions. In: ICTE 2013, pp. 2573–2580. https://doi.org/10.1061/9780784413159.374
12. Miller, P., De Barros, A., Kattan, L., Wirasinghe, S.: Public transportation and sustainability: a review. KSCE J. Civil Eng. **20**, 1076–1083 (2016). https://doi.org/10.1007/s12205-016-0705-0
13. Redman, L., Friman, M., Gärling, T., Hartig, T.: Quality attributes of public transport that attract car users: a research review. Transp. Policy **25**(C), 119–127 (2013). https://doi.org/10.1016/j.tranpol.2012.11.005
14. Ryus, P.: A summary of TCRP report 88: a guidebook for developing a transit performance-measurement system. TCRP Research Results Digest (56) (2003)
15. Sheth, C., Triantis, K., Teodorovic, D.: Performance evaluation of bus routes: a provider and passenger perspective. Transp. Res. Part E Logistics Transp. Rev. **43**, 453–478 (2007). https://doi.org/10.1016/j.tre.2005.09.010
16. Stewart, C., Bertini, R., El-Geneidy, A., Diab, E.: Perspectives on transit: potential benefits of visualizing transit data. Transp. Res. Rec. J. Transp. Res. Board **2544**(1), 90–101 (2016). https://doi.org/10.3141/2544-11
17. Suh, M., Günay, M., Jasper, W.: Prediction of surface uniformity in woven fabrics through 2-D anisotropy measures, part i: definitions and theoretical model. J. Text. Inst. **98**, 109–116 (2007a). https://doi.org/10.1533/joti.2005.0232
18. Suh, M.W., Gunay, M., Vangala, R.: Dynamic textile process and quality control system. NTC Annual Report (2007b)
19. Tan, P.N., Steinbach, M., Kumar, V.: Introduction to Data Mining. Addison Wesley (2005)
20. Valdés, D., Cruzado, I., Martínez, J., Taveras, Y.: Development of transit performance measures using big data (2017). https://trid.trb.org/view/1592142
21. Welch, T.F., Widita, A.: Big data in public transportation: a review of sources and methods. Transp. Rev. **39**(6), 795–818 (2019). https://doi.org/10.1080/01441647.2019.1616849
22. Zhu, L., Yu, F.R., Wang, Y., Ning, B., Tang, T.: Big data analytics in intelligent transportation systems: a survey. IEEE Trans. Intell. Transp. Syst. **20**(1), 383–398 (2019)

# Methods for Trajectory Prediction in Table Tennis

Mehmet Fatih Kaftanci[1(⊠)], Melih Günay[1], and Özge Öztimur Karadağ[2]

[1] Akdeniz University, Antalya, Turkey
fatih@kaftanci.com, mgunay@akdeniz.edu.tr
[2] Alanya Alaaddin Keykubat University, Alanya, Antalya, Turkey
ozge.karadag@alanya.edu.tr

**Abstract.** Training and skill improvement is totally about finding a valuable partner in table tennis. It is difficult to find a partner who has various of good technical skills, teaching capacity to improve your skills, and whose training hours are compatible. Thus, so many researchers improve several methods and systems to solve this deficiency with robot table tennis.

This paper provides a review of the common methods and systems then tries to explain how far these methods are from the solution that the professional/semiprofessional table tennis players expected. Some researchers are focused on the whole system, some study on a specific part of the problem. So this review provides a category-based approach to this subject and gives information about the last situation.

**Keywords:** Trajectory prediction · Machine vision · Object tracking · Object detection · Table tennis robot · Artificial intelligence · Artificial neural networks

## 1 Introduction

One of the most important problems for professional/semi-professional table tennis players is finding a trainer or a training partner who will contribute to their development. In order to develop skills against different styles of play, it is necessary to practice/match with players in these styles. Due to less popularity compared to other sports, it is very difficult to find this opportunity in this sport branch. In order to solve this problem, various studies have been carried out on the development of table tennis robot for years. Different methods have been proposed in these recent studies. Common issue for all methods is finding the proper hitting point, racket trajectory, hitting velocity and angle for table tennis robot that strikes the ball back well. The main goal for all methods is obtaining trajectory prediction at the proper time. For an attack player ball flight time can be under 200 ms and prediction should finish under 100 ms. Hence the variety of methods and there is no certain solution about this problem, both experts and newcomers would have need the bases of the methods. This paper provides a

J. Hemanth et al. (Eds.): ICAIAME 2020, LNDECT 76, pp. 696–704, 2021.
https://doi.org/10.1007/978-3-030-79357-9_64

detailed review about all methods for all studies about trajectory prediction in table tennis.

This paper is organized as follow. In Sect. 2, all methods obtained from literature review are overviewed. In Sect. 3, performance analysis of the methods are compared. Concluding remarks are made in Section IV.

# 2    An Overview of the Trajectory Prediction Methods

All of the methods we have reviewed have evolved over time with the experience of other researchs. Researchers tried to find the best method for trajectory prediction in proper time. Some researchers used advanced physical formulas on dynamic features of the ball, some used machine learning on training trajectory data.

- Some of the methods used in this research are listed below:
- Physics modelling by second order polynomial
- Regression Learning with experienced data
- Aerodynamics Model and Bouncing Model
- EKF (Extended Kalman Filter) supported physical model
- Fuzzy rectification for spinning ball
- UKF (Unscented Kalman Filter) on non-linear system and spin analyse with BP neural network
- ADM (Aerodynamics Model) and RRM (Racket Rebound Model) backward in time
- IELM (Improved Extreme Learning Machine) with non-massive data
- e-SVR (e Support Vector Regression) method
- Weighted least-square method

**Table 1.** Typical values of parameters in the physics model

| Parameter | Value |
|---|---|
| g | $9.802\,\mathrm{m/s^2}$ |
| m | $0.0027\,\mathrm{kg}$ |
| $\rho$ | $1.29 \times 10^3\,\mathrm{kg/m^3}$ |
| $C_D$ | $0.4 \sim 0.5$ |
| $\|\vec{V}\|$ | $3 \sim 10$ |
| $C_L$ | $0.2 \sim 0.6$ |
| $r_b$ | $0.02\,\mathrm{m}$ |
| $\omega$ | $0 \sim (10 \times 2\pi)\,rad/s$ |

## 2.1   Physics Modelling by Second Order Polynomial

Zhang et al. [1] has proposed a method based on physics modelling of the ball without rotation. Second-order polynomials (1) are used to compute the x,y and z points of estimated trajectory on a given time.

$$\begin{cases} x = a_1^2 + b_1 t + c_1 \\ y = a_2^2 + b_2 t + c_2 \\ z = a_3^2 + b_3 t + c_3 \end{cases} \tag{1}$$

The coefficients $a_1$, $a_2$, $a_3$, $b_1$, $b_2$, $b_3$, $c_1$, $c_2$, $c_3$ can be computed via the least square method (LSM) with 3-D positions that calculated from the images captured by stereo-vision system.

Velocities $(V_x, V_y, V_z)$ can be calculated via state vector:

$$\begin{bmatrix} \dot{x} \\ \dot{y} \\ \dot{z} \\ \dot{V_x} \\ \dot{V_y} \\ \dot{V_z} \end{bmatrix} = \begin{bmatrix} V_x \\ V_y \\ V_z \\ -K_m \|\vec{V}\| V_x \\ -K_m \|\vec{V}\| V_y \\ -K_m \|\vec{V}\| V_z - g \end{bmatrix} \tag{2}$$

$$K_m = \frac{1}{2m} \rho S C_D \tag{3}$$

Parameters used in physics model are in Table 1.

Time dependent positions and velocities can be predicted via iteration with

$$\begin{bmatrix} x_j \\ y_j \\ z_j \\ V_{xj} \\ V_{yj} \\ V_{zj} \end{bmatrix} = \begin{bmatrix} x_j - 1 \\ y_j - 1 \\ z_j - 1 \\ V_{xj} - 1 \\ V_{yj} - 1 \\ V_{zj} - 1 \end{bmatrix} + \begin{bmatrix} V_{xj} - 1 \\ V_{yj} - 1 \\ V_{zj} - 1 \\ -K_m \|\vec{V}_{j-1}\| V_{xj-1} \\ -K_m \|\vec{V}_{j-1}\| V_{yj-1} \\ -K_m \|\vec{V}_{j-1}\| V_{zj-1} - g \end{bmatrix} T_c \tag{4}$$

## 2.2   Regression Learning with Experienced Data

Huang et al. [2] proposed a method based on constructing the ball map for trajectory prediction. Initial state can be estimated after applying polinomial fitting [1] to the initial trajectory which came from vision system.

Rebound trajectory of the ball can be approximated by a parabolic curve and rebound trajectory pattern can be extracted using least squares method.

$$p_b(t) = \Phi(t)\omega_b, \tag{5}$$

$$\omega_b = \left(\Phi_b{}^T \Phi_b\right)^{-1} \omega_b{}^T p_b \tag{6}$$

With the regression learning method, entire ball trajectory can be estimated by experience data. How big is the experienced database, trajectory prediction by regression method improves.

## 2.3   Aerodynamics Model and Bouncing Model

Chen et al. [3] proposed aerodynamic model and bouncing model used together to predict ball trajectory. Aerodynamic model estimate the trajecory in flight time. Bouncing model used to calculate the velocity changes after bouncing.

In aerodynamic model, gravity (7), air resistance (8) and the magnus force (9) are formulated to obtain the trajectory for flying ball.

$$F_g = mg \tag{7}$$

$$F_d = -\frac{1}{8}C_D\rho_a\pi D^2\|v\|v \tag{8}$$

$$F_m = -\frac{1}{8}C_m\rho_a\pi D^3\omega \times v \tag{9}$$

In bouncing model, velocity and spin values after rebound can be computed if the incidence velocity, incidence spin, vertical restitution coefficient and friction coefficient of table are known.

## 2.4   EKF (Extended Kalman Filter) Supported Physical Model

Liu et al. [4] proposed an algorithm which combines the historical data and the real time information. Flying and rebounding model of the ball are formulated considered on spin.

EKF (extenden kalman filter) is used to estimate the status of the ball. Hitting point for robot can be calculated on a virtual hitting plane by iterations with theses estimated ball status.

In this architecture system would keep gathering information and do prediction continuous till motion ends. This method is applied on different trajectories and the results are considered as ground truth.

## 2.5   Fuzzy Rectification for Spinning Ball

Ren et al. [5] informed a new algorithm that analyzes how the spinning effects on a flying ball for trajectory prediction. This algorithm is based on both model and experience learning. Due to both model based learning and experience based learning methods needs large sample of data and has prediction errors, this method proposed an algorithm that is combining these 2 method based on the error of the prediction results for the non-spinning ball and the real collected trajectory.

This algorithm first predict the trajectory with non-spinning method then correct the errors with fuzzy rules like iterative method.

Fuzzy rules are standart triangle membership functions applied to X, Ym and rot variables

## 2.6   UKF (Unscented Kalman Filter) on Non-linear System and Spin Analyse with BP Neural Network

Wang et al. [6] proposed a method based on applying Unscented Kalman Filter (UKF) to the non-linear model. Because of the Jacobian matrix calculation like complex operations Extended Kalman Filter (EKF) has intensive computation and less efficiency on prediction then UKF.

UKF is more suitable for realtime table tennis robot systems because of its accuracy and less time usage on computing.

UKF uses time dependent 9 dimensional vectors to respresent state variables. Unscented Tranformation (UT) is used as a method to calculate the statistics of a random variable in a non-linear transformation.

By constructing a BP neural network it is possible to decide the spin type. Three-dimensional velocity values for 5 virtual plane that comes from UKF filter as input values for the BP neural network. After data normalization and a method proposed for preventing overfitting, spin type can be classified that comes from an output from BP neural network.

## 2.7   ADM (Aerodynamics Model) and RRM (Racket Rebound Model) Backward in Time

Nakashima et al. [7] proposed a method for trajectory prediction by solving ADM (Aerodynamics Model) and RRM (Racket Rebound Model) problems backward in time.

Finite Difference Method (FDM) is linearized from an approximated aerodynamics model (ADM) which has a Two Point Boundary Value Problem (TPBVP) that has to be solved. Also, TPBVP can be solved if striking point, landing points and angular velocity after striking is given.

In FDM time section is divided into subsections with the same interval and non-linear vectors are calculated by solving linearized ones iteratively.

## 2.8   IELM (Improved Extreme Learning Machine) with Non-massive Data

Wang et al. [8] proposed a new method for trajectory prediction named IELM (Improved Extreme Learning Machine). In this method trajectory prediction accuracy is mostly related with identification and classification of spinning ball. So IELM is improved to predict spin value and classification with a non-massive data.

Primitive ELM (Extreme Learning Machine) is powerful on big data but this increases computation time. IELM is approximately 40 times faster than primitive ELM on non-massive data.

## 2.9   E-SVR (e Support Vector Regression) Method

Liu et al. [9] proposed a learning algorithm based on e-Support Vector Regression (e-SVR) to learn hitting policy that enables the robot to return the incoming balls to a desired location.

e-SVR uses the position and velocity of the incoming ball when it is passing through the specific virtual plane to represent the dynamic features of the ball. Virtual plane (state vector) is computed by interpolation of neighborhood locations of the ball around the virtual plane using a first-order polynomial of time in both x and y directions and a second-order polynomial of time in z direction.

## 2.10   Weighted Least-Square Method

Liu et al. [10] proposed Weighted Least-Square method to predict the trajectory of the flying ball by set of 3D locations and velocities that obtained from vision system.

GPU based stereo vision system is constructed to use multithread technology and speed up the image processing on blurred images.

Spatial coordinates of the ball calculated by using iterative procedure provided by OpenCV. Then trajectory predicted by fitting positions and velocities of the ball obtained from these coordinates to (10) using weighted least square method.

$$
\begin{bmatrix} P_x(t) \\ P_y(t) \\ P_z(t) \\ v_x(t) \\ v_y(t) \\ v_z(t) \end{bmatrix} = \begin{bmatrix} P_x0\ v_x0\ \ 0 \\ P_y0\ v_y0\ a_y0 \\ P_z0\ v_z0\ \ g \\ v_x0\ \ 0\ \ \ 0 \\ v_y0\ a_y0\ \ 0 \\ v_z0\ \ g\ \ \ 0 \end{bmatrix} + \begin{bmatrix} 1 \\ t \\ \frac{t^2}{2} \end{bmatrix} \tag{10}
$$

# 3   Performance Analysis of Methods

In this section a comparative performance analysis of the methods based on accuracy, time usage and relevance for a table tennis robot system that is suitable for professional table tennis training. Results of performance and analyses are presented in Table 2. Accuracy and time usage is the most important parameters to analyse a method. Because if accuracy is not enough to predict the hit point for robot system, ball van not be returned. On the other hand all computations, prediction and robotic arms movement to return the ball have to finish in a flight time of a flying ball in table tennis.

**Table 2.** Accuracy, prediction time and solution results for all methods in this review

| Method | Accuracy | Time | Sol. |
|---|---|---|---|
| Physics modelling by second order polynomial | Average 6.5 mm | 2658 ms | No |
| Regression Learning with experienced data | %96 Successful return | – | No |
| Aerodynamics Model and Bouncing Model | Less than 1 cm | ~1 s | No |
| EKF supported physical model | Accuracy increases approximately %20 after filtered | Time decreases approximately %20 after filtered | No |
| Fuzzy rectification for spinning ball | Left Rotation: 1–6 cm Right Rotation: 0.1–3 cm | – | No |
| UKF on non-linear system and spin analyse with BP neural network | X < 30 mm Y < 35 mm Z < 40 mm | 1 s | No |
| ADM and RRM backward in time | ~0.05 cm | ~0.63 s | No |
| IELM with non-massive data (precision of identification on spin) | %83–%95 | 0.086–0.62 s | No |
| e-SVR Method | 2–26 cm | 0.44–0.53 s | No |
| Weighted Least-Square Method | 0.5–3 cm | – | No |

## 3.1   Accuracy

Some researchers suppose that hitting the ball and turning back it to the opponent side is a success and calculate the accuracy with returning percentage. Some suppose the distance from the predicted striking point to real striking point as accuracy.

## 3.2   Time Usage

Time passed for a successful prediction or flight time of the ball that has been predicted well.

**Acknowledgements.** Special thanks to the Akdeniz University Scientific Research Projects Coordination Unit for accepting my research project and supporting materials.

## References

1. Zhang, Z., Xu, D., Tan, M.: Visual measurement and prediction of ball trajectory for table tennis robot. IEEE Trans. Instrum. Meas. **59**(12), 3195–3205 (2010)

2. Huang, Y., Buchler, D., Koc, O., Scholkopf, B., Peters, J.: Jointly learning trajectory generation and hitting point prediction in robot table tennis. In: Proceedings of the IEEE-RAS 16th International Conference on Humanoid Robots, pp. 650–655 (2016)
3. Chen, X., Huang, Q., Zhang, W., Yu, Z., Li, R., Lv, P.: Ping-pong trajectory perception and prediction by a PC based high speed four-camera vision system. In: Proceedings of the 7th World Congress Intelligent Control Automation, Taipei, Taiwan, pp. 1087–1092 (2011)
4. Liu, J., Fang, Z., Zhang, K., Tan, M.: A new data processing architecture for table tennis robot. In: IEEE Conference on Robotics and Biomimetics 2015, pp. 1780–1785 (2010)
5. Ren, Y., Fang, Z., Xu, D., Tan, M.: A trajectory prediction algorithm based on fuzzy rectification for spinning ball. In: 9th Asian Control Conference (ASCC) (2013)
6. Wang, Q., Zhang, K., Wang, D.: The trajectory prediction and analysis of spinning ball for a table tennis robot application. In: The 4th Annual IEEE International Conference on Cyber Technology in Automation, Control and Intelligent Systems, pp. 496–501 (2016)
7. Nakashima, A., Ito, D., Hayakawa, Y.: An online trajectory planning of struck ball with spin by table tennis robot. In: IEEE/ASME International Conference on Advanced Intelligent Mechatronics (AIM), pp. 865–870 (2014)
8. Wang, Q., Sun, Z.: Trajectory identification of spinning ball using improved extreme learning machine in table tennis robot system. In: I The 5th Annual IEEE International Conference on Cyber Technology in Automation, Control and Intelligent Systems, pp. 551–554 (2015)
9. Liu, H., Li, Z., Wang, B., Zhou, Y., Zhang, Q.: Table tennis robot with stereo vision and humanoid manipulator I: system design and learning-based batting decision. In: IEEE International Conference on Robotics and Biomimetics (ROBIO), pp. 905–910 (2013)
10. Liu, H., Li, Z., Wang, B., Zhou, Y., Zhang, Q.: Table tennis robot with stereo vision and humanoid manipulator II: visual measurement of motion-blurred ball. In: IEEE International Conference on Robotics and Biomimetics (ROBIO), pp. 2430–2435 (2013)
11. Nakashima, A., Takayanagi, K., Hayakawa, Y.: A learning method for returning ball in robotic table tennis. In: International Conference on Multisensor Fusion and Information Integration for Intelligent Systems (MFI) (2014)
12. Zhao, Y., Wu, J., Zhu, Y., Yu, H., Xiong, R.: A learning framework towards real-time detection and localization of a ball for robotic table tennis system. In: IEEE International Conference on Real-time Computing and Robotics (RCAR), pp. 97–102 (2017)
13. Mülling, K., Kober, J., Peters, J.: A biomimetic approach to robot table tennis. In: IEEE/RSJ International Conference on Intelligent Robots and Systems, pp. 1921–1926 (2010)
14. Yang, P., Xu, D., Zhang, Z., Chen, G., Tan, M.: A vision system with multiple cameras designed for humanoid robots to play table tennis. In: IEEE International Conference on Automation Science and Engineering, pp. 737–742 (2011)
15. Chen, G., Xu, D., Yang, P.: High precision pose measurement for humanoid robot based on PnP and OI algorithms. In: IEEE International Conference on Robotics and Biomimetics, pp. 620–624 (2010)

16. Liu, C., Hayakawa, Y., Nakashima, A.: A registration algorithm for online measuring the rotational velocity of a table tennis ball. In: IEEE International Conference on Robotics and Biomimetics, pp. 2270–2275 (2010)
17. Koç, O., Maeda, G., Peters, J.: A new trajectory generation framework in robotic table tennis. In: IEEE/RSJ International Conference on Intelligent Robots and Systems (IROS), pp. 3750–3756 (2016)
18. Zhang, Z., Xu, D.: Design of high-speed vision system and algorithms based on distributed parallel processing architecture for target tracking. In: 7th Asian Control Conference, pp. 1638–1643 (2009)
19. Bao, H., Chen, X., Wang, Z.T., Pan, M., Meng, F.: Bouncing model for the table tennis trajectory prediction and the strategy of hitting the ball. In: IEEE International Conference on Mechatronics and Automation, pp. 2002–2006 (2012)
20. Sun, L., Liu, J., Wang, Y., Zhou, L., Yang, Q., He, S.: Ball's flight trajectory prediction for table-tennis game by humanoid robot. In: IEEE International Conference on Robotics and Biomimetics (ROBIO), pp. 2379–2384 (2009)
21. Zhang, K., Cao, Z., Liu, J., Fang, Z., Tan, M.: Real-time visual measurement with opponent hitting behavior for table tennis robot. IEEE Trans. Instrument. Measur. **67**(4), 811–820 (2018)

# Emotion Analysis Using Deep Learning Methods

Bekir Aksoy[✉] and İrem Sayin

Faculty of Technology Mechatronics Engineering, Isparta University of Applied Sciences,
Isparta, Turkey
bekiraksoy@isparta.edu.tr

**Abstract.** With the rapid advancement of technology, the importance of artificial intelligence methods has also increased. Artificial intelligence methods have started to be used frequently in many fields such as health, education, engineering, medicine and agriculture. In this study, the emotion image data set from the open-source access website (Kaggle.com) was classified as normal, angry and happy using VGG16, Resnet50 and DenseNet121 architectures. Among the deep learning architectures used, the best result was obtained with the DenseNet121 algorithm with an accuracy rate of 84.16% .

**Keywords:** Deep learning · Emotion analysis · Classification

## 1 Introduction

Today, with the development of computer technology, studies in the field of human-computer interaction are gaining momentum and becoming more and more important [1]. Thanks to these developments in human-machine interaction, the communication between user and machine are strengthened. One of the innovative studies conducted in this direction is the identification of human facial expressions from the obtained images and the analysis of these images for human emotions by computerized systems [2].

In the last two decades, the identification of human facial expressions plays an important role in human-computer interaction [3]. Facial expressions express human emotions as a non-verbal communication tool and make it possible to understand people's feelings. Similar to the communication between humans, it is aimed for computers to recognize human facial expressions [4]. The main goal of facial expression recognition by computers is to define the emotional state of people based on the given facial images. These facial expressions to be defined are classified as anger, disgust, fear, happiness, sadness and surprise [5]. The emotion analysis method can be used in many areas to examine people's emotions. In the lessons given by the distance education method, the facial expressions of the students who are bored in the lesson or have difficulty understanding the lesson can be identified, the way the lesson is carried out can be changed, the difficulty level of the lesson can be adjusted, and things that provide encouragement can be included into the environment to attract the student's attention to the lesson. In the field of security, emotions such as fear and anxiety can be detected by making emotion analysis in public areas. In the field of health, to understand the difficulties that patients

J. Hemanth et al. (Eds.): ICAIAME 2020, LNDECT 76, pp. 705–714, 2021.
https://doi.org/10.1007/978-3-030-79357-9_65

or individuals with autism experience in the treatment process emotion analysis can be used and the treatment process can be adjusted using emotion analysis results. In the automotive field, the emotional state of the driver can affect their driving performance. Therefore, emotion analysis can be used to support the driver's driving experience and encourage a better driving performance [2].

When academic studies on emotion analysis are examined; Soyel et al. (2010) presented a feature selection process for pose-invariant 3D facial expression recognition [3]. The process provides a lower-dimensional subspace representation, taken from the geometric localization of facial feature points to classify facial expressions, which is optimized to improve classification accuracy. A Fisher criterion-based approach is adopted for the most appropriate feature selection. The two-step probabilistic neural network architecture is used as a classifier to recognize human facial expressions. Facial expressions are classified into one of three expression groups created using seven basic facial expressions in the first stage. At the second classification stage, the expression is determined using within-group classification. Facial expressions were successfully recognized with an average recognition rate of 93.72 Kudiri et al. (2012) studied the problem of data loss in the feature extraction scheme based on the limited number of positions of facial muscles [6]. To improve the detection performance, the authors proposed relative sub-image based features. Classifications were made using a support vector machine to implement an automatic emotion detection system. According to the results, the relative sub-image-based features proposed by the authors increased the classification rates. Kumar et al. (2017) presented an approach to predict human emotions and how the intensity of emotions in the face changes from low to high emotion using the deep Convolution Neural Network [7]. In this study, authors have used the FERC-2013 database for training. They have achieved an accuracy of 90+% with the proposed CNN architecture and method. Ensar and Günay (2017) analyzed the algorithms used for face recognition and compared the performance of these algorithms [8]. Analyzed methods are k-nearest neighbor, Naive Bayes, eigenfaces, principal component analysis and k-mean algorithms. These methods have been implemented on the ORL face dataset. As a result of the analysis, the k-nearest neighbor algorithm and eigenfaces algorithms had the best performance, and the Naive Bayes algorithm had the worst performance result. The performance of the k-nearest neighbor algorithm, which gave the most successful result, dropped from 94% to 91.5% after principal component analysis. This difference increased by 7% in the Naive Bayes algorithm. In another study, Hossain and Muhammad proposed an emotion recognition system using a deep learning approach on big data structures consisting of speech and video data. In the proposed system, Mel-spectrogram was obtained by first switching the audio signal to the frequency domain. Features were extracted from Mel-spectrogram using a CNN network [9]. Kuo et al. (2018) created a CNN model using three data sets in their study and aimed to alleviate the problem of overfitting when training the model with images from hybrid data sources [10].

Emotion detection consists of three stages. These stages are; facial recognition, feature extraction and emotion classification. In the face recognition stage, it is ensured that human faces are detected in the image obtained. In the feature extraction phase, features such as eyes, nose and mouth are extracted from the image using artificial intelligence methods [11].

In this study, the emotion data set obtained from Kaggle, an open-access website, was modeled by using VGG16, ResNet50 and DenseNet121 deep learning architectures. Dataset images are classified into three classes as neutral angry and happy. Among the deep learning architectures used, the DenseNet121 model gave the most successful result with an accuracy rate of 84.16%.

## 2   Material Method

The "facial expression recognition data set" used in the study was taken from Kaggle, an open-source website [12]. The taken data set consists of a total of 20286 images belonging to the normal, angry and happy classes. Detailed information about the VGG16, ResNet50 and DenseNet121 architectures used in the study is given in the material section.

### 2.1   Material

#### 2.1.1   DenseNet121 Architecture

In DenseNet121 architecture, each convolution layer is connected forwardly to other layers and thus a deep architecture is built [13]. In this architecture, in addition to the layers being intensely connected, each layer uses the properties of all previous layers as input and gives its properties in the layer as input to the next layers [14]. Thus, in DenseNet architectures, the number of training is reduced and the feature is reused by providing feature diffusion [15]. The advantage of DenseNet architectures is that the number of parameters is reduced as well as the loss of gradient problem with skip connections [16]. An example of DenseNet-121 architecture is given in Fig. 1 [17]. This architecture consists of four dense blocks, three transition layers and a total of 121 layers (117 loops, 3 transitions and 1 classification) [17].

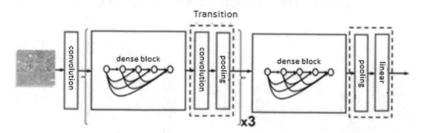

**Fig. 1.** DenseNet-121 architecture [17]

#### 2.1.2   VGG16 Architecture

VGG16, presented by Simonyan and Zisserman, is a large-scale image recognition architecture with approximately 138 million parameters that trains the fully connected network [18, 19]. Figure 2 shows the architecture of VGG16 [20]. This architecture consists of 16 layers (13 convolutional and 3 fully connected) with trainable weights [21]. The VGG16 architecture has an increasing network structure and fully connected layers in the last layers are used for feature extraction [22].

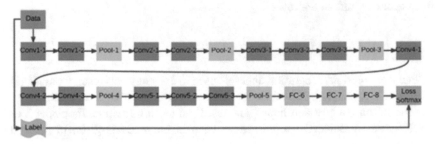

**Fig. 2.** VGG16 architecture [20]

### 2.1.3 ResNet50 Architecture

ResNet architecture, He et al. In 2015, it won first place in the ImageNet competition with an error rate of 3.57% [23]. The ResNet50 architecture has a good performance in image classification and can extract high-quality features of images [24]. In the ResNet50 construction, which is a residual neural network architecture with a depth of 50 times, there are convolution layers in the form of $(1 \times 1)$, $(3 \times 3)$, $(1 \times 1)$ [25]. ResNet architecture can classify images in 1000 object categories and the architecture consists of millions of parameters [26]. In the ResNet architecture, blocks containing many layers are used to reduce the training error and residual values are transferred to the next layers instead of learning the unreferenced functions. Thanks to the process of skipping between layers, as the network deepens, the distortion and disappearing gradient problems are solved [27, 28]. Figure 3 shows a ResNet connection example [29].

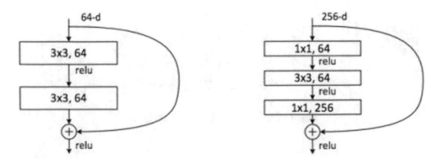

**Fig. 3.** ResNet connection example [29]

## 2.2  Method

The workflow diagram used in predicting emotion analysis in the study is given in Fig. 4. As can be seen in Fig. 4, in the first stage of the study, data collection and data preprocessing were carried out. In the data preprocessing phase, the emotion data set images taken from the open-access website Kaggle were labeled and resized. The labeled data were then normalized. In the second phase of the study, a total of 20286 data, which were labeled and normalized, were divided into two groups, 16285 being in the training and 4001 being in the testing. In the last stage of the study, the image data set was modeled in three classes as normal, angry and happy using VGG16, ResNet50 and DenseNet121 deep learning models. Learning rates obtained from all models were determined between 0.0001 and 0.00001. Using the learning rates obtained, the results obtained from three deep learning models were evaluated on the test data set and the most suitable model was tried to be found.

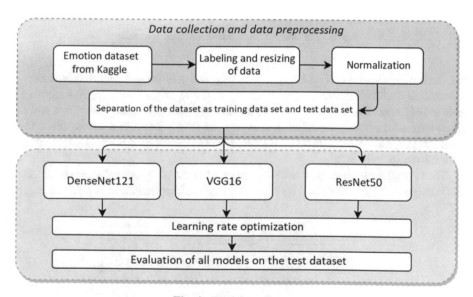

**Fig. 4.** Workflow diagram

# 3  Research Findings

The results of VGG16, ResNet50 and DenseNet121 architectures used in classifying the emotion data set used in the study are given. The first applied deep learning architecture is VGG16. The complexity matrix obtained by using the VGG16 architecture is given in Fig. 5.

When the VGG16 architecture in Fig. 5 is examined, it is seen that it correctly classified 730 images from a total of 960 test images for the angry class and classified 230 images incorrectly. For the happy class, it was determined that out of a total of 1825

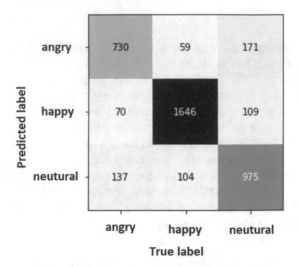

**Fig. 5.** Confusion matrix for VGG16 architecture

test images, it classified 1646 images correctly and 179 images incorrectly. Lastly, it is seen that it classified 975 images correctly and 241 images incorrectly within 1216 images for the normal class. As a result, 3351 images were correctly classified and 650 images were incorrectly classified among a total of 4001 test images for these three classes. Thus, it is seen that this architecture makes classification with 83.75% accuracy from the complexity matrix.

The second deep learning architecture applied in the study is ResNet50 architecture. The complexity matrix obtained by using the ResNet50 architecture is given in Fig. 6.

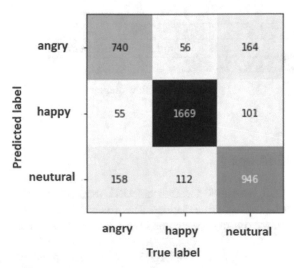

**Fig. 6.** Confusion matrix for ResNet50 architecture

When the Resnet50 architecture is examined in Fig. 6, it is seen that it classifies 740 images correctly and 220 images incorrectly out of 960 test images for the angry class. For the happy class, it was determined that out of a total of 1825 test images, it classified 1669 images correctly and 156 images incorrectly. Lastly, it is seen that it classifies 946 images correctly and 270 images incorrectly in 1216 images for the normal class. As a result, 3355 images were correctly classified and 646 images were incorrectly classified among a total of 4001 test images for these three classes. Thus, it is seen that this architecture makes classification with 83.85% accuracy from the complexity matrix.

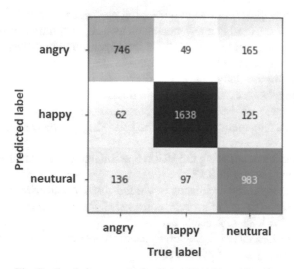

**Fig. 7.** Confusion matrix for DenseNet121 architecture

Lastly, the complexity matrix obtained by using DenseNet121 architecture is given in Fig. 7. When the DenseNet50 architecture is examined in Fig. 7, it is seen that it classifies 746 images correctly and 214 images incorrectly out of 960 test images for the angry class. For the happy class, it was determined that out of a total of 1825 test images, it classified 1638 images correctly and 187 images incorrectly. Lastly, it is seen that it classifies 983 images correctly and 233 images incorrectly in 1216 images for the normal class. As a result, 3367 images were correctly classified and 634 images were incorrectly classified among a total of 4001 test images for these three classes. Thus, it is seen that this architecture makes classification with 84.16% accuracy from the complexity matrix.

The results of the three deep learning methods used in the study are given in Table 1.

When Table 1 is examined, it is seen that the DenseNet121, VGG16 and ResNet50 deep learning models applied in the study give similar results. It is seen that DenseNet121 architecture gives the most accurate result among these three deep learning architectures with 84.16% accuracy.

**Table 1.** Accuracy rates achieved with deep learning architectures

| Model | DenseNet121 | VGG16 | ResNet50 |
|---|---|---|---|
| Accuracy rate (%) | 84.16 | 83.75 | 83.85 |

# 4 Results

With the rapid development of image processing methods in our age, image processing methods are frequently used in many areas such as face recognition systems, cybersecurity and education. In recent years, artificial intelligence methods have been applied to image processing methods, and meaningful results are obtained from images, providing great convenience in our daily life. In the study, the emotion data set taken from Kaggle, an open-access website, was divided into three classes as angry, happy and normal, and the learning rate was determined and classified using DenseNet121, VGG16 and ResNet50 deep learning architectures and the following results were obtained.

- Firstly, the model created using the VGG16 architecture performed the classification process with an accuracy of 83.75%.
- Secondly, in the Resnet50 architecture, which is used as a deep learning architecture, it was observed that the accuracy rate increased slightly and the classification process was performed with 83.85% accuracy.
- Lastly, in the DenseNet121 architecture used, the classification process was carried out with an accuracy rate of 84.16%.

It is thought that it will be possible to increase the accuracy rate by using more accurately labeled pictures in future studies.

**Acknowledgments.** I would like to thank Jonathan Oheix, who uploaded the data set to Kaggle, the open-source website used in the study.

# References

1. Akça, B.N., Çubukçu, B., Yüzgeç, U.: Detection of happiness emotion on images. Acad. Perspect. Procedia **2**(3), 324–333 (2019)
2. Demır, A., Atıla, O., Şengür, A.: Deep learning and audio based emotion recognition. In: 2019 International Artificial Intelligence and Data Processing Symposium (IDAP), pp. 1–6. IEEE (2019)
3. Soyel, H., Demirel, H.: Optimal feature selection for 3D facial expression recognition using coarse-to-fine classification. Turk. J. Electr. Eng. Comput. Sci. **18**(6), 1031–1040 (2010)
4. Tümen, V., Söylemez, Ö.F., Ergen, B.: Facial emotion recognition on a dataset using convolutional neural network. In: 2017 International Artificial Intelligence and Data Processing Symposium (IDAP), pp. 1–5. IEEE (2017)
5. Sağbaş, E.A., Gökalp, O., Uğur, A.: Feature extraction based on distance ratio and selection with genetic algorithms for facial expression recognition. Veri Bilimi **2**(1), 19–29 (2019)

6. Kudiri, K.M., Said, A.M., Nayan, M.Y.: Emotion detection using sub-image based features through human facial expressions. In: 2012 International Conference on Computer and Information Science (ICCIS), vol. 1, pp. 332–335. IEEE (2012)

7. Kumar, G.R., Kumar, R.K., Sanyal, G.: Facial emotion analysis using deep convolution neural network. In: 2017 International Conference on Signal Processing and Communication (ICSPC), pp. 369–374. IEEE (2017)

8. Ensar E.T., Günay, M.: Comparison of face recognition algorithms. In: 25th Signal Processing and Communications Applications Conference (SIU), Antalya, pp. 1–4 (2017). https://doi.org/10.1109/SIU.2017.7960469

9. Hossain, M.S., Muhammad, G.: Emotion recognition using deep learning approach from audio–visual emotional big data. Inf. Fusion **49**, 69–78 (2019)

10. Kuo, C.M., Lai, S.H., Sarkis, M.: A compact deep learning model for robust facial expression recognition. In: Proceedings of the IEEE Conference on Computer Vision and Pattern Recognition Workshops, pp. 2121–2129 (2018)

11. Pathak, A.R., Bhalsing, S., Desai, S., Gandhi, M., Patwardhan, P.: Deep learning model for facial emotion recognition. In: Singh, P.K., Panigrahi, B.K., Suryadevara, N.K., Sharma, S.K., Singh, A.P. (eds.) Proceedings of ICETIT 2019. LNEE, vol. 605, pp. 543–558. Springer, Cham (2020). https://doi.org/10.1007/978-3-030-30577-2_48

12. Kaggle. [Online]. Available: https://www.kaggle.com/jonathanoheix/face-expression-recognition-dataset

13. Hara, K., Kataoka, H., Satoh, Y.: Can spatiotemporal 3D CNNS retrace the history of 2D CNNS and ImageNet? In: Proceedings of the IEEE conference on Computer Vision and Pattern Recognition, pp. 6546–6555 (2018)

14. Kumar, R.: Adding binary search connections to improve DenseNet performance (2020). Available at SSRN 3545071

15. Huang, G., Liu, Z., Van Der Maaten, L., Weinberger, K.Q.: Densely connected convolutional networks. In: Proceedings of the IEEE Conference on Computer Vision and Pattern Recognition, pp. 4700–4708 (2017)

16. Leong, M.C., Prasad, D.K., Lee, Y.T., Lin, F.: Semi-CNN architecture for effective spatio-temporal learning in action recognition. Appl. Sci. **10**(2), 557 (2020)

17. Li, X., Shen, X., Zhou, Y., Wang, X., Li, T.Q.: Classification of breast cancer histopathological images using interleaved DenseNet with SENet (IDSNet). *PloS one*, **15**(5), e0232127 (2020)

18. Simonyan, K., Zisserman, A.: Very deep convolutional networks for large-scale image recognition. arXiv preprint arXiv:1409.1556 (2014)

19. Zeiler, M.D., Fergus, R.: Visualizing and understanding convolutional networks. In: Fleet, D., Pajdla, T., Schiele, B., Tuytelaars, T. (eds.) Computer Vision – ECCV 2014. LNCS, vol. 8689, pp. 818–833. Springer, Cham (2014). https://doi.org/10.1007/978-3-319-10590-1_53

20. Qassim, H., Verma, A., Feinzimer, D.: Compressed residual-VGG16 CNN model for big data places image recognition. In: 2018 IEEE 8th Annual Computing and Communication Workshop and Conference (CCWC), pp. 169–175. IEEE (2018)

21. Demir, U., Ünal, G.: Inpainting by deep autoencoders using an advisor network. In: 2017 25th Signal Processing and Communications Applications Conference (SIU), pp. 1–4. IEEE (2017)

22. Toğaçar, M., Ergen, B., Özyurt, F.: Classification of flower species by using feature selection methods in convolutional neural network models. Fırat Univ. J. Eng. Sci. **32**(1), 47–56 (2017)

23. He, K., Zhang, X., Ren, S., Sun, J.: Delving deep into rectifiers: surpassing human-level performance on ImageNet classification. In: Proceedings of the IEEE International Conference on Computer Vision, pp. 1026–1034. IEEE (2015)

24. Wen, L., Li, X., Gao, L.: A transfer convolutional neural network for fault diagnosis based on ResNet-50. Neural Comput. Appl. **32**(10), 6111–6124 (2019). https://doi.org/10.1007/s00521-019-04097-w

25. Talo, M.: Classification of histopathological breast cancer images using convolutional neural networks. Fırat Univ. J. Eng. Sci. **31**(2), 391–398 (2018)
26. Rezende, E., Ruppert, G., Carvalho, T., Ramos, F., De Geus, P.: Malicious software classification using transfer learning of resnet-50 deep neural network. In: 2017 16th IEEE International Conference on Machine Learning and Applications (ICMLA), pp. 1011–1014. IEEE (2017)
27. Korfiatis, P., Kline, T.L., Lachance, D.H., Parney, I.F., Buckner, J.C., Erickson, B.J.: Residual deep convolutional neural network predicts MGMT methylation status. J. Digit. Imaging **30**(5), 622–628 (2017)
28. Fu, Y., Aldrich, C.: Flotation froth image recognition with convolutional neural networks. Miner. Eng. **132**, 183–190 (2019)
29. Doğan, F., Türkoğlu, İ: Derin öğrenme algoritmalarının yaprak sınıflandırma başarımlarının karşılaştırılması. Sakarya Univ. J. Comput. Inf. Sci. **1**(1), 10–21 (2018)

# A Nested Unsupervised Learning Model for Classification of SKU's in a Transnational Company: A Big Data Model

Gabriel Loy-García[1], Román Rodríguez-Aguilar[2],
and Jose-Antonio Marmolejo-Saucedo[3]([⊠])

[1] Facultad de Ingeniería, Universidad Anáhuac, Av Universidad Anáhuac 46, Mexico, Mexico
[2] Escuela de Ciencias Económicas y Empresariales, Universidad Panamericana, Augusto Rodin 498, Mexico, 03920 Mexico City, Mexico
[3] Facultad de Ingeniería, Universidad Panamericana, Augusto Rodin 498, Mexico, 03920 Mexico City, Mexico
jmarmolejo@up.edu.mx

**Abstract.** This work seeks to develop a nested non-supervised model that allows a transnational soft drink company to improve its decision-making for the discontinuation of products from its portfolio with the use of unsupervised models from a database with commercial and financial information for all your product line in your most important operation. The integration of different cluster methodologies through a nested non-supervised model allowed to generate a correct identification of the products that should be refined from the catalog due to financial and operational factors. Given the magnitude of the information, a cluster was integrated into a platform for data processing as well as the generation of automatic reports that could be consulted automatically through the cloud. The products identified through the nested unsupervised model made it possible to identify products that had low demand and a low contribution to the utility of the company. Removing said products from the catalog will allow maximizing the profit of the business in addition to not incurring sunk costs related to the production and distribution of low-demand products. The platform developed will allow continuous monitoring of business performance in order to automatically identify the products likely to leave the catalog.

## 1 Introduction

A fundamental activity for the sales planning and operations (S&OP) cycles in large companies with value chains for the production, distribution and marketing of consumer goods, is the product discontinuation process. This product portfolio management process helps optimize the value chain and must be aligned with the financial and commercial objectives of the company. This process is generally driven by the commercial or marketing area of the companies and is accompanied by the financial and supply chain area. Together, these areas of the organization make decisions about the portfolio empirically and through data that allow them to act in a logical way, but without a formal and scientific mechanism to support the actions taken.

© The Author(s), under exclusive license to Springer Nature Switzerland AG 2021
J. Hemanth et al. (Eds.): ICAIAME 2020, LNDECT 76, pp. 715–731, 2021.
https://doi.org/10.1007/978-3-030-79357-9_66

In the analytics literature, there is a shortage of articles focused on solving the problem of discontinuing products and in most cases, they correspond to techniques for classifying inventory types with a classic supply chain approach in A, B, C and D. Works more focused on the application of machine learning methods to categorize products based on consumer behavior are identified. Transactional databases are generally used, and rules are general for decision making. [1] show an application to the retail sector on the type and quantity of products that they keep in their catalogs, through the application of a hybrid model based on k-means algorithm and the construction of association rule mining. [2] shows a case of innovation and discontinuation but using the firm as the unit of analysis. Estimating a production function comparing two sectors, the manufacturing sector and the service sector.

[3] shows the application of text mining methodologies to assert the voice of the customer by analyzing the case of a household and consumer products company. [4] analyze the classification of products in an e-commerce framework using a deep multi-modal architecture using text and images as inputs, showing the application of a deep learning model. [5] develops an intelligent platform for the classification of products in e-commerce based on natural language processing algorithms, vector space models, k-nearest neighbor, Naive-Bayes classifier and hierarchical classification. [6] address a classification problem with big data. The goal is to properly classify millions of products by type at Walmart-Labs. Applying large scales crowdsourcing, learning rules and in-house analysts.

Other studies such as the work of [7] use eigen color feature with ensemble machine learning combining artificial neural networks and support vector machines for the classification of product images. The assembly model is trained with eigen color feature for the classification of a catalog of 100 product classes. The little literature identified on the application of product classification does not address discontinuation and the vast majority focus on supervised models. However, the problem of this study lacks a dependent variable since what is sought is to identify underlying information that allows grouping products for later discontinuation based on a group of variables that address various areas of the business beyond the physical characteristics of the product.

Based on the above, this work seeks to develop a nested unsupervised model that allows a transnational soft drink company to improve its decision-making for the discontinuation of products from its portfolio. With the use of database with commercial and financial information for entire product line in Mexico operation. This international soft drink company serves more than 290 million consumers, selling 19 billion liters of beverage per year through 2 million points of sale. This company seeks to have an attractive portfolio of non-alcoholic beverage brands and presentations to address the unique dynamics of its markets and stimulate demand from a growing customer and consumer base. In this way, its customers have the opportunity to purchase one of the 131 brands of soft drinks and non-carbonated beverages that it offers.

The operations of this company cover certain territories in Mexico, Colombia, Brazil and Argentina, and at the national level in Guatemala, Nicaragua, Costa Rica, Panama, Uruguay and Venezuela. The company has 52 bottling plants and more than 300 distribution centers, its corporate headquarters are located in Mexico City. Depending on the country, the company manages a portfolio that includes around 250 and up to 536 different product codes (SKUs). With the constant changes in consumer habits, such as leading healthier lifestyles, and the demand from consumers for packaging that has a smaller carbon footprint to help the environment, the company is constantly innovating that consists of introducing products. Based on the above, the proliferation of new product codes is a constant that different areas must deal with when carrying out the quarterly sales and operations plan.

In order to optimize the capabilities of the company's supply chain and the profitability of the beverage portfolio, it is important that before carrying out the plan described above, those product codes that are not generating profits or the product are removed from the current portfolio. Sales volume defined as target in the plan of the previous period. This process is known as discontinuation of products in the beverage portfolio. Currently, there is no objective and standard methodology in the company to generate the list of product codes that must leave the portfolio each quarter. This list is a fundamental input in the work meetings between the commercial, finance and supply chain areas of the company, where the quarterly plan of sales and operations of the company is generated. Based on the results of the previous quarter, those product codes that have a poor financial performance are known quite precisely, and it is these products that should be the first candidates to be considered in the discontinuation process. Leaving them in the portfolio for the next quarter would only mean allocating productive resources to codes that reduce operating profit. The lack of this type of methodology is an especially important need for the Mexican operation. The application of an unsupervised model that allows integrating groups susceptible of SKUs to be discontinued will allow to have a standard methodology so that this process is carried out in the most objective way possible based on technical evidence.

The work is organized as follows, in the first section the theoretical framework and the sources of information to be used are presented. The second section presents the proposal of the nested model using different cluster methodologies, as well as the structure of the tool for treating big data. The third section presents the main results identified. Finally, the conclusions and future lines of research.

## 2  Materials and Methods

### 2.1  Description of the Data

The profitability base of the Mexico operation corresponding to the accumulated months from December 2019 to February 2020. The database contains 69,726 records with 16 mixed variables. The Table 1 contains the description of the variables:

The records that have a negative CV and that represent a loss of economic value for the company and that are directly candidates to be discontinued were eliminated. Additionally, presentations that correspond to specific consumption occasions and that are important for the brand image were eliminated; as well as products that according to their life cycle are in a phase of introduction to the market, or growth.

The resulting base has 49,508 records. The first step with this database is to review the number of levels of each categorical variable (Table 2). When considering the different combinations between the categories of the variables and the total of variables, the total number of variables increases significantly.

**Table 1.** Variables to consider in the model

| Variable | Descripción | Type |
|---|---|---|
| State | Geographic location where the product is sold | Categorical |
| Distributors (U.O) | Distributor from where the products are distributed to customers and that belongs to a particular State | Categorical |
| SKUs | Unique product code, also known as SKU for its acronym in English | Categorical |
| Category | Indicate if the product is a soft drink or a non-carbonated drink and what is its nature: Cola, flavor, isotonic… | Categorical |
| Type of consumption | Indicates whether the size of the product is for personal consumption (consumed by a single person) or family (consumed by two or more people) | Categorical |
| Brand | Brand of the product | Categorical |
| Product size | Number of milliliters with which the product is sold | Categorical |
| Presentation | Product packaging type | Categorical |
| Returnability | Indicates if the product is sold in returnable or non-returnable packaging | Categorical |
| CU | Sales volume of a product in packages of 5,678 L | Numerical |
| TR | Number of bottles of a product sold | Numerical |
| VN | Sales in Mexican pesos after taxes and without discounts generated by a product | Numerical |
| Desc | Discounts in Mexican pesos granted to a product resulting from a commercial condition granted to the customer | Numerical |
| CV | Variable Contribution in Mexican pesos generated by a product. The variable contribution is the result of subtracting from Sales, Variable Costs and Expenses, as well as the Marketing Expense of each product. It is the amount that remains per product to cover costs and fixed expenses | Numerical |

It is assumed that the discontinuation of any SKU will be for total channels and the Customer Business Model variable will not be taken into account. As we can see, the complexity of making an unsupervised model with this database lies in the number of levels or factors of each categorical variable, especially by the number of Distributors (133) and SKUs (536). Attempting to classify this base would be a huge effort in computing power and would be inefficient. Additionally, and from business knowledge, it is known that there are two variables with redundant information for classification. The first is the Type of Consumption and that is a classification that depends on the Product Size. The second is Category, which depends on the Brand. Finally, we also know that the Distributors variable has a high correlation with the State variable (each

**Table 2.** Levels of categorical variables

| Categorical variables | Levels |
|---|---|
| State | 11 |
| Distributors (U.O.) | 133 |
| SKUs | 536 |
| Product size | 30 |
| Type of consumption | 2 |
| Returnability | 2 |
| Category | 10 |
| Brand | 33 |
| Presentation | 4 |

Distributor, being a physical geographic location, belongs to a State) and that the SKU can be summarized with its attributes that correspond to the variables Size of Product, Returnability, Brand and Presentation.

Based on the above, we decided to group the base of 49,508 by Status, Product Size, Returnability, Brand and Presentation. We obtain a new base with those five categorical variables and the five numerical variables presented previously. The resulting database now contains 1,618 records. This step was vitally important as a starting point for the classification model. An exploratory analysis of the resulting database was carried out and the levels of each categorical variable were converted into factors. Mixed Principal Components analysis was applied to confirm whether the data are grouped according to the factors of the categorical variables.

## 2.2  Cluster Analysis

A set of unsupervised cluster model methodologies were applied in order to compare their performance. A methodological structure was followed for the elaboration of an unsupervised model based on [8]. Establishing a series of non-parametric tests prior to the application of the methodologies as well as internal validation methodologies of the clusters generated in order to select the best methodology to apply. The objective of comparing a series of methodologies is to be able to robustly select an ideal methodology for the treatment of the data used. The following steps were considered in the development of the unsupervised model:

1. Cluster trend validation
2. Selection of the number of partitions to be made
3. Selection of the best algorithm
4. Validation of the clusters generated
5. Implementation on a platform for big data

Randomness validation of the base to define if the data are subject to clustering. To carry out the validation, we carry out a hypothesis test that tells us whether the study database behaves as a uniform one or not. The Hopkins statistic is used [9, 10].

$$H = \frac{\sum_{i=1}^{n} y_i}{\sum_{i=1}^{n} x_i + \sum_{i=1}^{n} y_i} \tag{1}$$

Where $x_{i=dist(p_i,p_j)}$ is the distance for each observation from its closest neighbor and $y_{i=dist(q_i,q_j)}$ is the distance for each point from its closest neighbor in a data set simulated using a uniform distribution.

Since it is a database with mixed data, it is necessary to establish an ad hoc metric, the Gower distance was used [11, 12].

$$GOW_{jk} = \frac{\sum_{i=1}^{n} W_{ijk} S_{ijk}}{\sum_{i=1}^{n} W_{ijk}} \tag{2}$$

Where $W_{ijk} = 0$ if the objects $j$ and $k$ cannot be compared for variable $i$ either because $x_{ij}$ or $x_{ik}$ are not known. Additionally, for nominal variables:
$W_{ijk} = 1$ if $x_{ij}$ and $x_{ik}$ are known; so
$S_{ijk} = 0$ if $x_{ij} \neq x_{ik}$
$S_{ijk} = 1$ if $x_{ij} = x_{ik}$
To determine the number of partitions to be made, the Silhouette Coefficient was used, which allows an optimization criterion to be used to determine the number of partitions in a cluster using a measure of the quality of the cluster classification [13].

$$Silhouette = \frac{1}{N} \sum_{i=1}^{N} \frac{d_i - s_i}{\max\{d_i - s_i\}} \tag{3}$$

Where $s_i$ is the mean distance to objects in the same cluster and $d_i$ is the mean distance to objects in the next nearest cluster. Observations with a large Silhouette (almost 1) are well grouped. A small Silhouette (around 0) means that the observation is between two groups. Observations with a negative Silhouette are probably located in the wrong group.

A group of hard and soft cluster methodologies were selected to select the best methodology to apply in the analyzed database. The selected methodologies were k-means, k-medoids, hierarchical agglomerative and fuzzy cluster.

### K-means
Clustering by K means is an effective algorithm for dividing a data set into K groups according to the similarity of the observations. The value of K is defined a priori. The grouping generated seeks to maximize the homogeneity of the observations within each group. The objective will be to solve the following minimization problem [14].

$$min_{C_1,...,C_k} \left\{ \sum_{k=1}^{K} W(C_k) \right\} \tag{4}$$

Where, $W(C_k)$ is a measure of the intra-cluster variation $C_k$. To solve this problem, it is necessary to define the measure of intra-cluster variation. There are different alternatives for this, but the most common one considers the quadratic Euclidean distance.

### K-medoids
K-medoids is a partition clustering technique. In k-medoids each cluster is represented

by an element of the cluster, these points (elements) are the medoids. The term medoid refers to an object within a cluster for which the average dissimilarity between it and all other members of the cluster is minimal. It corresponds to the most central point of the group. K-medoids is less sensitive to noise and outliers, because it uses medoids as cluster centers rather than centroids (used in k means). The most common k-medoids grouping methods are the PAM algorithm (Partitioning Around Medoids, [13, 15]. The PAM algorithm is based on finding $k$ representative objects or medoids among the observations in the data set.

Hierarchical Agglomerative

Agglomerative grouping is the most common type of hierarchical grouping. Groups objects into clusters based on their similarity. The cluster pairs are then successively merged until all the clusters have been merged into one large group that contains all the objects. The result is a tree of representation of the objects (dendrogram) [16].

Fuzzy Clustering

Fuzzy grouping is considered soft grouping, in which each item has a probability of belonging to each group. In other words, each element has a set of membership coefficients corresponding to the degree of being in a given group. The degree, to which an element belongs to a given group, is a numerical value that varies from 0 to 1. The fuzzy c-means algorithm is one of the most widely used fuzzy clustering algorithms. Fuzzy c-means is a clustering method that allows a part of the data to belong to two or more clusters. This method is frequently used in pattern recognition. It is based on the minimization of the following objective function [17, 18]:

$$J_m = \sum_{i=1}^{N} \sum_{j=1}^{C} u_{ij}^m \| X_i - C_j \|^2 \qquad (5)$$

Where $m$ is any real number greater than 1, $u_{ij}$ is the degree of membership of $X$ in group $j$, $X_i$ is the i-th of the data measured in $d$ dimensions, $C_j$ is the center of dimension $d$ of the group and $\| * \|$ it is any norm that expresses the similarity between any measured data and the center.

In order to select the best algorithm for the classification and profiling of the groups, the validation of each model or of each stage of the model is carried out by generating an internal validation and the calculation of several stability means.

Internal Validation

To perform internal validation of clusters, the Average Silhouette statistic is calculated. This statistic is calculated for each cluster and indicates the quality of grouping of individuals within the cluster. The grouping quality will be better when the value of this statistic is close to 1. This statistic is calculated for each selected algorithm by varying the number of clusters. In this way, the algorithm and with the number of groups that the Silhouette statistic has closest to 1, should be chosen as the best model [13].

### 2.3 Architecture of the Proposed Model

The proposed classification model for discontinuation applies a nested methodology where, as the groups are structured in a first level, they are disaggregated at a higher level of detail in order to reach the presentation level of each product. The objective of proposing this strategy was that when carrying out the generalized models it was observed that it was not possible to identify relevant partitions. In order to achieve a greater level of detail in the selection of the products to be discontinued, it was decided to replicate the structure of the cluster methodology at different levels of disaggregation based on key variables such as the distributor, the state, type of consumption, category, brand, presentation and finally SKU (Fig. 1).

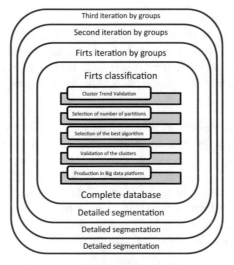

**Fig. 1.** Structure of the proposed nested cluster model

When taking into account the detail in the disaggregation of the categorical variables used in each iteration, the increase in the dimension of the problem grows exponentially, due to all the possible combinations of levels between the categorical variables considered. The code of the proposed model was coded on a free software platform. Based on the above, the Business Intelligence area will be able to generate at the beginning of each planning cycle, the list of products to be discontinued in an efficient and objective manner, significantly shortening the three weeks that this process currently takes. The Business Intelligence area use the Microsoft Azure suite as a work tool, which has the following architecture and installed components it was selected to run with the R program used through the interface provided by R Studio. R Studio was installed on a "notebook" in the Azure Machine Learning module called Data-Bricks (Fig. 2).

The data processing power in the Data-Bricks module comes from a dedicated processing cluster within the module with the following features: 140 GB of memory and 20 cores. Additionally, it was configured with a minimum of 6 "Workers" or parallel computing modules (Fig. 3).

Once the architecture of the model and the nested modeling strategy were structured, the model was implemented with the company's database for the period December 2019 to February 2020, in the following section the main results are presented.

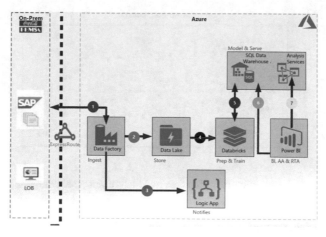

**Fig. 2.** Integration of the nested cluster model in the Microsoft Azure architecture.

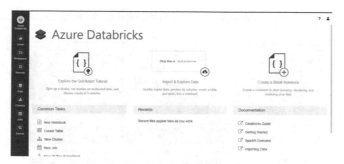

**Fig. 3.** Azure Databricks configuration.

## 3 Results

### 3.1 Cluster Analysis (Selection Between Hard and Soft Models)

We work with the base previously described in the exploratory data analysis with 1,618 records, from which we exclude 15 records that correspond to extreme values. Therefore, the database with which we began to work has 1,615 records. To verify that the data are subject to clustering, a graphic exploration was carried out and Hopkins Statistic. The representation of the data on a two-dimensional map using the first two mix principal components shows that the distribution of the data is not random [19] (Fig. 4).

Based on the Silhoutte method we obtain the graph with which we can decide the optimal number of clusters for $K = 2$ and up to $K = 5$. Note that having categorical variables it is necessary to previously calculate the Gower distance for the dataset and it must be used within any clustering algorithm. To select the appropriate cluster algorithm, a comparison was made between the different proposed methodologies: k-means, k-medoids, herarquical and fuzzy. The Silhoutte index for different methods in order to alternatively obtain the optimal number of clusters. Three hard algorithms and one soft

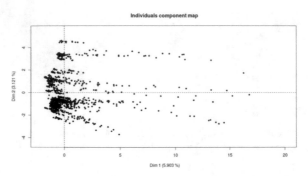

**Fig. 4.** Representation of the data in a map of principal components.

algorithm were run, varying the number of clusters for $Ki_{i=2,...,5}$. The results are shown in Table 3.

**Table 3.** Comparison of the proposed methodologies

| Método | K = 2 | | | | K = 3 | | | |
|---|---|---|---|---|---|---|---|---|
| | KMEANS | KMEDOIDS | HIERARCHICAL | FUZZY | KMEANS | KMEDOIDS | HIERARCHICAL | FUZZY |
| Índice Silhouette | 0.41 | 0.15 | 0.41 | 0.25 | 0.41 | 0.09 | 0.41 | |
| Tamaño de Clúster 1 | 977 | 638 | 601 | 581 | 406 | 541 | 195 | |
| Tamaño de Clúster 2 | 626 | 965 | 1,002 | 1,022 | 206 | 573 | 1,002 | |
| Tamaño de Clúster 3 | | | | | 991 | 489 | 406 | |
| Tamaño de Clúster 4 | | | | | | | | |
| Tamaño de Clúster 5 | | | | | | | | |
| Índice Silhouette Clúster 1 | 0.55 | 0.11 | 0.20 | 0.15 | 0.40 | 0.11 | 0.37 | |
| Índice Silhouette Clúster 2 | 0.18 | 0.18 | 0.53 | 0.30 | 0.34 | 0.17 | 0.42 | |
| Índice Silhouette Clúster 3 | | | | | 0.43 | (0.03) | 0.40 | |
| Índice Silhouette Clúster 4 | | | | | | | | |
| Índice Silhouette Clúster 5 | | | | | | | | |

| Método | K = 4 | | | | K = 5 | | | |
|---|---|---|---|---|---|---|---|---|
| | KMEANS | KMEDOIDS | HIERARCHICAL | FUZZY | KMEANS | KMEDOIDS | HIERARCHICAL | FUZZY |
| Índice Silhouette | 0.43 | 0.13 | 0.42 | | 0.39 | 0.13 | 0.39 | |
| Tamaño de Clúster 1 | 993 | 196 | 155 | | 908 | 192 | 155 | |
| Tamaño de Clúster 2 | 160 | 547 | 1,002 | | 157 | 375 | 917 | |
| Tamaño de Clúster 3 | 406 | 428 | 406 | | 405 | 338 | 406 | |
| Tamaño de Clúster 4 | 44 | 432 | 40 | | 40 | 354 | 40 | |
| Tamaño de Clúster 5 | | | | | 93 | 344 | 85 | |
| Índice Silhouette Clúster 1 | 0.43 | 0.34 | 0.57 | | 0.38 | 0.33 | 0.57 | |
| Índice Silhouette Clúster 2 | 0.55 | 0.14 | 0.42 | | 0.56 | 0.02 | 0.39 | |
| Índice Silhouette Clúster 3 | 0.39 | - | 0.39 | | 0.39 | 0.23 | 0.39 | |
| Índice Silhouette Clúster 4 | 0.27 | 0.15 | 0.31 | | 0.28 | 0.01 | 0.27 | |
| Índice Silhouette Clúster 5 | | | | | 0.19 | 0.16 | 0.18 | |

In this case the fuzzy clster algorithm can no longer be calculated for $K = 3$, $K = 4$ and $K = 5$. We observe that the highest Silhouette index is obtained for the K-Means and Hierarchical Agglomerative algorithms with $K = 4$ respectively and it is verified that the optimal number of groups is 4. Additionally, the number of records with a negative Silhouette index (misclassified) is 9 for the K-Means algorithm and 14 for the Hierarchical Agglomerative. Therefore, we selected the K-Means algorithm with 4 groups for the classification of the base, graphically the clusters are represented on a principal component map to observe the initial classification generated by the selected model (Fig. 5).

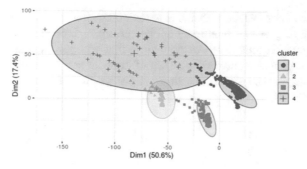

**Fig. 5.** Cluster plot using the two principal components.

## 3.2   Definition of Profiles by Iterations Within Each Group

The groups were summarized and characterize by the variable Contribution that each group generates per bottle and the percentage that it contributes to the total volume in Unit Boxes (CU) to the enterprise in Mexico. As reviewed in the part corresponding to the exploratory data analysis, a total of 20,218 records were excluded as they had a negative variable contribution or a special reason, which is equivalent to 16% of the volume in Unit Cases to the total of enterprise in Mexico. 82% of the remaining volume is what was classified with this first model. In this case the outliers correspond to the best products of the company (Fig. 6).

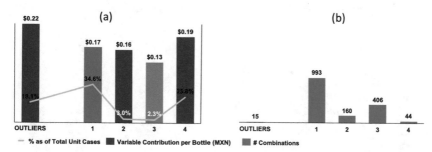

**Fig. 6.** (a) Characterization and (b) dimension of the clusters generated in the first stage.

Group 1 is mainly made up of Non-Returnable Pet products. Group 2 corresponds mostly to 355 and 500 mL Returnable Glass products. Group 3 corresponds mostly to non-returnable Can and Glass products. Group 4 corresponds to products of the most important brand of the company, mostly 500 mL Returnable Pet. Making a decision to discontinue products based on what was found in this first classification is impossible. More iterations of classification on specific groups need to be done. In this case we choose group 1, which still groups a considerable number of records and concentrates a large volume mix, as well as group 3, which has the lowest CV per bottle and still has a significant number of records.

Using the Silhouette method to select the optimal algorithm between K-means and Hierarchical Agglomerative, $n$ iterations will be generated to split groups 1 and 3 until we obtain a number of combinations that, due to their characteristics, correspond to product profiles that are subject to discontinuation. For group 1, three additional iterations are made where there is a group of 194 combinations that are equivalent to 0.4% of the sales volume and that generate only $0.10 MXN of contribution per unit. So is concluded that this group necessarily contains products that at the SKU-Distributor level should be discontinued (Fig. 7).

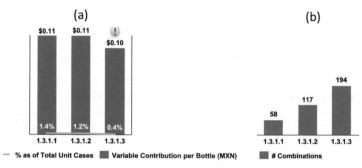

**Fig. 7.** (a) Characterization and (b) dimension of cluster 1 disaggregated.

In the case of group 3, only one iteration is made where there are two groups that are equivalent to 265 combinations that are equivalent to 1.1% of the sales volume and that generate only around $1.8 MXN, so we conclude that this group necessarily contains products that at the SKU-Distributor should be discontinued (Fig. 8).

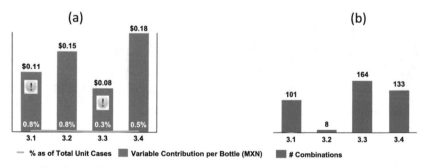

**Fig. 8.** (a) Characterization and (b) dimension of cluster 3 disaggregated.

## 4   Discontinuation SKU-Distributor

Based on the groups previously established with the grouped base, we set up a key with the variables State, Product Size, Returnability, Brand and Presentation. This key is then searched for in the original base in order to reach the original granularity level, whose lowest level of expression is SKU-Distributor. Based on the records of the original database, we will proceed to make new clusters on groups 1.3.1.3, 3.1 and 3.3 according to the findings of the previous sections. In the new cluster analysis, the only categorical variables that enter are SKU and Distributor, respectively and according to the fields of the original database. As observed in the previous section, after three iterations it was concluded that group 1.3.1.3 necessarily contains some SKU-Distributor combinations to be discontinued, this group has 194 records or combinations of the grouped database and 3227 records in the original database with the minimum level of granularity SKU-Distributor. The steps mentioned above are repeated to cluster the records of the group 1.3.1.3, 3.1 and 3.3. For group 1.3.1.3, four new sub-partitions were identified, of which group 1.3.1.3.3 was the final candidate group containing the SKU-Distributor selected to be discontinued (Fig. 9).

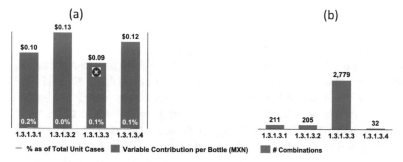

**Fig. 9.** (a) Characterization and (b) dimension of cluster 1.3.1.3 at SKU-Distributor level

According to what has been observed, the SKUs and Distributors that are in groups 3.1 and 3.4 must be part of the list of candidates to be discontinued. For group 3.1 was broken down into four more partitions of which partition 3.1.2 and 3.1.4 are the candidates to be discontinued at the SKU-Distributor level (Fig. 10).

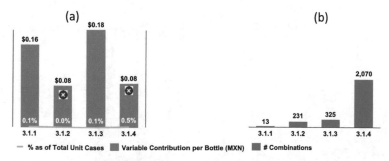

**Fig. 10.** (a) Characterization and (b) dimension of cluster 3.1 at SKU-Distributor level

In the case of the first partition, 3.3 was disaggregated into four subgroups at the SKU-Distributor level, of which groups 3.3.1 and 3.3.2 are the candidate products for discontinuation (Fig. 11).

By representing these groups (1.3.1.3.3; 3.1.2; 3.1.4; 3.3.1 and 3.3.2) 8,232 records of the original base and only 0.8% of the company's sales volume with a variable contribution per bottle of $0.1 (MXN), we can conclude that makes a great effort in the

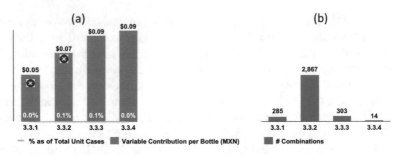

**Fig. 11.** (a) Characterization and (b) dimension of cluster 3.3 at SKU-Distributor level

value chain to guarantee the supply of these products. Leaving them in the portfolio would be inefficient. According to what was previously reviewed, the proposed list of SKU-Distributor combinations to be discontinued correspond to 8,232 records from the original database. This equates to 11.8% of the records from the original base. The SKUs and Distributors that are on the list with the candidates to be discontinued have a CV per bottle of $1.6 MXN. Being the average fixed (costs and expenses) per bottle of $1.59 MXN, this important amount of records does not generate any profit for the company, with that we confirm that the methodology proposed in this application project is correct to solve the problem silvered at the beginning of the form agile and methodologically robust.

As the list represents 8,232 records from the original base and only 0.8% of the company's sales volume, we can conclude that a great effort is made in the value chain to guarantee the supply of these products. Leaving them in the portfolio would be inefficient. With the results obtained and making use of technology-oriented Business Intelligence, a dashboard was made in "Tableau" where the profiles by group and the detail of the SKUs to be discontinued by Distributor and State can be reviewed interactively (Fig. 12).

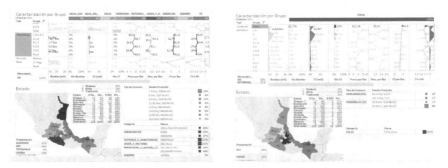

**Fig. 12.** Dashboard of SKUs to be discontinued by state.

These are mostly products that do not make commercial sense in the states of Northeast (Tamaulipas), Chiapas, Guerrero and Veracruz due to the type of size and presentation with which they are sold or because of their selling price that is too onerous for states with a low socioeconomic status. The most representative sizes belong to Personal

Presentations and most correspond to Tin and Non-Returnable Glass. These records are equivalent to discontinuing 195 SKUs in 133 Distributors, which do not represent any profit for the company and which only increase the complexity of the value chain as we have previously pointed out. To contrast the records that are candidates to be discontinued, in the same Dashboard (Fig. 12) we can see the profile of those SKU's-Distributor that are the Stars-products of the beverage portfolio in the enterprise. The detail of the records to be discontinued by SKU and Distributor can be consulted within the Dashboard in the tab called "Discontinued: Detail by SKU" as shown in Fig. 13.

**Fig. 13.** Detailed dashboard by SKUs to be discontinued.

The detail of the records that represent the "stars" of the company's beverage portfolio by SKU and Distributor can be consulted within the Dashboard in the tab called "Stars: Detail by SKU".

## 5 Conclusions and Recommendations

The implementation of a standard clustering methodology comparing various methodologies allowed to generate coherent and robust results for decision making. The approach of a nested cluster allowed us to carry out various segmentations and to apply this clustering methodology at different levels of disaggregation. The development of an analytical intelligence tool such as the proposal generates added value for companies, especially by supporting the discontinuation of products based on an unsupervised model which considers a comprehensive set of variables from the entire supply chain of the company and not the emphasis is only on financial aspects.

The way in which this project was approached is quite innovative applying a nested clustering approach and implemented on a big data platform. Since a complex problem was solved due to the number of levels or factors of each categorical variable, despite the fact that the base is apparently small with 69,726 records. Due to the number of levels or factors of each category, this base is quite complex to start if we do it with all its granularity from the beginning and it would require a lot of computing power, this would be inefficient. On the other hand, innovation is also found in the way of considering numerical variables, since their absolute value is always considered and without any transformation, which guarantees an impartial classification and purely based on statistics. This adds a lot of value against the traditional method offered by the financial area of the company where the variables are generally transformed, and they are considered units by bottles or percentages to net sales to classify the records empirically.

With the list obtained from SKU's to discontinued, the supply chain planning area will be able, through the origin-destination matrix, to know the primary origin of the plants that supply the Distributors and will be able to determine if the complexity of the supply chain value will drop as a result of the departure of those SKUs that will no longer be produced in certain plans, nor will they be chartered to the Distributors. This decrease in some production costs and some expenses should maximize the company's profit margin. The model designed and the reports generated to measure will allow to analyze in detail the decision-making or discontinuation of products. Making use of technology oriented Business Intelligence, a dashboard was made in "Tableau" where the profiles by group and the detail of the SKUs to be discontinued by Distributor and State can be reviewed interactively.

As future work it is necessary to evaluate the hypothesis of value having discontinued the SKU's and based on the new sales and operations plan for the new quarter, the above, not including the new launches and before incorporating the new pricing plan. This recommendation depends on the execution of the sales and operations plan carried out for the quarter from April to June 2020 and corresponds to an additional scope to this application project.

# References

1. Karki, D.: A hybrid approach for managing retail assortment by a categorizing products based on consumer behavior. Dublin, National College of Ireland, Ph.D. thesis (2018)
2. Crowley, F.: Product and service innovation and discontinuation in manufacturing and service firms in Europe. Eur. J. Innov. Manag. **20**(2), 250–268 (2017)
3. Gonzalez, R.A., Rodriguez-Aguilar, R., Marmolejo-Saucedo, J.A.: Text mining and statistical learning for the analysis of the voice of the customer. In: Hemanth, D.J., Kose, U. (eds.) ICAIAME 2019. LNDECT, vol. 43, pp. 191–199. Springer, Cham (2020). https://doi.org/10.1007/978-3-030-36178-5_16
4. Zahavy, T., Magnani, A., Krishnan, A., Mannor, S.: Is a picture worth a thousand words? A deep multi-modal fusion architecture for product classification in e-commerce. In: The Thirtieth Conference on Innovative Applications of Artificial Intelligence (IAAI) (2018)
5. Ding, Y., et al.: Goldenbullet: automated classification of product data in e-commerce. In: Proceedings of Business Information Systems Conference (BIS 2002), Poznan, Poland (2002)
6. Sun, C., Rampalli, N., Yang, F., Doan, A.: Chimera: large-scale classification using machine learning, rules, and crowdsourcing. PVLDB **7**(13), 1529–1540 (2014)
7. Oyewole, S.A., Olugbara, O.O.: Product image classification using Eigen Colour feature with ensemble machine learning. Egypt. Inform. J. **19**(2), 83–100 (2018)
8. Kassambara, A.: Practical guide to cluster analysis in R: unsupervised machine learning. CreateSpace Independent Publishing Platform (2017)
9. Hopkins, B., Skellam, J.G.: A new method for determining the type of distribution of plant individuals. Ann. Bot. **18**(2), 213–227 (1954)
10. Banerjee, A.: Validating clusters using the Hopkins statistic. In: IEEE International Conference on Fuzzy Systems, pp. 149–153 (2004)

11. Gower, J.: A general coefficient of similarity and some of its properties. Biometrics **27**, 857–872 (1971)

12. Tuerhong, G., Kim, S.B.: Gower distance-based multivariate control charts for a mixture of continuous and categorical values. Elsevier, South Korea (2013)

13. Kaufman, L., Rousseeuw, P.J.: Finding Groups in Data: An Introduction to Cluster Analysis, p. 1990. Wiley, Hoboken (1990)

14. Lloyd, S.P.: Least squares quantization in PCM. IEEE Trans. Inf. Theory **28**(1982), 129–137 (1982)

15. Park, H.-S., Jun, C.-H.: A simple and fast algorithm for K-medoids clustering. Expert Syst. Appl. **36**, 3336–3341 (2009)

16. Ward, J.H.: Hierarchical grouping to optimize an objective function. J. Am. Stat. Assoc. **58**(301), 236–244 (1963)

17. Dunn, J.C.: A fuzzy relative of the ISODATA process and its use in detecting compact well-separated clusters. J. Cybern. **3**(3), 32–57 (1973)

18. Bezdek, J.C.: Pattern Recognition with Fuzzy Objective Function Algorithms (1981)

19. Rodriguez-Aguilar, R.: Proposal for a comprehensive environmental key performance index of the green supply chain. Mob. Netw. Appl. (2020)

# Identification of Trading Strategies Using Markov Chains and Statistical Learning Tools

Román Rodríguez-Aguilar[1] and Jose-Antonio Marmolejo-Saucedo[2]($\boxtimes$)

[1] Facultad de Ciencias Económicas y Empresariales, Universidad Panamericana,
Augusto Rodin 498, 03920 Mexico City, Mexico
[2] Facultad de Ingeniería, Universidad Panamericana, Augusto Rodin 498,
03920 Mexico City, Mexico
jmarmolejo@up.edu.mx

**Abstract.** Technological advances have modified many operational and strategic areas in companies, the financial sector has been one of the sectors highly influenced by the methods of artificial intelligence and machine learning. The operation in the stock exchanges have used more technological tools to process information and be able to make investment decisions. The main objective is to be able to detect buying and selling opportunities at the right time. Stock markets have traditionally based their decisions on two major approaches, technical analysis and fundamental analysis, with new machine learning and artificial intelligence technologies, these paradigms have been updated making use of additional tools for their analysis. The present work is a proposal for the detection of trading signals in the markets through the use of Markov models and generalized additive models. In order to identify investment opportunities in the stock markets.

**Keywords:** Trading algorithms · Machine learning · Markov models · Generalized additive models (GAM)

## 1 Introduction

The current financial dynamics represent great challenges for the modeling and estimation of the prices of the shares of a company, the perception of the markets is not always sufficiently prepared to recognize the scope of the decisions and macroeconomic movements that generate volatility in the prices of a share.

The traditional approach to stock analysis has been based on two fundamental paradigms, the technical and fundamental analysis of company stocks. The first focuses on seeking to identify trends through the use of graphs with the aim of identifying future trends in stock prices. The Dow theory is the basis of technical analysis where it seeks to identify ascending, descending or lateral trends through the monitoring of prices and volumes of the stock market activity. In order to identify these trends, technical formations or patterns are used that will allow the analyst to know if a trend changes or remains. In addition to the graphical analysis, quantitative analysis tools are also used, among which we can mention, moving averages, weighted moving averages, oscillators, Bollinger bands, etc. (Brown 1989).

J. Hemanth et al. (Eds.): ICAIAME 2020, LNDECT 76, pp. 732–741, 2021.
https://doi.org/10.1007/978-3-030-79357-9_67

On the other hand, we have the fundamental analysis approach, where it is sought to determine the value of a share according to key information and to know if this value corresponds to the market price. The fundamental objective is to know what the intrinsic value of the shares is and consequently make investment decisions. It is a medium and long-term approach since it takes into account key information from the company's operation as well as the stock market. As tools to carry out fundamental analysis, the analysis of financial ratios, company valuation techniques, as well as the analysis of the economic situation are traditionally used (Arbanell 1997).

Until now there is no definitive verdict on which approach generates better results, in fact, mixed approaches are generally used combining technical and fundamental analysis tools. Generally, not allowing to obtain results above the average in most cases. One of the justifications in this regard is based on the efficient market's hypothesis, which states that a stock market is efficient when competition leads to an equilibrium situation in which the market price of a share constitutes a good estimate of its value intrinsic. In other words, prices in an efficient market reflect all the existing information and adjust quickly when new data emerges. All stocks are correctly valued so there are no undervalued or overvalued values (Fama 1970).

Currently, although the technical and fundamental analysis approach prevails as dominant, more sophisticated quantitative tools have been integrated that seek the same objective, to identify investment opportunities in the markets. The integration of high-frequency data in real time, as well as artificial intelligence and machine learning methodologies in the operation of markets has generated contributions in investment strategies, however the main approaches prevail. In other words, these new tools are integrated into the fundamental and technical analysis paradigm, and investors make their decisions based on these results as well as their experience and intuition in many cases.

Various methodologies have been applied to the estimation of share prices, such as Shen et al. (2012) show the application of support vector machines and deep learning using the existing correlation between the different markets in the world to predict the trend of stocks for the next day. Authors such as Milosevic (2016) focus on long-term prediction in the movement of stocks, for this they apply the intrinsic value of stocks through various machine learning algorithms such as random forest, logistic regression, naive bayes, Bayesian networks, etc.

Other studies such as Rundo (2019) propose an advanced Markov model to perform an adaptive trading system. Through the analysis of online trading information, where automatic trading has been a trend in recent years whereby significantly increasing the number of operations it is known as high frequency trading, which together with machine learning algorithms have allowed to improve performance of these systems. Yan (2007) analyzes the stock selection through the Prototype Ranking method, applying a competitive learning approach to predict the stock rankings. Yang (2018) propose a dynamic stock recommendation system based on acquiring 20% of the top stocks dynamically using machine learning methods such as linear regression, ridge regression, stepwise regression, random forest and generalized boosted regression to model the stocks indicators and quaterly log-return in a rolling window. Huck (2019) integrates the use of large historical databases into the analysis and treating a large data problem for statistical arbitrage, using deep belief networks and random forest. Xiong (2018) shows the use of a deep reinforcement learning to stock trading model to obtain an adaptive strategy.

This work seeks to address the application of statistical learning methods for the analysis of stock markets. However, the objective is to provide additional information for decision-making, the role of the analyst can hardly be replaced by an algorithm, but the fact that an analyst has additional information processed efficiently does add value to investment decisions. In this approach were explored the benefits of using Markov chains and generalized additive models (GAM), through which they will seek to determine the stocks most susceptible to sell or buy in a portfolio given the current state. As well as an expected price estimate for the next day based on the closing price of other stocks through the fitting of a non-linear GAM model. The work integrates as follows, section two shows the materials and methods used in the proposal to be developed, section three shows the main results obtained and finally the conclusions and future lines of research are shown.

## 2   Materials and Methods

Information on the prices of shares representative of different sectors of economic activity was used, as well as the Standard and Poor's index. The selection of assets was made based on the risk beta, using balanced stocks including a set with low risk and a set with higher risk, allowing diversification of the variability of the profitability of the selected stocks with respect to the average profitability of the market. Table 1 shows the stocks and index selected in total 12 assets were selected.

**Table 1.** Selected assets

| Name | Beta coefficient |
| --- | --- |
| Kellogg | 0.43 |
| AT&T | 0.46 |
| Walmart | 0.47 |
| Exxon Mobil | 0.69 |
| Pfizer | 0.73 |
| Honda | 0.86 |
| Alphabet | 1.30 |
| American Airlines | 1.43 |
| Tiffany | 1.71 |
| Bank of America | 1.73 |
| Amazon | 1.94 |
| Standard & Poor's 500 | 1.00 |

This portfolio has a series of companies that have very different levels of volatility due to the Beta, but together they show the entire volatility panorama, where at the extremes we have Kellogg and Amazon. The analysis period considers 2017–2018, these years were selected because they were years where high impact disruptive events were not observed in the markets, such as the COVID-19 pandemic.

Based on the available information, a hybrid model will be structured to identify investment opportunities by modeling the variations of the shares using a Markov model and based on the information of highly correlated actions with the selected ones, it will be estimated using a non-linear GAM model price for the next period. Both indicators could be used by analysts to base their investment decisions based on the expected trend of the stocks analyzed. It is a hybrid model in two stages, in the first, the change in trend will be estimated based on the defined states of a Markov process, and in the second stage, the expected point value for the price of each share will be estimated (Fig. 1).

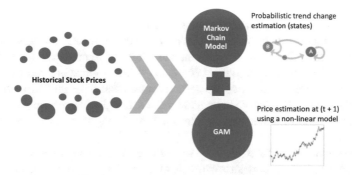

**Fig. 1.** Structure of the proposed model.

The objective is to have two measures about the expected trend of the asset, one based on probabilities and the other on the non-linear historical trend.

## 3   Markov Model

The Markov model estimation uses the fact that changes in stock price trends can be considered as changes of state in a stochastic Markov process. In order to estimate the different states, the selection of the concepts used in the technical analysis is used as a criterion, so the states to go through would be "go up", "go down" or "stay". These states will be determined by tolerance ranges in each share, that is, the analyst will define when he considers that a share has risen enough in price to think about buying (selling), or when a share has fallen enough to be wanting to sell (buy). Likewise, the analyst will define the range of variation in the price of the share that he considers to be maintaining its price trend. In order to construct the states that an action can take within the Markov process, we will model its behavior using a stochastic Markov process. Where the present state summarizes the relevant information of the process to describe its future state.

A Markov chain is a sequence $X_1$, $X_2$, $X_3$, ..., $X_n$ of random variables, whose domain is state space. The value of $X_n$ is the state of the process at time $n$. If the conditional probability distribution of $X_{n+1}$ in past states is a function of $X_n$, then $X_i$ is the state of the process at time $i$ and fulfills the Markov property.

$$P(X_{n+1} = x_{n+1} | X_n = x_n, X_{n-1} = x_{n-1}, \ldots, X_2 = x_2, X_1 = x_1) = P(X_{n+1} = x_{n+1} | X_n = x_n) \tag{1}$$

Using the Markov property, the probabilities of transition between states will be estimated according to the ranges of variations tolerated by the analyst. Based on the historical data, the matrix of transition probabilities will be calculated for each share.

## 4 Generalized Additive Model (GAM)

In a second stage of analysis, once the probabilistic estimate of the change in the status of the share is obtained, the point value of the share price is estimated using a non-linear GAM model using the information of the shares mostly correlated with the stock under evaluation. And the variables that are considered relevant in each action in particular. It is assumed that the price of the stock of the next day can be predicted from the price of the stocks of other companies today. Where we will recognize and start from the fact that for certain share prices, there are certain factors that have predictive power, from the perspective that a company is affected by the same global economic dynamics that affect other companies and that what is observed today may represent the future of other companies.

The complexity in the forecast of the share price has been documented, therefore this work only aims to have a short-term estimate of the variation in the price of the indicator action that together with the probability of change in trend obtained through the Model of Markov provide greater certainty to the analyst. It is clear that the behavior of the stock price does not follow a linear trend, despite this in many machine learning models this assumption is maintained. In this work, the use of a Generalized Additive Model is proposed that allows contemplating the non-linearity in the behavior of the share price.

The different methodologies for making predictions or forecasts using linear models have limitations when working with data that presents non-linear behavior. Recently the Generalized Additive Models (GAMs) approach has been developed. These models are an extension of the classical regression models allowing non-linear functions for the variables and maintaining the additivity of the model. GAMS models allow working with quantitative and qualitative response variables (James et al. 2013).

The linear regression model can be extended to capture nonlinear relationships between independent variables and the dependent variable. For this, it is necessary to replace each linear component $(\beta_j x_{ij})$ with a non-linear (smoothed) function $f_j(x_{ij})$. The model specification would be as follows (James et al. 2013).

$$\begin{aligned} y_i &= \beta_0 + \sum_{j=1}^{p} f_j(x_{ij}) + \varepsilon_i \\ &= \beta_0 + f_1(x_{i1}) + f_2(x_{i2}) + \ldots + f_p(x_{ip}) + \varepsilon_i \end{aligned} \tag{2}$$

Where $f_j$ are smooth no linear functions. The type of functions that can be used include natural splines, step functions, polynomial, smoothing splines, basis functions, etc. The use of different functions for non-linear treatment with each explanatory variable allows improving the efficiency of the estimates. Given the additive nature, it is feasible to analyze the effect of each independent variable separately keeping the rest of the variables constant.

# 5 Results

## 5.1 Markov Chain

It is a system that changes states over time, in which we know that the price of a stock does not depend on chance, we can say that each record is the result of a random variable for each time index, this collection of variables is defined as a stochastic process, which can be modeled as a Markov chain since presumably today's price only depends on yesterday's price. Before starting with the analysis, the state space was defined for the first analysis in which we will try to know the level of increase or decrease given the increase or decrease of the previous period or day. At this stage, the analyst will define the variation margins that he considers relevant to be able to estimate the transition probability matrices. In this case, variation ranges between [−10%, 10%] were considered.

For this, it was necessary to build a class depending on the percentage increase or decrease of the share with respect to the previous day (Fig. 2) The daily price variation ranges are estimated for each share in order to build a transition matrix for each stock.

**Fig. 2.** Intraday changes in Stock prices.

In the Fig. 2 we can see the actions that were evaluated, and the profit or loss percentages and the index related to each action, this information will be used to determine the Markov chain that corresponds to each Stock considered. It is observed that there are no absorbing states, but we do have transitory and recurring states. Despite the fact that the previous statements are totally empirical, we must say that this first analysis did not yield very good results, since due to the low probabilities and the wide ranges we cannot obtain a good fit for the prediction, these results led us to carry out a less precise analysis, but that could help us to determine or generate a better prediction of the actions based on the trend if it remains low or increases, at the cost of not knowing how much percentage. Given the previous results decided to change the state space limiting the analysis to whether the share price rises, does not change or falls and where the state of stability is limited to [−0.5%, 0.5%].

With this analysis we will be able to identify the states based on a random walk, since this is a stochastic process at discrete time that evolves through its transition to different states. The process starts in the zero state and can go to the next state [+1, 0, or −1] with a certain probability which uses the same rule for the following times. Since the price of a stock can be defined as a random walk, this series of records can be defined as a succession of independent and identically distributed random variables. For this we start from the assumption that the random walk starts at [−1, 0 or 1] and depends on each of the actions, which can perform $n$ steps where the probabilities of moving from one state to the next using the following transition diagrams. For reasons of space, only the results for two actions will be exemplified (Fig. 3).

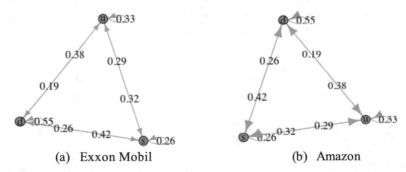

(a)  Exxon Mobil                    (b)  Amazon

**Fig. 3.** Transition diagrams based on the fitted transition probability matrices.

It is observed that there are no absorbing states, despite the fact that in some cases we have values close to 100%, this is because our confidence level is 95%, but we do have transitory and recurring states. This analysis, as mentioned above, is less robust, however, since the probabilities in the previous exercise are very low, we decided to apply this second option only to determine which actions are viable in order to determine the position that would presumably be to retain, sell or buy.

In general we can see that the results are not favorable, since the only action you could venture into, given the previous results, is Amazon, since it generates the most positive results (38% probability that the action will rise given that it is It rose in the previous period), however, it is not very sure that it should be invested in it (24% probability that the stock will go down given that in the previous period it rose). With this analysis we sought to determine which some stocks were the most profitable for an investment portfolio, since we can see that the transition matrix of some stocks gives us a certainty perspective and how the stocks behave, however, it does not give us sufficient precision to be able to venture investing unless the portfolio is much more diversified.

Despite the above, we can say that with these results we can better appreciate the dynamics of current markets where fluctuations presumably with a downward trend are represented by the reality of the share price for the Exxon Mobile company, since at the beginning of 2017 it had a price per share of almost $90 USD when in the last data it is almost $75 USD. Amazon presents a similar behavior but in a positive sense, indicating a trend towards the appreciation of the value of the Stock. What we can see in the following graph since, at the beginning of 2017 it had a price of almost $750 USD and in the last data it is at $1,502 USD per Stock.

## 5.2 Generalized Additive Model (GAM)

In the case of the use of the GAM, the case of the Walmart share price is shown, which will be estimated based on the prices of a set of shares (the most correlated) in addition to the behavior of the S&P 500 index. The approach of cross validation was used to analyze the Mean Square Error (MSE) which gives us an overview of the error generated since we will no longer be working with linear functions. We also do resample when changing the training and validation set. We will use resampling based on change of base, this to generate a greater number of MSE values by changing the training base, this implies repeatedly extracting samples from a training set and re-estimating a model in each sample to obtain additional information about the model tight. Three different non-linear models were proposed to perform the adjustment based on cross validation and select the best model.

Model 1

$$lm.fit1 = lm(Walmart \sim poly(BoA, 3) + poly(XOM, 3)$$
$$+ poly(Alphabet, 2) + poly(Tiffany, 3) \tag{3}$$

Where all the variables have the flexibility of a polynomial function that depends on each of the independent variables.
Model 2

$$gam.m1 = gam(Walmart \sim spline(BoA, 4) + spline(XOM, 4)$$
$$+ spline(Alphabet, 4) + spline(Tiffany, 4) + spline(AT.T, 4) \tag{4}$$

The degrees of freedom for the spline functions that allow modeling non-linearity were estimated by cross validation.
Model 3

$$gam.m2 = gam(Walmart \sim BoA + spline(XOM, 4) + spline(Alphabet, 4)$$
$$+ spline(Tiffany, 4) + spline(AT.T, 4) + spline(S\&P500, 4) \tag{5}$$

Where all the variables have the flexibility of 4 degrees of freedom for each of the independent variables, except Bank of America since this had come out as an important variable, but not at the level of the others given its probability value in the main analysis. To estimate the variability of a linear model we repeatedly generate different samples of the training data and fit each of the regressions to each of the new samples, and then examine the MSE measure with which we decide which regression to keep.

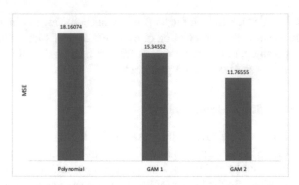

**Fig. 4.** Mean Square Error in three specifications of no lineal model.

The smallest MSE results regardless of the training and scoring base was obtained for the second specification of the GAM. This is due to the level of flexibility in each of the elements for the independent variables (Fig. 4).

The results of the GAM for the validation set are shown in Fig. 5, where we can observe an adequate fit of the model to the observed data, this due to its low mean square error that is between 11.7 and 12.5, which is very good for a Stock that fluctuates between 70 and 100 USD.

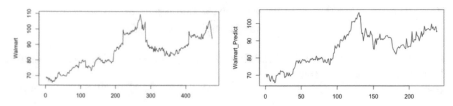

**Fig. 5.** Estimated vs observed values.

The predicted values show a correct fit of the non-linear model to the trend of the share price, it is important to note that the objective of the GAM is not to estimate long-term forecasts, but rather for short periods, for the analyst's decision making.

## 6 Conclusions

The proposed model allows obtaining relevant information on the behavior of shares in two dimensions, the Markov approach allows having a probability measure on the change in the trend in the prices of a share. For its part, the non-linear GAM model allows obtaining more precise information on the variation of the share price through a prediction for the next analysis period.

The integration of both approaches allows the generation of relevant information for decision-making, the role of the analyst is transcendental and can hardly be replaced by an algorithm. So far, the evidence shows that there is no methodology that exceeds the average of the market returns, however there are outstanding results by experienced analysts.

This proposal is an advance in the contribution of new technologies in stock market analysis, the integration of probabilistic approaches with specific estimates allows having a range of options when deciding on the rebalancing of an investment portfolio. As future lines of research, it would be relevant to explore hidden processes in the trend changes in stocks through hidden Markov chains in addition to integrating portfolio optimization in the trend change estimation.

# References

Brown, D.P., Jennings, R.H.: On technical analysis. Rev. Financ. Stud. **2**(4), 527–551 (1989)

Abarbanell, J., Bushee, B.: Fundamental analysis, future earnings, and stock prices. J. Account. Res. **35**(1), 1–24 (1997)

Fama, E.: Efficient capital markets: a review of theory and empirical work. J. Financ. **25**(2), 383–417 (1970)

Shen, S., Jiang, H., Zhang, T.: Stock market forecasting using machine learning algorithms. Department of Electrical Engineering, Stanford University, Stanford, CA (2012)

Milosevic, N.: Equity forecast: predicting long term stock price movement using machine learning. J. Econ. Libr. (2016)

Rundo, F., Trenta, F., Di Stallo, A.L., Battiato, S.: Advanced Markov-based machine learning framework for making adaptive trading system. Computation **7**, 4 (2019)

Yan, R.J., Ling, C.X.: Machine learning for stock. In: Proceedings of the 13th ACM SIGKDD International Conference on Knowledge Discovery and Data Mining Selection. ACM (2007)

Yang, H., Liu, X., Wu, Q.: A practical machine learning approach for dynamic stock recommendation. In: 2018 17th IEEE International Conference on Trust, Security and Privacy in Computing and Communications/12th IEEE International Conference on Big Data Science and Engineering (TrustCom/BigDataSE), New York, NY, pp. 1693–1697 (2018)

Huck, N.: Large data sets and machine learning: applications to statistical arbitrage. Eur. J. Oper. Res. **278**(1), 330–342 (2019)

Xiong, Z., Liu, X.-Y., Zhong, S., Yang, H., Walid, A.: Practical deep reinforcement learning approach for stock trading. arXiv preprint arXiv:1811.07522 (2018)

James, G., Witten, D., Hastie, T., Tibshirani, R.: An Introduction to Statistical Learning with Applications in R. Springer Text in Statistics, USA (2013)

# Estimation of the Stochastic Volatility of Oil Prices of the Mexican Basket: An Application of Boosting Monte Carlo Markov Chain Estimation

Román Rodríguez-Aguilar[1]([⊠]) and Jose-Antonio Marmolejo-Saucedo[2]

[1] Escuela de Ciencias Económicas y Empresariales, Universidad Panamericana,
Augusto Rodin 498, 03920 Mexico City, Mexico, Mexico
rrodrigueza@up.edu.mx
[2] Facultad de Ingeniería, Universidad Panamericana, Augusto Rodin 498, 03920 Mexico City,
Mexico, Mexico
jmarmolejo@up.edu.mx

**Abstract.** The volatility of the returns on financial assets is not a constant number over time as many valuation models, mainly derivatives, developed during the 80's, assume. The complexity of non-heteroscedasticity and the difference in results when estimated with different methodologies such as historical, implicit or stochastic calculation, make this subject too extensive a field to be covered in this work. However, stochastic volatility has been widely accepted in recent years. Monte Carlo Markov Chain (MCMC) method is explained and used to estimate the distribution of oil prices of Mexican basket as a stochastic variable. MCMC in the univariate case, supposes that we can estimate the distribution of a latent (hidden) variable through the behavior of another variable observed posteriori with the help of Bayesian inference; this method allows an efficient inference independent of the underlying process through an algorithm. The results show a correct adjustment of stochastic volatility to the behavior of the oil prices.

**Keywords:** Stochastic volatility · Non-heteroscedasticity · MCMC · Bayesian inference · Boosting

## 1 Introduction

Due to the complexity of the valuation models of some financial instruments over time, some constant values are assumed for many of them. The variation in exchange rate returns, for example, in most derivative pricing models is assumed constant when in reality it is a value that changes over time. In the specific case of the European exchange rate option valuation model proposed by [1], volatility is a crucial input that is assumed constant. In reality, the volatility of a currency cannot remain unchanged since there are times of more uncertainty than others in the markets. Over the years, countless more complex models have been proposed to estimate the volatility of exchange rates;

J. Hemanth et al. (Eds.): ICAIAME 2020, LNDECT 76, pp. 742–754, 2021.
https://doi.org/10.1007/978-3-030-79357-9_68

however, the reality is that on a day-to-day basis, options transactions are still made by calculating the price with implied volatility.

Models for estimating volatility such as Autoregressive conditional heterocedasticity (ARCH) or Generalized autoregressive conditional heterocedasticity (GARCH) assume that the observed volatility is estimated through past information. A more realistic improvement to this type of model proposes an equation that has a predictable component, which depends on past information, and a noise component. Then the volatility variable is an unobserved latent variable. This type of model is known as stochastic volatility [2, 3]. In a stochastic volatility model, both the asset price and the volatility behavior are given by Brownian movements which are possibly correlated. This is presented through a system of differential equations that when discretized using Euler's method results in the standard self-regressive stochastic volatility (SV) model [4, 5].

$$y_t = e^{\left(\frac{h_t}{2}\right)\sigma_\varepsilon \varepsilon_t} \tag{1}$$

Where:

$$\varepsilon_t \sim N(0, 1)\, iid$$

$$h_t = \mu + \phi h_{t-1} + \tau \eta_t$$

$$\eta_t \sim N(0, 1)\, iid$$

$y_t$ = logarithm of returns.
$h_t$ = logarithm of volatility.

The main objective of an SV model is to estimate the succession of unobserved volatilities and predict their values a certain number of periods ahead.

There are several stochastic volatility studies, where the main problem is the estimation of the parameters. Two main groups of applications are mainly distinguished: stochastic volatility models used in mathematical finance and models applied to financial econometrics [6]. Several studies have focused on the comparison between different classical volatility models (ARCH and GARCH) with respect to other alternatives, such as the estimation of stochastic volatility models through Markov Chain Monte Carlo sampling [5]. Despite the complexity in estimating the parameters in a stochastic volatility model, the main motivation in its recent boom is the limitation represented in various valuation models by assuming that volatility is constant, among the parameter estimation alternatives. Some authors such as [7] explore the efficiency of using simulation-intensive based methods, using the property that quadratic variation-like measures of activity in financial markets or realized volatility could derive the moments and the asymptotic distribution of the realized volatility error.

Works such as that of [8] propose the use of price range in estimating stochastic volatility models. Showing that range-based volatility proxies are highly efficient, approximately Gaussian and robust to microstructure noise, they applied a method that uses range-based Gaussian quasi-maximum likelihood estimation procedure. Other authors like [9] propose the use of the distribution of stock prices when they follow a diffusion process with a stochastic volatility parameter applying the Ornstein-Uhlenbeck arithmetic method. They also analyze the relationship between stochastic volatility and the presence of fat tails in stock price distributions. [10] shows a closed solution for the valuation of options with stochastic volatility using a proposed technique based on characteristic functions. In recent years the proposals for the application of stochastic volatility models focus on the use of simulation methods such as the MCMC method, based on the generalized stochastic volatility models defined by heavy-tailed Student-t distributions and the use of particle filtration methods [11]. Multivariate stochastic volatility models have also been developed, approached according to various categories, asymmetric models, factor models, time-varing correlation models an alternative MSV specifications. Among the estimation methods of multivariate models, the quasi-maximum likelihood method can be mentioned, simulated maximum likelihood and Markov Chain Monte Carlo methods [12].

Among the main parameter estimation methods for stochastic volatility models, we can mention estimators based on the Generalized Method of Moments principle, models based on the likelihood approximation, for example Monte Carlo-based estimators [13]. Markov Chain as the Gibss sampling [14] and models based on Simulated Maximum Likelihood proposed by [15] which is based on approximating the likelihood using simulation to integrate volatility. Other proposals are based on direct maximum likelihood estimation through the use of numerical integration [16]. Some other methods proposed for the estimation of parameters are based on the use of auxiliary models [17], estimators based on linear approximations [18], pseudo maximum likelihood using the logarithms of the squared returns [19]. Recently [20] proposed the application of ancillarity-sufficiency interwaving strategy (ASIS) for boosting MCMC estimation of stochastic volatility models. This method allows you to improve the sampling efficiency for all parameters in the same way it allows you to make inference about the constellations that have been infeasible previously to estimate without the need to select a particular parametrization beforehand. A review of the state of art shows that stochastic volatility models are a plausible alternative to GARCH models to analyze the evolution of volatility. They have greater flexibility to represent the inherent properties of financial returns and would of course be more attached to contemporary financial models. The empirical use has been limited due to the complexity of estimating the parameters of the stochastic volatility model, however, as the development of different methodologies has been documented in recent years, its use has increased empirically [21].

The objective of this work is to estimate volatility with an Stochastic Volatility model using MCMC version proposed by [20]. The aim is to model the stochastic volatility of Mexican oil prices that are highly relevant indicator for the performance of the Mexican economy. The series of prices of the Mexican oil basket expressed in dollars will be used, for the period 2010–2020, seeking to find a necessary input in the calculation of price options that is usually cleared from the formula Black-Scholes once the price is known. This issue is highly controversial in the derivatives market since without volatility there is no price and without price there is no volatility. The volatility is assumed constant for each trading term, which sometimes makes the estimates inconsistent. Also, finding volatility for intermediate terms where there is no liquidity is a problem.

## 2  Methodology

Given the behavior of financial series, one of the main concerns of investors and agents that participate in the financial market is to have knowledge of the risk presented by the investments they make. However, due to the complexity of the valuation models of some instruments over time, some constant values are assumed in many of them. From the daily prices $X_t$, profitability can be calculated and modeled as a random variable based on the volatility of its profitability. To measure market risk, the behavior of all those variables that influence the determination of its performance is investigated, or the series of market price returns is considered, which for small values of the price return, turns out to be a good approximation of the real profitability, and allows the sum of the profitability. These returns define the asset's variation rate, which is calculated depending on the time of its quotes. If assets are traded for a certain time $t$ discrete, the returns are of the form [21].

$$R_t = \frac{X_t - X_{t-1}}{X_{t-1}} \tag{2}$$

On the other hand, if they are quoted for a time t in a continuously way, then

$$R_t = Log\left(\frac{X_t}{X_{t-1}}\right) \tag{3}$$

Let $R_t$ be the return of an asset at time t that generally behaves like a leptokurtic, asymmetric distribution and a stationary process. The central idea behind the study of volatility is to test the conditional heteroscedasticity of the series $R_t^2$. To estimate these models, it is important to consider the expectation and conditional variance of $R_t^2$ given $F_{t-1}$ are [2]:

$$\mu_t = E(R_t|F_{t-1}) \tag{4}$$

$$\sigma_t^2 = Var(R_t|F_{t-1}) = E[\left(R_t - \mu_t\right)^2|F_{t-1})] \tag{5}$$

where $F_{t-1}$ is a filtration or the set of information available at time $t-1$ that expresses past returns. It is assumed that $\mu_t$ and $R_t$ follow an ARMA $(p, q)$ model, where [22]:

$$R_t = \mu_t + a_t \quad a_t \sim IID(0, 1) \tag{6}$$

$$\mu_t = \phi_0 + \sum_{i=1}^{p} \phi_i R_{t-i} + \sum_{j=1}^{q} \theta_j a_{t-j} \tag{7}$$

With $|\phi_i| < 1$ and $|\theta_j| < 1$.
From the above equation it can be deduced that

$$\sigma_t^2 = Var(R_t|F_{t-1}) = Var(a_t|F_{t-1}), \tag{8}$$

where $\sigma_t$ is the volatility. Conditional heteroscedastic models can be classified into two categories general. Those in the first category use an exact function to govern the evolution of $\sigma_t^2$, while those in the second category use a stochastic equation for $\sigma_t^2$. The GARCH model belongs to the first category, while the stochastic volatility model is in the second category. The expression $a_t$ refers to the shocks or innovations of a return at time $t$. The model of $\mu\_t$ in the above equation is called the mean equation for $R_t$ and the model $\sigma_t^2$ the volatility equation for $R_t$. For the construction of $\sigma_t^2$ models, the existence of significant correlations in the $R_t$ series is verified using the Ljung-Box test or if the residuals for the mean $a_t = \mu_t - R_t$ have statistically significant ARCH effects. To do this, either a Ljung-Box test with the $Q(k)$ statistic is performed on the $a_t^2$ series or a Lagrange multiplier test. This test is equivalent to using a Fisher distribution to prove that $\alpha_i = 0$ for all $i = 1, 2, \ldots, k$ in linear regression [23].

$$a_t^2 = \alpha_0 + \sum_{i=1}^{k} \alpha_i a_{t-i}^2 + \epsilon_i, \quad t = k + 1, \ldots, T \tag{9}$$

Where $\epsilon_i$ is the error term, $k$ is a positive integer and $T$ is the sample size. If $[\![ SSR_0 = \sum_{t=k+1}^{T}\left(a_t^2 - \varpi\right)^2$, with $\varpi = \frac{1}{T}\sum_{t=1}^{T} a_t^2$ and, $SSR_1 = \sum_{t=k+1}^{T}\widehat{\epsilon_t}^2$ where $\widehat{\epsilon_t}$ is the least squares residual from the previous linear regression, then F statistic is given by [23].

$$F = \frac{(SSR_0 - SSR_1)/k}{SSR_1/(T - 2k - 1)} \tag{11}$$

With $F$ distribution for $k$ and $T - 2k - 1$ degrees of freedom. The purpose of the test is to determine if there are ARCH effects in the series of returns, that is, $\alpha_i \neq 0$ *for some* $i = 1, 2, \ldots, k$. After proposing a model in conditional mean $\mu_t$ and a model of conditional or stochastic variance $\sigma_t^2$, the assumptions are verified in their standardized residuals [23]:

$$\tilde{a}_t = \frac{a_t}{\sigma_t} \tag{12}$$

The tests to verify the assumptions are:

- The Ljung-Box statistics of $\tilde{a}_t$ to verify the absence of significant autocorrelations and to validate the equation in mean.
- The Ljung-Box statistics of $\left(\tilde{a}_t\right)^2$ to test the validity of the equation for conditional or stochastic volatility.
- The adequate distribution of $\tilde{a}_t$, as the case may be, stated in the error terms $\epsilon_t$ of the conditional or stochastic volatility models.

### Stochastic Volatility Models

Conditional heteroscedasticity models such as ARCH or GARCH assume that volatility can be observed one step ahead with the information of the random variables $\epsilon_t$ or $a_t$. In these models, shocks are only incorporated through the innovations $\epsilon_t$, that is, there are only disturbances on the mean equation. In this way, volatility is estimated through past information, becoming predictable because for each moment $t$ the information in

$t - 1$ is already known. More realistic model for volatility can be based on a behavioral equation that has a predictable component that depends on past information, as in GARCH models, and an unexpected noise component. In this case, volatility is a latent unobserved variable. One interpretation of latent volatility is that it represents the arrival of new information in the market. This type of modeling corresponds to the so-called stochastic volatility models.

## Model SV-AR(1)

The autoregressive stochastic volatility model or SV-AR (1) is a state space model where the state variable is the log-volatility, given by [21, 22].

$$R_t = \sigma_t \epsilon_t \tag{13}$$

$$\sigma_t^2 = exp\left\{\frac{h_t}{2}\right\} \tag{14}$$

$$h_t = \mu + \phi(h_{t-1} - \mu) + \eta_t \tag{15}$$

Where $\mu \in \mathbb{R}$, $h_t$ is the logarithm of the volatility that follows a stationary process AR (1) with persistence parameter $|\phi| < 1$. It is assumed that $\epsilon_t \sim \mathcal{N}(0, 1)$ and $\eta_t \sim \mathcal{N}(0, \sigma_\eta^2)$ are the unexpected noise or disturbances for the returns and for the volatility, respectively, mutually independent for all $t$ and $s$. To complete the configuration of the model, an a priori distribution for the vector of parameters $\theta = (\mu, \phi, \sigma_\eta)$ must be specified in order to take the median, as a robust estimator, of each distribution as a point parameter of the model. For this, independent components are chosen for each parameter, that is, $p(\theta) = (p(\mu)p(\phi)p(\sigma_\eta))$. The parameter $\mu$ follows a normal a priori distribution $\mathcal{N}(b, B)$. According to (Kastner 2016a), $b = 0$ and $B \geq 100$ can be taken.

For the parameter $\phi \in (-1, 1)$, we choose $(\phi + 1) \sim 2\mathcal{B}(a_0, b_0)$, which implies that this distribution is in the interval $(-1, 1)$ and guarantees the stationarity of the autoregressive volatility process. Its expected value and variance are given through the expressions [21, 22].

$$E(\phi) = \frac{2a_0}{a_0 + b_0} - 1 \tag{16}$$

$$Var(\phi) = \frac{4a_0b_0}{(a_0 + b_0)^2(a_0 + b_0 + 1)} \tag{17}$$

The a priori distribution of $\phi$ depends only on the relationship between $a_0$ and $b_0$. This parameter is greater than zero if and only if $a_0 > b_0$ and less than zero if and only if $a_0 < b_0$. The uniform distribution in $(-1, 1)$ arises as a particular case of $a_0 = b_0 = 1$. When the underlying data generation process is (almost) homoscedastic, the volatility of the log-variance $\sigma_\eta$ is (very close to) zero and therefore the probability contains little or no information about $\varphi$. Consequently, the a priori distribution of $\phi$ is (almost) equal to its former, no matter how many data points are observed. According to (Kastner, 2016a), we can take $a_0 = 5$ and $b_0 = 1.5$, which implies a previous mean of 0.86 with a previous standard deviation of 0.11 and, therefore, very little mass for non-positive values of $\phi$.

For the volatility of the log-variance $\sigma$ $\sigma_\eta \in \mathbb{R}^+$, $\sigma_\eta^2 \sim \mathcal{B} \times \chi_1^2$ is chosen, where the choice of hyperparameter $B$ turns out to be of less importance in empirical applications, as long as it is not set too small [20].

**MCMC (Markov Chain Monte Carlo) Sampling**
The fundamental idea of MCMC methods is to design a Markov chain to generate independent samples based on the distribution of interest. This algorithm provides representations of the posterior distribution of the desired random variables, which in our case are the log-variances $h$ and the vector of parameters $\theta$. Numerical methods based on Monte Carlo Markov Chain have had a great boom in estimating stochastic volatility models. The fundamental problem in estimating a stochastic volatility model is the likelihood assessment.

$$f(Y|\theta) = \int f(Y|H, \theta)f(H|\theta)dH \tag{18}$$

Where $Y = (y_1, \ldots, y_T)$, $H = (log\sigma_1^2, \ldots, log\sigma_T^2)$ and $\theta$ is a vector of unknow parameters.

A novel and crucial feature of the algorithm is the use of the "Anciliarity-Sufficiency Interweaving Strategy" (ASIS) which has been presented in the general context of state-space models. ASIS exploits the fact that, for certain groupings of parameters, the sampling efficiency is substantially improved when considering a non-centered version of a state-space model. This fact is commonly known as a reparameterization problem [20]. In the case of the SV model, a movement of this type can be achieved by changing the level of the log-variance $\mu$ and/or the volatility $\sigma_\eta$ from the variation of the state process to the observation process through a simple reparameterization of the latent process $h$ [20]. In the case of the SV model, it turns out that there is no single parameterization. Rather, for some underlying processes, standard parameterization produces better results, while for other processes, non-centered versions are better. To overcome this problem, the vector of parameters $\theta$ is sampled twice: once in the centered and once in a non-centered parameterization. This method allows efficient inference regardless of the underlying process with an algorithm [20].

For this work, oil prices in USD for the period 2010–2020 was considered. They are daily observations. As only the univariate case is exposed. The source of this series is the Central Bank of Mexico information system and each observation corresponds to the close of the day. In the case of oil prices, it is important to highlight that a negative price was observed on April 20, 2020, for which it was decided to discard this price from the data since estimating logarithmic returns generated inconsistencies in the results.

## 3   Stochastic Volatility Estimation by MCMC Using ASIS Algorithm

Once the methods used in this work for the estimation of volatility have been exposed, the results of applying the proposal of [20] to the series of data of daily prices of Mexican oil basket (USD) between the dates 1/01/2010 – 14/08/2020. Oil prices is a relevant variable to monitor in the case of the Mexican economy. Since a large part of the public income comes from the sale of oil it is a crucial variable. Figure 1 shows the series of oil prices, the price of oil has shown a downward trend in recent years due to structural changes in the energy sector. April 2020 was the month that presented the biggest drop in prices and even on the 20th of that month negative prices were observed, a phenomenon never seen in the markets.

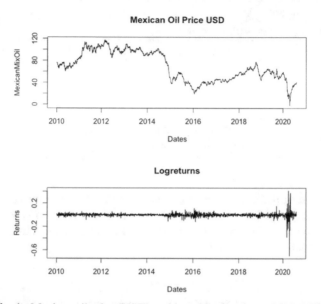

**Fig. 1.** Mexican oil price (USD) and logarithmic returns, 2010–2020.

The MCMC function used in this work contains the following initial parameters, the series has 2534 observations of log-variances:

- A priori distribution of $\mu$ as a Normal $N$ (0,10) Where $\mu$ is the log-variance.
- A priori distribution of the volatility persistence parameter $\varphi \in (-1, 1)$ where $\frac{(\varphi+1)}{2}$ is distributed as a $B(a_0 = 20, b_0 = 1.1)$. This parameter is that it provides information on the process in a previous time.
- The last parameter $\sigma_n = 0.1$ refers to the variance of the log-variance and is distributed as an $\chi_i^2$ as explained in the methodology section, the important thing is that the initial value is small.

The results of the algorithm are shown in Table 1.

**Table 1.** Prior and posterior parameter distributions

Prior distributions:

mu ~ Normal (mean = 0, sd = 10)

(phi+1)/2 ~ Beta (a0 = 20, b0 = 1.1)

sigma^2 ~ 0.1 * Chisq(df = 1)

Posterior draws of parameters (thinning = 1):

|          | mean    | sd     | 5%      | 50%     | 95%     | ESS  |
|----------|---------|--------|---------|---------|---------|------|
| mu       | − 8.128 | 0.2453 | − 8.513 | − 8.128 | − 7.736 | 6433 |
| phi      | 0.980   | 0.0058 | 0.970   | 0.981   | 0.989   | 237  |
| sigma    | 0.219   | 0.0244 | 0.182   | 0.218   | 0.261   | 145  |
| exp(mu/2)| 0.017   | 0.0021 | 0.014   | 0.017   | 0.021   | 6433 |
| sigma^2  | 0.049   | 0.0109 | 0.033   | 0.048   | 0.068   | 145  |

The selection of the initial values of the algorithm was carried out through cross validation, according to [20] the initial values tend to converge to stable values when increasing the number of iterations. Figure 2 shows the results of the estimated parameters in the 10,000 iterations carried out.

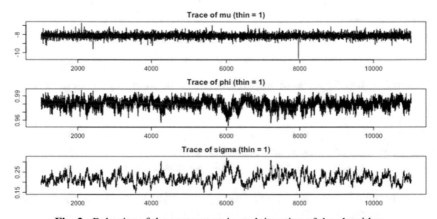

**Fig. 2.** Behavior of the parameters in each iteration of the algorithm.

As we can see, the algorithm is quite robust for estimating the necessary stable parameters, even when our initial values change a bit. Even as part of the results of the algorithm we can obtain the posterior distributions (Fig. 3).

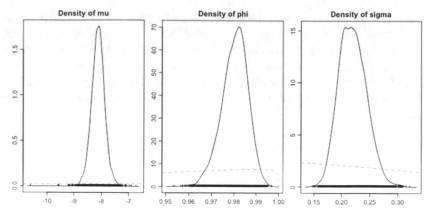

**Fig. 3.** Posterior distributions of the parameters.

Based on the estimated volatility model, a volatility forecast 60 days ahead was estimated. It is worth mentioning that the volatility estimated by the stochastic model considers the abrupt variations in prices that were observed in the month of April in international oil prices, except on April 20, where negative prices were observed. Figure 4 show the hidden variable, daily volatility, it looks quite coherent and we see that its behavior over time is within the limits established by the references of both the historical series of prices and the estimates made through the implicit data for different time horizons.

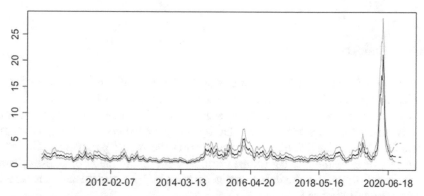

**Fig. 4.** Estimated volatilities in percent (5%, 50%, 95% posterior quantiles with 60 days ahead

One of the advantages of the adjustment of a stochastic volatility model using MCMC is the estimation of the posterior distribution of the parameters, in this case we can see how the parameters stabilize when increasing the number of iterations. When making a forecast 60 days ahead regarding the volatility of oil prices, an open interval is observed, but with a decreasing trend at the 50th percentile. The volatility window observed in the month of April is not replicated in the forecasts, however, it is considered in the estimation of the stochastic volatility model parameters.

Using the estimated stochastic volatility model, a forecast was generated based on an AR (5) -SV model for 60 days ahead. It is observed that the 50% percentile shows a trend of mean in oil prices around 40 dollars for the next sixty days. Again, it is important to emphasize that the stochastic volatility used contemplates the high volatility windows presented in April with very low oil prices. The dotted lines in green in Fig. 5 show the 10th and 90th percentiles of the forecast and the solid green lines represent the 50th percentile.

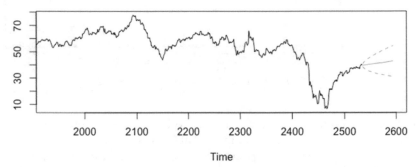

**Fig. 5.** Predicted Mexican oil prices in percent (10%, 50%, 90% posterior quan-tiles with 60 days ahead)

As shown in Fig. 5 the AR (5) structure for the average based on the SV model shows a slight recovery in the price of oil, keeping the mean at a value around 40 USD per barrel.

## 4   Conclusions

The modeling of financial series requires using models that adhere as closely as possible to reality. In the valuation of financial instruments, especially in the case of derivatives, on many occasions constant volatility is assumed, which implies an oversimplification of reality. In recent years, stochastic volatility models have been developed to a great extent using various methodologies for estimating the model parameters. One of the most widely used methods is the Monte Carlo Markov Chain method, whose approach is carried out through the simulation and generation of independent samples based on the distribution of interest.

As part of the MCMC methodologies, [20] proposes a methodology called anciliatiry-sufficiency interwaving strategy (ASIS), this methodology allows to significantly improve the sampling efficiency for all parameters and throughout the entire parameter range. Without the restriction of selection, a particular parameterization beforehand. The results show that the initiation of a parameterization of the MCMC allows convergence to obtain the model parameters for the estimation of stochastic volatility. Stochastic volatility, in addition to better modeling reality by construction, does not depend on a formula or given data to be able to calculate it. Furthermore, the Kastner (2014) algorithm is quite efficient even with large data samples. The algorithm, when you have large data samples, is quite robust, so even for initial values without much sense the convergence to the real parameters is good. This type of numerical methodologies presents a great alternative for estimating distribution functions that is very complex to integrate.

After reviewing the algorithm, we found that, when performing the exercise, daily volatilities are estimated with daily prices, which are not usually very different between different currency pairs because the time horizon is very short. Therefore, it is recommended to do the analysis over longer time horizons to check the consistency of the estimate. It is suggested to review several series of different time frames expressed in daily terms to check for inconsistencies. The outcomes of the application of the MCMC methodology based on the ASIS methodology shows a consistent adjustment of the volatility of oil prices. The estimates are not affected by the volatility window observed in April 2020, where there was great volatility in oil prices, even presenting negative prices on April 20. The estimated volatility and the predictions made with it through an AR (5) process show consistent results with respect to what was observed in the market in the analyzed time window, without greatly replicating the trend observed in prices in the month of April 2020.

# References

1. Black, F., Scholes, M.: The pricing of options and corporate liabilities. J. Polit.l Econ. 637–654 (1973)
2. Taylor, S.: Financial returns modelled by the product of two stochastic process, a study of daily sugar prices. In: Time Series Analysis: Theory and Practice, Amsterdam, North Holland (1982)
3. Wiley West, M., Harrison, J.: Bayesian Forecasting and Dynamic Models, 2nd edn. Springer, New York (2007). https://doi.org/10.1007/b98971
4. Lord, R., Koekkoek, R., Van Dijk, D.: A comparison of biased simulation schemes for stochastic volatility models. Quant. Financ. **10**(2), 177–194 (2010). https://doi.org/10.1080/146976 80802392496
5. Sangjoon, K., Shepard, N., Chib, S.: Stochastic volatility: likehood inference and comparison with ARCH models. Rev. Econ. Studies **65**(3), 361 (1998)
6. Ghysels, E., Harvey, A.C., Renault, E.: Stochastic Volatility. Handbook of Statistics, vol.14, pp. 119–191. Elsevier (1996)
7. Barndorff-Nielsen, O.E., Shephard, N.: Econometric analysis of realized volatility and its use in estimating stochastic volatility models. J. Roy. Stat. Soc. Ser. B (Stat. Methodol.) **64**, 253–280 (2002). https://doi.org/10.1111/1467-9868.00336
8. Alizadeh, S., Brandt, M.W., Diebold, F.X.: Range-based estimation of stochastic volatility models. J. Financ. **57**, 1047–1091 (2002). https://doi.org/10.1111/1540-6261.00454
9. Stein, E.M., Stein, J.C.: Stock price distributions with stochastic volatility: an analytic approach. Rev. Financ. Studies 4(4), 727–752 (1991). https://doi.org/10.1093/rfs/4.4.727
10. Heston, S.L.: A closed-form solution for options with stochastic volatility with applications to bond and currency options. Rev. Financ. Studies **6**(2), 327–343 (1993). https://doi.org/10.1093/rfs/6.2.327
11. Chib, S., Nardari, F., Shephard, N.: Markov chain Monte Carlo methods for stochastic volatility models. J. Econom. **108**(2), 281–316 (2002)
12. Asai, M., McAleer, M., Yu, J.: Multivariate stochastic volatility: a review. Econom. Rev. **25**(2–3), 145–175 (2006). https://doi.org/10.1080/07474930600713564

13. Melino, A., Turnbull, S.: Pricing foreing currency options with stochastic volatility. J. Econom. **45**, 239–265 (1990)
14. Jacquier, E., Polson, N., Rossi, P.: Bayesian analysis of stochastic volatility models. J. Bus. Econ. Stat. **12**, 314–417 (1994)
15. Danielsson, J.: Stochastic volatility in asset prices: estimation with simulated maximum likelihood. J. Econom. **61**, 375–400 (1994)
16. Friedman, M., Harris, L.: A maximum likelihood approach for non Gaussian stochastic volatility models. J. Bus. Econ. Stat. **16**, 284–291 (1998)
17. Gourieroux, C., Monfort, A., Renault E.: Indirect inference. J. Appl. Econom. **8**, 85–118 (1993)
18. Koopman, S.J., Uspensky, E.H.: The stochastic volatility in mean model: empirical evidence from international stock markets. J. Appl. Econom. **17**, 667–689 (2002)
19. Harvey, A.C., Ruiz, E., Shepard, N.G.: Multivariate stochastic variance models. Rev. Econ. Stud. **61**, 247–264 (1994)
20. Kastner, G., Frühwirth-Schnatter, S.: Ancillarity-sufficiency interweaving strategy (ASIS) for boosting MCMC estimation of stochastic volatility models. Comput. Stat. Data Anal. **76**(2014), 408–423 (2014)
21. Ruiz, E., Veiga, H.: Modelos de volatilidad estocástica: una alternativa atractiva y factible para modelizer la evolución de la volatilidad. Anales de Estudios Económicos y Empresariales **18**, 9–68 (2008)
22. Gamerman, D., Lopes, H.F.: MCMC: Stochastic Simulation for Bayesian Inference, 2nd edn. Chapman & Hall/CRC, Boca Raton (2006).
23. Ljung, G.M., Box, G.E.P.: On a measure of a lack of fit in time series models. Biometrika **65**(2), 297–303 (1978)

# Optimization of the Input/Output Linearization Feedback Controller with Simulated Annealing and Designing of a Novel Stator Flux-Based Model Reference Adaptive System Speed Estimator with Least Mean Square Adaptation Mechanism

Remzi Inan[1]($\boxtimes$), Mevlut Ersoy[2], and Cem Deniz Kumral[3]

[1] Faculty of Technology Department of Electrical and Electronics Engineering, Isparta University of Applied Sciences, Isparta, Turkey
remziinan@isparta.edu.tr

[2] Faculty of Engineering Department of Computer Engineering, Süleyman Demirel University, Isparta, Turkey
mevlutersoy@sdu.edu.tr

[3] Distance Learning Vocational School Department of Computer Technologies, Isparta University of Applied Sciences, Isparta, Turkey
cemkumral@isparta.edu.tr

**Abstract.** In this study, the positive coefficients of the auxiliary control inputs of the input/output feedback linearization controller (IOFLC) are optimized with simulated annealing (SA) and Particle Swarm Optimization (PSO) algorithm. The IOFLC produces the $-\alpha\beta$ stator stationary axis components of the reference voltage of the space vector pulse width modulation (SVPWM) and is used instead of the proportional and integral type (PI) controller in direct torque control (DTC) of the induction motor (IM). In addition to optimization with SA, a novel stator flux-based model reference adaptive system (MRAS) speed estimator is proposed for the speed-sensorless IOFLC-based DTC (IOFLC-DTC) of IM. The least mean square (LMS) algorithm is used as the adaptation mechanism in the novel MRAS speed estimator. In the optimization process, the mean square errors between the estimated and the reference values of the states which are affected by the optimized coefficients are based on. Due to this, the mean square error between the estimated and reference mechanical velocity and the mean square error between the estimated and reference stator flux are used as the control boundaries of optimized coefficients. The proposed study includes implementation of the novel MRAS speed estimator-based optimized IOFLC-DTC drive system of the IM in simulation and also testing the estimation and the control performance under load torque variations in a wide speed range consists of zero speed to rated speed at forward and reverse direction. The simulation results show clearly the effectiveness of the proposed IM drive system.

**Keywords:** Simulating annealing · Particle Swarm Optimization · Induction motor · Input/output feedback linearization control · Model reference adaptive system · Least mean square

© The Author(s), under exclusive license to Springer Nature Switzerland AG 2021
J. Hemanth et al. (Eds.): ICAIAME 2020, LNDECT 76, pp. 755–769, 2021.
https://doi.org/10.1007/978-3-030-79357-9_69

# 1 Introduction

The dynamic control of Induction Motor (IM) is difficult because of having nonlinear multivariable coupled time varying system and the rotor variables are not measurable [1]. The Field Oriented Control (FOC), Direct Torque Control (DTC), Sliding Mode Control (SMC), Model Predictive Control (MPC), adaptive control, artificial intelligence-based control, and observer-based nonlinear control methods are proposed for the fast dynamic control of IM [2]. The DTC has fast torque response and a very simple and less sensitive structure to parameter changes than the other control methods, making it widely preferred in industrial applications of IM dynamic control. But in conventional DTC, variable switching frequency due to the use of the switching table causes high torque ripples and switching losses. The use of Space Vector Pulse Width Modulation (SVPWM) in DTC gives the ability to switch with a fixed switching frequency, as well as using the DC voltage with high efficiency and reducing harmonic distortions. Thus, in SVPWM-based DTC (SVPWM-DTC), torque ripples and switching losses are reduced [3]. However, in SVPWM-DTC, the use of PI-type controller and passive reference-frame transformations are required. These passive reference-frame transformations increase the complexity of the system and determining the coefficients of PI-type controller requires expertise. In addition, the dynamic structure of the coefficients of PI-type controller against operating conditions and parameter changes negatively affects control performance [4].

The most common method used to solve this problem is model-based Input/Output Feedback Linearization Control (IOFLC) which has a nonlinear control structure. The IOFLC increases the control performance by decoupling the correlation between variables. Thus a simple control design is obtained by converting the nonlinear system to linear one. With IOFLC, decoupled control of torque and stator flux can be achieved exactly, and the design of the control system is simplified by eliminating the passive reference-frame transformations [5, 6]. However, in IOFLC the determination of the exact values of the coefficients which are used during the derivation of the $-\alpha\beta$ stator stationary axis components of auxiliary control inputs and defined as positive constants in the literature affects the control performance of drive system directly. In the literature, these coefficients are determined by trial-and-error method. However, it is obvious that a higher control performance can be achieved if these coefficients are determined by heuristic optimization methods.

Meta-heuristic approaches are recognized as effective methods used in many optimization problems. They are widely used to solve complex problems in industry and service areas, especially finance problems, production management and engineering problems. Almost all meta-heuristic approaches provide the best solution by taking inspiration from nature, using random variables, and using various parameters to fit the problem at hand [7]. Simulated Annealing (SA) and Particle Swarm Optimization (PSO) are meta-heuristic approaches that are frequently used in optimization problems. The use of SA and PSO algorithms is preferred in the realized study.

SA works efficiently with random systems and cost functions, guaranteeing finding the most statistically appropriate solution. It is relatively easier to encode than other meta-intuition algorithms, and often offers a good solution even for complex problems. The disadvantages of SA are that Repeatedly annealing with a $1/\log k$ schedule is very slow, especially if the cost function is expensive to compute, for problems where the energy landscape is smooth, or there are few local minima.

PSO is one of the modern heuristic algorithms that can be used to solve nonlinear and non-continuous optimization problems. The main advantages of the PSO algorithm can be summarized as simple concept, easy implementation, providing of robustness to control parameters, better computational efficiency compared to mathematical algorithms and other heuristic optimization techniques. The disadvantages of the PSO algorithm are that it is easy to fall to the local optimum level in high-dimensional space and has a low convergence rate in the iterative process.

Although both approaches have disadvantages they are use in solving the problem in the study carried out brings successful results. For the reasons presented, SA and PSO algorithms have been used in order to optimize the positive coefficients of the auxiliary control inputs of the IOFLC control parameters.

As a result of realizated work it is proved from the Mean Square Error (MSE) between the reference and measured speed of IM under both transient and steady-state condition that SA algorithm gives most effective optimization results against to PSO algorithm. However, in the SA algorithm, assigning the initial conditions of the parameters that need to be optimized is a problem, and it is observed that this problem directly combines the global minimums and the inability to improve the optimization process, causing the algorithm to get stuck, while the PSO algorithm has no problem in assigning the initial conditions for the parameters that need to be optimized.

In this study, the specified coefficients are determined by using two different heuristic optimization methods as SA and PSO. The optimization performances of the SA and PSO are tested in the IOFLC-based DTC (IOFLC-DTC) drive system of IM by using the heuristically determined coefficients. Thus, the performance of IOFLC-DTC drive system of IM is carried to a very advanced level.

Moreover, the SVPWM-IOFLC drive system of IM needs the exact knowledge of rotor mechanical angular speed ($\omega_m$) and $-\alpha\beta$ components of stator flux ($\varphi_{s\alpha}$ and $\varphi_{s\beta}$). Measurement of $\omega_m$, $\varphi_{s\alpha}$, and $\varphi_{s\beta}$ is possible with subsequent interventions to the IM. This increases the cost of the control system and adversely affects its robustness. For the estimation of $\omega_m$, a novel stator flux-based Model Reference Adaptive System (MRAS) speed estimator with least mean square (LMS) adaptation mechanism is proposed for the speed-sensorless IOFLC-DTC of IM. Where $\varphi_{s\alpha}$, and $\varphi_{s\beta}$ take place of the reference voltage model of MRAS. The proposed novel stator flux-based MRAS with LMS adaptation mechanism is adopted to the IOFLC-DTC drive system in order to avoid the complexity of observer and improve control performance of the drive system.

The main contribution of this study is the optimization of the positive coefficients of the auxiliary control inputs of the IOFLC with SA and PSO algorithm in order to increase the control performance. And also a novel stator flux-based MRAS speed estimator with LMS adaptation mechanism is proposed for the speed-sensorless IOFLC-DTC of IM.

## 2  Method and Material

### 2.1  Stator Flux-Based IM Model in Stator Stationary Reference Frame

The differential expressions of the dynamic stator flux-based IM model in $-\alpha\beta$ stator stationary reference frame and the motion equation are given in the following form:

$$\frac{di_{s\alpha}}{dt} = -\frac{1}{\sigma}\left(\frac{R_r}{L_r} + \frac{R_s}{L_s}\right)i_{s\alpha} - \omega_m i_{s\alpha} + \frac{R_r}{L_\sigma L_r}\varphi_{s\alpha} + \frac{\omega_m}{L_\sigma}\varphi_{s\beta} + \frac{1}{L_\sigma}v_{s\alpha} \tag{1}$$

$$\frac{di_{s\beta}}{dt} = -\frac{1}{\sigma}\left(\frac{R_r}{L_r} + \frac{R_s}{L_s}\right)i_{s\beta} + \omega_m i_{s\beta} + \frac{R_r}{L_\sigma L_r}\varphi_{s\beta} - \frac{\omega_m}{L_\sigma}\varphi_{s\alpha} + \frac{1}{L_\sigma}v_{s\beta} \tag{2}$$

$$\frac{d\varphi_{s\alpha}}{dt} = v_{s\alpha} - R_s i_{s\alpha} \tag{3}$$

$$\frac{d\varphi_{s\beta}}{dt} = v_{s\beta} - R_s i_{s\beta} \tag{4}$$

$$\frac{d\omega_m}{dt} = \frac{1}{j_L}\left(\frac{3}{2}p_p\left(\varphi_{s\alpha}i_{s\beta} - \varphi_{s\beta}i_{s\alpha}\right) - \beta_L\omega_m - t_L\right) \tag{5}$$

where, $i_{s\alpha}$ and $i_{s\beta}$ are $-\alpha\beta$ components of stator current. $v_{s\alpha}$ and $v_{s\beta}$ are $-\alpha\beta$ components of stator voltage. $R_s$ and $R_r$ are stator and rotor resistance, respectively. $L_s$ and $L_r$ are stator and rotor self inductance, respectively. $L_\sigma = \sigma L_s$ is the stator transient inductance. $\sigma = 1 - L_m^2/(L_s L_r)$ is the leakage or coupling factor. $L_m$ is the mutual inductance. $t_L$ is load torque. $j_L$ and $\beta_L$ are total inertia and viscous friction of IM and load, respectively. $p_p$ is pole pair of IM.

The generalized state space equations of IM model is given in (6) and (7). In state space representation, the IM model equations are discretized by using Forward Euler method with a sampling time $T$. Thus, the IM model can be implemented on digital platform in discrete-time. Also the parameters of IM model is given in Table 1.

$$\dot{x} = f(x, u) + w \tag{6}$$

$$= A(x)x + Bu + w$$

$$z = h(x) + v \ (\text{Measurement Equation}) \tag{7}$$

$$= Hx + v$$

where, $x$ is the extended state vector. $f$ is the nonlinear function of states and inputs. $u$ is the control input vector. $A$ is the system matrix. $B$ is the input matrix. $z$ is the measurement vector. $h$ is the nonlinear functions of the outputs. $H$ is the measurement matrix. $w$ and $v$ are process and measurement noise, respectively.

**Table 1.** IM parameters

| $P$[kW] | $f$[Hz] | $j_L\left[\text{kg}\cdot\text{m}^2\right]$ | $\beta_L$[N · m · s/rad] | $p_p$ | $V$[V] | $I$[A] |
|---|---|---|---|---|---|---|
| 2.2 | 50 | 0.055 | 0.001 | 3 | 380 | 5.5 |
| $R_s$[Ω] | $R_r$[Ω] | $L_s$[H] | $L_r$[H] | $L_m$[H] | $n_m$[r/min] | $t_L$[N · m] |
| 3.03 | 2.53 | 0.1385 | 0.1443 | 0.1269 | 950 | 20 |

## 2.2 Overwiev Simulated Annealing Algorithm

Simulated Annealing algorithm is a local search method that performs a probabilistic operation in order to obtain better heuristic solutions in combinatorial optimization problems proposed by Kirkpatrick et al. in 1983. Taboo search algorithm, genetic algorithms etc. as in the approaches, it aims to find the best solution to the problem as soon as possible. Simulation Annealing algorithm is based on Metropolis Monte Carlo method, which is used to collect atoms be in equilibrium at a certain temperature. The standard Monte Carlo method accepts only a move to a lower-energy state, while the Metropolis method also allows a higher-energy state to be accepted within the possibilities. Simulated annealing algorithm; image processing applications, path finding problems and travel, electronic and logical circuit design, simulations, Materials Physics, flow scheduling, job scheduling, cutting and packing gave positive results it was observed that in the solution of problems [8, 9].

Annealing simulation is an algorithm that attempts to compensate for the shortfall caused by other local search methods stopping searching to find the global low after reaching the local low. In annealing simulation, the same basic processes in other local search methods are used, except in cases where there are exceptions [10]. Decisions that must be made when performing an annealing simulation algorithm on a problem are divided into two groups:

1. Decisions based on algorithm:

   - Determining the first temperature
   - Determining the cooling parameter
   - Algorithm's halt condition

2. Decisions depending on the applied problem:

   - How solutions are represented
   - Determining the objective function
   - Choosing the starting solution
   - Determining the neighbor breeding plan

The name Simulation Annealing comes from the algorithm's similarity with the physical annealing processes of solids. If a solid substance is heated to its melting point and then rapidly cooled, the molecular structure of the substance changes according to the speed of the cooling process. If the optimization problem is represented depending on the states of matter, each state of the matter will correspond to different solutions and the available energy will correspond to the cost function. In the annealing process, the material is cooled slowly until it turns into a solid, that is, to the lowest energy level. If the global lowest energy is reached, the solution to the optimization problem is provided. The initial temperature (T) should be set high enough when the algorithm starts to accept bad solutions in line with the possibilities and ensure that the new solution to be reached is independent of the initial solution. In the simulated annealing algorithm, the temperature value is not constant and decreases depending on a certain iteration. Reducing the temperature parameter is an important factor in reaching the best solution [11]. The pseudo code of the simulated annealing algorithm is given in Algorithm 1 [11].

1. Determine the initial temperature (T)
2. Determine the cooling parameter ($\alpha$)
3. Generate the first solution ($\pi$) and calculate the function value ($f(\pi)$)
4. **while** T > 0.1 **do**
5. Generate a new solution ($\pi$') and calculate the function value $f(\pi')$
6. $\Delta f = f(\pi') - f(\pi)$
7. **if** $\Delta f < 0$ **then**
8. $\pi = \pi'$
9. **else if** random[0,1] $< \exp(-\frac{\Delta f}{T})$ **then**
10. $\pi = \pi'$
11. **else** cancel new solution ($\pi'$)
12. T = T x $\alpha$                //update temperature
13. return Best Solution

**Algorithm 1.** Simulated Annealing pseudo code [11]

## 2.3   Overview of Particle Swarm Optimization

Particle Swarm Optimization (PSO) is a population-based heuristic optimization method that was first proposed by Eberhart and Kennnedy in 1995 to solve combinatorial problems. It can be said that PSO is a method inspired by sociology, because the basis of the algorithm is based on social information sharing, that is, herd intelligence, that animals gather in herds. It has been observed that the random behaviors of animals moving in herds, such as finding food and ensuring their safety, make it easier for them to reach their goals. Herds follow a collaborative method to find food and ensure their safety, and each animal in the herd continues to change the way they search based on their learning experiences and those of other animals in the herd [12–14].

Searching in PSO is done according to the number of individuals as in genetic algorithms. The community of particles and particles for each individual is also called a flock. Each particle directs its position to the best position in the herd, drawing on previous experience. The PSO is basically based on the movement of the particles in the herd towards the part with the best position in the herd. This diverting movement is a random occurrence and usually the particles in the herd reach a better position in their new movement than the previous position. These orientation processes continue until the global best point is reached. Particle Swarm Optimization; it has been observed to bring successful results in optimization problems such as supply selection problems, order quantity analysis, scheduling problems, sequencing problems, voltage and power control, determination of motor control parameters [14, 15].

In the algorithm, each particle represents a bird in the flock and offers a solution. All particles are passed through the fitness function to find their suitability values. Particles have a speed information that affects the flight of birds in the flock. The algorithm is started with a certain number of randomly generated particles. These particles are updated as the algorithm steps progresses and the best solution value is sought. Each particle is updated based on its own best solution (local Best) and best solution (global best) values among all particles. These values are kept in memory and when the algorithm is terminated, the global best value in solving the problem is obtained. The pseudo code of Particle Swarm Optimization is given in Algorithm 2 [16].

1. Define the particle value (X)
2. Define the rate of change (V) of the particle
3. Define constants (c1, c2)
4. Define random values (r1, r2)
5. Generate the flock by giving random initial values for Xi,j and Vi,j
6. **Do**
7. **For** i = 1 **to** Particle count (Pc)
8. **if** f(Xi) < f(yi) **then** yi = Xi             // update local best
9. yi' = min(Xneighbors)                      // update global best
10. **For** j = 1 **to** Number of sizes optimized
11. Vi,j = Vi,j + c1r1,j*|yi,j – Xi,j| + c2r2,j*|yi' - Xi,j|     //update speed vector
12. Xi,j = Xi,j + Vi,j                         // update position vector
13. **Next** j
14. **Next** i
15. **Until** Termination condition

**Algorithm 2.** Particle Swarm Optimization pseudo code[16]

### 2.4 IOFLC Method

The torque and square of the stator flux modulus of the IM which are taken as the system outputs in order to develope the IOFLC. In IOFLC, $v_{s\alpha}$ and $v_{s\beta}$ are defined as control inputs and $i_{s\alpha}$ and $i_{s\beta}$ are measured directly from the IM. Expressions and details regarding the derivation of IOFLC are given as follows [17, 18].

The system output in IOLC is given as follows:

$$y_1 = t_e = \frac{3}{2}p_p\left(\hat{\varphi}_{s\alpha}i_{s\beta} - \hat{\varphi}_{s\beta}i_{s\alpha}\right) \tag{8}$$

$$y_2 = \left|\hat{\varphi}_s\right|^2 = \hat{\varphi}_{s\alpha}^2 + \hat{\varphi}_{s\beta}^2 \tag{9}$$

Controller objectives $e_1$ and $e_2$ are defined as follows:

$$e_1 = \Delta t_e = t_e - t_e^r \tag{10}$$

$$e_2 = \Delta |\varphi_s|^2 = |\hat{\varphi}_s|^2 - |\varphi_s^r|^2 \tag{11}$$

where $t_e^r$ and $|\varphi_s^r|$ are induced torque and stator flux references, respectively.

The time derivatives of $e_1$ and $e_2$ are obtained by using the stator flux-based model of IM as given below:

$$\begin{bmatrix} \dot{e}_1 \\ \dot{e}_2 \end{bmatrix} = \begin{bmatrix} g_1 \\ g_2 \end{bmatrix} + D \begin{bmatrix} v_{s\alpha}^r \\ v_{s\beta}^r \end{bmatrix} \tag{12}$$

where $v_{s\alpha}^r$ and $v_{s\beta}^r$ are the $-\alpha\beta$ components of stator voltage reference applied to the SVPWM as the reference inputs in order to determine the switching conditions of inverter. The expressions of the coefficients in (12) are as follows:

$$g_1 = \frac{3}{2} p_p \left[ -\left( \frac{R_s}{L_\sigma} + \frac{R_r}{\sigma L_r} \right) \left( \hat{\varphi}_{s\alpha} i_{s\beta} - \hat{\varphi}_{s\beta} i_{s\alpha} \right) + \hat{\omega}_m \left( \hat{\varphi}_{s\alpha} i_{s\alpha} - \hat{\varphi}_{s\beta} i_{s\beta} \right) \right.$$
$$\left. - \frac{p_p \hat{\omega}_m}{L_\sigma} |\hat{\varphi}_s|^2 \right] \tag{13}$$

$$g_2 = -2R_s \left( \hat{\varphi}_{s\alpha} i_{s\beta} + \hat{\varphi}_{s\beta} i_{s\alpha} \right) \tag{14}$$

$$D = \begin{bmatrix} \frac{3}{2} p_p \left( i_{s\beta} - \frac{\hat{\varphi}_{s\beta}}{L_\sigma} \right) & -\frac{3}{2} p_p \left( i_{s\alpha} - \frac{\hat{\varphi}_{s\alpha}}{L_\sigma} \right) \\ 2\hat{\varphi}_{s\alpha} & 2\hat{\varphi}_{s\beta} \end{bmatrix} \tag{15}$$

Due to $i_{s\alpha} = \left( \hat{\varphi}_{s\alpha} / L_\sigma \right) - \left( L_m / (L_\sigma L_r) \right) \hat{\varphi}_r$, the determination of $D$ is obtained as follows:

$$\det(D) = \frac{3 L_m}{\sigma L_r} p_p |\hat{\varphi}_s| |\hat{\varphi}_r| \cos(\hat{\varphi}_r, \hat{\varphi}_s) \tag{16}$$

In (16), product of the $|\hat{\varphi}_s|$ and the magnitude of rotor flux vector ($|\hat{\varphi}_r|$) is different from zero. Therefore it is certain that $D$ is a nonsingular matrix. According to input/output feedback linearization, the contorl inputs are defined as below:

$$\begin{bmatrix} v_{s\alpha}^r \\ v_{s\beta}^r \end{bmatrix} = D^{-1} \begin{bmatrix} g_1 + v_x \\ g_2 + v_y \end{bmatrix} \tag{17}$$

where $v_x$ and $v_y$ are auxiliary control inputs and they can be taken as below by taking into account pole placement concept:

$$v_x = -c_1 e_1 \tag{18}$$

$$v_y = -c_2 e_2 \tag{19}$$

Where $c_1$ and $c_2$ are positive constant and they are determined by trial-and-error method in the literature. In this study, a $c_1$ and $c_2$ are determined by optimizing with SA and PSO. And also for the speed and flux estimations of IM, a novel stator flux-based MRAS with LMS adaptation mechanism is proposed in this study. Thus, the control performance of speed-sensorless DTC of IM can be improved.

### 2.5 Novel Stator Flux-Based MRAS Speed Observer with LMS Adaptation Mechanism

In this section, development and design of a novel speed observer which is derived from stator flux-based model reference adaptive system (MRAS). The proposed MRAS-based speed observer is derived by using two different stator flux-based IM models given in the stator stationary axis $(-\alpha\beta)$ reference frame. As shown in Fig. 1, the rotor speed is observed by an adaptation mechanism which uses the difference between two different $-\alpha\beta$ components of stator fluxes which one is obtained from the reference model and the other one is obtained from adaptive model as given in (20) and (21):

$$\frac{d}{dt}\vec{\varphi}_{s,\alpha\beta}^{r} = \vec{v}_{s,\alpha\beta} - R_s\vec{i}_{s,\alpha\beta} \tag{20}$$

$$\frac{d}{dt}\vec{\varphi}_{s,\alpha\beta}^{a} = -\frac{R_r}{L_r}\vec{\varphi}_{s,\alpha\beta}^{a} + jp_p\omega_m\left(\vec{\varphi}_{s,\alpha\beta}^{a} - L_\sigma\vec{i}_{s,\alpha\beta}\right) + \left(\frac{L_sR_r}{L_r}\right)\vec{i}_{s,\alpha\beta} + L_\sigma\frac{d}{dt}\vec{i}_{s,\alpha\beta} \tag{21}$$

Here, $(*)^r$ and $(*)^a$ represent the reference and adaptive model, respectively. The IM speed is estimated as the output of the adaptation mechanism which utilizes the difference between the outputs of the reference and adaptive models which is defined as error input in the literature and this error is applied to the adaptation mechanism as input.

In conventional MRAS-based observers, the PI-type controller-based adaptation mechanism [19], sliding mode and fuzzy logic-based adaptation machanism are generally used [20]. But the performance of the MRAS which utilizes PI-type adaptation mechanism is affected by the changes of the IM parameters and operating conditions [20]. Furthermore, the sliding mode-based MRAS has chattering problem which is also referred as high frequency oscillations due to the improper system modelling around the sliding surface. And this chattering problem decreases robustness of the system against the disturbances [21]. Also the fuzzy logic-based MRAS requires expertise to implement specific complex system. In [22], a LMS algorithm is proposed as an adaptation mechanism with active power-based MRAS because it provides simple recursive algorithm and requires determination of a single coefficient. Due to this a LMS algorithm is preferred as an adaptation mechanism for speed estimation with stator flux-based MRAS. In the literature it is the first time that a LMS algorithm is used as an adaptation mechanism with stator flux-based MRAS for speed estimation of IM.

The standard LMS algorithm is described as follows [22]:

$$y(k) = \mathbf{W}^T(k)\mathbf{X}(k) \tag{22}$$

$$e(k) = d(k) - y(k) \tag{23}$$

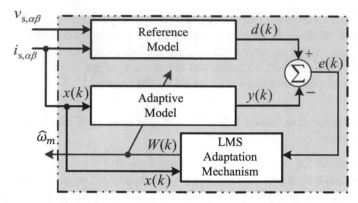

**Fig. 1.** Novel stator flux-based MRAS with LMS adaptation mechanism.

$$W(k + 1) = W(k) + \mu e(k)X(k) \tag{24}$$

where $d(k)$ is the reference model output. $y(k)$ is the adaptive model output. $e(k)$ is the error signal. $\mu$ is step size considered as constant. $X(k)$ and $W(k)$ are control input vector and filter weigth matrix, respectively. For $\omega_m$ estimation $(\hat{\omega}_m)$ $X(k)$ and $W(k)$ are derived from (21) as follows:

$$X = \begin{bmatrix} 1 - \frac{R_r T}{L_r} * \varphi_{s\alpha}^a(k + 1) & 1 - \frac{R_r T}{L_r} * \varphi_{s\beta}^a(k + 1) \\ \frac{R_r T L_s i_{s\alpha}(k)}{L_r} + L_\sigma (i_{s\alpha}(k + 1) - i_{s\alpha}(k)) & \frac{R_r T L_s i_{s\beta}(k)}{L_r} + L_\sigma (i_{s\beta}(k + 1) - i_{s\beta}(k)) \\ L_\sigma i_{s\beta}(k) - \varphi_{s\beta}^a(k + 1) & \varphi_{s\alpha}^a(k + 1) - L_\sigma i_{s\alpha}(k) \end{bmatrix} \tag{25}$$

$$\mathbf{W^T} = \begin{bmatrix} \hat{\omega}_m p_p T \\ 1 \\ 1 \end{bmatrix} \tag{26}$$

## 3    Optimization Results and Estimation and Control Performance Simulation of MRAS with LMS Adaptation-Based Speed-Sensorles IOFLC-DTC of IM

In this study, firstly the coefficients of auxiliary control inputs of IOFLC ($c_1$ and $c_2$) are optimized with PSO and SA. The IOFLC-DTC of IM with speed sensor and open-loop stator flux observer is shown in Fig. 2. While the optimization of the coefficients, $\omega_m$, $i_{s\alpha}$ and $i_{s\beta}$ are assumed as to be measured and $\varphi_{s\alpha}$ and $\varphi_{s\beta}$ are observed with an classical open-loop stator flux observer which expressions is given as follows:

$$\varphi_{s\alpha}(k + 1) = \frac{L_m^2 R_r T i_{s\alpha}(k + 1)}{L_r^2} + \left( \frac{L_m}{L_r} - \frac{L_m R_r T}{L_r^2} \right) \varphi_{s\alpha}(k) - \frac{L_m p_p \omega_m(k + 1) T \varphi_{s\beta}(k)}{L_r}$$
$$+ L_\sigma i_{s\alpha}(k + 1) \tag{27}$$

$$\varphi_{s\beta}(k+1) = \frac{L_m^2 R_r T i_{s\beta}(k+1)}{L_r^2} + \left(\frac{L_m}{L_r} - \frac{L_m R_r T}{L_r^2}\right)\varphi_{s\beta}(k) + \frac{L_m p_p \omega_m (k+1) T \varphi_{s\alpha}(k)}{L_r}$$
$$+ L_\sigma i_{s\beta}(k+1) \tag{28}$$

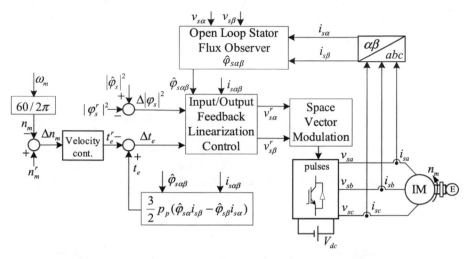

**Fig. 2.** The IOFLC-DTC of IM with speed sensor and open-loop stator flux observer.

The coefficients of auxiliary control inputs of IOFLC ($c_1$ and $c_2$) are optimized on the control system given in Fig. 2 by using the mean square errors (MSE) between the reference speed and the measured speed given in (29) with PSO and SA.

$$f(\cdot) = \frac{1}{N} \sum_{n=1}^{N} \left(\omega_m^r - \omega_m\right)^2 \tag{29}$$

While the optimization of $c_1$ and $c_2$, the control system runs for 4 $s$ in steady-state with $t_L$ variation. Under these conditions, optimization results of PSO and SA are given in Tables 2 and 3, respectively.

**Table 2.** MSE on speed control of SA with different initial temperature under steady-state condition

| Init. temp. | Final temp. | # Func. calls | $c_1^{IOFLC}$ | $c_2^{IOFLC}$ | MSE |
|---|---|---|---|---|---|
| 1 | 0.0687195 | 241 | 5.7209e+03 | 5.0280e+03 | 0.5616873 |
| 10 | 0.0377789 | 501 | 1.1186e+04 | 1.1170e+04 | 0.5900013 |
| 100 | 0.193428 | 561 | 1.2211e+04 | 1.3593e+04 | 0.6110063 |

**Table 3.** MSE on speed control of PSO with different swarm numbers (population number) under steady-state condition

| #Iteration | #Swarm | $c_1^{PSO}$ | $c_2^{PSO}$ | $c_1^{IOFLC}$ | $c_2^{IOFLC}$ | MSE |
|---|---|---|---|---|---|---|
| 50 | 20 | 2.2 | 1.8 | 10,000 | 8.8322e+03 | 0.5563012 |
| **50** | **30** | **2.2** | **1.8** | **7.4877e+03** | **6.4644e+03** | **0.5526747** |
| 50 | 40 | 2.2 | 1.8 | 9.0777e+03 | 8.2879e+03 | 0.5543446 |

The best optimization performances of SA and PSO algorithm are shown in Tables 2 and 3, respectively. The best optimization performances are indicated in red text in the tables. Their optimization performances are nearly same. However, it is seen that, PSO is more successful than SA in the optimization made under steady-state operation condition of IM.

After the optimization of $c_1$ and $c_2$, these values are used in the MRAS with LMS adaptation-based speed-sensorles IOFLC-DTC of IM given in Fig. 3 for testing the estimation and control performance of the proposed IM drive system.

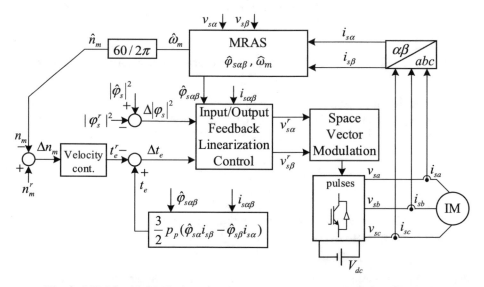

**Fig. 3.** MRAS with LMS adaptation-based speed-sensorless IOFLC-DTC of IM.

Estimation and control performances are tested under $t_L$ variations in a wide speed range which contains transient and steady-state conditions in simulation. The simulation results are shown in Fig. 4. Furthermore, the MSE between the $\omega_m$ and $\hat{\omega}_m$ which are obtained when the optimized values of $c_1$ and $c_2$ are used in the proposed speed-sensorless drive system of IM are given in Tables 4 and 5, respectively.

The best MSE results are indicated in red text in tables. From these results, SA is more successful than PSO in the optimization made under both transient and steady-state

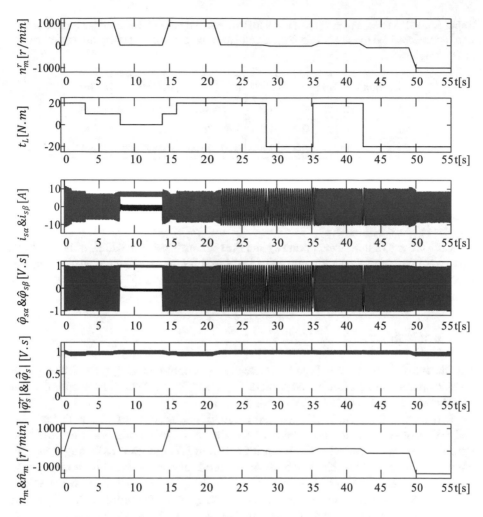

**Fig. 4.** Simulation results of MRAS with LMS adaptation-based speed-sensorless IOFLC-DTC of IM.

**Table 4.** MSE on speed estimation of novel MRAS-based speed estimator with optimized values of $c_1$ and $c_2$ obtained form SA with different initial temperature under both transient and steady-state conditions

| Init. temp. | Final temp. | #Func. calls | $c_1^{IOFLC}$ | $c_2^{IOFLC}$ | MSE |
|---|---|---|---|---|---|
| 1 | 0.0687195 | 241 | 5.7209e+03 | 5.0280e+03 | 6.3864 |
| 10 | 0.0377789 | 501 | 1.1186e+04 | 1.1170e+04 | 6.3529 |
| **100** | **0.193428** | **561** | **1.2211e+04** | **1.3593e+04** | **6.3256** |

**Table 5.** MSE on speed estimation of novel MRAS-based speed estimator with optimized values of $c_1$ and $c_2$ obtained form PSO with different initial temperature under both transient and steady-state conditions

| #Iteration | #Swarm | $c_1^{PSO}$ | $c_2^{PSO}$ | $c_1^{IOFLC}$ | $c_2^{IOFLC}$ | MSE |
|---|---|---|---|---|---|---|
| **50** | **20** | **2.2** | **1.8** | **10000** | **8.8322e+03** | **6.3937** |
| 50 | 30 | 2.2 | 1.8 | 7.4877e+03 | 6.4644e+03 | 6.3946 |
| 50 | 40 | 2.2 | 1.8 | 9.0777e+03 | 8.2879e+03 | 6.3949 |

operation conditions of IM. Also, it is clearly seen from Fig. 4 that, the proposed novel stator flux-based MRAS speed estimator with LMS adaptation mechanism has a high success in speed estimation.

Both the use of optimized $c_1$ and $c_2$ and the high estimation accuracy of novel stator flux-based MRAS speed estimator with LMS adaptation mechanism ensure that the control performance of the proposed speed-sensorless IOFLC-DTC drive system of IM is quite high.

## 4   Conclusions

Speed control performance of the IM is accurately accomplished by novel MRAS with LMS adaptation mechanism-based speed-sensorless IOFLC-DTC with the optimization of unknown constants of IOFLC with SA and PSO.

It is proved from the MSE on speed estimation and control under both transient and steady-state operation conditions of IM that SA algorithm gives most effective optimization results against to PSO algorithm. But in SA algorithm, assigning the initial conditions of paramaters that need to be optimized is a problem and this problem cause an another problem that the algorithm converges the global minima directly and sticks that point. And also it can not be able to improve the optimization process. PSO algorithm has not a problem about assigning the initial conditions of paramaters that need to be optimized.

The proposed study is the first time in the literature with two different features. The first feature is a LMS algorithm is used as an adaptation mechanism with stator flux-based MRAS for speed estimation of IM and the second contains the optimization of the unknown coefficients of the auxiliary control inputs of IOFLC with SA and PSO.

## References

1. Zeghib, O., Allag, A., Hamidani, B., Allag, A.: Input-output linearizing control of induction motor based on a newly extended MVT observer design. In: Proceedings of the IEEE International Conference on Communications and Electrical Engineering (ICCEE), El Oued, Algeria, pp. 1–6 (2018)
2. Devanshu, A., Singh, M., Kumar, N.: Sliding mode control of induction motor drive based on feedback linearization. IETE J. Res. **66**(2), 256–269 (2020)

3. Ammar, A., Kheldoun, A., Metidji, B., Ameid, T., Azzoug, Y.: Feedback linearization based sensorless direct torque control using stator flux MRAS-sliding mode observer for induction motor drive. ISA Trans. **98**, 382–392 (2020)
4. Liu, X., Yu, H., Yu, J., Zhao, L.: Combined speed and current terminal sliding mode control with nonlinear disturbance observer for PMSM drive. IEEE Access **6**, 29594–29601 (2018)
5. Accetta, A., et al.: Robust control for high performance induction motor drives based on partial state-feedback linearization. IEEE Trans. Ind. Appl. **55**(1), 490–503 (2019)
6. Young-Sik, C., Han Ho, C., Jin-Woo, J.: Feedback linearization direct torque control with reduced torque and flux ripples for IPMSM drives. IEEE Trans. Power Electron. **31**(5), 3728–3737 (2016)
7. BoussaïD, I., Lepagnot, J., Siarry, P.: A survey on optimization metaheuristics. Inf. Sci. **237**, 82–117 (2013)
8. Kirkpatrick, S., Gelatt, C.D., Vecchi, M.P.: Optimization by simulated annealing. Science **220**(4598), 671–680 (1983)
9. Metropolis, N., Rosenbluth, A.W., Rosenbluth, M.N., Teller, A.H., Teller, E.: Equation of state calculations by fast computing machines. J. Chem. Phys. **21**(6), 1087–1092 (1953)
10. Avci, Y.: Sonlu eleman modeli güncellemesi tekniğinde benzetilmiş tavlama algoritması kullanılarak mekanik sistemlerde hasar tespiti. Doctoral Dissertation, Institute of Science (2008)
11. Kutucu, H., Durgut, R.: Silah hedef atama problemi için tavlama benzetimli bir hibrit yapay arı kolonisi algoritması. Süleyman Demirel Univ. J. Inst. Sci. **22**, 263–269 (2018)
12. Kennedy, J., Eberhart, R.: Particle swarm optimization. In: Proceedings of the ICNN 1995-International Conference on Neural Networks, vol. 4, pp. 1942–1948. IEEE (1995)
13. Wang, D., Tan, D., Liu, L.: Particle swarm optimization algorithm: an overview. Soft. Comput. **22**(2), 387–408 (2017). https://doi.org/10.1007/s00500-016-2474-6
14. Özsağlam, M.Y., Çunkaş, M.: Particle swarm optimization algorithm for solving optimization problems. Polytech. J. **11**(4), 299–305 (2008)
15. Poli, R., Kennedy, J., Blackwell, T.: Particle swarm optimization. Swarm Intell. **1**(1), 33–57 (2007)
16. Bansal, J.C.: Particle swarm optimization. In: Bansal, J.C., Singh, P.K., Pal, N.R. (eds.) Evolutionary and Swarm İntelligence Algorithms, pp. 11–23. Springer, Cham (2019). https://doi.org/10.1007/978-3-319-91341-4_2
17. Zheng, Z., Xue, X., Xu, B.: Direct torque control of asynchronous motor based on space vector modulation. In: Proceedings of the International Conference on Electronics, Communications and Control (ICECC), Ningbo, China, pp. 1440–1443 (2011)
18. Soltani, J., Markadeh, G.R.A., Abjadi, N.R., Ping, H.W.: A new adaptive direct torque control (DTC) scheme based-on SVM for adjustable speed sensorless induction motor drive. In: Proceedings of the International Conference on Electrical Machines and Systems (ICEMS), Busan, South Korea, pp. 497–502 (2007)
19. Das, S., Kumar, R., Pal, A.: MRAS-based speed estimation of induction motor drive utilizing machines' d- and q-circuit impedances. IEEE Trans. Industr. Electron. **66**(6), 4286–4295 (2019)
20. Gadoue, S.M., Giaouris, D., Finch, J.W.: MRAS sensorless vector control of an induction motor using new sliding-mode and fuzzy-logic adaptation mechanisms. IEEE Trans. Energy Convers. **25**(2), 394–402 (2010). https://doi.org/10.1109/TEC.2009.2036445
21. Holakooie, M., Ojaghi, M., Taheri, A.: Modified DTC of a six-phase induction motor with a second-order sliding-mode MRAS-based speed estimator. IEEE Trans. Power Electron. **34**(1), 600–611 (2019)
22. Demir, R., Barut, M.: Novel hybrid estimator based on model reference adaptive system and extended Kalman filter for speed-sensorless induction motor control. Trans. Inst. Meas. Control. **40**(13), 3884–3898 (2018)

# Author Index

**A**

Adegboye, Oluwatayomi, 400
Akhmet, Marat, 540, 625
Akram, Vahid, 369
Akram, Vahid Khalilpour, 381
Aksoy, Bekir, 429, 705
Aksoy, Sencer, 654
Alaadin, Asmaa, 360
Aldağ, Mehmet, 400
Alkan, Taha Yigit, 125
Alkım, Erdem, 360
Allahverdiev, Bilender P., 10
Altun, Adem Alpaslan, 93
Aruğaslan Çinçin, Duygu, 524, 540
Atagün, Ercan, 220, 246, 256, 344
Ateş, Fatmanur, 429
Atıcı, Birkan, 211
Aydın, Zafer, 455
Az, Şevval, 76

**B**

Baba, A. Fevzi, 390
Bakir, Huseyin, 136, 150
Başaran, Barış Doruk, 680
Basarslan, M. Sinan, 237
Bayıroğlu, Hacer, 246, 256, 344
Baysan, Fatih, 680
Beqa, E., 601
Bilen, Mehmet, 315
Bilge, Alper, 280
Bilgin, Cahit, 42
Bilgin, Metin, 409
Biroğul, Serdar, 220
Bozkurt, Armagan, 390

Bozkurt, Ferda, 59
Bozkurt, Mehmet Recep, 42, 59
Bulut, Batuhan, 680
Bulut, Nisanur, 409
Burunkaya, Mustafa, 446
Bytyçi, E., 601

**C**

Çabuk, Umut Can, 369, 381
Cengiz, Nur, 524, 540
Çetin, Gürcan, 465
Chehbi Gamoura, Samia, 325
Çolak, Recep, 669
Curukoglu, Nur, 82

**D**

Dagdeviren, Orhan, 369, 381
Dagdeviren, Zuleyha Akusta, 553
Daldır, Irmak, 165
Danisman, Taner, 350
Demir, Emre, 190
Demirci, Sercan, 360
Demirhan, Mervenur, 643
Derin, Bayram, 590
Duman, Mehmet, 183
Duman, Serhat, 136, 150

**E**

Eken, Süleyman, 484
Erçelebi Ayyıldız, Tülin, 265
Erdoğan, Hande, 174
Ersoy, Mevlüt, 1, 231, 491, 669, 755
Ersoz, Metin, 390

J. Hemanth et al. (Eds.): ICAIAME 2020, LNDECT 76, pp. 771–773, 2021.
https://doi.org/10.1007/978-3-030-79357-9

**G**
Genç, Burkay, 113
Gerek, Ömer Nezih, 280
Gögebakan, Maruf, 502
Gökçen, Ahmet, 190
Gönen, Serkan, 418
Görmez, Yasin, 455
Gülmez, Esra, 325
Günay, Melih, 33, 125, 680, 696
Gündüz, Hakan, 246, 256, 344
Gurfidan, Remzi, 1
Güvenç, Ercüment, 643
Guvenc, Ugur, 136, 150

**H**
Hadiwidodo, Yoyok Setyo, 613
Halis, Helin Diyar, 567
Hazir, Uğur, 125, 350

**I**
İlhan Omurca, Sevinç, 211
İlhan, Özcan, 265
Inan, Remzi, 755
İnce, Muhammed Numan, 33
İnce, Murat, 439
Işık, Ali Hakan, 106, 315, 669
Islamiyah, Wardah Rahmatul, 198

**J**
Juniani, Anda I., 198

**K**
Kaftanci, Mehmet Fatih, 696
Kalemsiz, Melisa, 680
Kara, Kemal Can, 15
Karacayılmaz, Gökçe, 418
Karagül Yıldız, Tuba, 246
Karakaş, Aşkın, 15, 76
Kayaalp, F., 237
Kayakuş, Mehmet, 590
Khelil, Imane, 325
Kilinçarslan, Şemsettin, 439
Kocabıçak, Ümit, 42
Kocaer, Emine Rümeysa, 325
Koçak, Hilal, 465
Koruca, Halil İbrahim, 325
Köse, Utku, 491, 669
Kuçi, Rifai, 106
Kumral, Cem Deniz, 755
Kurt, Zühal, 280

**L**
Ledet, Joseph William, 125
Loy-García, Gabriel, 715

**M**
Marichelvam, M. K., 174
Marmolejo-Saucedo, Jose-Antonio, 715, 732, 742
Melekoğlu, Engin, 42
Melenli, Sadettin, 578
Mertoğlu, Uğur, 113
Michail, Seth, 125

**N**
Naufal, Rizqullah Yusuf, 613
Nugayeva, Zakhira, 625
Nuraisyah, Triajeng, 198

**O**
Ocak, Cemil, 514
Oner, Yusuf, 390
Oturanç, Galip, 304
Özgün, Kamer, 680
Özkan, Kemal, 280
Özkaraca, Osman, 643, 654
Ozkaya, Burcin, 136
Ozpinar, Alper, 82
Özsoy, Koray, 567
Öztimur Karadağ, Özge, 696

**P**
Purnami, Santi Wulan, 198

**R**
Rochani, Imam, 613
Rodríguez-Aguilar, Román, 715, 732, 742
Rogova, E., 601
Rosyid, Daniel Mohammad, 613

**S**
Salman, Osamah, 429
Şanlıalp, Elif, 25
Şanlıalp, İbrahim, 25
Şansal, Uğur, 231
Sayan, H. Hüseyin, 418
Sayin, İrem, 705
Şenol, Ramazan, 429
Servı, Sema, 304
Sever, Hayri, 113
Şimşek Türker, Yasemin, 439
Sönmez, Yusuf, 514

**T**
Takazoğlu, Uygar, 76
Taşdemir, Şakir, 93
Timuçin, Tunahan, 246, 256, 344
Tleubergenova, Madina, 625
Topkaya, Aylin, 578
Tosun, Mustafa, 369

Tosun, Nedret, 165
Tosun, Ömür, 165, 174
Tuna, Hüseyin, 10
Tunç, Ali, 93
Tuncer, Işılay, 15
Türkçetin, Ayşen Özün, 491, 669

**U**
Uçar, Ali, 484
Uçar, Muhammed Kürşad, 42, 59
Ülker, Ezgi Deniz, 400
Üncü, İsmail Serkan, 590
Üstünsoy, Furkan, 418

**W**
Wulandari, Diah P., 198

**Y**
Yagci, Mustafa, 292
Yalçinkaya, Yüksel, 10
Yenipinar, Burak, 514
Yiğit, Tuncay, 25, 315
Yıldırım, Kılıçarslan, 59
Yildirim, Mehmet, 93
Yıldız, Sadık, 446
Yildiz, Tuba Karagül, 256, 344
Yilmaz, Cemal, 514
Yılmaz, Ercan Nurcan, 418
Yürüten, Ege, 680

Printed in the United States
by Baker & Taylor Publisher Services